Lecture Notes in Computer Science **10665**

Commenced Publication in 1973
Founding and Former Series Editors:
Gerhard Goos, Juris Hartmanis, and Jan van Leeuwen

More information about this series at http://www.springer.com/series/7407

Ivan Lirkov · Svetozar Margenov (Eds.)

Large-Scale Scientific Computing

11th International Conference, LSSC 2017
Sozopol, Bulgaria, June 5–9, 2017
Revised Selected Papers

Editors
Ivan Lirkov
Institute of Information and Communication Technologies
Bulgarian Academy of Sciences
Sofia
Bulgaria

Svetozar Margenov
Institute of Information and Communication Technologies
Bulgarian Academy of Sciences
Sofia
Bulgaria

ISSN 0302-9743 ISSN 1611-3349 (electronic)
Lecture Notes in Computer Science
ISBN 978-3-319-73440-8 ISBN 978-3-319-73441-5 (eBook)
https://doi.org/10.1007/978-3-319-73441-5

Library of Congress Control Number: 2017962887

LNCS Sublibrary: SL1 – Theoretical Computer Science and General Issues

Printed on acid-free paper

This Springer imprint is published by Springer Nature
The registered company is Springer International Publishing AG
The registered company address is: Gewerbestrasse 11, 6330 Cham, Switzerland

Preface

The 11th International Conference on Large-Scale Scientific Computations (LSSC 2017) was held in Sozopol, Bulgaria, June 5–9, 2017. The conference was organized by the Institute of Information and Communication Technologies at the Bulgarian Academy of Sciences in cooperation with Society for Industrial and Applied Mathematics (SIAM) and Sozopol municipality.

The plenary invited speakers and lectures were:

- J. Adler, "Energy-Minimization, Finite Elements, and Multilevel Methods for Liquid Crystals"
- J. Gopalakrishnan, "Theoretical and Practical Aspects of DPG Methods"
- U. Langer, "Multi-Patch Discontinuous Galerkin Space and Space-Time Isogeometric Analysis"
- J. Pasciak, "Numerical Approximation of a Space-Time Fractional Parabolic Equation"
- G. G. Yin, "Numerical Methods for Consensus of Networked Systems and Stochastic Control with Soft Constraints"

The success of the conference and the present volume are the outcome of the joint efforts of many partners from various institutions and organizations. First, we would like to thank all the members of the Scientific Committee for their valuable contribution forming the scientific face of the conference, as well as for their help in reviewing contributed papers. We especially thank the organizers of the special sessions. We are also grateful to the staff involved in the local organization.

Traditionally, the purpose of the conference is to bring together scientists working with large-scale computational models in natural sciences and environmental and industrial applications, and specialists in the field of numerical methods and algorithms for modern high-performance computers. The invited lectures reviewed some of the most advanced achievements in the field of numerical methods and their efficient applications. The conference talks were presented by researchers from academic institutions and practical industry engineers including applied mathematicians, numerical analysts, and computer experts. The general theme for LSSC 2017 was "Large-Scale Scientific Computing" with a particular focus on the organized special sessions.

The special sessions and organizers were:

- Space-Time Methods for Solving Time-Dependent PDEs — U. Langer
- Advanced Discretizations and Solvers for Coupled Systems of Partial Differential Equations — J. Adler, X. Hu, L. Zikatanov
- Least-Squares Finite Element Methods — F. Bertrand and P. Bochev
- Advances in Heterogeneous Numerical Methods for Multi Physics Problems — P. Bochev, M. Perego

- Advanced Numerical Methods for Nonlinear Elliptic Partial Differential Equations — J. Kraus
- Solvers and Error Estimators for Mixed Finite Elements in Solid Mechanics — G. Starke
- Control and Optimization of Dynamical Systems — M. Krastanov, V. Veliov
- HPC and Big Data: Algorithms and Applications — A. Karaivanova, T. Gurov, E. Atanassov
- Toward Exascale Computation — O. Iliev
- Monte Carlo Methods: Theory, Applications and Distributed Computing — I. Dimov, R. Georgieva, M. Nedjalkov
- Application of Metaheuristics to Large-Scale Problems — S. Fidanova, G. Luque
- Large-Scale Models: Numerical Methods, Parallel Computations and Applications — K. Georgiev, Z. Zlatev
- Large-Scale Numerical Computations for Sustainable Energy Production and Storage — P. D'Ambra, D. di Serafino, S. Filippone

About 150 participants from all over the world attended the conference representing some of the strongest research groups in the field of advanced large-scale scientific computing. This volume contains 66 papers by authors from 21 countries.

The next international conference LSSC will be organized in June 2019.

November 2017

Ivan Lirkov
Svetozar Margenov

Organization

Scientific Committee

James Adler	Tufts University, USA
Pasqua D'Ambra	Istituto per le Applicazioni del Calcolo Mauro Picone, CNR, Naples, Italy
Emanouil Atanassov	Institute of Information and Communication Technologies, BAS, Bulgaria
Fleurianne Bertrand	Universität Duisburg-Essen, Germany
Pavel Bochev	Sandia National Laboratories, USA
Ivan Dimov	Institute of Information and Communication Technologies, BAS, Bulgaria
Stefka Dimova	Sofia University, Bulgaria
Stefka Fidanova	Institute of Information and Communication Technologies, BAS, Bulgaria
Salvatore Filippone	Cranfield University, UK
Krassimir Georgiev	Institute of Information and Communication Technologies, BAS, Bulgaria
Jay Gopalakrishnan	Portland State University, USA
Todor Gurov	Institute of Information and Communication Technologies, BAS, Bulgaria
Xiaozhe Hu	Tufts University, USA
Oleg Iliev	ITWM, Germany
Aneta Karaivanova	Institute of Information and Communication Technologies, BAS, Bulgaria
Mikhail Krastanov	Sofia University, Bulgaria
Johannes Kraus	University of Duisburg-Essen, Germany
Ulrich Langer	Johannes Kepler University Linz, Austria
Raytcho Lazarov	Texas A&M University, USA
Ivan Lirkov	Institute of Information and Communication Technologies, BAS, Bulgaria
Gabriel Luque	University of Málaga, Spain
Svetozar Margenov	Institute of Information and Communication Technologies, BAS, Bulgaria
Joseph Pasciak	Texas A&M University, College Station, USA
Mauro Perego	Sandia National Laboratories, USA
Daniela di Serafino	Università degli Studi della Campania Luigi Vanvitelli, Italy
Gerhard Starke	University of Duisburg-Essen, Germany
Vladimir Veliov	TU Vienna, Austria

Gang George Yin	Wayne State University, USA
Zahari Zlatev	Aarhus University, Denmark
Ludmil Zikatanov	Pennsylvania State University, USA

Contents

Least-Squares Finite Element Methods

Advances in Heterogeneous Numerical Methods for Multi Physics Problems

Advanced Numerical Methods for Nonlinear Elliptic Partial Differential Equations

Control and Optimization of Dynamical Systems

HPC and Big Data: Algorithms and Applications

Toward Exascale Computation

Application of Metaheuristics to Large-Scale Problems

Large-Scale Models: Numerical Methods, Parallel Computations and Applications

Contributed Papers

Invited Papers

Discrete Energy Laws for the First-Order System Least-Squares Finite-Element Approach

J. H. Adler[1(✉)], I. Lashuk[1], S. P. MacLachlan[2], and L. T. Zikatanov[3]

[1] Department of Mathematics, Tufts University, Medford, MA 02155, USA
james.adler@tufts.edu, ilya.lashuk@gmail.com
[2] Department of Mathematics and Statistics, Memorial University of Newfoundland, St. John's, NL A1C 5S7, Canada
smaclachlan@mun.ca
[3] Department of Mathematics, Penn State, University Park, PA 16802, USA
ludmil@psu.edu

Abstract. This paper analyzes the discrete energy laws associated with first-order system least-squares (FOSLS) discretizations of time-dependent partial differential equations. Using the heat equation and the time-dependent Stokes' equation as examples, we discuss how accurately a FOSLS finite-element formulation adheres to the underlying energy law associated with the physical system. Using regularity arguments involving the initial condition of the system, we are able to give bounds on the convergence of the discrete energy law to its expected value (zero in the examples presented here). Numerical experiments are performed, showing that the discrete energy laws hold with order $\mathcal{O}\left(h^{2p}\right)$, where h is the mesh spacing and p is the order of the finite-element space. Thus, the energy law conformance is held with a higher order than the expected, $\mathcal{O}\left(h^{p}\right)$, convergence of the finite-element approximation. Finally, we introduce an abstract framework for analyzing the energy laws of general FOSLS discretizations.

1 Introduction

First-order system least squares (FOSLS) is a finite-element methodology that aims to reformulate a set of partial differential equations (PDEs) as a system of first-order equations [8,9]. The problem is posed as a minimization of a functional in which the first-order differential terms appear quadratically, so that the functional norm is equivalent to a norm meaningful for the given problem. In equations of elliptic type, this is usually a product H^1 norm. Some of the compelling features of the FOSLS methodology include: self-adjoint discrete equations stemming from the minimization principle; good operator conditioning stemming from the use of first-order formulations of the PDE; and finite-element and multigrid performance that is optimal and uniform in certain parameters (e.g., Reynolds number for the Navier-Stokes equations), stemming from uniform product-norm equivalence.

I. Lirkov and S. Margenov (Eds.): LSSC 2017, LNCS 10665, pp. 3–20, 2018.
https://doi.org/10.1007/978-3-319-73441-5_1

Successful FOSLS formulations have been developed for a variety of applications [5,7]. One example of a large-scale physical application is in magnetohydrodynamics (MHD) [1–3]. These numerical methods have led to substantial improvements in MHD simulation technology; however, several important estimates remain to be analyzed to confirm their quantitive accuracy. One of these is the energy of the system. Using an energetic-variational approach [11–13,15], energy laws of the MHD system can be derived that show that the total energy should decay as a direct result of the dissipation in the system. Initial computations show that the FOSLS method indeed captures this energy law, but it remains to be shown why it should.

In this paper, we describe the *discrete* energy laws associated with FOSLS discretizations of time-dependent PDEs, such as the heat equation or Stokes' equation, and show quantitatively how they are related to the continuous physical law. While we only show results for these "simple" linear systems, the results appear generalizable to more complicated systems, such as MHD. Getting the correct energy law is not only important for numerical stability, but it is crucial for capturing the correct physics, especially if singularities or high contrasts in the solution are present.

The paper is outlined as follows. In Sect. 2, we discuss the energy laws of a given system and describe their discrete analogues. Section 3 analyzes the energy laws associated with the FOSLS discretizations of the heat equation, and the same is done for Stokes' equations in Sect. 4. For both examples, we present numerical simulations in Sect. 5. Finally, we discuss a generalization of the concepts presented here in Sect. 6, and some concluding remarks in Sect. 7.

2 Energy Laws

The energetic-variational approach (EVA) [11–13,15] of hydrodynamic systems in complex fluids is based on the second law of thermodynamics and relies on the fundamental principle that the change in the total energy of a system over time must equal the total dissipation of the system. This energy principle plays a crucial role in understanding the interactions and coupling between different scales or phases in a complex fluid. In general, any set of equations that describe the system can be derived from the underlying energy laws. The energetic variational principle is based on the energy dissipation law for the whole coupled system:

$$\frac{\partial E_{total}}{\partial t} = -\mathcal{D}, \tag{1}$$

where E_{total} is the total energy of the system, and $\mathcal{D}$ is the dissipation.

Simple fluids, where we assume no internal (or elastic) energies, can also be described in this setting and yield the following energy law:

$$\frac{\partial}{\partial t}\left(\frac{1}{2}\int_{\Omega}|\mathbf{u}|^2\,d\mathbf{x}\right) = -\int_{\Omega}\nu|\nabla\mathbf{u}|^2\,d\mathbf{x}, \tag{2}$$

where $\mathbf{u}$ represents the fluid velocity and ν is the fluid viscosity, accounting for the dissipation in the system. Applying the so-called least-action principle results in the integral equation,

$$\left\langle \frac{\partial \mathbf{u}}{\partial t} + \nabla p, \mathbf{y} \right\rangle = \langle \nabla \cdot \nu \nabla \mathbf{u}, \mathbf{y} \rangle, \forall \mathbf{y} \in \mathcal{V},$$

where we assume an incompressible fluid, $\nabla \cdot u = 0$, and an appropriate Hilbert space, $\mathcal{V}$. Here, we use $\langle \cdot, \cdot \rangle$ to denote the $L^2(\Omega)$ inner product. In strong form, we obtain the time-dependent Stokes' equations (assuming appropriate boundary conditions):

$$\frac{\partial \mathbf{u}}{\partial t} + \nabla p - \nabla \cdot \nu \nabla \mathbf{u} = 0, \tag{3}$$

$$\nabla \cdot \mathbf{u} = 0. \tag{4}$$

Note that the energy law can also be derived directly from the PDE itself. First, we consider the weak form of (3)–(4), multiplying (3) by $\mathbf{u}$ and (4) by p and integrate over Ω. After integration by parts we obtain the following relations:

$$\begin{aligned} 0 &= \left\langle \frac{\partial \mathbf{u}}{\partial t} + \nabla p - \nabla \cdot \nu \nabla \mathbf{u}, \mathbf{u} \right\rangle + \langle \nabla \cdot \mathbf{u}, p \rangle \\ &= \left\langle \frac{\partial \mathbf{u}}{\partial t}, \mathbf{u} \right\rangle + \langle \nabla p, \mathbf{u} \rangle - \langle \nabla \cdot \nu \nabla \mathbf{u}, \mathbf{u} \rangle + \langle \nabla \cdot \mathbf{u}, p \rangle \\ &= \frac{1}{2} \frac{\partial}{\partial t} \langle \mathbf{u}, \mathbf{u} \rangle + \langle \nabla p, \mathbf{u} \rangle + \langle \nu \nabla \mathbf{u}, \nabla \mathbf{u} \rangle - \langle \mathbf{u}, \nabla p \rangle . \end{aligned}$$

Here, we have assumed that the boundary conditions are such that the boundary terms, resulting from the integration by parts, vanish. Hence, we have

$$\frac{1}{2} \frac{\partial}{\partial t} \langle \mathbf{u}, \mathbf{u} \rangle = - \langle \nu \nabla \mathbf{u}, \nabla \mathbf{u} \rangle .$$

This approach can also be applied to other PDEs, such as the heat equation, to show similar energy dissipation relations. Let ν be the thermal diffusivity of the body Ω, and u its temperature. Then the PDE describing the temperature distribution in Ω is as follows,

$$\frac{\partial u}{\partial t} - \nabla \cdot \nu \nabla u = 0, \quad \text{on } \Omega, \quad u = 0, \quad \text{on } \partial\Omega. \tag{5}$$

As before, we multiply (5) by u and integrate over Ω to obtain that

$$\begin{aligned} 0 &= \left\langle \frac{\partial u}{\partial t} - \nabla \cdot \nu \nabla u, u \right\rangle = \left\langle \frac{\partial u}{\partial t}, u \right\rangle - \langle \nabla \cdot \nu \nabla u, u \rangle \\ &= \frac{1}{2} \frac{\partial}{\partial t} \langle u, u \rangle + \langle \nu \nabla u, \nabla u \rangle \end{aligned}$$

Hence,

$$\frac{1}{2} \frac{\partial}{\partial t} \langle u, u \rangle = - \langle \nu \nabla u, \nabla u \rangle ,$$

which is the scalar version of (2).

For the remainder of the paper, we analyze (2), specifically how closely the FOSLS method can approximate the energy law discretely. We will consider both the scalar (heat equation) and the vector version (Stokes' equation) in the numerical results, as the form of the energy law is identical. First, we discuss how moving to a finite-dimensional space affects the energy law.

3 Heat Equation

First, we consider the heat equation, assuming a constant diffusion coefficient $\nu = 1$ for simplicity, homogeneous Dirichlet boundary conditions, and a given initial condition:

$$\frac{\partial u(\boldsymbol{x},t)}{\partial t} = \Delta u(\boldsymbol{x},t) \quad \forall \boldsymbol{x} \in \Omega,\ \forall t > 0 \tag{6}$$

$$u(\boldsymbol{x},t) = 0 \quad \forall \boldsymbol{x} \in \partial\Omega,\ \forall t \geq 0 \tag{7}$$

$$u(\boldsymbol{x},0) = u_0(\boldsymbol{x}) \quad \forall \boldsymbol{x} \in \bar{\Omega}. \tag{8}$$

To discretize the problem in time, we consider a symplectic, or energy-conserving, time-stepping scheme such as Crank-Nicolson. Given a time step size, τ, and time $t_n = \tau n$, we approximate $u_n = u(\mathbf{x}, t_n)$ with the following semi-discrete version of (6),

$$\frac{u_{n+1} - u_n}{\tau} = \frac{\Delta u_{n+1} + \Delta u_n}{2}$$

To simplify the calculations later, we introduce an intermediate approximation, $u_{n+\frac{1}{2}}$, and re-write the semi-discrete problem as

$$\begin{aligned} \frac{u_{n+\frac{1}{2}} - u_n}{\left(\frac{\tau}{2}\right)} &= \Delta u_{n+\frac{1}{2}} \\ u_{n+\frac{1}{2}}(\boldsymbol{x}) &= 0 \quad \forall \boldsymbol{x} \in \partial\Omega,\ n = 0,1,2,\ldots \\ u_{n+1} &= 2u_{n+\frac{1}{2}} - u_n \end{aligned} \tag{9}$$

Remark 1. To obtain the *semi-discrete* energy law for (9), we perform a similar procedure as done in Sect. 2, where we multiply the first equation in (9) by $u_{n+\frac{1}{2}}$ and integrate over the domain. After some simple calculations, we obtain the corresponding energy law, using L^2–norm notation:

$$\frac{||u_{n+1}||^2 - ||u_n||^2}{2\tau} = -||\nabla u_{n+\frac{1}{2}}||^2 \tag{10}$$

To use the FOSLS method, we now put the operator into a first-order system. Since we have reduced the problem to a reaction-diffusion type problem, we introduce a new vector $\mathbf{V} = \nabla u$, and use the H^1-elliptic equivalent system [8,9]:

$$L_\tau \begin{pmatrix} u_{n+\frac{1}{2}} \\ \boldsymbol{V}_{n+\frac{1}{2}} \end{pmatrix} = \begin{pmatrix} -\nabla \cdot \boldsymbol{V}_{n+\frac{1}{2}} + \frac{2}{\tau} u_{n+\frac{1}{2}} \\ \boldsymbol{V}_{n+\frac{1}{2}} - \nabla u_{n+\frac{1}{2}} \\ \nabla \times \boldsymbol{V}_{n+\frac{1}{2}} \end{pmatrix} = \begin{pmatrix} \frac{2}{\tau} u_n \\ \mathbf{0} \\ \mathbf{0} \end{pmatrix}. \tag{11}$$

Note that Dirichlet boundary conditions on the continuous solution, u, gives rise to tangential boundary conditions on $\mathbf{V}$, $\mathbf{V} \times \mathbf{n} = \mathbf{0}$, where $\mathbf{n}$ is the normal vector to the boundary.

Next, we consider a finite-dimensional subspace of a product H^1 space, $\mathcal{V}^h$, and perform the FOSLS minimization of (11) over $\mathcal{V}^h$:

$$\left(u^h_{n+\frac{1}{2}}, \boldsymbol{V}^h_{n+\frac{1}{2}}\right) = \underset{(u,\boldsymbol{V})\in\mathcal{V}^h}{\arg\min} \left\| L_\tau \begin{pmatrix} u \\ \mathbf{V} \end{pmatrix} - \begin{pmatrix} \frac{2}{\tau} u^h_n \\ \mathbf{0} \\ \mathbf{0} \end{pmatrix} \right\|,$$

$$u^h_{n+1} = 2u^h_{n+\frac{1}{2}} - u^h_n.$$

For each n, the above minimization results in the following weak set of equations:

$$\left\langle L_\tau \begin{pmatrix} u^h_{n+\frac{1}{2}} \\ \mathbf{V}^h_{n+\frac{1}{2}} \end{pmatrix} - \begin{pmatrix} \frac{2}{\tau} u^h_n \\ \mathbf{0} \\ \mathbf{0} \end{pmatrix}, L_\tau \phi^h \right\rangle = 0 \quad \forall \phi^h \in \mathcal{V}^h, \tag{12}$$

where the inner products and norms are all in L^2 (scalar or vector, depending on context), unless otherwise noted.

Note, that with the introduction of $\mathbf{V}$, the discrete form of the FOSLS energy law can now be written,

$$\frac{||u^h_{n+1}||^2 - ||u^h_n||^2}{2\tau} - ||\mathbf{V}^h_{n+\frac{1}{2}}||^2 \to 0, \quad \text{as} \quad h \to 0. \tag{13}$$

The goal of the remainder of this Section is to show how well this energy law is satisfied. To do so, we make use of the following assumption.

Assumption 1. *Assume that the initial condition is smooth enough and the projection onto the finite-element space has the following property,*

$$\left\| u_0 - u^h_0 \right\|_{H^1} \leq Ch^p \left\| u_0 \right\|_{H^{p+1}},$$

where p is the order of the finite-element space being considered.

Then, using standard regularity estimates we obtain the following Lemma.

Lemma 1. *Let $\{u_i\}_{i=0,1,\ldots}$ be a sequence of semi-discrete solutions to* (9). *Then, for any successive time steps, there exists a constant $C > 0$, such that*

$$\left\| u_{n+1} \right\|_{H^p} \leq C \left\| u_n \right\|_{H^p}$$

A consequence of this regularity estimate is a bound on the error in the approximation.

Lemma 2. *Let $f \in H^p \cap H^1_0$ and let the pair $(u^h, \boldsymbol{V}^h) \in \mathcal{V}^h$ solve*

$$\begin{pmatrix} u^h \\ \boldsymbol{V}^h \end{pmatrix} = \underset{(u,\boldsymbol{V})\in\mathcal{V}^h}{\arg\min} \left\| L_\tau \begin{pmatrix} u \\ \boldsymbol{V} \end{pmatrix} - \begin{pmatrix} \frac{2}{\tau} f \\ \mathbf{0} \\ \mathbf{0} \end{pmatrix} \right\|^2.$$

Let $\hat{u}$ be the exact solution of the corresponding PDE, i.e.,

$$-\Delta \hat{u} + \frac{2}{\tau}\hat{u} = \frac{2}{\tau} f \quad \text{in } \partial\Omega,$$
$$\hat{u} = 0 \quad \text{on } \partial\Omega.$$

Then,

$$\left\|u^h - \hat{u}\right\|_{H^1} \leq \frac{C(\tau)h^p}{\tau} \left\|f\right\|_{H^{p-1}},$$

where the constant $C(\tau)$ may also depend on τ.

Proof. For a fixed τ, the PDE is a reaction-diffusion equation. Therefore, standard results from the FOSLS discretization of reaction-diffusion can be used [8,9]. Note that for a standard FOSLS approach, $C(\tau) = \mathcal{O}\left(\frac{1}{\tau^2}\right)$, but a rescaling of the equations may ameliorate this "worst-case scenario."

Next, we make the following observation, which follows from the well-posedness of the FOSLS formulation [8,9].

Lemma 3. *Let $(u_1, \boldsymbol{V}_1) \in \mathcal{V}^h$ and $(u_2, \boldsymbol{V}_2) \in \mathcal{V}^h$ be two solutions to the following FOSLS weak forms with different right-hand sides,*

$$\left\langle L_\tau \begin{pmatrix} u_1 \\ \boldsymbol{V}_1 \end{pmatrix} - F_1, L_\tau \boldsymbol{\phi}^h \right\rangle = 0, \quad \left\langle L_\tau \begin{pmatrix} u_2 \\ \boldsymbol{V}_2 \end{pmatrix} - F_2, L_\tau \boldsymbol{\phi}^h \right\rangle = 0 \quad \forall \boldsymbol{\phi}^h \in \mathcal{V}.$$

Then,

$$\left\|u_1 - u_2\right\|_{H^1} + \left\|\boldsymbol{V}_1 - \boldsymbol{V}_2\right\|_{H^1} \leq C(\tau) \left\|F_1 - F_2\right\|.$$

This, then, yields the following result.

Lemma 4. *Given the solution to the semi-discrete Eq. (9), and the fully discrete solution, we can bound the error in the L^2 norm:*

$$\left\|u^h_{n+\frac{1}{2}} - u_{n+\frac{1}{2}}\right\| \leq \frac{C_1(\tau)}{\tau} h^p \left\|u_n\right\|_{H^p} + C_2(\tau) \left\|u^h_n - u_n\right\|. \tag{14}$$

Proof. Let $\tilde{u}^h_{n+\frac{1}{2}}$ be the scalar part of the FOSLS solution $\left(\tilde{u}^h_{n+\frac{1}{2}}, \tilde{\boldsymbol{V}}^h_{n+\frac{1}{2}}\right)$ of

$$-\Delta u + \frac{2}{\tau} u = \frac{2}{\tau} u_n, \text{ in } \Omega, \quad u = 0 \text{ on } \partial\Omega, \tag{15}$$

where the exact semi-discrete solution u_n, at the previous time step, is used in the right-hand side. By the triangle inequality,

$$\left\|u^h_{n+\frac{1}{2}} - u_{n+\frac{1}{2}}\right\| \leq \left\|u^h_{n+\frac{1}{2}} - \tilde{u}^h_{n+\frac{1}{2}}\right\| + \left\|\tilde{u}^h_{n+\frac{1}{2}} - u_{n+\frac{1}{2}}\right\|. \tag{16}$$

By Lemma 3, we have

$$\begin{aligned} \left\|u^h_{n+\frac{1}{2}} - \tilde{u}^h_{n+\frac{1}{2}}\right\| &\leq \left\|u^h_{n+\frac{1}{2}} - \tilde{u}^h_{n+\frac{1}{2}}\right\|_{H^1} + \left\|\boldsymbol{V}^h_{n+\frac{1}{2}} - \tilde{\boldsymbol{V}}^h_{n+\frac{1}{2}}\right\|_{H^1} \\ &\leq C_2(\tau) \left\|u^h_n - u_n\right\|. \end{aligned} \tag{17}$$

The functions $\tilde{u}^h_{n+\frac{1}{2}}$ and $u_{n+\frac{1}{2}}$ are, respectively, FOSLS and exact solutions of the same boundary value problem (15). Hence, from Lemma 2, we have

$$\left\|\tilde{u}^h_{n+\frac{1}{2}} - u_{n+\frac{1}{2}}\right\| \leq \frac{C(\tau)}{\tau} h^p \left\|u_n\right\|_{H^{p-1}} \leq \frac{C(\tau)}{\tau} h^p \left\|u_n\right\|_{H^p} \tag{18}$$

Combining (16), (17) and (18), we obtain (14).

Finally, we have the following result on the approximation of the exact energy law (13).

Theorem 1. *Let* $\begin{pmatrix} u^h_n \\ \mathbf{V}^h_n \end{pmatrix}$ *be the solution to the FOSLS system,* (12), *at time step* n *(with* $u^h_{n+\frac{1}{2}}$ *and* $V^h_{n+\frac{1}{2}}$ *defined as before). There exists* $C(\tau) > 0$ *such that*

$$\left|\frac{\|u^h_{n+1}\|^2 - \|u^h_n\|^2}{2\tau} + \|\mathbf{V}^h_{n+\frac{1}{2}}\|^2\right| \leq C(\tau)\frac{2}{\tau}\left\|u^h_n - u_n\right\| \min_{\phi^h \in \mathcal{V}^h}\left\|\begin{pmatrix} u^h_{n+\frac{1}{2}} \\ \mathbf{V}^h_{n+\frac{1}{2}} \\ \mathbf{0} \end{pmatrix} - L_\tau \phi^h\right\|.$$

Proof. To simplify the notation, define the energy law we wish to bound as

$$E^h_n := \frac{\|u^h_{n+1}\|^2 - \|u^h_n\|^2}{2\tau} + \|\mathbf{V}^h_{n+\frac{1}{2}}\|^2.$$

Note that

$$\frac{1}{2}\frac{\left\|u^h_{n+1}\right\|^2 - \left\|u^h_n\right\|^2}{\tau} = \left\langle \frac{u^h_{n+1} - u^h_n}{\tau}, u^h_{n+\frac{1}{2}} \right\rangle = \left\langle \frac{u^h_{n+\frac{1}{2}} - u^h_n}{\frac{\tau}{2}}, u^h_{n+\frac{1}{2}} \right\rangle,$$

and

$$\left\|\mathbf{V}^h_{n+\frac{1}{2}}\right\|^2 = \left\langle -\nabla\cdot\mathbf{V}^h_{n+\frac{1}{2}}, u^h_{n+\frac{1}{2}}\right\rangle + \left\langle \mathbf{V}^h_{n+\frac{1}{2}} - \nabla u^h_{n+\frac{1}{2}}, \mathbf{V}^h_{n+\frac{1}{2}}\right\rangle,$$

where the latter equation is obtained by integration by parts, continuity of the spaces, and appropriate boundary conditions. Thus,

$$\begin{aligned} E^h_n &= \left\langle -\nabla\cdot\mathbf{V}^h_{n+\frac{1}{2}} + \frac{u^h_{n+\frac{1}{2}} - u^h_n}{\frac{\tau}{2}}, u^h_{n+\frac{1}{2}}\right\rangle + \left\langle \mathbf{V}^h_{n+\frac{1}{2}} - \nabla u^h_{n+\frac{1}{2}}, \mathbf{V}^h_{n+\frac{1}{2}}\right\rangle \\ &= \left\langle L_\tau \begin{pmatrix} u^h_{n+\frac{1}{2}} \\ \mathbf{V}^h_{n+\frac{1}{2}} \end{pmatrix} - \begin{pmatrix} \frac{2}{\tau}u^h_n \\ \mathbf{0} \\ 0 \end{pmatrix}, \begin{pmatrix} u^h_{n+\frac{1}{2}} \\ \mathbf{V}^h_{n+\frac{1}{2}} \\ \mathbf{0} \end{pmatrix}\right\rangle. \end{aligned}$$

Using (12), for any $\phi^h \in \mathcal{V}^h$,

$$E^h_n = \left\langle L_\tau \begin{pmatrix} u^h_{n+\frac{1}{2}} \\ \mathbf{V}^h_{n+\frac{1}{2}} \end{pmatrix} - \begin{pmatrix} \frac{2}{\tau}u^h_n \\ \mathbf{0} \\ 0 \end{pmatrix}, \begin{pmatrix} u^h_{n+\frac{1}{2}} \\ \mathbf{V}^h_{n+\frac{1}{2}} \\ \mathbf{0} \end{pmatrix} - L_\tau\phi^h\right\rangle.$$

Next, consider adding and subtracting the solutions to the semi-discrete, (11), and fully discrete, (12), FOSLS system from the previous time step,

$$
\begin{aligned}
E_n^h = &\left\langle L_\tau \begin{pmatrix} u_{n+\frac{1}{2}}^h \\ \boldsymbol{V}_{n+\frac{1}{2}}^h \end{pmatrix} - \begin{pmatrix} \frac{2}{\tau} u_n^h \\ \mathbf{0} \\ \mathbf{0} \end{pmatrix} + \frac{2}{\tau} \begin{pmatrix} u_n^h - u_n \\ \mathbf{0} \\ \mathbf{0} \end{pmatrix}, \begin{pmatrix} u_{n+\frac{1}{2}}^h \\ \boldsymbol{V}_{n+\frac{1}{2}}^h \\ \mathbf{0} \end{pmatrix} - L_\tau \boldsymbol{\phi}^h \right\rangle \\
&- \frac{2}{\tau} \left\langle \begin{pmatrix} u_n^h - u_n \\ \mathbf{0} \\ \mathbf{0} \end{pmatrix}, \begin{pmatrix} u_{n+\frac{1}{2}}^h \\ \boldsymbol{V}_{n+\frac{1}{2}}^h \\ \mathbf{0} \end{pmatrix} - L_\tau \boldsymbol{\phi}^h \right\rangle \\
= &\left\langle L_\tau \begin{pmatrix} u_{n+\frac{1}{2}}^h \\ \boldsymbol{V}_{n+\frac{1}{2}}^h \end{pmatrix} - \frac{2}{\tau} \begin{pmatrix} u_n \\ \mathbf{0} \\ \mathbf{0} \end{pmatrix}, \begin{pmatrix} u_{n+\frac{1}{2}}^h \\ \boldsymbol{V}_{n+\frac{1}{2}}^h \\ \mathbf{0} \end{pmatrix} - L_\tau \boldsymbol{\phi}^h \right\rangle \\
&- \frac{2}{\tau} \left\langle \begin{pmatrix} u_n^h - u_n \\ \mathbf{0} \\ \mathbf{0} \end{pmatrix}, \begin{pmatrix} u_{n+\frac{1}{2}}^h \\ \boldsymbol{V}_{n+\frac{1}{2}}^h \\ \mathbf{0} \end{pmatrix} - L_\tau \boldsymbol{\phi}^h \right\rangle \\
\leq &\left\| L_\tau \begin{pmatrix} u_{n+\frac{1}{2}}^h \\ \boldsymbol{V}_{n+\frac{1}{2}}^h \end{pmatrix} - \frac{2}{\tau} \begin{pmatrix} u_n \\ \mathbf{0} \\ \mathbf{0} \end{pmatrix} \right\| M_n^h + \frac{2}{\tau} M_n^h \left\| u_n^h - u_n \right\|,
\end{aligned}
$$

where we have defined $M_n^h := \min_{\boldsymbol{\phi}^h \in \mathcal{V}^h} \left\| \begin{pmatrix} u_{n+\frac{1}{2}}^h \\ \boldsymbol{V}_{n+\frac{1}{2}}^h \\ \mathbf{0} \end{pmatrix} - L_\tau \boldsymbol{\phi}^h \right\|$. Then, adding and subtracting $L_\tau \begin{pmatrix} u_{n+\frac{1}{2}} \\ \boldsymbol{V}_{n+\frac{1}{2}} \end{pmatrix}$ yields

$$
E_n^h \leq \left\| L_\tau \begin{pmatrix} u_{n+\frac{1}{2}}^h - u_{n+\frac{1}{2}} \\ \boldsymbol{V}_{n+\frac{1}{2}}^h - \boldsymbol{V}_{n+\frac{1}{2}} \end{pmatrix} + \underbrace{L_\tau \begin{pmatrix} u_{n+\frac{1}{2}} \\ \boldsymbol{V}_{n+\frac{1}{2}} \end{pmatrix} - \frac{2}{\tau} \begin{pmatrix} u_n \\ \mathbf{0} \\ \mathbf{0} \end{pmatrix}}_{\to 0} \right\| M_n^h + \frac{2}{\tau} M_n^h \left\| u_n^h - u_n \right\|.
$$

Using the continuity of L_τ, followed by Lemma 3, gives

$$
\begin{aligned}
|E_n^h| &\leq C(\tau) M_n^h \left\| \begin{pmatrix} u_{n+\frac{1}{2}}^h - u_{n+\frac{1}{2}} \\ \boldsymbol{V}_{n+\frac{1}{2}}^h - \boldsymbol{V}_{n+\frac{1}{2}} \end{pmatrix} \right\|_{H^1} + \frac{2}{\tau} M_n^h \left\| u_n^h - u_n \right\| \\
&\leq C(\tau) \frac{2}{\tau} M_n^h \left\| u_n^h - u_n \right\| + \frac{2}{\tau} M_n^h \left\| u_n^h - u_n \right\|.
\end{aligned}
$$

Combining the two terms completes the proof.

To provide a better bound for the FOSLS energy law (13), we introduce a measure for the truncation error defined as

$$
\delta_n = \max_{v \in H^{p+1}(\Omega)} \frac{1}{\|v\|_{H^{p+1}}} \min_{\boldsymbol{\phi}^h \in \mathcal{V}^h} \left\| \begin{pmatrix} u_{n+\frac{1}{2}}^h(v) \\ \boldsymbol{V}_{n+\frac{1}{2}}^h(v) \\ \mathbf{0} \end{pmatrix} - \mathcal{L} \boldsymbol{\phi}^h \right\|, \tag{19}
$$

where $u^h_{n+\frac{1}{2}}(v)$ and $\mathbf{V}^h_{n+\frac{1}{2}}(v)$ are the corresponding solutions to the fully discrete problem with $u_0 = v$ as the initial condition.

Corollary 1. *Using the same assumptions as Theorem 1 and Assumption 1,*

$$\left| \frac{\|u^h_{n+1}\|^2 - \|u^h_n\|^2}{2\tau} + \|\mathbf{V}^h_{n+\frac{1}{2}}\|^2 \right| \leq \frac{C(\tau)\delta}{\tau} h^p \|u_0\|^2_{H^{p+1}}, \quad \delta = \max_n \delta_n. \tag{20}$$

Proof. Using the definitions of $u^h_{n+\frac{1}{2}}$ and $u_{n+\frac{1}{2}}$, the triangle inequality, and Lemma 4,

$$\begin{aligned} \|u^h_{n+1} - u_{n+1}\| &\leq 2\|u^h_{n+\frac{1}{2}} - u_{n+\frac{1}{2}}\| + \|u^h_n - u_n\| \\ &\leq \frac{2}{\tau} C_1(\tau) h^p \|u_n\|_{H^p} + (C_2(\tau) + 1) \|u^h_n - u_n\|. \end{aligned}$$

An induction argument then gives

$$\|u^h_n - u_n\| \leq \frac{2}{\tau} C_1(\tau) h^p \sum_{j=1}^{n} (C_2(\tau) + 1)^{j-1} \|u_{n-j}\|_{H^p} + (C_2(\tau) + 1)^n \|u^h_0 - u_0\|.$$

With Assumption 1,

$$\|u^h_n - u_n\| \leq \frac{2}{\tau} C_1(\tau) h^p \sum_{j=1}^{n} (C_2(\tau) + 1)^{j-1} \|u_{n-j}\|_{H^p} + (C_2(\tau) + 1)^n \|u_0\|_{H^{p+1}}.$$

Using some regularity arguments for each u_i, we get,

$$\|u^h_n - u_n\| \leq C(n) h^p \|u_0\|_{H^{p+1}}.$$

Then, with the definition of δ and the result from Theorem 1, the proof is complete.

We note that the bound in Corollary 1 is a rather pessimistic one. At a fixed time, t, we expect the quality of both the fully discrete and semi-discrete approximations to the true solution to improve as $\tau \to 0$ and more time-steps are used to reach time t; thus, $\|u^h_n - u_n\|$ should decrease as $\tau \to 0$ for $n = t/\tau$. Furthermore, for the unforced heat equation, we expect both u^h_n and u_n to decrease in magnitude with n, but this is not accounted for in the bound in Corollary 1. The bound above worsens with smaller τ and bigger n, showing the limitations of bounding $\|u^h_n - u_n\|$ by terms depending only on u_0 and the finite-element space.

Remark 2. As shown in the numerical experiments, Sect. 5, the constant δ defined in (19) is of order h^p for a smooth solution. This indicates that the energy law (13) holds with order $\mathcal{O}\left(h^{2p}\right)$. While the theoretical justification of such statement may be plausible, it is nontrivial as the discrete quantities involved in the definition of δ do not possess enough regularity (they are just finite-element functions, only in H^1).

4 Stokes' Equations

Next, we return to the time-dependent Stokes' equations, (3)–(4). For simplicity, we again assume $\nu = 1$, and rewrite the equations using Dirichlet boundary conditions for the normal components of the velocity field, and zero-mean average for the pressure field,

$$\frac{\partial \boldsymbol{u}(\boldsymbol{x},t)}{\partial t} - \Delta \boldsymbol{u}(\boldsymbol{x},t) + \nabla p(\boldsymbol{x},t) = \mathbf{0} \quad \forall \boldsymbol{x} \in \Omega,\ \forall t > 0 \tag{21}$$

$$\nabla \cdot \boldsymbol{u}(\boldsymbol{x},t) = 0 \quad \forall \boldsymbol{x} \in \Omega,\ \forall t > 0 \tag{22}$$

$$\mathbf{n} \cdot \boldsymbol{u}(\boldsymbol{x},t) = 0 \quad \forall \boldsymbol{x} \in \partial\Omega,\ \forall t \geq 0 \tag{23}$$

$$\boldsymbol{u}(\boldsymbol{x},0) = \mathbf{g}(\boldsymbol{x}) \quad \forall \boldsymbol{x} \in \bar{\Omega}, \tag{24}$$

$$\int_\Omega p(\boldsymbol{x},t)\, \mathrm{d}V = 0 \quad \forall t \geq 0. \tag{25}$$

Using a similar semi-discretization in time with Crank-Nicolson that was done in (9) yields

$$\begin{aligned}
\frac{\boldsymbol{u}_{n+\frac{1}{2}} - \boldsymbol{u}_n}{\left(\frac{\tau}{2}\right)} - \Delta \boldsymbol{u}_{n+\frac{1}{2}} + \nabla p_{n+\frac{1}{2}} &= \mathbf{0}, \\
\nabla \cdot \boldsymbol{u}_{n+\frac{1}{2}} &= 0, \\
\mathbf{n} \cdot \boldsymbol{u}_{n+\frac{1}{2}}(\boldsymbol{x}) &= 0 \quad \forall \boldsymbol{x} \in \partial\Omega, \\
\int_\Omega p_{n+\frac{1}{2}}\, \mathrm{d}V &= 0 \quad \forall n \geq 0, \\
\boldsymbol{u}_{n+1} &= 2\boldsymbol{u}_{n+\frac{1}{2}} - \boldsymbol{u}_n, \\
p_{n+\frac{1}{2}} &= 2p_{n+\frac{1}{2}} - p_n.
\end{aligned} \tag{26}$$

To use the FOSLS method, we put the operator into a first-order system in a similar fashion to the heat equation. Least-squares formulations are well-studied for Stokes' system and we consider a simple, velocity-gradient-pressure formulation, where a new gradient tensor, $\mathbf{V} = \nabla \boldsymbol{u}$, is used to obtain an H^1-elliptic equivalent system [4,6,14]:

$$L_\tau \begin{pmatrix} \boldsymbol{u}_{n+\frac{1}{2}} \\ \boldsymbol{V}_{n+\frac{1}{2}} \\ p_{n+\frac{1}{2}} \end{pmatrix} = \begin{pmatrix} -\nabla \cdot \boldsymbol{V}_{n+\frac{1}{2}} + \nabla p_{n+\frac{1}{2}} + \frac{2}{\tau}\boldsymbol{u}_{n+\frac{1}{2}} \\ \nabla \cdot \boldsymbol{u}_{n+\frac{1}{2}} \\ \boldsymbol{V}_{n+\frac{1}{2}} - \nabla \boldsymbol{u}_{n+\frac{1}{2}} \\ \nabla \times \boldsymbol{V}_{n+\frac{1}{2}} \\ \nabla \mathbf{tr} \boldsymbol{V}_{n+\frac{1}{2}} \end{pmatrix} = \begin{pmatrix} \frac{2}{\tau}\boldsymbol{u}_n \\ 0 \\ \mathbf{0} \\ \mathbf{0} \\ \mathbf{0} \end{pmatrix}. \tag{27}$$

Appropriate boundary conditions on the continuous solution, such as $\mathbf{n} \cdot \boldsymbol{u} = 0$, gives rise to tangential boundary conditions on $\mathbf{V}$, $\mathbf{V} \times \mathbf{n} = \mathbf{0}$, where $\mathbf{n}$ is the normal vector to the boundary. Ultimately, the corresponding semi-discrete energy law is

$$\frac{||\boldsymbol{u}_{n+1}||^2 - ||\boldsymbol{u}_n||^2}{2\tau} = -||\mathbf{V}_{n+\frac{1}{2}}||^2. \tag{28}$$

Finally, we minimize the residual of (27) over a finite-dimensional subspace of the product H^1 Sobolev space in the L^2 norm obtaining the weak equations,

$$\left\langle L_\tau \begin{pmatrix} \boldsymbol{u}^h_{n+\frac{1}{2}} \\ \mathbf{V}^h_{n+\frac{1}{2}} \\ p^h_{n+\frac{1}{2}} \end{pmatrix} - \begin{pmatrix} \frac{2}{\tau}\boldsymbol{u}^h_n \\ 0 \\ \mathbf{0} \\ \mathbf{0} \\ 0 \end{pmatrix}, L_\tau \phi^h \right\rangle = 0 \quad \forall \phi^h \in \mathcal{V}^h. \tag{29}$$

Note that the weak system is similar to (12) and the energy law is identical to (13) in vector form. Thus, all the above theory still holds subject to enough regularity of the solution to the time-dependent Stokes' equations [21,22] and a suitable generalization of the definition of δ.

5 Numerical Experiments

For the numerical results presented here, we use a C++ implementation of the FOSLS algorithm, using the modular finite-element library MFEM [20] for managing the discretization, mesh, and timestepping. The linear systems are solved by direct method using the UMFPACK package [10].

5.1 Heat Equation

First, we consider the heat Eq. (6), and its discrete FOSLS formulation, (12), on a triangulation of $\Omega = (0,1) \times (0,1)$. The data is chosen so that the true solution is $u(x,y,t) = \sin(\pi x)\sin(\pi y)e^{-2\pi^2 t}$. Note that this solution satisfies the boundary conditions and other assumptions discussed above.

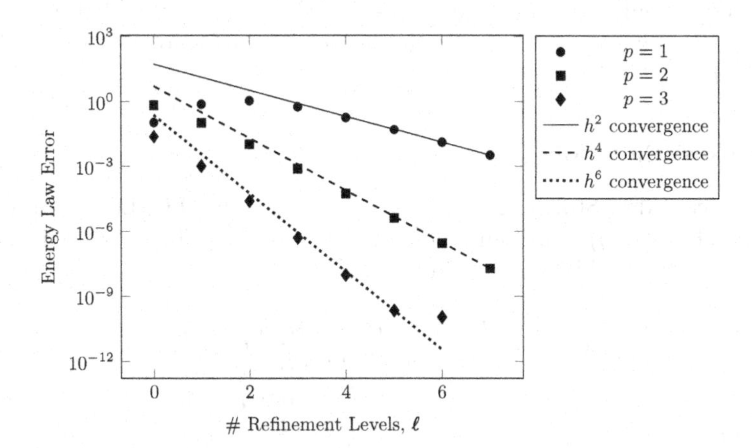

Fig. 1. Energy law error, (13), vs. number of mesh refinements, ℓ ($h = \frac{1}{2^\ell}$), for the FOSLS discretization of the heat equation, (12), using various orders of the finite-element space ($p = 1$ - linear; $p = 2$ - quadratic; and $p = 3$ - cubic). One time step is performed with $\tau = 0.005$.

Figure 1 displays the convergence of the energy law to zero as the mesh is refined for a fixed time step. The convergence is $\mathcal{O}\left(h^{2p}\right)$, where p is the order of the finite-element space being considered, confirming Theorem 1. It also suggests that the constant δ is $\mathcal{O}\left(h^{p}\right)$, as is remarked above.

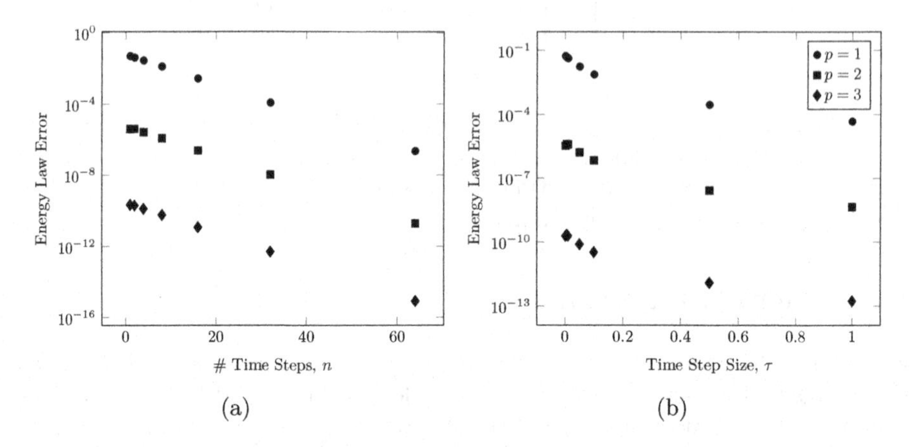

Fig. 2. Energy law error, (13), vs. (a) number of time steps, n (with fixed $\tau = 0.005$), and (b) time step size, τ, for the FOSLS discretization of the heat equation, (12), using various orders of the finite-element space ($p = 1$ - linear; $p = 2$ - quadratic; and $p = 3$ - cubic). Mesh spacing is $h = \frac{1}{32}$.

Figure 2 indicates how the timestepping affects the convergence of the energy law. As discussed above, taking more time steps decreases the error in the energy law, showing that we can improve the results on the bound, $\|u_n - u_n^h\|$. On the other hand, if only one time step is taken, the convergence slightly worsens for small τ, which is consistent with the constants found in Theorem 1 and Corollary 1.

5.2 Stokes' Equations

Next, we consider Stokes' Equations, (21), and the FOSLS discretization described above, (29). The same domain, $\Omega = (0,1) \times (0,1)$, is used, and we assume data that yields the exact solution,

$$\mathbf{u}(\mathbf{x},t) = \begin{pmatrix} \sin(\pi x)\cos(\pi y) \\ -\cos(\pi x)\sin(\pi y) \end{pmatrix} e^{-2\pi^2 t},$$

$$p(\mathbf{x},t) = 0.$$

This produces a C^{∞} solution that satisfies the appropriate boundary conditions and regularity arguments needed for the bounds on the energy law described above.

Similarly to the heat equation, Fig. 3 compares the convergence of the energy law to zero as the mesh is refined for a fixed time step. Again, we see that the convergence is $\mathcal{O}\left(h^{2p}\right)$, where p is the order of the finite-element space being considered, confirming that Theorem 1 can also be applied to the time-dependent Stokes' equations. Thus, the FOSLS discretization can adhere to the energy law for fluid-type systems, and has the potential for capturing the relevant physics of other complex fluids.

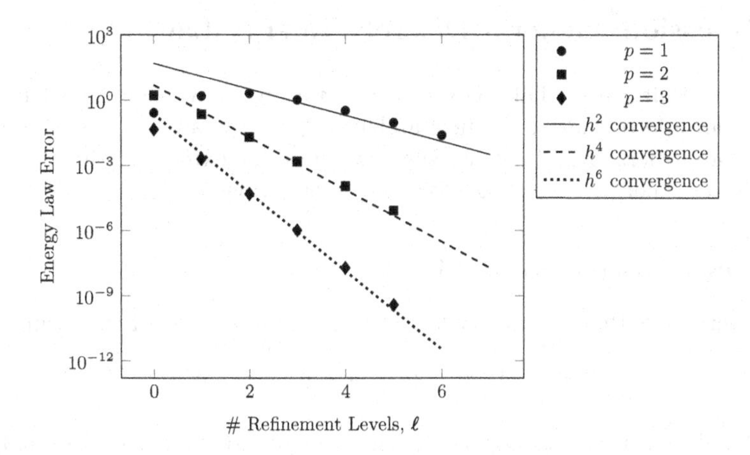

Fig. 3. Energy law error, (28), vs. number of mesh refinements, ℓ ($h = \frac{1}{2^\ell}$), for the FOSLS discretization of the Stokes' equation, (27), using various orders of the finite-element space ($p = 1$ - linear; $p = 2$ - quadratic; and $p = 3$ - cubic). One time step is performed with $\tau = 0.005$.

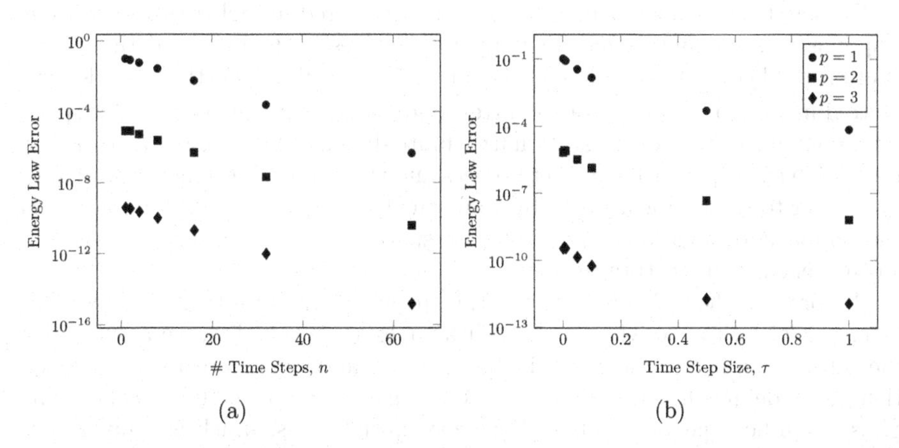

Fig. 4. Energy law error, (28), vs. (a) number of time steps, n (with fixed $\tau = 0.005$), and (b) time step size, τ, for the FOSLS discretization of the Stokes' equation, (27), using various orders of the finite-element space ($p = 1$ - linear; $p = 2$ - quadratic; and $p = 3$ - cubic). Mesh spacing is $h = \frac{1}{32}$.

Figure 4 again confirms how we expect the timestepping to affect the convergence of the energy law. Taking more time steps decreases the error in the energy law, while the convergence slightly worsens for small τ. These results also highlight the similarities between the energy laws of the heat equation and the time-dependent Stokes equations. Since both have underlying energy laws that are similar, the FOSLS discretization is capable of capturing both with high accuracy.

6 Discussion: General Discrete Energy Laws

The above results show that FOSLS discretizations of two specific PDEs yield higher-order approximation of their underlying energy laws. In this section, we give a more general result, which suggests ideas for extending this theory for other discrete energy laws using FOSLS discretizations.

6.1 FOSLS Discrete Energy Laws

As encountered earlier, an energy law is an integral relation of the form:

$$\langle \mathcal{L}u, u \rangle = 0, \quad \text{for} \quad u(x,0) = u_0(x), \tag{30}$$

where $\mathcal{L} : \widetilde{\mathcal{V}} \to \widetilde{\mathcal{V}}$ is a linear operator (that involves boundary conditions), $\widetilde{\mathcal{V}}$ is a function space, and $u \in \widetilde{\mathcal{V}}$ is the solution to

$$\mathcal{L}u = 0, \quad u(x,0) = u_0, \quad \text{for example:} \quad \mathcal{L} = \partial_t - \Delta. \tag{31}$$

To match the time-dependent problems considered in earlier sections, $\widetilde{\mathcal{V}}$ corresponds to a computational domain that involves both space and time, or as is often dubbed, a "space-time" domain: $\widetilde{\Omega} = \Omega \times [0,T]$. Further, we define a finite-dimensional space, $\widetilde{V}_h$ on $\widetilde{\Omega}$ corresponding to a triangulation of this space-time domain, as well as a "stationary" finite-dimensional space, V_h, for $t = 0$. Regarding such space-time discrete spaces and the related constructions, we refer the reader to the classical works by Johnson et al. [16,17], to [19] for space-time least squares formulations, and to [18] for space-time iso-geometric analysis and a comprehensive literature review.

To present the FOSLS discretization in an abstract setting, we define an extension of u_0 to the whole of $\widetilde{\Omega}$. Without loss of generality, we assume that the initial condition is a piecewise polynomial and, more precisely, $u_0 \in V_h$. Hence, we define the extension $w_h \in \widetilde{V}_h$ of u_0 so that $w_h(x,0) = u_0(x)$. This gives a non-homogenous problem with zero initial guess, which is equivalent to (31). Its weak form is: Find $u \in \widetilde{\mathcal{V}}$ such that for all $v \in \widetilde{\mathcal{V}}_o$ there holds

$$u = \varphi + w_h, \quad \text{where} \quad \langle \mathcal{L}\varphi, v \rangle = -\langle \mathcal{L}w_h, v \rangle, \tag{32}$$

Here, the space, $\widetilde{\mathcal{V}}_o$, is the subspace of $\widetilde{\mathcal{V}}$ of functions with vanishing trace at $t = 0$ (zero initial condition). In a typical FOSLS setting, for the heat equation, u is a vector-valued function and the extension w_h needs to be modified accordingly. We then have the following space-time FOSLS discrete problem: Find $u_h \in \widetilde{\mathcal{V}}_h$ such that for all $v_h \in \widetilde{\mathcal{V}}_{h,o}$ there holds

$$u_h = w_h + \varphi_h, \quad \text{where}, \quad \langle \mathcal{L}\varphi_h, \mathcal{L}v_h \rangle = -\langle \mathcal{L}w_h, \mathcal{L}v_h \rangle . \tag{33}$$

Restricting $\widetilde{\mathcal{V}}$ to a finite-element space-time space, $\widetilde{\mathcal{V}}_h \subset \widetilde{\mathcal{V}}$, results in a restriction of $\mathcal{L}$ on $\widetilde{\mathcal{V}}_h$, which is often called the "discrete operator".

In the following, we keep $\langle \mathcal{L}u, u \rangle$ in all estimates allowing for a non-homogenous right-hand side in (30). We now estimate the error in the energy law, namely the difference $\langle \mathcal{L}u, u \rangle - \langle \mathcal{L}u_h, u_h \rangle$.

Theorem 2. *If $u_h \in \widetilde{\mathcal{V}}_h$ is the FOSLS solution of* (33). *Then, the following estimate holds:*

$$|\langle \mathcal{L}u, u \rangle - \langle \mathcal{L}u_h, u_h \rangle| \le Ch^p \|u\|_{H^{p+1}}. \tag{34}$$

Proof. For the left side of (34) we have

$$\begin{aligned}\langle \mathcal{L}u, u \rangle - \langle \mathcal{L}u_h, u_h \rangle &= \langle \mathcal{L}u, u \rangle - \langle \mathcal{L}u, u_h \rangle + \langle \mathcal{L}(u - u_h), u_h \rangle \\ &= \langle \mathcal{L}u, u - u_h \rangle + \langle \mathcal{L}(u - u_h), u_h \rangle .\end{aligned}$$

Using the continuity of $\mathcal{L}$ and the standard error estimates for the FOSLS discretization,

$$\begin{aligned}|\langle \mathcal{L}u, u \rangle - \langle \mathcal{L}u_h, u_h \rangle| &\le |\langle \mathcal{L}u, u - u_h \rangle| + |\langle \mathcal{L}(u - u_h), u_h \rangle| \\ &\le C\|u - u_h\|_{H^1} (\|u\|_{H^1} + \|u_h\|_{H^1}) \\ &\le Ch^p \|u\|_{H^{p+1}}.\end{aligned}$$

This concludes the proof.

6.2 Exact Discrete Energy Law

Next, we provide a necessary and sufficient condition for the FOSLS discretization to *exactly* satisfy an energy law, namely conditions under which we have $\langle \mathcal{L}u_h, u_h \rangle = \langle \mathcal{L}u, u \rangle$. Recall the assumption that $u_0 \in \mathcal{V}_h$. Consider two standard projections on the finite-element space, $\widetilde{\mathcal{V}}_{h,o}$: (1) the Galerkin projection $\Pi_h : \widetilde{\mathcal{V}} \mapsto \widetilde{\mathcal{V}}_{h,o}$; and (2) the $L^2(\widetilde{\Omega})$-orthogonal projection, $Q_h : L^2(\widetilde{\Omega}) \mapsto \widetilde{\mathcal{V}}_{h,o}$. These operators are defined in a standard fashion:

$$\begin{aligned}\langle \mathcal{L}\Pi_h u, v_h \rangle &:= \langle \mathcal{L}u, v_h \rangle , \quad \text{for all } v_h \in \widetilde{\mathcal{V}}_{h,o} \text{ and } u \in \widetilde{\mathcal{V}}, \\ \langle Q_h u, v_h \rangle &:= \langle u, v_h \rangle , \quad \text{for all } v_h \in \widetilde{\mathcal{V}}_{h,o} \text{ and } u \in L^2(\widetilde{\Omega}).\end{aligned}$$

Consider a well-known identity (see for example [23] for the case of symmetric $\mathcal{L}$) relating Π_h and Q_h, which is used in the later proof of Theorem 3.

Lemma 5. *The projections Q_h and Π_h satisfy the relation*

$$\mathcal{L}_h \Pi_h = Q_h \mathcal{L}, \tag{35}$$

where $\mathcal{L}_h : \widetilde{\mathcal{V}}_h \mapsto \widetilde{\mathcal{V}}_h$ is the restriction of $\mathcal{L}$ on $\widetilde{\mathcal{V}}_h$, namely,

$$\langle \mathcal{L}_h v_h, w_h \rangle = \langle \mathcal{L} v_h, w_h \rangle, \quad \textit{for all} \quad v_h,\ w_h \in \widetilde{\mathcal{V}}_h.$$

Proof. The result easily follows from the definitions of Q_h, Π_h, $\mathcal{L}_h$, and the fact that $\mathcal{L}_h \Pi_h v \in \widetilde{\mathcal{V}}_h$. For $v \in \widetilde{\mathcal{V}}$, and $w \in \widetilde{\mathcal{V}}$ we have

$$\begin{aligned}\langle \mathcal{L}_h \Pi_h v, w \rangle &= \langle \mathcal{L}_h \Pi_h v, Q_h w \rangle = \langle \mathcal{L} \Pi_h v, Q_h w \rangle \\ &= \langle \mathcal{L} v, Q_h w \rangle = \langle Q_h \mathcal{L} v, Q_h w \rangle = \langle Q_h \mathcal{L} v, w \rangle .\end{aligned}$$

This completes the proof.

Note that we use $Q_h \chi_h = \Pi_h \chi_h = \chi_h$ for all $\chi_h \in \widetilde{\mathcal{V}}_{h,o}$. In general, such an identity is not true for $\chi_h \in \widetilde{\mathcal{V}}_h$. However, we can relate the solution to (33) to a discrete analogue of the energy law (30) using Lemma 5. Further, notice that the FOSLS solution, u_h, satisfies $\langle \mathcal{L} u_h, \mathcal{L} \chi_h \rangle = 0$ only for $\chi_h \in \widetilde{\mathcal{V}}_{h,o}$ corresponding to a zero initial guess. Thus, it is not obvious how to estimate $\langle \mathcal{L} u_h, u_h \rangle - \langle \mathcal{L} u, u \rangle$.

Theorem 3. *The solution u_h of (33) satisfies the discrete energy law $\langle \mathcal{L} u_h, u_h \rangle = \langle \mathcal{L} u, u \rangle$ if and only if there exists a $w_h \in \widetilde{\mathcal{V}}_h$ satisfying the initial condition $w_h(x,0) = u_0(x)$ and if $\langle \mathcal{L} u_h, w_h \rangle = \langle \mathcal{L} u, u \rangle$.*

Proof. Let $w_h \in \widetilde{\mathcal{V}}_h$ be any extension of $u_0 \in V_h$ in $\widetilde{\Omega}$, that is, w_h satisfies the initial condition. The following relations follow directly from the definitions given earlier, Eq. (33), and Lemma 5.

$$\begin{aligned}\langle \mathcal{L} u_h, u_h \rangle &= \Big\langle \mathcal{L} u_h, \underbrace{(u_h - w_h)}_{\in \widetilde{\mathcal{V}}_{h,o}} \Big\rangle + \langle \mathcal{L} u_h, w_h \rangle \\ &= \langle \mathcal{L} u_h, Q_h (u_h - w_h) \rangle + \langle \mathcal{L} u_h, w_h \rangle \\ &= \langle \mathcal{L} u_h, Q_h \mathcal{L} \mathcal{L}^{-1} (u_h - w_h) \rangle + \langle \mathcal{L} u_h, w_h \rangle \\ &= \Big\langle \mathcal{L} u_h, \mathcal{L} \underbrace{\Pi_h \mathcal{L}^{-1} (u_h - w_h)}_{v_h \in \widetilde{\mathcal{V}}_{h,o}} \Big\rangle + \langle \mathcal{L} u_h, w_h \rangle = \langle \mathcal{L} u_h, w_h \rangle .\end{aligned}$$

In the last identity, we use the fact that $v_h = \Pi_h \mathcal{L}^{-1} (u_h - w_h)$ is an element of $\widetilde{\mathcal{V}}_{h,o}$ and the first term on the right side vanishes (by Eq. (33)). As a result, we have

$$\langle \mathcal{L} u, u \rangle - \langle \mathcal{L} u_h, u_h \rangle = \langle \mathcal{L} u, u \rangle - \langle \mathcal{L} u_h, w_h \rangle .$$

which gives the desired necessary and sufficient condition.

From the proof, we immediately obtain the following relation,

$$|\langle \mathcal{L}u, u\rangle - \langle \mathcal{L}u_h, u_h\rangle| = \inf_{w_h} \{|\langle \mathcal{L}u, u\rangle - \langle \mathcal{L}u_h, w_h\rangle|,\ w_h(\cdot, 0) = u_0\}. \quad (36)$$

In addition to the estimate in Theorem 2, it is plausible that one can use the right side of (36) to obtain a sharper result. While this is beyond the scope of this paper, some comments are in order. The difficulties associated with each particular case in hand (heat equation, Stokes' equation, etc.) amount to estimating the quantity on the right side of (36) and such estimates depend on the spaces chosen for discretization and how well the timestepping approximates the space-time formulation. Sharper estimates on the error in discrete energy law, which uses (36), can lead to sharper bounds on the constant defined in (19).

7 Conclusions

In this work, we have shown numerically that convergence of the discrete energy law is of order higher than the finite-element approximation order for two typical transient problems. Thus, while it is known that the FOSLS method may have issues with adherence to some conservation laws (i.e., mass conservation), energy conservation is not such an issue, and can be satisfied with high accuracy. The rigorous theoretical justification of such claims are topics of current and future research.

Acknowledgements. The work of J. H. Adler was supported in part by NSF DMS-1216972. I. V. Lashuk was supported in part by NSF DMS-1216972 (Tufts University) and DMS-1418843 (Penn State). S. P. MacLachlan was partially supported by an NSERC Discovery Grant. The research of L. T. Zikatanov was supported in part by NSF DMS-1720114 and the Department of Mathematics at Tufts University.

References

1. Adler, J.H., Manteuffel, T.A., McCormick, S.F., Nolting, J.W., Ruge, J.W., Tang, L.: Efficiency based adaptive local refinement for first-order system least-squares formulations. SIAM J. Sci. Comput. **33**(1), 1–24 (2011). http://dx.doi.org/10.1137/100786897
2. Adler, J.H., Manteuffel, T.A., McCormick, S.F., Ruge, J.W.: First-order system least squares for incompressible resistive magnetohydrodynamics. SIAM J. Sci. Comput. **32**(1), 229–248 (2010). http://dx.doi.org/10.1137/080727282
3. Adler, J.H., Manteuffel, T.A., McCormick, S.F., Ruge, J.W., Sanders, G.D.: Nested iteration and first-order system least squares for incompressible, resistive magnetohydrodynamics. SIAM J. Sci. Comput. **32**(3), 1506–1526 (2010). http://dx.doi.org/10.1137/090766905
4. Bochev, P., Cai, Z., Manteuffel, T.A., McCormick, S.F.: Analysis of velocity-flux first-order system least-squares principles for the Navier-Stokes equations. I. SIAM J. Numer. Anal. **35**(3), 990–1009 (1998). http://dx.doi.org/10.1137/S0036142996313592
5. Bochev, P., Gunzburger, M.: Analysis of least-squares finite element mehtods for the Stokes equations. Math. Comput. **63**(208), 479–506 (1994)

6. Bochev, P., Manteuffel, T.A., McCormick, S.F.: Analysis of velocity-flux least-squares principles for the Navier-Stokes equations. II. SIAM J. Numer. Anal. **36**(4), 1125–1144 (1999). (Electronic). http://dx.doi.org/10.1137/S0036142997324976
7. Bramble, J.H., Kolev, T.V., Pasciak, J.: A least-squares approximation method for the time-harmonic Maxwell equations. J. Numer. Math. **13**, 237–263 (2005)
8. Cai, Z., Lazarov, R., Manteuffel, T.A., McCormick, S.F.: First-order system least squares for second-order partial differential equations. I. SIAM J. Numer. Anal. **31**(6), 1785–1799 (1994). http://dx.doi.org/10.1137/0731091
9. Cai, Z., Manteuffel, T.A., McCormick, S.F.: First-order system least squares for second-order partial differential equations. II. SIAM J. Numer. Anal. **34**(2), 425–454 (1997). http://dx.doi.org/10.1137/S0036142994266066
10. Davis, T.A.: Algorithm 832: Umfpack v4.3–an unsymmetric-pattern multifrontal method. ACM Trans. Math. Softw. **30**(2), 196–199 (2004). http://doi.acm.org/10.1145/992200.992206
11. Feng, J., Liu, C., Shen, J., Yue, P.: A energetic variational formulation with phase field methods for interfacial dynamics of complex fluids: advantages and challenges. In: Calderer, M.C.T., Terentjev, E.M. (eds.) Modeling of Soft Matter. The IMA Volumes in Mathematics and its Applications, vol. 141, pp. 1–26. Springer, New York (2005). https://doi.org/10.1007/0-387-32153-5_1
12. Gelfand, I.M., Fomin, S.V.: Calculus of Variations. Prentice-Hall Inc., Englewood Cliffs (1963). Revised English edition translated and edited by R.A. Silverman
13. Girault, V., Raviart, P.A.: Finite Element Approximation of the Navier-Stokes Equations. LNM, vol. 749. Springer, Berlin (1979). https://doi.org/10.1007/BFb0063447
14. Heys, J.J., Lee, E., Manteuffel, T.A., McCormick, S.F.: An alternative least-squares formulation of the Navier-Stokes equations with improved mass conservation. J. Comput. Phys. **226**(1), 994–1006 (2007). http://dx.doi.org/10.1016/j.jcp.2007.05.005
15. Hyon, Y., Kwak, D.Y., Liu, C.: Energetic variational approach in complex fluids: maximum dissipation principle. Discret. Contin. Dyn. Syst. **26**(4), 1291–1304 (2010). http://dx.doi.org/10.3934/dcds.2010.26.1291
16. Johnson, C.: Numerical Solution of Partial Differential Equations by the Finite Element Method. Dover Publications Inc., Mineola (2009). Reprint of the 1987 edition
17. Johnson, C., Nävert, U., Pitkäranta, J.: Finite element methods for linear hyperbolic problems. Comput. Methods Appl. Mech. Eng. **45**(1–3), 285–312 (1984). http://dx.doi.org/10.1016/0045-7825(84)90158-0
18. Langer, U., Moore, S.E., Neumüller, M.: Space-time isogeometric analysis of parabolic evolution problems. Comput. Methods Appl. Mech. Eng. **306**, 342–363 (2016). http://dx.doi.org/10.1016/j.cma.2016.03.042
19. Masud, A., Hughes, T.J.R.: A space-time Galerkin/least-squares finite element formulation of the Navier-Stokes equations for moving domain problems. Comput. Methods Appl. Mech. Eng. **146**(1–2), 91–126 (1997). http://dx.doi.org/10.1016/S0045-7825(96)01222-4
20. MFEM: Modular finite element methods library (2016). http://mfem.org
21. Solonnikov, V.A.: Estimates for solutions of a non-stationary linearized system of Navier-Stokes equations. Trudy Mat. Inst. Steklov. **70**, 213–317 (1964)
22. Solonnikov, V.A.: On boundary value problems for linear parabolic systems of differential equations of general form. Trudy Mat. Inst. Steklov. **83**, 3–163 (1965)
23. Xu, J.: Iterative methods by space decomposition and subspace correction. SIAM Rev. **34**(4), 581–613 (1992). http://dx.doi.org/10.1137/1034116

Multipatch Space-Time Isogeometric Analysis of Parabolic Diffusion Problems

U. Langer[1,2(✉)], M. Neumüller[1], and I. Toulopoulos[2]

[1] Institute of Computational Mathematics, Johannes Kepler University Linz, Altenberger Str. 69, 4040 Linz, Austria
{ulanger,neumueller}@numa.uni-linz.ac.at

[2] RICAM, Austrian Academy of Sciences, Altenberger Str. 69, 4040 Linz, Austria
{ulrich.langer,ioannis.toulopoulos}@ricam.oeaw.ac.at

Abstract. We present and analyze a new stable multi-patch space-time Isogeometric Analyis (IgA) method for the numerical solution of parabolic diffusion problems. The discrete bilinear form is elliptic on the IgA space with respect to a mesh-dependent energy norm. This property together with a corresponding boundedness property, consistency and approximation results for the IgA spaces yields a priori discretization error estimates. We propose an efficient implementation technique via tensor product representation, and fast space-time parallel solvers. We present numerical results confirming the efficiency of the space-time solvers on massively parallel computers using more than 100.000 cores.

Keywords: Parabolic initial-boundary value problems
Space-time isogeometric analysis
A priori discretization error estimates · Parallel solvers

1 Introduction

The standard discretization methods for parabolic initial-boundary value problems (IBVP) are based on a separation of the discretizations in space and time, i.e., first space, then time, or, vice versa, first time, then space. The former one is called vertical method of lines, whereas the latter one is called horizontal method of lines or Rothe's method. Both methods use some kind of time-stepping method for time discretization. This is a sequential procedure that needs some smart ideas for the parallelization with respect to time, see [4] for a historical overview of time-parallel methods. Other disadvantages of these approaches are connected with a separation of adaptivity with respect to space and time, and with difficulties in the numerical treatment of moving interfaces and spatial domains. To overcome this curse of sequentiality of time-stepping methods, one should look at the time variable t as just another variable, say, x_{d+1} if $x_1, \dots, x_d$ are the spatial variable, and at the time derivative as a strong convection in the direction x_{d+1}. In [10], we were inspired by this view at parabolic problems, and proposed upwind-stabilized single-patch space-time IgA schemes for parabolic

I. Lirkov and S. Margenov (Eds.): LSSC 2017, LNCS 10665, pp. 21–32, 2018.
https://doi.org/10.1007/978-3-319-73441-5_2

evolution problems. For comprehensive overview on the literature on different space-time methods for solving parabolic IPVP, we also refer to [10].

In this paper, we generalize the results of [10] from the single-patch to the time dG multi-patch IgA case. As in [10], we consider the linear parabolic IBVP: find $u : \overline{Q} \to \mathbb{R}$ such that

$$\partial_t u - \Delta u = f \text{ in } Q,\ u = 0 \text{ on } \Sigma, \text{ and } u = u_0 \text{ on } \Sigma_0, \tag{1}$$

as a typical model problem posed in the space-time cylinder $\overline{Q} = \overline{\Omega} \times [0, T] = Q \cup \Sigma \cup \overline{\Sigma}_0 \cup \overline{\Sigma}_T$ where ∂_t denotes the partial time derivative, Δ is the Laplace operator, f is a given source function, u_0 are the given initial data, T is the final time, $Q = \Omega \times (0, T)$, $\Sigma = \partial\Omega \times (0, T)$, $\Sigma_0 := \Omega \times \{0\}$, $\Sigma_T := \Omega \times \{T\}$, and $\Omega \subset \mathbb{R}^d$ $(d = 1, 2, 3)$ denotes the spatial computational domain with the boundary $\partial\Omega$. The spatial domain Ω is supposed to be bounded and Lipschitz. Later we will assume that Ω has a single- or multipatch NURBS representation as is used in CAD respectively IgA.

2 Space-Time Variational Formulation

Using the standard procedure and integration by parts with respect to both x and t, we can easily derive the following space-time variational formulation of (1): find $u \in H_0^{1,0}(Q) = \{u \in L_2(Q) : \nabla_x u \in [L_2(Q)]^d, u = 0 \text{ on } \Sigma\}$ such that

$$a(u, v) = l(v), \quad \forall v \in H^{1,1}_{0,\overline{0}}(Q), \tag{2}$$

with the bilinear form

$$a(u, v) = -\int_Q u(x, t)\partial_t v(x, t)\, dx\, dt + \int_Q \nabla_x u(x, t) \cdot \nabla_x v(x, t)\, dx\, dt \tag{3}$$

and the linear form

$$l(v) = \int_Q f(x, t)v(x, t)\, dx\, dt + \int_\Omega u_0(x)v(x, 0)\, dx, \tag{4}$$

where $H^{1,1}_{0,\overline{0}}(Q) = \{u \in L_2(Q) : \nabla_x u \in [L_2(Q)]^d, \partial_t u \in L_2(Q), u = 0 \text{ on } \Sigma,$ and $u = 0$ on $\Sigma_T\}$. The space-time variational formulation (2) has a unique solution, see, e.g., [8,9]. In these monographs, beside existence and uniqueness results, one can also find useful a priori estimates and regularity results. For simplicity, we below assume that $u_0 = 0$.

3 Stable Multi-patch Space-Time IgA Discretization

Let us now assume that the space-time cylinder $\overline{Q} = \cup_{n=1}^N \overline{Q}_n$ consists of N subcylinders (patches or time slices) $Q_n = \Omega \times (t_{n-1}, t_n)$, $n = 1, \ldots, N$, where $0 = t_0 < t_1 < \ldots < t_N = T$ is some subdivision of time interval $[0, T]$. The time faces between the time patches are denoted by $\overline{\Sigma}_n = \overline{Q}_{n+1} \cap \overline{Q}_n = \overline{\Omega} \times \{t_n\}$. We obviously have $\Sigma_N = \Sigma_T$. Every space-time patch $Q_n = \mathbf{\Phi}_n(\widehat{Q})$ in the physical domain Q can be represented as the image of the parameter domain $\widehat{Q} = (0,1)^{d+1}$ by means of a sufficiently regular IgA (B-Spline, NURBS etc.) map $\mathbf{\Phi}_n : \widehat{Q} \to Q_n$, i.e.,

$$\mathbf{\Phi}_n(\xi) = \sum_{i \in \mathcal{I}_n} \mathbf{P}_{n,i} \widehat{\varphi}_{n,i}(\xi), \tag{5}$$

where $\{\widehat{\varphi}_{n,i}\}_{i \in \mathcal{I}_n}$ are the IgA basis functions, and $\{\mathbf{P}_{n,i}\}_{i \in \mathcal{I}_n} \subset \mathbb{R}^{d+1}$ are the control points for the patch Q_n. The IgA basis functions are usually multivariant B-Splines or NURBS defined on a mesh given by the knot vector wrt to each direction in the parameter domain $\widehat{Q}$, and the underlying polynomial degrees and multiplicities of the knots defining the smoothnesses of the basis functions, see, e.g., [2] or [11] for more detailed information.

Now, we can construct our finite-dimensional IgA (B-Spline, NURBS etc.) space $V_{0h} = \{v_h : v_n = v_h|_{Q_n} \in V_{0n}, n = 1, \ldots, N\}$, the functions of which are smooth in each time patch Q_n in correspondence to the smoothness of the splines, but in general discontinuous across the time faces Σ_n, $n = 1, \ldots, N-1$. The smooth IgA spaces $V_{0n} = V_{0h_n} = \mathrm{span}\{\varphi_{n,i}\}_{i \in \mathcal{I}_n} \subset H_0^{1,1}(Q_n)$ are spanned by IgA basis functions $\{\varphi_{n,i}\}_{i \in \mathcal{I}_n}$ that are nothing but the images of the basis functions $\{\widehat{\varphi}_{n,i}\}_{i \in \mathcal{I}_n}$, which were already used for defining the patch Q_n, by the map $\mathbf{\Phi}_n$, i.e., $\varphi_{n,i} = \widehat{\varphi}_{n,i} \circ \mathbf{\Phi}_n^{-1}$. The basis functions $\varphi_{1,i}$ should vanish on Σ_0 for all $i \in \mathcal{I}_1$. Therefore, all functions v_h from V_{0h} fulfil homogeneous boundary and initial conditions. The discretization parameter h_n denotes the average mesh-size of the mesh induced by the corresponding mesh in the parameter domain $\widehat{Q}$ via the map $\mathbf{\Phi}_n$. The IgA technology of using the same basis functions for describing the patches of the computational domain (geometry) and for defining the approximation spaces V_{0h} was introduced by Hughes, Cottrell and Bazilevs in 2005 [7] and analyzed in [1], see also monograph [2] for more comprehensive information.

In order to derive our dG IgA scheme for defining the IgA solution $u_h \in V_{0h}$, we multiply the parabolic PDE (1) by a time-upwind test function of the form $v_n + \theta_n h_n \partial_t v_n$ with an arbitrary $v_n \in V_{0n}$ and a positive, sufficiently small constant θ_n, and integrate over the space-time subcylinder Q_n. After integration by parts wrt x, we get

$$\begin{aligned} &\int_{Q_n} (\partial_t u(v_n + \theta_n h_n \partial_t v_n) + \nabla_x u \cdot \nabla_x v_n + \theta_n h_n \nabla_x u \cdot \nabla_x \partial_t v_n)\, dxdt \\ &\quad - \int_{\partial Q_n} \mathbf{n}_x \cdot \nabla_x u\, (v_n + \theta_n h_n \partial_t v_n)\, ds = \int_{Q_n} f(v_n + \theta_n h_n \partial_t v_n)\, dxdt. \end{aligned} \tag{6}$$

We mention that $\partial_t v_n$ is differentiable wrt x due to the special tensor product structure of V_{0n}. Using the facts that v_n and $\partial_t v_n$ are always zero on Σ, and the x-components $\mathbf{n}_x = (n_1, \ldots, n_d)^\top$ of the normal $\mathbf{n} = (n_1, \ldots, n_d, n_{d+1})^\top = (\mathbf{n}_x, n_t)^\top$ are zero on Σ_{n-1} and Σ_n, we observe that the integral over ∂Q_n is always zero. Now, adding to the left-hand side of (6) a consistent time-upwind term for stabilization, and summing over all time patches, we get the identity

$$\sum_{n=1}^{N} \int_{Q_n} (\partial_t u(v_n + \theta_n h_n \partial_t v_n) + \nabla_x u \cdot \nabla_x v_n + \theta_n h_n \nabla_x u \cdot \nabla_x \partial_t v_h)\, dxdt + \sum_{n=1}^{N} \int_{\Sigma_{n-1}} [|u|]\, v_n \, dx = \sum_{n=1}^{N} \int_{Q_n} f(v_n + \theta_n h_n \partial_t v_n)\, dxdt \tag{7}$$

that holds for a sufficiently smooth solution u of our parabolic IBVP, where $[|u|] := u|_{Q_n} - u|_{Q_{n-1}}$ on Σ_{n-1} denotes the jump of u across Σ_{n-1} that is obviously zero.

The time multipach space-time IgA scheme for solving the parabolic IBVP (1) respectively (2) can now be formulated as follows: find $u_h \in V_{0h}$ such that

$$a_h(u_h, v_h) = l_h(v_h), \quad \forall v_h \in V_{0h}, \tag{8}$$

where

$$a_h(u_h, v_h) = \sum_{n=1}^{N} a_n(u_h, v_h) = \sum_{n=1}^{N} \Big(\int_{Q_n} (\partial_t u_n (v_n + \theta_n h_n \partial_t v_n) + \nabla_x u_n \cdot \nabla_x v_n + \theta_n h_n \nabla_x u_n \cdot \nabla_x \partial_t v_n)\, dxdt + \int_{\Sigma_{n-1}} [|u_h|]\, v_n \, dx \Big), \tag{9}$$

$$l_h(v_h) = \sum_{n=1}^{N} l_n(v_h) = \sum_{n=1}^{N} \int_{Q_n} f(v_n + \theta_n h_n \partial_t v_n)\, dxdt. \tag{10}$$

Here and below we formally set $[|u_1|]$ on Σ_0 to zero since we assumed homogeneous initial conditions. It is clear that this jump term can be used to include inhomogeneous initial conditions in a weak sense. In this case, the test functions v_n are not forced to be zero on Σ_0. The derivation of the IgA scheme given above immediately yields that this scheme is consistent for sufficiently smooth solution, cf. identity (7). Indeed, if the solution $u \in H_0^{1,0}(Q)$ of (2) belongs to $H_0^{1,1}(Q)$, then it satisfies the consistency identity

$$a_h(u, v_h) = l_h(v_h), \quad \forall v_h \in V_{0h}, \tag{11}$$

yielding Galerkin orthogonality

$$a_h(u - u_h, v_h) = 0, \quad \forall v_h \in V_{0h}. \tag{12}$$

Now we will show that the bilinear form $a_h(\cdot,\cdot)$ is V_{0h}-elliptic wrt the norm $\|v\|_h$ defined by

$$\|v\|_h^2 = \sum_{n=1}^{N} \Big(\frac{1}{2} \|\nabla_x v\|^2_{L_2(Q_n)} + \theta_n\, h_n\, \|\partial_t v\|^2_{L_2(Q_n)} + \frac{1}{2} \|\, [|v|]\, \|^2_{L_2(\Sigma_{n-1})} \Big) + \frac{1}{2} \|v\|^2_{L_2(\Sigma_N)}.$$

In order to show the V_{0h}-ellipticity of $a_h(\cdot,\cdot)$, we need the inverse inequality

$$\|\nabla_x v_n\|^2_{L_2(\Sigma_{n-1})} \leq c^2_{inv,0} h_n^{-1} \|\nabla_x v_n\|^2_{L_2(Q_n)} \tag{13}$$

that is valid for all $v_n \in V_{0n}$ and $n = 1, \ldots, N$, see [1,3].

Lemma 1. *The bilinear form $a_h(\cdot,\cdot)$ defined by (9) is V_{0h}-elliptic, i.e., there exist a generic positive constant μ_e such that*

$$a_h(v_h, v_h) \geq \mu_e \|v_h\|_h^2, \quad \forall v_h \in V_{0h}. \tag{14}$$

provided that the parameters θ_n are sufficiently small. More precisely, $\mu_e = 1$ if $0 < \theta_n \leq c_{inv,0}^{-2}$ for all $n = 1, 2, \ldots, N$, where $c_{inv,0}$ is the constant from the inverse inequality (13).

Proof. Using integration by parts with respect to t and the inverse inequality (13), we can derive the following estimates:

$$\begin{aligned}
a_n(v_h, v_h) &= \int_{Q_n} \left(\frac{1}{2}\partial_t v_n^2 + \theta_n h_n (\partial_t v_n)^2 + |\nabla_x v_n|^2 + \frac{\theta_n h_n}{2} \partial_t |\nabla_x v_n|^2 \right) dx\,dt \\
&\quad + \int_{\Sigma_{n-1}} [|v_h|]\, v_n \, dx \\
&= \frac{1}{2} \int_{\Sigma_n} v_n^2 dx - \frac{1}{2} \int_{\Sigma_{n-1}} v_n^2 dx + \int_{Q_n} \left(\theta_n h_n (\partial_t v_n)^2 + |\nabla_x v_n|^2 \right) dx\,dt \\
&\quad + \frac{\theta_n h_n}{2} \int_{\Sigma_n} |\nabla_x v_n|^2 dx - \frac{\theta_n h_n}{2} \int_{\Sigma_{n-1}} |\nabla_x v_n|^2 dx + \int_{\Sigma_{n-1}} [|v_h|]\, v_n \, dx \\
&\geq \theta_n h_n \|\partial_t v_n\|^2_{L_2(Q_n)} + \|\nabla_x v_n\|^2_{L_2(Q_n)} - 0.5\, \theta_n h_n \|\nabla_x v_n\|^2_{L_2(\Sigma_{n-1})} \\
&\quad + \frac{1}{2} \int_{\Sigma_n} v_n^2 dx - \frac{1}{2} \int_{\Sigma_{n-1}} v_n^2 dx + \int_{\Sigma_{n-1}} v_n^2 \, dx - \int_{\Sigma_{n-1}} v_{n-1} v_n \, dx \\
&\geq \theta_n h_n \|\partial_t v_n\|^2_{L_2(Q_n)} + (1 - 0.5\, \theta_n c^2_{inv,0}) \|\nabla_x v_n\|^2_{L_2(Q_n)} \\
&\quad + \int_{\Sigma_{n-1}} \left(\frac{1}{2} v_n^2 - v_{n-1} v_n \right) dx + \frac{1}{2} \int_{\Sigma_n} v_n^2 dx
\end{aligned}$$

Summing over all $n = 1, \ldots, N$, we obtain

$$\begin{aligned}
a_h(v_h, v_h) &= \sum_{n=1}^N a_n(v_h, v_h) \\
&\geq \sum_{n=1}^N \theta_n h_n \|\partial_t v_h\|^2_{L_2(Q_n)} + \left(1 - 0.5\, \theta_n\, c^2_{inv,0}\right) \|\nabla_x v_h\|^2_{L_2(Q_n)} \\
&\quad + \sum_{n=1}^N \frac{1}{2} \|\, [|v_h|]\, \|^2_{L_2(\Sigma_{n-1})} + \frac{1}{2} \|v_N\|^2_{L_2(\Sigma_N)}.
\end{aligned}$$

Choosing $0 < \theta_n \leq c_{inv,0}^{-2}$ for all $n = 1, 2, \ldots, N$, we immediately arrive at (14) with $\mu_e = 1$. ■

Lemma 1 immediately implies that the solution $u_h \in V_{0h}$ of (8) is unique. Since the IgA scheme (8) is posed in the finite dimensional space V_{0h}, the uniqueness yields existence of the solution $u_h \in V_{0h}$ of (8).

Once the basis is chosen, the IgA scheme (8) can be rewritten as a huge linear system of algebraic equations of the form

$$\mathbf{L}_h \mathbf{u}_h = \mathbf{f}_h \tag{15}$$

for determining the vector $\mathbf{u}_h = ((u_{1,i})_{i \in \mathcal{I}_1}, \ldots, (u_{N,i})_{i \in \mathcal{I}_N}) \in \mathbb{R}^{N_h}$ of the control points of the IgA solution

$$u_h(x,t) = \sum_{i \in \mathcal{I}_n} u_{n,i} \varphi_{n,i}(x,t), \ (x,t) \in \overline{Q}_n, \ n = 1, \ldots, N, \tag{16}$$

solving the IgA scheme (8). The system matrix $\mathbf{L}_h$ is the usual Galerkin (stiffness) matrix, and $\mathbf{f}_h$ is the corresponding right-hand side (load) vector.

4 A Priori Discretization Error Estimates

In order to derive a priori discretization error estimates, we will first show that the IgA bilinear form $a_h(\cdot,\cdot)$ is bounded on $V_{0h,*} \times V_{0h}$, where the space $V_{0h,*} = V + V_{0h}$ is equipped with the norm $\|\cdot\|_{h,*}$ defined by the relation

$$\|v\|_{h,*}^2 = \|v\|_h^2 + \sum_{n=1}^{N} (\theta_n h_n)^{-1} \|v\|_{L_2(Q_n)}^2 + \sum_{n=2}^{N} \|v|_{Q_n}\|_{L_2(\Sigma_{n-1})}^2, \tag{17}$$

and V is a suitable infinite-dimensional space containing the solution u, e.g., we can choose $V = H_{0,\underline{0}}^{1,1}(Q) = \{u \in L_2(Q) : \nabla_x u \in [L_2(Q)]^d, \partial_t u \in L_2(Q), u = 0 \text{ on } \Sigma, \text{ and } u = 0 \text{ on } \Sigma_0\}$ assuming that the solution u belongs to this space. In order to prove the boundedness of $a_h(\cdot,\cdot)$, we need the inverse inequality

$$\|\nabla_x \partial_t v_n\|_{L_2(Q_n)}^2 \leq c_{inv,1}^2 h_n^{-2} \|\nabla_x v_n\|_{L_2(Q_n)}^2 \tag{18}$$

that is valid for all $v_n \in V_{0n}$ and $n = 1, \ldots, N$, see [1,3].

Lemma 2. *The bilinear form $a_h(\cdot,\cdot)$ defined by (9) is bounded on the space $V_{0h,*} \times V_{0h}$, i.e., there exists a generic positive constant μ_b such that*

$$|a_h(u, v_h)| \leq \mu_b \|u\|_{h,*} \|v_h\|_h, \quad \forall u \in V_{0h,*}, \forall v_h \in V_{0h}. \tag{19}$$

with the boundedness constant $\mu_b = 2 \max\{\sqrt{1 + c_{inv,1}/c_{inv,0}^2}, \sqrt{2}\}$, where $c_{inv,0}$ and $c_{inv,1}$ are the constants from inequalities (13) and (18). We always assume that the parameters θ_n are chosen as in Lemma 1.

Proof. For the first and the interface jump terms of a_h, we use Green's formula and the Cauchy inequality to derive the following estimates:

$$\begin{aligned}
&\sum_{n=1}^{N}\left(\int_{Q_n} \partial_t u\, v_h \,dx\,dt + \int_{\Sigma_{n-1}} [|u|]\ v_n\, ds\right) \\
&= \sum_{n=1}^{N}\left(-\int_{Q_n} u\,\partial_t v_h \,dx\,dt + \int_{\Sigma_n} u\,v_n\,ds - \int_{\Sigma_{n-1}} u\,v_n\,ds + \int_{\Sigma_{n-1}} [|u|]\ v_n\,ds\right) \\
&\le \left(\sum_{n=1}^{N}(\theta_n h_n)^{-1}\Big(\int_{Q_n} u^2\,dx\,dt\Big)^2\right)^{\frac{1}{2}} \left(\sum_{n=1}^{N}\theta_n h_n\Big(\int_{Q_n}(\partial_t v_n)^2\,dx\,dt\Big)^2\right)^{\frac{1}{2}} \\
&\quad + \sum_{n=1}^{N}\int_{\Sigma_{n-1}}(v_{n-1}-v_n)u\,ds + \int_{\Sigma_N} v_n u\,ds \\
&\le \left(\sum_{n=1}^{N}(\theta_n h_n)^{-1}\Big(\int_{Q_n} u^2\,dx\,dt\Big)^2\right)^{\frac{1}{2}} \left(\sum_{n=1}^{N}\theta_n h_n\Big(\int_{Q_n}(\partial_t v_n)^2\,dx\,dt\Big)^2\right)^{\frac{1}{2}} \\
&\quad + \sqrt{2}\Big(\frac{1}{2}\sum_{n=1}^{N}[|v_h|]^2_{L_2(\Sigma_{n-1})} + \frac{1}{2}\|v_N\|^2_{L_2(\Sigma_N)}\Big)^{\frac{1}{2}} \sqrt{2}\Big(\sum_{n=1}^{N}\|u\|^2_{L_2(\Sigma_{n-1})} + \frac{1}{2}\|u\|^2_{L_2(\Sigma_N)}\Big)^{\frac{1}{2}}.
\end{aligned}$$

Using again Cauchy's inequality, we get the estimates

$$\begin{aligned}
&\sum_{n=1}^{N}\int_{Q_n}(\theta_n h_n)^{\frac{1}{2}}\partial_t u\,(\theta_n h_n)^{\frac{1}{2}}\partial_t v_n\,dx\,dt + \sum_{n=1}^{N}\int_{Q_n}\nabla_x u\cdot\nabla_x v_n\,dx\,dt \\
&\le \Big(\sum_{n=1}^{N}\theta_n h_n\|\partial_t u\|^2_{L_2(Q_n)}\Big)^{\frac{1}{2}}\Big(\sum_{n=1}^{N}\theta_n h_n\|\partial_t v_n\|^2_{L_2(Q_n)}\Big)^{\frac{1}{2}} \\
&\quad + \sqrt{2}\Big(\frac{1}{2}\sum_{n=1}^{N}\|\nabla_x u\|^2_{L_2(Q_n)}\Big)^{\frac{1}{2}}\sqrt{2}\Big(\frac{1}{2}\sum_{n=1}^{N}\|\nabla_x v_n\|^2_{L_2(Q_n)}\Big)^{\frac{1}{2}}
\end{aligned}$$

for the second and third terms. Finally, for the last but one term, we apply Cauchy's and inverse inequalities to show

$$\begin{aligned}
&\sum_{n=1}^{N}\int_{Q_n}\nabla_x u\cdot(\theta_n h)\nabla_x\partial_t v_n\,dx\,dt \\
&\le \left(\sum_{n=1}^{N}\|\nabla_x u\|^2_{L_2(Q_n)}\right)^{\frac{1}{2}}\left(\sum_{n=1}^{N}(\theta_n h_n)^2\|\nabla_x\partial_t v_n\|^2_{L_2(Q_n)}\right)^{\frac{1}{2}} \\
&\le \left(\sum_{n=1}^{N}\|\nabla_x u\|^2_{L_2(Q_n)}\right)^{\frac{1}{2}}\left(\sum_{n=1}^{N}(\theta_n h_n)^2 c_{inv,1}^2 h_n^{-2}\|\nabla_x v_n\|^2_{L_2(Q_n)}\right)^{\frac{1}{2}} \\
&\le c_{inv,1}\theta_n\sqrt{2}\Big(\frac{1}{2}\sum_{n=1}^{N}\|\nabla_x u\|^2_{L^2(Q_n)}\Big)^{\frac{1}{2}}\sqrt{2}\Big(\frac{1}{2}\sum_{n=1}^{N}\|\nabla_x v_n\|_{L^2(Q_n)}\Big)^{\frac{1}{2}}.
\end{aligned}$$

Gathering together the bounds obtained above yields estimate (19) with $\mu_b = 2\max\{\sqrt{1+c_{inv,1}\theta}, \sqrt{2}\}$, where $\theta = \max_{n=1,\ldots,N}\{\theta_n\} \le c_{inv,0}^{-2}$. ∎

Let v_h be an arbitrary IgA function from V_{h0}. Using the fact that $v_h - u_h \in V_{h0}$, the V_{h0}-ellipticity of the bilinear form $a_h(\cdot,\cdot)$ as was shown in Lemma 1, the Galerkin orthogonality (12), and the boundedness (19) of $a_h(\cdot,\cdot)$ on $V_{0h,*} \times V_{0h}$, we can derive the following estimate

$$\mu_e \|v_h - u_h\|_h^2 \leq a_h(v_h - u_h, v_h - u_h) = a_h(v_h - u, v_h - u_h)$$
$$\leq \mu_b \|v_h - u\|_{h,*} \|v_h - u_h\|_h.$$

Therefore, we can proceed as follows:

$$\begin{aligned} \|u - u_h\|_h &\leq \|u - v_h\|_h + \|v_h - u_h\|_h \\ &\leq \|u - v_h\|_h + (\mu_b/\mu_e)\|v_h - u\|_{h,*} \\ &\leq (1 + \mu_b/\mu_c)\|v_h - u\|_{h,*}, \end{aligned}$$

which proves the following Cea-like Lemma providing an estimate of the discretization error wrt the norm $\|\cdot\|_h$ by the best approximation error wrt to the $\|\cdot\|_{h,*}$ norm.

Lemma 3. *Under the assumption made above, the discretization error wrt the $\|\cdot\|_h$ norm can be estimated from above by the best approximation error wrt to the $\|\cdot\|_{h,*}$ norm as follows:*

$$\|u - u_h\|_h \leq (1 + \frac{\mu_b}{\mu_c}) \inf_{v_h \in V_{0h}} \|u - v_h\|_{h,*}. \tag{20}$$

Theorem 1. *Let the solution $u \in H_0^{1,0}(Q)$ of the parabolic initial-boundary value model problem (2) belong to $V = H_{0,\underline{0}}^{1,1}(Q)$ globally, and patch-wise to $H^{s_n}(Q_n)$ with some $s_n \geq 2$ for $n = 1, \ldots, N$, and let $u_h \in V_{0h}$ be the solution to the IgA scheme (8) with fixed positive θ_n, $n = 1, \ldots, N$, defined as in Lemma 1. Then the discretization error estimate*

$$\|u - u_h\|_h \leq (1 + \frac{\mu_b}{\mu_c}) \sum_{n=1}^{N} c_n h_n^{r_n - 1} \|u\|_{H^{r_n}(Q_n)} \tag{21}$$

holds, where c_n are generic positive constants, $r_n = \min\{s_n, p_n + 1\}$, and p_n denotes the underlying polynomial degree of the B-splines or NURBS used in patch Q_n with $n = 1, \ldots, N$.

Proof. Let Π_n be a projective operator from $L_2(Q_n)$ to V_{0n} that delivers optimal approximation error estimates in the $L_2(Q_n)$ and $H^1(Q_n)$ norms, see, e.g., [1] or [12]. We define the multi-patch projective operator $(\Pi_h u)|_{Q_n} = \Pi_n(u|_{Q_n})$ for all $n = 1, \ldots, N$. Employing the approximation results given in [1] or [12], we can easily derive the approximation error estimates

$$\|\nabla_x(u - \Pi_n u)\|_{L^2(Q_n)}^2 + \theta_n h_n \|\partial_t(u - \Pi_n u)\|_{L^2(Q_n)}^2 \leq C_1 h_n^{2(r_n - 1)} \|u\|_{H^{r_n}(Q_n)}^2 \tag{22}$$

and

$$\theta_n h_n^{-1} \|u - \Pi_n u\|_{L^2(Q_n)}^2 \leq C_2 h_n^{2r_n - 1} \|u\|_{H^{r_n}(Q_n)}^2, \tag{23}$$

with positive generic constants C_1 and C_2. Based on the previous estimates and the trace inequality

$$\|u\|^2_{L^2(\partial Q_n)} \leq C_{tr} h_n^{-1} \big(\|u\|^2_{L^2(Q_n)} + h_n^2 \, |u|^2_{H^1(Q_n)}\big),$$

we can further show the approximation error estimate

$$\|u - \Pi_n u\|^2_{L^2(\partial Q_n)} \leq C_3 h^{2r_n - 1} \|u\|^2_{H^{r_n}(Q_n)}$$

that in turn implies

$$\|u - \Pi_n u\|^2_{L^2(\Sigma_n)} \leq C_4 h^{2r_n - 1} \|u\|^2_{H^{r_n}(Q_n)} \tag{24}$$

and

$$\begin{aligned} \frac{1}{2} \| \, [[u - \Pi_h u]] \, \|^2_{L_2(\Sigma_{n-1})} &\leq \|u_n - \Pi_n u\|^2_{L^2(\Sigma_{n-1})} + \|u_{n-1} - \Pi_{n-1} u\|^2_{L^2(\Sigma_{n-1})} \\ &\leq C_4 h_n^{2r_n - 1} \|u\|^2_{H^{r_n}(Q_n)} + C_5 h_{n-1}^{2r_{n-1} - 1} \|u\|^2_{H^{r_{n-1}}(Q_{n-1})}, \end{aligned} \tag{25}$$

with positive generic constants C_4 and C_5. Finally, gathering together (22), (23), (24) and (25), summing over all space-time patches Q_n, and recalling definition (17), we get the approximation error estimate

$$\|u - \Pi_h u\|_{h,*} \leq \sum_{n=1}^{N} c_n h_n^{r_n - 1} \|u\|_{H^{r_n}(Q_n)}. \tag{26}$$

Inserting (26) into (20) yields the desired result. ■

Remark 1. The above estimate has been derived under the isotropic assumption $u \in H^{s_n}(Q_n)$ for the patch-wise regularity of the solution. In the forthcoming work [6], we will present a discretization error analysis for the case when the solution can have anisotropic regularity behavior with respect to time and space.

5 Matrix Representation and Space-Time Multigrid Solvers

We now assume that the IgA map $\mathbf{\Phi}_n : \widehat{Q} \to Q_n$ preserves the tensor product structure of the IgA basis functions $\varphi_{n,i} = \widehat{\varphi}_{n,i} \circ \mathbf{\Phi}_n^{-1}$. Hence, for each time slice Q_n, $n = 1, \ldots, N$, the basis functions $\varphi_{n,i}$, $i \in \mathcal{I}_n$, can be rewritten in the form

$$\varphi_{n,i}(x,t) = \phi_{n,i_x}(x) \psi_{n,i_t}(t), \text{ with } i_x \in \{1, \ldots, N_{n,x}\} \text{ and } i_t \in \{1, \ldots, N_{n,t}\},$$

where $\dim(V_{0n}) = N_{n,x}N_{n,t}$. Using this representation in the definition of the bilinear form $a_n(\cdot,\cdot)$, we obtain

$$\int_{Q_n} \Big(\partial_t\varphi_{n,j}(\varphi_{n,i} + \theta_n h_n \partial_t\varphi_{n,i}) + \nabla_x\varphi_{n,j}\cdot\nabla_x\varphi_{n,i} + \theta_n h_n \nabla_x\varphi_{n,j}\cdot\nabla_x\partial_t\varphi_{n,i}\Big)\,dxdt$$
$$+ \int_{\Sigma_{n-1}} [|\varphi_{n,j}|]\,\varphi_{n,i}\,dx$$
$$= \Big[\int_\Omega \phi_{n,j_x}\phi_{n,i_x}\,dx\Big]\Big[\int_{t_{n-1}}^{t_n} \partial_t\psi_{n,j_t}(\psi_{n,i_t} + \theta_n h_n\partial_t\psi_{n,i_t})\,dt\Big]$$
$$+ \Big[\int_\Omega \nabla_x\phi_{n,j_x}\cdot\nabla_x\phi_{n,i_x}\,dx\Big]\Big[\int_{t_{n-1}}^{t_n} \psi_{n,j_t}(\psi_{n,i_t} + \theta_n h_n\partial_t\psi_{n,i_t})\,dt\Big]$$
$$+ \Big[\int_\Omega \phi_{n-1,k_x}\phi_{n,i_x}\,dx\Big]\Big[\big(\psi_{n,j_t}(t_{n-1}) - \psi_{n-1,k_t}(t_{n-1})\big)\psi_{n,i_t}(t_{n-1})\Big]$$
$$= \mathbf{M}_{n,x}[i_x,j_x]\mathbf{K}_{n,t}[i_t,j_t] + \mathbf{K}_{n,x}[i_x,j_x]\mathbf{M}_{n,t}[i_t,j_t] - \widetilde{\mathbf{M}}_{n,x}[i_x,k_x]\mathbf{N}_{n,t}[i_t,k_t],$$

with the standard mass and stiffness matrices wrt to space

$$\mathbf{M}_{n,x}[i_x,j_x] := \int_\Omega \phi_{n,j_x}\phi_{n,i_x}\,dx, \qquad \mathbf{K}_x[i_x,j_x] := \int_\Omega \nabla_x\phi_{n,j_x}\cdot\nabla_x\phi_{n,i_x}\,dx,$$
$$\widetilde{\mathbf{M}}_{n,x}[i_x,k_x] := \int_\Omega \phi_{n-1,k_x}\phi_{n,i_x}\,dx,$$

and the corresponding matrices wrt to time

$$\mathbf{K}_{n,t}[i_t,j_t] := \int_{t_{n-1}}^{t_n} \partial_t\psi_{n,j_t}(\psi_{n,i_t} + \theta_n h_n\partial_t\psi_{n,i_t})\,dt + \psi_{n,j_t}(t_{n-1})\psi_{n,i_t}(t_{n-1}),$$
$$\mathbf{M}_{n,t}[i_t,j_t] := \int_{t_{n-1}}^{t_n} \psi_{n,j_t}(\psi_{n,i_t} + \theta_n h_n\partial_t\psi_{n,i_t})\,dt,$$
$$\mathbf{N}_{n,t}[i_t,k_t] := \psi_{n-1,k_t}(t_{n-1})\psi_{n,i_t}(t_{n-1}).$$

With this computations, we have shown that the Galerkin matrix $\mathbf{L}_h$ can be rewritten in the block form

$$\mathbf{L}_h = \begin{pmatrix} \mathbf{A}_1 & & & \\ -\mathbf{B}_2 & \mathbf{A}_2 & & \\ & \ddots & \ddots & \\ & & -\mathbf{B}_N & \mathbf{A}_N \end{pmatrix},$$

with the matrices $\mathbf{A}_n := \mathbf{M}_{n,x}\otimes\mathbf{K}_{n,t} + \mathbf{K}_{n,x}\otimes\mathbf{M}_{n,t}$ for $n = 1,\ldots,N$, and $\mathbf{B}_n := \widetilde{\mathbf{M}}_{n,x}\otimes\mathbf{N}_{n,t}$ for $n = 2,\ldots,N$.

Thus, the linear system (15) can sequentially be solved from one time slice Q_{n-1} to the next time slice Q_n, where a linear system with the system matrix $\mathbf{A}_n$ has to be solved. This can be done, for example, by means of an algebraic multigrid method, which was already successfully used for the single patch case in [10]. More advanced solvers for the linear system (15) are given by space-time multigrid methods, which allow parallelization wrt to space and time. The problem given in (15) perfectly fits into the framework of space-time multigrid methods introduced in [5].

6 Numerical Results

In this section, we demonstrate the proposed method for the spatial computational domain $\Omega = (0,1)^3$ and $T = 1$, i.e., $Q = (0,1)^4$. We consider the manufactured solution $u(x,t) = \sin(\pi x_1)\sin(\pi x_2)\sin(\pi x_3)\sin(\pi t)$ for problem (1). Here, we only show results for the case $p_n = 1$, $n = 1,\dots,N$, i.e., for lowest order splines. We start with an initial space-time mesh consisting of 64 elements in space and one time slice ($N = 1$) which is subdivided into 8 elements. We then apply uniform refinement wrt space, and increase the number of time slices by a factor of two. At the same time, we keep the number of subdivision per time slice constant. For each time slice, we always use the same parameter $\theta_n = 0.2$. Using the results of Sect. 5, we can generate the linear system (15) very fast. Moreover, we can apply the solver technology given in [10] to solve the linear system in parallel wrt space and time. In detail, we use the space-time multigrid method (1 V-cycle in time and space, and 1 *hypre* algebraic multigrid (AMG) V-cycle in space) as a preconditioner for the GMRES method, and we stop the iterations until a relative residual error of 10^{-8} is reached. In Table 1, we show the convergence of this approach with respect to the $L_2(Q)$-norm. We observe the optimal convergence rate of 2. The number of cores used for the *hypre* AMG is denoted by c_x, whereas c_t gives the number of cores with respect to time. Overall, we use $c_x c_t$ cores, which is also listed in this table. We also observe quite small iteration numbers. Finally, we can solve the global linear system with 9 777 365 568 unknowns in less than 5 min on a massively parallel machine with 131 072 cores. The weak parallel efficiency corresponding to the last two rows of Table 1 is about 50%. This is due to the massive space parallelization of the AMG that is not especially adapted to the problem under consideration. All computations have been performed on the Vulcan BlueGene/Q at Livermore, U.S.A, MFEM.

Table 1. Convergence results for the space-time IgA as well as iteration numbers and solving times for the parallel space-time multigrid preconditioned GMRES method.

N	dof per slice	Overall dof	$\|\|u - u_h\|\|_{L_2(Q)}$	eoc	c_x	c_t	Cores	Iter	Time [s]
1	1 125	1 125	1.8815E−02	-	1	1	1	1	0.04
2	6 561	13 122	4.8619E−03	1.95	1	2	2	9	1.35
4	44 217	176 868	1.2294E−03	1.98	1	4	4	12	18.72
8	323 433	2 587 464	3.0834E−04	2.00	4	8	32	13	58.81
16	2 471 625	39 546 000	7.7092E−05	2.00	32	16	512	15	109.79
32	19 320 201	618 246 432	1.92621E−05	2.00	256	32	8192	16	133.31
64	152 771 337	9 777 365 568	4.81335E−06	2.00	2048	64	131072	18	273.27

7 Summary and Conclusion

We presented new time-upwind stabilized multi-patch space-time IgA schemes for parabolic IBVP, derived a priori discretization error estimates, and provided fast generation and solution methods, which can be efficiently implemented on massively parallel computers as the first numerical results show. This space-time method can be generalized to more general parabolic evolution problems.

Acknowledgments. The authors gratefully acknowledge the financial support by the Austrian Science Fund (FWF) under the grants NFN S117-03. We also want to thank the Lawrence Livermore National Laboratory for the possibility to perform numerical test on the Vulcan Cluster. In particular, the second author wants to thank P. Vassilevski for the support and the fruitful discussions during his visit at the Lawrence Livermore National Laboratory.

References

1. Bazilevs, Y., Beirão da Veiga, L., Cottrell, J., Hughes, T., Sangalli, G.: Isogeometric analysis: approximation, stability and error estimates for h-refined meshes. Comput. Methods Appl. Mech. Eng. **194**, 4135–4195 (2006)
2. Cottrell, J.A., Hughes, T.J.R., Bazilevs, Y.: Isogeometric Analysis: Toward Integration of CAD and FEA. Wiley, Chichester (2009)
3. Evans, J., Hughes, T.: Explicit trace inequalities for isogeometric analysis and parametric hexahedral finite elements. Numer. Math. **123**(2), 259–290 (2013)
4. Gander, M.J.: 50 years of time parallel time integration. In: Carraro, T., Geiger, M., Körkel, S., Rannacher, R. (eds.) Multiple Shooting and Time Domain Decomposition Methods. CMCS, vol. 9, pp. 69–114. Springer, Cham (2015). https://doi.org/10.1007/978-3-319-23321-5_3. http://www.unige.ch/~gander/Preprints/50YearsTimeParallel.pdf
5. Gander, M., Neumüller, M.: Analysis of a new space-time parallel multigrid algorithm for parabolic problems. SIAM J. Sci. Comput. **38**(4), A2173–A2208 (2016). https://doi.org/10.1137/15M1046605
6. Hofer, C., Langer, U., Neumüller, M., Toulopoulos, I.: Multipatch time discontinuous Galerkin space-time isogeometric analysis of parabolic evolution problems. Under preperation (2017)
7. Hughes, T.J.R., Cottrell, J.A., Bazilevs, Y.: Isogeometric analysis: CAD, finite elements, NURBS, exact geometry and mesh refinement. Comput. Methods Appl. Mech. Eng. **194**, 4135–4195 (2005)
8. Ladyzhenskaya, O.A.: The Boundary Value Problems of Mathematical Physics. Springer, New York (1985). https://doi.org/10.1007/978-1-4757-4317-3
9. Ladyzhenskaya, O.A., Solonnikov, V.A., Uraltseva, N.N.: Linear and Quasilinear Equations of Parabolic Type. AMS, Providence (1968)
10. Langer, U., Moore, S., Neumüller, M.: Space-time isogeometric analysis of parabolic evolution equations. Comput. Methods Appl. Mech. Eng. **306**, 342–363 (2016)
11. Piegl, L., Tiller, W.: The NURBS Book. Springer, Heidelberg (1997). https://doi.org/10.1007/978-3-642-97385-7
12. da Veiga, L.B., Buffa, A., Sangalli, G., Vázquez, R.: Mathematical analysis of variational isogeometric methods. Acta Numer. **23**, 157–287 (2014)

Numerical Methods for Controlled Switching Diffusions

G. Yin(✉), C. Zhang, and L. Y. Wang

Wayne State University, Detroit, USA
{gyin,czhang,lywang}@wayne.edu

Abstract. This work presents a survey on some of the recent results on numerical methods for controlled switching diffusions. Before presenting the numerical parts, the basics of switching diffusions are recalled. Finally, some numerical examples are presented for demonstration.

1 Introduction: Switching Diffusions and Controlled Switching Diffusions

The objective of this work is to provide a survey of some recent progress on numerical methods for controlled switching diffusions. Although studies of controlled diffusions have been around for many years, the investigation of controlled switching diffusions is relatively recent. Here, we choose a simple setting to convey the main ideas without going through technical details. The rest of the paper is arranged as follows. Section 1 introduces the concepts of switching diffusions and controlled switching diffusions. Section 2 presents the numerical methods for controlled switching diffusions. Section 3 demonstrates the methods by a numerical example. Finally, Sect. 4 gives some concluding remarks.

Switching Diffusions. In recent years, there have been growing and resurgent interests devoted to the study of switching diffusions or hybrid switching diffusions, which has substantially enlarged the applicability of diffusion systems. These systems are called hybrid systems because the continuous dynamics and discrete events coexist and interact. The rationale is that the traditional formulation of differential equations or stochastic differential equations is no longer adequate for many applications. Discrete-event processes are added to reflect the random environment changes that cannot be represented in the usual differential equation setup. Comprehensive development and survey of the progress in switching diffusions can be found in [13,27]; see also [14–16,23–26] and references therein. For the development, we refer to [17,21] for related numerical methods for solutions of differential equations with switching, [7,18–20] for numerical methods for control and game problems, and [1,2,8,9] and references therein for applications to real options and insurance, among others.

This work was supported in part by the Air Force Office of Scientific Research under FA9550-15-1-0131.

I. Lirkov and S. Margenov (Eds.): LSSC 2017, LNCS 10665, pp. 33–44, 2018.
https://doi.org/10.1007/978-3-319-73441-5_3

We work with a complete filtered probability space $(\Omega, \mathcal{F}, \{\mathcal{F}_t\}, P)$, where $(\Omega, \mathcal{F}, P)$ is the probability space, and $\{\mathcal{F}_t\}$ is the filtration satisfying the usual conditions (it is increasing and $\mathcal{F}_0$ contains all null sets). On this probability space, there are a standard r-dimensional Brownian motion $W(\cdot)$, and a continuous-time Markov chain $\alpha(\cdot)$ independent of the Brownian motion with a finite state space $\mathcal{M} = \{1, \dots, m_0\}$.[1] The generator of the Markov chain $Q = (q_{ij})$ satisfies the usual conditions $q_{ij} \geq 0$ for $i, j = 1, 2, \dots, m_0$ with $j \neq i$ and $\sum_{j=1}^{m_0} q_{ij} = 0$ for each $i = 1, 2, \dots, m_0$. For $b(\cdot, \cdot) : \mathbb{R}^r \times \mathcal{M} \mapsto \mathbb{R}^r$ and $\sigma(\cdot, \cdot) : \mathbb{R}^r \times \mathcal{M} \mapsto \mathbb{R}^{r \times r}$, consider

$$dX(t) = b(X(t), \alpha(t))dt + \sigma(X(t), \alpha(t))dW(t), \;\; (X(0), \alpha(0)) = (x, \alpha). \tag{1}$$

The two-component process $(X(t), \alpha(t))$ is a switching diffusion process. Similar to diffusion processes, $b(\cdot)$ and $\sigma(\cdot)$ are referred to as the drift and diffusion coefficients, respectively. Note that in contrast to a usual diffusion process, both the drift and diffusion depend on the switching process $\alpha(t)$. For example, to model the price of a stock in a financial market, we use $dS = \mu(\alpha(t))Sdt + \sigma(\alpha(t))Sdw$, where $S(\cdot)$ represents the stock price, $\mu(\cdot)$ and $\sigma(\cdot)$ the appreciation and volatility rates, and $w(\cdot)$ a real-valued standard Brownian motion. The use of the Markov chain $\alpha(\cdot)$ is an effort to represent the random environment, the market trend, as well as other economic factors. Under suitable conditions such as certain growth conditions and local Lipschitz conditions, it can be shown that there exists a unique solution to (1). For further properties such as well-posedness, Feller properties, recurrence, ergodicity, and stability, we refer to [27].

Controlled Switching Diffusions. Next we consider controlled processes. Redefine $b(\cdot, \cdot, \cdot) : \mathbb{R}^r \times \mathcal{M} \times U \mapsto \mathbb{R}^r$, where $U \subset \mathbb{R}^d$, and consider controlled switching diffusion

$$dX(t) = b(X(t), \alpha(t), u(t))dt + \sigma(X(t), \alpha(t))dW(t), \;\; X(0) = x, \;\; \alpha(0) = \alpha, \tag{2}$$

where $\sigma(\cdot), W(\cdot)$, and $\alpha(\cdot)$ are the same as before. Let $\{\mathcal{F}_t\}$ be a filtration that measures at least $\{W(s), \alpha(s) : s \leq t\}$, and $u(t) \in U$ a compact subset of $\mathbb{R}^d$. A control $u(t) = u(x(t))$ is an admissible feedback control, if $u(\cdot)$ is an $\{\mathcal{F}_t\}$ adapted process, $u(t) \in U$, and (2) has a unique solution with given (x, α) and $u(\cdot)$ used. In this paper, $\sigma(\cdot)$ is not controlled for simplicity.

To proceed, let τ be the first exit time of the switching diffusion from a compact set $G \in \mathbb{R}^r$, i.e.,

$$\tau = \min\{t : x(t) \notin G^o\}, \tag{3}$$

and consider the cost function

$$\begin{aligned} J(x, \iota, u) &= \mathbb{E}^u_{x,\iota}\Big[\int_0^\tau \widetilde{k}(x(s), \alpha(s), u(s))ds + g(x(\tau), \alpha(\tau))\Big] \\ J(x, \iota, u) &= g(x, \iota), \quad \text{for} \quad (x, \iota) \in G^o \times \mathcal{M}, \end{aligned} \tag{4}$$

[1] Note that in [27], the switching process $\alpha(\cdot)$ is not a Markov chain but depends on the switching process. In this paper, we consider the simpler case of Markovian switching diffusions for simplicity.

where $\widetilde{k}(\cdot)$ and $g(\cdot)$ are appropriate real-valued functions representing the running cost and terminal cost, respectively. In the above, the notation $\mathbb{E}^u_{x,\iota}$ denotes the expectation taken with the initial data $x(0) = x$ and $\alpha(0) = \iota$ and given control process $u(\cdot)$. For any $u \in U$, $(x, \iota) \in G \times \mathcal{M}$, and $\phi(\cdot, \iota) \in C^2(\mathbb{R}^r)$, define a control-dependent operator $\mathcal{L}^u$ by

$$\mathcal{L}^u \phi(x, \iota) = (\phi_x(x, \iota))' b(x, \iota, u) + \frac{1}{2}\mathrm{tr}[\phi_{xx}(x, \iota) a(x, \iota)] + Q\phi(x, \cdot)(\iota), \quad (5)$$

where $\phi_x(\cdot, \iota)$ and $\phi_{xx}(\cdot, \iota)$ denote the gradient vector and the Hessian matrix with respect to x, $a(x, \iota) = \sigma(x, \iota)\sigma'(x, \iota)$, z' denotes the transpose of z, and

$$Q\phi(x, \cdot)(\iota) = \sum_{\ell=1}^{m_0} q_{\iota\ell}\phi(x, \ell) = \sum_{\ell \neq \iota} q_{\iota\ell}(\phi(x, \ell) - \phi(x, \iota)).$$

Let $\mathcal{U}$ be the collection of admissible controls. For each $\iota \in \mathcal{M}$, let $V(x, \iota)$ be the value function

$$V(x, \iota) = \inf_{u \in \mathcal{U}} J(x, \iota, u). \quad (6)$$

The value functions are solutions to the following system of Hamilton-Jacobi-Bellman (HJB) equations

$$\begin{aligned} &\inf_{u \in U} [\mathcal{L}^u V(x, \iota) + \widetilde{k}(x, \iota, u)] = 0, \ (x, \iota) \in G^o \times \mathcal{M}, \\ &V(x, \iota) = g(x, \iota), \ \text{for } (x, \iota) \in \partial G \times \mathcal{M}, \end{aligned} \quad (7)$$

where G^o and ∂G denote the interior and the boundary of G, respectively. That is, the value function is a solution of a system of partial differential equations with Dirichlet boundary conditions; see also [5]. Note that the random stopping time can be replaced by a finite time T or ∞ (if one works on an infinite horizon problem, a certain discount or long-run average cost may be used). Because of the equations are highly nonlinear together with the inf operation used, analytic solutions are virtually impossible. Thus, numerical solutions are sorely needed. Our next task is to construct a numerical procedure for solving the optimal control problem. It is possible to solve (7) by using discretization of the system of PDEs, but such a direct approach normally requires extensive *a prior* information of the system and regularity of the system of PDEs. Because of the use of the controls, such regularity may not be readily available. As a viable alternative, Kushner initiated the so-called Markov chain approximation methods in the 1960s for controlled diffusions, which was summarized in his book [10]. This method was developed further in Kushner and Dupuis [12] for controlled diffusions; see also related reference [6]. The idea is that one constructs a controlled Markov chain in discrete time. This controlled Markov chain serves as an approximation to the controlled diffusion process and is locally consistent in that the local mean and covariance coincide with that of the controlled diffusion. Then by taking appropriate interpolations, one can show that the interpolated process converges to that of the controlled diffusion. One of the main advantages of this approach is that not much regularity or prior information of the system is needed. Taking this into consideration, we design a Markov chain approximation method for the switching diffusion. In our setup, the approximating Markov chain has two components. One of them delineates the diffusive behavior (as in the controlled diffusions) and the other represents the switching process.

2 Numerical Methods

We shall construct a discrete-time controlled Markov chain with a finite state space to approximate the controlled switching diffusion. Since we are dealing with controlled switching diffusions now, the approximating controlled Markov chain in discrete time has two components (two chains, one of them is an approximation to the switching process and the other is an approximation to the controlled diffusion for each $\iota \in \mathcal{M}$). The methods are based on our work [19,20]. To proceed, choose the step size $h > 0$ and define $S_h = \{x : x = kh, k = 0, \pm 1, \pm 2, \ldots\}$. Construct $\{(\xi_n^h, \alpha_n^h), n < \infty\}$, a controlled discrete-time Markov chain on a discrete state space $S_h \times \mathcal{M}$. Construct the transition probabilities from state $(x, \iota) \in \mathcal{M}$ to state $(y, \ell) \in \mathcal{M}$ as $p^h((x, \iota), (y, \ell)|u)$ for $u \in U$. Denote by u_n^h the random variable that is the control action for the chain at discrete time n. To approximate the continuous-time $(x(\cdot), \alpha(\cdot))$, we use a suitable continuous-time interpolation. Select $\Delta t^h(\cdot, \cdot, \cdot) > 0$ on $S_h \times \mathcal{M} \times U$, and denote $\Delta t_n^h = \Delta t^h(\xi_n^h, \alpha_n^h, u_n^h)$. Define the interpolation time $t_n^h = \sum_{k=0}^{n-1} \Delta t_k^h(\xi_k^h, \alpha_k^h, u_k^h)$. Define piecewise constant interpolations $(\xi^h(\cdot), \alpha^h(\cdot)), u^h(\cdot)$, and $z^h(\cdot)$, as follows. For $t \in [t_n^h, t_{n+1}^h)$, define

$$\xi^h(t) = \xi_n^h, \ \alpha^h(t) = \alpha_n^h, \ u^h(t) = u_n^h, \ z^h(t) = n. \tag{8}$$

Denote $\Delta \xi_n^h = \xi_{n+1}^h - \xi_n^h$ and assume $\inf_{x,\iota,r} \Delta t^h(x, \iota, r) > 0$ for each $h > 0$ and $\lim_{h\to 0} \sup_{x,\iota,r} \Delta t^h(x, \iota, r) \to 0$. Let $\mathbb{E}_{x,\iota,n}^{r,h}$, $\mathbb{V}_{x,\iota,n}^{r,h}$, and $\mathbb{P}_{x,\iota,n}^{r,h}$ denote the conditional expectation, covariance, and marginal probability given $\{\xi_k^h, \alpha_k^h, u_k^h, k \le n, \xi_n^h = x, \alpha_n^h = \iota, u_n^h = r\}$, respectively. So for example,

$$\mathbb{V}_{x,\iota,n}^{u,h} \Delta \xi_n^h = \mathbb{E}_{x,\iota,n}^{u,h} [\Delta \xi_n^h - \mathbb{E}_{x,\iota,n}^{u,h}][\Delta \xi_n^h - \mathbb{E}_{x,\iota,n}^{u,h}]'.$$

The sequence $\{(\xi_n^h, \alpha_n^h)\}$ is locally consistent with (2), if it satisfies, for $\varepsilon^h = o(\Delta t^h(x, \iota, r))$,

$$\begin{aligned}
&\mathbb{E}_{x,\iota,n}^{u,h} \Delta \xi_n^h = b(x, \iota, u) \Delta t^h(x, \iota, u) + \varepsilon^h, \\
&\mathbb{V}_{x,\iota,n}^{u,h} \Delta \xi_n^h = a(x, \iota) \Delta t^h(x, \iota, u) + \varepsilon^h, \\
&\mathbb{P}_{x,\iota,n}^{u,h} \{\alpha_{n+1}^h = \ell\} = \Delta t^h(x, \iota, u) q_{\iota\ell} + \varepsilon^h, \ \text{for } \ell \neq \iota, \\
&\mathbb{P}_{x,\iota,n}^{r,h} \{\alpha_{n+1}^h = \iota\} = \Delta t^h(x, \iota, u)(1 + q_{\iota\iota}) + \varepsilon^h, \\
&\sup_{n,\omega\in\Omega} |\Delta \xi_n^h| \to 0 \text{ as } h \to 0.
\end{aligned} \tag{9}$$

Roughly, the local consistency means that the local mean and covariance of the approximating controlled Markov chain "match" those of the controlled switching diffusion. Let $G_h^o = S_h \cap G^o$. Thus $G_h^o \times \mathcal{M}$ is a finite state space.

Let N_h denote the first time that $\{\xi_n^h\}$ leaves G_h^o. Natural cost functions for the chain that approximates (4) are, for $(x,\iota) \in G_h^o \times \mathcal{M}$,

$$J^h(x,\iota,u^h) = \mathbb{E}_{x,\iota}^{u^h} \sum_{n=0}^{N_h-1} \widetilde{k}(\xi_n^h, \alpha_n^h, u_n^h)\Delta t_n^h + \mathbb{E}_{x,\iota}^{u^h} g(\xi_{N_h}^h, \alpha_{N_h}^h). \tag{10}$$

Corresponding to the continuous-time problems, the first term on the right-hand side of (10) represents the running cost and the last term gives the terminal cost. Using $\mathcal{U}^h$ to denote the collection of controls, which are determined by a sequence of measurable functions $F_n^h(\cdot)$ such that $u_n^h = F_n^h(\xi_k^h, \alpha_k^h, k \le n; u_k^h, k < n)$. We can find approximation of $V(x,\iota)$ by

$$V^h(x,\iota) = \inf_{u^h \in \mathcal{U}^h} J^h(x,\iota,u^h). \tag{11}$$

Practically, we can compute $V^h(x,\iota)$ by solving the corresponding dynamic programming equation using iteration method. That is, for $(x,\iota) \in G_h^o \times \mathcal{M}$,

$$V^h(x,\iota) = \min_{u\in U}\{ \sum_{(y,\ell)\in G_h^o \times \mathcal{M}} \mathbb{P}^h((x,\iota),(y,\ell)|u)V^h(y,\ell) + \widetilde{k}(x,\iota,r)\Delta t^h(x,\iota,r)\}, \tag{12}$$

with the boundary condition $V^h(x,\iota) = g(x,\iota)$ for $(x,\iota) \in G_h^o \times \mathcal{M}$. Now we proceed to find the transition probabilities and interpolation intervals for the controlled Markov chain $\{(\xi_n^h, \alpha_n^h)\}$. To find a reasonable Markov chain that is locally consistent, we first consider a special case, in which the control space has unique admissible control $u^h \in \mathcal{U}^h$. In this case, min in (12) can be dropped. That is,

$$V^h(x,\iota) = \sum_{(y,\ell)\in G_h^o \times \mathcal{M}} p^h((x,\iota),(y,\ell)|u)V^h(y,\ell) + \widetilde{k}(x,\iota,u)\Delta t^h(x,\iota,u). \tag{13}$$

If we assume $\mathcal{U}$ has a single admissible control $u(\cdot)$, we can drop inf in (7), and apply $u = u(0)$ in $\mathcal{L}^u$. That is,

$$V_x(x,\iota)b(x,\iota,u) + \frac{1}{2}V_{xx}(x,\iota)a(x,\iota) + \sum_{\ell} V(x,\ell)q_{\iota\ell} + \widetilde{k}(x,\iota,u) = 0. \tag{14}$$

To proceed, denote the standard unit vector by $e_i \in \mathbb{R}^r$ that is the vector with the ith component being 1 and all other components being 0. Denote also $b_i = b'e_i$, the ith component of b. Discretize (14) using upwind finite difference approximation with step-size $h > 0$ by

$$\begin{aligned}
&V(x,\iota) \to V^h(x,\iota)\\
&V_{x_i}(x,\iota) \to \frac{V^h(x+he_i,\iota) - V^h(x,\iota)}{h} \text{ for } b_i(x,\iota,u) > 0,\\
&V_{x_i}(x,\iota) \to \frac{V^h(x,\iota) - V^h(x-he_i,\iota)}{h} \text{ for } b_i(x,\iota,u) < 0,\\
&V_{x_ix_i}(x,\iota) \to \frac{V^h(x+he_i,\iota) - 2V^h(x,\iota) + V^h(x-he_i,\iota)}{h^2}.
\end{aligned}$$

For off diagonal elements of $V_{xx}(x,\iota)$, we use

$$\begin{aligned}
V_{x_ix_j}(x,\iota) \mapsto & [2V^h(x,\iota) + V^h(x+he_i+he_j,\iota) + V^h(x-he_i-he_j,\iota)]/(2h^2) \\
& -[V^h(x+he_i,\iota) + V^h(x-he_j,\iota) + V^h(x+he_j,\iota) \\
& \quad + V^h(x-he_j,\iota)]/(2h^2) \text{ if } a_{ij}(x,\iota) \geq 0, \text{ and} \\
V_{x_ix_j}(x,\iota) \mapsto & -[2V^h(x,\iota) + V^h(x+he_i+he_j,\iota) + V^h(x-he_i-he_j,\iota)]/(2h^2) \\
& +[V^h(x+he_i,\iota) + V^h(x-he_j,\iota) + V^h(x+he_j,\iota) \\
& \quad + V^h(x-he_j,\iota)]/(2h^2) \text{ if } a_{ij}(x,\iota) < 0.
\end{aligned}$$

Assume that $a_{ii}(x,\iota) - \sum_{j:j\neq i} |a_{ij}(x,\iota)| \geq 0$. For relaxation of the above condition, we refer the reader to [12, Chap. 5] for controlled diffusions.

Denote $\widetilde{D} = \sum_i a_{ii}(x,\iota) - \sum_{i,j:i\neq j} |a_{ij}(x,\iota)|/2 + h\sum_i |b(x,\iota,u)| - h^2 q_{\iota\iota}$, define $\Delta t^h(x,\iota,u) = (h^2/\widetilde{D})$, and suppose $\Delta t^h(x,\iota,u) = O(h)$. We define the transition probabilities as

$$\begin{aligned}
\mathbb{P}^h((x,\iota),(x \pm he_i,\iota)|u) &= \frac{\frac{a_{ii}(x,\iota)}{2} - \sum\limits_{j:j\neq i} \frac{|a_{ij}(x,\iota)|}{2} + hb_i^{\pm}(x,\iota,u)}{\widetilde{D}}, \\
\mathbb{P}^h((x,\iota),(x+he_i+he_j,\iota)|u) &= \frac{a_{ij}^{+}(x,\iota)}{2\widetilde{D}}, \\
\mathbb{P}^h((x,\iota),(x-he_i+he_j,\iota)|u) &= \frac{a_{ij}^{-}(x,\iota)}{2\widetilde{D}}, \\
\mathbb{P}^h((x,\iota),(x,\ell)|u) &= \frac{h^2}{\widetilde{D}} q_{\iota\ell} \\
\mathbb{P}^h(\cdot) &= 0, \quad \text{otherwise},
\end{aligned} \tag{15}$$

with $\mathbb{P}^h((x,\iota),(x+he_i+he_j,\iota)|u) = \mathbb{P}^h((x,\iota),(x-he_i-he_j,\iota|u))$ and $\mathbb{P}^h((x,\iota),(x-he_i+he_j,\iota)|u) = \mathbb{P}^h((x,\iota),(x+he_i-he_j,\iota|u))$. Moreover, we can verify that with the transition probabilities so specified, the corresponding approximating Markov chain is locally consistent. In the above, ψ^+ and ψ^- denote the positive and negative parts of ψ for $\psi = a$ and $\psi = b$.

To facilitate the analysis, we introduce the relaxed control representation; see [12, Sect. 4.6]. The sequence of ordinary controls may not converge in a traditional sense, and the use of the relaxed control terminology enables us to obtain and appropriately characterize the weak limit.

Let $\mathcal{B}(U \times [0,\infty))$ be the σ-algebra of Borel subsets of $U \times [0,\infty)$. An *admissible relaxed control* (or deterministic relaxed control) $m(\cdot)$ is a measure on $\mathcal{B}(U \times [0,\infty))$ such that $m(U \times [0,t]) = t$ for each $t \geq 0$. Given a relaxed control $m(\cdot)$, there is an $m_t(\cdot)$ such that $m(drdt) = m_t(dr)dt$. In fact, we can define $m_t(B) = \lim_{\delta\to 0} \frac{m(B\times[t-\delta,t])}{\delta}$ for $B \in \mathcal{B}(U)$. With the given probability space, we say that $m(\cdot)$ is an admissible relaxed (stochastic) control for $(W(\cdot),\alpha(\cdot))$, or $(m(\cdot),W(\cdot),\alpha(\cdot))$ is admissible if $m(\cdot,\omega)$ is a deterministic relaxed control with probability 1 and if $m(A \times [0,t])$ is $\mathcal{F}_t$-adapted for all $A \in \mathcal{B}(U)$. There is a derivative $m_t(\cdot)$ such that $m_t(\cdot)$ is $\mathcal{F}_t$-adapted for all $A \in \mathcal{B}(U)$. Note that

$m_t(\cdot)$ is a probability measure on $\mathcal{B}(U)$. Loosely, it is the time derivative of $m(\cdot)$. It is natural to define the relaxed control representation $m^h(\cdot)$ of $u^h(\cdot)$ by $m_t^h(A) = I_{\{u^h(t)\in A\}}, \forall A \in \mathcal{B}(U)$. Let $\mathcal{F}_t^h$ denote the minimal σ-algebra that measures $\{\xi^h(s), \alpha^h(\cdot), m_s^h(\cdot), w^h(s), z^h(s), s \le t\}$. Use Γ^h to denote the set of admissible relaxed controls $m^h(\cdot)$ with respect to $(\alpha^h(\cdot), w^h(\cdot))$ such that $m_t^h(\cdot)$ is a fixed probability measure in the interval $[t_n^h, t_{n+1}^h)$ given $\mathcal{F}_t^h$. Then Γ^h is a larger control space containing $\mathcal{U}^h$.

With the notion of relaxed control given above, we can write the system, the cost, and the value functions as

$$\xi^h(t) = x + \int_0^t \int_U b(\xi^h(s), \alpha^h(s), c) m_s^h(dc) ds + \int_0^t \sigma(\xi^h(s), \alpha^h(s)) dw^h(s) + \varepsilon^h(t), \tag{16}$$

$$\begin{aligned} J^h(x,\iota,m^h) &= \mathbb{E}_{x,\iota}^{m^h} \int_0^{\tau^h} \int_U \widetilde{k}(\xi^h(s), \alpha^h(s), c) m_s^h(dc) ds + \mathbb{E}_{x,\iota}^{m^h} g(\xi^h(\tau_h), \alpha^h(\tau_h)), \\ V^h(x,\iota) &= \inf_{m^h \in \Gamma^h} J^h(x,\iota,m^h). \end{aligned} \tag{17}$$

The use of the relaxed controls makes the control appears essentially linearly in the dynamics and cost function. We can rewrite (2) and (4) as

$$x(t) = x + \int_0^t \int_U b(x(s), \alpha(s), c) m_s(dc) ds + \int_0^t \sigma(x(s), \alpha(s)) dw(s), \tag{18}$$

$$J(x,\iota,m) = \mathbb{E}_{x,\iota}^m \Big[\int_0^\tau \int_U \widetilde{k}(x(s), \alpha(s), c) m_s(dc) ds + g(x(\tau), \alpha(\tau)) \Big]. \tag{19}$$

By a weak solution of (18), we mean that there exist a probability space $(\Omega, \mathcal{F}, P)$, a filtration $\mathcal{F}_t$, and processes $(x(\cdot), \alpha(\cdot), m(\cdot), w(\cdot))$ such that $w(\cdot)$ is a standard $\mathcal{F}_t$-Wiener process, $\alpha(\cdot)$ is a Markov chain with generator Q and state space $\mathcal{M}$, $m(\cdot)$ is admissible with respect to $(w(\cdot), \alpha(\cdot)), x(\cdot)$ is $\mathcal{F}_t$-adapted, and (18) is satisfied. For an initial condition (x, ι), by weak sense uniqueness, we mean that the probability law of admissible process $(\alpha(\cdot), m(\cdot), w(\cdot))$ determines the probability law of solution $(x(\cdot), \alpha(\cdot), m(\cdot), w(\cdot))$ to (18), irrespective of the probability space. To proceed, some conditions are needed.

(A1) For each $\iota \in \mathcal{M}$ and each $r \in U$, the functions $b(\cdot, \iota, r)$ and $\sigma(\cdot, \iota)$ are continuous in G.

(A2) For each $\iota \in \mathcal{M}$, denote $a(x,\iota) = \sigma(x,\iota)\sigma'(x,\iota) = (a_{ij}(x,\iota))$ and assume that $a_{ii}(x,\iota) - \sum_{j:j\neq i} |a_{ij}(x,\iota)| \ge 0 \ \forall (x,\iota) \in G \times \mathcal{M}$.

(A3) For each $\iota \in \mathcal{M}$ and each $r \in U$, the functions $\widetilde{k}(\cdot, \iota, r)$ and $g(\cdot, \iota)$ are continuous in G.

(A4) Let $u(\cdot)$ be an admissible ordinary control with respect to $(w(\cdot), \alpha(\cdot))$; suppose $u(\cdot)$ is piecewise constant and takes only a finite number of values. Then for each initial condition, there exists a solution to (18) unique in the weak sense, where $m(\cdot)$ is the relaxed control representation of $u(\cdot)$.

(A5) Let $\hat{\tau}(\phi) = \infty$, if $\phi(t) \in G^o$, for all $t < \infty$, otherwise, define $\hat{\tau}(\phi) = \inf\{t : \phi(t) \notin G^o\}$. The function $\hat{\tau}(\cdot)$ is continuous (as a map from $D[0,\infty)$, the space of functions that are right continuous and have left limits endowed with the Skorohod topology) with probability one relative to the measure induced by any solution to (18) with initial condition (x, ι).

Note that if condition (A2) is not satisfied, we can modify it similar to that of the controlled diffusions as in [12]. Because of the page limitation, the proofs of the theorems below are not presented here. We refer the readers to [19] for related treatment of systems whose states are one dimensional.

Theorem 1. *Assume* (A1) *and* (A2). *Let the approximating chain* $\{\xi_n^h, \alpha_n^h, n < \infty\}$ *be constructed with transition probabilities defined in* (15), $\{u_n^h, n < \infty\}$ *be a sequence of admissible controls,* $(\xi^h(\cdot), \alpha^h(\cdot))$ *be the continuous-time interpolation defined in* (8), $m^h(\cdot)$ *be the relaxed control representation of* $\{u_n^h, n < \infty\}$, *and* $\{\tilde{\tau}_h\}$ *be a sequence of* $\mathcal{F}_t^h$*-stopping times. Then* $\{\xi^h(\cdot), \alpha^h(\cdot), m^h(\cdot), w^h(\cdot), \tilde{\tau}_h\}$ *is tight. Denote by* $(x(\cdot), \alpha(\cdot), m(\cdot), w(\cdot), \tilde{\tau})$ *the limit of a weakly convergent subsequence, and denote by* $\mathcal{F}_t$ *the* σ*-algebra generated by* $\{x(s), \alpha(s), m(s), W(s), s \leq t, \tilde{\tau} I_{\{\tilde{\tau} \leq t\}}\}$. *Then* $W(\cdot)$ *is a standard* $\mathcal{F}_t$*-Wiener process,* $\tilde{\tau}$ *is an* $\mathcal{F}_t$*-stopping time, and* $m(\cdot)$ *is an admissible control. Moreover,* (18) *is satisfied.*

Theorem 2. *Assume* (A1)–(A5). *Then for the approximation of the value function* $V^h(x, \iota)$ *and the value function in the original problem* $V(x, \iota)$, *we have* $V^h(x, \iota) \to V(x, \iota)$ *as* $h \to 0$.

3 Examples

This section presents a couple of examples for demonstration.

Example 3. Consider a linear quadratic regulator with regime switching

$$dX(t) = U(t) - \Xi(\alpha(t))X(t)dt + \sigma(\alpha(t))X(t)dW(t), \tag{20}$$

with $X(t) = (x_1(t), x_2(t))', x_i(t) \in \mathbb{R}$, respectively, admissible control $U(t) = (0.1u_1(t), 0.3u_2(t))'$, the continuous-time Markov chain $\alpha(t)$ takes values in $\{1, 2\}$, and $W(t) = (w_1(t), w_2(t))'$ is a 2-dimension Brownian motion. We put

$$\Xi(1) = \begin{pmatrix} 0.2 & 0.3 \\ 0.6 & 0.1 \end{pmatrix}, \ \Xi(2) = \begin{pmatrix} 1 & 3 \\ 6 & 2 \end{pmatrix}, \ a(1) = \begin{pmatrix} 0.09 & 0.03 \\ 0.03 & 0.01 \end{pmatrix}, \ a(2) = \begin{pmatrix} 0.36 & 0.3 \\ 0.3 & 0.25 \end{pmatrix}$$

Assume that the generator of the Markov chain is $Q = \begin{pmatrix} -1.5 & 1.5 \\ 1.5 & -1.5 \end{pmatrix}$

For simplicity, assume there are only 4 possible control actions $\mathcal{U} = (u_1, u_2) = \{(0, 0), (0, 1), (1, 0), (1, 1)\}$, indexed by $0, 1, 2, 3$, respectively, which corresponds to no action, only taking actions on process 1, only taking actions on 2, and taking actions on both, respectively.

We use the cost function

$$J(x_1, x_2, i, u) = E_{x,i}^u \int_0^\infty e^{-\rho t}[x_1^2(t) + x_2^2(t) + u_1^2(t) + u_2^2(t)]dt, \quad g(x, i) = 0$$

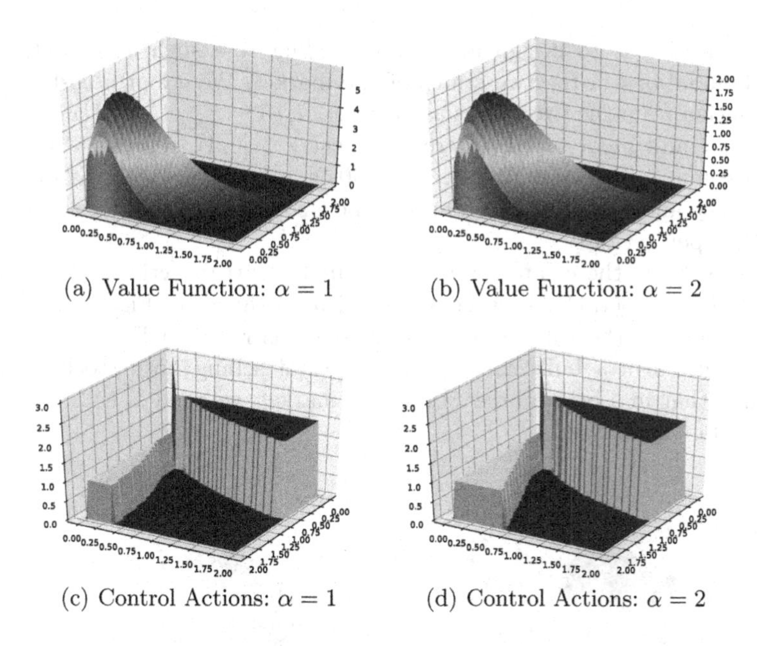

(a) Value Function: $\alpha = 1$ (b) Value Function: $\alpha = 2$

(c) Control Actions: $\alpha = 1$ (d) Control Actions: $\alpha = 2$

Fig. 1. The control actions and value functions for Example 3.

where $u = (u_1, u_2)$. Our objective is to design an optimal strategy to maximize the cost. Then the value function is $V(x,i) = \max_{u \in (u_1,u_2)} J(x,i,u)$. We implemented our numerical algorithm using PYTHON 2.7 and obtained the approximation of value functions and optimal control actions as in Fig. 1.

Example 4. Consider an application to the optimal liquidation rule for trading two related stocks. The underlying dynamic system is

$$dX(t) = -U(t) + \Xi(\alpha(t))X(t)dt + \sigma(\alpha(t))X(t)dW(t), \tag{21}$$

where $X(t) = (x_1(t), x_2(t))', x_i(t) \in \mathbb{R}$, denotes the prices of two stocks, respectively, $U(t) = (0.01u_1(t), 0.2u_2(t))'$ is the depressing effects of selling on each stock's price, the continuous-time Markov chain $\alpha(t)$ takes values in $\{1, 2\}$, and $W(t) = (w_1(t), w_2(t))'$ represents two mutually independent real-valued Brownian motions. $\Xi(i)$ for $i = 1, 2$ represent the appreciation rates of the two stocks in the two states. $\sigma(i) i = 1, 2$ characterize the volatilities and covariances of the two processes, respectively.

We use the simulated data $\Xi(1) = \begin{pmatrix} 0.2 \; 0.3 \\ 0.6 \; 0.1 \end{pmatrix}, \Xi(2) = \begin{pmatrix} -0.1 \; -3 \\ -0.6 \; -2 \end{pmatrix}$, to represent a bull market and a bear one with variances satisfying: $a(1) = \begin{pmatrix} 0.087 \; 0.032 \\ 0.032 \; 0.015 \end{pmatrix}$, $a(2) = \begin{pmatrix} 0.48 \; 0.32 \\ 0.32 \; 0.25 \end{pmatrix}$, and $Q = \begin{pmatrix} -.5 & .5 \\ .5 & -.5 \end{pmatrix}$. Note that in practice the parameter estimations given real data can be obtained using statistical inference. We refer the interested readers to [28].

Our objective is to design an optimal selling rule such as to obtain the biggest liquidations in the two states of the market. We adopt a cost function for liquidation from the two stocks as: $J(x_1, x_2, i, u) = E^u_{x,i} \int_0^\tau e^{-\rho\tau}[x_1(t)u_1(t) + x_2(t)u_2(t)]dt$ and $g(x, i) = 0$, where $u_1(\cdot), u_2(\cdot) \in [0, 1]$, represent the selling rates, discount rate $\rho = 0.01, G_1, G_2 = [0, 2]$ as compact subsets of $\mathbb{R}$. Let $G = G_1 \times G_2$, and $\tau = \min\{t : X(t) \in G^0\}$ be a stopping time of the investment, for G_0 the open set of G.

Suppose that there are only 4 optimal control actions, $(u_1, u_2) = \{(0, 0), (1, 0), (0, 1), (1, 1)\}$, indexed by 0, 1, 2, 3 as shown in Fig. 2(c), (d). The approximations of the value functions are demonstrated in Fig. 2(a), (b). For optimal controls in a bull market, the region of selling both stocks is much larger than that in a bear market. In a bear market, we tend to sell only one of the stocks for liquidation.

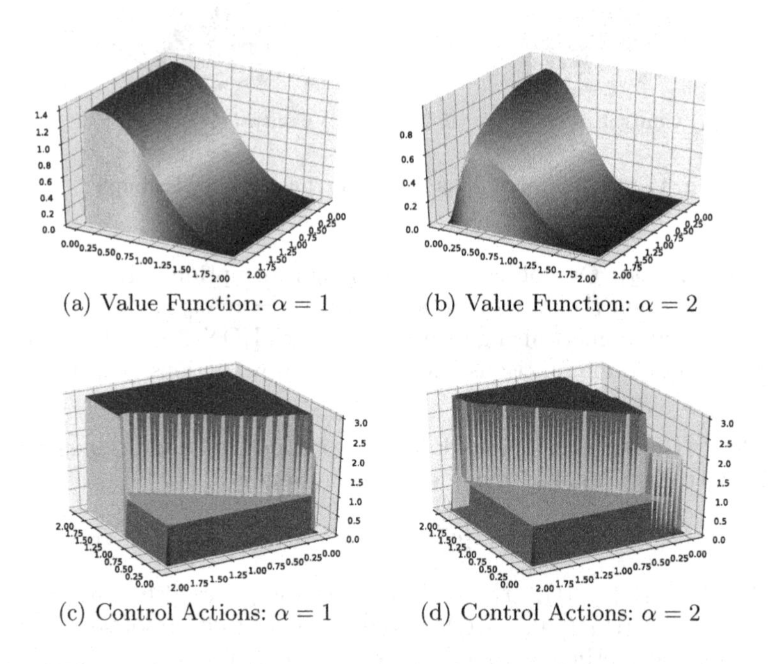

(a) Value Function: $\alpha = 1$ (b) Value Function: $\alpha = 2$

(c) Control Actions: $\alpha = 1$ (d) Control Actions: $\alpha = 2$

Fig. 2. The control actions and value functions for Example 4.

4 Concluding Remarks

This paper presents a survey of some of our recent work on controlled switching diffusions. Simple examples are presented. The numerical methods are rather versatile and can be used for a wide range of applications.

Very recently, we have begun the study of a more general class of hybrid systems. Let $Y(t) = (X(t), \Lambda(t))$ be a two component Markov process such that X is an $\mathbb{R}^d$-valued process, and Λ is a switching process taking values in a finite set $\mathcal{M} = \{1, 2, \ldots, m_0\}$. Let $b(\cdot,\cdot) : \mathbb{R}^d \times \mathcal{M} \mapsto \mathbb{R}^d, \sigma(\cdot,\cdot) : \mathbb{R}^d \times \mathcal{M} \mapsto \mathbb{R}^d \times \mathbb{R}^d$, and for each $x \in \mathbb{R}^d, \pi_i(x, dz)$ is a σ-finite measure on $\mathbb{R}^d$ satisfying $\int_{\mathbb{R}^d}(1 \wedge |z|^2)\pi_i(x, dz) < \infty$. Let $Q(x) = (q_{ij}(x))$ be an $m_0 \times m_0$ matrix depending on x such that $q_{ij}(x) \geq 0$ for $i \neq j, \sum_{j\in\mathcal{M}} q_{ij}(x) \leq 0$. Define $Q(x)f(x,\cdot)(i) := \sum_{j\in\mathcal{M}} q_{ij}(x)f(x,j)$. The generator $\mathcal{G}$ of the process $(X(t), \Lambda(t))$ is given as follows. For $f : \mathbb{R}^d \times \mathcal{M} \mapsto \mathbb{R}$ with $f(\cdot, i) \in C^2(\mathbb{R}^d)$, and for $i \in \mathcal{M}$, define

$$\begin{aligned}
\mathcal{G}f(x,i) &= \mathcal{L}_i f(x,i) + Q(x)f(x,\cdot)(i), \quad (x,i) \in \mathbb{R}^d \times \mathcal{M}, \text{ where} \\
\mathcal{L}_i f(x,i) &= \sum_{k,l=1}^{d} a_{kl}(x,i)\frac{\partial^2 f(x,i)}{\partial x_k \partial x_l} + \sum_{k=1}^{d} b_k(x,i)\frac{\partial f(x,i)}{\partial x_k} \\
&\quad + \int_{\mathbb{R}^d} \left(f(x+z,i) - f(x,i) - \nabla f(x,i)\cdot z\mathbf{1}_{\{|z|<1\}}\right)\pi_i(x,dz),
\end{aligned}$$

where $a(x,i) = \sigma(x,i)\sigma'(x,i), \nabla f(\cdot,i)$ denotes the gradient of $f(\cdot,i)$. In contrast to the switching diffusions considered in this paper, such switching jump diffusions are even more difficult since the associated operator is non-local; see [3,4] and references therein. For systems without switching, controlled jump diffusions were considered for example, in [11]. Nevertheless, in [11] (see also [19]), jumps are confined to a compact set. In the setup outlined above, we are in fact looking at a non-compact set. The corresponding control problems and the associated numerical methods deserve careful thoughts and consideration.

References

1. Bensoussan, A., Hoe, S., Yan, Z., Yin, G.: Real options with competition and regime switching. Math. Financ. **27**, 224–250 (2017)
2. Bensoussan, A., Yan, Z., Yin, G.: Threshold-type policies for real options using regime-switching models. SIAM J. Financ. Math. **3**, 667–689 (2012)
3. Chen, X., Chen, Z.-Q., Tran, K., Yin, G.: Properties of switching jump diffusions: maximum principles and Harnack inequalities, to appear in Bernouli
4. Chen, X., Chen, Z.-Q., Tran, K., Yin, G.: Recurrence and ergodicity for a class of regime-switching jump diffusions (2017, preprint)
5. Eidelman, S.D.: Parabolic Systems. North-Holland, New York (1969)
6. Fleming, W.H., Soner, H.M.: Controlled Markov Processes and Viscosity Solutions. Springer, New York (1992). https://doi.org/10.1007/0-387-31071-1
7. Higham, D., Roj, M., Mao, X., Song, Q.S., Yin, G.: Mean exit times and the multilevel Monte Carlo method. SIAM/ASA J. Uncertain. Quantif. **1**, 2–18 (2013)
8. Jin, Z., Yang, H.L., Yin, G.: Numerical methods for optimal dividend payment and investment strategies of regime-switching jump diffusion models with capital injections. Automatica **49**, 2317–2329 (2013)
9. Jin, Z., Yin, G., Zhu, C.: Numerical solutions of optimal risk control and dividend optimization policies under a generalized singular control formulation. Automatica **48**, 1489–1501 (2012)
10. Kushner, H.J.: Probability Methods for Approximations in Stochastic Control and for Elliptic Equations. Academic Press, New York (1977)

11. Kushner, H.J.: Approximation and Weak Convergence Methods for Random Processes with Applications to Stochastic Systems Theory. MIT Press, Cambridge (1984)
12. Kushner, H.J., Dupuis, P.: Numerical Methods for Stochastic Control Problems in Continuous Time, 2nd edn. Springer, New York (2001). https://doi.org/10.1007/978-1-4613-0007-6
13. Mao, X., Yuan, C.: Stochastic Differential Equations with Markovian Switching. Imperial College Press, London (2006)
14. Nguyen, D.H., Yin, G.: Modeling and analysis of switching diffusion systems: past dependent switching with a countable state space. SIAM J. Control Optim. **54**, 2450–2477 (2016)
15. Nguyen, D.H., Yin, G.: Recurrence and ergodicity of switching diffusions with past-dependent switching having a countable state space. Potential Anal. https://doi.org/10.1007/s11118-017-9641-y
16. Nguyen, S.L., Yin, G.: Pathwise convergence rate for numerical solutions of stochastic differential equations. IMA J. Numer. Anal. **32**, 701–723 (2012)
17. Nguyen, S.L., Hoang, T., Nguyen, D., Yin, G.: Milstein-type procedures for numerical solutions of stochastic differential equations with Markovian switching. SIAM J. Numer. Anal. **55**, 953–979 (2017)
18. Song, Q.S., Yin, G.: Rates of convergence of numerical methods for controlled regime-switching diffusions with stopping times in the costs. SIAM J. Control Optim. **48**, 1831–1857 (2009)
19. Song, Q.S., Yin, G., Zhang, Z.: Numerical method for controlled regime-switching diffusions and regime-switching jump diffusions. Automatica **42**, 1147–1157 (2006)
20. Song, Q.S., Yin, G., Zhang, Z.: Numerical solutions for stochastic differential games with regime switching. IEEE Trans. Autom. Control **53**, 509–521 (2008)
21. Yin, G., Mao, X.R., Yuan, C., Cao, D.: Approximation methods for hybrid diffusion systems with state-dependent switching processes: numerical algorithms and existence and uniqueness of solutions. SIAM J. Math. Anal. **41**, 2335–2352 (2010)
22. Yin, G., Wang, L.Y., Sun, Y.: Stochastic recursive algorithms for networked systems with delay and random switching: multiscale formulations and asymptotic properties. Multiscale Model. Simul.: SIAM J. **9**, 1087–1112 (2011)
23. Yin, G., Xi, F.: Stability of regime-switching jump diffusions. SIAM J. Control Optim. **48**, 4525–4549 (2010)
24. Yin, G., Yuan, Q., Wang, L.Y.: Asynchronous stochastic approximation algorithms for networked systems: regime-switching topologies and multi-scale structure. Multiscale Model. Simul.: SIAM J. **11**, 813–839 (2013)
25. Yin, G., Zhang, Q.: Continuous-Time Markov Chains and Applications: A Two-Time-Scale Approach, 2nd edn. Springer, New York (2013). https://doi.org/10.1007/978-1-4614-4346-9
26. Yin, G., Zhao, G., Wu, F.: Regularization and stabilization of randomly switching dynamic systems. SIAM J. Appl. Math. **72**, 1361–1382 (2012)
27. Yin, G., Zhu, C.: Hybrid Switching Diffusions: Properties and Applications. Springer, New York (2010). https://doi.org/10.1007/978-1-4419-1105-6
28. Zhang, Q., Zhang, C., Yin, G.: Optimal stopping of two-time scale Markovian systems: analysis, numerical methods, and applications. Nonlinear Anal.: Hybrid Syst. **26**, 151–167 (2017)

Space-Time Methods for Solving Time-Dependent PDEs

Preconditioners for Time-Harmonic Optimal Control Eddy-Current Problems

Owe Axelsson[1] and Dalibor Lukáš[2](✉)

[1] Institute of Geonics, Czech Academy of Sciences, Studentská 1768, 708 00 Ostrava-Poruba, Czech Republic
owe.axelsson@it.uu.se

[2] VŠB-Technical University of Ostrava, 17. listopadu 15, 708 33 Ostrava-Poruba, Czech Republic
dalibor.lukas@vsb.cz

Abstract. Time-harmonic formulations enable solution of time-dependent PDEs without use of normally slow time-stepping methods. Two efficient preconditioners for the discretized parabolic and eddy current electromagnetic optimal control problems, one on block diagonal form and one utilizing the two by two block structure of the resulting matrix, are presented with simplified analysis and numerical illustrations. Both methods result in tight eigenvalue bounds for the preconditioned matrix and very few iterations that hold uniformly with respect to the mesh, problem and method parameters, with the exception of the dependence on reluctivity for the block diagonal preconditioner.

Keywords: Preconditioning · Krylov subspace methods
Optimal control · Eddy currents · Time-harmonic

1 Introduction

Time dependent partial differential equations are normally solved numerically with some time stepping method. Due to reasons of numerical stability, if an explicit method is used, one must choose very small time steps or use a stable implicit method which requires the solution of a large scale linear system at each step, both of which can be computationally costly.

However, for time-harmonic problems one can approximate the solution on the whole given global time interval by a truncated Fourier series of trigonometric functions and use a multiharmonic approach. For linear problems the arising problems for each frequency separate, which makes it possible to solve for all frequencies, $\omega = k\pi/T$, $k = 0, 1, 2, \cdots$, where T is the end time, in parallel.

This work was supported by The Ministry of Education, Youth and Sports from the National Programme of Sustainability (NPU II) project "IT4Innovations excellence in science - LQ1602". The second author was supported by the Czech Science Foundation under the project 17-22615S.

I. Lirkov and S. Margenov (Eds.): LSSC 2017, LNCS 10665, pp. 47–54, 2018.
https://doi.org/10.1007/978-3-319-73441-5_4

Furthermore, to achieve a sufficient numerical accuracy it suffices normally with the use of few terms in the Fourier series expansion.

For nonlinear problems one can often use a two-grid method, which enables the solution of the nonlinear equation only on a coarse grid, see e.g. [4]. To illustrate the ideas, in this talk we mainly consider an optimal control problem for a linear parabolic heat equation and only shortly describe the application for an eddy current electromagnetic problem.

The main contribution of the paper is the presentation of two preconditioners for the arising two-by-two and four-by-four block matrix systems. The analysis and numerical tests show that they are very efficient, leading to few, mostly single digits iterations. One of the preconditioners has previously been presented by Kolmbauer and Langer [1] and the other is based on previously presented preconditioners for optimal control problems by Axelsson et al. [2]. A simplified analysis of the methods will appear in [7].

We show first the derivation of the two-by-two block matrix system and show then that the two preconditioners lead to tight eigenvalue bounds which for the first type preconditioner hold uniformly under some restrictions and for the second one without any restrictions with respect to all parameters. To illustrate the methods, the paper ends with numerical tests. It is found that they need very few iterations, in particular for practical values of the problem and method parameters.

2 The Optimal Control Problem

Following [1], consider the optimal control problem of finding the state $y(x,t)$ and control $u(x,t)$ that minimizes the functional,

$$J(y,u) = \frac{1}{2}\int_{\Omega\times(0,T)} |y(x,t) - y_d(x,t)|^2 dx\, dt + \frac{1}{2}\beta \int_{\Omega\times(0,T)} |u(x,t)|^2 dx\, dt,$$

subject to the time periodic parabolic, heat equation problem,

$$\frac{\partial y(x,t)}{\partial t} - \Delta y(x,t) = u(x,t) \text{ in } \Omega \times (0,T),$$
$$y(x,t) = 0 \text{ in } \Gamma \times (0,T), \quad y(x,0) = y(x,T) \text{ and } u(x,0) = u(x,T) \text{ in } \Omega.$$

Here $\Gamma = \partial\Omega$, y_d is the desired state and $\beta > 0$ is the cost regularization parameter for the control function $u(x,t)$. The target function is assumed to be time-harmonic, $y_d(x,t) = y_d(x)e^{i\omega t}$ with frequency $\omega = 2\pi k/T$ for some non-negative integer k.

Remark 1. If the target function is not time-harmonic, one can approximate it by a truncated Fourier series of the form $y_d = \sum_{k=0}^{N}\left(y_{d,k}^c \cos(k\omega t) + y_{d,k}^s \sin(k\omega t)\right)$, where the Fourier coefficients are given by classical expressions. Since the equation is linear, the solution and the control are also time-harmonic, $y(x,t) = y(x)e^{i\omega t}$ and $u(x,t) = u(x)e^{i\omega t}$ and the problem separates for the different frequencies.

Therefore it suffices to consider the single frequency problem,

$$\text{minimize}_{y,u} \ \frac{1}{2}\int_{\Omega}|y(x)-y_d(x)|^2dx+\frac{1}{2}\beta\int_{\Omega}|u(x)|^2dx$$

subject to $i\omega y(x)-\Delta y(x)=u(x)$ in Ω. We assume that $y(x)$ and $y_d(x)$ are real valued but the control $u(x)=u_0(x)+iu_1(x)$ must be complex valued.

The state equation and hence also the minimization problem, has a unique solution. Using an appropriate finite element subspace V_h for both y and u and a complex-valued Lagrange multiplier vector $\underline{\zeta}$, the corresponding Lagrangian functional for the discretized constrained optimization problem becomes

$$L(y,\underline{u},\underline{\zeta})=\frac{1}{2}(y-y_d)^TM(y-y_d)+\frac{1}{2}\beta\underline{u}^*M\underline{u}+Re\{\underline{\zeta}^*(i\omega My+Ky-M\underline{u})\},$$

where M is the mass matrix and K is the negative discrete Laplacian. One of the first order necessary condition, $\nabla L(y,\underline{u},\underline{\zeta})=0$, shows that $\beta M\underline{u}=-M\underline{\zeta}$ and lead to the reduced system, $\begin{bmatrix} M & \beta(K-i\omega M)\\ K+i\omega M & -M\end{bmatrix}\begin{bmatrix} y\\ \underline{u}\end{bmatrix}=\begin{bmatrix} My_d\\ 0\end{bmatrix}$, using the relation $\underline{\zeta}=\beta\underline{u}$. Here we multiply the second equation with $\sqrt{\beta}$ and introduce $\tilde{u}=\sqrt{\beta}u$, which gives

$$\begin{bmatrix} M & \sqrt{\beta}(K-i\omega M)\\ \sqrt{\beta}(K+i\omega M) & -M\end{bmatrix}\begin{bmatrix} y\\ \underline{\tilde{u}}\end{bmatrix}=\begin{bmatrix} My_d\\ 0\end{bmatrix}. \tag{1}$$

We present now two preconditioners for this block matrix.

3 A Block Diagonal Preconditioner

We consider now a general form of two-by-two block matrices, for which the matrix in (1) is a special case. Let then $\mathcal{A}=\begin{bmatrix} A & E-iF\\ E+iF & -A\end{bmatrix}$, where A is spd and E and F are symmetric and positive semidefinite (spsd). Following [1], but with a simplified analysis, we consider block diagonal preconditioner. The following eigenvalue bounds hold for the preconditioned matrix.

Proposition 1. *Let* $\mathcal{D}=\begin{bmatrix} D & 0\\ 0 & D\end{bmatrix}$, $D=A+E+F$ *and assume that* $ED^{-1}F=FD^{-1}E$. *This holds if* $F=\omega(A+\delta E)$, $\omega>0$, $0\le\delta\le 1$. *Then the matrix* $(\mathcal{D}^{-1}\mathcal{A})^2$ *is block diagonal, its eigenvalues are real and contained in the interval* $\frac{1}{4}\le\lambda\left((\mathcal{D}^{-1}\mathcal{A})^2\right)\le 1$. *If* $F=\omega A$, *then* $\frac{1}{3}\le\frac{1}{2(1+\omega/(1+\omega^2))}\le\lambda((\mathcal{D}^{-1}\mathcal{A})^2)\le 1$.

Proof. A proof is presented in [7]. Note that for ω small or large, the lower bound is close to its value $1/2$ taken for $\omega=0$.

To solve a system with $\mathcal{A}$ using the block diagonal preconditioner in a Krylov subspace iteration method, as well known requires then $2m$ iterations so that the residual r^{2m} satisfies $\|r^{2m}\|/\|r^0\|\le\frac{2q^m}{1+q^{2m}}$, where $q=\frac{\sqrt{3}-1}{\sqrt{3}+1}=\frac{1}{2+\sqrt{3}}$. This is guaranteed for $2m$ proportional to the logarithm of the relative precision.

4 A Preconditioner for a Two-by-Two Block Matrix of Special Form with Square Matrix Blocks

Consider now a matrix $\mathcal{A} = \begin{bmatrix} A & B_2 \\ B_1 & -A \end{bmatrix}$ and preconditioner $\mathcal{C} = \begin{bmatrix} A + B_1 + B_2 & B_2 \\ B_1 & -A \end{bmatrix}$, where A, of order $n \times n$, is spd and $A + B_i$, $i = 1, 2$ are nonsingular. The preconditioner and the given matrix will be used in a Krylov subspace type of iteration method. We show first that linear systems $\mathcal{C} \begin{bmatrix} x \\ y \end{bmatrix} = \begin{bmatrix} f \\ g \end{bmatrix}$ can be readily solved. For this reason, change the sign of the second equation and add the first, which results in $\begin{bmatrix} A + B_1 & B_2 \\ 0 & A + B_2 \end{bmatrix} \begin{bmatrix} x \\ z \end{bmatrix} = \begin{bmatrix} f \\ f - g \end{bmatrix}$, where $z = x + y$. Hence the algorithm to compute the solution (x, y) can be written: (1) Solve $(A + B_2)z = f - g$, compute $\tilde{f} = f - B_2 z$. (2) Solve $(A + B_1)x = \tilde{f}$, compute $y = z - x$. Therefore, besides a matrix vector multiplication with matrix B_2 and some vector additions, the algorithm involves a solution with matrix $A + B_2$ and with $A + B_1$. In our problems, they will be elliptic type of matrices.

It is seen that the above algorithm is equivalent to computing the action of the following form of the inverse of $\mathcal{C}$,

$$\mathcal{C}^{-1} = \begin{bmatrix} I & 0 \\ -I & I \end{bmatrix} \begin{bmatrix} (A + B_1)^{-1} & 0 \\ 0 & I \end{bmatrix} \begin{bmatrix} I & -B_2 \\ 0 & I \end{bmatrix} \begin{bmatrix} I & 0 \\ 0 & -(A + B_2)^{-1} \end{bmatrix} \begin{bmatrix} I & 0 \\ -I & I \end{bmatrix}.$$

This form was already given in [2].

For the computation of the rate of convergence of the iteration method we need information about the eigenvalue distribution of $\mathcal{C}^{-1}\mathcal{A}$.

Proposition 2. *Let* $\mathcal{A} = \begin{bmatrix} A & B_2 \\ -B_1 & A \end{bmatrix}$, $\mathcal{C} = \mathcal{A} + \begin{bmatrix} B_1 + B_2 & 0 \\ 0 & 0 \end{bmatrix}$, *where* A, *of order* $n \times n$ *is spd and* $B_2 = B_1^*$, $B_1 = B$, $B + B^*$ *is positive semidefinite and* $A + B$ *is nonsingular. Then the eigenvalues* λ *of* $\mathcal{C}^{-1}\mathcal{A}$ *are real and satisfy* $\frac{1}{2} \le \frac{1}{1+\alpha} \le \lambda \le 1$, *where* $\alpha = \max_{\mu} \{Re(\mu)/|\mu|\}$, *and* μ *are eigenvalues of* $Bz = \mu Az$, $z \neq 0$. *The eigenvector space is complete, so* $\mathcal{C}^{-1}\mathcal{A}$ *is a normal matrix. Here* $\lambda = 1$ *is an eigenvalue of dimension* $n + n_0$, *where* $n_0 = dim\{\mathcal{N}(B + B^*)\}$.

Proof. For a proof, see [7].

This proposition shows that the relative size, $Re(\mu)/|\mu|$ of the real part of the eigenvalues of $\mu Az = Bz$, $\|z\| \neq 0$, determines the lower bound of $\mathcal{C}^{-1}\mathcal{A}$.

For the matrix in (1) it follows that

$$\mathcal{C} = \begin{bmatrix} M + 2\sqrt{\beta}K & \sqrt{\beta}(K - i\omega M) \\ \sqrt{\beta}(K + i\omega M) & -M \end{bmatrix},$$

and $\sqrt{\beta}(K - i\omega M)z = \mu Mz$, $\|z\| \neq 0$, so $\alpha = \frac{Re(\mu)}{\sqrt{Re(\mu)^2 + \omega^2}} \le 1$. The correspondingly preconditioned Krylov subspace iteration method converges therefore fast with a rate determined by the narrow eigenvalue bounds $\frac{1}{2} \le \frac{1}{1+\alpha} \le \lambda \le 1$, and will be particularly fast for large values of ω where α gets small.

5 A Double Two-by-Two Block Matrix Arising in Eddy Current Electromagnetic Problems

Following [1], consider now the multiharmonic method to numerically solve an eddy current problem. Here the vector Laplacian operator in Sect. 2 is replaced by a curl curl operator. It is assumed that Ω is a bounded Lipschitz domain in $\mathbb{R}^3$. The reluctivity $\nu \in L^\infty(\Omega)$ is uniformly positive and we assume that it does not depend on the solution y, so the problem is linear. The conductivity $\sigma \in L^\infty(\Omega)$ is piecewise constant, positive in conducting and zero in nonconducting subdomains. Due to the discontinuity of σ and to obtain uniqueness in the nonconducting domains, the state equation must be regularized. This is done here by adding a positive term εy, $\varepsilon > 0$ to the state equation. However, for the case of divergence free vector solutions, this is not needed. The regularized optimal control problem takes then the form,

$$\text{minimize}_{(y,u)} \frac{1}{2} \int_{\Omega \times (0,T)} |y - y_d|^2 dx\, dt + \frac{\beta}{2} \int_{\Omega \times (0,T)} |u|^2 dx\, dt,$$

subject to the state equation,

$$\begin{cases} \sigma \frac{\partial y}{\partial t} + \operatorname{curl}(\nu \operatorname{curl} y) + \varepsilon y = u \text{ in } \Omega \times (0,T) \\ y \times \underline{n} = 0 \text{ on } \Gamma \times (0,T), \qquad y = y_0 \text{ on } \Gamma \times \{0\}. \end{cases}$$

For a time-harmonic problem, the initial condition is replaced by the periodicity equation, $y(0) = y(T)$, in Ω. Applying a Lagrange multiplier w to impose the state equation, the Lagrangian functional becomes

$$\mathcal{L}(y,u,w) = J(y,u) + \int_{\Omega \times (0,T)} \left(\sigma \frac{\partial y}{\partial t} + \operatorname{curl}(\nu \operatorname{curl} y) + \varepsilon y - u \right) w\, dx\, dt.$$

The first order necessary condition $\nabla_w \mathcal{L}(y,u,w) = 0$ gives the relation $\beta u = w$ in $\Omega \times (0,T)$, which enables elimination of the control variable. As before we use a truncated Fourier series expansion for y and u, which decouple the equations so that it suffices to consider only one frequency.

For the finite element discretization we use the lowest order tetrahedral edge elements, originally introduced in Nèdèlec [3]. After a reordering of the equations, this yields the following system of linear equations,

$$\begin{bmatrix} M & 0 & K & -M_\omega \\ 0 & M & M_\omega & K \\ K & M_\omega & -\beta^{-1} M & 0 \\ -M_\omega & K & 0 & -\beta^{-1} M \end{bmatrix} \begin{bmatrix} y^c \\ y^s \\ w^c \\ w^s \end{bmatrix} = \begin{bmatrix} y_d^c \\ y_d^s \\ 0 \\ 0 \end{bmatrix},$$

where $M = [M_{ij}]$, $M_{ij} = \int_\Omega u_j v_i dx$, $(M_\omega)_{ij} = \int_\Omega \sigma \omega u_j v_i dx$, $i,j = 1,2,\cdots,n$, $K = [K_{ij}]$, $K_{ij} = \int_\Omega \nu \operatorname{curl} u_j \operatorname{curl} v_i + \varepsilon \int_\Omega u_j v_i dx$. Further u_i, v_i are taken from the set of finite element basis functions in $H_0(\text{curl}) = \{v \in L^2(\Omega) : \operatorname{curl} v \in$

$L^2(\Omega), v \times \underline{n} = 0$ on $\Gamma\}$ on edges i, j. The values on the right hand side are given by $(y_d^c)_i = \int_\Omega y_d^c v_i dx, \quad (y_d^s)_i = \int_\Omega y_d^s v_i dx$.

In a similar way as was done before, we modify the system by multiplying the last two equations with $\sqrt{\beta}$ and scale the multiplier variable to $\begin{bmatrix} \tilde{w}^c \\ \tilde{w}^s \end{bmatrix} = \frac{1}{\sqrt{\beta}} \begin{bmatrix} w^c \\ w^s \end{bmatrix}$. Using the same type of preconditioning as in Sect. 4, obtained by adding the off-diagonal blocks to the primary diagonal block, we get

$$\mathcal{C} = \begin{bmatrix} M + 2\tilde{K} & 0 & \tilde{K} & -\tilde{M}_\omega \\ 0 & M + 2\tilde{K} & \tilde{M}_\omega & \tilde{K} \\ \tilde{K} & \tilde{M} & -M & 0 \\ -\tilde{M} & \tilde{K} & 0 & -M \end{bmatrix},$$

where $\tilde{K} = \sqrt{\beta} K$ and $\tilde{M}_\omega = \sqrt{\beta} M_\omega$. To get the same form of the matrix as in Sect. 4, let $A = \begin{bmatrix} M & 0 \\ 0 & M \end{bmatrix}$, $B = \begin{bmatrix} \tilde{K} & \tilde{M}_\omega \\ -\tilde{M}_\omega & \tilde{K} \end{bmatrix}$. Then $\mathcal{A} = \begin{bmatrix} A & B^* \\ B & -A \end{bmatrix}$ and $C = \begin{bmatrix} A + 2\tilde{K}' & B^* \\ B & -A \end{bmatrix}$, where $\tilde{K}' = \begin{bmatrix} \tilde{K} & 0 \\ 0 & \tilde{K} \end{bmatrix}$. Then $B + B^* = 2\tilde{K}'$, which is spd and it follows from Proposition 2 that the eigenvalues λ of $\mathcal{C}^{-1}\mathcal{A}$ satisfy $\frac{1}{1+\alpha} \le \lambda \le 1$, where α is the ratio, $\alpha = Re(\mu)/|\mu|$ and μ is eigenvalue of the generalized eigenvalue problem,

$$\begin{bmatrix} \tilde{K} & \tilde{M}_\omega \\ -\tilde{M}_\omega & \tilde{K} \end{bmatrix} \begin{bmatrix} x \\ y \end{bmatrix} = \mu \begin{bmatrix} M & 0 \\ 0 & M \end{bmatrix} \begin{bmatrix} x \\ y \end{bmatrix} \text{ or } \begin{bmatrix} \hat{K} & \hat{M}_\omega \\ -\hat{M}_\omega & \hat{K} \end{bmatrix} \begin{bmatrix} M^{1/2} x \\ M^{1/2} y \end{bmatrix} = \mu \begin{bmatrix} M^{1/2} x \\ M^{1/2} y \end{bmatrix},$$

where $\hat{K} = M^{-\frac{1}{2}} \tilde{K} M^{-\frac{1}{2}}$, $\hat{M}_\omega = M^{-\frac{1}{2}} \tilde{M}_\omega M^{-\frac{1}{2}}$. Hence $\alpha = \frac{\|\hat{K}^{-1/2} \hat{M}_\omega \hat{K}^{-1/2}\|}{1 + \|\hat{K}^{-1/2} \hat{M}_0 \hat{K}^{-1/2}\|}$.

The arising inner systems with the block matrix $\begin{bmatrix} M + \tilde{K} & \tilde{M}_\omega \\ -\tilde{M}_\omega & M + \tilde{K} \end{bmatrix}$ can also be solved by iteration using the same type of preconditioner as for the outer system, i.e. with $\begin{bmatrix} M + \tilde{K} + 2\tilde{M}_\omega & \tilde{M}_\omega \\ -\tilde{M}_\omega & M + \tilde{K} \end{bmatrix}$.

The corresponding eigenvalues $\tilde{\lambda}$ satisfy $(\tilde{\lambda} - 1) \begin{bmatrix} M + \tilde{K} + 2\tilde{M}_\omega & \tilde{M}_\omega \\ -\tilde{M}_\omega & M + \tilde{K} \end{bmatrix} \begin{bmatrix} x \\ y \end{bmatrix} = - \begin{bmatrix} 2\tilde{M}_\omega & 0 \\ 0 & 0 \end{bmatrix} \begin{bmatrix} x \\ y \end{bmatrix}$, from which it follows that $\tilde{\lambda} \le 1$ and $(\tilde{\lambda} - 1) x^T (M + \tilde{K} + 2\tilde{M}_\omega + \tilde{M}_\omega (M + \tilde{K})^{-1} \tilde{M}_\omega) x = -2x^T \tilde{M}_\omega x$, i.e. $(\tilde{\lambda} - 1)\hat{x}^T (I + 2\hat{\hat{M}}_\omega + \hat{\hat{M}}_\omega^2)\hat{x} = -2\hat{x}^T \hat{\hat{M}}_\omega \hat{x}$, where $\hat{\hat{M}}_\omega = (M + \tilde{K})^{-1/2} \tilde{M}_\omega (M + \tilde{K})^{-1/2}$ and $\hat{x} = (M + \tilde{K})^{1/2} x$. It follows that $\tilde{\lambda} - 1 \ge -\frac{1}{2}$, i.e. $\tilde{\lambda} \ge \frac{1}{2}$, so $\frac{1}{2} \le \tilde{\lambda} \le 1$.

In practice mostly the control and observation are restricted to subdomains of Ω. Due to limitation of space we do not consider this here, but it can be shown that our preconditioner performs as well for these problems also. However, as reported in [5], the performance of the block diagonal preconditioner deteriorates for small values of ν and β.

6 Numerical Illustrations

In order to demonstrate the efficiency of our method, which holds uniformly with respect to all model and method parameters involved, some numerical tests were done. To enable a comparison with the block diagonal preconditioner used in the thesis by Kolmbauer [6], we test our method on the same problems as done there. Due to limitations we thereby choose only a subset of the problems.

We demonstrate this only on the eddy current electromagnetic problem with constant and with jump of the conductivity coefficient. In Table 1 we show how the uniformly bounded and low number of flexible GMRES (FGMRES) iterations varies for different values of the frequency ω and of the control cost parameter β. This is done for two values of the mesh size parameter h. In Table 2 (left) it is shown how the number of iterations vary with respect to the values of reluctivity and ω, and in Table 2 (right) how they vary with respect to conductivity. Thereby the conductivity is fixed in the domain $\Omega = [0,1]^3$, respectively takes a constant positive value σ_2 in the subcube $\Omega_2 = [1/4, 3/4]^3$ and $\sigma_1 = 1$ in the rest of the domain $\Omega_1 = \Omega \setminus \Omega_2$. The listed numbers of degrees of freedom (DOFs) for the lowest order Nédélec elements on tetrahedron is equal to one per edge.

All tests demonstrate a remarkable, uniformly low number of iterations. For the problem with no jump in the conductivity coefficient the number of iteration decreases for large values of the frequency, which is in accordance with the theoretical bound of the condition number. It can be seen that the number of iterations demonstrate a more favourable performance of our method as compared to the block diagonal preconditioner. Furthermore, the choice of elliptic

Table 1. Robustness of outer and total inner (in brackets) FGMRES iterations with respect to β, ω, and h, while fixing $\nu = \sigma_2 = 1$ and outer rel. prec. 10^{-8}.

h	β	Inner rel. prec. 10^{-2}					Inner rel. prec. 10^{-6}				
		ω					ω				
		10^{-8}	10^{-4}	10^{0}	10^{4}	10^{8}	10^{-8}	10^{-4}	10^{0}	10^{4}	10^{8}
1/16	10^{-10}	10(20)	10(20)	10(20)	10(40)	3(11)	10(20)	10(20)	10(29)	10(87)	2(12)
	10^{-8}	11(22)	11(22)	11(22)	10(57)	3(11)	11(22)	11(22)	11(33)	9(125)	2(12)
	10^{-6}	11(22)	11(22)	11(22)	6(48)	3(11)	11(22)	11(22)	11(36)	5(80)	2(12)
	10^{-4}	9(18)	9(18)	9(18)	6(48)	3(11)	9(18)	9(18)	9(45)	4(72)	2(12)
	10^{-2}	5(10)	5(10)	6(23)	7(56)	3(11)	5(10)	5(15)	5(30)	3(55)	2(12)
	10^{0}	4(8)	4(8)	5(19)	7(56)	3(11)	4(8)	4(12)	4(24)	3(56)	2(12)
1/32	10^{-10}	10(20)	10(20)	10(20)	11(42)	4(15)	10(20)	10(20)	10(29)	10(89)	2(14)
	10^{-8}	11(22)	11(22)	11(22)	10(58)	4(15)	11(22)	11(22)	11(33)	10(144)	2(13)
	10^{-6}	11(22)	11(22)	11(22)	6(48)	4(15)	11(22)	11(22)	11(36)	5(80)	2(14)
	10^{-4}	9(18)	9(18)	9(18)	7(56)	4(15)	9(18)	9(18)	9(45)	4(72)	2(14)
	10^{-2}	5(10)	5(10)	6(23)	7(56)	4(15)	5(10)	5(14)	5(30)	3(55)	2(14)
	10^{0}	4(8)	4(8)	5(19)	7(56)	4(15)	4(8)	4(12)	4(24)	3(55)	2(14)

Table 2. Robustness of outer and total inner (in brackets) FGMRES iterations with respect to: (a) β, ν, and h, while fixing $\omega = \sigma_2 = 1$ (left); (b) β, σ_2, and h, while fixing $\omega = \nu = 1$ (right). In both cases we fix outer rel. prec. 10^{-8} and inner rel. prec. 10^{-2}.

h	β	ν					σ_2				
		10^{-8}	10^{-4}	10^{0}	10^{4}	10^{8}	10^{-8}	10^{-4}	10^{0}	10^{4}	10^{8}
1/16	10^{-10}	1(2)	2(4)	10(20)	5(10)	2(4)	10(20)	10(20)	10(20)	10(38)	6(24)
	10^{-8}	2(4)	3(6)	11(22)	4(8)	3(6)	11(22)	11(22)	11(22)	11(65)	6(23)
	10^{-6}	2(4)	4(8)	11(22)	3(6)	3(6)	11(22)	11(22)	11(22)	12(75)	6(23)
	10^{-4}	3(6)	6(12)	9(18)	4(8)	4(8)	10(21)	10(21)	9(18)	10(80)	5(19)
	10^{-2}	2(8)	10(40)	6(23)	4(13)	5(17)	7(24)	7(24)	6(23)	7(56)	3(12)
	10^{0}	2(8)	10(57)	5(19)	3(12)	5(20)	7(32)	7(32)	5(19)	6(46)	2(8)
1/32	10^{-10}	1(2)	2(4)	10(20)	5(10)	3(6)	10(20)	10(20)	10(20)	10(38)	7(30)
	10^{-8}	2(4)	3(6)	11(22)	4(8)	3(6)	11(22)	11(22)	11(22)	12(71)	6(24)
	10^{-6}	2(4)	5(10)	11(22)	3(6)	3(6)	11(22)	11(22)	11(22)	12(76)	6(25)
	10^{-4}	3(6)	9(18)	9(18)	4(8)	4(8)	10(21)	10(21)	9(18)	10(80)	5(22)
	10^{-2}	2(8)	11(42)	6(23)	4(13)	5(16)	8(25)	8(25)	6(23)	7(56)	4(18)
	10^{0}	3(12)	10(58)	5(19)	3(12)	5(20)	8(43)	8(43)	5(19)	6(46)	2(9)

operator problem to be solved on the innermost level is straightforward in our method while it is somewhat more involved for the block diagonal preconditioner used in [5].

Tables 1 and 2 correspond to Tables 7.1–7.5 in [6] with the same stopping tolerance. It is seen that our method never gives more iterations than those reported there but mostly fewer and in some cases significantly smaller.

References

1. Kolmbauer, M., Langer, U.: A robust preconditioned MINRES solver for distributed time-periodic eddy current optimal control problems. SIAM J. Sci. Comput. **34**, B785–B809 (2012)
2. Axelsson, O., Farouq, S., Neytcheva, M.: A preconditioner for optimal control problems constrained by Stokes equation with a time-harmonic control. J. Comp. Appl. Math. **310**, 5–18 (2017)
3. Nédélec, J.C.: Mixed finite elements in $\mathbb{R}^3$. Numer. Math. **35**, 315–341 (1980)
4. Axelsson, O., Layton, W.: A two-level method for the discretization of nonlinear boundary value problems. SIAM J. Numer. Anal. **33**, 2359–2374 (1996)
5. Kollmann, M., Kolmbauer, M.: A preconditioned MinRes solver for time-periodic parabolic optimal control problems. Numer. Linear Algebra Appl. **20**, 761–784 (2013)
6. Kolmbauer, M.: The Multiharmonic finite element and boundary element method for simulation and control of eddy current problems. Ph.D. thesis, Johannes Kepler Universität, Linz, Austria (2012)
7. Axelsson, O., Lukáš, D.: Preconditioning methods for eddy current optimally controlled time-harmonic electromagnetic problems. J. Numer. Math. (to appear)

Functional Type Error Control for Stabilised Space-Time IgA Approximations to Parabolic Problems

Ulrich Langer[1], Svetlana Matculevich[1](✉), and Sergey Repin[2,3]

[1] RICAM Linz, Johann Radon Institute, Linz, Austria
{ulanger,smatculevich}@ricam.oeaw.ac.at
[2] St. Petersburg Department of V.A. Steklov Institute of Mathematics RAS, Saint Petersburg, Russia
serepin@pdmi.ras.ru
[3] University of Jyvaskyla, Jyväskylä, Finland
sergey.s.repin@jyu.fi

Abstract. The paper is concerned with reliable space-time IgA schemes for parabolic initial-boundary value problems. We deduce a posteriori error estimates and investigate their applicability to space-time IgA approximations. Since the derivation is based on purely functional arguments, the estimates do not contain mesh dependent constants and are valid for any approximation from the admissible (energy) class. In particular, they imply estimates for discrete norms associated with stabilised space-time IgA approximations. Finally, we illustrate the reliability and efficiency of presented error estimates for the approximate solutions recovered with IgA techniques on a model example.

Keywords: Error control · Functional error estimates
Stabilised space-time IgA schemes · Fully-adaptive space-time schemes

Countless usage of the time-dependent systems governed by parabolic partial differential equations (PDEs) in scientific and engineering applications trigger their active investigation in mathematical and numerical modelling. By virtue of the fast development of parallel computers, treating time in the evolutionary equations as yet another dimension in space became quite natural. The so-called *space-time approach* is not restricted with pitfalls of time-marching schemes. On the contrary, it becomes quite useful when efficient parallel methods and their implementation on massively parallel computers are considered (rather than attempt to reiterate all prior work, we refer the reader to [17], whose introductory section contains an extensive overview of various space-time techniques).

Investigation of effective adaptive refinement methods is crucial for the construction of fast and efficient solvers for PDEs. In the same time, the aspect of scheme localisation is strongly linked with reliable and quantitatively efficient a posteriori error estimation tools. The latter one is expected to identify the

I. Lirkov and S. Margenov (Eds.): LSSC 2017, LNCS 10665, pp. 55–65, 2018.
https://doi.org/10.1007/978-3-319-73441-5_5

areas of the considered computational domain with relatively high discretization errors and provide an automated refinement strategy in order to reach the desired accuracy level for the current reconstruction. Local refinement tools of IgA (e.g., T-splines, THB-splines, and LR-splines) have been combined with various a posteriori error estimation techniques, e.g., error estimates (EEs) using the hierarchical basis [4,26], residual-based [2,9,13], and goal-oriented EEs [3,14,27]. Below, we use a different (functional) method providing fully guaranteed EEs in the various weighted norms equivalent to the global energy norm. These estimates include only global constants (independent of the mesh characteristic h) and are valid for any approximation from the admissible functional space. Functional EEs (so-called majorants and minorants) were introduced in [24] and later applied to different mathematical models [18,22]. They provide guaranteed, sharp, and fully computable upper and lower bounds of errors. This approach in combination with IgA approximations generated by tensor-product splines was investigated in [11] for elliptic boundary value problems (BVPs).

In this paper, we derive functional-type a posteriori EEs for time-dependent problems in the context of the space-time IgA scheme introduced in [17]. The latter one exploits the time-upwind test function motivated by the space-time streamline diffusion method (see, e.g., [8,10]) and approximations provided by IgA framework. By exploiting the universality and efficiency of the considered EEs as well as the smoothness of the IgA approximations, we aim at the construction of fully-adaptive, fast and efficient parallel space-time methods that could tackle complicated problems inspired by industrial applications.

This work has the following structure: Sect. 1 defines the problem and discusses its solvability, whereas Sect. 2 presents the stabilised space-time IgA scheme with its main properties. An overview of main ideas and definitions used in the IgA framework can be found in the same section. In Sect. 3, we introduce new functional type a posteriori EEs using the stabilised formulation of parabolic initial BVPs (I-BVPs). Finally, Sect. 4 presents numerical results demonstrating the efficiency of the majorants in the elliptic case.

1 Model Problem

Let $\overline{Q} := Q \cup \partial Q, Q := \Omega \times (0,T)$, denote the space-time cylinder, where $\Omega \subset \mathbb{R}^d, d \in \{1,2,3\}$, is a bounded Lipschitz domain with boundary $\partial\Omega$, and $(0,T)$ is a given time interval, $0 < T < +\infty$. Here, the cylindrical surface is defined as $\partial Q := \Sigma \cup \overline{\Sigma}_0 \cup \overline{\Sigma}_T$ with $\Sigma = \partial\Omega \times (0,T), \Sigma_0 = \Omega \times \{0\}$, and $\Sigma_T = \Omega \times \{T\}$. We discuss our approach to guaranteed error control of space-time approximations with the paradigm of the classical *linear parabolic I-BVP*: find $u : \overline{Q} \to \mathbb{R}$ satisfying the system

$$\partial_t u - \Delta_x u = f \quad \text{in} \quad Q, \qquad u = 0 \quad \text{on} \quad \Sigma, \qquad u = u_0 \quad \text{on} \quad \overline{\Sigma}_0, \tag{1}$$

where ∂_t is the time derivative, Δ_x denotes the Laplace operator in space, $f \in L^2(Q)$, and $u_0 \in H_0^1(\Sigma_0)$ are the given source function and initial data, respectively. Here, $L^2(Q)$ is the space of square-integrable functions over

Q quipped with the usual norm and scalar product denoted respectively by $\| v \|_Q := \| v \|_{L^2(Q)}$ and $(v, w)_Q := \int_Q v(x,t) w(x,t) \, dxdt, \forall v, w \in L^2(Q)$.

By $H^k(Q), k \geq 1$, we denote spaces of functions having generalised square-summable derivatives of the order k with respect to (w.r.t.) space and time. Next, we introduce the Sobolev spaces $H_0^1(Q) := \{ w \in H^1(Q) : w|_\Sigma = 0 \}$, $H^1_{0,\bar{0}}(Q) := \{ w \in H_0^1(Q) : w_{\Sigma_T} = 0 \}, V_0 := H^1_{0,\underline{0}}(Q) := \{ w \in H_0^1(Q) : w_{\Sigma_0} = 0 \}$, and $V_0^{\Delta_x} := H_0^{\Delta_x,1}(Q) := \{ w \in H_0^1(Q) : \Delta_x w \in L^2(Q) \}$. Moreover, we use auxiliary Hilbert spaces for vector-valued functions

$$H^{\mathrm{div}_x,0}(Q) := \{ \boldsymbol{y} \in [L^2(Q)]^d \ : \ \mathrm{div}_x \boldsymbol{y} \in L^2(Q) \} \text{ and}$$
$$H^{\mathrm{div}_x,1}(Q) := \{ \boldsymbol{y} \in H^{\mathrm{div}_x,0}(Q) \ : \ \partial_t \boldsymbol{y} \in [L^2(Q)]^d \}$$

equipped with respective semi-norms $\|\boldsymbol{y}\|^2_{H^{\mathrm{div}_x,0}} := \|\mathrm{div}_x \boldsymbol{y}\|^2_Q$ and $\|\boldsymbol{y}\|^2_{H^{\mathrm{div}_x,1}} := \|\mathrm{div}_x \boldsymbol{y}\|^2_Q + \|\partial_t \boldsymbol{y}\|^2_Q$.

Further in the paper, C_{F} stands for the constant in the Friedrichs inequality $\|w\|_Q \leq C_{\mathrm{F}} \|\nabla_x w\|_Q, \forall w \in H_0^{1,0}(Q) := \{ w \in L^2(Q) : \nabla_x w \in [L^2(Q)]^2, w|_\Sigma = 0 \}$. From [15, Theorem 2.1] it follows that, if $f \in L^2(Q)$ and $u_0 \in H_0^1(\Sigma_0)$, the problem (1) is uniquely solvable in $V_0^{\Delta_x}$, and the solution u depends continuously on t in the $H_0^1(\Omega)$-norm. Moreover, according to [15, Remark 2.2], $\| \nabla_x u(\cdot, t) \|^2_\Omega$ is an absolutely continuous function of $t \in [0, T]$ for any $u \in V_0^{\Delta_x}$. If $u_0 \in L^2(\Sigma_0)$, then the problem has a unique solution u in the wider class $H_0^{1,0}(Q)$, and it satisfies the generalised formulation

$$(\nabla_x u, \nabla_x w)_Q - (u, \partial_t w)_Q =: a(u, w) = l(w) := (f, w)_Q + (u_0, w)_{\Sigma_0} \tag{2}$$

for all $w \in H^1_{0,\bar{0}}(Q)$, where $(u_0, w)_{\Sigma_0} := \int_{\Sigma_0} u_0(x) \, w(x, 0) dx = \int_\Omega u_0(x) \, w(x, 0) dx$. According to the well-established arguments (see [15,28]), without loss of generality, we can 'homogenise' the problem, i.e., consider (2) with $u_0 = 0$.

Our main goal is to derive fully computable estimates for space-time IgA approximations of this class of problems. For this purpose, we use the functional approach to a posteriori EE. Initially, their simplest form has been obtained for a heat equation in [23]. Numerical properties of above-mentioned EE w.r.t. the time-marching and space-time method are discussed in [6,20,21].

2 Stabilized Formulation of the Problem and Its Discretization

For the convenience of the reader, we first recall the general concept of the IgA approach, the definition of B-splines (NURBS) and their use in the geometrical representation of the space-time cylinder Q, as well as in the construction of the IgA trial spaces, used to approximate solutions satisfying (2).

Let $p \geq 2$ denote a degree of polynomials used for the IgA approximations and n denote the number of basis functions used to construct a B-spline curve. A *Knot-vector* is a non-decreasing set of coordinates in a parameter domain, written as $\Xi = \{\xi_1, \ldots, \xi_{n+p+1}\}, \xi_i \in \mathbb{R}$, where $\xi_1 = 0$ and $\xi_{n+p+1} = 1$. The knots can be repeated, and the multiplicity of the i-th knot is indicated by m_i. Throughout the paper, we consider only so-called open knot vectors, i.e., $m_1 = m_{n+p+1} = p+1$. For $\widehat{Q} := (0,1)$, $\widehat{\mathcal{K}}_h$ denotes a locally quasi-uniform mesh, where each element $\widehat{K} \in \widehat{\mathcal{K}}_h$ is constructed by the distinct neighbouring knots. The global size of $\widehat{\mathcal{K}}_h$ is denoted by $\hat{h} := \max_{\widehat{K} \in \widehat{\mathcal{K}}_h} \{\hat{h}_{\widehat{K}}\}$, where $\hat{h}_{\widehat{K}} := \operatorname{diam}(\widehat{K})$.

The *univariate B-spline basis functions* $\widehat{B}_{i,p} : \widehat{Q} \to \mathbb{R}$ are defined by means of Cox-de Boor recursion formula and are $(p - m_i)$-times continuously differentiable across the i-th knot with multiplicity m_i. The scope of this paper is limited to a single-patch domain. The *multivariate B-splines* on $\widehat{Q} := (0,1)^{d+1}, d = \{1,2,3\}$, is defined as a tensor-product of the univariate ones. In multidimensional case, we define the knot-vector dependent on the coordinate direction $\Xi^\alpha = \{\xi_1^\alpha, \ldots, \xi_{n^\alpha+p^\alpha+1}^\alpha\}, \xi_i^\alpha \in \mathbb{R}$, where $\alpha = 1, \ldots, d+1$ indicates the direction (in space or time). Furthermore, we introduce set of multi-indices $\mathcal{I} = \{i = (i_1, \ldots, i_{d+1}) : i_\alpha = 1, \ldots, n_\alpha,\ \alpha = 1, \ldots, d+1\}$ and multi-index $p := (p_1, \ldots, p_{d+1})$ indicating the order of polynomials. Then, multivariate B-spline basis functions are defined as $\widehat{B}_{i,p}(\boldsymbol{\xi}) := \prod_{\alpha=1}^{d+1} \widehat{B}_{i_\alpha,p_\alpha}(\xi^\alpha)$, where $\boldsymbol{\xi} = (\xi^1, \ldots, \xi^{d+1}) \in \widehat{Q}$. The *univariate and multivariate NURBS basis* functions are defined in $\widehat{Q}$ by means of B-spine basis functions, i.e., for given p and any $i \in \mathcal{I}$ $\widehat{R}_{i,p} : \widehat{Q} \to \mathbb{R}$ is generated as $\widehat{R}_{i,p}(\boldsymbol{\xi}) := \frac{w_i\, \widehat{B}_{i,p}(\boldsymbol{\xi})}{W(\boldsymbol{\xi})}$. Here, $W(\boldsymbol{\xi})$ is a weighting function $W(\boldsymbol{\xi}) := \sum_{i \in \mathcal{I}} w_i\, \widehat{B}_{i,p}(\boldsymbol{\xi})$, where $w_i \in \mathbb{R}^+$.

The physical space-time domain $Q \subset \mathbb{R}^{d+1}$ is defined by the geometrical mapping of the parametric domain $\widehat{Q} := (0,1)^{d+1}$:

$$\Phi : \widehat{Q} \to Q := \Phi(\widehat{Q}) \subset \mathbb{R}^{d+1}, \quad \Phi(\boldsymbol{\xi}) := \sum_{i \in \mathcal{I}} \widehat{R}_{i,p}(\boldsymbol{\xi})\, \mathbf{P}_i, \tag{3}$$

where $\{\mathbf{P}_i\}_{i \in \mathcal{I}} \in \mathbb{R}^{d+1}$ are the control points. For simplicity, we assume the same polynomial degree for all coordinate directions, i.e., $p_\alpha = p$ for all $\alpha = 1, \ldots, d+1$. By means of geometrical mapping (3), the mesh $\mathcal{K}_h$ discretising Q is defined as $\mathcal{K}_h := \{K = \Phi(\widehat{K}) : \widehat{K} \in \widehat{\mathcal{K}}_h\}$. The global mesh size is denoted by

$$h := \max_{K \in \mathcal{K}_h} \{\, h_K \,\}, \quad h_K := \|\nabla \Phi\|_{L^\infty(K)} \hat{h}_{\widehat{K}}. \tag{4}$$

Moreover, we assume that $\mathcal{K}_h$ is quasi-uniform mesh, i.e., there exists a positive constant C_u independent of h, such that $h_K \le h \le C_u\, h_K$.

The finite dimensional spaces on Q are constructed by a push-forward of the NURBS basis functions $V_h := \operatorname{span}\{\phi_{h,i} := \widehat{R}_{i,p} \circ \Phi^{-1}\}_{i \in \mathcal{I}}$, where the geometrical mapping Φ is invertible in Q, with smooth inverse on each element $K \in \mathcal{K}_h$ (see [1,25]). The subspace $V_{0h} := V_h \cap V_{0,\underline{0}}(Q)$, where $V_{0,\underline{0}} := V_0 \cap H^1_{0,\underline{0}}(Q)$ is introduced for the functions satisfying homogeneous boundary condition (BC).

In order to provide efficient discretization method, we test (1) with the time-upwind test-function

$$\lambda w + \mu\, \partial_t w, \quad w \in V_{0,\underline{0}}^{\nabla_x \partial_t} := \{w \in V_{0,\underline{0}}^{\Delta_x} : \nabla_x \partial_t w \in L^2(Q)\}, \quad \lambda, \mu \ge 0. \tag{5}$$

and arrive at the stabilised weak formulation for $u \in V_0$, i.e.,

$$\begin{aligned}\big(\partial_t u, \lambda\, w + \mu\, \partial_t w\big)_Q + \big(\nabla_x u, \nabla_x (\lambda\, w + \mu\, \partial_t w)\big)_Q \\ =: a_s(u, w) = l_s(w) := (f, \lambda\, w + \mu\, \partial_t w)_Q, \quad \forall w \in V_{0,\underline{0}}^{\nabla_x \partial_t}. \end{aligned} \tag{6}$$

In [17], it was shown that *stable discrete space-time IgA scheme* corresponds to the case, when $\lambda = 1$ and $\mu = \delta_h = \theta h$ in (5) with $\theta > 0$ and global mesh-size h (cf. (4)) both for the fixed and moving spatial computational domains. Hence, (6) implies the discrete stabiliaed space-time problem: find $u_h \in V_{0h}$ satisfying

$$\begin{aligned}(\partial_t u_h, w_h + \delta_h \partial_t w_h)_Q + \big(\nabla_x u_h, \nabla_x (w_h + \delta_h \partial_t w_h)\big)_Q \\ =: a_{s,h}(u_h, w_h) = l_{s,h}(w_h) := (f, w_h + \delta_h\, \partial_t w_h)_Q, \quad \forall w_h \in V_{0h}. \end{aligned} \tag{7}$$

The V_{0h}-coercivity of $a_h(\cdot,\cdot) : V_{0h} \times V_{0h} \rightarrow \mathbb{R}$ w.r.t. the norm

$$|\!|\!| w_h |\!|\!|_{s,h}^2 := \|\nabla_x w_h\|_Q^2 + \delta_h\, \|\partial_t w_h\|_Q^2 + \|w_h\|_{\Sigma_T}^2 + \delta_h\, \|\nabla_x w_h\|_{\Sigma_T}^2 \tag{8}$$

follows from [17, Lemma 1] or [16, Lemma 3]. Moreover, one can show a boundedness property of the bilinear form $a_{h,s}(\cdot,\cdot)$ in appropriately chosen norms. Combining these coercivity and boundedness properties of $a_{h,s}(\cdot,\cdot)$ with the consistency of the scheme (7) and approximation results for the IgA spaces implies a corresponding a priori EE presented in Theorem 1 below.

Theorem 1. *Let* $u \in H_0^s(Q) := H^s(Q) \cap H_0^{1,0}(Q), s \in \mathbb{N}, s \geq 2,$ *be the exact solution of* (2) *and* $u_h \in V_{0h}$ *be the solution of* (7) *with some fixed parameter* θ. *Then, the following a priori EE*

$$\|u - u_h\|_{s,h} \leq C\, h^{r-1}\, \|u\|_{H^r(Q)} \tag{9}$$

holds, where $r = \min\{s, p+1\}, C > 0$ *is a generic constant independent of* h.

Proof: See, e.g., [17, Theorem 8]. □

3 Error Majorant

In this section, we derive error majorants for stabilised weak formulation of parabolic I-BVPs. The functional nature of these majorants allows obtaining a posteriori EEs for $u \in V_{0,\underline{0}}^{\Delta_x}$ and any $v \in V_{0,\underline{0}}^{\Delta_x}$. The error $e = u - v$ is measured in terms of

$$|\!|\!| e |\!|\!|_{s,\nu_i}^2 := \nu_1\, \|\nabla_x e\|_Q^2 + \nu_2\, \|\partial_t e\|_Q^2 + \nu_3\, \|\nabla_x e\|_{\Sigma_T}^2 + \nu_4\, \|e\|_{\Sigma_T}^2, \tag{10}$$

where $\{\nu_i\}_{i=1,\dots,4}$ are the positive weights introduced in the derivation process.

To obtain guaranteed error bounds of $|\!|\!| e |\!|\!|_{s,\nu_i}^2$, we apply a method similar to the one developed in [21,23] for parabolic I-BVPs. For the derivation process, we consider space of smoother functions $V_{0,\underline{0}}^{\nabla_x \partial_t}$ (cf. (5)) equipped with the norm

$\|w\|_{V_{0,\underline{0}}^{\nabla_x \partial_t}} := \sup_{t\in[0,T]} \|\nabla_x w(\cdot,t)\|^2_Q + \|w\|^2_{V_{0,\underline{0}}^{\Delta_x}}$, where $\|w\|^2_{V_{0,\underline{0}}^{\Delta_x}} := \|\Delta_x w\|^2_Q + \|\partial_t w\|^2_Q$, which is dense in $V_{0,\underline{0}}^{\Delta_x}$. According to [15, Remark 2.2], norms $\|\cdot\|_{V_{0,\underline{0}}^{\nabla_x \partial_t}} \approx \|\cdot\|_{V_{0,\underline{0}}^{\Delta_x}}$.

Let u_n be a sequence in $V_{0,\underline{0}}^{\nabla_x \partial_t}$. We consider the corresponding stabilised identity

$$a_s(u_n, w) = (f_n, \lambda\, w + \mu\, \partial_t w)_Q, \text{ where } f_n = (u_n)_t - \Delta_x u_n \in L^2(Q). \tag{11}$$

By subtracting $a_s(v_n, w), v_n \in V_{0,\underline{0}}^{\nabla_x \partial_t}$, from (11), and by setting $w = e_n = u_n - v_n \in V_{0,\underline{0}}^{\nabla_x \partial_t}$, we arrive at the so-called 'error-identity'

$$\begin{aligned}
&\lambda\, \|\nabla_x e_n\|^2_Q + \mu\, \|\, \partial_t e_n\|^2_Q + \tfrac{1}{2}\, (\mu\, \|\nabla_x e_n\|^2_{\Sigma_T} + \lambda \|e_n\|^2_{\Sigma_T}) \\
&= \lambda\Big((f_n - \partial_t v_n, e_n)_Q - (\nabla_x v_n, \nabla_x e_n)_Q\Big) \\
&\quad + \mu\Big((f_n - \partial_t v_n, \partial_t e_n)_Q - (\nabla_x v_n, \nabla_x\, \partial_t e_n)_Q\Big),
\end{aligned}$$

which is used in the derivation of the majorants of (10) in Theorems 2 and 3.

Theorem 2. *For any $v \in V_{0,\underline{0}}^{\Delta_x}$ and $\boldsymbol{y} \in H^{\mathrm{div}_x, 0}(Q)$, the following estimate holds:*

$$\begin{aligned}
\|e\|^2_{s,\nu_i} \leq \overline{\mathrm{M}}(v, \boldsymbol{y}; \gamma, \alpha_i) := \gamma\Big\{\lambda\big((1 + \alpha_1)\, \|\mathbf{r}_\mathrm{d}\|^2_Q \\
+ (1 + \tfrac{1}{\alpha_1})\, C_\mathrm{F}^2\, \|\mathbf{r}_\mathrm{eq}\|^2_Q\big) + \mu\big((1 + \alpha_2)\, \|\mathrm{div}_x \mathbf{r}_\mathrm{d}\|^2_Q + (1 + \tfrac{1}{\alpha_2})\, \|\mathbf{r}_\mathrm{eq}\|^2_Q\big)\Big\},
\end{aligned} \tag{12}$$

where $\nu_1 = (2 - \frac{1}{\gamma})\, \lambda, \nu_2 = (2 - \frac{1}{\gamma})\, \mu, \nu_3 = \mu, \nu_4 = \lambda$, C_F is the Friedrichs constant, $\mathbf{r}_\mathrm{eq}$ and $\mathbf{r}_\mathrm{d}$ are residuals defined by relations

$$\mathbf{r}_\mathrm{eq}(v, \boldsymbol{y}) := f - \partial_t v + \mathrm{div}_x\, \boldsymbol{y} \text{ and } \mathbf{r}_\mathrm{d}(v, \boldsymbol{y}) := \boldsymbol{y} - \nabla_x v, \tag{13}$$

$\lambda, \mu > 0$ are weights introduced in (5), $\gamma \in [\frac{1}{2}, +\infty)$, and $\alpha_i > 0, i = 1, 2$.

Proof: The detailed proof can be found in [16, Theorem 2], where we use the 'error-identity' and the density of space $V_{0,\underline{0}}^{\nabla_x \partial_t}$ in $V_{0,\underline{0}}^{\Delta_x}$ to obtain (12). □

The next theorem assumes higher regularity on the approximations v and $\boldsymbol{y}$.

Theorem 3. *For any $v \in V_{0,\underline{0}}^{\nabla_x \partial_t}$ and $\boldsymbol{y} \in H^{\mathrm{div}_x, 1}(Q)$, we have the estimate*

$$\begin{aligned}
\|e\|^2_{s,\nu_i} \leq \overline{\mathrm{M}}^{\mathrm{II}}(v, \boldsymbol{y}; \zeta, \beta_i, \epsilon) := \epsilon\, \mu \|\mathbf{r}_\mathrm{d}\|^2_{\Sigma_T} + \zeta\, \Big(\lambda\big((1 + \beta_1)\big((1 + \beta_2)\, \|\mathbf{r}_\mathrm{d}\|^2_Q \\
+ (1 + \tfrac{1}{\beta_2}) C_\mathrm{F}^2\, \|\mathbf{r}_\mathrm{eq}\|^2_Q\big) + (1 + \tfrac{1}{\beta_1})\, \tfrac{\mu^2}{\lambda^2}\, \|\partial_t \mathbf{r}_\mathrm{d}\|^2_Q\big) + \mu\, \|\mathbf{r}_\mathrm{eq}\|^2_Q\Big),
\end{aligned} \tag{14}$$

where $\nu_1 = (2 - \frac{1}{\zeta})\, \lambda, \nu_2 = (2 - \frac{1}{\zeta})\, \mu, \nu_3 = \mu\, (1 - \frac{1}{\epsilon}), \nu_4 = \lambda$, where C_F is the Friedrichs constant, $\mathbf{r}_\mathrm{eq}(v, y)$ and $\mathbf{r}_\mathrm{d}(v, y)$ are residuals in (13), $\lambda, \mu > 0$ are parameters in (5), $\zeta \in [\frac{1}{2}, +\infty), \epsilon \in [1, +\infty)$, and $\beta_i > 0, i = 1, 2$.

Proof: By using analogous density arguments and integral manipulations with the 'error-identity', we obtain (14) (see also [16, Theorem 3]). □

Corollary 1 presents majorants for $\lambda = 1$ and $\mu = \delta_h$, where $\delta_h = \theta\, h$, $\theta > 0$.

Corollary 1

(i) If $v \in V_{0,\underline{0}}^{\Delta_x}$ *and* $\boldsymbol{y} \in H^{\mathrm{div}_x,0}(Q)$*, Theorem 2 yields the estimate*

$$\|e\|^2_{s,\nu_i} \leq \overline{\mathrm{M}}_{\delta_h}(v, \boldsymbol{y}; \gamma, \alpha_i) := \gamma \Big((1+\alpha_1)\, \|\mathbf{r}_{\mathrm{d}}\|^2_Q + (1 + \tfrac{1}{\alpha_1})\, C_{\mathrm{F}}^2\, \|\mathbf{r}_{\mathrm{eq}}\|^2_Q + \delta_h \left((1+\alpha_2)\, \|\mathrm{div}_x \mathbf{r}_{\mathrm{d}}\|^2_Q + (1 + \tfrac{1}{\alpha_2})\, \|\mathbf{r}_{\mathrm{eq}}\|^2_Q \right) \Big), \quad (15)$$

where $\nu_1 = (2 - \frac{1}{\gamma}), \nu_2 = (2 - \frac{1}{\gamma})\, \delta_h, \nu_3 = \delta_h\ \nu_4 = 1.$

(ii) If $v \in V_{0,\underline{0}}^{\nabla_x \partial_t}$ *and* $\boldsymbol{y} \in H^{\mathrm{div}_x,1}(Q)$*, then Theorem 3 yields*

$$\|e\|^2_{s,\nu_i} \leq \overline{\mathrm{M}}^{\mathrm{II}}_{\delta_h}(v, \boldsymbol{y}; \zeta, \beta_i, \epsilon) := \epsilon\, \delta_h\, \|\mathbf{r}_{\mathrm{d}}\|^2_{\Sigma_T} + \zeta \Big((1+\beta_1)\big((1+\beta_2)\, \|\mathbf{r}_{\mathrm{d}}\|^2_Q + (1 + \tfrac{1}{\beta_2}) C_{\mathrm{F}}^2\, \|\mathbf{r}_{\mathrm{eq}}\|^2_Q\big) + (1 + \tfrac{1}{\beta_1}) \delta_h^2\, \|\partial_t \mathbf{r}_{\mathrm{d}}\|^2_Q + \delta_h\, \|\mathbf{r}_{\mathrm{eq}}\|^2_Q \Big), \quad (16)$$

where $\nu_1 = (2 - \frac{1}{\zeta}), \nu_2 = (2 - \frac{1}{\zeta})\, \delta_h, \nu_3 = \delta_h$*, and* $\nu_4 = 4$*. In (i) and (ii),* $\mathbf{r}_{\mathrm{d}}$ *and* $\mathbf{r}_{\mathrm{eq}}$ *are defined in* (13)*,* C_{F} *is the Friedrichs constant,* δ_h *is discretisation parameter,* $\gamma, \zeta \in [\frac{1}{2}, +\infty), \epsilon \in [1, +\infty)$*, and* $\beta_i > 0, i = 1, 2.$

4 Numerical Example

In the final section of this work, we present an example demonstrating the numerical behaviour of the derived majorants for the static case of the parabolic I-BVP. In fact, the space-time approach treats the parabolic problem as yet another elliptic problem in $\mathbb{R}^{d+1}$ with strong convection in $(d+1)$-th direction. Therefore, for the simplicity of presentation, we consider the Poisson Dirichlet problem

$$-\Delta_x u = f \quad \text{in} \quad Q := (0,1)^2 \in \mathbb{R}^2, \qquad u = 0 \quad \text{on} \quad \partial\Omega. \quad (17)$$

Let $u_h \in V_{0h}$, where $V_h \equiv \mathcal{S}_h^{p,p} := \{ \hat{V}_h \circ \Phi^{-1} \}$ and $\hat{V}_h \equiv \hat{\mathcal{S}}_h^{p,p}$, be generated with NURBS of degree $p = 2$. Due to the restriction on the knots-multiplicity of $\hat{\mathcal{S}}_h^{p,p}$, we have $u_h \in C^{p-1}$. Then, $u_h(x) := \sum_{i \in \mathcal{I}} \underline{\mathrm{u}}_{h,i}\, \phi_{h,i}$, where $\underline{\mathrm{u}}_h := \left[\underline{\mathrm{u}}_{h,i}\right]_{i \in \mathcal{I}} \in \mathbb{R}^{|\mathcal{I}|}$ is a vector of degrees of freedom (DOFs) defined by a system

$$\mathrm{K}_h\, \underline{\mathrm{u}}_h = \mathrm{f}_h, \quad \mathrm{K}_h := \left[(\nabla_x \phi_{h,i}, \nabla_x \phi_{h,j})_Q \right]_{i,j \in \mathcal{I}}, \quad \mathrm{f}_h := \left[(f, \phi_{h,i})_Q \right]_{i \in \mathcal{I}}. \quad (18)$$

The majorant corresponding to (17) can be presented as

$$\overline{\mathrm{M}}(u_h, \boldsymbol{y}_h) := (1+\beta)\, \|\boldsymbol{y}_h - \nabla_x u\|^2_Q + (1 + \tfrac{1}{\beta})\, C_{\mathrm{F}}^2\, \|\mathrm{div}_x \boldsymbol{y}_h + f\|^2_Q, \quad (19)$$

where $\beta > 0$ and $\boldsymbol{y} \in H^{\mathrm{div}_x,0}(Q)$. The approximation space for $\boldsymbol{y}_h \in Y_h \equiv \mathcal{S}_h^{q,q} := \{\hat{Y}_h \circ \Phi^{-1}\}$ is generated by the push-forward of $\hat{Y}_h := \hat{\mathcal{S}}_h^{q,q} \oplus \hat{\mathcal{S}}_h^{q,q}$, where $\mathcal{S}_h^{\hat{q},q}$ is the space of NURBS functions of degree q for each of the components of $\boldsymbol{y}_h = (y_h^{(1)}, y_h^{(2)})^{\mathrm{T}}$. The best EE is obtained by optimisation of $\overline{\mathrm{M}}(u_h, \boldsymbol{y}_h)$ w.r.t. $\boldsymbol{y}_h := \sum_{i \in \mathcal{I}} \underline{\mathrm{y}}_{h,i} \, \boldsymbol{\psi}_{h,i}$. Here $\boldsymbol{\psi}_{h,i}$ is the basis function of the space Y_h, and $\underline{\mathrm{y}}_h := \left[\underline{\mathrm{y}}_{h,i}\right]_{i \in \mathcal{I}} \in \mathbb{R}^{2|\mathcal{I}|}$ is a vector of DOFs of $\boldsymbol{y}_h$ defined by a system

$$\left(C_{\mathrm{F}}^2 \, \mathrm{Div}_h + \beta \, \mathrm{M}_h\right) \underline{\mathrm{y}}_h = -C_{\mathrm{F}}^2 \, \mathrm{z}_h + \beta \, \mathrm{g}_h, \tag{20}$$

where

$$\mathrm{Div_h} := \left[(\mathrm{div}_x \boldsymbol{\psi}_i, \mathrm{div}_x \boldsymbol{\psi}_j)_Q\right]_{i,j=1}^{2|\mathcal{I}|}, \qquad \mathrm{z}_h := \left[\left(f, \mathrm{div}_x \boldsymbol{\psi}_j\right)_Q\right]_{j=1}^{2|\mathcal{I}|},$$

$$\mathrm{M_h} := \left[(\boldsymbol{\psi}_i, \boldsymbol{\psi}_j)_Q\right]_{i,j=1}^{2|\mathcal{I}|}, \qquad \mathrm{g}_h := \left[\left(\nabla_x v, \boldsymbol{\psi}_j\right)_Q\right]_{j=1}^{2|\mathcal{I}|}.$$

According to [11], the most effective results for the majorant reconstruction (with uniform refinement) is obtained, when q is set substantially higher than p. We assume that $q = p + k, k \in \mathbb{N}^+$. In the same time, when u_h is reconstructed on the mesh $\mathcal{T}_h$, we use a coarser one $\mathcal{T}_{Kh}, K \in \mathbb{N}^+$, to recover the flux $\boldsymbol{y}_{Kh}$.

Example 1. We consider a basic example with $u = (1 - x_1)\, x_1^2\, (1 - x_2)\, x_2$, $f = -(2\,(1 - 3\,x_1)\,(1 - x_2)\,x_2 - 2\,(1 - x_1)\,x_1^2)$, and homogenous Dirichlet BC. For the uniform refinement, we set $p = 2$, i.e., $u_h \in S_h^{2,2}$, and compare two different settings: (a) $\boldsymbol{y}_h \in S_{Kh}^{q,q} \oplus S_{Kh}^{q,q}, q = 5, k = 3, K = 3$ and (b) $\boldsymbol{y}_h \in S_{Kh}^{q,q} \oplus S_{Kh}^{q,q}, q = 9, k = 7$, $K = 7$. The upper and lower parts of Tables 1 and 2, correspond to the cases (a) and (b), respectively. In the case (a), the time spent on the reconstruction of $\boldsymbol{y}_h$ (i.e., $t_{\mathrm{as}}(\boldsymbol{y}_h) + t_{\mathrm{sol}}(\boldsymbol{y}_h)$) is about 3 times higher than the time $t_{\mathrm{as}}(u_h) + t_{\mathrm{sol}}(u_h)$. However, for the case (b), the assembling time of the systems $\mathrm{Div_h}$ and $\mathrm{M_h}$ (denoted by $t_{\mathrm{as}}(\boldsymbol{y}_h)$) takes approximately 1/10-th of the assembling time for $\mathrm{K_h}$ ($t_{\mathrm{as}}(u_h)$). Moreover, solving the system (20) ($t_{\mathrm{sol}}(\boldsymbol{y}_h)$) takes only 1/500-th part of the time spent on solving (18) ($t_{\mathrm{sol}}(u_h)$). The efficiency of the obtained functional majorant is illustrated by $I_{\mathrm{eff}}(\overline{\mathrm{M}}) = 1.0936$ (see the fifth column of Table 1).

Table 1. Assembling and solving time for systems (18) and (20) w.r.t. the last 2 refinement steps.

DOFs(u_h)	DOFs($\boldsymbol{y}_h$)	$t_{\mathrm{as}}(u_h)$	$t_{\mathrm{as}}(\boldsymbol{y}_h)$	$t_{\mathrm{sol}}(u_h)$	$t_{\mathrm{sol}}(\boldsymbol{y}_h)$
$\boldsymbol{y} \in S^{5,5} \oplus S^{5,5}, k = 3, K = 3$					
1 004 004	42 431	4.2770	**10.9889**	17.0640	**43.3740**
4 010 004	205 031	17.1461	**45.0032**	143.2929	**328.5911**
$\boldsymbol{y} \in S^{9,9} \oplus S^{9,9}, k = 7, K = 7$					
1 004 004	441	4.3506	**0.4213**	17.1396	**0.0456**
4 010 004	1 161	17.4620	**1.7268**	142.9116	**0.2667**

Table 2. The error, the majorant, the corresponding efficiency index, and the e.o.c. (error order or convergence) p w.r.t. the last 2 refinement steps.

$\|\nabla_x(u-u_h)\|_Q^2$	$\overline{\mathrm{M}}$	I_{eff}	p
$\boldsymbol{y} \in S^{5,5} \oplus S^{5,5}, k=3, K=3$			
6.229382e−07	7.450215e−07	**1.0936**	2.0113
1.557344e−07	1.862552e−07	**1.0936**	2.0056
$\boldsymbol{y} \in S^{9,9} \oplus S^{9,9}, k=7, K=7$			
6.229382e−07	6.363897e−07	**1.0107**	2.0113
1.557344e − 07	1.557499e − 07	**1.0000**	2.0056

We now consider an adaptive refinement strategy, i.e., THB-Splines [7,12,26] in combination with the functional EE (19). We use the so-called Dörfler's marking [5] with a parameter $\theta = 0.6$. We start with the following setting: $u_h \in S_h^{2,2}$ is THB-Splines basis (with one level and 36 basis functions of degree 2), and $\boldsymbol{y}_h \in S_{3h}^{5,5} \oplus S_{3h}^{5,5}$ is THB-Splines basis (with one level and 81 basis functions of degree 5). We execute 16 refinement steps to obtain the error illustrated in Table 3 (where only the last two refinement steps are shown). The time spent on the assembling, solving, and generating corresponding EEs is illustrated in Table 4. By using 3 times courser mesh in the refinement of the basis for $\boldsymbol{y}_h$, we have managed to spare the effort of reconstructing the optimal $\boldsymbol{y}_h$ and speed up the over-all reconstruction of the majorant. In the current configuration, we obtain the following ratios of the times spent on reconstruction of $\boldsymbol{y}_h$ and u_h, i.e., $\frac{t_{\mathrm{as}}(u_h)}{t_{\mathrm{as}}(\boldsymbol{y}_h)} \approx 19$ and $\frac{t_{\mathrm{sol}}(u_h)}{t_{\mathrm{sol}}(\boldsymbol{y}_h)} \approx 658$.

Table 3. Assembling and solving time for systems (18) and (20) w.r.t. the last 2 refinements of total 16 steps.

DOFs(u_h)	DOFs($\boldsymbol{y}_h$)	$t_{\mathrm{as}}(u_h)$	$t_{\mathrm{as}}(\boldsymbol{y}_h)$	$t_{\mathrm{sol}}(u_h)$	$t_{\mathrm{sol}}(\boldsymbol{y}_h)$
55 005	145	12.1002	**0.9302**	0.6105	**0.0022**
107 444	132	17.7635	**0.9269**	1.1858	**0.0018**

Table 4. Error, majorant, its efficiency index, and e.o.c. w.r.t. the last 2 refinements of total 16 steps.

$\|\nabla_x(u-u_h)\|_Q^2$	$\overline{\mathrm{M}}$	I_{eff}	p
3.263617e−06	3.373925e−06	**1.0338**	1.4511
2.187249e−06	2.202933e−06	**1.0072**	1.6345

Acknowledgements. The research is supported by the Austrian Science Fund (FWF) through the NFN S117-03 project. Implementation was carried out using the open-source C++ library *G+smo* [19] developed at RICAM.

References

1. Bazilevs, Y., Beirão da Veiga, L., Cottrell, J.A., Hughes, T.J.R., Sangalli, G.: Isogeometric analysis: approximation, stability and error estimates for h-refined meshes. Math. Models Methods Appl. Sci. **16**(7), 1031–1090 (2006). http://dx.doi.org/10.1142/S0218202506001455
2. Buffa, A., Giannelli, C.: Adaptive isogeometric methods with hierarchical splines: error estimator and convergence. Technical report arXiv:1502.00565 [math.NA] (2015)
3. Dedè, L., Santos, H.A.F.A.: B-spline goal-oriented error estimators for geometrically nonlinear rods. Comput. Mech. **49**(1), 35–52 (2012). http://dx.doi.org/10.1007/s00466-011-0625-2
4. Dörfel, M.R., Jüttler, B., Simeon, B.: Adaptive isogeometric analysis by local h-refinement with T-splines. Comput. Methods Appl. Mech. Eng. **199**(5–8), 264–275 (2010). http://dx.doi.org/10.1016/j.cma.2008.07.012
5. Dörfler, W.: A convergent adaptive algorithm for Poisson's equation. SIAM J. Numer. Anal. **33**(3), 1106–1124 (1996)
6. Gaevskaya, A.V., Repin, S.I.: A posteriori error estimates for approximate solutions of linear parabolic problems. Differ. Equ. **41**(7), 970–983 (2005). Springer
7. Giannelli, C., Jüttler, B., Speleers, H.: THB-splines: the truncated basis for hierarchical splines. Comput. Aided Geom. Des. **29**(7), 485–498 (2012). http://dx.doi.org/10.1016/j.cagd.2012.03.025
8. Hansbo, P.: Space-time oriented streamline diffusion methods for nonlinear conservation laws in one dimension. Comm. Numer. Meth. Eng. **10**(3), 203–215 (1994)
9. Johannessen, K.: An adaptive isogeometric finite element analysis. Technical report, Master's thesis, Norwegian University of Science and Technology (2009)
10. Johnson, C., Saranen, J.: Streamline diffusion methods for the incompressible Euler and Navier-Stokes equations. Math. Comput. **47**(175), 1–18 (1986)
11. Kleiss, S.K., Tomar, S.K.: Guaranteed and sharp a posteriori error estimates in isogeometric analysis. Comput. Math. Appl. **70**(3), 167–190 (2015). http://dx.doi.org/10.1016/j.camwa.2015.04.011
12. Kraft, R.: Adaptive and linearly independent multilevel B-splines. In: Surface Fitting and Multiresolution Methods (Chamonix-Mont-Blanc, 1996), pp. 209–218. Vanderbilt University Press, Nashville (1997)
13. Kumar, M., Kvamsdal, T., Johannessen, K.A.: Simple a posteriori error estimators in adaptive isogeometric analysis. Comput. Math. Appl. **70**(7), 1555–1582 (2015). http://dx.doi.org/10.1016/j.camwa.2015.05.031
14. Kuru, G., Verhoosel, C.V., van der Zee, K.G., van Brummelen, E.H.: Goal-adaptive isogeometric analysis with hierarchical splines. Comput. Methods Appl. Mech. Eng. **270**, 270–292 (2014). https://doi.org/10.1016/j.cma.2013.11.026
15. Ladyzhenskaya, O.A.: The Boundary Value Problems of Mathematical Physics. Springer, New York (1985). https://doi.org/10.1007/978-1-4757-4317-3
16. Langer, U., Matculevich, S., Repin, S.: A posteriori error estimates for space-time IgA approximations to parabolic initial boundary value problems. arXiv.org/1612.08998 [math.NA] (2016)

17. Langer, U., Moore, S., Neumüller, M.: Space-time isogeometric analysis of parabolic evolution equations. Comput. Methods Appl. Mech. Eng. **306**, 342–363 (2016)
18. Mali, O., Neittaanmäki, P., Repin, S.: Accuracy Verification Methods. Computational Methods in Applied Sciences, vol. 32. Springer, Dordrecht (2014). https://doi.org/10.1007/978-94-007-7581-7
19. Mantzaflaris, A., et al.: G+Smo (geometry plus simulation modules) v0.8.1 (2015). http://gs.jku.at/gismo
20. Matculevich, S.: Fully reliable a posteriori error control for evolutionary problems. Ph.D. thesis, Jyväskylä Studies in Computing, University of Jyväskylä (2015)
21. Matculevich, S., Repin, S.: Computable estimates of the distance to the exact solution of the evolutionary reaction-diffusion equation. Appl. Math. Comput. **247**, 329–347 (2014). http://dx.doi.org/10.1016/j.amc.2014.08.055
22. Repin, S.: A Posteriori Estimates for Partial Differential Equations. Radon Series on Computational and Applied Mathematics, vol. 4. Walter de Gruyter GmbH & Co. KG, Berlin (2008)
23. Repin, S.I.: Estimates of deviations from exact solutions of initial-boundary value problem for the heat equation. Rend. Mat. Acc. Lincei **13**(9), 121–133 (2002)
24. Repin, S.: A posteriori error estimation for nonlinear variational problems by duality theory. Zapiski Nauchnych Seminarov POMIs **243**, 201–214 (1997)
25. Tagliabue, A., Dedè, L., Quarteroni, A.: Isogeometric analysis and error estimates for high order partial differential equations in fluid dynamics. Comput. Fluids **102**, 277–303 (2014)
26. Vuong, A.V., Giannelli, C., Jüttler, B., Simeon, B.: A hierarchical approach to adaptive local refinement in isogeometric analysis. Comput. Methods Appl. Mech. Eng. **200**(49–52), 3554–3567 (2011). http://dx.doi.org/10.1016/j.cma.2011.09.004
27. van der Zee, K.G., Verhoosel, C.V.: Isogeometric analysis-based goal-oriented error estimation for free-boundary problems. Finite Elem. Anal. Des. **47**(6), 600–609 (2011). http://dx.doi.org/10.1016/j.finel.2010.12.013
28. Zeidler, E.: Nonlinear Functional Analysis and Its Applications. II/A. Springer, New York (1990). https://doi.org/10.1007/978-1-4612-0985-0

An Algebraic Multigrid Method for an Adaptive Space–Time Finite Element Discretization

Olaf Steinbach(✉) and Huidong Yang

Institut für Numerische Mathematik, Technische Universität Graz, Steyrergasse 30, 8010 Graz, Austria
o.steinbach@tugraz.at, hyang@math.tugraz.at

Abstract. This work is devoted to numerical studies on an algebraic multigrid preconditioned GMRES method for solving the linear algebraic equations arising from a space–time finite element discretization of the heat equation using h–adaptivity on tetrahedral meshes. The finite element discretization is based on a Galerkin–Petrov variational formulation using piecewise linear finite elements simultaneously in space and time. In this work, we focus on h–adaptivity relying on a residual based a posteriori error estimation, and study some important components in the algebraic multigrid method for solving the space–time finite element equations.

Keywords: Adaptive space–time finite element · Algebraic multigrid

1 Introduction

The space–time finite element method has a long history since the starting work for the application in elastodynamics [11]. Recently, in [10,15,16], the authors have proposed a discontinuous Galerkin space–time finite element approach and robust multigrid methods for parabolic problems. In [12], space–time isogeometric analysis discretization methods for parabolic evolution equations in fixed and moving spatial computational domains have been investigated. In [3], the classical streamline–diffusion and edge averaged finite element methods for time–dependent convection–diffusion problems are considered. In [7,8], space–time discontinuous Petrov–Galerkin finite elements with optimal test functions for fluid problems have been exploited. Further, a class of methods based on well–known space–time tensor product ansatz spaces can be found in, e.g., [1,19,22].

In this work, we follow the Galerkin–Petrov space–time finite element method recently proposed and analyzed in [20] for solving the model heat equation

$$\alpha\partial_t u(x,t) - \Delta_x u(x,t) = f(x,t) \quad \text{for } (x,t) \in Q := \Omega \times (0,T), \tag{1}$$

with the boundary and initial conditions $u(x,t) = 0$ for $(x,t) \in \Sigma := \partial\Omega \times (0,T)$ and $u(x,0) = u_0$ for $x \in \Omega$, respectively. Here, $\Omega \subset \mathbb{R}^2$ is a bounded Lipschitz domain, and $\alpha \in \mathbb{R}_+$ is the heat capacity constant.

I. Lirkov and S. Margenov (Eds.): LSSC 2017, LNCS 10665, pp. 66–73, 2018.
https://doi.org/10.1007/978-3-319-73441-5_6

The Galerkin–Petrov variational formulation for the heat Eq. (1) is to find $\overline{u} \in X := \{v \in L^2(0,T;H^1_0(\Omega)) \cap H^1(0,T;H^{-1}(\Omega)), v(x,0) = 0 \text{ for } x \in \Omega\}$ such that

$$a(\overline{u}, v) = \langle f, v\rangle - a(\overline{u}_0, v) \tag{2}$$

is satisfied for all $v \in Y := L^2(0,T;H^1_0(\Omega))$, where

$$a(u,v) := \int_0^T \int_\Omega \Big[\alpha\partial_t u(x,t)v(x,t) + \nabla_x u(x,t)\cdot\nabla_x v(x,t)\Big]\,dx\,dt,$$

$$\langle f, v\rangle := \int_0^T \int_\Omega f(x,t)v(x,t)\,dx\,dt,$$

and $\overline{u}_0 \in L^2(0,T;H^1_0(\Omega)) \cap H^1(0,T;H^{-1}(\Omega))$ denotes an arbitrary but fixed extension of the initial datum $u_0 \in H^1_0(\Omega)$. Existence and uniqueness of the solution to (2) is provided in [20], see also [19,22].

The related discrete Galerkin–Petrov problem is to find $\overline{u}_h \in X_h \subset X$ such that

$$a(\overline{u}_h, v_h) = \langle f, v_h\rangle - a(\overline{u}_0, v_h) \tag{3}$$

is satisfied for all $v_h \in Y_h \subset Y$ where we assume $X_h \subset Y_h$. Then, a discrete inf–sup condition was shown in [20], from which we conclude a standard stability and error analysis.

In particular, the space–time cylinder Q is decomposed into finite elements $Q_h = \cup_{\ell=1}^N \overline{q}_\ell$. For simplicity, we assume that Ω is polygonal bounded, $\overline{Q} = Q_h$. The finite element spaces are given by $X_h = S^1_h(Q_h) \cap X$ and $Y_h = X_h$ with $S^1_h(Q_h) = \text{span}\{\varphi_i\}_{i=1}^M$ being the span of piecewise linear and continuous basis functions φ_i. The following energy error estimate is shown in [20],

$$\|\overline{u} - \overline{u}_h\|_{L^2(0,T;H^1_0(\Omega))} \leq c\,h\,|\overline{u}|_{H^2(Q)}, \tag{4}$$

where $\overline{u} \in X$ and $\overline{u}_h \in X_h$ denote the unique solutions of the variational problems (2) and (3), respectively, $c > 0$ is a constant independent of the mesh size h, and we assume $\overline{u} \in H^2(Q)$, i.e., $u_0 \in H^3(\Omega)$ and $f \in H^{2,1}(Q)$.

In comparison with other space–time methods, this approach is very suitable for the development of h–adaptivity simultaneously in space and time. We may further pose the question how to solve the arising linear system of equations. In fact, this will be tackled by an algebraic multigrid (AMG) preconditioned GMRES method.

The remainder of this paper is organized as follows: In Sect. 2, we discuss the space–time adaptive approach while Sect. 3 deals with the algebraic multigrid method for the finite element equations. Some numerical experiments are prescribed in Sect. 4. Finally, some conclusions are drawn in Sect. 5.

2 Space–Time Adaptivity

2.1 Local Error Indicator

Let $\overline{u}_h \in X_h$ be the space–time finite element solution of the variational problem (3) implying $u_h := \overline{u}_0 + \overline{u}_h$ for which we can define the local residuals

$$R_{q_\ell}(u_h) := f + \Delta_x u_h - \alpha\partial_t u_h$$

on each tetrahedral element q_ℓ and the jumps

$$J_\gamma(u_h) := [n_x \cdot \nabla_x u_h]_{|\gamma}$$

of the normal flux in the spatial direction across the inner boundaries γ between q_ℓ and its neighbouring elements. Then, the local error indicator on each element q_ℓ is given as

$$\eta_{q_\ell} = \left\{ c_1 h_{q_\ell}^2 \|R_{q_\ell}\|_{L_2(q_\ell)}^2 + c_2 h_{q_\ell} \|J_\gamma\|_{L_2(\partial q_\ell)}^2 \right\}^{\frac{1}{2}}, \tag{5}$$

with suitably chosen positive constants c_1, c_2. For more details, we refer to our recent work [21]. In comparison with more conventional adaptive methods for time–dependent problems, see, e.g., [9,14,18], our method allows to perform the spatial and temporal adaptivity simultaneously.

2.2 Adaptive Mesh Refinement

Two local mesh refinement methods on the tetrahedral meshes Q_h have been employed in order to perform the space–time adaptivity, namely, the octasection based method [4], and the newest vertex bisection based method [2]. In the octasection based method, the marked tetrahedral elements are refined using a regular refinement [24], the hanging nodes are closed by the so–called irregular refinement with 62 possible cases. It is important that the irregular elements will never be further refined on the next refinement levels in order to keep shape regularity of the tetrahedral elements. If such irregular elements are marked, we have to perform the regular refinement on their parental elements, and irregular refinement on those affected neighbouring elements. In the newest vertex bisection based method, the local refinement pattern for each tetrahedral element is fixed a priori following certain rules. In [2], there exist 5 refinement patterns defined for tetrahedral elements. The local refinement strictly follows the natural rules from one pattern to another. The closure of hanging nodes is realized by calling such a local refinement recursively until no more hanging nodes exist.

2.3 The Adaptive Space–Time Finite Element Loop

The adaptive loop in the space–time finite element method follows the one in the standard adaptive finite element approach, see, e.g., [23], that consists of the following four main steps: Given a conforming decomposition Q_0 on the initial mesh level $k = 0$,

1. SOLVE: *Solve the discrete problem* (3) *on the adaptive mesh level* k.
2. ESTIMATE: *Compute the local error indicators* (5) *on each element* q_ℓ.
3. MARK: *Mark the elements for refinement using a proper marking strategy.*
4. REFINE: *Perform the local mesh refinement using octasection or bisection, increase level* $k := k + 1$, *obtain the conforming decomposition* Q^k, *and go to Step 1 if the solution is not accurate enough.*

In particular, for the module MARK, we use the Dörfler marking strategy [6]: For a given parameter $\theta \in [0,1]$, find N_k such that

$$\theta \sum_{\ell=1}^{N_k} \eta_{q_\ell} \geq \sum_{\ell=1}^{M_k} \eta_{q_\ell},$$

where M_k denotes the total number of tetrahedral elements on the current level k. In order to mark as few elements as possible, it is desirable to hold $q_\ell \geq q_m$ if $\ell < m$. The tetrahedral elements with index from 1 to N_k will be marked for the refinement on the next level $k+1$. In our numerical experiments we used $\theta = 0.5$ which seems to be almost optimal compared with other values.

3 The Algebraic Multigrid Method

The remaining task is to solve the linear system of algebraic equations arising from the space–time finite element discretization of the Galerkin–Petrov problem (3). It is clear that the stiffness matrix is not symmetric but positive definite. Hence we develop an AMG preconditioner for the GMRES method, that requires special care for its components, namely, the coarsening and smoother. So far, we have considered a greedy strategy for coarse–grid selection [13]. After the selection of coarse and fine grids the interpolation matrix is defined as in classical AMG [5]. As a smoother, we employ the ω-Kaczmarz relaxation which satisfies the algebraic smoothing property [17]. More details on the development of the robust AMG method for such space–time finite element equations will be provided in a near future report.

4 Numerical Experiments

4.1 Convergence History

As a numerical example we consider $\Omega = (0,1)^2$ and $T = 1$, i.e. $Q = (0,1)^3$, and we chose a sufficiently smooth solution u given as

$$u(x,t) = (x_1^2 - x_1)(x_2^2 - x_2)(t^2 - t)e^{-100.0((x_1-0.25)^2+(x_2-0.25)^2+(t-0.25)^2)}. \quad (6)$$

To verify the estimated order of convergence (eoc), the $L^2(0,T;H_0^1(\Omega))$–norm of the error $e_h := u - u_h$ between the given and the discrete solution is calculated on five uniformly refined mesh levels L_1–L_5 with increasing number of degrees of freedom (#Dof) and tetrahedral elements (#Tet), and decreasing mesh size h. The mesh information is prescribed in Table 1.

The convergence results are illustrated in Table 2 for varying heat conductivity parameters $\alpha = 1$, $\alpha = 10$, and $\alpha = 100$. From the numerical results, we observe a linear order of convergence as predicted. The convergence history of the space–time finite element method using an adaptive mesh refinement in comparison with the uniform one is plotted in Fig. 1 for $\alpha = 1$. The adaptive mesh always starts from the L_1 mesh level. To reach the same accuracy as the uniform refinement, the two adaptive methods require much fewer degrees of freedom. Further, they provide a linear order of convergence.

Table 1. Mesh information on five uniformly refined levels.

Level	L_1	L_2	L_3	L_4	L_5
#Dof	125	729	4913	35937	274625
#Tet	384	3072	24586	196608	1572864
h	0.25	0.125	0.0625	0.03125	0.015625

Table 2. The estimated order of convergence (eoc) on the mesh levels L_1–L_5 with different values for the heat conductivity: $\alpha = 1$, $\alpha = 10$, and $\alpha = 100$.

Level	$\alpha = 1$		$\alpha = 10$		$\alpha = 100$	
	$\Vert e_h\Vert_{L_2(0,T;H_0^1(\Omega))}$	eoc	$\Vert e_h\Vert_{L_2(0,T;H_0^1(\Omega))}$	eoc	$\Vert e_h\Vert_{L_2(0,T;H_0^1(\Omega))}$	eoc
L_1	$3.77 \cdot 10^{-3}$	–	$3.78 \cdot 10^{-3}$	–	$4.10 \cdot 10^{-3}$	–
L_2	$2.93 \cdot 10^{-3}$	0.36	$3.01 \cdot 10^{-3}$	0.33	$4.21 \cdot 10^{-3}$	–
L_3	$2.00 \cdot 10^{-3}$	0.55	$2.04 \cdot 10^{-3}$	0.56	$2.36 \cdot 10^{-3}$	0.84
L_4	$1.07 \cdot 10^{-3}$	0.89	$1.08 \cdot 10^{-3}$	0.91	$1.13 \cdot 10^{-3}$	1.06
L_5	$5.47 \cdot 10^{-4}$	0.97	$5.49 \cdot 10^{-4}$	0.98	$5.56 \cdot 10^{-4}$	1.02

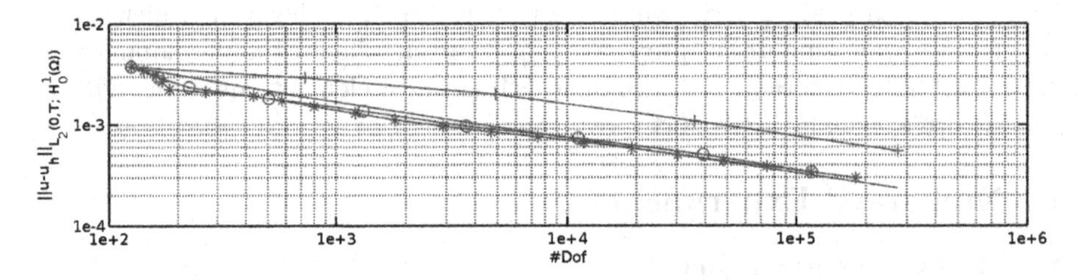

Fig. 1. Convergence history of the space–time finite element method using uniform and adaptive refinements: Uniform ($-+-$), octasection ($-\circ-$), bisection ($-*-$) and linear ($-$), for $\alpha = 1.0$.

4.2 AMG Performance

To study the AMG performance, in Table 3 we show the number of AMG preconditioned GMRES iterations and the cost in seconds (s) for solving the model heat equation on uniformly refined five mesh levels (L_1–L_5). We set the relative residual error $\varepsilon = 10^{-7}$ as the stopping criterion for the GMRES iteration, and run 1–2 AMG iterations in each preconditioning step. We observe a relatively fair robustness with respect to the mesh size h and heat capacity α.

In Figs. 2 and 3, respectively, we compare the computational time in seconds (s) and the number of AMG preconditioned GMRES iterations to reach the same accuracy of the space–time finite element solution between the adaptive and

Table 3. Performance of the AMG preconditioned GMRES method on five uniform mesh levels: Number of GMRES iterations (left) and time in seconds (right).

α	L_1	L_2	L_3	L_4	L_5
1	3	4	7	14	32
10	2	4	6	14	24
100	3	4	6	9	15

α	L_1	L_2	L_3	L_4	L_5
1	0.001 s	0.032 s	0.373 s	7.437 s	143.190 s
10	0.001 s	0.033 s	0.667 s	9.171 s	190.070 s
100	0.001 s	0.060 s	1.383 s	27.060 s	448.590 s

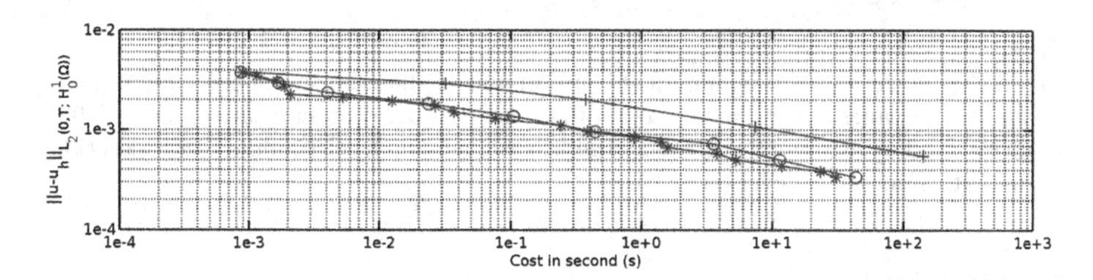

Fig. 2. Comparison of the computational cost in seconds (s) to reach the same accuracy of the space–time finite element solution using uniform and adaptive refinements: Uniform (− + −), octasection (− ∘ −) and bisection (− ∗ −), for $\alpha = 1.0$.

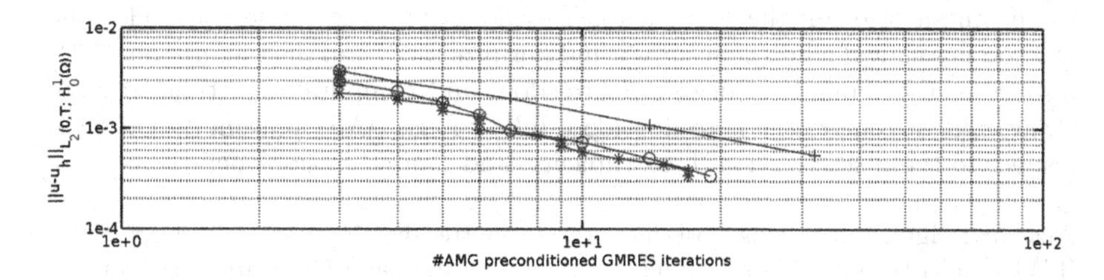

Fig. 3. Comparison of the AMG preconditioned GMRES iterations to reach the same accuracy of the space–time finite element solution using the uniform and adaptive refinements: Uniform (− + −), octasection (− ∘ −) and bisection (− ∗ −), for $\alpha = 1.0$.

uniform refinements. In comparison with the uniform refinement, the adaptive one shows more efficiency in saving the number of AMG preconditioned GMRES iterations as well as in the computational time.

4.3 Visualization

The visualization of the numerical solution and of the adaptive space–time meshes at three planes, in particular for $x_1 = 0.5$, $x_2 = 0.5$, and $t = 0.25$, is shown in Fig. 4. It is easy to see that our adaptive methods capture the interest in the space–time domain effectively and make the adaptive mesh refinement in space and time simultaneously.

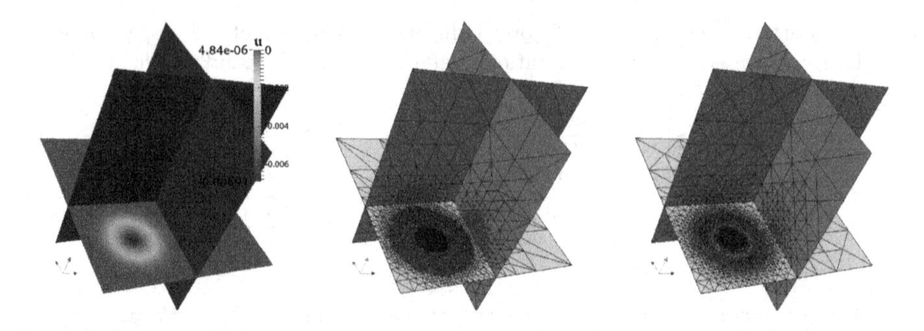

Fig. 4. Visualization of the numerical solution and of the adaptive space–time meshes at the three planes for $x_1 = 0.5$, $x_2 = 0.5$, and $t = 0.25$: Numerical solution (left), adaptive meshes using octasection at the 9th refinement level (middle) and bisection at the 19th refinement level (right).

5 Conclusions

In this work, we have developed an AMG preconditioned GMRES method for solving the adaptive space–time finite element discretized heat equation. The proposed method has demonstrated a relatively good performance with respect to the mesh size and heat capacity. The adaptive space–time finite element method has shown a better accuracy and performance than the uniform one with respect to the number of degrees of freedom and computational time, respectively. The ongoing work is to develop a fully robust AMG preconditioner with respect to the discretization, heat capacity, and more diffusion coefficients.

Acknowledgements. This work has been supported by the Austrian Science Fund (FWF) under the Grant SFB Mathematical Optimisation and Applications in Biomedical Sciences.

References

1. Andreev, R.: Wavelet-in-time multigrid-in-space preconditioning of parabolic evolution equations. SIAM J. Sci. Comput. **38**(1), A216–A242 (2016)
2. Arnold, D.N., Mukherjee, A., Pouly, L.: Locally adapted tetrahedral meshes using bisection. SIAM J. Sci. Comput. **22**(2), 431–448 (2000)
3. Bank, R.E., Vassilevski, P.S., Zikatanov, L.T.: Arbitrary dimension convection-diffusion schemes for space-time discretizations. J. Comput. Appl. Math. **310**, 19–31 (2017)
4. Bey, J.: Tetrahedral grid refinement. Computing **55**(4), 355–378 (1995)
5. Briggs, W.L., Henson, V.E., McCormick, S.F.: A multigrid tutorial. SIAM, Philadelphia (2000)
6. Dörfler, W.: A convergent adaptive algorithm for Poisson's equation. SIAM J. Numer. Anal. **33**(3), 1106–1124 (1996)
7. Ellis, T.E.: Space-time discontinuous Petrov-Galerkin finite elements for transient fluid mechanics. Ph.D. thesis. University of Texas at Austin (2016)

8. Ellis, T.E., Demkowicz, L., Chan, J.: Locally conservative discontinuous Petrov-Galerkin finite elements for fluid problems. Comput. Math. Appl. **68**(11), 1530–1549 (2014)
9. Eriksson, K., Johnson, C.: Adaptive finite element methods for parabolic problems I: a linear model problem. SIAM J. Numer. Anal. **28**(1), 43–77 (1991)
10. Gander, M.J., Neumüller, M.: Analysis of a new space-time parallel multigrid algorithm for parabolic problems. SIAM J. Sci. Comput. **38**(4), A2173–A2208 (2016)
11. Hughes, T.J.R., Hulbert, G.M.: Space-time finite element methods for elastodynamics: formulations and error estimates. Comput. Methods Appl. Math. **66**(3), 339–363 (1988)
12. Langer, U., Moore, S.E., Neumüller, M.: Space-time isogeometric analysis of parabolic evolution problems. Comput. Methods Appl. Math. **306**, 342–363 (2016)
13. MacLachlan, S., Saad, Y.: A greedy strategy for coarse-grid selection. SIAM J. Sci. Comput. **29**(5), 1825–1853 (2007)
14. Moore, P.K.: A posteriori error estimation with finite element semi-and fully discrete methods for nonlinear parabolic equations in one space dimension. SIAM J. Numer. Anal. **31**(1), 149–169 (1994)
15. Neumüller, M.: Space-time methods: fast solvers and applications. Ph.D. thesis. TU Graz (2013)
16. Neumüller, M., Steinbach, O.: Refinement of flexible space-time finite element meshes and discontinuous Galerkin methods. Comput. Vis. Sci. **14**, 189–205 (2011)
17. Popa, C.: Algebraic multigrid smoothing property of Kaczmarz's relaxation for general rectangular linear systems. Electron. Trans. Numer. Anal. **29**, 150–162 (2007)
18. Schmich, M., Vexler, B.: Adaptivity with dynamic meshes for space-time finite element discretizations of parabolic equations. SIAM J. Sci. Comput. **30**(1), 369–393 (2008)
19. Schwab, C., Stevenson, R.: Space-time adaptive wavelet methods for parabolic evolution problems. Math. Comput. **78**(267), 1293–1318 (2009)
20. Steinbach, O.: Space-time finite element methods for parabolic problems. Comput. Methods Appl. Math. **15**, 551–566 (2015)
21. Steinbach, O., Yang, H.: An adaptive space-time finite element method for solving the heat equation, Technical report. TU Graz (2017, in preparation)
22. Urban, K., Patera, A.T.: An improved error bound for reduced basis approximation of linear parabolic problems. Math. Comput. **83**(288), 1599–1615 (2014)
23. Verfürth, R.: A Posteriori Error Estimation Techniques for Finite Element Methods. Oxford Unversity Press, Oxford (2013)
24. Zhang, S.: Multi-level iterative techniques. Ph.D. thesis. Penn State University (1988)

Advanced Discretizations and Solvers for Coupled Systems of Partial Differential Equations

Splitting Schemes for Mixtures of Nematic-Isotropic Flows with Anchoring Effects

Giordano Tierra[1(✉)], Francisco Guillén-González[2], and María Ángeles Rodríguez-Bellido[2]

[1] Temple University, Philadelphia, PA 19122, USA
gtierra@temple.edu
[2] Universidad de Sevilla and IMUS, C/Tarfia, S/N, 41012 Seville, Spain
{guillen,angeles}@us.es
https://gtierra.wordpress.com, http://personal.us.es/guillen,
http://personal.us.es/angeles

Abstract. This work is devoted to the study of complex fluids composed by the mixture between isotropic (newtonian fluid) and nematic (liquid crystal) flows, taking into account how the liquid crystal molecules behave on the interface between both fluids (anchoring effects) and the influence of the shape of the liquid crystal molecules on the dynamics of the system (stretching effects).

First, we present the PDE system to model Nematic-Isotropic mixtures, taking into account viscous, mixing, nematic, anchoring and stretching effects. Then, we provide a new linear unconditionally energy-stable splitting scheme. Moreover, we present numerical simulations to show the efficiency of the proposed numerical scheme and the influence of the different types of anchoring effects in the dynamics of the system.

Keywords: Liquid crystal · Phase field · Finite elements
Multiphase flows · Energy stability · Stretching effects
Anchoring effects

1 The Model

The dynamics of the interface between two different materials is a key role to understand the behavior of many interesting systems in science, engineering and industry. In this work we are interested in the diffuse interface approach to represent mixtures of two immiscible incompressible fluids: one isotropic and one nematic liquid crystal.

We consider a bounded domain $\Omega \subset \mathbb{R}^M, (M = 2, 3)$, whose boundary will be represented by Γ (i.e., $\Gamma := \partial\Omega$) and we introduce a phase field function $\phi(\mathbf{x}, t)$ that will be used to localize the components along the domain Ω, such that

$$\phi(\mathbf{x}, t) = \begin{cases} -1 & \text{Newtonian Fluid,} \\ 1 & \text{Nematic Liquid Crystal.} \end{cases}$$

I. Lirkov and S. Margenov (Eds.): LSSC 2017, LNCS 10665, pp. 77–84, 2018.
https://doi.org/10.1007/978-3-319-73441-5_7

Moreover, the physical variables are: $(\mathbf{u}, p)$ the velocity field and the pressure, and $\mathbf{d}$ the director field related to the orientation of the nematic liquid crystal molecules. We introduce the total energy of the system as the addition of the energies related to each component plus the energy associated to the mixture in the interface between both components:

$$\begin{aligned} E_{tot}(\mathbf{u}, \phi, \mathbf{d}) &= E_{kin}(\mathbf{u}) + \lambda_{mix} E_{mix}(\phi) \\ &\quad + \lambda_{nem} E_{nem}(\mathbf{d}, \phi) + \lambda_{anch} E_{anch}(\mathbf{d}, \phi), \end{aligned} \tag{1}$$

where $E_{kin}(\mathbf{u})$ denotes the kinetic energy of the system, $E_{mix}(\phi)$ denotes the mixing energy associated to the mixture process, $E_{nem}(\mathbf{d}, \phi)$ denotes the elastic energy due to the nematic liquid crystal (which is the simplest elastic energy considered in the literature) and $E_{anch}(\mathbf{d}, \phi)$ denotes the anchoring energy that represents the influence of the interfacial effects on the orientation of the nematic liquid crystal molecules in the interface between both components. Moreover, positive parameters λ_{mix}, λ_{nem} and λ_{anch} are introduced to balance the effect of each energy in the system. In particular, the energy terms read:

$$\begin{aligned} E_{kin}(\mathbf{u}) &= \frac{1}{2}\int_\Omega |\mathbf{u}|^2 d\mathbf{x}, \quad E_{mix}(\phi) = \int_\Omega \left(\frac{1}{2}|\nabla \phi|^2 + F(\phi)\right) d\mathbf{x}, \\ E_{nem}(\mathbf{d}, \phi) &= \int_\Omega I(\phi) \left(\frac{1}{2}|\nabla \mathbf{d}|^2 + G(\mathbf{d})\right) d\mathbf{x}, \\ E_{anch}(\mathbf{d}, \phi) &= \frac{1}{2}\int_\Omega \left(\delta_1 |\mathbf{d}|^2 |\nabla \phi|^2 + \delta_2 \left|\mathbf{d} \cdot \nabla \phi\right|^2\right) d\mathbf{x} \end{aligned}$$

and the anchoring energy will take different forms depending on the anchoring effect considered:

$$(\delta_1, \delta_2) = \begin{cases} (0,0) & \text{no anchoring,} \\ (0,1) & \text{parallel anchoring,} \\ (1,-1) & \text{homeotropic anchoring.} \end{cases} \tag{2}$$

For the functionals $F(\phi)$ and $G(\mathbf{d})$ we assume the following double-well potentials which in both cases have their minimums (and consequently their stable equilibrium states) at ± 1:

$$F(\phi) = \frac{1}{4\varepsilon^2}(\phi^2 - 1)^2 \quad \text{and} \quad G(\mathbf{d}) = \frac{1}{4\eta^2}(|\mathbf{d}|^2 - 1)^2,$$

where ε and η are small parameters related to the interface width and the constraint $|\mathbf{d}| = 1$, respectively. We denote $f(\phi) := F'(\phi)$ and $\mathbf{g}(\mathbf{d}) := G'(\mathbf{d})$. The functional $I(\phi)$ is an interpolation function that represents the volume fraction of liquid crystal and it is used to localize the nematic energy only where necessary, that is, this function is introduced to assure that the contributions from the nematic energy comes only from the regions where the nematic liquid crystal is contained. Here, we choose $I(\phi) \in C^2(\mathbb{R})$ with $I(\phi) \in [0, 1]$ for any $\phi \in \mathbb{R}$ defined by [3]:

$$I(\phi) := \begin{cases} \frac{1}{16}(\phi + 1)^3 (3\phi^2 - 9\phi + 8) & \text{if } \phi \in (-1, 1) \\ 1 & \text{if } \phi \geq 1 \\ 0 & \text{if } \phi \leq -1. \end{cases}$$

We denote by $i(\phi) := I'(\phi)$. The model that represents the mixture between a nematic liquid crystal and an isotropic flow can be derived phenomenologically: starting from the total free energy (1) and combining the ideas from the least action and the maximum dissipation principles, we can arrive at a thermodynamically consistent PDE system with respect to this free energy. This PDE system is an extension of the models proposed in [1–3], with the difference that in this case we consider a more realistic model for the nematic liquid crystal that takes into account the contribution of the shape of the molecules to the dynamics of the system (stretching effects). Precisely,

$$\begin{cases} \mathbf{u}_t + \mathbf{u}\cdot\nabla\mathbf{u} + \nabla p - \nabla\cdot(2\nu(\phi)\mathbf{D}\mathbf{u}) + \phi\nabla\mu - (\nabla\mathbf{d})^t\mathbf{w} \\ \qquad + \nabla\cdot\Big(\beta\mathbf{w}\mathbf{d}^t + (1+\beta)\mathbf{d}\mathbf{w}^t\Big) = \mathbf{0}, \\ \nabla\cdot\mathbf{u} = 0, \\ \mathbf{d}_t + (\mathbf{u}\cdot\nabla)\mathbf{d} + \beta(\nabla\mathbf{u})\mathbf{d} + (1+\beta)(\nabla\mathbf{u})^t\mathbf{d} + \gamma_{nem}\mathbf{w} = \mathbf{0}, \\ \lambda_{nem}\Big[-\nabla\cdot\big(I(\phi)\nabla\mathbf{d}\big) + I(\phi)\mathbf{g}(\mathbf{d})\Big] + \lambda_{anch}\Lambda_{\mathbf{d}}(\mathbf{d},\phi) = \mathbf{w}, \\ \phi_t + \nabla\cdot(\phi\mathbf{u}) - \gamma_{mix}\Delta\mu = 0, \\ \lambda_{mix}[-\Delta\phi + f(\phi)] + \lambda_{nem}I'(\phi)\left(\frac{1}{2}|\nabla\mathbf{d}|^2 + G(\mathbf{d})\right) \\ \qquad - \lambda_{anch}\nabla\cdot\Lambda_{\phi}(\mathbf{d},\phi) = \mu, \end{cases} \tag{3}$$

where $\mathbf{D}\mathbf{u} := (1/2)(\nabla\mathbf{u} + (\nabla\mathbf{u})^t)$, $\Lambda_{\mathbf{d}}(\mathbf{d},\phi) = \delta_1|\nabla\phi|^2\,\mathbf{d} + \delta_2\,(\mathbf{d}\cdot\nabla\phi)\,\nabla\phi$, and $\Lambda_{\phi}(\mathbf{d},\phi) = (\delta_1\,|\mathbf{d}|^2\,\nabla\phi + \delta_2\,(\mathbf{d}\cdot\nabla\phi)\,\mathbf{d})$, with (δ_1,δ_2) defined in (2). The geometrical parameter $\beta\in[-1,0]$ is a constant associated with the aspect ratio of the liquid crystal particles. For instance, $\beta = -1/2, -1$ and 0, corresponds to spherical, rod-like and disk-like liquid crystal molecules [4], respectively. Also, γ_{nem} and γ_{mix} are time relaxation parameters and $\nu = \nu(\phi)$ is the fluid viscosity (because the viscosities of each component of the mixture could be different). The PDE system (3) is written using the auxiliary variables $\mathbf{w}$ and μ (which are the variational derivatives of the total energy with respect to $\mathbf{d}$ and ϕ respectively) and is supplemented with the initial and boundary conditions:

$$\begin{aligned} \mathbf{u}|_{t=0} = \mathbf{u},\quad \mathbf{d}|_{t=0} = \mathbf{d},\quad \phi|_{t=0} = \phi_0 \text{ in } \Omega, \\ \mathbf{u}|_{\partial\Omega} = \big(I(\phi)\nabla\mathbf{d}\big)\mathbf{n}\big|_{\partial\Omega} = \mathbf{0} \quad\text{and}\quad \partial_{\mathbf{n}}\phi|_{\partial\Omega} = \partial_{\mathbf{n}}\mu|_{\partial\Omega} = 0 \quad \text{in } (0,T), \end{aligned} \tag{4}$$

where $\mathbf{n}$ denotes the outwards normal vector to the boundary $\partial\Omega$.

Remark 1. By integrating Eq. $(3)_5$ over Ω, we recover the conservation of volume (standard for Cahn-Hilliard models) $\frac{d}{dt}\int_\Omega \phi(t,\mathbf{x})\,d\mathbf{x} = 0$.

Theorem 1. *System (3) is dissipative. In fact, the following (dissipative) energy law holds:*

$$\frac{d}{dt}E_{tot}(\mathbf{u},\mathbf{d},\phi) + 2\int_\Omega \nu|\mathbf{D}\mathbf{u}|^2 d\mathbf{x} + \gamma_{nem}\int_\Omega |\mathbf{w}|^2 d\mathbf{x} + \gamma_{mix}\int_\Omega |\nabla\mu|^2 d\mathbf{x} = 0\,. \tag{5}$$

Proof. Testing $(3)_1$ by $\mathbf{u}$, $(3)_2$ by p, $(3)_3$ by $\mathbf{w}$, $(3)_4$ by $\mathbf{d}_t$, $(3)_5$ by μ and $(3)_6$ by ϕ_t, integrating over Ω and using the boundary conditions (4), then the energy law (5) holds.

2 Numerical Scheme

Let $\mathbf{V}_h \times P_h \times \mathbf{D}_h \times \mathbf{W}_h \times C_h \times M_h$ be conformed finite element spaces in $\mathbf{H}_0^1(\Omega) \times L_0^2(\Omega) \times \mathbf{H}^1(\Omega) \times \mathbf{L}^2(\Omega) \times H^1(\Omega) \times H^1(\Omega)$ corresponding to a regular and quasi-uniform triangulation $\mathcal{T}_h$ of the domain Ω with polyhedric boundary Γ. For simplicity, we describe the time discretization using a uniform partition of the time interval $[0,T]$: $t_n = n\Delta t$, where $\Delta t = T/N$ is the time step.

Let $(\mathbf{u}^n, p^n, \mathbf{d}^n, \mathbf{w}^n, \phi^n, \mu^n) \in \mathbf{V}_h \times P_h \times \mathbf{D}_h \times \mathbf{W}_h \times \Phi_h \times M_h$ be known and we denote $\delta_t a^{n+1} := (1/\Delta t)(a^{n+1} - a^n)$.
Step 1: Find $(\mathbf{d}^{n+1}, \mathbf{w}^{n+1}) \in \mathbf{D}_h \times \mathbf{W}_h$ such that, for each $(\bar{\mathbf{d}}, \bar{\mathbf{w}}) \in \mathbf{D}_h \times \mathbf{W}_h$

$$\left\{\begin{aligned} &(\delta_t \mathbf{d}^{n+1}, \bar{\mathbf{w}}) + (\mathbf{u}^\star, (\nabla \mathbf{d}^n)^t \bar{\mathbf{w}}) - \beta(\mathbf{u}^{\star\star}, \nabla\cdot(\bar{\mathbf{w}}(\mathbf{d}^n)^t)) \\ &\qquad -(1+\beta)(\mathbf{u}^{\star\star\star}, \nabla\cdot(\mathbf{d}^n \bar{\mathbf{w}}^t)) + \gamma_{nem}(\mathbf{w}^{n+1}, \bar{\mathbf{w}}) = 0, \\ &\lambda_{nem}\Big(I(\phi^n)\nabla\mathbf{d}^{n+1}, \nabla\bar{\mathbf{d}}\Big) + \lambda_{nem}\Big(I(\phi^n)\mathbf{g}_k(\mathbf{d}^{n+1}, \mathbf{d}^n), \bar{\mathbf{d}}\Big) \\ &\qquad + \lambda_{anch}\Big(\Lambda_{\mathbf{d}}(\mathbf{d}^{n+1}, \phi^n), \bar{\mathbf{d}}\Big) - (\mathbf{w}^{n+1}, \bar{\mathbf{d}}) = 0, \end{aligned}\right. \tag{6}$$

where

$$\begin{aligned} \mathbf{u}^\star &:= \mathbf{u}^n + 4\,\Delta t\,(\nabla\mathbf{d}^n)^t \mathbf{w}^{n+1}, \\ \mathbf{u}^{\star\star} &:= \mathbf{u}^n - 4\,\Delta t\,\beta\,\nabla\cdot(\mathbf{w}^{n+1}(\mathbf{d}^n)^t), \\ \mathbf{u}^{\star\star\star} &:= \mathbf{u}^n - 4\,\Delta t\,(1+\beta)\,\nabla\cdot(\mathbf{d}^n(\mathbf{w}^{n+1})^t) \end{aligned}$$

and $\mathbf{g}_k(\mathbf{d}^{n+1}, \mathbf{d}^n)$ denotes a first order approximation of $\mathbf{g}(\mathbf{d}(t_{n+1}))$.
Step 2: Find $(\phi^{n+1}, \mu^{n+1}) \in \Phi_h \times M_h$ such that, for each $(\bar{\phi}, \bar{\mu}) \in \Phi_h \times M_h$

$$\left\{\begin{aligned} &(\delta_t \phi^{n+1}, \bar{\mu}) - (\phi^n \mathbf{u}^\triangle, \nabla\bar{\mu}) + \gamma_{mix}(\nabla\mu^{n+1}, \nabla\bar{\mu}) = 0, \\ &\lambda_{mix}(\nabla\phi^{n+1}, \nabla\bar{\phi}) + \lambda_{mix}(f_k(\phi^{n+1}, \phi^n), \bar{\phi}) \\ &+ \lambda_{nem}\left(i_k(\phi^{n+1}, \phi^n)\left[\tfrac{1}{2}|\nabla\mathbf{d}^{n+1}|^2 + G(\mathbf{d}^{n+1})\right], \bar{\phi}\right) \\ &\qquad + \lambda_{anch}\Big(\Lambda_\phi(\mathbf{d}^{n+1}, \phi^{n+1}), \nabla\bar{\phi}\Big) - (\mu^{n+1}, \bar{\phi}) = 0, \end{aligned}\right. \tag{7}$$

where

$$\mathbf{u}^\triangle := \mathbf{u}^n - 4\,\Delta t\,\phi^n \nabla\mu^{n+1}$$

and $f_k(\phi^{n+1}, \phi^n)$ denotes a first order approximation of $f(\phi(t_{n+1}))$.
Step 3: Find $(\mathbf{u}^{n+1}, p^{n+1}) \in \mathbf{V}_h \times P_h$ such that, for each $(\bar{\mathbf{u}}, \bar{p}) \in \mathbf{u}_h \times P_h$

$$\left\{\begin{aligned} &\left(\frac{\mathbf{u}^{n+1} - \widehat{\mathbf{u}}}{\Delta t}, \bar{\mathbf{u}}\right) + c(\mathbf{u}^n, \mathbf{u}^{n+1}, \bar{\mathbf{u}}) - (p^{n+1}, \nabla\cdot\bar{\mathbf{u}}) \\ &\qquad\qquad + 2\,(\nu(\phi^{n+1})\mathbf{D}\mathbf{u}^{n+1}, \mathbf{D}\bar{\mathbf{u}}) = 0, \\ &\qquad\qquad (\nabla\cdot\mathbf{u}^{n+1}, \bar{p}) = 0, \end{aligned}\right. \tag{8}$$

where $c(\mathbf{u}, \mathbf{v}, \mathbf{w}) := \Big((\mathbf{u}\cdot\nabla)\mathbf{v}, \mathbf{w}\Big) + \frac{1}{2}\Big(\nabla\cdot\mathbf{u}, \mathbf{v}\cdot\mathbf{w}\Big)$ and $\widehat{\mathbf{u}} := \dfrac{\mathbf{u}^\star + \mathbf{u}^{\star\star} + \mathbf{u}^{\star\star\star} + \mathbf{u}^\triangle}{4}$.

Remark 2. It is easy to deduce, taking $\bar{\mu} = 1$ in (7), the mass conservation property of the scheme $\int_\Omega \phi^{n+1}\, d\mathbf{x} = \int_\Omega \phi^n\, d\mathbf{x}$.

A local in time discrete energy law is deduced in the next theorem, which is the main step to provide the (unconditional) energy stability of this linear scheme.

Theorem 2. *Scheme (6)–(8) satisfies the following local in time discrete energy law:*

$$\begin{array}{c}\delta_t E(\mathbf{d}^{n+1},\phi^{n+1},\mathbf{u}^{n+1})+\gamma_{nem}\left\|\mathbf{w}^{n+1}\right\|_{L^2}^2+\gamma_{mix}\|\nabla\mu^{n+1}\|_{L^2}^2\\ +2\left\|\nu(\phi^{n+1})^{1/2}\mathbf{D}\mathbf{u}^{n+1}\right\|_{L^2}^2+ND_{\mathbf{u}}^{n+1}+ND_{elast}^{n+1}(\phi^n)+ND_{penal}^{n+1}(\phi^n)\\ +ND_{philic}^{n+1}+ND_{phobic}^{n+1}+ND_{interp}^{n+1}(\mathbf{d}^{n+1})+ND_{anch}^{n+1}=0,\end{array}\tag{9}$$

where the numerical dissipation terms are:

$$\begin{aligned}
ND_{\mathbf{u}}^{n+1} &= \frac{1}{4\Delta t}\Big(2\|\mathbf{u}^{n+1}-\mathbf{u}^n\|_{L^2}^2+\|\mathbf{u}^{\star}-\mathbf{u}^n\|_{L^2}^2+\|\mathbf{u}^{\star\star}-\mathbf{u}^n\|_{L^2}^2\\
&\qquad+\|\mathbf{u}^{\star\star\star}-\mathbf{u}^n\|_{L^2}^2+\|\mathbf{u}^{\triangle}-\mathbf{u}^n\|_{L^2}^2\Big),\\
ND_{elast}^{n+1}(\phi) &= \lambda_{nem}\frac{\Delta t}{2}\int_\Omega I(\phi)\left|\delta_t\nabla\mathbf{d}^{n+1}\right|^2 d\mathbf{x},\\
ND_{penal}^{n+1}(\phi) &= \lambda_{nem}\int_\Omega I(\phi)\left(\mathbf{g}_k(\mathbf{d}^{n+1},\mathbf{d}^n)\cdot\delta_t\mathbf{d}^{n+1}-\delta_t G(\mathbf{d}^{n+1})\right)d\mathbf{x},\\
ND_{philic}^{n+1} &= \lambda_{mix}\frac{\Delta t}{2}\int_\Omega\left|\delta_t\nabla\phi^{n+1}\right|^2 d\mathbf{x},\\
ND_{phobic}^{n+1} &= \lambda_{mix}\int_\Omega\left(f_k(\phi^{n+1},\phi^n)\,\delta_t\phi^{n+1}-\delta_t F(\phi^{n+1})\right)d\mathbf{x},\\
ND_{interp}^{n+1}(\mathbf{d}) &= \lambda_{nem}\int_\Omega\left(\frac{|\nabla\mathbf{d}|^2}{2}+G(\mathbf{d})\right)\left(i_k(\phi^{n+1},\phi^n)\delta_t\phi^{n+1}-\delta_t I(\phi^{n+1})\right)d\mathbf{x},\\
ND_{anch}^{n+1} &= \lambda_{anch}\frac{\Delta t}{2}\int_\Omega\Big[\delta_1\left(|\delta_t\mathbf{d}^{n+1}|^2|\nabla\phi^n|^2+|\mathbf{d}^{n+1}|^2|\delta_t\nabla\phi^{n+1}|^2\right)\\
&\quad+\delta_2\left(|\delta_t\mathbf{d}^{n+1}\cdot\nabla\phi^n|^2+|\mathbf{d}^{n+1}\cdot\nabla\delta_t\phi^{n+1}|^2\right)\Big]d\mathbf{x},
\end{aligned}$$

with the values of (δ_1,δ_2) depending on the type of anchoring defined in (2).

Proof. Taking $(\bar{\mathbf{w}},\bar{\mathbf{d}})=(\mathbf{w}^{n+1},\delta_t\mathbf{d}^{n+1})$ in (6), $(\bar{\mu},\bar{\phi})=(\mu^{n+1},\delta_t\phi^{n+1})$ in (7) and $(\bar{\mathbf{u}},\bar{p})=(\mathbf{u}^{n+1},p^{n+1})$ in (8) and following similar arguments to the ones presented in [3] and [2], then (9) is derived.

Remark 3. In practice, there is no need to introduce the extra unknowns $\mathbf{u}^\star$, $\mathbf{u}^{\star\star}$, $\mathbf{u}^{\star\star\star}$, $\mathbf{u}^\triangle$ and $\widehat{\mathbf{u}}$ for carrying out the simulations, they are only used as a tool to show the energy-stability of the scheme.

Lemma 1. *Let us consider the linear approximations for the potential terms $f_k(\phi^{n+1},\phi^n)$, $i_k(\phi^{n+1},\phi^n)$ and $\mathbf{g}_k(\mathbf{d}^{n+1},\mathbf{d}^n)$ presented in* [3]. *Assuming that $1\in C_h$, $\mathbf{D}_h\subseteq\mathbf{W}_h$ and the pair of FE spaces $(\mathbf{V}_h,P_h)$ satisfies the discrete inf-sup condition: there exists $\widetilde{C}>0$ such that $\|p\|_{L^2}\le\widetilde{C}\sup_{\bar{\mathbf{u}}\in\mathbf{V}_h\setminus\{\Theta\}}\frac{(p,\nabla\cdot\bar{\mathbf{u}})}{\|\bar{\mathbf{u}}\|_{H^1}}$ for all $p\in P_h$, then there exist a unique solution of the scheme (6)–(8); a unique solution $(\mathbf{d}^{n+1},\mathbf{w}^{n+1})$ of (6), (ϕ^{n+1},μ^{n+1}) of (7) and $(\mathbf{u}^{n+1},p^{n+1})$ of (8).*

3 Numerical Simulations

In this section we present numerical experiments to show the effectiveness of the numerical scheme. In particular, the potential terms $f_k(\phi^{n+1},\phi^n)$, $\mathbf{g}_k(\mathbf{d}^{n+1},\mathbf{d}^n)$

Table 1. Parameters

Ω	$[0,T]$	h	Δt	ν_0	λ_{nem}	λ_{mix}	λ_{anch}	γ_{nem}	γ_{mix}	ε	η	β
$[-1,1]^2$	$[0,6]$	$2/90$	0.001	1.0	0.1	0.01	0.1	0.5	0.01	0.025	0.075	-1.0

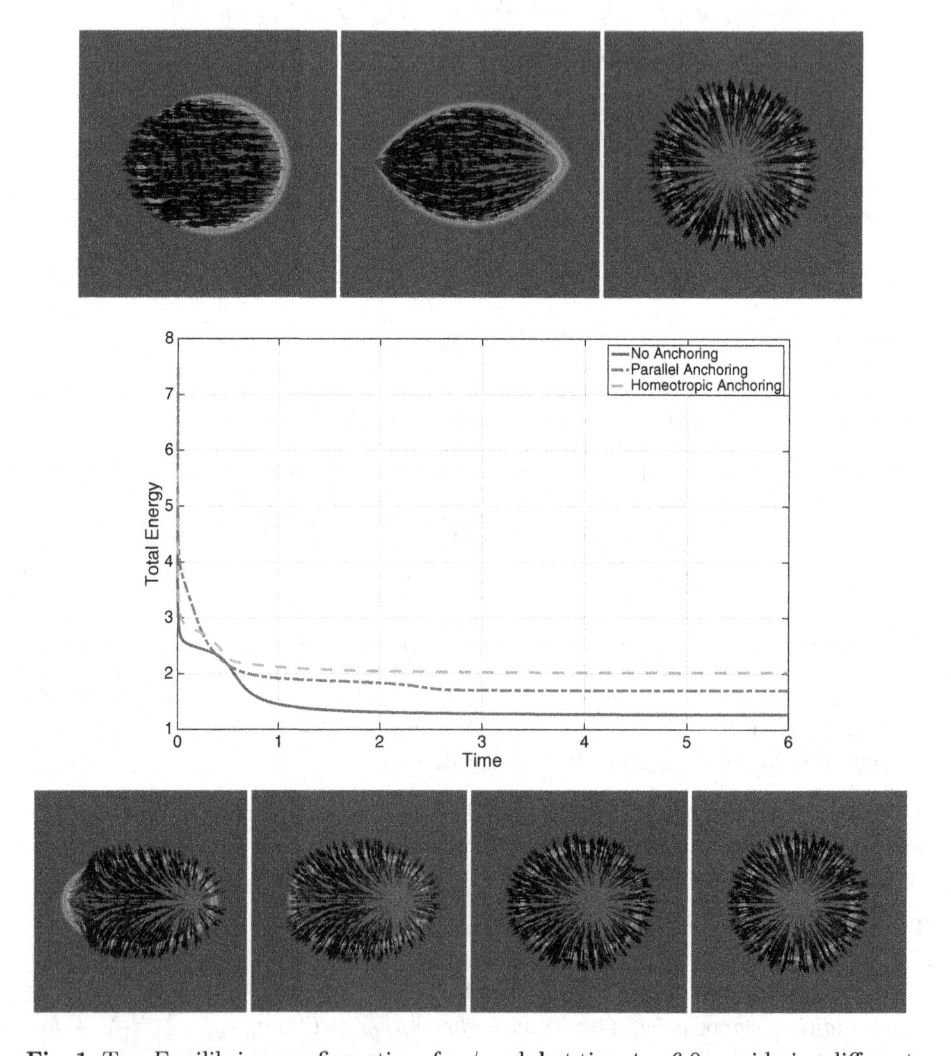

Fig. 1. Top: Equilibrium configurations for ϕ and $\mathbf{d}$ at time $t = 6.0$ considering different anchoring energies (Left: No anchoring. Center: Parallel anchoring. Right: homeotropic anchoring). Center: Evolution in time of the total energy for the three anchoring cases. Bottom: evolution of ϕ and $\mathbf{d}$ at times $t = 0, 0.5, 2.0, 6.0$ in the homeotropic case.

and $i_k(\phi^{n+1}, \phi^n)$ have been approximated considering the ideas introduced in [3] to design linear and energy-stable approximations. All the simulations have been carried out in $2D$ using *FreeFem++* software, and we consider the choice for the discrete spaces

$$(\mathbf{u}, p) \sim P_2 \times P_1, \qquad (\phi, \mu) \sim P_1 \times P_1 \qquad \text{and} \qquad (\mathbf{d}, \mathbf{w}) \sim P_1 \times P_1,$$

that satisfy the assumptions from Lemma 1 assuring the well-posedness of each of the sub-steps of the numerical scheme (6)–(8). The discrete and physical parameters are presented in Table 1, where for simplicity we are considering constant viscosity $\nu(\phi) = \nu_0$. The physical parameters have been taken such that the kinetic energy is not playing a main role in the behavior of the system, in order to be able to identify the influence of the rest of the terms in the total energy (1) over the dynamics of the system.

We have considered an elliptic nematic droplet with two defect points at $(\pm 0.5, 0)$, a Hedgehog defect at $(0.5, 0)$ and an anti-Hedgehog defect at $(-0.5, 0)$. This configuration is a classical benchmark for numerical schemes to approximate nematic liquid crystals and is generated by using the function

$$\mathbf{d}_0(x) = I(\phi)\,\widehat{\mathbf{d}}/\sqrt{|\widehat{\mathbf{d}}|^2 + 0.05^2}, \qquad \text{with} \qquad \widehat{\mathbf{d}} = (x^2 + y^2 - 0.25, y).$$

In the dynamics of the system, there are three main processes that are competing: to minimize the mixing energy, the system would like to arrive to a circular configuration, while to minimize the elastic energy the system would like to annihilate the possible defects. Moreover, the system should rearrange the director field through the interface to minimize as much as possible the anchoring energy. We have compared the behavior of the three possible choices of the anchoring energy (no anchoring, parallel and homeotropic) when the stretching parameter is fixed to represent the case of rod-like molecules ($\beta = -1.0$). It is clear how the anchoring energy influences the dynamic of the system, arriving at three different equilibrium configurations (Fig. 1-Top). In all the cases the total energy is decreasing until it reaches an equilibrium state (Fig. 1-Center). Moreover we present the details of the dynamics of the Homeotropic case in Fig. 1-Bottom.

4 Conclusion

We have described a new linear unconditionally stable scheme for a nematic-isotropic mixture whose dynamics is given by the interactions of several effects, including hydrodynamics and mixing effects, the preferred orientation of the molecules in the interface (anchoring) as in [3], adding as novelty the effect of the shape of the molecules (stretching). The proposed scheme is efficient due to its linearity and the fact of decoupling into smaller sub-steps. Moreover, it maintains the energy stability and achieves equilibrium configurations predicted by experimental groups [5,6].

References

1. Yue, P., Feng, J.J., Liu, C., Shen, J.: A diffuse-interface method for simulating two-phase flows of complex fluids. J. Fluid Mech. **515**, 293–317 (2004)

2. Cabrales, R.C., Guillén-González, F., Gutiérrez-Santacreu, J.V.: A projection-based time-splitting algorithm for approximating nematic liquid crystal flows with stretching. ZAMM **97**(10), 1204–1219
3. Guillén-González, F., Rodríguez-Bellido, M.A., Tierra, G.: Linear unconditional energy-stable splitting schemes for a phase-field model for nematic-isotropic flows with anchoring effects Int. J. Numer. Methods Eng. **108**, 535–567 (2016)
4. Jeffery, G.B.: The motion of ellipsoidal particles immersed in a viscous fluid. R. Soc. Proc. **102**, 161–179 (1922)
5. van Bijnen, R.M.W., Otten, R.H.J., van der Schoot, P.: Texture and shape of two-dimensional domains of nematic liquid crystals. Phys. Rev. E **86**, 051703 (2012)
6. Kim, Y.K., Shiyanovskii, S.V., Lavrentovich, O.D.: Morphogenesis of defects and tactoids during isotropic-nematic phase transition in self-assembled lyotropic chromonic liquid crystals. J. Phys. Condens. Matter **25**, 404202 (2013)

Two Classes of Vector Domain Decomposition Schemes for Time-Dependent Problems with Overlapping Subdomains

Petr N. Vabishchevich[1,2](✉)

[1] Nuclear Safety Institute, 52, B. Tulskaya, 115191 Moscow, Russia
vabishchevich@gmail.com

[2] North-Eastern Federal University, 58, Belinskogo, 677000 Yakutsk, Russia

Abstract. The domain decomposition methods for time-dependent problems are based on special schemes of splitting into subdomains. To construct homogeneous numerical algorithms, overlapping subdomain methods are preferable. The domain decomposition is associated with corresponding additive representation of the problem operator. Such regionally-additive schemes are based on the general theory of additive operator-difference schemes. There are variants of decomposition operators differing by distinct types of data exchanges on interfaces.

New classes of domain decomposition schemes for transient problems based on subdomains overlapping are constructed. The boundary value problem for the parabolic equation of second order is considered as a model problem. We propose a general approach to construct vector domain decomposition schemes for time-dependent systems of equations. Using a partition of unity for a computational domain we perform a transition to finding the individual components of the solution in the subdomains. General stability conditions are obtained for vector regionally-additive schemes with first and second order accuracy.

1 Introduction

The domain decomposition methods for time-dependent problems can be classified according to (I) the method of domain decomposition, (II) the selection of decomposition operators (exchange boundary conditions), and (III) the used splitting schemes (approximation in time). For multi-dimensional boundary value problems, it is possible to construct domain decomposition methods with or without overlapping of subdomains [2,5]. Methods without overlapping of subdomains are associated with the explicit formulation of exchange conditions on the boundaries of subdomains. These methods are particularly useful for solving problems where each individual subdomain has its own computational grid (triangulation). To construct homogeneous numerical algorithms, methods with overlapping subdomains are more preferable. In the case of minimal overlapping, where the width of overlapping is equal to one step of spatial grid, the methods can be treated as non-overlapping with special exchange boundary conditions on the common boundaries.

I. Lirkov and S. Margenov (Eds.): LSSC 2017, LNCS 10665, pp. 85–92, 2018.
https://doi.org/10.1007/978-3-319-73441-5_8

To construct decomposition operators for solving boundary value problems, it is convenient to apply a partition of unity for the computational domain (see, e.g., [1]). In the methods of decomposition with overlapping of subdomains, the functions are associated with individual subdomains with values between zero and one. Domain decomposition methods for the Cauchy problem for partial differential equations are summarized in the book [4]. Among more recent researches we note the work [6], which discusses domain decomposition methods more suitable for numerical implementation.

In the present work, we construct regionally-additive schemes and study their convergence on the basis of the general theory of splitting schemes [3,7]. We start with the simplest case of two-component splitting. In this case, we construct unconditionally stable factorized splitting schemes, such as the classical alternating direction methods, predictor-corrector schemes and so on. The most interesting for practice is the situation, where the problem operator is split into a sum of three or more non-commutative non-self-adjoint operators. For such multi-component representation, the splitting schemes are based on the concept of summarized approximation. Additively averaged splitting schemes are of particular interest for parallel computers. In the class of splitting schemes with full approximation [7], we highlight the vector additive schemes, where the original equation is transformed into a system of similar equations. The most convenient approach for construction of additive operator-difference schemes of multi-component splitting is based on regularization of the difference schemes. In this approach, stability is achieved by perturbing a difference scheme operator.

To solve approximately the Cauchy problem for an evolutionary equation, the standard approach with a partition of unity involves an additive representation of the problem operator, where individual operator terms are associated with subproblems in the related subdomains. A transition to a new time-level is performed using various additive schemes (splitting schemes) for the original scalar problem and involves the solution of problems in separate subdomains. We investigate a new class of domain decomposition schemes that is based on an explicit transition from the original scalar problem to a vector one for solving problems in separate subdomains.

2 Problem Formulation

Let Ω be a bounded domain ($\Omega \subset \boldsymbol{R}^m, m = 2,3$) with a piecewise smooth boundary $\partial\Omega$. Let $\mathcal{A}$ be an elliptic operator such that

$$\mathcal{A}u = -\nabla(k(\boldsymbol{x})\nabla u) + c(\boldsymbol{x})u, \quad \boldsymbol{x} \in \Omega \tag{1}$$

on the set of functions

$$u(\boldsymbol{x}) = 0, \quad \boldsymbol{x} \in \partial\Omega. \tag{2}$$

Assume that the coefficients $k(\boldsymbol{x})$ and $c(\boldsymbol{x})$ are smooth functions in $\overline{\Omega}$ and

$$k(\boldsymbol{x}) \geq \kappa > 0, \quad c(\boldsymbol{x}) \geq 0, \quad \boldsymbol{x} \in \Omega.$$

We consider the Cauchy problem

$$\frac{du}{dt} + \mathcal{A}u = f(t), \quad 0 < t \leq T, \tag{3}$$

$$u(0) = u^0, \tag{4}$$

where, for instance, $f(\boldsymbol{x}, t) \in L_2(\Omega)$, $u^0(\boldsymbol{x}) \in L_2(\Omega)$. Here we use the notation $u(t) = u(\boldsymbol{x}, t)$.

Let $(\cdot, \cdot)$, $\|\cdot\|$ be the scalar product and norm in $H = L_2(\Omega)$, respectively:

$$(u, v) = \int_\Omega u(\boldsymbol{x})v(\boldsymbol{x})d\boldsymbol{x}, \quad \|u\| = (u, u)^{1/2}.$$

The symmetric positive definite bilinear form $d(u, v)$ such that

$$d(u, v) = d(v, u), \quad d(u, u) \geq \delta\|u\|^2, \quad \delta > 0,$$

is associated with the Hilbert space H_d, where the scalar product and norm are defined as

$$(u, v)_d = d(u, v), \quad \|u\|_d = (d(u, u))^{1/2}.$$

Define $H_0^1(\Omega)$ as a subspace of $H^1(\Omega)$:

$$H_0^1(\Omega) = \{v(\boldsymbol{x}) \in H^1(\Omega) : \ v(\boldsymbol{x}) = 0, \quad \boldsymbol{x} \in \partial\Omega\}.$$

Multiplying Eq. (3) by $v(\boldsymbol{x}) \in H_0^1(\Omega)$ and integrating over the domain Ω, we get the equality:

$$\left(\frac{du}{dt}, v\right) + a(u, v) = (f, v), \quad \forall v \in H_0^1(\Omega), \quad 0 < t \leq T. \tag{5}$$

Here $a(\cdot, \cdot)$ is a bilinear form, which has the form

$$a(u, v) = \int_\Omega (k\nabla u \cdot \nabla v + c\,u\,v)d\boldsymbol{x},$$

and

$$a(u, v) = a(v, u), \quad a(u, u) \geq \delta\|u\|^2, \quad \delta > 0.$$

In view of (4), we put

$$(u(0), v) = (u^0, v), \quad \forall v \in H_0^1(\Omega). \tag{6}$$

The variational (weak) formulation of the problem (1)–(4) is to find $u(\boldsymbol{x}, t) \in H_0^1(\Omega)$, $0 < t \leq T$ satisfying the boundary conditions (2) and Eqs. (5) and (6). The solution of the problems (5) and (6) satisfies the following a priori estimate:

$$\|u(t)\|_a^2 \leq \|u^0\|_a^2 + \frac{1}{2}\int_0^t \|f(\theta)\|^2 d\theta. \tag{7}$$

To solve numerically the initial-boundary value problems (3) and (4), approximation in space is constructed using the finite element method. The weak formulation (5) and (6) is applied.

Let us define a subspace of finite elements $V_h \subset H_0^1(\Omega)$. The domain Ω is divided into non-overlapping triangles (tetrahedra) using first-order Lagrangian finite elements (piecewise-linear approximation). In view of (5), the approximate solution is determined by

$$\left(\frac{dw}{dt}, v\right) + a(w, v) = (f, v), \quad \forall v \in V_h, \quad 0 < t \leq T. \tag{8}$$

For $t = 0$ (see (6)), we have

$$(w(0), v) = (u^0, v), \quad \forall v \in V_h. \tag{9}$$

To solve numerically the problem (8) and (9), we apply the implicit two-level scheme [3]. Let τ be a step of the uniform grid in time $y^n = y(t^n)$, $t^n = n\tau$, $n = 0, 1, ..., N$, $N\tau = T$. We approximate equation (8) by the following two-level scheme:

$$\left(\frac{y^{n+1} - y^n}{\tau}, v\right) + a(\sigma y^{n+1} + (1 - \sigma) y^n, v) = (f(\sigma t^{n+1} + (1 - \sigma) t^n), v), \tag{10}$$

$$(y^0, v) = (u^0, v), \quad \forall v \in V_h, \quad n = 0, 1, ..., N - 1, \tag{11}$$

where σ is a numerical parameter (weight). If $\sigma = 0$, then (10) and (11) is the explicit scheme. For $\sigma = 1$, we obtain the fully implicit scheme. The value $\sigma = 0.5$ corresponds to the symmetric scheme (the Crank-Nicolson scheme). The condition

$$(v, v) + \tau \left(\sigma - \frac{1}{2}\right) a(v, v) \geq 0, \quad \forall v \in V_h$$

is the necessary and sufficient condition for stability in the space H_a [3,4]. In particular, for $\sigma \geq 0.5$, the solution of (10) and (11) satisfies the a priori estimate

$$\|y^{n+1}\|_a^2 \leq \|u^0\|_a^2 + \frac{1}{2} \sum_{k=0}^{n} \tau \|f^{k+\sigma}\|^2, \tag{12}$$

with the following notation:

$$t^{n+\sigma} = \sigma t^{n+1} + (1 - \sigma) t^n, \quad f^{n+\sigma} = f(t^{n+\sigma}).$$

3 Standard Vector Domain Decomposition Schemes

For the above differential problem, we consider the following domain decomposition:

$$\overline{\Omega} = \bigcup_{\alpha=1}^{p} \overline{\Omega}_\alpha, \quad \overline{\Omega}_\alpha = \Omega_\alpha \cup \partial\Omega_\alpha, \quad \alpha = 1, 2, ..., p \tag{13}$$

with overlapping ($\Omega_{\alpha\beta} \equiv \Omega_\alpha \cap \Omega_\beta \neq \varnothing$) or without overlapping of subdomains ($\Omega_{\alpha\beta} = \varnothing$) [2,5].

To construct domain decomposition schemes, we use a partition of unity for the computational domain Ω [1]. The individual subdomains Ω_α, $\alpha = 1, 2, ..., p$ are associated with functions $\eta_\alpha(\boldsymbol{x})$, $\alpha = 1, 2, ..., p$ such that

$$\eta_\alpha(\boldsymbol{x}) = \begin{cases} > 0, \ \boldsymbol{x} \in \Omega_\alpha, \\ \ \ 0, \ \ \boldsymbol{x} \notin \Omega_\alpha, \end{cases} \quad \alpha = 1, 2, ..., p, \quad \sum_{\alpha=1}^{p} \eta_\alpha(\boldsymbol{x}) = 1, \quad \boldsymbol{x} \in \Omega.$$

The standard approach [4] involves the following additive representation of the operator of the problems (3) and (4):

$$\mathcal{A} = \sum_{\alpha=1}^{p} \mathcal{A}_\alpha, \tag{14}$$

where the individual operator term $\mathcal{A}_\alpha$ is associated with the subdomain Ω_α, $\alpha = 1, 2, ..., p$. For example, taking into account (1), it seems natural to put

$$\mathcal{A}_\alpha = -\nabla(\eta_\alpha(\boldsymbol{x}) k(\boldsymbol{x}) \nabla u) + \eta_\alpha(\boldsymbol{x}) c(\boldsymbol{x}) u, \quad \alpha = 1, 2, ..., p, \quad \boldsymbol{x} \in \Omega.$$

In this case, for the representation (14), we have

$$\mathcal{A}_\alpha = \mathcal{A}_\alpha^* \geq 0, \quad \alpha = 1, 2, ..., p.$$

The construction and investigation of domain decomposition schemes for time-dependent problems (3), (4) and (14) is based on the corresponding splitting schemes [7]. Separately, we can highlight the case of two-component splitting ($p = 2$). For multi-component splitting ($p > 2$), we use vector additive schemes.

For the additive representation of the problem operator in (14), applying vector schemes, we search (see, e.g., [7]) the vector $\boldsymbol{u} = \{u_1, u_2, ..., u_p\}$. The individual components are obtained from the similar problems:

$$\frac{du_\alpha}{dt} + \sum_{\beta=1}^{p} \mathcal{A}_\beta u_\beta = f,$$

$$u_\alpha(0) = u^0, \quad \alpha = 1, 2, ..., p.$$

Below, for this problem, we consider various approximations in time associated with solving problems in separate subdomains.

4 Vector Schemes with an Additive Representation of the Solution

A new class of domain decomposition schemes is based on the use of functions

$$\xi_\alpha(\boldsymbol{x}) = \eta_\alpha^{1/2}(\boldsymbol{x}), \quad \alpha = 1, 2, ..., p.$$

Thus

$$\sum_{\alpha=1}^{p} \xi_\alpha^2(\boldsymbol{x}) = 1, \quad \boldsymbol{x} \in \Omega.$$

To solve the problem (3) and (4), we apply the decomposition

$$u(\boldsymbol{x},t) = \sum_{\alpha=1}^{p} \xi_\alpha(\boldsymbol{x}) u_\alpha(\boldsymbol{x},t), \quad u_\alpha(\boldsymbol{x},t) = \xi_\alpha(\boldsymbol{x}) u(\boldsymbol{x},t), \quad \alpha = 1,2,...,p. \quad (15)$$

Let us formulate the appropriate problem for the individual components of the solution $u_\alpha(\boldsymbol{x},t)$, $\alpha = 1,2,...,p$.

Multiplying Eq. (3) by $\xi_\alpha(\boldsymbol{x},t)$, $\alpha = 1,2,...,p$, in view of (15), we obtain the following system of equations:

$$\frac{du_\alpha}{dt} + \sum_{\beta=1}^{p} \mathcal{A}_{\alpha\beta} u_\beta = f_\alpha, \quad \alpha = 1,2,...,p. \quad (16)$$

In (16), we have

$$\mathcal{A}_{\alpha\beta} u_\beta = -\xi_\alpha(\boldsymbol{x}) \nabla (k(\boldsymbol{x}) \nabla \xi_\beta(\boldsymbol{x}) u_\beta) + \delta_{\alpha\beta} c(\boldsymbol{x}) \xi_\alpha^2(\boldsymbol{x}) u_\beta, \quad \beta = 1,2,...,p,$$

$$f_\alpha(\boldsymbol{x},t) = \xi_\alpha(\boldsymbol{x}) f(\boldsymbol{x},t), \quad \alpha = 1,2,...,p,$$

where

$$\delta_{\alpha\beta} = \begin{cases} 1, \ \alpha = \beta, \\ 0, \ \alpha \neq \beta, \end{cases}$$

is the Kronecker's symbol. The Eq. (16) are supplemented by the initial conditions

$$u_\alpha(0) = u_\alpha^0, \quad \alpha = 1,2,...,p, \quad (17)$$

where $u_\alpha^0(\boldsymbol{x}) = \xi_\alpha(\boldsymbol{x}) u^0(\boldsymbol{x})$, $\alpha = 1,2,...,p$.

The Cauchy problems (16) and (17) is associated with a variational problem. We search $u_\alpha(\boldsymbol{x},t) \in H_0^1(\Omega)$, $\alpha = 1,2,...,p$, $0 < t \leq T$ from the conditions

$$\left(\frac{du_\alpha}{dt}, v_\alpha \right) + \sum_{\beta=1}^{p} a_{\alpha\beta}(u_\beta, v_\alpha) = (f_\alpha, v_\alpha), \quad 0 < t \leq T, \quad (18)$$

$$(u_\alpha(0), v_\alpha) = (u_\alpha^0, v_\alpha), \quad \forall v_\alpha \in H_0^1(\Omega), \quad \alpha = 1,2,...,p. \quad (19)$$

For the bilinear forms $a_{\alpha\beta}(\cdot,\cdot)$, we have

$$a_{\alpha\beta}(u_\beta, v_\alpha) = \int_\Omega (k \nabla (\xi_\beta u_\beta) \cdot \nabla (\xi_\alpha v_\alpha) + \delta_{\alpha\beta} \xi_\alpha^2 c \, u_\alpha v_\alpha) d\boldsymbol{x},$$

and

$$a_{\alpha\beta}(u,v) = a_{\alpha\beta}(v,u), \quad \alpha, \beta = 1,2,...,p.$$

Theorem 1. *For the solution of problems (18) and (19), the following a priori estimate holds:*

$$\|u(t)\|_a^2 \leq \|u^0\|_a^2 + \frac{1}{2}\int_0^t \sum_{\alpha=1}^{p} \|f_\alpha(\theta)\|^2 d\theta, \tag{20}$$

where

$$u(t) = \sum_{\alpha=1}^{p} \xi_\alpha u_\alpha(t). \tag{21}$$

Applying finite element approximations to (18) and (19), we arrive at the problem of searching $w_\alpha(\boldsymbol{x}, t) \in V_h$, $\alpha = 1, 2, ..., p$, $0 < t \leq T$ from the conditions

$$\left(\frac{dw_\alpha}{dt}, v_\alpha\right) + \sum_{\beta=1}^{p} a_{\alpha\beta}(w_\beta, v_\alpha) = (f_\alpha, v_\alpha), \quad 0 < t \leq T, \tag{22}$$

$$(w_\alpha(0), v_\alpha) = (u_\alpha^0, v_\alpha), \quad \forall v_\alpha \in V_h, \quad \alpha = 1, 2, ..., p. \tag{23}$$

Using the standard approximations in time, we search the individual components of the solution from the coupled system of equations. For example, for two-level schemes with weights

$$\left(\frac{y_\alpha^{n+1} - y_\alpha^n}{\tau}, v_\alpha\right) + \sum_{\beta=1}^{p} a_{\alpha\beta}(y_\beta^{n+\sigma}, v_\alpha) = (f_\alpha^{n+\sigma}, v_\alpha), \quad n = 0, 1, ..., N-1, \tag{24}$$

$$(y_\alpha^0(0), v_\alpha) = (u_\alpha^0, v_\alpha), \quad \forall v_\alpha \in V_h, \quad \alpha = 1, 2, ..., p, \tag{25}$$

the solution at a new time-level is obtained from the system of equations:

$$(y_\alpha^{n+1}, v_\alpha) + \tau\sigma \sum_{\beta=1}^{p} a_{\alpha\beta}(y_\beta^{n+1}, v_\alpha) = (\varphi_\alpha^n, v_\alpha), \quad \forall v_\alpha \in V_h, \quad \alpha = 1, 2, ..., p.$$

We get the rhs φ_α^n, $\alpha = 1, 2, ..., p$ from (24), first moving over all *known* functions (y_α^n, $y_\beta^n, \beta = 1, 2, ..., p$) to the rhs, and then multiplying by τ. Consider domain decomposition schemes, where at the new time-level we solve the problems

$$(y_\alpha^{n+1}, v_\alpha) + \tau\gamma a_{\alpha\beta}(y_\alpha^{n+1}, v_\alpha) = (\varphi_\alpha^n, v_\alpha), \quad \forall v_\alpha \in V_h, \quad \alpha = 1, 2, ..., p,$$

for the individual components y_α^{n+1}, $\alpha = 1, 2, ..., p$. We distinguish the asynchronous (parallel) version of domain decomposition schemes, namely, an analogue of the block Jacobi method, and synchronous (serial) variant, which is an analogue of the block Seidel method.

Instead of (24), we apply the explicit-implicit scheme:

$$\begin{aligned} \left(\frac{y_\alpha^{n+1} - y_\alpha^n}{\tau}, v_\alpha\right) + \sum_{\beta=1}^{\alpha} a_{\alpha\beta}(y_\beta^{n+1}, v_\alpha) + \sum_{\beta=\alpha+1}^{p} a_{\alpha\beta}(y_\beta^n, v_\alpha) \\ = (f_\alpha^{n+1}, v_\alpha), \quad n = 0, 1, ..., N-1. \end{aligned} \tag{26}$$

The implementation of the scheme (25) and (26) is serial. First, we search y_1^{n+1}, then y_2^{n+1} and so on. The scheme (25) and (26) has first-order approximation in time.

Theorem 2. *The difference scheme (25) and (26) is unconditionally stable and the following a priori estimate holds:*

$$\|y^{n+1}\|_a^2 \leq \|u^0\|_a^2 + \frac{\tau}{2}\sum_{k=0}^{n}\sum_{\alpha=1}^{p}\|f_\alpha^{k+1}\|^2, \tag{27}$$

$$y^{n+1} = \sum_{\alpha=1}^{p}\xi_\alpha y_\alpha^{n+1}. \tag{28}$$

In the scheme (25) and (26), the transition to a new time-level involves the lower triangular part of the operator matrix (16) and we have an analogue of the Seidel method. It is possible to construct parallel domain decomposition schemes, where the new time-level includes the diagonal part of the operator in (16), and we obtain an analogue of the Jacobi method. In this case, the explicit-implicit scheme has the form:

$$\begin{aligned}\left(\frac{y_\alpha^{n+1} - y_\alpha^n}{\tau}, v_\alpha\right) + a_{\alpha\alpha}(\sigma y_\alpha^{n+1} + (1-\sigma)y_\alpha^n, v_\alpha) + \sum_{\beta=1,\beta\neq\alpha}^{p} a_{\alpha\beta}(y_\beta^n, v_\alpha) \\ = (f_\alpha^{n+1}, v_\alpha), \quad n = 0, 1, ..., N-1,\end{aligned} \tag{29}$$

The implementation of the scheme (25) and (29) allows to evaluate y_α^{n+1}, $\alpha = 1, 2, ..., p$ independently of each other.

Theorem 3. *The explicit-implicit scheme (25) and (29) is unconditionally stable for $\sigma \geq p/2$ and the a priori estimate (27) and (28) holds.*

References

1. Mathew, T.: Domain Decomposition Methods for the Numerical Solution of Partial Differential Equations. Springer, Heidelberg (2008). https://doi.org/10.1007/978-3-540-77209-5
2. Quarteroni, A., Valli, A.: Domain Decomposition Methods for Partial Differential Equations. Clarendon Press, Oxford (1999). https://doi.org/10.1007/978-3-0348-7885-2
3. Samarskii, A.A.: The Theory of Difference Schemes. Marcel Dekker, New York (2001). https://doi.org/10.1201/9780203908518
4. Samarskii, A.A., Matus, P.P., Vabishchevich, P.N.: Difference Schemes with Operator Factors. Kluwer Academic Pub, Heidelberg (2002). https://doi.org/10.1007/978-94-015-9874-3
5. Toselli, A., Widlund, O.: Domain Decomposition Methods - Algorithms and Theory. Springer, Heidelberg (2005). https://doi.org/10.1007/b137868
6. Vabishchevich, P.N.: A substructuring domain decomposition scheme for unsteady problems. Comput. Methods Appl. Math. **11**(2), 241–268 (2011). https://doi.org/10.2478/cmam-2011-0013
7. Vabishchevich, P.N.: Additive Operator-Difference Schemes: Splitting Schemes. de Gruyter, Berlin (2014). https://doi.org/10.1515/9783110321463

Least-Squares Finite Element Methods

An Alternative Proof of a Strip Estimate for First-Order System Least-Squares for Interface Problems

Fleurianne Bertrand(✉)

Fakultät für Mathematik, Universität Duisburg-Essen, 45117 Essen, Germany
fleurianne.bertrand@uni-due.de

Abstract. The purpose of this paper is an alternative proof of a strip estimate, used in Least-Squares methods for interface problems, as in [4] for a two-phase flow problem with incompressible flow in the subdomains. The Stokes flow problems in the subdomains are treated as first-order systems and a combination of $H(\text{div})$-conforming Raviart-Thomas and standard H^1-conforming elements were used for the discretization. The interface condition is built directly in the $H(\text{div})$-conforming space. Using the strip estimate, the homogeneous Least-Squares functional is shown to be equivalent to an appropriate norm allowing the use of standard finite element approximation estimates.

1 Introduction

Least-Squares finite element methods (see e.g. [10]) minimizing the L^2 residuals in the partial differential equations combine the advantages of the mixed finite element methods (see e.g. [11]) with the production of symmetric and positive definite discrete systems and an inherent a posteriori error indicator (see e.g. [3]). Interface conditions can be handled by an auxiliary Least-Squares functional (as in [22]) or built directly in the finite element spaces. In this case, a closer look has to be taken at the error that appears using this approach on phases with curved boundaries approximated by a triangulation. This relies crucially on an estimate for the discrepancy of the normal flux associated with the parametric Raviart-Thomas spaces on the curved boundary if it is set to zero on its piecewise polynomial approximation ([6,7]). Implementation details are given in [5].

For incompressible Newtonian fluid flow with homogeneous density, the primitive physical equations are the conservation of momentum and the constitutive law, and correspond to different first-order systems depending on the variables used ([9,13,16]). In [4], the formulation developed in [15] is used for each phase in a least-squares context building the interface condition directly in the spaces. The main theorem of [4] states that the optimal order of convergence is retained for the least-squares formulation using the (iso)parametric elements on the approximated domain and uses the strip estimate of [21], Lemma 2.1 in order to estimate the norm of the flux on a strip.

I. Lirkov and S. Margenov (Eds.): LSSC 2017, LNCS 10665, pp. 95–102, 2018.
https://doi.org/10.1007/978-3-319-73441-5_9

Since in our case, this estimate reduces to one-dimensional considerations, a proof involving differential calculus of scalar functions can be given. In the following section the problem and its formulation as a first-order system is presented and the second section is concerned with the introduction of the finite-element spaces. The least squares formulation is introduced in Sect. 4 and the main ingredients of [4] for the proof of optimal order of convergence are recalled, before the alternative one-dimensional proof of the strip estimate is given.

2 The Two-Phase Incompressible Flow Model

Similarly to [18,19] the domain $\Omega \subset \mathbb{R}^2$, is assumed to be completely covered by Ω_1 and Ω_2, i.e. $\Omega = \Omega_1 \cup \Omega_2$. Further, assume that $\Omega_1 \subset \Omega_2$ (i.e. $\Omega = \Omega_2$), let Γ denote the boundary of Ω_1, $\mathbf{n}$ be the unit normal on Γ that is pointing from Ω_1 to Ω_2 and κ denote the curvature of Γ. The case $\Omega_1 \not\subset \Omega_2$ usually involves incompatibilities between boundary and interface condition. Using the inherent error estimator of the least-squares method for adaptiv refinement for this case is subject of future work. Assume that Γ is a Lipschitz continuous and piecewise C^{k+2} curve. Since no phase transition takes place and the two phases are viscous, the two-phase model reduces to governing equations in each phase and coupling conditions at the interface. For the equations in each phase, the flows are assumed to be incompressible and the velocity $\mathbf{u}$ is continuous over the whole domain Ω, i.e. in the context of variational formulations $\mathbf{u}$ is sought in $H^1(\Omega)$. In contrast to the velocity, the pressure is discontinuous over the interface. Therefore $p_{\Omega_i} \in H^2(\Omega_i)$ denotes the pressure in each phase Ω_i. However, in order to simplify the notation the index i is skipped whenever the restriction on each phase is not needed. Moreover, the flows in each phase are assumed to be Newtonian, such that for the stress tensor, $\boldsymbol{\sigma}_{\Omega_i}$, there holds $\boldsymbol{\sigma}_{\Omega_i} = -p_{\Omega_i}\mathbf{I} + \mu_i \mathbf{D}(\mathbf{u})$ in each phase Ω_i, with a constant dynamic viscosity $\mu_i > 0$ and the deformation tensor $\mathbf{D}(\mathbf{u}) = \nabla \mathbf{u} + (\nabla \mathbf{u})^\top$. Due to the fact that on both sides of Γ there are different molecules with different attractive forces, a surface tension force acts at the interface, and this leads to the coupling condition

$$(\boldsymbol{\sigma}_{\Omega_2} - \boldsymbol{\sigma}_{\Omega_1}) \cdot \mathbf{n} = -\tau\kappa\mathbf{n} \text{ on } \Gamma, \tag{1}$$

where τ denotes the constant surface tension coefficient. For the further analysis in this work, τ is assumed to be one to simplify the exposition and the following simplified stationary Stokes two-phase model where the density ρ_i in each phase is assumed to be constant, is considered.

$$\left.\begin{aligned} \operatorname{div} \boldsymbol{\sigma}_{\Omega_i} &= \mathbf{0} \\ \operatorname{dev} \boldsymbol{\sigma} - \mu_i \mathbf{D}(\mathbf{u}) &= \mathbf{0} \end{aligned}\right\} \quad \text{in } \Omega,\ i = 1,2,$$
$$\begin{aligned} (\boldsymbol{\sigma}_{\Omega_2} - \boldsymbol{\sigma}_{\Omega_1}) &= -\kappa\mathbf{n} \qquad \text{on } \Gamma \\ (\operatorname{tr} \boldsymbol{\sigma}, 1)_\Omega &= 0 \end{aligned} \tag{2}$$

As the results for isoparametric approximation combined with Dirichlet boundary are well known (see e.g. [14]), the boundary of Ω is assumed to be polygonal and Dirichlet boundary conditions are set on $\partial\Omega$. The Least-Squares functional associated with the problem (2) is

$$\mathcal{F}(\boldsymbol{\sigma},\mathbf{u}) = \sum_{i=1}^{2} \|\frac{1}{\sqrt{\mu_i}}\text{dev }\boldsymbol{\sigma} - \sqrt{\mu_i}\,\mathbf{D}(\mathbf{u})\|^2_{0,\Omega_i} + \|\text{div }\boldsymbol{\sigma}_{\Omega_i}\|^2_{0,\Omega_i} \tag{3}$$

for $\mathbf{u} \in \mathcal{W} = \left(H^1_0(\Omega)\right)^2$ and

$$\begin{aligned}\boldsymbol{\sigma} \in \boldsymbol{\Sigma} = \{\boldsymbol{\sigma} = (\boldsymbol{\sigma}_{\Omega_1}, \boldsymbol{\sigma}_{\Omega_2}) \ :\boldsymbol{\sigma}_{\Omega_i} &\in (H(\text{div}, \Omega_i))^2,\ i = 1,2, \\ [\boldsymbol{\sigma}\cdot\mathbf{n}]_\Gamma &:= (\boldsymbol{\sigma}_{\Omega_2} - \boldsymbol{\sigma}_{\Omega_1})\cdot\mathbf{n} = -\kappa\mathbf{n} \text{ on } \Gamma \\ &\text{and } (\text{tr }\boldsymbol{\sigma}, 1)_{0,\Omega} = 0\}.\end{aligned} \tag{4}$$

3 Finite-Element Spaces

The purpose of this section is to recall the notation used for the Finite-Element Spaces. Consider the interior domain Ω_1 and construct $\hat{\Omega}_1$ and a regular triangulation $\hat{\mathcal{T}}_{h,1}$ as in [7]. Note that therefore, Γ is linearly interpolated by $\hat{\Gamma} = \partial\hat{\Omega}_1$. Define $\hat{\Omega}_2 = \Omega\backslash\hat{\Omega}_1$ and construct a triangulation $\hat{\mathcal{T}}_{h,2}$. Moreover, in the higher-order case, construct as in [5] the approximated domain $\Omega_{h,1} = F_h(\hat{\Omega}_1)$ with a piecewise polynomial mapping F_h such that the distance between $\Gamma_h = \partial\Omega_{h,1}$ and Γ is proportional to h^{k+2}, and define $\Omega_{h,2} = \Omega\backslash\Omega_{h,1}$. Note that then, it also holds $\Omega_{h,2} = F_h(\hat{\Omega}_2)$ and

$$\|F_{h,i}\|_{W^s_\infty(\hat{T}_j)} \lesssim h^s\ , \quad \|F^{-1}_{h,i}\|_{W^s_\infty(\hat{T}_j)} \lesssim h^{-s} \tag{5}$$

for a positive integer s with $s \leq k+2$.

The triangulation $\mathcal{T}_{h,i}$ with curved triangles is defined from $\hat{\mathcal{T}}_{h,i}$ by replacing each straight boundary segment by the approximated one. Note that then Ω is completely covered by both of the triangulations $\mathcal{T}_h = \mathcal{T}_{h,1} \cup \mathcal{T}_{h,2}$ and $\hat{\mathcal{T}}_h = \hat{\mathcal{T}}_{h,1} \cup \hat{\mathcal{T}}_{h,2}$. Then, in both domains Ω_i, the injective mapping $\hat{\boldsymbol{\Phi}}_{h,i} : \hat{\Omega}_i \to \Omega_i$, $\boldsymbol{\Phi}_{h,i} = \hat{\boldsymbol{\Phi}}_{h,i} \circ F_h^{-1} : \Omega_{h,i} \to \Omega_i$, $\hat{\boldsymbol{\Psi}}_{h,i} = \hat{\boldsymbol{\Phi}}^{-1}_{h,i}$ and $\boldsymbol{\Psi}_{h,i} = \boldsymbol{\Phi}^{-1}_{h,i}$ can be computed. As the velocity is continuous over the whole domain, standard (isoparametric) conforming elements are used for its approximation $\mathbf{u}_h$:

$$\mathbf{u}_h \in \mathbf{W}^{k+1}_{h,0} = \{\mathbf{w}_h = (w_{h,1}, w_{h,2}) \ :\ w_{h,i} \in Q^{k+1}_h(\Omega_i) \text{ and } w_{h,2} = 0 \text{ on } \partial\Omega_2\}.$$

For the approximation of the stress tensor, parametric Raviart-Thomas elements are used as each row $[\boldsymbol{\sigma}_h]_j$ of the stress tensor belongs to $H_{\text{div}}(\Omega_1) \times H_{\text{div}}(\Omega_2)$:

$$\underline{\underline{RT}}_k(\Omega) = \{(([\boldsymbol{\sigma}_{h,\Omega_1}]_1, [\boldsymbol{\sigma}_{h,\Omega_1}]_2), ([\boldsymbol{\sigma}_{h,\Omega_2}]_1, [\boldsymbol{\sigma}_{h,\Omega_2}]_2)) \in (RT_k(\Omega_1))^2 \times (RT_k(\Omega_2))^2\},$$

with the parametric Raviart-Thomas space

$$RT_k(\Omega_i) = \{\mathbf{v}_h : \Omega \to \mathbb{R}^2 : \mathbf{v}_h = \left(\frac{1}{\det J_{F_{h,i}}} J_{F_{h,i}}\hat{\mathbf{v}}_h\right) \circ F^{-1}_{h,i} \text{ with } \hat{\mathbf{v}}_h \in \hat{RT}_k(\Omega_i)\}. \tag{6}$$

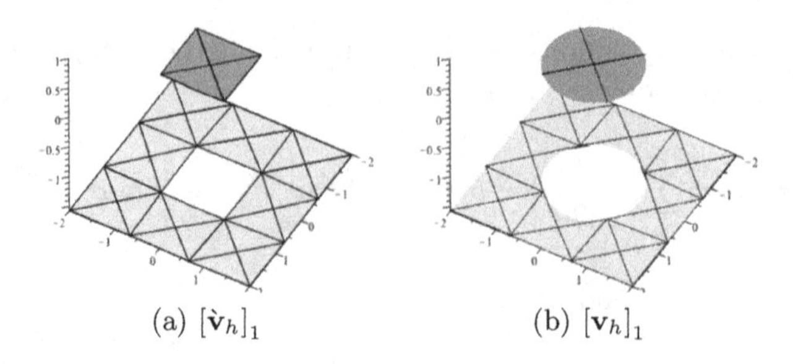

(a) $[\grave{\mathbf{v}}_h]_1$ (b) $[\mathbf{v}_h]_1$

Fig. 1. Location of the jump on examplary triangulation

Note that in this definition, the polynomial definition of the finite-element on $\Omega_{h,i}$ is extended in order to define those on Ω_i, similarly to [7]. Hence, $\mathbf{v}_h$ jumps over Γ as illustrated in Fig. 1. For better distinction, let $\grave{\mathbf{v}}_h$ denote the standard parametric Raviart-Thomas function that jumps on Γ_h. Further, for $[\boldsymbol{\sigma}_{h,\Omega_i}]_j \in RT_k(\Omega_i)$, $[\hat{\boldsymbol{\sigma}}_{h,\Omega_i}]_j \in \hat{RT}_k(\Omega_i)$ denotes the standard Raviart-Thomas function such that

$$[\boldsymbol{\sigma}_{h,\Omega_i}]_j = \left(\frac{1}{\det J_{F_{h,i}}} J_{F_{h,i}}[\hat{\boldsymbol{\sigma}}_{h,\Omega_i}]_j\right) \circ F_{h,i}^{-1} \tag{7}$$

Moreover, define

$$\hat{\boldsymbol{\sigma}}_h = (\hat{\boldsymbol{\sigma}}_{h,\Omega_1}, \hat{\boldsymbol{\sigma}}_{h,\Omega_2}) = (([\hat{\boldsymbol{\sigma}}_{h,\Omega_1}]_1, [\hat{\boldsymbol{\sigma}}_{h,\Omega_1}]_2), [\hat{\boldsymbol{\sigma}}_{h,\Omega_2}]_1, [\hat{\boldsymbol{\sigma}}_{h,\Omega_2}]_2)) . \tag{8}$$

In order to add the interface condition into the space $\underline{\underline{RT}}_k(\Omega)$, $\mathbf{g} = -\kappa\mathbf{n}$ has to be approximated by $\mathbf{g}_h$. Since the Piola transform preserves the normal direction, for $\boldsymbol{\sigma}_h \in \underline{\underline{RT}}_k(\Omega)$ it holds

$$\grave{\boldsymbol{\sigma}}_{h,\Omega_1} \cdot \mathbf{n}|_{\Gamma_h} = \left.\frac{\hat{\boldsymbol{\sigma}}_{h,\Omega_1} \cdot \mathbf{n}}{|J_{F_{h,1}}^{-\top}\mathbf{n}|\ (\det J_{F_{h,1}})}\right|_{\Gamma_h} . \tag{9}$$

This motivates the definition $\mathbf{g}_h = \mathcal{Q}_h(\omega_h(\mathbf{g} \circ \hat{\boldsymbol{\Phi}}_{h,1}))$ with $\omega_h = |J_{\boldsymbol{\Psi}_{h,1}}^{\top}\mathbf{n}|\,(\det J_{F_{h,1}})$ on Γ_h, where $\mathcal{Q}_h$ is the orthogonal projection in $L^2(F_{h,1}^{-1}(\Gamma))$ onto the piecewise polynomials of degree k. Now, define

$$\boldsymbol{\Sigma}_h^k = \{\boldsymbol{\sigma}_h \in \underline{\underline{RT}}_k(\Omega)\ :\ [\grave{\boldsymbol{\sigma}}_h \cdot \mathbf{n}]_{\Gamma_h} = \mathbf{g}_h \text{ and } (\operatorname{tr} \boldsymbol{\sigma}_h, 1)_\Omega = 0\} \tag{10}$$

The minimizing problem corresponding to (2) is now given by

$$\begin{aligned} &\text{Find } (\boldsymbol{\sigma}_h, \mathbf{u}_h) \in \boldsymbol{\Sigma}_h^k \times \mathbf{W}_{h,0}^{k+1} \text{ such that} \\ &\mathcal{F}(\boldsymbol{\sigma}_h, \mathbf{u}_h) \le \mathcal{F}(\boldsymbol{\tau}_h, \mathbf{w}_h) \quad \forall\ (\boldsymbol{\tau}_h, \mathbf{w}_h) \in \boldsymbol{\Sigma}_h^k \times \mathbf{W}_{h,0}^{k+1}. \end{aligned} \tag{11}$$

Recall that the following estimate for the jump of the normal flux on the interpolated boundary similar to the crucial estimate of [7] was derived in [4], involving the strip S_h which consists of all triangles in $\mathcal{T}_h$ whose intersection with Γ is not empty.

Theorem 1. *Let Ω be the two-phase domain having the properties described in Sect. 2. In particular, assume that the interface Γ is a piecewise C^{k+2} curve, $k \geq 0$. Then,*

$$|\langle [\boldsymbol{\sigma}_h \cdot \mathbf{n}]_\Gamma - \mathbf{g}, \mathbf{q}\rangle_{0,\Gamma}| \lesssim \left(h^{2k+2}\|\mathbf{g}\|_{k+1,\Gamma} + h^{2k+1}\|\boldsymbol{\sigma}_h\|^2_{0,S_h(\Gamma)}\right)^{\frac{1}{2}} \|\mathbf{q}\|_{0,\Gamma} \quad (12)$$

holds for all $\boldsymbol{\sigma}_h \in \boldsymbol{\Sigma}_h^k$ and $\mathbf{q} \in (H^{\frac{1}{2}}(\Gamma))^2$.

4 Least-Squares Functional and Ellipticity

Recall the Korn inequality (see e.g. [12])

$$\exists\, C_K > 0 \text{ with } \|\mathbf{D}(\mathbf{w})\|_{0,\Omega} + \|\mathbf{w}\|_{0,\Omega} \geq C_K \|\mathbf{w}\|_{1,\Omega} \qquad \forall \mathbf{w} \in (H^1(\Omega))^2, \quad (13)$$

and the following Lemma, (two-phase version of Lemma 3.1 in [1] or Proposition 9.1.1 in [14], see also [2,17]),

Lemma 1. *There exists a constant $C_D > 0$ such that*

$$\|\text{tr}\ \boldsymbol{\tau}\|_{0,\Omega} \leq C_D \left(\|[\boldsymbol{\tau} \cdot \mathbf{n}]\|_{-\frac{1}{2},\Gamma} + \|\text{dev}\ \boldsymbol{\tau}\|_{0,\Omega} + \sum_{i=1}^{2} \|\mathit{div}\ \boldsymbol{\tau}\|_{0,\Omega_i} \right). \quad (14)$$

holds for all $\boldsymbol{\tau} = (\boldsymbol{\tau}_1, \boldsymbol{\tau}_2) \in (H(div, \Omega_1))^d \times (H(div, \Omega_2))^d$ with $(\text{tr}\ \boldsymbol{\tau}, 1)_{0,\Omega} = 0$.

This leads to a lower bound for the Least-Squares functional:

Lemma 2. *Let $(\boldsymbol{\sigma}, \mathbf{u}) \in \boldsymbol{\Sigma} \times \mathcal{W}$ be the exact solution of (2). Then, it holds*

$$\mathcal{F}(\boldsymbol{\sigma}_h, \mathbf{u}_h) + \left(h^{2k+2}\|\mathbf{g}\|_{k+1,\Gamma} + h^{2k+1}\|\boldsymbol{\sigma}_h\|^2_{0,S_h(\Gamma)}\right)^{\frac{1}{2}} \|\mathbf{u} - \mathbf{u}_h\|_{0,\Gamma}$$
$$\gtrsim \|\text{dev}(\boldsymbol{\sigma} - \boldsymbol{\sigma}_h)\|^2_{0,\Omega} + \|(\mathbf{u} - \mathbf{u}_h)\|^2_{1,\Omega} + \sum_{i=1}^{2} \|\mathit{div}(\boldsymbol{\sigma}_{\Omega_i} - \boldsymbol{\sigma}_{h,\Omega_i})\|^2_{0,\Omega_i}$$

for all $(\boldsymbol{\sigma}_h, \mathbf{u}_h) \in \boldsymbol{\Sigma}_h^k \times \mathbf{W}_{h,0}^{k+1}$.

The optimal order of convergence is retained using the approximated functional

$$\mathcal{F}_h(\boldsymbol{\sigma}, \mathbf{u}) = \sum_{i=1}^{2} \|\text{div}\ \boldsymbol{\sigma}_{\Omega_i}\|^2_{0,\Omega_{h,i}} + \|\frac{1}{\sqrt{\mu_i}}\text{dev}\ \boldsymbol{\sigma} - \sqrt{\mu_i}\mathbf{D}(\mathbf{u})\|^2_{0,\Omega_{h,i}}. \quad (15)$$

The proof in [4] use the following estimate (see [21], Lemma 2.1)

$$\|\boldsymbol{\sigma}\|_{0,S_h(\Gamma)\cap\Omega_i} \leq h^{\frac{1}{2}} \|\boldsymbol{\sigma}\|_{1,\Omega_i} \quad (16)$$

and can be replaced by a one-dimensional consideration, i.e.:

Lemma 3. *Let g be continuously differentialble on $[0,1]$. Then, it holds*

$$\int_0^h g^2(x) \leq h \int_0^1 (g'(t))^2 + g^2(t)\ dt \tag{17}$$

Proof. *Let ξ be such that $g(\xi) = \int_0^1 g(t)$. Then, it holds*

$$\begin{aligned} g(x)^2 = g(\xi)^2 + \int_\xi^x 2g'(t)g(t)dt &\leq \int_\xi^x (g'(t))^2 + g^2(t)dt + \int_0^1 g(t)^2 dt \\ &\leq 2\int_0^1 (g'(t))^2 + g^2(t)dt. \end{aligned} \tag{18}$$

Since the right hand side does not depend on x, integration leads to the result.

Theorem 2. *Let $(\boldsymbol{\sigma}, \mathbf{u}) \in \boldsymbol{\Sigma}_h^k \times \mathcal{W}$ denote the exact solution of the system (2) and assume that it satisfies $\mathbf{u} \in (H^{k+2}(\Omega))^2$ and $div\ \boldsymbol{\sigma}_i \in (H^{k+1}(\Omega_i))^4$. Further, let $(\boldsymbol{\sigma}_h, \mathbf{u}_h) \in \boldsymbol{\Sigma}_h^k \times \mathbf{W}_{h,0}^{k+1}$ denote the (parametric) finite-element approximation minimizing $\mathcal{F}_h(\boldsymbol{\sigma}_h,\ \mathbf{u}_h)$ under all $(\boldsymbol{\tau}_h, \mathbf{w}_h) \in \boldsymbol{\Sigma}_h^k \times \mathbf{W}_{h,0}^{k+1}$. Then it holds*

$$\|\boldsymbol{\sigma} - \boldsymbol{\sigma}_h\|_{div,\Omega} + \|\mathbf{u} - \mathbf{u}_h\|_{1,\Omega} \lesssim h^{k+1} \left(\|\mathbf{u}\|_{k+2,\Omega} + \sum_{i=1}^{2} \|\boldsymbol{\sigma}_i\|_{k+1,\Omega_i} + |div\ \boldsymbol{\sigma}_i|_{k+1,\Omega_i} \right).$$

The proof of the above theorem in [4] is similar to the proof of Theorem 3 in [7]. In a first step, using the fact that $(\boldsymbol{\sigma}_h, \mathbf{u}_h) \in \boldsymbol{\Sigma}_h^k \times \mathbf{W}_{h,0}^{k+1}$ minimizes $\mathcal{F}_h(\boldsymbol{\tau}_h,\ \mathbf{w}_h)$ under all $(\boldsymbol{\tau}_h, \mathbf{w}_h) \in \boldsymbol{\Sigma}_h^k \times \mathbf{W}_{h,0}^{k+1}$, leads to

$$\mathcal{F}_h(\boldsymbol{\sigma}_h, \mathbf{u}_h) \lesssim h^{2k+2} \left(\|\mathbf{u}\|_{2,\Omega}^2 + \sum_{i=1}^{2} \|\boldsymbol{\sigma}_{\Omega_i}\|_{k+1,\Omega_i}^2 + |\text{div}\ \boldsymbol{\sigma}_{\Omega_i}|_{k+1,\Omega_i}^2 \right). \tag{19}$$

Further, combining this with Lemmas 1 and 2, this leads to

$$\|\boldsymbol{\sigma} - \boldsymbol{\sigma}_h\|_{0,\Omega}^2 \lesssim h \left(\|\mathbf{g}\|_{k+1,\Gamma} + \|\boldsymbol{\sigma}_h\|_{0,S_h(\Gamma)}^2 \right). \tag{20}$$

Here the strip estimate (16) is applied in [4] to retain the convergence order:

$$\frac{1}{h}\|\boldsymbol{\sigma}_h\|_{0,S_h(\Gamma)\cap\Omega_i}^2 \leq \frac{1}{h}\|\boldsymbol{\sigma}\|_{0,S_h(\Gamma)\cap\Omega_i}^2 + \frac{1}{h}\|\boldsymbol{\sigma}_h - \boldsymbol{\sigma}\|_{0,S_h(\Gamma)\cap\Omega_i}^2 \tag{21}$$

$$\lesssim \|\boldsymbol{\sigma}\|_{1,\Omega_i}^2 + \frac{1}{h}\|\boldsymbol{\sigma}_h - \boldsymbol{\sigma}\|_{0,S_h(\Gamma)\cap\Omega_i}^2. \tag{22}$$

and the previous one-dimensional estimate can be used as an alternative proof. Combine this with (20) leads to

$$\frac{1}{h}\|\boldsymbol{\sigma}_h\|_{0,S_h(\Gamma)\cap\Omega_i}^2 \lesssim \|\mathbf{g}\|_{k+1,\Gamma} + \|\boldsymbol{\sigma}_h\|_{0,S_h(\Gamma)}^2. \tag{23}$$

Combine this again with Lemma 2 and with (19) in the same way as in [7] leads to the result.

References

1. Arnold, D.N., Douglas, J., Gupta, C.P.: A family of higher order mixed finite element methods for plane elasticity. Numerische Mathematik **45**(1), 1–22 (1984)
2. Bauer, S., Neff, P., Pauly, D., Starke, G.: Dev-Div-and DevSym-DevCurl-inequalities for incompatible square tensor fields with mixed boundary conditions. ESAIM: Control Optimisation Calc. Var. **22**(1), 112–133 (2016)
3. Cai, Z., Manteuffel, T., McCormick, S.: First-order system least squares for second-order partial differential equations: part II. SIAM J. Num. Anal. **34**, 425–454 (1997)
4. Bertrand, F.: Approximated Flux Boundary Conditions for Raviart-Thomas Finite Elements on Domains with Curved Boundaries and Applications to First-Order System Least Squares, Thesis (2014)
5. Bertrand, F.: Considerations for the finite element approximation of three-dimensional domains. In: Équations aux dérivées partielles et leurs applications Actes du colloque Edp-Normandie, Le Havre, pp. 185–195 (2015)
6. Bertrand, F., Münzenmaier, S., Starke, G.: First-order system least squares on curved boundaries: Lowest-order Raviart-Thomas elements. SIAM J. Num. Anal. **52**(2), 880–894 (2014)
7. Bertrand, F., Münzenmaier, S., Starke, G.: First-order system least squares on curved boundaries: higher-order Raviart-Thomas elements. SIAM J. Num. Anal. **52**(6), 3165–3180
8. Bertrand, F., Starke, G.: Parametric Raviart-Thomas elements for mixed methods on domains with curved surfaces. SIAM J. Num. Anal. **54**(6), 3648–3667
9. Bochev, P., Gunzburger, M.: Analysis of least squares finite element methods for the Stokes equations. Math. Comput. **63**(208), 479–506 (1994)
10. Bochev, P., Gunzburger, M.: Least-Squares Finite Element Methods. Springer, New York (2009). https://doi.org/10.1007/b13382
11. Boffi, D., Brezzi, F., Fortin, M.: Mixed Finite Element Methods and Applications. Springer, Heidelberg (2013). https://doi.org/10.1007/978-3-642-36519-5
12. Braess, D.: Finite Elements: Theory, Fast Solvers, and Applications in Solid Mechanics, 2nd edn. Cambridge University Press, Cambridge (2001)
13. Bramble, J., Lazarov, R., Pasciak, J.: A least-squares approach based on a discrete minus one inner product for first order systems. Math. Comput. Am. Math. Soc. **66**(219), 935–955 (1997)
14. Brenner, S.C., Scott, L.R.: The Mathematical Theory of Finite Element Methods, 3rd edn. Springer, New York (2008). https://doi.org/10.1007/978-0-387-75934-0
15. Cai, Z., Barry, L., Ping, W.: Least-squares methods for incompressible Newtonian fluid flow: linear stationary problems. SIAM J. Num. Anal. **42**(2), 843–859 (2004)
16. Cai, Z., Manteuffel, T.A., McCormick, S.F.: First-order system least squares for the Stokes equations, with application to linear elasticity. SIAM J. Num. Anal. **34**(5), 1727–1741 (1997)
17. Carstensen, C., Dolzmann, D.: A posteriori error estimates for mixed FEM in elasticity. Numerische Mathematik **81**(2), 187–209 (1998)
18. Gross, S., Reichelt, V., Reusken, A.: A finite element based level set method for two-phase incompressible flows. Comput. Vis. Sci. **9**(4), 239–257 (2006)
19. Gross, S., Reusken, A.: Numerical Methods for Two-Phase Incompressible Flows. Springer, Heidelberg (2011). https://doi.org/10.1007/978-3-642-19686-7

20. Lenoir, M.: Optimal isoparametric finite elements and error estimates for domains involving curved boundaries. SIAM J. Num. Anal. **23**, 562–580 (1986)
21. Li, J., Melenk, J.M., Wohlmuth, B., Zou, J.: Optimal a priori estimates for higher order finite elements for elliptic interface problems. Appl. Num. Math. **60**, 19–37 (2010)
22. Münzenmaier, S., Starke, G.: First-order system least squares for coupled Stokes-Darcy flow. SIAM J. Num. Anal. **49**, 387–404 (2011)

Spectral Mimetic Least-Squares Method for Div-curl Systems

Marc Gerritsma[1(✉)] and Artur Palha[1,2]

[1] TU Delft, Kluyverweg 1, 2629 HS Delft, The Netherlands
M.I.Gerritsma@TUDelft.nl
[2] Department of Mechanical Engineering, Eindhoven University of Technology, P.O. Box 513, 5600 MB Eindhoven, The Netherlands
a.palha@tue.nl

Abstract. In this paper the spectral mimetic least-squares method is applied to a two-dimensional div-curl system. A test problem is solved on orthogonal and curvilinear meshes and both h- and p-convergence results are presented. The resulting solutions will be pointwise divergence-free for these test problems. For $N > 1$ optimal convergence rates on an orthogonal and a curvilinear mesh are observed. For $N = 1$ the method does not converge.

Keywords: Div-curl system · Spectral element method
Mimetic methods

1 Introduction

Div-curl systems play an important role in static electromagnetic fields, [4,8] and incompressible viscous flows, [8, Chap. 5]. One of the first papers where mimetic discretization for div-curl problems is described, is by Nicolaides, [9]. Nicolaides introduces geometric degrees of freedom and incidence matrices for metric-free derivatives on dual grids. When homogeneous tangential boundary conditions, $\boldsymbol{n} \times \boldsymbol{u} = 0$ (or homogeneous normal boundary conditions, $\boldsymbol{u} \cdot \boldsymbol{n} = 0$), are prescribed we have that $\mathcal{N}_0(\nabla\times) \perp \mathcal{N}(\nabla\cdot)$, where $\mathcal{N}(\mathsf{A})$ denotes the null space of the operator A. This orthogonality property is important for well-posedness of div-curl systems. Mimetic discretizations preserve this property at the finite dimensional level. In this paper mimetic spectral element methods are used in a *conforming least-squares formulation* as described in [2, Chap. 6]. Application of the non-conforming approach described in [2, Chap. 6] can be found in [3].

2 Div-curl System

Let Ω be a contractible domain $\mathbb{R}^d$, $d = 2, 3$ with Lipschitz continuous boundary $\partial\Omega$. The div-curl problem consists of finding $\boldsymbol{u} \in H_0(\nabla\times, \boldsymbol{\Theta}_1, \Omega) \cap H(\nabla\cdot, \boldsymbol{\Theta}_1^{-1}, \Omega)$ which satisfies

$$\begin{cases} \nabla \times \boldsymbol{u} = \boldsymbol{g} & \text{in } \Omega \\ \Theta_0^{-1} \nabla \cdot \boldsymbol{\Theta}_1 \boldsymbol{u} = 0 & \text{in } \Omega \end{cases} \quad \text{and} \quad \boldsymbol{n} \times \boldsymbol{u} = \boldsymbol{0} \text{ along } \partial\Omega. \tag{1}$$

I. Lirkov and S. Margenov (Eds.): LSSC 2017, LNCS 10665, pp. 103–110, 2018.
https://doi.org/10.1007/978-3-319-73441-5_10

A very brief explanation of the function spaces is given in Appendix A of this paper. See [2, Appendix A] for a full discussion.

The construction of conforming finite dimensional subspaces for $H_0(\nabla\times, \boldsymbol{\Theta}_1, \Omega) \cap H(\nabla\cdot, \boldsymbol{\Theta}_1^{-1}, \Omega)$ is non-trivial on arbitrary domains, therefore a formulation in terms of $H_0(\nabla\times, \Theta_1, \Omega) \times H(\nabla\cdot, \Theta_1^{-1}, \Omega)$ is preferred. See [5] for weak formulations based on (1).

Following the derivation in [2, Chap. 6] the first order div-curl system is given by

$$\begin{cases} \nabla \times \boldsymbol{u} = \boldsymbol{g} & \text{in } \Omega \\ \boldsymbol{v} - \Theta_1 \boldsymbol{u} = \boldsymbol{0} & \text{in } \Omega \\ \nabla \cdot \boldsymbol{v} = 0 & \text{in } \Omega \end{cases} \quad \text{and} \quad \boldsymbol{n} \times \boldsymbol{u} = \boldsymbol{0} \text{ along } \partial\Omega. \tag{2}$$

There exists a solution if $\boldsymbol{g} \in \mathcal{R}(\nabla\times)$, which, due to Poincaré's Lemma, is equal to $\nabla \cdot \boldsymbol{g} = 0$. Uniqueness follows from: Let $(\boldsymbol{u}_1, \boldsymbol{v}_1)$ and $(\boldsymbol{u}_2, \boldsymbol{v}_2)$ be two solutions of (2), then $(\boldsymbol{u}_2 - \boldsymbol{u}_1, \boldsymbol{v}_2 - \boldsymbol{v}_1)$ satisfies (2) with $\boldsymbol{g} = \boldsymbol{0}$, therefore $\boldsymbol{u}_2 - \boldsymbol{u}_1 \in \mathcal{N}_0(\nabla\times)$ and $\boldsymbol{v}_2 - \boldsymbol{v}_1 \in \mathcal{N}(\nabla\cdot)$. But since $\mathcal{N}_0(\nabla\times) \perp \mathcal{N}(\nabla\cdot)$, the second equation in (2) implies that $\boldsymbol{u}_1 = \boldsymbol{u}_2$ and $\boldsymbol{v}_1 = \boldsymbol{v}_2$, which proves uniqueness.

Consider the least-squares functional

$$\begin{cases} \mathcal{J}(\boldsymbol{u}, \boldsymbol{v}; \boldsymbol{g}) = \|\nabla \times \boldsymbol{u} - \boldsymbol{g}\|^2_{0,\Theta_2} + \|\nabla \cdot \boldsymbol{v}\|^2_{0,\Theta_0^{-1}} + \|\boldsymbol{v} - \Theta_1 \boldsymbol{u}\|^2_{0,\Theta_1^{-1}} \\ X = H_0(\nabla\times, \Theta_1, \Omega) \times H(\nabla\cdot, \Theta_1^{-1}, \Omega) \end{cases}, \tag{3}$$

The functional setting in terms of a two-dimensional double DeRham complex for the variables $(\boldsymbol{u}, \boldsymbol{v})$ and the data $\boldsymbol{g}$ is shown in (4), the three-dimensional double DeRham complex is given in Appendix B

$$\begin{array}{ccccc} H_0(\nabla\times, \Theta_2^{-1}, \Omega) & \xrightarrow{\nabla\times} & \boldsymbol{v} \in H_0(\nabla\cdot, \Theta_1^{-1}, \Omega) & \xrightarrow{\nabla\cdot} & L_0^2(\Theta_0^{-1}, \Omega) \\ \Theta_2^{-1} \Big\Uparrow \Big\Downarrow \Theta_2 & & \Theta_1^{-1} \Big\Uparrow \Big\Downarrow \Theta_1 & & \Theta_0^{-1} \Big\Uparrow \Big\Downarrow \Theta_0 \\ \boldsymbol{g} \in H(\nabla\cdot, \Theta_2, \Omega) & \xleftarrow[\nabla\times]{} & \boldsymbol{u} \in H(\nabla\times, \Theta_1, \Omega) & \xleftarrow[\nabla]{} & H^1(\Theta_0, \Omega) \end{array} \tag{4}$$

Theorem 6.5 in [2] asserts that the least-squares functional (3) is coercive with respect to the natural norm on X. This property is inherited on conforming subspaces of $H_0(\nabla\times, \Theta_1, \Omega) \times H(\nabla\cdot, \Theta_1^{-1}, \Omega)$.

3 Spectral Mimetic Basis Functions

On contractible domains, the horizontal operators in (4) form an exact sequence. The aim of mimetic spectral methods is to form a sequence of finite dimensional subspaces which also forms an exact sequence, see for instance [1,6,14]. Higher order methods for div-curl systems are also described in [10].

Let $L_N(\xi)$ the Legendre polynomial of degree N with derivative $L'_N(\xi)$. The $N+1$ roots, ξ_i, of $(1-\xi^2)L'_N(\xi)$ satisfy $-1 = \xi_0 < \xi_1 < \ldots < \xi_{N-1} < \xi_N = 1$

and are called the Gauss-Lobatto-Legendre (GLL) points. Next construct the Lagrange polynomials, $h_i(\xi)$ through the GLL points with

$$h_i(\xi_j) = \begin{cases} 1 & \text{if } i = j \\ 0 & \text{if } i \neq j \end{cases}, \quad i,j = 0,\ldots,N.$$

From the Lagrange polynomials, we can construct the so-called edge polynomials, [6], as

$$e_i(\xi) = -\sum_{k=0}^{i-1} \mathrm{d}h_k = -\sum_{k=0}^{i-1} \frac{dh_k}{d\xi}\mathrm{d}\xi\,, \quad i = 1,\ldots,N.$$

The edge polynomials have the property that

$$\int_{\xi_{j-1}}^{\xi_j} e_i(\xi) = \begin{cases} 1 & \text{if } i = j \\ 0 & \text{if } i \neq j \end{cases}, \quad i,j = 1,\ldots,N.$$

These polynomials were presented for the first time at 7th International Conference on Large-Scale Scientific Computations in Sozopol, 2009, [7,12]. If we expand a function in terms of Lagrange polynomials, then its derivative is naturally expanded in terms of edge polynomials

$$f(\xi) = \sum_{i=0}^{N} f_i h_i(\xi) \quad \Longrightarrow \quad f'(\xi) = \sum_{i=1}^{N} (f_i - f_{i-1})\, e_i(\xi). \tag{5}$$

In multiple dimensions we use tensor products of Lagrange and edge functions. For instance, on $I^2 = [-1,1]^2$ vector fields $\boldsymbol{v} \in H(\nabla\cdot, I^2)$ are expanded as

$$\boldsymbol{v} = (p,q) = \left(\sum_{i=0}^{N}\sum_{j=1}^{N} p_{i,j} h_i(\xi) e_j(\eta), \sum_{i=1}^{N}\sum_{j=0}^{N} q_{i,j} e_i(\xi) h_j(\eta) \right). \tag{6}$$

Then, using (5) we have

$$\nabla \cdot \boldsymbol{v} = \sum_{i=1}^{N}\sum_{j=1}^{N} \left[p_{i,j} - p_{i-1,j} + q_{i,j} - q_{i,j-1}\right] e_i(\xi) e_j(\eta).$$

Since the $e_i(\xi)e_j(\eta)$ form a basis for $\mathbb{P}^{N-1,N-1}$, we have that

$$\nabla \cdot \boldsymbol{v} = 0 \quad \Longleftrightarrow \quad p_{i,j} - p_{i-1,j} + q_{i,j} - q_{i,j-1} = 0. \tag{7}$$

Note that $\nabla \cdot \boldsymbol{v} = 0$ can be completely expressed in terms of the expansion coefficients $p_{i,j}$ and $q_{i,j}$ and the basis functions cancel from this equation. Secondly, the signs $(+1)$ and (-1) in the discrete divergence, (7), correspond to the incidence matrices used in [1,9,13].

For $\boldsymbol{u} \in H(\nabla\times; I^2)$ we will use the expansion

$$\boldsymbol{u} = (u,v) = \left(\sum_{i=1}^{M}\sum_{j=0}^{M} u_{i,j} e_i(\xi) h_j(\eta), \sum_{i=0}^{M}\sum_{j=1}^{M} v_{i,j} h_i(\xi) e_j(\eta) \right). \tag{8}$$

Using (5) again, we have

$$\nabla \times \boldsymbol{u} = \sum_{i=1}^{M} \sum_{j=1}^{M} \left[v_{i,j} - v_{i-1,j} - u_{i,j} + u_{i,j-1} \right] e_i(\xi) e_j(\eta).$$

If the right hand side function $\boldsymbol{g}$ in (2) is projected onto $e_i(\xi)e_j(\eta)$ as

$$\boldsymbol{g}^h = \sum_{i=1}^{M} \sum_{j=1}^{M} g_{i,j} e_i(\xi) e_j(\eta),$$

then $\nabla \times \boldsymbol{u} = \boldsymbol{g}$ can be represented on the grid by the difference equation

$$v_{i,j} - v_{i-1,j} - u_{i,j} + u_{i,j-1} = g_{i,j}. \tag{9}$$

Note, that although we use high order polynomial expansions, the discrete Eqs. (7) and (9) are very sparse. In fact, the sparsity of these two equations is independent of the polynomial degree.

It is in the equation $\boldsymbol{v} - \boldsymbol{\Theta}_1 \boldsymbol{u} = \boldsymbol{0}$ that the two different expansions are equated. Even when $\boldsymbol{\Theta}_1$ is the identity map, this will give a full matrix. The div and curl equations can be discretized independent of the particular choice of basis functions. The dependence on the basis functions only appears in the constitutive equation $\boldsymbol{v} - \boldsymbol{\Theta}_1 \boldsymbol{u} = \boldsymbol{0}$.

The variables $\boldsymbol{u}$ and $\boldsymbol{v}$ will be treated as contravariant vectors. If we transform the equation to curvilinear coordinates only the equation $\boldsymbol{v} - \boldsymbol{\Theta} \boldsymbol{u} = \boldsymbol{0}$ is affected, the div and curl equations remain unchanged. In Sect. 5 the performance of this discretization in curvilinear coordinates is demonstrated.

4 Mapping to Curvilinear Coordinates

In Sect. 3 the expansions are given on the square $(\xi, \eta) \in I^2$. Consider the map

$$x = x(\xi, \eta), \quad y = y(\xi, \eta),$$

then the components of $\boldsymbol{u}$ and $\boldsymbol{v}$ transform as

$$\boldsymbol{u}(x,y) = (p(x,y), q(x,y)), \qquad \begin{cases} p(x,y) = \frac{1}{detJ} \left[p(\xi,\eta) \frac{\partial x}{\partial \xi} + q(\xi,\eta) \frac{\partial x}{\partial \eta} \right] \\ q(x,y) = \frac{1}{detJ} \left[p(\xi,\eta) \frac{\partial y}{\partial \xi} + q(\xi,\eta) \frac{\partial y}{\partial \eta} \right] \end{cases},$$

and

$$\boldsymbol{v}(x,y) = (u(x,y), v(x,y)), \qquad \begin{cases} u(x,y) = \frac{1}{detJ} \left[u(\xi,\eta) \frac{\partial y}{\partial \eta} - v(\xi,\eta) \frac{\partial y}{\partial \xi} \right] \\ v(x,y) = \frac{1}{detJ} \left[-u(\xi,\eta) \frac{\partial x}{\partial \eta} + v(\xi,\eta) \frac{\partial x}{\partial \xi} \right] \end{cases},$$

where $detJ = \frac{\partial x}{\partial \xi}\frac{\partial y}{\partial \eta} - \frac{\partial x}{\partial \eta}\frac{\partial y}{\partial \xi}$. We use the expansions from Sect. 3 for $p(\xi,\eta)$, $q(\xi,\eta)$, $u(\xi,\eta)$ and $v(\xi,\eta)$.

5 Numerical Results

Consider problem (2) on $\Omega = I^2 = [-1,1]^2 \subset \mathbb{R}^2$ with right hand side function $\boldsymbol{g} = 2\pi^2 \cos(2\pi x)\cos(2\pi y)$. For $\boldsymbol{\Theta}_1 = \mathbb{I}$ the exact solution $\boldsymbol{u} = (u, v)$ for this test case is

$$\begin{cases} u = -\pi \cos(\pi x)\sin(\pi y) \\ v = \pi \sin(\pi x)\cos(\pi y) \end{cases},$$

which resembles the test case used in [10]. For the expansions of $\boldsymbol{u}$ and $\boldsymbol{v}$ we use $N = M$ in (8) and (6), respectively. Consider the map $\Phi : \Omega \to \Omega$ given by

$$\begin{cases} x = \xi + c\sin(\pi\xi)\sin(\pi\eta) \\ y = \eta + c\sin(\pi\xi)\sin(\pi\eta) \end{cases}.$$

For $c = 0.0$ this mapping maps the orthogonal coordinate system (ξ, η) in the orthogonal coordinate system (x, y), see the grid on the left in Fig. 1, while for $c = 0.2$ the orthogonal coordinates (ξ, η) are mapped on the curvilinear coordinates (x, y) on the grid grid in Fig. 1. Figure 2 displays h-convergence on

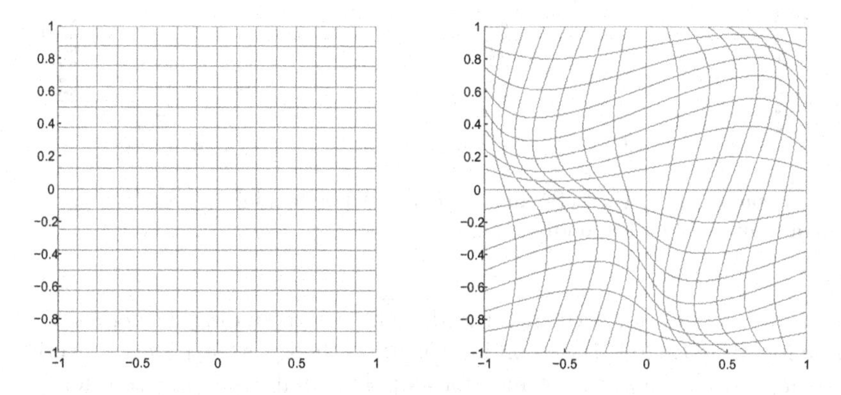

Fig. 1. A 16×16 grid for $c = 0.0$ (left) and $c = 0.2$ (right).

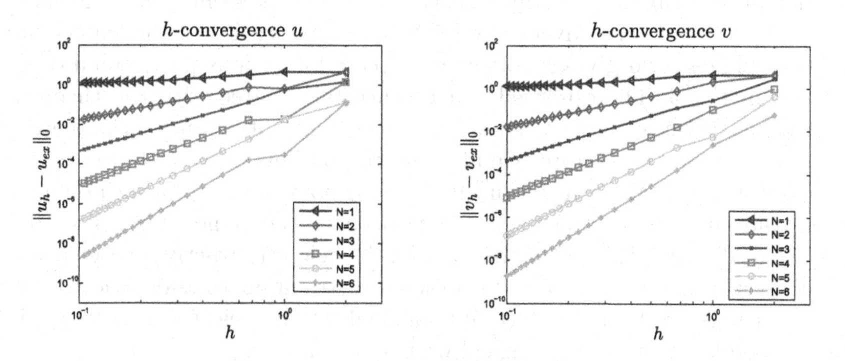

Fig. 2. h-convergence of $\boldsymbol{u}$ (left) and $\boldsymbol{v}$ (right) for polynomial degrees $N = 1, \ldots, 6$ on the orthogonal grid corresponding to $c = 0.0$.

Table 1. Convergence rates for the div-curl least-squares solution on orthogonal ($c = 0.0$) and curvilinear grids ($c = 0.2$).

	$c = 0.0$			$c = 0.2$		
N	$\boldsymbol{u}$	$\boldsymbol{v}$	$\Vert\nabla\cdot\boldsymbol{v}\Vert_\infty$	$\boldsymbol{u}$	$\boldsymbol{v}$	$\Vert\nabla\cdot\boldsymbol{v}\Vert_\infty$
1	0.2	0.2	0.0	0.1	0.1	0.0
2	2.0	2.0	0.0	2.0	2.0	0.0
3	3.0	3.0	0.0	3.0	3.0	0.0
4	4.0	4.0	0.0	4.0	4.0	0.0
5	5.0	5.0	0.0	5.0	5.0	0.0
6	6.0	6.0	0.0	6.0	6.0	0.0

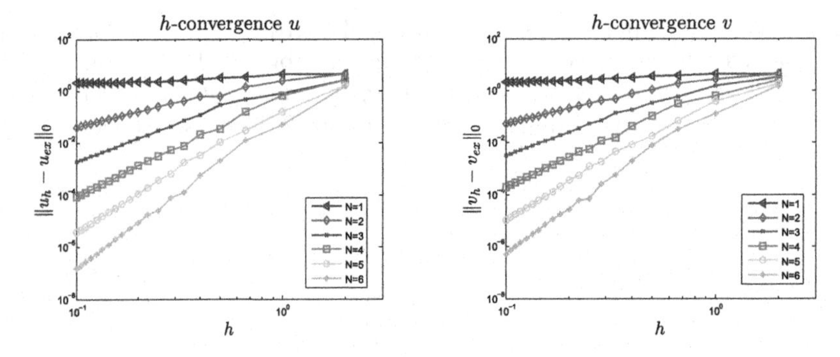

Fig. 3. h-convergence of $\boldsymbol{u}$ (left) and $\boldsymbol{v}$ (right) for polynomial degrees $N = 1, \ldots, 6$ on the orthogonal grid corresponding to $c = 0.2$.

a sequence of uniform, orthogonal grids. The corresponding convergence rates can be found in Table 1. Based on interpolation theory, we expect a convergence rate equal to N, which is confirmed for all polynomial degrees, except for $N = 1$ which does not seem to converge at all. Although the least-squares formulation does not necessarily require dual grids, see [9,10], it seems that the lack of proper duality hampers convergence for $N = 1$. Application of the least-squares to the curvilinear grid gives h-convergence plots for various polynomial degrees as shown in Fig. 3. The observed convergence rates agree with the theoretical expected convergence rates, as shown in Table 1, except again for the piecewise linear-piecewise constant approximation corresponding to $N = 1$.

Table 1 also contains the L^∞-norm of the divergence of $\boldsymbol{v}$ for all polynomial degrees and all number of element counts on both the orthogonal grid, $c = 0.0$ and the curvilinear grid, $c = 0.2$. In all cases the field $\boldsymbol{v}$ is exactly divergence-free. This conservation property (or involution constraint in time-dependent problems), which is essential for incompressible flows and electromagnetism, is a direct consequence of the topological property (7).

Acknowledgements. The authors want to thank the reviewers for the valuable feedback.

A Weighted Sobolev Spaces

Weighted Sobolev spaces are discussed in [2, Appendix A]. The space $H_0(\nabla\times, \boldsymbol{\Theta}_1, \Omega)$ is the Hilbert space of vector-valued functions

$$H_0(\nabla\times, \boldsymbol{\Theta}_1, \Omega) := \left\{ \boldsymbol{u} \in [L^2(\boldsymbol{\Theta}_1, \Omega)]^d \,\middle|\, \nabla \times \boldsymbol{u} \in [L^2(\boldsymbol{\Theta}_2, \Omega)]^d \text{ and } \boldsymbol{u} \times \boldsymbol{n} = 0 \text{ along } \partial\Omega \right\}, \tag{10}$$

where

$$\boldsymbol{u}, \boldsymbol{v} \in [L^2(\boldsymbol{\Theta}, \Omega)]^d \quad (\boldsymbol{u}, \boldsymbol{v})_{0,\Omega,\boldsymbol{\Theta}} = \int_\Omega \boldsymbol{u} \cdot \boldsymbol{\Theta} \cdot \boldsymbol{v} \, \mathrm{d}\Omega$$

and associated norm

$$\|\boldsymbol{u}\|^2_{0,\Omega,\boldsymbol{\Theta}} = (\boldsymbol{u}, \boldsymbol{u})_{0,\Omega,\boldsymbol{\Theta}}.$$

The space $H(\nabla\cdot, \Omega, \boldsymbol{\Theta}_1^{-1})$ is defined by

$$H_0(\nabla\cdot, \boldsymbol{\Theta}_1^{-1}, \Omega) := \left\{ \boldsymbol{u} \in [L^2(\boldsymbol{\Theta}_1^{-1}, \Omega)]^d \,\middle|\, \nabla \cdot \boldsymbol{u} \in L^2(\Theta_0^{-1}, \Omega) \right\}. \tag{11}$$

If $\boldsymbol{u} \in L^2(\boldsymbol{\Theta}, \Omega)]^d$, then $\boldsymbol{\Theta}\boldsymbol{u} \in L^2(\boldsymbol{\Theta}^{-1}, \Omega)]^d$, therefore the second equation in (2) therefore equates two functions in $L^2(\Omega, \boldsymbol{\Theta}_1^{-1})]^d$.

Weighted Sobolev spaces incorporate material parameters in the functional setting, thus allowing for inhomogeneous and anisotropic relations, see for example Remark A.4 in [2, p.542]. If a description in curvilinear coordinates is obtained from a mapping as described in Sect. 4 then the weight functions naturally arise as a consequence of the pullbacks of those maps.

B Three-Dimensional Double DeRham Complex

In (4) the two-dimensional DeRham complex is given. For $d = 3$ the double DeRham setting is given by

$$\begin{array}{ccccccc}
H_0(\nabla, \Theta_3^{-1}, \Omega) & \xrightarrow{\nabla} & H_0(\nabla\times, \Theta_2^{-1}, \Omega) & \xrightarrow{\nabla\times} & \boldsymbol{v} \in H_0(\nabla\cdot, \Theta_1^{-1}, \Omega) & \xrightarrow{\nabla\cdot} & L_0^2(\Theta_0^{-1}, \Omega) \\
\Theta_3^{-1} \uparrow \downarrow \Theta_3 & & \Theta_2^{-1} \uparrow \downarrow \Theta_2 & & \Theta_1^{-1} \uparrow \downarrow \Theta_1 & & \Theta_0^{-1} \uparrow \downarrow \Theta_0 \\
L^2(\Theta_3, \Omega) & \xleftarrow[\nabla\times]{} & g \in H(\nabla\cdot, \Theta_2, \Omega) & \xleftarrow[\nabla\times]{} & \boldsymbol{u} \in H(\nabla\times, \Theta_1, \Omega) & \xleftarrow[\nabla]{} & H^1(\Theta_0, \Omega)
\end{array}$$

Although the current paper focused on the two-dimensional div-curl system, the three dimensional analogue of (1) is much more challenging, because it constitutes a system of 4 partial differential equations for 3 unknown vector components of $\boldsymbol{u}$, [11].

References

1. Bochev, P.B., Gerritsma, M.I.: A spectral mimetic least-squares method. Comput. Math. Appl. **68**, 1480–1502 (2014)
2. Bochev, P.B., Gunzburger, M.D.: Least-Squares Finite Element Methods. Spinger, New York (2009). https://doi.org/10.1007/b13382
3. Bochev, P.B., Peterson, K., Siefert, C.: Analysis and computation of compatible least-squares methods for div-curl systems. SIAM J. Numer. Anal. **49**(1), 159–181 (2011)
4. Boulmezaoud, T.Z., Kaliche, K., Kerdid, N.: Explicit div-curl inequalities in bounded and unbounded domains of $\mathbb{R}^3$. Ann. Univ. Ferrara **63**(2), 249–276 (2017). https://doi.org/10.1007/s11565-016-0266-7
5. Bramble, J.H., Pasciak, J.E.: A new approximation technique for div-curl systems. Math. Comput. **73**(248), 1739–1762 (2004)
6. Gerritsma, M.: Edge functions for spectral element methods. In: Hesthaven, J., Rønquist, E. (eds.) Spectral and High Order Methods for Partial Differential Equations. LNCS, pp. 199–208. Springer, Heidelberg (2011). https://doi.org/10.1007/978-3-642-15337-2_17
7. Gerritsma, M., Bouman, M., Palha, A.: Least-squares spectral element method on a staggered grid. In: Lirkov, I., Margenov, S., Waśniewski, J. (eds.) LSSC 2009. LNCS, vol. 5910, pp. 653–661. Springer, Heidelberg (2010). https://doi.org/10.1007/978-3-642-12535-5_78
8. Jiang, B.-N.: The Least-Squares Finite Element Method. Springer, Heidelberg (1998). https://doi.org/10.1007/978-3-662-03740-9
9. Nicolaides, R.A.: Direct discretization of planar div-curl problems. SIAM J. Numer. Anal. **29**(1), 32–56 (1992)
10. Nicolaides, R.A., Wang, D.-Q.: A higher-order covolume method for planar div-curl problems. Int. J. Numer. Meth. Fluid **31**, 299–308 (1999)
11. Nicolaides, R.A., Wu, X.: Covolume solutions of three-dimensional div-curl equations. SIAM J. Numer. Aanl. **34**(6), 2195–2203 (1997)
12. Palha, A., Gerritsma, M.: Mimetic least-squares spectral/*hp* finite element method for the poisson equation. In: Lirkov, I., Margenov, S., Waśniewski, J. (eds.) LSSC 2009. LNCS, vol. 5910, pp. 662–670. Springer, Heidelberg (2010). https://doi.org/10.1007/978-3-642-12535-5_79
13. Palha, A., Rebelo, P.P., Hiemstra, R., Kreeft, J., Gerritsma, M.I.: Physics-compatible discretization techniques on single and dual grids, with application to the Poisson equation of volume forms. J. Comput. Phys. **257**, 1394–1422 (2014)
14. Palha, A., Gerritsma, M.I.: Spectral mimetic least-squares method for curl-curl problems. In: Lirkov, I., Margenov, S. (eds.) LSSC 2017. LNCS, vol. 10665, pp. 119–127. Springer, Cham (2017)

Spectral Mimetic Least-Squares Methods on Curvilinear Grids

R. O. Hjort and B. Gervang(✉)

Department of Engineering, Aarhus University,
Inge Lehmanns Gade 10, 8000 Aarhus C, Denmark
bge@ase.au.dk

Abstract. We present a spectral mimetic least-squares method on curvilinear grids, which conserves important invariants. The method is developed using differential forms where the topological part and the constitutive part have been separated. It is shown that the topological part is solved exactly, independent of the order of the spectral expansion. The method is applied to a model convection-diffusion problem, where we show that conservation of a potential is satisfied up to machine precision. The convective term is represented using the Lie derivative, by means of Cartans homotopy formula. The spectral mimetic least-squares method is compared to a standard spectral least-squares method. It is shown that both schemes lead to spectral convergence.

Keywords: Mimetic · Least-squares · Spectral · Convection-diffusion
Curvilinear grid

1 Introduction

We first consider general convection-diffusion of a scalar in 2D:

$$\nabla \cdot (\mathbf{u}\phi) + \nabla \cdot D\nabla\phi = f \qquad in\ \Omega, \tag{1}$$

where ϕ is the unknown potential, f the source term, D is the diffusion coefficient tensor of the system, and $\mathbf{u}$ a known divergence free vector field. Equation 1 is subjected to a homogeneous Dirichlet boundary condition:

$$\phi = 0 \qquad on\ \partial\Omega. \tag{2}$$

The method presented is based on a combination of mimetic methods, presented in [2,4,12] and least-squares spectral element methods, [1,14]. Recent work combining the two methods include [6]. The method is derived using basic components from differential geometry, which leads to conservation of invariants of the system. Using the least-squares principles lead to a symmetric positive definite matrix for the discretized problem.

I. Lirkov and S. Margenov (Eds.): LSSC 2017, LNCS 10665, pp. 111–118, 2018.
https://doi.org/10.1007/978-3-319-73441-5_11

2 Differential Geometry

In differential geometry the unknowns are presented by forms instead of vector and scalar fields, as in vector calculus. Variables associated with points, such as the temperature, are represented by a 0-form while variables associated with a volume are represented by 3-forms, e.g. the density. 1-forms and 2-forms can likewise represent variables associated with lines and surfaces. Furthermore forms have inner and outer orientation. Outer oriented 2-forms represent variables working through surfaces, such as a flux, while inner oriented 2-forms represent variables working on a surface e.g. describing vorticity in a plane.

Generalising the definitions of 0-forms, 1-forms, 2-forms and 3-forms, the general k-form is denoted $\omega^{(k)} \in \Lambda^k(\Omega_n)$ on the n-dimensional domain Ω_n, for $0 \leq k \leq n$. $\Lambda^k(\Omega_n)$ is the space of k-forms on Ω_n, i.e. the collection of all k-linear, antisymmetric mappings of vectors belonging to the n-dimensional tangent vector space V:

$$\omega^{(k)} : \underbrace{V \times ... \times V}_{k} \rightarrow \mathbb{R}. \tag{3}$$

Differential geometry also introduces the wedge product between k-forms and m-forms, which produces a $(k+m)$-form: $\wedge : \Lambda^k(\Omega_n) \times \Lambda^m(\Omega_n) \rightarrow \Lambda^{k+m}(\Omega_n)$. The wedge product, also called a skew-symmetric product, has the property: $\alpha^{(k)} \wedge \beta^{(m)} = (-1)^{km}\beta^{(m)} \wedge \alpha^{(k)}$.

Instead of using three different operators to represent curl, divergence and gradient, differential forms are equipped with an operator representing all three operators; the exterior derivative, d. The exterior derivative operates on k-forms and maps them into $(k+1)$-forms: $d : \Lambda^k(\Omega_n) \rightarrow \Lambda^{k+1}(\Omega_n)$. The exterior derivative can be defined by means of the Stokes theorem [3]:

$$\int_{\Omega_{k+1}} d\omega^{(k)} = \int_{\partial\Omega_{k+1}} \omega^{(k)}. \tag{4}$$

Since the exterior derivative is constructed using only the boundary of the domain of interest, the discrete version of the exterior derivative can be performed exactly.

The interior product is the inverse operation of the exterior derivative and is the mapping: $\iota_Y : \Lambda^k(\Omega_n) \rightarrow \Lambda^{k-1}(\Omega_n)$ for some vector field $Y \in \Omega_n$ and $1 \leq k \leq n$, defined as:

$$\iota_Y \alpha^{(k)}(X_2, \cdots, X_k) = \alpha^{(k)}(Y, X_2, \cdots, X_k) \quad \forall X_i, Y \in V \tag{5}$$

The Lie-derivative represents how forms change when they are altered by the flow of some vector field $Y \in \Omega_n$ and is the mapping: $\mathcal{L}_Y : \Lambda^k(\Omega_n) \rightarrow \Lambda^k(\Omega_n)$, see [13,15]. The Lie-derivative can be seen as the convection operator for differential geometry and is defined by applying Cartan's formula:

$$\mathcal{L}_Y \alpha^{(k)} = \iota_Y d\alpha^{(k)} + d\iota_Y \alpha^{(k)}. \tag{6}$$

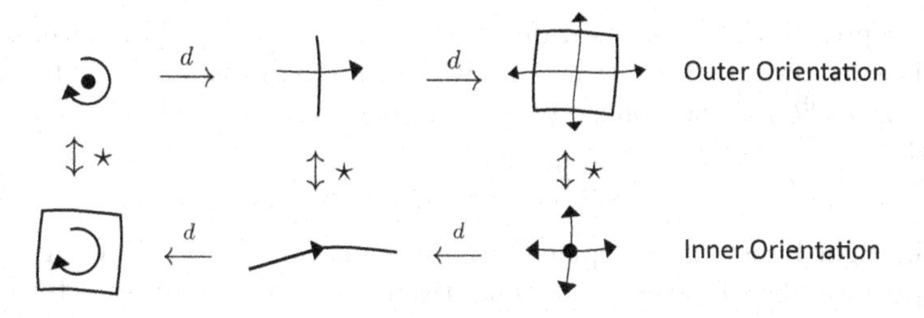

Fig. 1. The six different forms in the two dimensional space. The upper row is the inner oriented forms while the bottom shows the outer oriented forms.

The inner and outer oriented forms are connected using the Hodge-star operator denoted with a $\star$, see [3,10]. The Hodge-star operator, $\star$, is a map between k-forms and $(n-k)$-forms of opposite orientation in a n-dimensional domain: $\star : \Lambda^k(\Omega^n) \rightarrow \Lambda^{n-k}(\Omega^n)$ see for example Fig. 1. In this report the $\sim$ denotes inner oriented forms. The Hodge-star operator is defined using the following relation:

$$\alpha^{(k)} \wedge \star\beta^{(k)} = (\alpha^{(k)}, \beta^{(k)})\omega^{(n)}, \tag{7}$$

where $\omega^{(n)}$ is a unit n-form and the brackets $(\cdot, \cdot)$ denote an inner product, which computes a scalar field from the vector proxy of the forms. This inner product results in a 0-form defined such that on the n-dimensional Euclidean domain: $(\alpha^{(1)}, \beta^{(1)}) = \sum_{i=1}^{n}\sum_{j=1}^{n} A_i B_j g^{ij}$, where $(A_1, ..., A_n)$ and $(B_1, ..., B_n)$ define the vector proxies of $\alpha^{(1)}$ and $\beta^{(1)}$, respectively, and g^{ij} is the inverse of the metric tensor.

The spaces of forms are also equipped with an integral inner product or also referred to as an L^2 inner product defined as:

$$(\alpha^{(k)}, \beta^{(k)})_\Omega = \int_\Omega (\alpha^{(k)}, \beta^{(k)})\omega^{(n)}. \tag{8}$$

When working with multiple domains, differential forms are equipped with an operator, which transforms forms defined on the codomain to the domain; the pullback operator. Consider the mapping of coordinates $\Phi : \widehat{\Omega} \rightarrow \Omega$, then the pullback operator is the mapping: $\Phi^* : \Lambda(\Omega) \rightarrow \Lambda(\widehat{\Omega})$. For the k-form $a^{(k)}$ defined on Ω, the following relation can then be constructed:

$$\int_\Omega a^{(k)} = \int_{\Phi(\widehat{\Omega})} a^{(k)} = \int_{\widehat{\Omega}} \Phi^* a^{(k)}. \tag{9}$$

3 Mimetic Least-Squares Formulation

The variable ϕ in (1), is represented by the inner oriented 0-form $\tilde{\phi}^{(0)}$. The Laplace operator working on a 0-form is constructed using the exterior derivative and Hodge star operator $\Delta \rightarrow \star d \star d$, which results in a 0-form. The source function in (1) can be represented by an inner oriented 0-form, $\tilde{f}^{(0)}$. The term

$\nabla \cdot (\mathbf{u}\phi)$ represents the conservation of convective flux, which is naturally constructed using the Lie-derivative. One way of implementing this is to consider the 2-form $\star\tilde{\phi}^{(0)}$ for the convective term, such that the following equation is obtained:

$$\mathcal{L}_{\mathbf{u}} \star \tilde{\phi}^{(0)} + d \star d\tilde{\phi}^{(0)} = \star\tilde{f}^{(0)}, \tag{10}$$

where the right hand side is replaced by the 2-form $f^{(2)} = \star\tilde{f}^{(0)}$. Using Cartans homotopy formula, (6), the convective term reduces to only one term, since $d \circ d \equiv 0$. This leads to the following equation:

$$d\iota_{\mathbf{u}} \star \tilde{\phi}^{(0)} + d \star d\tilde{\phi}^{(0)} = f^{(2)}. \tag{11}$$

This allows defining the outer oriented 1-form $q^{(1)} = \iota_{\mathbf{u}} \star \tilde{\phi}^{(0)} + \star d\tilde{\phi}^{(0)}$, which can be interpreted as the total flux of the potential, i.e. the sum of convective and diffusive fluxes. A solution to the problem in (1), can then be obtained by solving a conservation equation and a constitutive relation:

$$\nabla \cdot (\mathbf{u}\phi) + \Delta\phi = f \Leftrightarrow \begin{cases} dq^{(1)} = f^{(2)} \\ q^{(1)} = \iota_{\mathbf{u}} \star \tilde{\phi}^{(0)} + \star d\tilde{\phi}^{(0)}. \end{cases} \tag{12}$$

The conservation equation can be solved exactly, while the approximation is introduced in the constitutive equation. A least-squares functional is established by integrating the squared residual over the domain,

$$\mathcal{J}(\tilde{\phi}^{(0)}, q^{(1)}; f^{(2)}) = \frac{1}{2}\left(\left|\left| dq^{(1)} - f^{(2)} \right|\right|_0^2 + \left|\left| q^{(1)} - \iota_{\mathbf{u}} \star \tilde{\phi}^{(0)} - \star d\tilde{\phi}^{(0)} \right|\right|_0^2 \right).$$

where $\left|\left|\alpha^{(k)}\right|\right|_0^2 = (\alpha^{(k)}, \alpha^{(k)})_\Omega$. The least-squares method is a minimisation problem where the functional $\mathcal{J}$ is minimised by setting the gradient of $\mathcal{J}$ to zero, [8]. If we define $\tilde{\Lambda}_0^0(\Omega)$ as the space of all inner oriented 0-forms, satisfying the boundary conditions in (2), and $\Lambda^1(\Omega)$ as the space of all outer oriented 1-forms, then the variational formulation is obtained as: Find $\tilde{\phi}^{(0)} \in \tilde{\Lambda}_0^0(\Omega)$ and $q^{(1)} \in \Lambda^1(\Omega)$ such that:

$$\begin{aligned} (dp^{(1)}, dq^{(1)} - f^{(2)}) &= 0 \quad \forall \tilde{\varsigma}^{(0)} \in \tilde{\Lambda}_0^0(\Omega) \\ (p^{(1)} - \iota_{\mathbf{u}} \star \tilde{\varsigma}^{(0)} - \star d\tilde{\varsigma}^{(0)}, q^{(1)} - \iota_{\mathbf{u}} \star \tilde{\phi}^{(0)} - \star d\tilde{\phi}^{(0)}) &= 0 \quad \forall p^{(1)} \in \Lambda^1(\Omega). \end{aligned} \tag{13}$$

4 Mimetic Spectral Discretization

The unknowns in the system are expanded using Lagrange polynomials [9] and edge polynomials [5]. Consider the one dimensional domain $\Omega_1 = [-1, 1]$ on which $N+1$ Gauss-Lobatto-Legendre (GLL) nodes are defined: $-1 = x_0 < \cdots < x_N = 1$. Using these nodes we define $N+1$ Lagrange polynomials $h_i(x)$, such that $h_i(x_j) = \delta_{ij}$. The expansion coefficients are then equal to the 0-form evaluated in the nodes: $a_i = a^{(0)}(x_i)$. The 0-form $\phi^{(0)} = \phi(x,y)$ is expanded using Lagrange polynomials in both coordinate directions:

$$\phi_h^{(0)} = \sum_{i=0}^{N}\sum_{j=0}^{N} \phi_{ij} h_i(x) h_j(y). \tag{14}$$

For 1-forms and 2-forms edge polynomials, presented in [5], are used to construct the approximation of the form. Consider the 1-form $q^{(1)} = q^x(x,y)dx + q^y(x,y)dy$. Then the approximated form is represented as:

$$q^{(1)} \approx q_h^{(1)} = \sum_{i=0}^{N}\sum_{j=1}^{N} q_{ij}^x h_i(x) e_j(y) dx + \sum_{i=1}^{N}\sum_{j=0}^{N} q_{ij}^y e_i(x) h_j(y) dy \tag{15}$$

where $e_i(x)$ and $e_j(y)$ are edge polynomials defined from derivative of the Lagrange polynomials. From the N+1 Lagrange polynomials it is possible to define N edge polynomials:

$$e_i(x) = -\sum_{j=0}^{i-1} \frac{dh_j(x)}{dx}, \qquad \text{for } i = 1 : N \tag{16}$$

which are connected to line segments between the nodes by the following relation: $\int_{x_{j-1}}^{x_j} e_i(x)dx = \delta_{ij}$ for $i,j = 1 : N$. For the 2-form $\rho^{(2)} = P(x,y)dx \wedge dy$ the approximate form is constructed as:

$$\rho^{(2)} \approx \rho_h^{(2)} = \sum_{i=1}^{N}\sum_{j=1}^{N} \rho_{ij} e_i(x) e_j(y). \tag{17}$$

5 Numerical Results

Choosing a solution for the convection-diffusion problem as:

$$\phi_{sol}(x,y) = (x^2-1)(y^2-1)\sin\left(\frac{1}{2}\pi x\right), \tag{18}$$

and a known divergence free velocity field as $\mathbf{u} = u^x \frac{\partial}{\partial x} + u^y \frac{\partial}{\partial y}$ where $u^x = \sin(\pi x)\cos(\pi y)$ and $u^y = -\sin(\pi y)\cos(\pi x)$ we can calculate the source term.

The grid is constructed using the mappings; $x(\xi,\eta) = \xi + c\sin(\pi\xi)\sin(\pi\eta)$ and $y(\xi,\eta) = \eta + c\sin(\pi\xi)\sin(\pi\eta)$, where c is a skewness parameter, see Fig. 2. The results are shown in Fig. 3. It is observed that we obtain exponential convergence for the unknown potential (Fig. 3b) as well as for the accuracy of the constitutive equation (Fig. 3c). Both convergence plots show lower accuracy for large skewness of the mesh (the c parameter). In Fig. 3d we plot the invariant, $q^{(1)}$, as function of the polynomial order and it is solved to machine accuracy.

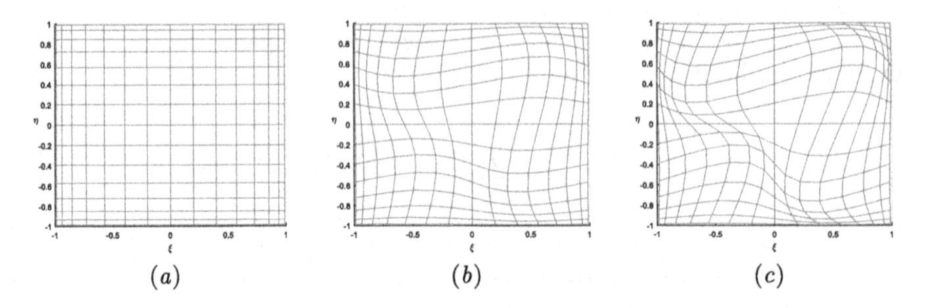

Fig. 2. Grid shown for $N = 15$ using mapping. (a) $c = 0$. (b) $c = 0.1$. (c) $c = 0.2$.

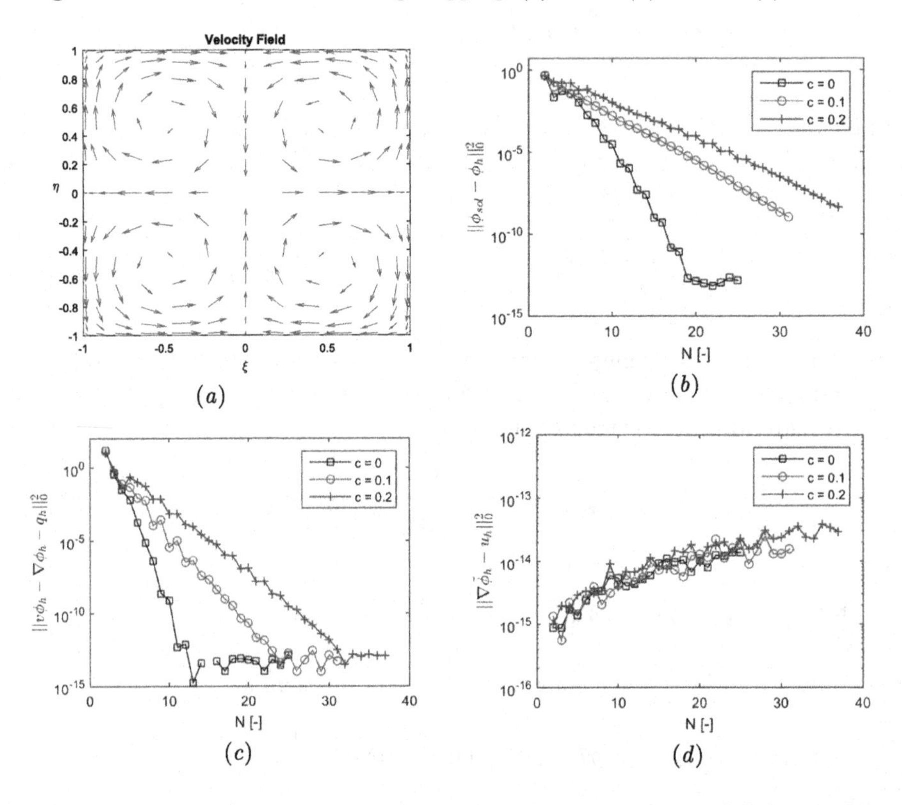

Fig. 3. (a) The divergence free vector field defined for the convection-diffusion problem in (18). (b) Convergence plot showing the error calculated using the analytical solution for the convection-diffusion problem. (c) Accuracy of the constitutive equation. (d) Conservation of $q^{(1)}$ shown for increasing polynomial order.

In order to study the method on convection-dominated problems we consider the slightly different problem:

$$\nabla\phi + \epsilon\nabla^2\phi = f, \tag{19}$$

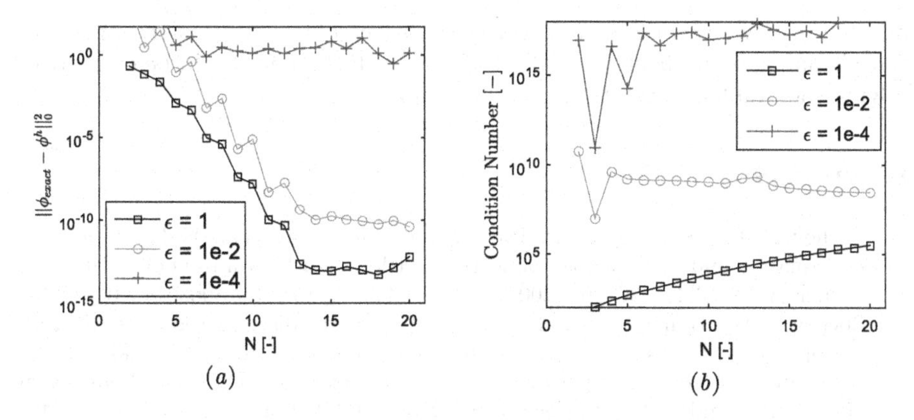

Fig. 4. (a) Convergence plot showing the error calculated for the convection-dominated problem in (19). (b) Condition number of the coefficient matrix.

with the solution:

$$\phi_{sol}(x, y) = (x^2 - 1)(y^2 - 1)\cos(x),$$

In Fig. 4 we present the results for the second problem. We observe that for small values of ϵ it is not possible to obtain an accurate solution for the problem. The condition number has increased to a critical level which causes problems. In [11] a stabilization term is introduced in the form as 'upwinding' flux for a Least-Squares finite element method. This introduces stabilization, however a slightly modified convection-diffusion equation system is solved. In [7] it is argued that all least-squares methods do not give reasonable results for convection-dominated problems possessing both interior and boundary layer structures in the solution. Solving convection-dominated problems with the least-squares method require further investigation. This introduces stabilization, however a slightly modified convection-diffusion equation system is solved for.

6 Conclusion

In this paper we present a spectral mimetic least-squares method for convection-diffusion problems. We show that by encapsulating the underlying geometric properties in the problem, we are able to discretize the convection-diffusion problem such that the invariant is conserved. The topological part of the problem can be satisfied to machine precision for moderate low values of the diffusion coefficient. However, for highly convection-dominated problems, i.e. for low values of the diffusion coefficient, the condition number of the associated matrix is huge and it is not possible to obtain the correct solution of the matrix system. We have in the present problem used a direct solver, however one of the main features of the Least-Squares method is that we obtain symmetric coefficient matrices and we can therefore use fast iterative solvers such as the preconditioned conjugate gradient method. Stabilization techniques could be one way of solving the problems with high condition numbers.

References

1. Bochev, P., Gunzburger, M.: On least-squares finite element methods for the Poisson equation and their connection to the Dirichlet and Kelvin principles. SIAM J. Numer. Anal. **43**(1), 340–362 (2005). https://doi.org/10.1137/S003614290443353X
2. Bochev, P.B., Hyman, J.M.: Principles of mimetic discretizations of differential operators. In: Arnold, D.N., Bochev, P.B., Lehoucq, R.B., Nicolaides, R.A., Shashkov, M. (eds.) Compatible Spatial Discretizations. The IMA Volumes in Mathematics and its Applications, vol. 142, pp. 89–119. Springer, New York (2006). https://doi.org/10.1007/0-387-38034-5_5
3. Burke, W.L.: Applied Differential Geometry. Cambridge University Press, Cambridge (1985)
4. Desbrun, M., Hirani, A.N., Leok, M., Marsden, J.E.: Discrete Exterior Calculus. ArXiv Mathematics e-prints, August 2005
5. Gerritsma, M.: Edge functions for spectral element methods. In: Hesthaven, J., Rønquist, E. (eds.) Spectral and High Order Methods for Partial Differential Equations, vol. 76, pp. 199–207. Springer, Heidelberg (2011). https://doi.org/10.1007/978-3-642-15337-2_17
6. Gerritsma, M., Bochev, P.: A spectral mimetic least-squares method for the stokes equations with no-slip boundary condition. Comput. Math. Appl. **71**(11), 2285–2300 (2016)
7. Hsieh, P.W., Yang, S.Y.: On efficient least-squares finite element methods for convection-dominated problems. Comput. Methods Appl. Mech. Eng. **199**(1), 183–196 (2009)
8. Jiang, B.: The Least-Squares Finite Element Method: Theory and Applications in Computational Fluide Dynamics and Electromagnetics. Scientific Computation. Springer, New York (1998). http://opac.inria.fr/record=b1094463
9. Karniadakis, G., Sherwin, S.: Spectral/hp Element Methods for Computational Fluid Dynamics, 2nd edn. Oxford University Press, Oxford (2005)
10. Kreeft, J., Palha, A., Gerritsma, M.: Mimetic framework on curvilinear quadrilaterals of arbitrary order. ArXiv e-prints, November 2011
11. Lazarov, R., Tobiska, L., Vassilevski, P.: Streamline-diffusion least-squares mixed finite element methods for convection-diffusion problems. East-West J. Numer. Math **5**(4), 249–264 (1997)
12. Mattiussi, C.: A reference discretization strategy for the numerical solution of physical field problems. In: Hawkes, P.W. (ed.) Electron Microscopy and Holography, Advances in Imaging and Electron Physics, vol. 121, pp. 143–279. Elsevier (2002). http://www.sciencedirect.com/science/article/pii/S1076567002800271
13. McInerney, A.: First Steps in Differential Geometry: Riemannian, Contact, Symplectic. Undergraduate Texts in Mathematics, 1st edn. Springer, New York (2013). https://doi.org/10.1007/978-1-4614-7732-7
14. Proot, M.M.J., Gerritsma, M.: A least-squares spectral element formulation for the stokes problem. J. Sci. Comput. **17** (2002)
15. Tu, L.W.: An Introduction to Manifolds. Universitext, 2nd edn. Springer, New York (2011)

Spectral Mimetic Least-Squares Method for Curl-curl Systems

Artur Palha[1,2](✉) and Marc Gerritsma[1]

[1] TU Delft, Kluyverweg 1, 2629 HS Delft, The Netherlands
{a.palha,M.I.Gerritsma}@tudelft.nl
[2] Department of Mechanical Engineering, Eindhoven University of Technology, P.O. Box 513, 5600 MB Eindhoven, The Netherlands

Abstract. One of the most cited disadvantages of least-squares formulations is its lack of conservation. By a suitable choice of least-squares functional and the use of appropriate conforming finite dimensional function spaces, this drawback can be completely removed. Such a mimetic least-squares method is applied to a curl-curl system. Conservation properties will be proved and demonstrated by test results on two-dimensional curvilinear grids.

1 Curl-curl Systems

Let $\Omega \subset \mathbb{R}^d$, $d = 2, 3$ a contractible, bounded domain with Lipschitz-continuous boundary $\partial\Omega$. In [2, Sect. 6.2.2] curl-curl systems were introduced as the vector Laplacian of divergence-free vector fields

$$\begin{cases} \gamma \boldsymbol{w} + \nabla \times \boldsymbol{v} = \Theta_1 \boldsymbol{f} & \text{in } \Omega \\ \boldsymbol{y} - \nabla \times \boldsymbol{u} = \boldsymbol{0} & \text{in } \Omega \\ \boldsymbol{w} - \Theta_1 \boldsymbol{u} = \boldsymbol{0} & \text{in } \Omega \\ \boldsymbol{y} - \Theta_2^{-1} \boldsymbol{v} = \boldsymbol{0} & \text{in } \Omega \end{cases} \quad \text{and} \quad \boldsymbol{n} \times \boldsymbol{u} = \boldsymbol{0} \ \text{ along } \partial\Omega. \tag{1}$$

The symmetric positive, positive definite weights Θ_1 and Θ_2 satisfy

$$\alpha \boldsymbol{\xi}^T \boldsymbol{\xi} \leq \boldsymbol{\xi}^T \Theta_i \boldsymbol{\xi} \leq \frac{1}{\alpha} \boldsymbol{\xi}^T \boldsymbol{\xi}, \quad \alpha > 0. \tag{2}$$

Such curl-curl structures appear everywhere in physics. The Maxwell equations which describe electromagnetic fields for example, [2, Sect. 6.2.2]. For our mimetic least-squares formulation, we are going to rewrite the system slightly by introducing $\boldsymbol{z} = \gamma \boldsymbol{w} - \Theta_1 \boldsymbol{f}$ which then gives

$$\begin{cases} \boldsymbol{z} + \nabla \times \boldsymbol{v} = 0 & \text{in } \Omega \\ \boldsymbol{y} - \nabla \times \boldsymbol{u} = \boldsymbol{0} & \text{in } \Omega \\ \nabla \times \boldsymbol{v} + \gamma \Theta_1 \boldsymbol{u} = \Theta_1 \boldsymbol{f} & \text{in } \Omega \\ \nabla \times \boldsymbol{u} - \Theta_2^{-1} \boldsymbol{v} = \boldsymbol{0} & \text{in } \Omega \end{cases} \quad \text{and} \quad \boldsymbol{n} \times \boldsymbol{u} = \boldsymbol{0} \ \text{ along } \partial\Omega. \tag{3}$$

I. Lirkov and S. Margenov (Eds.): LSSC 2017, LNCS 10665, pp. 119–127, 2018.
https://doi.org/10.1007/978-3-319-73441-5_12

The least-squares functional corresponding to this system is given by

$$\begin{cases} \mathcal{J}(\boldsymbol{u},\, \boldsymbol{z},\, \boldsymbol{v},\, \boldsymbol{y};\, \boldsymbol{f}) = \|\boldsymbol{z} + \nabla\times\boldsymbol{v}\|^2_{0,(\gamma\Theta_1)^{-1}} + \|\boldsymbol{y} - \nabla\times\boldsymbol{u}\|^2_{0,\,\Theta_2} \\ \qquad\qquad + \|\nabla\times\boldsymbol{v} + \gamma\Theta_1\boldsymbol{u} - \Theta_1\boldsymbol{f}\|^2_{0,(\gamma\Theta_1)^{-1}} + \|\nabla\times\boldsymbol{u} - \Theta_2^{-1}\boldsymbol{v}\|^2_{0,\Theta_2} \\ (\boldsymbol{u},\boldsymbol{z},\boldsymbol{v},\boldsymbol{y}) \in X, \end{cases} \tag{4}$$

with

$$X = H_0(\nabla\times,\Theta_1,\Omega)\times H(\nabla\cdot,(\gamma\Theta_1)^{-1},\Omega)\times H(\nabla\times,\Theta_2^{-1},\Omega)\times H(\nabla\cdot,\Theta_2,\Omega) \tag{5}$$

where the function spaces $H(\mathbf{D},\Theta,\Omega)$ with $\Theta \in \{\Theta_1,\Theta_2\}$, and $H(\mathbf{D},\Theta^{-1},\Omega)$ with $\Theta^{-1} \in \{(\gamma\Theta_1)^{-1},\Theta_2^{-1}\}$, are defined as

$$\begin{cases} H(\mathbf{D},\Theta_i,\Omega) := \{\boldsymbol{u}\,|\,\int_\Omega \boldsymbol{u}\cdot\Theta_i\boldsymbol{u} \leq \infty \text{ and } \int_\Omega \mathbf{D}\boldsymbol{u}\cdot\Theta_{i+1}\mathbf{D}\boldsymbol{u} \leq \infty\}, \\ H(\mathbf{D},\Theta_i^{-1},\Omega) := \{\boldsymbol{u}\,|\,\int_\Omega \boldsymbol{u}\cdot\Theta_i^{-1}\boldsymbol{u} \leq \infty \text{ and } \int_\Omega \mathbf{D}\boldsymbol{u}\cdot\Theta_{i-1}^{-1}\mathbf{D}\boldsymbol{u} \leq \infty\}, \end{cases} \tag{6}$$

where $\mathbf{D} \in \{\nabla\times,\nabla\cdot\}$. For a more detailed definition of the functions spaces see [3].

Theorem 1. *There exists a constant $C > 0$ such that*

$$C\big(\|\boldsymbol{u}\|^2_{H_0(\nabla\times,\Theta_1,\Omega)} + \|\boldsymbol{y}\|^2_{H(\nabla\cdot,\Theta_2,\Omega)} + \|\boldsymbol{v}\|^2_{H(\nabla\times,\Theta_2^{-1},\Omega)} + \|\boldsymbol{z}\|^2_{H(\nabla\cdot,(\gamma\Theta_1)^{-1},\Omega)}\big) \leq \mathcal{J}(\boldsymbol{u},\boldsymbol{z},\boldsymbol{v},\boldsymbol{y};\mathbf{0})$$

Proof. Expanding $\mathcal{J}(\boldsymbol{u},\boldsymbol{z},\boldsymbol{v},\boldsymbol{y};\mathbf{0})$ and using $(\nabla\times\boldsymbol{v},\boldsymbol{u})_0 - (\nabla\times\boldsymbol{u},\boldsymbol{v})_0 = 0$ for $\boldsymbol{n}\times\boldsymbol{u} = 0$ along $\partial\Omega$ gives

$$\begin{aligned} \mathcal{J}(\boldsymbol{u},\boldsymbol{z},\boldsymbol{v},\boldsymbol{y};\mathbf{0}) &= 2\left[\|\nabla\times\boldsymbol{v}\|^2_{0,(\gamma\Theta_1)^{-1}} + \|\nabla\times\boldsymbol{u}\|^2_{0,\Theta_2}\right] + \|\boldsymbol{z}\|^2_{0,(\gamma\Theta_1)^{-1}} \\ &\quad + \|\boldsymbol{y}\|^2_{0,\Theta_2} + \|\boldsymbol{u}\|^2_{0,(\gamma\Theta_1)} + \|\boldsymbol{v}\|^2_{0,\Theta_2^{-1}} \\ &\quad + 2(\boldsymbol{z},\nabla\times\boldsymbol{v})_{0,(\gamma\Theta_1)^{-1}} + 2(\boldsymbol{y},\nabla\times\boldsymbol{u})_{0,\Theta_2} \\ &\geq \frac{1}{2}\left[\|\nabla\times\boldsymbol{v}\|^2_{0,(\gamma\Theta_1)^{-1}} + \|\nabla\times\boldsymbol{u}\|^2_{0,\Theta_2}\right] \\ &\quad + \frac{1}{3}\left[\|\boldsymbol{z}\|^2_{0,(\gamma\Theta_1)^{-1}} + \|\boldsymbol{y}\|^2_{0,\Theta_2}\right] + \|\boldsymbol{u}\|^2_{0,(\gamma\Theta_1)} + \|\boldsymbol{v}\|^2_{0,\Theta_2^{-1}}, \end{aligned}$$

where we used the ϵ-inequality with $\epsilon = 2/3$ for the inner products. Using (2) gives

$$\mathcal{J}(\boldsymbol{u},\boldsymbol{z},\boldsymbol{v},\boldsymbol{y};\mathbf{0}) \geq \frac{1}{2\alpha}\left[\|\boldsymbol{v}\|^2_{H(\nabla\times)} + \|\boldsymbol{u}\|^2_{H(\nabla\times)}\right] + \frac{1}{3\alpha}\left[\|\boldsymbol{z}\|^2_{H(\nabla\cdot)} + \|\boldsymbol{y}\|^2_{H(\nabla\cdot)}\right],$$

which concludes the proof with $C = 1/(3\alpha)$. □

Proposition 1. *The minimizers of (4) satisfy $\boldsymbol{z} + \nabla\times\boldsymbol{v} = \mathbf{0}$ and $\boldsymbol{y} - \nabla\times\boldsymbol{u} = \mathbf{0}$ in $L^2(\Omega)$.*

Proof. The variations of the least-squares functional in (4) with respect to $\boldsymbol{z}$ and $\boldsymbol{y}$ yield

$$\int_{\Omega}(\boldsymbol{z}+\nabla\times\boldsymbol{v})\cdot\tilde{\boldsymbol{z}}\,\mathrm{d}x=0\quad\text{and}\quad\int_{\Omega}(\boldsymbol{y}-\nabla\times\boldsymbol{u})\cdot\tilde{\boldsymbol{y}}\,\mathrm{d}x=0,$$

respectively. This implies that the conservation laws are satisfied in the L^2-sense. □

Remark 1. When this formulation is applied to the Maxwell equations it states that Faraday's law of induction and Ampere's law are satisfied for minimizers of the least-squares functional (4).

Corollary 1. *For sufficiently smooth solutions Proposition 1 implies that the 'redundant' equations $\nabla\cdot\boldsymbol{z}=0$ and $\nabla\cdot\boldsymbol{y}=0$ are also satisfied. For electromagnetic fields this translates to conservation of charge and magnetic monopoles.*

2 Two-Dimensional Case

2.1 Equations in 2D

In this work we consider the two-dimensional case, $\Omega=\mathbb{R}^2$, such that u, f, z are scalar fields and $\boldsymbol{v},\boldsymbol{y}$ are vector fields. Under these conditions (3) becomes

$$\begin{cases} z+\nabla\times\boldsymbol{v}=0, & \text{in } \Omega,\\ \boldsymbol{y}-\nabla\times u=\boldsymbol{0}, & \text{in } \Omega,\\ \nabla\times\boldsymbol{v}+\gamma\Theta_1 u=\Theta_1 f, & \text{in } \Omega,\\ \nabla\times u-\Theta_2^{-1}\boldsymbol{v}=\boldsymbol{0}, & \text{in } \Omega, \end{cases}\quad\text{and}\quad u=0\ \text{along}\ \partial\Omega. \tag{7}$$

Note that for scalar functions we have $\nabla\times u:=\frac{\partial u}{\partial y}\boldsymbol{e}_x-\frac{\partial u}{\partial x}\boldsymbol{e}_y$, and for vector fields the standard definition is used $\nabla\times\boldsymbol{v}:=\frac{\partial v^y}{\partial x}-\frac{\partial v^x}{\partial y}$.

2.2 Double DeRham Complex in 2D

The natural solution space for the two-dimensional problem (7) is a combination of two complementary pairs given by $\{\boldsymbol{v},\boldsymbol{y}\}\in H(\nabla\times,\Theta_2^{-1},\Omega)\times H_0(\nabla\cdot,\Theta_2,\Omega)$ and $\{u,z\}\in H_0(\nabla\times,\gamma\Theta_1,\Omega)\times H(\nabla\cdot,(\gamma\Theta_1)^{-1},\Omega)$. The first two equations in (7) are topological equations and the last two equations are constitutive equations. The relations established by this set of equations can be visualized in the following segment of the two-dimensional double DeRham complex[1]

$$\begin{array}{ccccc} u\in H_0(\nabla\times,\gamma\Theta_1,\Omega) & \xrightarrow{\nabla\times} & \boldsymbol{y}\in H_0(\nabla\cdot,\Theta_2,\Omega) & \xrightarrow{\nabla\cdot} & L^2_0(\Theta_3,\Omega)\\ \Theta_1\Big\downarrow\Big\uparrow\Theta_1^{-1} & & \Theta_2\Big\downarrow\Big\uparrow\Theta_2^{-1} & & \Theta_3\Big\downarrow\Big\uparrow\Theta_3^{-1}\\ z\in H(\nabla\cdot,(\gamma\Theta_1)^{-1},\Omega) & \xleftarrow[\nabla\times]{} & \boldsymbol{v}\in H(\nabla\times,\Theta_2^{-1},\Omega) & \xleftarrow[\nabla]{} & H^1(\Theta_3^{-1},\Omega) \end{array}. \tag{8}$$

[1] Note that these function spaces are the two-dimensional versions of the associated three-dimensional ones. For example, in two dimensions $H(\nabla\cdot,(\gamma\Theta_1)^{-1},\Omega)$ is solely the out-of-plane component of the associated full three-dimensional vector space.

In grey we represent the segments of the double DeRham complex not present in this problem. The topological relations are represented by the horizontal arrows and the constitutive relations are represented by the vertical arrows.

The goal of this work is to construct a least-squares finite element formulation that exactly satisfies the topological relations at the discrete level and includes all numerical approximations in the constitutive relations. The rationale behind this objective derives from the intimate relation between conservation properties and topological relations (e.g., in electromagnetics these topological state Faraday's law of induction and Ampere's law). Therefore, a scheme that exactly satisfies topological relations displays important conservation properties.

3 Numerical Discretization

3.1 Mimetic Spectral Element Spaces

For the numerical solution of (7) we introduce the finite dimensional function spaces for the discrete solutions: $u_h \in U_h \subset H_0(\nabla\times, \gamma\Theta_1, \Omega)$, $z_h \in Z_h \subset H(\nabla\cdot, (\gamma\Theta_1)^{-1}, \Omega)$, $\boldsymbol{v}_h \in V_h \subset H(\nabla\times, \Theta_2^{-1}, \Omega)$, $\boldsymbol{y}_h \in Y_h \subset H_0(\nabla\cdot, \Theta_2, \Omega)$. To exactly satisfy the topological equations, these function spaces must be part of a discrete DeRham complex, such that

$$\{\nabla \times u_h \,|\, u_h \in U_h\} \subseteq Y_h \quad \text{and} \quad \{\nabla \times \boldsymbol{v}_h \,|\, \boldsymbol{v}_h \in V_h\} \subseteq Z_h. \tag{9}$$

Each of these finite dimensional function spaces, U_h, Z_h, V_h, Y_h, has an associated finite set of basis functions ϵ_i^U, ϵ_i^Z, $\boldsymbol{\epsilon}_i^V$, $\boldsymbol{\epsilon}_i^Y$, such that

$$U_h = \text{span}\left\{\epsilon_1^U, \dots, \epsilon_{d_U}^U\right\}, \qquad Z_h = \text{span}\left\{\epsilon_1^V, \dots, \epsilon_{d_Z}^Z\right\},$$
$$V_h = \text{span}\left\{\boldsymbol{\epsilon}_1^V, \dots, \boldsymbol{\epsilon}_{d_V}^V\right\}, \qquad Y_h = \text{span}\left\{\boldsymbol{\epsilon}_1^Y, \dots, \boldsymbol{\epsilon}_{d_Y}^Y\right\},$$

where d_U, d_Z, d_V and d_Y denote the dimension of the discrete function spaces.

Such family of discrete function spaces can be constructed from tensor products of Lagrange and edge functions, as presented in [3, Sect. 3] and in more detail in [4].

As a consequence, we can express the approximate solutions as a linear combination of these basis functions

$$u_h := \sum_{i=1}^{d_U} u_i \epsilon_i^U, \quad z_h := \sum_{i=1}^{d_Z} z_i \epsilon_i^Z, \quad \boldsymbol{v}_h := \sum_{i=1}^{d_V} v_i \boldsymbol{\epsilon}_i^V \quad \text{and} \quad \boldsymbol{y}_h := \sum_{i=1}^{d_Y} y_i \boldsymbol{\epsilon}_i^Y.$$

The special feature of this family of basis functions is that

$$\nabla \times \boldsymbol{\epsilon}_i^V = \sum_{k=1}^{d_Z} \mathbb{E}_{k,i}^{2,1} \epsilon_k^Z \quad \text{and} \quad \nabla \times \epsilon_i^U = \sum_{k=1}^{d_Y} \mathbb{E}_{k,i}^{1,0} \boldsymbol{\epsilon}_k^Y. \tag{10}$$

where $\mathbb{E}^{1,0}$ and $\mathbb{E}^{2,1}$ are incidence matrices containing only 1's, 0's and -1's and express the topology of the mesh. For more details on the incidence matrices see [4]. Therefore at the discrete level the first and second equations in (7) become

$$\nabla \times \boldsymbol{v}_h = \sum_{i=1}^{d_V} v_i \nabla \times \boldsymbol{\epsilon}_i^V = \sum_{i=1}^{d_Z} \left(\sum_{k=1}^{d_V} \mathbb{E}_{i,k}^{2,1} v_k \right) \epsilon_i^Z = - \sum_{i=1}^{d_Z} z_i \epsilon_i^Z = -z_h, \tag{11}$$

$$\nabla \times u_h = \sum_{i=1}^{d_U} u_k \nabla \times \epsilon_i^U = \sum_{i=1}^{d_Y} \left(\sum_{k=1}^{d_U} \mathbb{E}_{i,k}^{1,0} u_k \right) \boldsymbol{\epsilon}_i^Y = \sum_{i=1}^{d_Y} y_i \boldsymbol{\epsilon}_i^Y = \boldsymbol{y}_h. \tag{12}$$

This shows that, at the discrete level, the topological relations are indeed purely topological since the coefficients z_i and y_i depend only on the incidence matrices $\mathbb{E}^{1,0}$ and $\mathbb{E}^{2,1}$ and on the coefficients v_i and u_i. There is no dependence on the particular basis function representation or on the geometry of the mesh.

Equation (9) imposes the fundamental constraints on the function spaces at the horizontal level of the discrete double DeRham complex (8). At the vertical level there is no strict relation between the function spaces since the relation introduced is an approximation relation. The discrepancy between dual variables is minimized by the least-squares formulation.

3.2 Transformation on a Curvilinear Mesh

In the case of a curvilinear mesh the transformation relations between fields in the computational domain $(\xi, \eta) \in \tilde{\Omega} = [-1, 1]^2$ and the physical domain $(x, y) \in \Omega$ require special attention.

In this work the unknown fields are contravariant tensors and if we consider a smooth injective map $\Phi: (\xi, \eta) \in \tilde{\Omega} \mapsto (x, y) \in \Omega$ the unknown fields transform in the following way

$$\tilde{u}(\xi,\eta) = \Phi^*[u(x,y)] := (u \circ \Phi)(\xi,\eta), \quad \tilde{z}(\xi,\eta) = \Phi^*[z(x,y)] := \det(\mathsf{J})(z \circ \Phi)(\xi,\eta)$$
$$\tilde{\boldsymbol{v}}(\xi,\eta) = \Phi^*[\boldsymbol{v}(x,y)] := \mathsf{J}(\boldsymbol{v} \circ \Phi)(\xi,\eta), \quad \tilde{\boldsymbol{y}}(\xi,\eta) = \Phi^*[\boldsymbol{y}(x,y)] := \mathsf{J}(\boldsymbol{y} \circ \Phi)(\xi,\eta).$$

$$u(x,y) = \left(\tilde{u} \circ \Phi^{-1}\right)(x,y), \qquad z(x,y) = \frac{1}{\det(\mathsf{J})}(\tilde{z} \circ \Phi^{-1})(x,y),$$
$$\boldsymbol{v}(x,y) = \mathsf{J}^{-1}(\tilde{\boldsymbol{v}} \circ \Phi^{-1})(x,y), \quad \boldsymbol{y}(x,y) = \frac{\mathsf{J}^{\mathsf{T}}}{\det(\mathsf{J})}(\boldsymbol{y} \circ \Phi^{-1})(x,y).$$

where J is the Jacobian matrix given by

$$\mathsf{J} := \begin{bmatrix} \frac{\partial \Phi^x}{\partial \xi} & \frac{\partial \Phi^y}{\partial \xi} \\ \frac{\partial \Phi^x}{\partial \eta} & \frac{\partial \Phi^y}{\partial \eta} \end{bmatrix}.$$

Another important aspect of these basis functions is that under this transformation properties (9), (11), and (12) are preserved. This results in a numerical method that exactly satisfies the topological relations even on highly deformed meshes as the ones presented in Sect. 4.

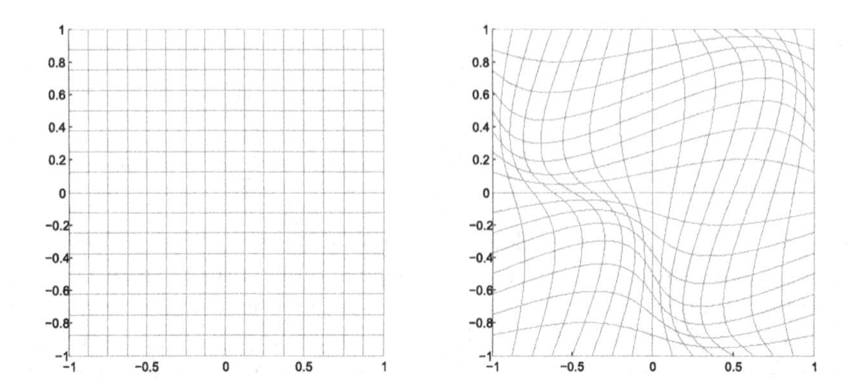

Fig. 1. A 16×16 grid for $c = 0.0$ (left) and $c = 0.2$ (right).

4 Numerical Test Case

Consider the two-dimensional domain $\Omega = [-1,1]^2$ and solve (1) with $f = (1+2\pi^2)\sin(\pi x)\sin(\pi y)$, $\gamma = \Theta_0 = 1$ and $\Theta_1 = \mathbb{I}$. We will solve (1) on sequences of orthogonal $K \times K$ grids and curvilinear $K \times K$ grids, where the curvilinear grid lines are given by

$$\begin{cases} x = \xi + c\sin(\pi\xi)\sin(\pi\eta) \\ y = \eta + c\sin(\pi\xi)\sin(\pi\eta) \end{cases}$$

An example of the orthogonal grid and curvilinear grid for $K = 16$ is shown in Fig. 1.

As can be seen in Figs. 2, 3, 4 and 5, and in more detail in Table 1, the method proposed in this work presents optimal convergence rates for all variables, independently of the mesh deformation.

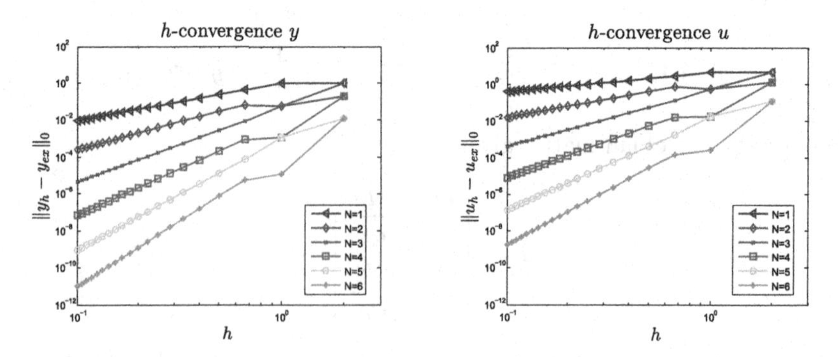

Fig. 2. h-convergence of $\boldsymbol{y}$ (left) and u (right) for polynomial degrees $N = 1, \ldots, 6$ on the orthogonal grid corresponding to $c = 0.0$.

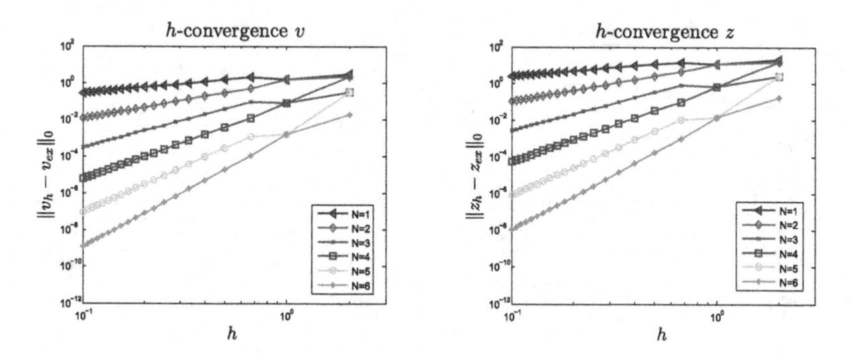

Fig. 3. h-convergence of $\boldsymbol{v}$ (left) and z (right) for polynomial degrees $N = 1, \ldots, 6$ on the orthogonal grid corresponding to $c = 0.0$.

Table 1. Convergence rates for the curl-curl least-squares solution on orthogonal ($c = 0.0$) and curvilinear grids ($c = 0.2$).

	$c = 0.0$				$c = 0.2$			
N	$\boldsymbol{y}$	u	$\boldsymbol{v}$	z	$\boldsymbol{y}$	u	$\boldsymbol{v}$	z
1	2.0	1.0	1.0	1.0	2.0	1.0	1.0	1.0
2	3.0	2.0	2.0	2.0	3.0	2.0	2.0	2.0
3	4.0	3.0	3.0	3.0	4.0	3.0	3.0	3.0
4	5.0	4.0	4.0	4.0	5.0	4.0	4.0	4.0
5	6.0	5.0	5.0	5.0	6.0	5.0	5.0	5.0
6	6.9	6.0	6.0	6.0	7.0	6.0	6.0	6.0

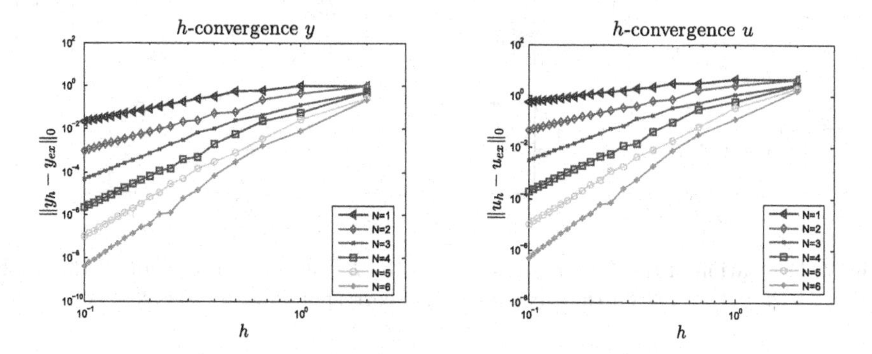

Fig. 4. h-convergence of $\boldsymbol{y}$ (left) and u (right) for polynomial degrees $N = 1, \ldots, 6$ on the curvilinear grid corresponding to $c = 0.2$.

According to Proposition 1 the least-squares solution should satisfy the conservation laws in the L^2-sense. Figures 6 and 7 show the conservation laws for various polynomial approximations as a function of the mesh size h in the L^∞-norm, confirming that the conservation laws are satisfied up to machine precision.

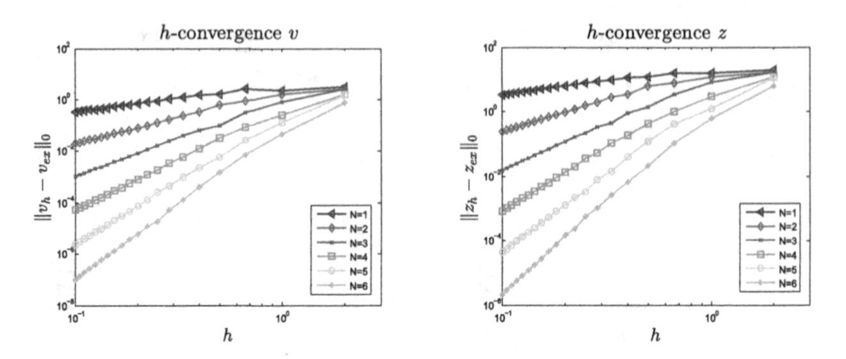

Fig. 5. h-convergence of $\boldsymbol{v}$ (left) and z (right) for polynomial degrees $N = 1, \ldots, 6$ on the curvilinear grid corresponding to $c = 0.2$.

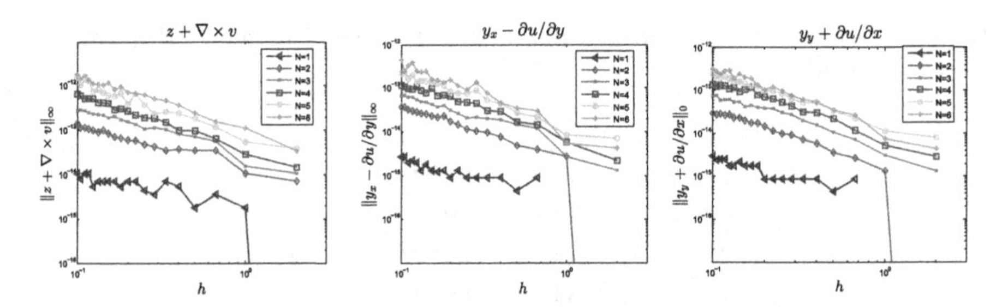

Fig. 6. Evaluation of the conservation laws in the L^{∞}-norm as a function of the mesh size h for $N = 1, \ldots, N$ on the orthogonal grid corresponding to $c = 0.0$.

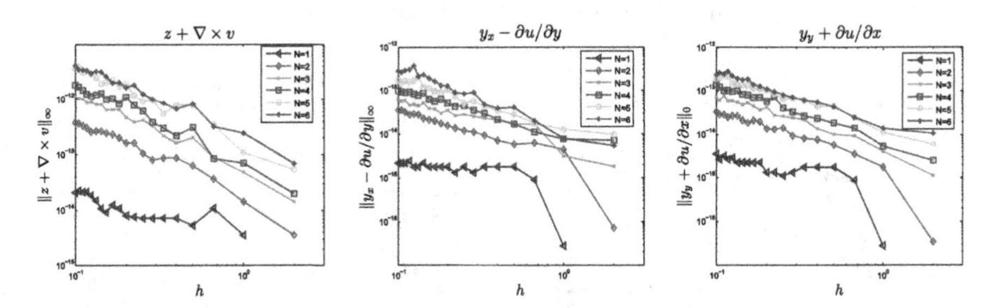

Fig. 7. Evaluation of the conservation laws in the L^{∞}-norm as a function of the mesh size h for $N = 1, \ldots, N$ on the curvilinear grid corresponding to $c = 0.2$.

5 Conclusions

We presented a least-squares spectral element formulation for the curl-curl problem that is capable of exactly satisfying the topological relations $\boldsymbol{y} - \nabla \times u = 0$ and $z + \nabla \times \boldsymbol{v} = 0$ even on highly deformed meshes. Additionally, this method is capable of optimal convergence rates on highly deformed meshes, improving upon the drawbacks of other families of finite elements identified in [5,6].

References

1. Bochev, P.B., Gerritsma, M.I.: A spectral mimetic least-squares method. Comput. Math. Appl. **68**, 1480–1502 (2014)
2. Bochev, P.B., Gunzburger, M.D.: Least-Squares Finite Element Methods. Spinger Verlag, Berlin (2009)
3. Gerritsma, M., Palha, A.: Spectral mimetic least-squares method for div-curl systems. In: Lirkov, I., Margenov, S. (eds.) LSSC 2017. LNCS, vol. 10665, pp. 103–110. Springer, Cham (2017)
4. Palha, A., Rebelo, P.P., Hiemstra, R., Kreeft, J., Gerritsma, M.: Physics-compatible discretization techniques on single and dual grids, with application to the Poisson equation of volume forms. J. Comput. Phys. **257**, 1394–1422 (2014)
5. Arnold, D., Boffi, D., Falk, R.S.: Quadrilateral $H(\text{div})$ finite elements. SIAM J. Numer. Anal. **42**, 2429–2451 (2005)
6. Falk, R.S., Gatto, P., Monk, P.: Hexahedral $H(\text{div})$ and $H(\text{curl})$ finite elements. ESAIM. Math. Model. Numer. Anal. **45**, 115–143 (2011)

Numerical Solution of Cahn-Hilliard System by Adaptive Least-Squares Spectral Element Method

Keunsoo Park[1,2(✉)], Marc Gerritsma[2], and Maria Fernandino[1]

[1] Department of Energy and Process Engineering, Norwegian University of Science and Technology, 7491 Trondheim, Norway
keunsoo.park@ntnu.no, whwr0428@gmail.com

[2] Aerospace Engineering, Delft University of Technology, Kluyverweg 1, Delft, The Netherlands

Abstract. There is a growing interest in the phase-field approach to numerically handle the interface dynamics in multiphase flow phenomena because of its accuracy. The numerical solution of phase-field models has difficulties in dealing with non-self-adjoint operators and the resolution of high gradients within thin interface regions. We present an h-adaptive mesh refinement technique for the least-squares spectral element method for the phase-field models. C^1 Hermite polynomials are used to give global differentiability in the approximated solution, and a space-time coupled formulation and the element-by-element technique are implemented. Two benchmark problems are presented in order to compare two refinement criteria based on the gradient of the solution and the local residual.

Keywords: Adaptive · Least-square · Phase-field · Cahn-Hilliard Parallel computation

1 Introduction

A phase-field model can avoid the problems of interface smearing and of the inaccurate computation of surface tension, which arise in the interface tracking methods such as volume-of-fluids or level-set method. The phase-field models have been widely used to simulate the flow of two or more fluids [1,2]. However, the phase-field method yields non-self-adjoint operators and the solution contains high gradients within thin interfacial regions. In this study, the Cahn-Hilliard equation is selected as a representative of the phase-field models.

The least-squares formulation with C^1 Hermite approximation is used in this study as a main setup. The advantage of this method in the phase-field methods is (1) it always provides a symmetric positive definite system, (2) the LBB condition is circumvented, and (3) the higher order global differentiability improves the approximation accuracy of the solution within the interface. Additionally, in

I. Lirkov and S. Margenov (Eds.): LSSC 2017, LNCS 10665, pp. 128–136, 2018.
https://doi.org/10.1007/978-3-319-73441-5_13

our solver a space-time coupled formulation is used, and the element-by-element technique is implemented.

We also present a high-order h-adaptive mesh refinement technique. The adaptive mesh makes our solver more efficiently by assigning finer elements within narrow interfaces and coarser elements in pure phases. Dealing with drastic topological changes requires tracking the interface movements. We compare the performance of two refinement criteria based on the solution gradient and the local residual.

2 The Mathematical Formulation

2.1 The Cahn-Hilliard Equation

We define the space-time set $\Omega := \Omega_{\mathbf{x}} \times (0,T), T > 0$, for a two-dimensional open domain $\Omega_{\mathbf{x}} \in \mathbb{R}^2$. The boundary of Ω is denoted as $\Gamma := \partial\Omega_{\mathbf{x}} \times (0,T)$. For a flow of two immiscible fluids, the dimensionless Cahn-Hilliard equation is stated as follows: find the unknowns $C = C(\boldsymbol{x},t) : \Omega \rightarrow (0,1)$, $\omega = \omega(\boldsymbol{x},t) : \Omega \rightarrow \mathbb{R}$ such that

$$\frac{\partial C}{\partial t} - \frac{1}{Pe}\nabla^2\omega = 0 \qquad \text{in } \Omega, \tag{1}$$

$$\omega = C^3 - 1.5C^2 + 0.5C - Cn^2\nabla^2 C \qquad \text{in } \Omega, \tag{2}$$

$$C(\boldsymbol{x},0) = C_0(\boldsymbol{x}) \qquad \text{in } \Omega_{\mathbf{x}}, \tag{3}$$

$$\nabla C \cdot \mathbf{n} = 0, \quad \nabla\omega \cdot \mathbf{n} = 0 \qquad \text{on } \Gamma. \tag{4}$$

Here C is the concentration, and w is the chemical potential. Peculet number $Pe = L_0U_0/M$ and Cahn number $Cn = \epsilon/L_0$ are used, where U_0 and L_0 are the reference velocity and length, M is the mobility, and ϵ is the interfacial parameter. The derivation and the physical meaning of the Cahn-Hilliard are explained in our previous study [3] in more detail.

2.2 Least-Squares Method

We use the Newton linearization method to handle with nonlinear terms. We use a subscript l to denote the terms from previous linearization step. For a two-dimensional spatial domain, the Cahn-Hilliard equation with the unknowns $\mathbf{u}^T = [C\ \omega]$ can be represented as

$$\frac{\partial}{\partial t}C - \frac{1}{Pe}\left(\frac{\partial^2}{\partial x^2} + \frac{\partial^2}{\partial y^2}\right)\omega = 0, \tag{5}$$

$$\left[3C_l^2 - 3C_l + 0.5 - Cn^2\left(\frac{\partial^2}{\partial x^2} + \frac{\partial^2}{\partial y^2}\right)\right]C - \omega = 2C_l^3 - 1.5C_l^2. \tag{6}$$

The final system with the boundary conditions is expressed in general as

$$\mathcal{L}\mathbf{u} = \mathcal{G} \qquad \text{in } \Omega, \tag{7}$$

$$\mathcal{B}\mathbf{u} = \mathbf{u}_\Gamma \qquad \text{on } \Gamma, \tag{8}$$

where $\mathcal{L}$ represents the partial differential operator, $\mathcal{G}$ is the corresponding source term, $\mathcal{B}$ is the boundary conditions operator, and $\mathbf{u}_\Gamma$ is the specified value on the boundaries. In this work, the boundary conditions are incorporated into the least-squares functional so that they are also a part of the minimization problem, namely

$$\mathcal{J}(\mathbf{u}) = \frac{1}{2}\|\mathcal{L}\mathbf{u} - \mathcal{G}\|_{0,\Omega}^2 + \frac{1}{2}\|\mathcal{B}\mathbf{u} - \mathbf{u}_\Gamma\|_{0,\Gamma}^2, \tag{9}$$

or equivalently,

Find $\mathbf{u} \in X(\Omega)$ such that

$$\mathcal{A}(\mathbf{u},\mathbf{v}) = \mathcal{F}(\mathbf{v}) \qquad \forall \mathbf{v} \in X(\Omega), \tag{10}$$

with

$$\mathcal{A}(\mathbf{u},\mathbf{v}) = (\mathcal{L}\mathbf{u}, \mathcal{L}\mathbf{v})_{0,\Omega} + (\mathcal{B}\mathbf{u}, \mathcal{B}\mathbf{v})_{0,\Gamma}, \tag{11}$$

$$\mathcal{F}(\mathbf{v}) = (\mathcal{G}, \mathcal{L}\mathbf{v})_{0,\Omega} + (\mathbf{u}_\Gamma, \mathcal{B}\mathbf{v})_{0,\Gamma}, \tag{12}$$

where $\mathcal{A} : X \times X \to \mathbb{R}$ is a symmetric, positive definite bilinear form, $\mathcal{F} : X \to \mathbb{R}$ a continuous linear form, and $X(\Omega)$ is a solution space.

2.3 Spectral Element Discretization

The computational domain Ω is divided into Ne subdomains Ω_e such that

$$\Omega = \sum_{e=1}^{Ne} \Omega_e, \qquad \Omega_i \cap \Omega_j = \emptyset, \quad i \neq j. \tag{13}$$

The discretization is based on a space-time coupled formulation. Space-time strips are consecutively aligned, and a strip is composed of only one element in time, $\Omega_e = \Omega_e^{\mathbf{x}} \times \Omega_e^t = (\mathbf{x}_e, \mathbf{x}_{e+1}) \times (t_n, t_{n+1})$ with the time step size $\Delta t = t_{n+1} - t_n$. Each subdomain is mapped onto the unit cube $(\xi, \sigma, \eta) \in [-1, 1]^3$ for a two-dimensional spatial domain, by an invertible mapping.

The transition region in the phase-field model is preferred to be close to a sharp interface, with the consequent impact on the numerical solution and need for higher spatial resolution. In this article, the local solution in each element Ω_e, $\mathbf{u}_e^h$, is approximated by C^1 p-version hierarchical approximation functions, so-called Hermite polynomials. A basis function for a two-dimensional space and time domain can be written as the tensor product of one-dimensional basis functions with same order, i.e., $\mathbf{\Phi}_m(\xi, \varsigma, \eta) = \phi_i(\xi) \otimes \phi_j(\varsigma) \otimes \phi_k(\eta)$, with $m = i + j(p+1) + k(p+1)^2$ where $0 \leq i, j, k \leq p$. Thus, the local approximation $\mathbf{u}_e^h$ is expanded in $\mathbf{\Phi}$ continuous basis functions as

$$\mathbf{u}_e^h = \sum_{m=1}^{(p+1)^3} \mathbf{U}_e^m \mathbf{\Phi}_e^m. \tag{14}$$

The same basis functions and construction approach have been used in our previous study [4]. For more details we also refer to [5,6].

Together with integration by the Gaussian quadrature based on the GLL-roots, the discretization of the least-squares formulation (10) can be expressed on an element-level as

$$\mathbf{L}_e^T \mathbf{W}_e \mathbf{L}_e \mathbf{U}_e = \mathbf{L}_e^T \mathbf{W}_e \mathbf{F}_e, \tag{15}$$

where $\mathbf{L}$ is a matrix whose components are the evaluation of $\mathcal{L}$ with the Hermite polynomials at the quadrature points, and $\mathbf{F}$ is a vector of the evaluation of $\mathcal{G}$. $\mathbf{W}$ is a diagonal matrix of the quadrature weights, and in this article, the number of quadrature points Q are fixed at the same number of polynomials of one dimensional basis function as $Q = p + 1$.

The discretized algebraic equation is solved element-by-element with the conjugated gradient method with the Jacobi preconditioner. Matlab code and Matlab MPI developed at our group are used. The local solutions in each elements $\mathbf{u}_e^h$, are glued to construct the global approximation of the solution $\mathbf{u}^h$, i.e.,

$$\mathbf{u}^h = \bigcup_{e=1}^{Ne} \mathbf{u}_e^h. \tag{16}$$

3 Adaptive Mesh Refinement

3.1 C^1 Continuous h-refinement

When the refinement level of neighboring elements is different, among the nodal basis of the coarser element, ones which have non-zero values on the element interface are shared with the finer element. To ensure the global C^1 continuity over a non-conformal element interface, we introduce two L^2-norm least-squares functionals to be minimized for the value of solution, $\mathcal{J}_0^r$, and for the derivative of solution, $\mathcal{J}_1^r$, respectively, over the inter-element interface γ between the finer element F and the coarser element C:

$$\mathcal{J}_0^r \left(\mathbf{u}_b^F; \mathbf{u}_b^C\right) = \int_\gamma \left(\mathbf{u}_b^F - \mathbf{u}_b^C\right)^2 ds, \tag{17}$$

$$\mathcal{J}_1^r \left(\mathbf{u}_b^F; \mathbf{u}_b^C\right) = \int_\gamma \left(\nabla \mathbf{u}_b^F \cdot \mathbf{n} - \nabla \mathbf{u}_b^C \cdot \mathbf{n}\right)^2 ds, \tag{18}$$

where $\mathbf{u}_b$ is the solution on the inter-element interface. With the expansion coefficients related to the solution value on the inter-element interface $\mathbf{U}_{b,0}$, the minimization statement of $\mathcal{J}_0^r$ can be written in an algebraic form as

$$\nabla \mathcal{J}_0^r = 0; \quad \mathbf{U}_{b,0}^F = \mathbf{H}_F^{-1} \mathbf{H}_C \mathbf{U}_{b,0}^C \equiv \widetilde{\mathbf{Z}}_0 \mathbf{U}_{b,0}^C, \tag{19}$$

where $\mathbf{H}$ is a matrix of Hermite polynomials at the quadrature points, and $\widetilde{\mathbf{Z}}_0$ is the projection matrix for the solution value. Similarly, with the expansion coefficients related to the derivative of solution on the inter-element interface $\mathbf{U}_{b,1}$, the minimization statement of $\mathcal{J}_1^r$ is expressed as

$$\nabla \mathcal{J}_1^r = 0; \quad \mathbf{U}_{b,1}^F = \mathbf{D}_F^{-1}\mathbf{D}_C\mathbf{U}_{b,1}^C \equiv \widetilde{\mathbf{Z}}_1\mathbf{U}_{b,1}^C, \tag{20}$$

with $\mathbf{D}$ a matrix of derivative of Hermite polynomials at the quadrature points, and $\widetilde{\mathbf{Z}}_1$ is the projection matrix for the derivative of solution.

With the relations (19) and (20), we can express all unknowns of finer element $\mathbf{U}^F$ in terms of $\mathbf{U}^{F'}$ composed of the unknowns of coarser element on the boarder and inner element unknowns $\mathbf{U}_i^F$ only:

$$\mathbf{U}^F = \begin{bmatrix} \mathbf{U}_{b,0}^F \\ \mathbf{U}_{b,1}^F \\ \mathbf{U}_i^F \end{bmatrix} = \begin{bmatrix} \widetilde{\mathbf{Z}}_0 & \mathbf{0} & \mathbf{0} \\ \mathbf{0} & \widetilde{\mathbf{Z}}_1 & \mathbf{0} \\ \mathbf{0} & \mathbf{0} & \mathbf{I} \end{bmatrix} \begin{bmatrix} \mathbf{U}_{b,0}^C \\ \mathbf{U}_{b,1}^C \\ \mathbf{U}_i^F \end{bmatrix} = \mathbf{Z}\mathbf{U}^{F'}, \tag{21}$$

where $\mathbf{Z}$ is the total projection matrix.

These constraints are implemented into the least-squares method by replacing $\mathbf{U}$ as $\mathbf{U}^{'}$ using Eq. (21). The formulation at an element-level becomes

$$\mathbf{Z}_e^T\mathbf{L}_e^T\mathbf{W}_e\mathbf{L}_e\mathbf{Z}_e\mathbf{U}_e^{'} = \mathbf{Z}_e^T\mathbf{L}_e^T\mathbf{W}_e\mathbf{F}_e. \tag{22}$$

$\mathbf{Z}_e^T$ is multiplied to maintain the symmetricity of the least-squares system.

3.2 Refinement Strategy

For a transient problem, the decision should be made on the elements in the original unrefined grid to be refined or retrieved at each time step. During the refinement, an element is split into four daughter elements. An element with the refinement level k can be made by k-th mesh refinements from the reference element. In this study, the maximum refinement level is set to 2, and we confine the irregularity up to 1-level, i.e., the difference in the refinement levels of neighboring elements is no larger than 1. In addition to the advantage of implementation, the 1-level irregularity can improve the accuracy in describing the interface [7].

Regarding the decision on refinement, we consider two refinement criteria. With the first criterion (gradient), the elements where the solution gradient exceed a certain tolerance are refined, and it requires that

$$\|\nabla \mathbf{C}\|_{0,\Omega_e}^2 \leq tol_g, \tag{23}$$

with tol_g the discretization tolerance for the gradient. The gradient criterion does not use any error estimator but it intensively targets to the interface.

The second criterion (residual) is based on the local residual in each element, and it is defined as

$$\|\mathcal{R}\|_{0,\Omega_e}^2 = \int_{\Omega_e} \left(\mathcal{L}\mathbf{u}^h - \mathcal{G}\right)^2 d\Omega_e. \tag{24}$$

The conforming elements with top certain percent of local residual are refined.

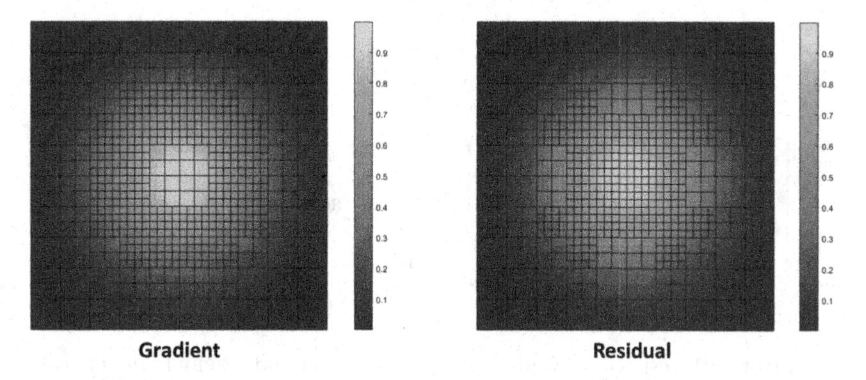

Fig. 1. Solution of Poisson problem and adaptive mesh with $Ne = 10 \times 10$ and $p = 4$ generated by gradient criterion (left) and by residual criterion (right).

4 Numerical Examples

4.1 Poisson Problem

We first solve the Poisson problem with exact solution and compare two refinement criteria. Consider the equation

$$\nabla^2 \mathbf{u} = \left(4r^2/\sigma^4 - 4/\sigma^2\right) \exp(-r^2/\sigma^2), \tag{25}$$

with the exact solution $\mathbf{u} = \exp(-r^2/\sigma^2)$, where $r^2 = x^2 + y^2$ and $\sigma = \sqrt{2}/5$.

The spatial domain is $\Omega = [-0.5, 0.5]^2$. In this example, tol_g is set to 2.8, and the percentage of refined elements with the residual criterion change by cases to have similar number of degrees of freedom (Ndofs) with the gradient criterion. Figure 1 shows the solution on the adaptive meshes by two refinement criteria with $Ne = 10 \times 10$ and $p = 4$. The hilltop area is refined into level 2 with the residual criterion, but in level 1 with the gradient criterion.

The estimated error, defined as $\|\mathbf{u} - \mathbf{u}_{ex}\|_{0,\Omega}^2$, with respect to the Ndofs for both refinement criteria is illustrated in Fig. 2. The error exponentially decays with increasing Ndofs for all expansion order cases. To compare the results from two refinement criteria, we define the following index, called refinement efficiency R_{ef}, composed of ratios of the error and Ndofs from the conforming grid and non-conforming grid, denoted as subscript c and n, respectively:

$$R_{ef} = \frac{\|\mathbf{u}_c - \mathbf{u}_{c,ex}\|_{0,\Omega}^2}{\|\mathbf{u}_n - \mathbf{u}_{n,ex}\|_{0,\Omega}^2} \times \frac{Ndofs_c}{Ndofs_n}. \tag{26}$$

The refinement efficiency is presented in Fig. 3. We can conclude that the efficiency is higher with the gradient criterion in lower Ndofs, while the residual criterion is more efficient in higher Ndofs. With the gradient criterion the efficiency decreases more monotonously with respect to the Ndofs. This conclusion is also applicable to the solution from the phase-field method because of similar contour of solution - having a plateau on the hilltop with steep hill.

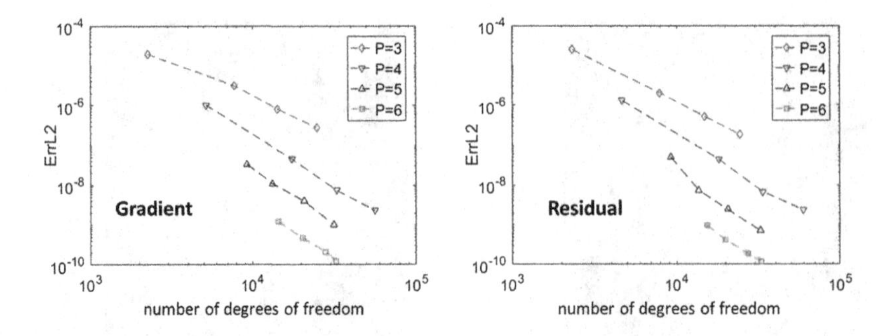

Fig. 2. Error with respect to Ndofs with gradient (left) and residual (right) criteria.

4.2 Benchmark Cross

The equilibrium state induced by the Cahn-Hilliard equation has the minimum local free energy and surface tension energy [8]. A cross-shaped droplet is initially located at the center of the domain $[0, 0.8]^2$, filled with another phase. The original unrefined grid in space is 10^2, and a single time-element of $\Delta t = 0.08$ is used. Polynomials with order of $p = 4$ are used to approximate the solution. The parameters used are $M = 1, Re = 1, Ca = 1, Pe = 100$ and $Cn = 0.01$. The tolerance in the gradient refinement criterion is set to 20, and for the residual refinement criterion 28% and 20% of elements in the initial coarse grid are refined, which correspond to the numbers of refined elements in the gradient criterion at the first and the last time step, respectively.

Among three cases, with the gradient criterion and with the residual criterion of 28% and 20%, no significant difference in the concentration is found. Figure 4 presents the evolution of the concentration and the local residuals of the three cases on the refined grids. With the gradient criterion, the only elements containing the interfaces are refined, so the distribution of the local residuals is symmetric. On the other hand, with the residual criterion, the refinement is performed rather assymetrically. Note that here the local residual from the

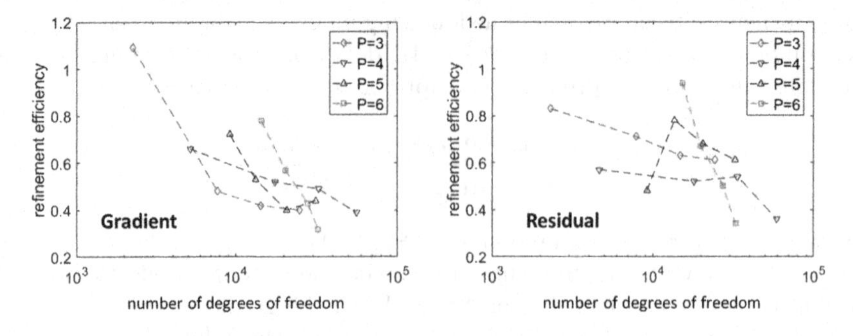

Fig. 3. Refinement efficiency with respect to Ndofs with gradient (left) and residual (right) criteria.

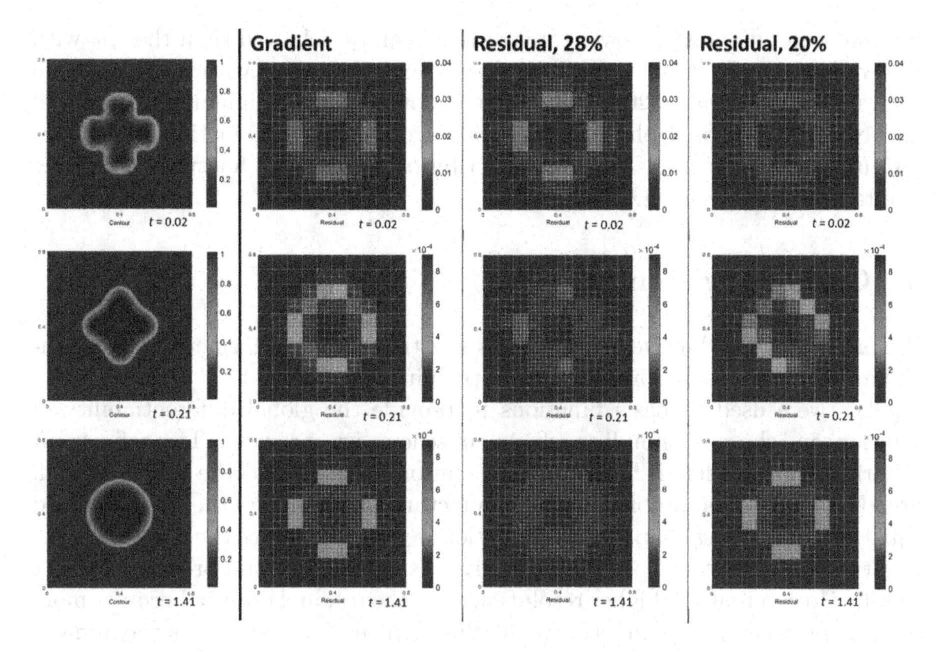

Fig. 4. Evolution of concentration (1st row) and local residuals on refined grid with gradient criterion (2nd row) and with residual criterion of 28% (3rd row) and of 20% (4th row) at $t = 0.02, 0.21$ and 1.41.

Navier-Stokes equation is negligible, of order under 10^{-6}, compared with the one from the Cahn-Hilliard equation over the entire domain. Figure 5 shows the total residual and the Ndofs in time in log scale for the three cases. With the

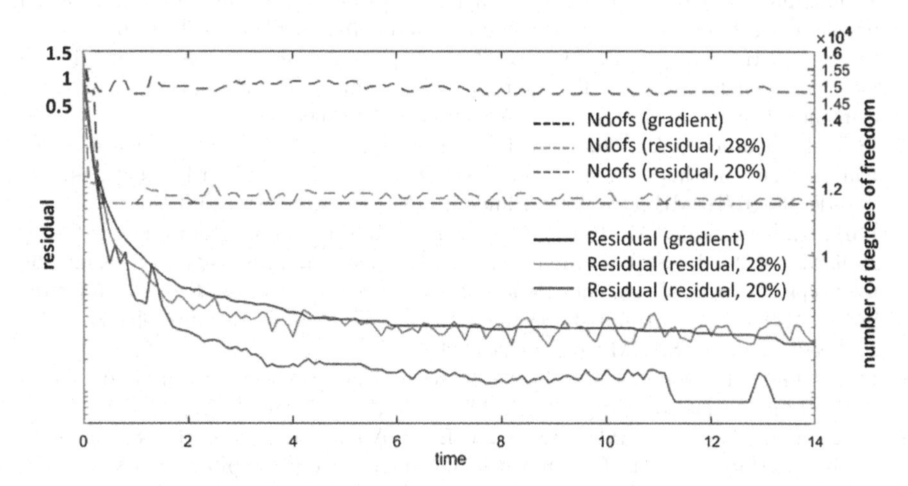

Fig. 5. Total residual (solid line) and Ndofs (dotted line) in time with the gradient criterion (black) and the residual criterion of 28% (blue) and of 20% (red) until $t = 14.0$. (Color figure online)

residual criterion of 28%, lower total residual can be achieved than the one with the gradient criterion, but it requires more Ndofs. Compared with the result from the residual criterion of 20%, the gradient criterion has a similar total residual and Ndofs at the final phase but yields more stable result in time. The total residual and Ndofs from the residual criterion have many fluctuation in time due to assymetric refinement of grid.

5 Concluding Remarks

We presented an adaptive least-squares spectral element scheme for the Cahn-Hilliard equation as a representative of the phase-field model. C^1 Hermite polynomials were used as basis functions to provide the global differentiability of solution, and the corresponding refinement scheme was provided. Two refinement criteria were considered, based on the solution gradient and the local residual. Steady-state Poisson problem with manufactured solution and the Cahn-Hilliard equation with a cross shaped initial solution were solved. Since the gradient criterion targets only interface elements, it gives us more stable and predictable error results. However, at the higher resolution case, the residual criteria becomes more efficient because the asymmetricity in refinement becomes subtle as the number of degrees of freedom increases. Therefore, we recommend to use the gradient criterion for the phase-field method, but it is worth considering the residual criterion if the initial grid already has a larger number of degrees of freedom.

References

1. Fernandino, M., La Forgia, N.: Refinement strategies for the Cahn-Hilliard equation using the least squares method. Comput. Fluids **35**(10), 1384–1399 (2006)
2. Stogner, R.H., Carey, G.F., Murray, B.T.: Approximation of Cahn-Hilliard diffuse interface models using parallel adaptive mesh refinement and coarsening with C1 elements. J. Numer. Methods Eng. **76**(5), 636–661 (2008)
3. Park, K., Dorao, C.A., Chiapero, E.M., Fernandino, M.: The least squares spectral element method for the Navier-Stokes and Cahn-Hilliard equations. In: ASME/JSME/KSME 2015 Joint Fluids Engineering Conference (2015)
4. Park, K., Fernandino, M., Dorao, C.A.: Numerical solution of incompressible Cahn-Hilliard and Navier-Stokes system with large density and viscosity ratio using the least-squares spectral element method. J. Fluid Flow Heat Mass **3**(1), 73–85 (2016)
5. Solin, P.: Towards optimal shape functions for hierarchical Hermite elements. In: Proceedings of the SANM Conference (2005)
6. Fernandino, M., Dorao, C.A.: The least squares spectral element method for the Cahn-Hilliard equation. Appl. Math. Model. **35**(2), 797–806 (2011)
7. Barosan, I., Anderson, P.D., Meijer, H.E.H.: Application of mortar elements to diffuse-interface methods. In: 7th National Conference on Computational Mechanics (2013)
8. Cahn, J.W., Hilliard, J.E.: Free energy of a nonuniform system. 1. Interfacial free energy. J. Chem. Phys. **28**, 258–267 (1958)

Stress-Velocity Mixed Least-Squares FEMs for the Time-Dependent Incompressible Navier-Stokes Equations

Alexander Schwarz(✉), Carina Nisters, Solveigh Averweg, and Jörg Schröder

Faculty of Engineering, Institute of Mechanics, University Duisburg-Essen, Universitätsstr. 15, 45141 Essen, Germany
alexander.schwarz@uni-due.de

Abstract. In this article a mixed least-squares finite element method (LSFEM) for the time-dependent incompressible Navier-Stokes equations is proposed and investigated. The formulation is based on the incompressible Navier-Stokes equations consisting of the balance of momentum and the continuity equations. In order to obtain a first-order system the Cauchy stress tensor is introduced as an additional variable to the system of equations. From this stress-velocity-pressure approach a stress-velocity formulation is derived by adding a redundant residual to the functional without additional variables in order to strengthen specific physical relations, e.g. mass conservation. We account for implementation aspects of triangular mixed finite elements especially regarding the approximation used for $H(\text{div}) \times H^1$ and the discretization in time using the Newmark method. Finally, we present the flow past a cylinder benchmark problem in order to demonstrate the derived stress-velocity least-squares formulation.

Keywords: Least-squares mixed finite element method
Stress-velocity formulation
Time-dependent incompressible Navier-Stokes

1 Introduction

In the last decades the least-squares finite element method (LSFEM) has increasingly gained attention as an alternative variational approach compared to the well-known mixed Galerkin method in constructing finite element formulations. LSFEM offer some theoretical benefits, such as rendering the LBB stability condition unnecessary. This allows a more flexible choice of the polynomial degree in the finite element spaces. Furthermore, the least-squares FEM is suited for the formulation of mixed methods based on different solution variables such as stresses, pressure, displacements or velocities, as they can directly be included in the variational formulation. Another advantage is that LSFEMs result in positive definite and symmetric system matrices, also for differential equations with not self-adjoint operators, see e.g. [1]. Besides that, the LSFEM provides an a

I. Lirkov and S. Margenov (Eds.): LSSC 2017, LNCS 10665, pp. 137–144, 2018.
https://doi.org/10.1007/978-3-319-73441-5_14

posteriori error estimator, which is available without additional costs and useful for adaptive meshing strategies.

These benefits led to various investigations of the LSFEM especially in the field of fluid dynamics in recent years. Thereby, the focus of extensive research was drawn to three mixed least-squares formulations, which differ in the combination of the solution variables: The VVP formulation in terms of vorticity, velocity and pressure (see e.g. [2–4]), the SVP formulation in terms of stress, velocity and pressure (see e.g. [5–7]) and the UVP formulation including the velocity gradient, velocity and pressure (see e.g. [8–10]).

SVP and stress-velocity (SV) formulations are promising approaches when the stresses are of special interest, e.g. for non-Newtonian, multiphase or turbulent flows. The SVP formulation is the basis of the discussed time-dependent SV formulation in the scope of the present work, as an extension of the SV formulation for stationary fluid flow investigated in [11]. The formulation in terms of stress and velocity is constructed by manipulation of the residual terms based on the findings in [12]. A central motivation of constructing SV formulations is to reduce the size of the system matrices, in order to solve a given boundary value problem at less cost. Furthermore, SV formulations are advantageous for fluid-structure interaction approaches, since the coupling conditions impose for the equality of the tractions and velocities on the interface of the fluid and the solid, see [13].

In the following section the stress-velocity least-squares formulation for the incompressible instationary Navier-Stokes equations is introduced. Next, the essential discretization and implementation aspects are discussed and finally the proposed formulation is numerically evaluated for the benchmark problem of a flow around a cylinder, see e.g. [14].

2 Least-Squares Finite Element Method for the Navier-Stokes Equations

This article is concerned with a least-squares mixed finite element method for first-order systems. Before constructing the LS functional some definitions and notations are given: The scalar multiplication of two matrices $\boldsymbol{B}, \boldsymbol{C} \in \mathbb{R}^{d \times d}$ and two vectors $\boldsymbol{b}, \boldsymbol{c} \in \mathbb{R}^d$ is given by the sum of component-wise multiplications, i.e. $(\boldsymbol{B}, \boldsymbol{C}) = \boldsymbol{B} \cdot \boldsymbol{C} = \text{tr}(\boldsymbol{B}\boldsymbol{C}^T)$ and $(\boldsymbol{b}, \boldsymbol{c}) = \boldsymbol{b} \cdot \boldsymbol{c} = \boldsymbol{b}^T \boldsymbol{c}$, respectively. Based on this, we define the L^2-norm of a matrix or a vector by the volume integral of the corresponding scalar multiplication, e.g.

$$||\boldsymbol{b}||_0 = \left\{ \int_{\mathcal{B}} \boldsymbol{b} \cdot \boldsymbol{b} \, \mathrm{d}V \right\}^{\frac{1}{2}} \quad \text{and} \quad ||\boldsymbol{B}||_0 = \left\{ \int_{\mathcal{B}} \boldsymbol{B} \cdot \boldsymbol{B} \, \mathrm{d}V \right\}^{\frac{1}{2}}. \tag{1}$$

In general, the least-squares method is to find the minimizer $\boldsymbol{b}_{LS}$ of a quadratic functional $\mathcal{F}(\boldsymbol{b})$, in order to obtain an equivalent variational statement, by solving the resulting optimization problem

$$\boldsymbol{b}_{LS} = \underset{\boldsymbol{b} \in \boldsymbol{X}}{\text{argmin}} \, \mathcal{F}(\boldsymbol{b}), \tag{2}$$

where $\boldsymbol{X}$ denotes the minimization space. With this at hand, we can construct the functional $\mathcal{F}(\boldsymbol{b})$ in terms of the square of the L^2-norm, such that

$$\mathcal{F}(\boldsymbol{b}) = \frac{1}{2}||\mathcal{R}(\boldsymbol{b})||_0^2. \tag{3}$$

The problem of minimization is solved using the calculus of variations with the condition that the first variation of the functional

$$\delta_{\boldsymbol{b}}\mathcal{F}(\boldsymbol{b};\boldsymbol{v}) = \lim_{\epsilon\to 0}\frac{\mathcal{F}(\boldsymbol{b}+\epsilon\,\boldsymbol{v})-\mathcal{F}(\boldsymbol{b})}{\epsilon} = (\mathcal{R}(\boldsymbol{b}),\mathcal{D}\mathcal{R}(\boldsymbol{b})[\boldsymbol{v}])_0 \quad \forall \boldsymbol{v}\in\boldsymbol{X} \tag{4}$$

equals to zero. All quantities $\boldsymbol{b}$ are defined in d dimensions on a domain $\mathcal{B}$, which is parameterized in $\boldsymbol{x}\in\mathbb{R}^d$. In order to find the minimizer $\boldsymbol{b}_{LS}$, we need a linearization step, such that the solution can be found with help of the Newton method, for instance.

The least-squares finite element formulation, which is investigated in this article, is based on the incompressible Navier-Stokes equations consisting of the balance of momentum, given by

$$-\rho\boldsymbol{a} - \rho\nabla\boldsymbol{v}\boldsymbol{v} + 2\rho\nu\,\mathrm{div}(\nabla^s\boldsymbol{v}) - \nabla p = \boldsymbol{0}, \tag{5}$$

and the continuity equation, which is

$$\mathrm{div}\ \boldsymbol{v} = 0. \tag{6}$$

In here, $\boldsymbol{v}$ denotes the velocity vector, $\boldsymbol{a} = \dot{\boldsymbol{v}}$ the acceleration vector, p the pressure, ρ the density and ν the kinematic viscosity of a medium flowing through a domain $\mathcal{B}$. The definition of the symmetric gradient of the velocity field is given by

$$\nabla^s\boldsymbol{v} = \frac{1}{2}(\nabla\boldsymbol{v} + (\nabla\boldsymbol{v})^T). \tag{7}$$

These governing equations need to be reformulated in order to obtain the well-known stress-velocity-pressure formulation according to [12]. Therefore, the Cauchy stress tensor $\boldsymbol{\sigma}$ is introduced as an additional variable to the system of equations, leading to the three residual forms

$$\mathcal{R}_1 := \mathrm{div}\ \boldsymbol{\sigma} - \rho\boldsymbol{a} - \rho\nabla\boldsymbol{v}\boldsymbol{v}, \quad \mathcal{R}_2 := \boldsymbol{\sigma} - 2\rho\nu\nabla^s\boldsymbol{v} + p\boldsymbol{1}, \quad \mathcal{R}_3 := \mathrm{div}\ \boldsymbol{v}. \tag{8}$$

Application of the square of the L^2-norm leads to

$$\begin{aligned}\mathcal{F}_{SVP}(\boldsymbol{\sigma},\boldsymbol{v},p) := \frac{1}{2}\Bigg(&\left|\left|\frac{1}{\sqrt{\rho}}(\mathrm{div}\ \boldsymbol{\sigma} - \rho\boldsymbol{a} - \rho\nabla\boldsymbol{v}\,\boldsymbol{v})\right|\right|_0^2 \\ &+ \left|\left|\frac{1}{\sqrt{\rho\nu}}(\boldsymbol{\sigma} - 2\rho\nu\nabla^s\boldsymbol{v} + p\boldsymbol{1})\right|\right|_0^2 + ||\,\mathrm{div}\ \boldsymbol{v}||_0^2\Bigg)\end{aligned} \tag{9}$$

with the additional weightings according to the elaborations in [15].

2.1 Stress-Velocity Formulation

In the following a stress-velocity formulation is derived based on the findings in [12]. Here, a redundant residual is added to the functional but without additional variables in order to strengthen specific physical relations, see e.g. [11,16]. For the derivation of the SV approach the equations given in (8) are considered as the starting point. Taking the trace of the second residual in (8) and using the continuity Eq. (6) yields $p = -1/3\,\mathrm{tr}(\boldsymbol{\sigma})$. We can reduce the system of equations by inserting this definition of the pressure into Eq. $(8)_2$. Furthermore, assuming the definition of the deviatoric part of the Cauchy stresses $\mathrm{dev}\,\boldsymbol{\sigma} = \boldsymbol{\sigma} - 1/3\,\mathrm{tr}(\boldsymbol{\sigma})\,\mathbf{1}$ we obtain the remaining residuals of the stress-velocity two-field formulation with

$$\mathcal{R}_1 := \mathrm{div}\,\boldsymbol{\sigma} - \rho\boldsymbol{a} - \rho\nabla\boldsymbol{v}\,\boldsymbol{v} \qquad \text{and} \qquad \mathcal{R}_2 := \mathrm{dev}\,\boldsymbol{\sigma} - 2\rho\nu\nabla^s\boldsymbol{v}, \tag{10}$$

see e.g. [12]. Applying the square of the L^2-norm, we obtain the related functional

$$\tilde{\mathcal{F}}_{SV}(\boldsymbol{\sigma},\boldsymbol{v}) = \frac{1}{2}\left(\left|\left|\frac{1}{\sqrt{\rho}}(\mathrm{div}\,\boldsymbol{\sigma} - \rho\boldsymbol{a} - \rho\nabla\boldsymbol{v}\boldsymbol{v})\right|\right|_0^2 + \left|\left|\frac{1}{\sqrt{\rho\nu}}(\mathrm{dev}\,\boldsymbol{\sigma} - 2\rho\nu\nabla^s\boldsymbol{v})\right|\right|_0^2\right). \tag{11}$$

This formulation has the drawback of a poor approximation quality especially with respect to mass conservation, as in [15]. Hence, [11] developed an enhanced formulation, taking a further residual into account. Based on that we introduce the continuity equation again as an additional condition, such that we end up with a system of the form

$$\mathcal{R}_1 := \mathrm{div}\,\boldsymbol{\sigma} - \rho\boldsymbol{a} - \rho\nabla\boldsymbol{v}\,\boldsymbol{v}, \quad \mathcal{R}_2 := \mathrm{dev}\,\boldsymbol{\sigma} - 2\rho\nu\nabla^s\boldsymbol{v}, \quad \mathcal{R}_3 := \mathrm{div}\,\boldsymbol{v}. \tag{12}$$

Analogous considerations were made by [17] for the Stokes equations, but without giving a detailed study on the influence on the numerical results of the third residual in contrast to the derived SV formulation. For more information about the influence of weighting the third residual we refer to [18].

Furthermore, before minimizing we also have to discretize the time-dependent variables in (12). Here, we apply the well-known Newmark method. The index n defines a value of the previous step, the index $n+1$ of the actual time step. For the sake of completeness the v-form of the Newmark method is given as

$$\begin{aligned} \boldsymbol{u}_{n+1} &= \boldsymbol{u}_n + \boldsymbol{v}_n\Delta t + \Delta t^2(\frac{1}{2} - \beta)\boldsymbol{a}_n + \frac{\beta\Delta t}{\gamma}(\boldsymbol{v}_{n+1} - \boldsymbol{v}_n - (1-\gamma)\boldsymbol{a}_n\Delta t) \\ \boldsymbol{a}_{n+1} &= \frac{1}{\gamma\Delta t}\boldsymbol{v}_{n+1} - \frac{1}{\gamma\Delta t}(\boldsymbol{v}_n + (1-\gamma)\boldsymbol{a}_n\Delta t), \end{aligned} \tag{13}$$

where the actual acceleration $\boldsymbol{a}_{n+1}$ and displacement $\boldsymbol{u}_{n+1}$ only depends on values of the previous time step and the actual velocity. In the framework of this work we choose the two Newmark parameters as $\gamma = 0.5$ and $\beta = 0.25$. Finally, the related two-field functional then reads

$$\mathcal{F}_{SV}(\boldsymbol{\sigma},\boldsymbol{v}) := \frac{1}{2}\left(\left|\left|\frac{\sqrt{\Delta t}}{\sqrt{\rho}}(\text{div }\boldsymbol{\sigma} - \rho\boldsymbol{a} - \rho\nabla\boldsymbol{v}\boldsymbol{v})\right|\right|_0^2 + \left|\left|\frac{1}{\sqrt{\rho\nu}}(\text{dev }\boldsymbol{\sigma} - 2\rho\nu\nabla^s\boldsymbol{v})\right|\right|_0^2 + ||\text{div }\boldsymbol{v}||_0^2\right), \tag{14}$$

where all quantities are evaluated at the the actual time step $(n+1)$. Here, we introduce the weightings in analogy to (9) and (11) and in addition to that a weighting with respect to the time dependency using Δt for the first residual.

3 Discretization in Space

The discretization in space for the mixed least-squares formulation in terms of stresses and velocities follows the assumptions in [15]. The approximation spaces are chosen as $\mathbf{W} = \{\boldsymbol{\sigma} \in H(\text{div}, \mathcal{B})^2\}$ and $\mathbf{V} = \{\boldsymbol{v} \in H^1(\mathcal{B})^2\}$, with the velocities $\boldsymbol{v}$ being approximated with standard Lagrange interpolation functions in H^1 and the stresses $\boldsymbol{\sigma}$ in $H(\text{div})$ being interpolated with vector-valued Raviart-Thomas interpolation functions, see [19]. This leads to the discretized least-squares finite element description RT_mP_k for the discretization of the SV two-field problem with

$$\mathbf{W}_m^h = \{\boldsymbol{\sigma} \in H(\text{div}, \mathcal{B})^2 : \boldsymbol{\sigma}|_{\mathcal{B}_e} \in RT_m(\mathcal{B}_e)^2 \;\forall\, \mathcal{B}_e\} \subseteq \mathbf{W}. \tag{15}$$

$$\mathbf{V}_k^h = \{\boldsymbol{v} \in H^1(\mathcal{B})^2 : \boldsymbol{v}|_{\mathcal{B}_e} \in P_k(\mathcal{B}_e)^2 \;\forall\, \mathcal{B}_e\} \subseteq \mathbf{V}, \tag{16}$$

The stress-velocity formulation is investigated for the finite elements of polynomial order RT_1P_3 in two dimensions.

As mentioned before, the minimization of the time-discretized least-squares functional requires the first variation to be zero. Hence, we seek the solution in $\mathbf{W}_m^h \times \mathbf{V}_k^h$ by means of $\delta\mathcal{F} = 0$. Here, the dependency of $\boldsymbol{a}_{n+1}$ on $\boldsymbol{v}_{n+1}$ in $(13)_2$ leads to $\delta\boldsymbol{a}_{n+1} = 1/(\gamma\Delta t)\delta\boldsymbol{v}_{n+1}$. The first variation of the functional with respect to each variable is given by

$$\begin{aligned}\delta_\sigma\mathcal{F}_{SV}(\boldsymbol{\sigma},\boldsymbol{v};\delta\boldsymbol{\sigma}) = &\int_{\mathcal{B}} \frac{\Delta t}{\rho}\text{div }\delta\boldsymbol{\sigma}\cdot(\text{div }\boldsymbol{\sigma} - \rho\boldsymbol{a} - \rho\nabla\boldsymbol{v}\boldsymbol{v})\text{d}V \\ &+ \int_{\mathcal{B}} \frac{1}{\rho\nu}\text{dev }\delta\boldsymbol{\sigma}\cdot(\text{dev }\boldsymbol{\sigma} - 2\rho\nu\nabla^s\boldsymbol{v})\text{d}V = 0,\end{aligned} \tag{17}$$

$$\begin{aligned}\delta_v\mathcal{F}_{SV}(\boldsymbol{\sigma},\boldsymbol{v};\delta\boldsymbol{v}) = &-\int_{\mathcal{B}} \frac{\Delta t}{\rho}(\rho\nabla\delta\boldsymbol{v}\boldsymbol{v} + \rho\nabla\boldsymbol{v}\delta\boldsymbol{v} + \rho\delta\boldsymbol{a})\cdot(\text{div }\boldsymbol{\sigma} - \rho\boldsymbol{a} - \rho\nabla\boldsymbol{v}\boldsymbol{v})\text{d}V \\ &- \int_{\mathcal{B}} \frac{1}{\rho\nu}2\rho\nu\nabla^s\delta\boldsymbol{v}\cdot(\text{dev }\boldsymbol{\sigma} - 2\rho\nu\nabla^s\boldsymbol{v})\text{d}V \\ &+ \int_{\mathcal{B}} \text{div }\delta\boldsymbol{v}\cdot\text{div }\boldsymbol{v}\text{d}V = 0.\end{aligned} \tag{18}$$

In order to solve the nonlinear problem with the Newton method we have to linearize the equations. The resulting Newton tangent can be established analytically, by an automated differentiation approach or numerically, e.g. in form of a standard difference quotient procedure (used in this work). Finally, the complete system of algebraic equations is obtained by a standard assembly operation and it should be noted that there is no local elimination of unknowns on the element level involved.

4 Numerical Example—Flow Around a Cylinder

The numerical investigation of the proposed mixed least-squares formulation in terms of stresses and velocities for a time-dependent incompressible fluid flow is carried out by solving the flow around a cylinder problem in two dimensions, see [14]. In the following all units are SI-units (kg, m, s). The geometry of the channel and the boundary conditions are shown in Fig. 1. The fluid density is defined as $\rho = 1.0$ and the kinematic viscosity as $\nu = 0.001$ yielding the Reynolds number $Re = \bar{v}d/\nu = 100$, with the mean velocity $\bar{v} = 2/3v_1(h/2)$ and the cylinder diameter $d = 0.1$. The evaluation of the problem includes the drag coefficient $c_D = (2F_D)/(\rho\bar{v}^2 d)$ and the lift coefficient $c_L = (2F_L)/(\rho\bar{v}^2 d)$ with the drag force F_D and the lift force F_L acting on the cylinder wall.

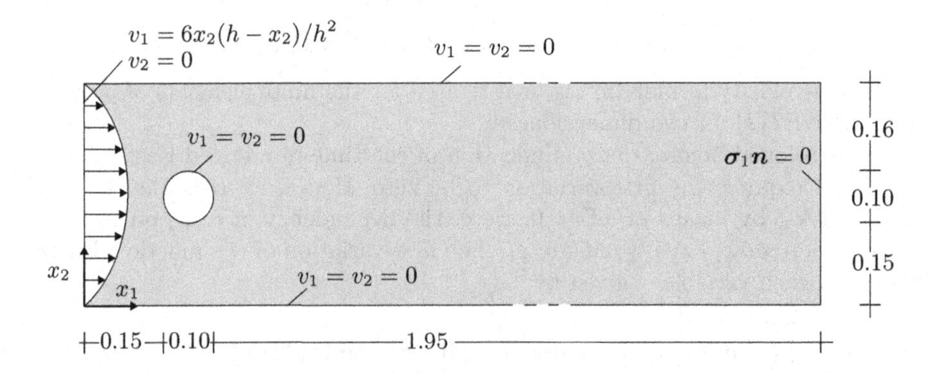

Fig. 1. Geometry and boundary conditions of flow around a cylinder in 2D

We use RT_1P_3 elements with a time step size $\Delta t = 0.005$ and the domain is discretized using an unstructured mesh with 4631 elements. To validate the formulation, the temporal evolutions of the drag and lift coefficients are compared to a reference solution by [14], who present the maximum values of the coefficients. The results of the drag and lift coefficients are in line with the reference solutions and are plotted over time in Fig. 2. The lift coefficient oscillates between −1 and 1 with a mean value very close to zero, whereas the drag coefficient starts at around 2.95, slightly rises and then oscillates between 3.16 and 3.22.

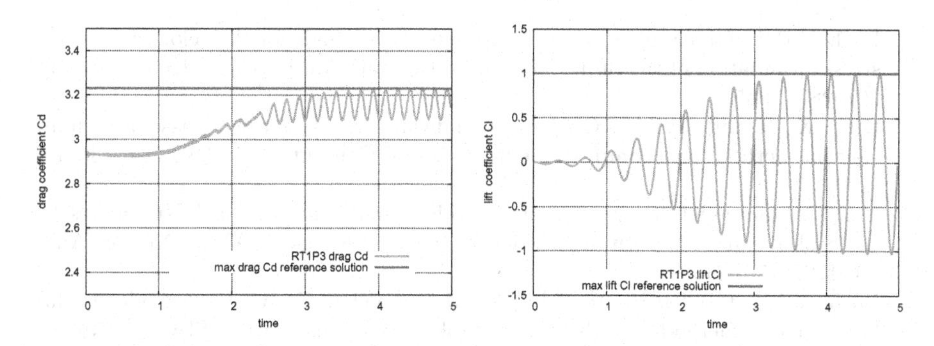

Fig. 2. Maximal drag and lift coefficient related to the cylinder for $Re = 100$

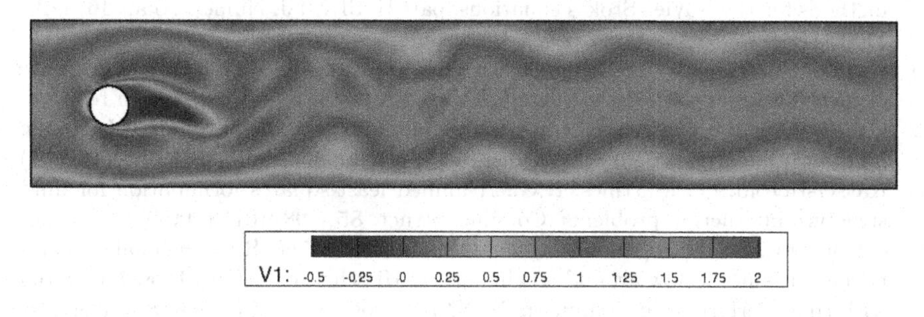

Fig. 3. Horizontal velocity v_1 at time $t = 5\,\mathrm{s}$ for $Re = 100$

The time-dependent behavior of both coefficients reflects the beginning of instability behind the cylinder with increasing velocity, leading to a periodic flow pattern that is characterized by a certain frequency. Furthermore, the velocity field at $t = 5\,\mathrm{s}$ is plotted in Fig. 3, visualizing the typical Von Kármán vortex street.

Acknowledgement. This work was supported by the German Research Foundation (DFG) under grant SCHW1355/3-1 and SCHR570/31-1.

References

1. Bochev, P.B., Gunzburger, M.D.: Least-Squares Finite Element Methods, vol. 166. Springer, New York (2009). https://doi.org/10.1007/b13382
2. Jiang, B.-N., Chang, C.L.: Least-squares finite elements for the Stokes problem. Comput. Meth. Appl. Mech. Eng. **78**, 297–311 (1990)
3. Bochev, P.B.: Analysis of least-squares finite element methods for the Navier-Stokes equations. SIAM J. Numer. Anal. **34**, 1817–1844 (1997)
4. Jiang, B.-N.: The Least-Squares Finite Element Method. Scientific Computation. Springer, Berlin (1998). https://doi.org/10.1007/978-3-662-03740-9
5. Bell, B.C., Surana, K.S.: p-version least squares finite element formulation for two-dimensional, incompressible, non-Newtonian isothermal and non-isothermal fluid flow. Int. J. Numer. Meth. Fluids **18**, 127–162 (1994)

6. Bochev, P.B., Gunzburger, M.D.: Least-squares methods for the velocity-pressure-stress formulation of the Stokes equations. Comput. Meth. Appl. Mech. Eng. **126**, 267–287 (1995)
7. Ding, X., Tsang, T.T.H.: On first-order formulations of the least-squares finite element method for incompressible flows. Int. J. Comput. Fluid Dyn. **17**, 183–197 (2003)
8. Cai, Z., Manteuffel, T.A., McCormick, S.F.: First-order system least squares for the Stokes equation, with application to linear elasticity. SIAM J. Numer. Anal. **34**, 1727–1741 (1997)
9. Bochev, P.B., Cai, Z., Manteuffel, T.A., McCormick, S.F.: Analysis of velocity-flux least-squares methods for the Navier-Stokes equations, part I. SIAM J. Numer. Anal. **35**, 990–1009 (1998)
10. Bochev, P.B., Manteuffel, T., McCormick, S.: Analysis of velocity-flux least squares methods for the Navier-Stokes equations, part II. SIAM J. Numer. Anal. **36**, 1125–1144 (1999)
11. Nisters, C., Schwarz, A., Schröder, J.: Efficient stress-velocity least-squares finite element formulations for the incompressible Navier-Stokes equations (in revision)
12. Cai, Z., Lee, B., Wang, P.: Least-squares methods for incompressible Newtonian fluid flow: linear stationary problems. SIAM J. Numer. Anal. **42**, 843–859 (2004)
13. Kayser-Herold, O., Matthies, H.G.: A unified least-squares formulation for fluid-structure interaction problems. Comput. Struct. **85**, 998–1011 (2007)
14. Schäfer, M., Turek, S., Durst, F., Krause, E., Rannacher, R.: Benchmark computations of laminar flow around a cylinder. In: Hirschel, E.H. (ed.) Flow Simulation with High-Performance Computers II. NNFM, vol. 48, pp. 547–566. Springer, Heidelberg (1996). https://doi.org/10.1007/978-3-322-89849-4_39
15. Schwarz, A., Nickaeen, M., Serdas, S., Nisters, C., Ouazzi, A., Schröder, J., Turek, S.: A comparative study of mixed least-squares FEMs for the incompressible Navier-Stokes equations. Int. J. Comput. Sci. Eng. (2017, in press)
16. Schwarz, A., Steeger, K., Schröder, J.: Weighted overconstrained least-squares mixed finite elements for static and dynamic problems in quasi-incompressible elasticity. Comput. Mech. **54**, 603–612 (2014)
17. Münzenmaier, S., Starke, G.: First-order system least squares for coupled Stokes-Darcy flow. SIAM J. Numer. Anal. **49**, 387–404 (2011)
18. Deang, J.M., Gunzburger, M.D.: Issues related to least-squares finite element methods for the Stokes equations. SIAM J. Sci. Comput. **20**, 878–906 (1998)
19. Raviart, P.A., Thomas, J.M.: A mixed finite element method for 2-nd order elliptic problems. In: Galligani, I., Magenes, E. (eds.) Mathematical Aspects of Finite Element Methods. LNM, vol. 606, pp. 292–315. Springer, Heidelberg (1977). https://doi.org/10.1007/BFb0064470

Advances in Heterogeneous Numerical Methods for Multi Physics Problems

A Virtual Control Coupling Approach for Problems with Non-coincident Discrete Interfaces

Pavel Bochev(✉), Paul Kuberry, and Kara Peterson

Center for Computational Research, Sandia National Laboratories, Albuquerque, NM 87125, USA
{pbboche,pakuber,kjpeter}@sandia.gov

Abstract. Independent meshing of subdomains separated by an interface can lead to spatially non-coincident discrete interfaces. We present an optimization-based coupling method for such problems, which does not require a common mesh refinement of the interface, has optimal H^1 convergence rates, and passes a patch test. The method minimizes the mismatch of the state and normal stress extensions on discrete interfaces subject to the subdomain equations, while interface "fluxes" provide virtual Neumann controls.

Keywords: PDE constrained optimization · Mesh tying
Transmission · Non-coincident interfaces · Optimal control
Virtual Neumann controls

1 Introduction

Solution of elliptic problems on two or more non-overlapping subdomains, subject to coupling conditions, occurs in multiple contexts. Independent meshing of these subdomains induces independent mesh partitions of the interface. In the more benign case the interface grids are non-matching but spatially coincident. However, when the interface is curved the induced interface grids may be spatially non-coincident, leading to gaps and/or overlaps between them. This complicates the accurate numerical solution of the coupled problem [1,2]. We present a new, optimization-based formulation, which avoids some difficulties associated with the application of domain decomposition methods [3,4] to such problems. Following [5,6], we switch the roles of the coupling conditions and the subdomain equations by couching the interface problem into a virtual control formulation in which the former define the objective, the latter define the constraints, and the interface flux serves as a Neumann control. Section 2 summarizes the

Sandia National Laboratories is a multimission laboratory managed and operated by National Technology and Engineering Solutions of Sandia, LLC., a wholly owned subsidiary of Honeywell International, Inc., for the U.S. Department of Energy's National Nuclear Security Administration under contract DE-NA-0003525.

I. Lirkov and S. Margenov (Eds.): LSSC 2017, LNCS 10665, pp. 147–155, 2018.
https://doi.org/10.1007/978-3-319-73441-5_15

germane notation and states the model interface problem. The optimization-based formulation, including the necessary state and flux extension operators are presented in Sect. 3, while Sect. 4 contains several representative numerical examples. Section 5 summarizes our findings.

2 Notation and Statement of the Problem

Consider a bounded open region $\Omega \subset \mathbf{R}^d$, $d = 2, 3$ with a Lipschitz continuous boundary Γ. An interface σ splits Ω into two non-overlapping subdomains Ω_1 and Ω_2 with Dirichlet boundaries $\Gamma_i = \partial\Omega_i \backslash \sigma$, $i = 1, 2$. We assume that each subdomain is endowed with an independently defined conforming finite element mesh Ω_i^h, $i = 1, 2$ with elements $\boldsymbol{k}_i^n$. These meshes induce finite element partitions σ_1^h and σ_2^h of the interface σ, containing the element sides $\boldsymbol{s}_i^n$ that have all their vertices in σ. The geometrical entities described by σ_1^h and σ_2^h are two different "versions" of the interface, denoted by σ_1 and σ_2, respectively. These entities and their associated finite element partitions are not required to match or to be spatially coincident; see Fig. 1. Given a mesh entity μ we denote the sets of all mesh vertices in μ by $V(\mu)$. For example $V(\sigma_i^h)$ are the vertices in the interface mesh σ_i^h and $V(\Omega_i^h)$ is the set of all vertices in the subdomain mesh Ω_i^h. If μ is a finite set, then $|\mu|$ is its dimension, e.g., $|\sigma_i^h|$ is the number of elements in σ_i^h. If μ is a geometric entity, then $|\mu|$ is its measure, e.g., $|\boldsymbol{k}_i^n|$ is the volume (or area) of an element $\boldsymbol{k}_i^n$.

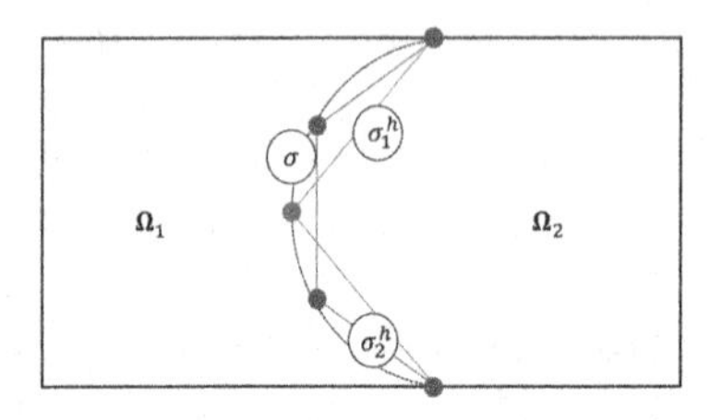

Fig. 1. Independent meshing of two subdomains separated by a curved interface σ results in two spatially non-coincident interface grids σ_1^h and σ_2^h.

We denote the standard Sobolev space of order one on Ω_i, $i = 1, 2$, and its subspace of functions with vanishing trace on Γ_i by $H^1(\Omega_i)$ and $H^1_{\Gamma_i}(\Omega_i)$, respectively. H_i^h is a conforming finite element subspace of $H^1(\Omega_i^h)$ with Lagrangian basis $\{N_i^k\}$, $H_{i,\Gamma}^h$ is a conforming subspace of $H^1_{\Gamma_i}(\Omega_i^h)$, $H_{i,\sigma}^h$ is the span of all basis functions associated with vertices on σ_i^h, and $T_i^h = H_{i,\sigma}^h\big|_{\sigma_i}$. The coefficient vector of $u_i^h \in H_i^h$ is $\boldsymbol{u}_i \in \mathbf{R}^{n_i}$, where $n_i = |H_i^h|$, the dimension of H_i^h.

In this paper we consider the model transmission problem

$$\begin{cases} -\nabla \cdot (\kappa_i \nabla u_i) = f_i \text{ in } \Omega_i, \ \ \mathrm{i} = 1, 2 & u_1 = u_2 \quad \text{on } \sigma \\ u_i = 0 \text{ on } \Gamma_i, \ \ \mathrm{i} = 1, 2 & \kappa_1 \nabla u_1 \cdot \mathbf{n} = \kappa_2 \nabla u_2 \cdot \mathbf{n} \ \text{ on } \sigma \end{cases} \tag{1}$$

where $\mathbf{n}$ is unit normal on σ and, for simplicity, κ_i is a positive constant on Ω_i. In this paper we develop stable and accurate methods for (1) that can handle spatially non-coincident interfaces $\sigma_1 \neq \sigma_2$. Our approach is based on the reformulation of (1) into a PDE-constrained optimization problem with virtual Neumann controls. We start by splitting (1) into a pair of subdomain equations with mixed Dirichlet and Neumann boundary conditions, and weak forms given by *seek* $u_i \in H^1_{\Gamma_i}(\Omega_i)$ *such that*

$$\kappa_i(\nabla u_i, \nabla v_i)_{\Omega_i} = (f_i, v_i)_{\Omega_i} + \langle g_i, v_i\rangle_\sigma \quad \forall v_i \in H^1_{\Gamma_i}(\Omega_i),\ i = 1,2. \tag{2}$$

We treat the Neumann data g_i as a virtual control and introduce the objective

$$\begin{aligned} J_\delta(u_1,u_2,g_1,g_2) &= \frac{1}{2}\int_\sigma (u_1-u_2)^2 dS + \frac{1}{2}\int_\sigma ((\kappa_1\nabla u_1 - \kappa_2 \nabla u_2)\cdot \mathbf{n})^2 dS \\ &+ \frac{\rho}{2}\left(\int_\sigma g_1 dS + \int_\sigma g_2 dS\right)^2 + \frac{\delta_1}{2}\int_\sigma g_1^2 dS + \frac{\delta_2}{2}\int_\sigma g_2^2 dS \end{aligned} \tag{3}$$

The reformulation of (4) is then given by the following optimization problem

$$\text{minimize } J_\delta(u_1,u_2,g_1,g_2) \text{ over } H^1_{\Gamma_1}(\Omega_1)\times H^1_{\Gamma_2}(\Omega_2)\times L^2(\sigma) \text{ subject to (2)}. \tag{4}$$

This problem provides the basis for our new method.

3 Virtual Control Formulation

For simplicity we consider C^0 piecewise linear elements on affine grids. When $\sigma_1 \neq \sigma_2$ we cannot discretize (4) directly because the interface integrals in (3) and (2) are undefined. We resolve this issue by using extension operators

$$E_{i,\gamma} : T_i^h \mapsto L^2(\sigma_\gamma) \text{ and } G_{i,\gamma} : \nabla T_i^h \mapsto [L^2(\sigma_\gamma)]^d, \quad \gamma \in \{1,2\}, \gamma \neq i. \tag{5}$$

to compare finite element fields defined on σ_i^h and their gradients to fields and gradients defined on σ_γ. The only requirement for these operators is consistency for linear and constant fields, respectively, i.e., $E_{i,\gamma}(p(\boldsymbol{x})) = p(\boldsymbol{x})\big|_{\sigma_\gamma^h}$ for all $p \in P_1(\mathbf{R}^d)$ and $G_{i,\gamma}(\boldsymbol{q}(\boldsymbol{x})) = \boldsymbol{q}(\boldsymbol{x})\big|_{\sigma_\gamma^h}$ for all $\boldsymbol{q} \in [P_0(\mathbf{R}^d)]^d$. A simple definition of $E_{i,\gamma}$, which satisfies this requirement is the linear extension

$$(E_{i,\gamma}u_i^h)(\boldsymbol{x}_\gamma) = \begin{cases} u_i^h(\boldsymbol{x}_\gamma^\perp) + \nabla u_i^h(\boldsymbol{x}_\gamma^\perp)\cdot(\boldsymbol{x}_\gamma - \boldsymbol{x}_\gamma^\perp) & \text{if } \boldsymbol{x}_\gamma \notin \Omega_i \\ u_i^h(\boldsymbol{x}_\gamma) & \text{if } \boldsymbol{x}_\gamma \in \Omega_i \end{cases} \tag{6}$$

where $\boldsymbol{x}_\gamma \in \sigma_\gamma$ is a given point and $\boldsymbol{x}_\gamma^\perp \in \sigma_i$ is the "closest" point on σ_i. Similarly, we define $G_{i,\gamma}$ to be an extension by a constant, i.e., given $\boldsymbol{x}_\gamma \in \sigma_\gamma$ we define

$$(G_{i,\gamma}\nabla u_i^h)(\boldsymbol{x}_\gamma) = \nabla u_i^h(\boldsymbol{x}_\gamma^\perp). \tag{7}$$

Finally, we note that although g_1 and g_2 belong in the same space $L^2(\sigma)$, their discretization requires two separate discrete control spaces $L^{2,h}_{1,\sigma}$ and $L^{2,h}_{2,\sigma}$, defined on σ_1 and σ_2, respectively. Here we choose $L^{2,h}_{i,\sigma}$ to be a piecewise constant space on σ_i^h, which is consistent with the piecewise linear discretization in Ω_i. These considerations yield the following extension of (3) to non-coincident interfaces:

$$
\begin{aligned}
J_\delta^h(u_1^h, u_2^h, g_1^h, g_2^h) &= \frac{\beta_1}{2}\int_{\sigma_1}(u_1^h - E_{2,1}u_2^h)^2 dS + \frac{\beta_2}{2}\int_{\sigma_2}(u_2^h - E_{1,2}u_1^h)^2 dS \\
&+ \frac{\gamma_1}{2}\int_{\sigma_1}((\kappa_1\nabla u_1^h - \kappa_2 G_{2,1}\nabla u_2^h)\cdot \mathbf{n}_1)^2 dS + \frac{\gamma_2}{2}\int_{\sigma_2}((\kappa_1 G_{1,2}\nabla u_1^h - \kappa_2\nabla u_2^h)\cdot \mathbf{n}_2)^2 dS \\
&+ \frac{\rho}{2}\left(\int_{\sigma_1} g_1^h dS + \int_{\sigma_2} g_2^h dS\right)^2 + \frac{\delta_1}{2}\int_{\sigma_1}(g_1^h)^2 dS + \frac{\delta_2}{2}\int_{\sigma_2}(g_2^h)^2 dS.
\end{aligned}
\tag{8}
$$

The first two pairs of terms in (8) generalize the state misfit and the flux misfit terms in (3), and the fifth term controls the total flux misfit between the interfaces. The last two terms generalize the control penalties necessary for the well-posedness of the optimization problem. The discretization of (4) on non-coincident interfaces is thus given by the following problem:

$$
\begin{array}{c}
\text{minimize} \quad J_\delta^h(u_1^h, u_2^h, g_1^h, g_2^h) \ \text{over} H_{1,\Gamma}^h \times H_{2,\Gamma}^h \times L_{1,\sigma}^{2,h} \times L_{2,\sigma}^{2,h} \\
\text{subject to a discretized form of the weak Eq. (2).}
\end{array}
\tag{9}
$$

Recovery of globally linear fields is desirable for any numerical method for (1). However, in order to pass this linear "patch test", methods based on Lagrange multipliers require carefully constructed multiplier spaces [7] and/or additional modifications of the interface grids [4,8]. An attractive property of (9) is that it does not require any additional considerations to pass a patch test: recovery of globally linear fields is built into the virtual control formulation.

Theorem 1. *Assume that $\kappa_1 = \kappa_2$ and that the discrete interfaces have matching boundaries, i.e., $\partial\sigma_1 = \partial\sigma_2$. Then, in the limit $\delta_i \to 0$, (9) recovers exactly any globally linear solution u_ℓ of (1).*

Proof. We show that $u_{i,\ell}^h = u_\ell\big|_{\sigma_i}$ and $g_{i,\ell}^h = \mathbf{n}_i \cdot \nabla u_{i,\ell}^h = \mathbf{n}_i \cdot \nabla u_\ell\big|_{\sigma_i}$, $i = 1, 2$ is an optimal solution of (9). Since any conforming discretization of (2) recovers linear solutions, $u_{i,\ell}^h$ is feasible. By construction $E_{i,\gamma}$ and $G_{i,\gamma}$ are exact for linear and constant fields, respectively and so, the first four terms in (8) vanish, i.e.,

$$
\begin{aligned}
&J_\delta^h(u_{1,\ell}^h, u_{2,\ell}^h, g_{1,\ell}^h, g_{2,\ell}^h) = \\
&\frac{\rho}{2}\left(\int_{\sigma_1} g_{1,\ell}^h dS + \int_{\sigma_2} g_{2,\ell}^h dS\right)^2 + \frac{\delta_1}{2}\int_{\sigma_1}(g_{1,\ell}^h)^2 dS + \frac{\delta_2}{2}\int_{\sigma_2}(g_{2,\ell}^h)^2 dS.
\end{aligned}
$$

Since u_ℓ is linear $\nabla u_{i,\ell}^h = \boldsymbol{c}$ for some $\boldsymbol{c} \in \mathbf{R}^d$, $d = 2, 3$. Let $\boldsymbol{u}_\ell^\perp \in \mathbf{R}^3$ be a linear vector field such that $\nabla \times \boldsymbol{u}_\ell^\perp = \boldsymbol{c}$. Stokes' theorem and $\partial\sigma_1 = \partial\sigma_2$ imply that

$$
\begin{aligned}
\int_{\sigma_1} \mathbf{n}_1 \cdot \nabla u_{2,\ell}^h dS &= \int_{\sigma_1} \mathbf{n}_1 \cdot \nabla \times \boldsymbol{u}_\ell^\perp dS = \int_{\partial\sigma_1} \boldsymbol{u}_\ell^\perp \cdot dl \\
&= -\int_{\partial\sigma_2} \boldsymbol{u}_\ell^\perp \cdot dl = -\int_{\sigma_2} \mathbf{n}_2 \cdot \nabla \times \boldsymbol{u}_\ell^\perp dS = -\int_{\sigma_2} \mathbf{n}_2 \cdot \nabla u_{2,\ell}^h dS,
\end{aligned}
$$

In two-dimensions the same identity follows by choosing a linear function $u_\ell^\perp$ such that $\nabla u_\ell^\perp = \mathbf{c}^\perp = (-c_2, c_1)$. Thus, we have that

$$J_\delta^h(u_{1,\ell}^h, u_{2,\ell}^h, g_{1,\ell}^h, g_{2,\ell}^h) = \frac{\delta_1}{2}\int_{\sigma_1}(g_{1,\ell}^h)^2 dS + \frac{\delta_2}{2}\int_{\sigma_2}(g_{2,\ell}^h)^2 dS.$$

The theorem follows by taking the limit $\delta_i \to 0$.

3.1 Solution of the Discrete Optimization Problem

Let $\boldsymbol{u}_i$, $\boldsymbol{g}_i$ denote the coefficient vectors of the states u_i^h and controls g_i^h, respectively. Setting $\vec{\boldsymbol{u}} = (\boldsymbol{u}_1, \boldsymbol{u}_2)$ and $\vec{\boldsymbol{g}} = (\boldsymbol{g}_1, \boldsymbol{g}_2)$, the virtual control formulation (9) is equivalent to the Quadratic Programming problem (QP)

$$\underset{\vec{\boldsymbol{u}},\vec{\boldsymbol{g}}}{\text{minimize}} \quad \mathbf{J}_\delta(\vec{\boldsymbol{u}}, \vec{\boldsymbol{g}}) \quad \text{subject to} \quad \begin{cases} K_1\boldsymbol{u}_1 = \boldsymbol{f}_1 - G_1\boldsymbol{g}_1 \\ K_2\boldsymbol{u}_2 = \boldsymbol{f}_2 + G_2\boldsymbol{g}_2 \end{cases}, \tag{10}$$

where K_i is the finite element stiffness matrix, $\boldsymbol{f}_i$ is the finite element load vector, $\boldsymbol{g}_i$ is the external load vector induced by the control g_i, and

$$\mathbf{J}_\delta(\vec{\boldsymbol{u}}, \vec{\boldsymbol{g}}) = \frac{1}{2}\vec{\boldsymbol{u}}^T H \vec{\boldsymbol{u}} + \vec{\boldsymbol{g}}^T M \vec{\boldsymbol{g}}$$

with suitable H and M. For clarity we have subsumed the weights β_i, γ_i and the penalty coefficients δ_i into the matrices H and M.

Because K_1 and K_2 are discretizations of mixed Dirichlet-Neumann boundary value problems they are invertible. Thus, we solve (10) by a reduced space approach, i.e., we eliminate the states by solving the constraint equations:

$$\boldsymbol{u}_i = K_i^{-1}(\boldsymbol{f}_i + \boldsymbol{g}_i), \quad i = 1, 2. \tag{11}$$

This yields an equivalent *unconstrained* optimization problem

$$\underset{\vec{\boldsymbol{g}}}{\text{minimize}} \qquad \frac{1}{2}\vec{\boldsymbol{g}}^T H_{\text{red}}\,\vec{\boldsymbol{g}} + \vec{\boldsymbol{g}}^T \boldsymbol{f}_{\text{red}}\,, \tag{12}$$

in terms of the virtual Neumann controls only. Setting the first variation of (12) to zero yields the following necessary condition

$$H_{\text{red}}\,\vec{\boldsymbol{g}} = \boldsymbol{f}_{\text{red}} \tag{13}$$

for the optimal virtual Neumann control. Since the dimensions of H_{red} and $\boldsymbol{f}_{\text{red}}$ equal the dimension of the virtual control vector $\vec{\boldsymbol{g}} = (\boldsymbol{g}_1, \boldsymbol{g}_2)$, the size of (13) is much smaller than the size of the optimality system of the original QP (10).

We solve (13) iteratively using GMRES, which requires the application of the reduced Hessian H_{red}. The latter involves multiple inversions of the stiffness matrices K_i. In our case, these matrices correspond to discretizations of second-order elliptic operators and so, they can be preconditioned by a number of algebraic and geometric multigrid preconditioners. Once the solution $\vec{\boldsymbol{g}} = (\boldsymbol{g}_1, \boldsymbol{g}_2)$ has been computed, one can recover the state variables from (11).

4 Numerical Results

We present three preliminary numerical studies of the virtual control formulation (9). These studies verify Theorem 1 and examine the convergence rates of the virtual control formulation for different interface configurations. In all cases we discretize the subdomain Eq. (2) using independently defined partitions Ω_i^h of Ω_i into affine triangles and standard C^0 piecewise linear nodal elements. Then we solve the QP (10) using the equivalent reduced-space formulation (12). This involves solving the optimality system (13) for the two Neumann controls by GMRES and then recovering the optimal states. We solve the reduced Hessian system to a relative residual of 1e−15. The optimization-based method is implemented in FreeFem++ [9].

Linear patch test. The first study confirms numerically Theorem 1, i.e., the ability of (9) to recover globally linear solutions. To this end, we set $u = 3x + 2y$ and define the Dirichlet boundary condition data and the right hand side by inserting this solution in (1). Then we set $\delta_1 = \delta_2 = 0$ and solve (9) for several different interface configurations. In general the well-posedness of (9) may require positive penalty parameters. However, in the case of the linear patch test, the optimization problem remained well-posed with $\delta_1 = \delta_2 = 0$. We note that in some related contexts, such as optimization-based additive operator splitting [10], one can prove that the associated optimization formulation is well-posed without control penalties. In all cases (9) recovers the exact solution to machine precision. Figure 2 shows this solution when the induced interface grids have a 2:3 ratio of elements. Despite the obvious gaps and overlaps between the interface grids we see a perfect recovery of the linear function.

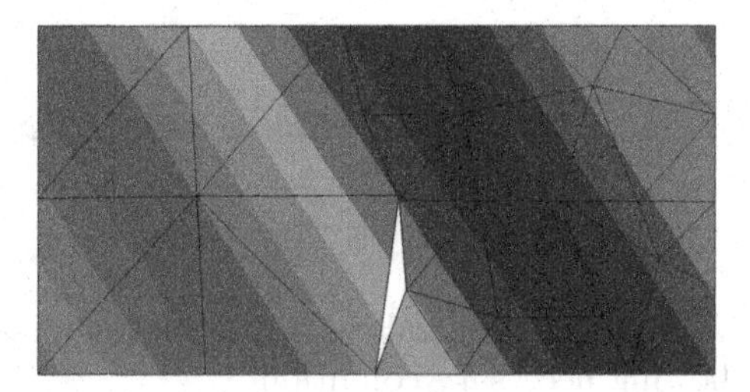

Fig. 2. Globally linear solution for an S-curve interface containing large gaps and overlaps.

Convergence study. To study the convergence of (9) we use the method of manufactured solutions on a domain with an S-curve interface; see Fig. 3. We set the exact solution of (1) to be the following function:

$$u = x^2(y-2)^3 \sin(2\pi x) - (x-3)^3 \cos(2\pi x - y). \tag{14}$$

Substitution of (14) into the interface problem (1) defines the right hand sides and Dirichlet boundary conditions for the subdomain problems. We measure the errors of the optimal finite element state variables u_i^h against the exact solution u_{ex} of (1) using sums of L^2 and H^1 norms on the discretized subdomains, i.e., we consider the following compound error norms:

$$\left\|u_i^h - u_{ex}\right\|_0^2 := \sum_{i=1}^{2} \left\|u_i^h - u_{ex}\right\|_{0,\Omega_i^h}^2 ; \quad \left\|u_i^h - u_{ex}\right\|_1^2 := \sum_{i=1}^{2} \left\|u_i^h - u_{ex}\right\|_{1,\Omega_i^h}^2 . \tag{15}$$

This study investigates the accuracy of the method when subdomain meshes have different resolutions. We consider several combinations of Ω_i^h providing a representative range of ratios $|\sigma_1^h|{:}|\sigma_2^h|$ between the numbers of elements in the discrete interfaces σ_1^h and σ_2^h. We compute the optimal finite element states on a sequence of six successively refined grids on Ω_1 and Ω_2. While the grids are defined independently on each subdomain by using the FreeFem++ mesh generator, the ratio of their interface segments $|\sigma_1^h|{:}|\sigma_2^h|$ is kept constant. This is accomplished by starting with an initial vertex distribution along $\partial\Omega_1$ and $\partial\Omega_2$, which produces the desired ratio $|\sigma_1^h|{:}|\sigma_2^h|$, and then driving the mesh refinement through doubling the number of vertices on the subdomain boundaries. We consider a total of eight different ratios in this study. For all interface ratios in this study we set $\beta_1 = \beta_2 = \gamma_1 = \gamma_2 = \rho = 1$, and $\delta_1 = \delta_2 = 1\text{e}{-10}$ in the objective (8).

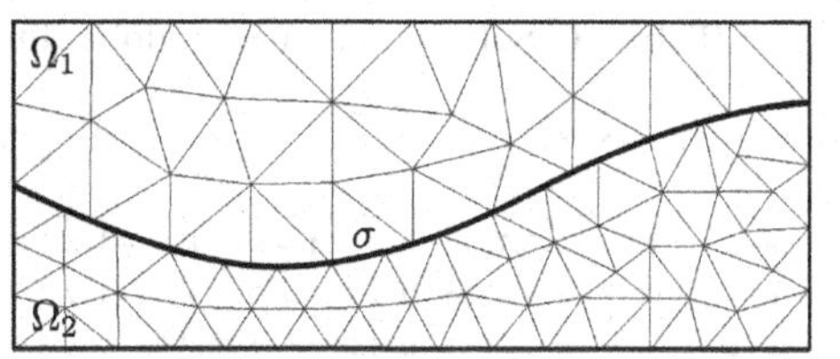

Fig. 3. An S-curve interface containing small gaps and overlaps. This is an example of grids having a 2:3 ratio of elements on the interface.

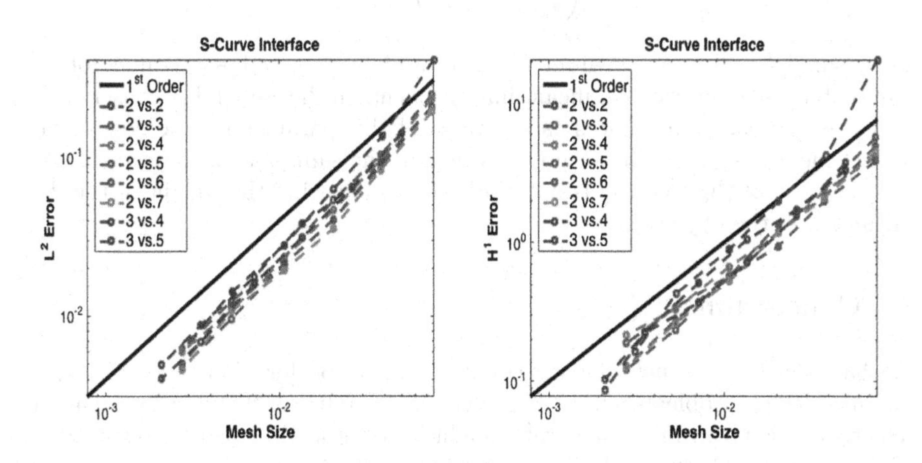

Fig. 4. Convergence rates of (9) for interface grids having different element ratios. In each case the interface element ratio $|\sigma_1^h|{:}|\sigma_2^h|$ is preserved throughout the grid refinement process.

Results in Fig. 4 reveal first order convergence in both compound norms. We believe that the suboptimal L^2 rate is due to the choice of piecewise constant controls g_i^h. Although this choice is enough to pass a linear patch test (see Theorem 1), it limits the accuracy of the finite element solution in the subdomain equations. In future work we will investigate a variant of the algorithm, which uses more accurate control representations.

Table 1. Solution error and global flux conservation as functions of ρ.

Parameter	$\rho=$ 1e−3			$\rho=1$			$\rho=$ 1e+3		
h	L^2	H^1	Δg	L^2	H^1	Δg	L^2	H^1	Δg
7.512e−2	2.495e−1	5.727e−0	3.264e−9	2.495e−1	5.727e−0	3.264e−15	2.495e−1	5.727e−0	3.267e−21
3.801e−2	1.036e−1	2.435e−0	1.408e−7	1.036e−1	2.435e−0	1.408e−13	1.036e−1	2.435e−0	1.408e−19
2.023e−2	5.337e−2	1.301e−0	4.072e−1	5.370e−2	1.349e−0	5.513e−7	5.370e−2	1.349e−0	5.515e−13
1.070e−2	2.356e−2	5.711e−1	1.118e−2	2.353e−2	5.751e−1	1.233e−8	2.353e−2	5.751e−1	1.234e−14
5.251e−3	1.240e−2	3.422e−1	3.154e−7	1.240e−2	3.422e−1	3.159e−13	1.240e−2	3.422e−1	3.159e−19
2.770e−3	6.236e−3	1.896e−1	2.772e−6	6.236e−3	1.896e−1	2.788e−12	6.236e−3	1.896e−1	2.788e−18

Flux conservation. Our last example examines global flux conservation across the interface as a function of the parameter ρ in the objective functional (8). We set $\delta_1 = \delta_2 =$ 1e−10 in (8) and use the S-curve interface in Fig. 3 with a sequence of refined grid from the convergence study with $|\sigma_1^h|{:}|\sigma_2^h|$ ratio of 2:3. Results in Table 1 compare the compound norm errors and global flux conservation, as measured by the global flux mismatch

$$\Delta g = \left(\int_{\sigma_1} g_1 \, dS - \int_{\sigma_2} g_2 \, dS \right)^2 ,$$

for a small ($\rho=$1e−3), medium ($\rho = 1$) and large ($\rho=$1e+3) values of the parameter ρ. We observe significant improvements in the global flux conservation over *non-coincident* interfaces as the value of this parameter increases. At the same time, the compound norm errors remain the same for all three cases, i.e., the accuracy of the solution is not affected by increasing the weight of the flux mismatch in the objective.

5 Conclusions

We have developed a new, virtual control formulation for discrete transmission and mesh tying problems with non-coincident discrete interfaces. The method is linearly consistent, while a moderate weight in the objective ensures conservation of the global flux between the subdomains to machine precision and *without any additional interface manipulations.* Preliminary results reveal first-order accuracy in compound L^2 and H^1 norms. Future work will consider more accurate choices for the virtual controls to improve the L^2 convergence rates.

Acknowledgments. This material is based upon work supported by the U.S. Department of Energy, Office of Science, Office of Advanced Scientific Computing Research, and the Laboratory Directed Research and Development program at Sandia National Laboratories.

References

1. Farhat, C., Lesoinne, M., Tallec, P.L.: Load and motion transfer algorithms for fluid/structure interaction problems with non-matching discrete interfaces: momentum and energy conservation, optimal discretization and application to aeroelasticity. Comput. Methods Appl. Mech. Engrg. **157**(1–2), 95–114 (1998)
2. de Boer, A., van Zuijlen, A., Bijl, H.: Review of coupling methods for non-matching meshes. Comput. Methods Appl. Mech. Engrg. **196**(8), 1515–1525 (2007)
3. Dohrmann, C.R., Key, S.W., Heinstein, M.W.: A method for connecting dissimilar finite element meshes in two dimensions. Int. J. Numer. Meth. Eng. **48**(5), 655–678 (2000)
4. Dohrmann, C.R., Key, S.W., Heinstein, M.W.: Methods for connecting dissimilar three-dimensional finite element meshes. Int. J. Numer. Meth. Eng. **47**(5), 1057–1080 (2000)
5. Gunzburger, M.D., Peterson, J.S., Kwon, H.: An optimization based domain decomposition method for partial differential equations. Comput. Math. Appl. **37**(10), 77–93 (1999)
6. Gunzburger, M.D., Lee, H.K.: An optimization-based domain decomposition method for the Navier-Stokes equations. SIAM J. Numer. Anal. **37**(5), 1455–1480 (2000)
7. Laursen, T.A., Heinstein, M.W.: Consistent mesh tying methods for topologically distinct discretized surfaces in non-linear solid mechanics. Int. J. Numer. Meth. Eng. **57**(9), 1197–1242 (2003)
8. Parks, M., Romero, L., Bochev, P.: A novel lagrange-multiplier based method for consistent mesh tying. Comput. Methods Appl. Mech. Engrg. **196**(35–36), 3335–3347 (2007)
9. Hecht, F.: New development in FreeFem++. J. Numer. Math. **20**(3–4), 251–265 (2012)
10. Bochev, P., Ridzal, D.: Optimization-based additive decomposition of weakly coercive problems with applications. Comput. Math. with Appl. **71**(11), 2140–2154 (2016)

Towards a Scalable Multifidelity Simulation Approach for Electrokinetic Problems at the Mesoscale

Brian D. Hong[1], Mauro Perego[2], Pavel Bochev[2](✉), Amalie L. Frischknecht[2], and Edward G. Phillips[2]

[1] University of Arizona, Tucson, USA
bhong@math.arizona.edu
[2] Sandia National Laboratories, Albuquerque, USA
{mperego,pbboche,alfrisc,egphill}@sandia.gov

Abstract. In this work we present a computational capability featuring a hierarchy of models with different fidelities for the solution of electrokinetics problems at the micro-/nano-scale. A multifidelity approach allows the selection of the most appropriate model, in terms of accuracy and computational cost, for the particular application at hand. We demonstrate the proposed multifidelity approach by studying the mobility of a colloid in a micro-channel as a function of the colloid charge and of the size of the ions dissolved in the fluid.

Keywords: Multifidelity modeling · Electrokinetics · Colloid mobility

1 Introduction

Accurate modeling of electrokinetics phenomena is essential for studying applications at meso- (micro/nano) scale, from understanding biological processes to designing nano-devices. Electrokinetic phenomena originate from the interplay between the polarization of a fluid exposed to an electric field and the consequent movement of the fluid itself. For many meso-scale applications the latter can be satisfactorily described by the incompressible Stokes equations. However, the choice of a model for the charge distribution is more involved and requires careful consideration of the relevant physics. This has led to a hierarchy of models that differ primarily in the manner in which the ions are treated. At the one end of this spectrum are local models such as Poisson-Boltzmann (PB) or

M. Perego, P. Bochev, A. L. Frischknecht and E. G. Phillips—Sandia National Laboratories is a multimission laboratory managed and operated by National Technology and Engineering Solutions of Sandia, LLC., a wholly owned subsidiary of Honeywell International, Inc., for the U.S. Department of Energy's National Nuclear Security Administration under contract DE-NA-0003525.

I. Lirkov and S. Margenov (Eds.): LSSC 2017, LNCS 10665, pp. 156–164, 2018.
https://doi.org/10.1007/978-3-319-73441-5_16

Poisson-Nernst-Planck (PNP), which treat the ions as idealized point charges. At the other end are nonlocal models based on classic density functional theory (cDFT) [2], which treat the ions as hard spheres with finite radii. Such models are desirable in the presence of highly charged surfaces or high ion concentrations where correlations between the ions become important and the local PB and PNP models often fail to adequately represent charge distribution.

Although accurate, the computational cost of cDFT can be prohibitive. This has prompted the development of phenomenonlogical models, which incorporate correlation effects into PB and PNP with relatively little additional computational cost and occupy a middle ground in our model hierarchy. We denote the enriched versions of these models by PB*and PNP*, respectively.

The main goal of this paper is to present a scalable computational capability for electrokinetic phenomena built on the lower two rungs of our model hierarchy, i.e., the set (PB, PB*, PNP, PNP*), where PB*and PNP* are obtained by modification of the chemical potential to account for ion crowding [1] (see [5] for a preliminary comparison between PB* and cDFT). Such a capability is of a significant practical interest because it enables a *multifidelity simulation* approach, which allows one to choose a charge distribution model that is the most appropriate and economical for a given application.

As a demonstration of the proposed multifidelity approach we study the mobility of a colloid in a micro-channel as a function of the size of the ions dissolved in the fluid and of the colloid charge, and investigate the computational cost and the accuracy for methods of different fidelity. Section 2 presents the hierarchy of models comprising the foundation of our multifidelity capability. Section 3 describes the numerical experiment to compute the mobility of a colloid in a nano-channel and compares the results obtained using the different models.

2 Models of Electrokinetics Phenomena

We consider a model hierarchy describing the steady state of a fluid with M charged ionic species $\alpha, \alpha = 1, \ldots, M$, dissolved within it. The flow of the bulk is determined by the electric field, which in turn depends on the net charge of the fluid. As such, each model is comprised of the same equations for fluid flow and electric potential, and models of varying levels of fidelity for the ion flux.

2.1 Electric Potential

In the quasi-static limit, the electric potential in the fluid is governed by the *Poisson* equation:

$$-\nabla \cdot (\epsilon_0 \epsilon \nabla \psi) = q. \tag{1}$$

Here ϵ_0 is the permittivity of free space, ϵ is the relative permittivity of the solvent, and q is the net charge density of the fluid, i.e. the sum of the charge densities of the individual ion species

$$q = \sum_{\alpha} e z_\alpha \rho_\alpha, \qquad \alpha = 1, \ldots, M. \tag{2}$$

In this expression z_α is the valence, ρ_α the number of ions per unit volume of the α species and e is the fundamental electron charge. Equation (1) is typically paired with a Dirichlet boundary condition, specifying a boundary potential ψ^{bd} or a Neumann boundary condition for a surface with fixed charge q_s, i.e.,

$$\psi = \psi^{bd} \text{ on } \Gamma_D, \qquad \nabla\psi \cdot \mathbf{n} = -\frac{q_s}{\epsilon\epsilon_0} \text{ on } \Gamma_N. \tag{3}$$

Here $\Gamma_D \neq \emptyset$ and Γ_N are the Dirichlet and Neumann parts of the boundary Γ and $\mathbf{n}$ is the outer unit normal to the boundary.

2.2 Ion Flux: Poisson-Nernst-Planck

The steady-state *Nernst-Planck* equations are conservation laws for the fluxes $\mathbf{J}_\alpha$ of each ion species:

$$\nabla \cdot \mathbf{J}_\alpha = 0, \tag{4}$$

where the fluxes are given by

$$\mathbf{J}_\alpha = \rho_\alpha \mathbf{u} - d_\alpha \rho_\alpha \nabla \mu_\alpha. \tag{5}$$

Here $\mathbf{u}$ is the fluid velocity, d_α is the ion diffusivity of the α species and μ_α is the chemical potential

$$\mu_\alpha = \ln \rho_\alpha + \frac{ez_\alpha}{kT}\psi, \tag{6}$$

where k is the Boltzmann constant and T is the temperature (in Kelvin). Substitution of (6) into (5) yields the following expression for the flux:

$$\mathbf{J}_\alpha = \rho_\alpha \mathbf{u} - d_\alpha \nabla\rho_\alpha - \frac{d_\alpha e z_\alpha}{kT}\rho_\alpha \nabla\psi. \tag{7}$$

The first term describes the transport of ions due to the fluid velocity, the second term is the diffusion down the concentration gradient. The last term describes transport due to the electric field $\mathbf{E} := -\nabla\psi$. The combination of (1) and (4) with (2) and (7) gives the PNP model. In addition to boundary conditions (3), boundary conditions for the Nernst-Planck equation (4) are required, e.g. one can prescribe boundary values ρ_α^b for the ion densities, or no-flux conditions $\mathbf{J}_\alpha \cdot \mathbf{n} = 0$ at a wall.

2.3 Ion Flux: Poisson Boltzmann

The PB equation simplifies the PNP equations by assuming that the ion densities follow the Boltzmann distribution (thermal equilibrium)

$$\rho_\alpha = \rho_\alpha^b \exp\left(-\frac{z_\alpha e}{kT}\psi\right), \tag{8}$$

where ρ_α^b is the density of the α species in the bulk (where ψ is assumed to be close to zero). This choice for ρ_α corresponds to a zero flux $\mathbf{J}_\alpha$ when the fluid velocity $\mathbf{u}$ is zero. Combining (1), (2), and (8) results in the classical PB equation:

$$-\nabla\cdot(\epsilon_0\epsilon\nabla\psi)=\sum_\alpha ez_\alpha\rho_\alpha^b\exp\left(-\frac{z_\alpha e}{kT}\psi\right). \tag{9}$$

As for Eq. (1), the PB model is closed with boundary conditions (3).

2.4 Ion Flux: Modified PNP and PB Models

In this section we relax the assumption that the ions are point charges and consider two extensions to the classic PB and PNP equations that accounts for ion crowding effect. The basic idea behind the crowding model [1] is that the energy associated with the system increases proportional to the ion density. The result is that in areas where the density would be unphysically large (e.g. near a charged surface) the distribution will instead saturate and spread out into the medium. This is obtained by adding the following excess chemical potential term μ_α^{ex} to the chemical potential (6):

$$\mu_\alpha=\ln\rho_\alpha+\frac{ez_\alpha}{kT}\psi+\mu_\alpha^{ex},\quad \mu_\alpha^{ex}:=-\ln(1-\Phi),$$

where $\Phi:=\sum_i a_i^3\rho_i$ is the local volume fraction of ions and a_α is the effective diameter of the ion species α. The ion flux becomes (see [3])

$$\mathbf{J}_\alpha=\rho_\alpha\mathbf{u}-d_\alpha\nabla\rho_\alpha-\frac{d_\alpha ez_\alpha}{kT}\rho_\alpha\nabla\psi-d_\alpha\rho_\alpha\frac{\sum\limits_i a_i^3\nabla\rho_i}{1-\sum\limits_i a_i^3\rho_i}. \tag{10}$$

We refer to (1), (4), (2) and (10) as the PNP* model. Analogously to the classic Boltzmann densities (8), the modified Boltzmann densities imply $\mathbf{J}_\alpha=\mathbf{0}$ (assuming $\mathbf{u}=\mathbf{0}$) and are given by

$$\rho_\alpha=\frac{\rho_\alpha^b\exp(-\frac{ez_\alpha}{kT}\psi)}{1+\sum_j a_j^3\rho_j^b\left(\exp(-\frac{ez_j}{kT}\psi)-1\right)}. \tag{11}$$

We refer to (1), (2), and (11) as the PB* model. When $a_\alpha\to 0$, PB* and PNP* reduce to the PB and PNP models, respectively. In cases where the ion diameters differ substantially the presented crowding model lacks the ability to describe the packing of the smaller ions between the larger ions [8].

2.5 Fluid Flow: Stokes Equations

All models in our hierarchy use the same fluid model given by incompressible Stokes equations

$$\begin{cases}-\eta\Delta\mathbf{u}+\nabla p=q\left(-\nabla\psi+\mathbf{E}_{app}\right)\\ \nabla\cdot\mathbf{u}=0,\end{cases} \tag{12}$$

where $\mathbf{u}$ is the fluid velocity, p is the fluid pressure, η the fluid viscosity, and $\mathbf{E}_{app}$ is an applied electric field, i.e., the fluid is driven by the electric field acting on the exposed charge. The use of the Stokes equations is justified by the low

Reynolds number regime typical of electrokinetics applications. Combination of (12) with the PB/PB* equations yields a one-way PB-Stokes coupled model in which the potential and charge density affect the velocity through the body force in (12), but they are not affected by the velocity. On the other hand, combining (12) with the PNP/PNP* equations leads to a fully coupled PNP-Stokes model in which the charge density is transported by the fluid velocity.

2.6 Numerical Implementation

The coupled system of PDEs comprising each model is discretized on a non-uniform tetrahedral grid using C^0 piecewise linear finite elements for all fields. The resulting equal order discretization of the Stokes problem is stabilized using the Pressure Stabilized Petrov-Galerkin method [7]. Implementation of the discretized problem is by Drekar, a multiphysics application code built on the Trilinos software library [4]. We solve the resulting nonlinear system in a monolithic fashion by a Newton-Krylov method in which the linear solves are performed by the AztecOO implementation of GMRES. Implementation of a fully coupled algebraic multigrid method [9] in the ML package provides the preconditioner. We generate the grids using Sandia's mesh generation toolkit Cubit [11]. The mesh is refined near the colloid surface to provide suitable resolution.

3 Numerical Studies

This section illustrates various aspects of the models in our hierarchy by computing the mobility of a spherical colloid placed in the middle of an infinite cylindrical nano-channel saturated with a solution of monovalent ions of opposite charge and the same size. The channel is represented by a finite domain with periodic boundary conditions at its inlet and outlet.

To compute the mobility with a steady-state Eulerian simulation, we exploit the symmetry of the problem and use the reference frame of the colloid. Therefore the colloid is fixed but the channel walls are moving with the same velocity v_c as the colloid (in the original reference frame) but in the opposite direction. To model this configuration we set the fluid velocity at the channel wall equal to $-v_c\,\mathbf{i}_a$, where $\mathbf{i}_a$ is the unit vector directed as the channel axis, and prescribe no-slip condition at the colloid surface. We compute v_c by an iterative procedure ensuring that the total (mechanical and electric) force $\mathbf{F} = \int_{\Sigma_c}(\sigma + \sigma_E)\mathbf{n}\, da$ acting on the colloid is zero. Here σ is the fluid stress tensor, $\sigma_E = \epsilon\epsilon_0\left(\mathbf{E}\mathbf{E}^\top - \frac{1}{2}(\mathbf{E}^\top\mathbf{E})I\right)$ is the Maxwell stress tensor, and Σ_c is the colloid surface. We set $\psi = 0$ at the channel wall and $-\nabla\psi\cdot\mathbf{n} = \frac{q_s}{\epsilon\epsilon_0}$ at the colloid surface, where q_s is the colloid surface charge density (total colloid charge divided by the colloid surface area). In the PNP case, we prescribe the charge densities to be equal to the bulk densities ρ_α^b at the channel inlet and outlet, and we prescribe no-flux boundary conditions at the channel wall and at the colloid surface. Unless specified otherwise, we use the parameters given in Table 1.

Table 1. Parameters used in simulations.

Bulk density (ρ_α^b)	2.5×10^{23} ions/m^3	Colloid radius	10 nm
Relative permittivity (ϵ)	80	Channel length	60 nm
Ionic charges (z_α)	$+1, -1$	Channel radius	20 nm
Temperature (T)	298 K	Viscosity	0.856 Pa · s
Ion diffusivity	10^{-10} m^2/s	Fluid density	10^3 kg/m^3
Colloid charge	200 e	Applied electric field (E_{app})	10^6 V/m

The role of the crowding model. We start by demonstrating the ability of the crowding model, described in Sect. 2.4, to mitigate unphysical effects in the PB and PNP equations. Figure 1 shows the solution of the PNP model. The fluid is polarized by the electric field, exposing charge near the colloid surface (Fig. 1, right) and the applied electric field induces a flow (Fig. 1, left) by acting on the exposed charge. Figure 2, left shows that the charge density computed using PNP (or PB) suffers from unphysical accumulation of ions near the colloid. In contrast, the right plot in Fig. 2 reveals that incorporation of the crowding model in the PNP* (or PB*) equations mitigates this undesirable effect. In this example we have assumed an effective ion diameter $a_\alpha = 5$ nm.

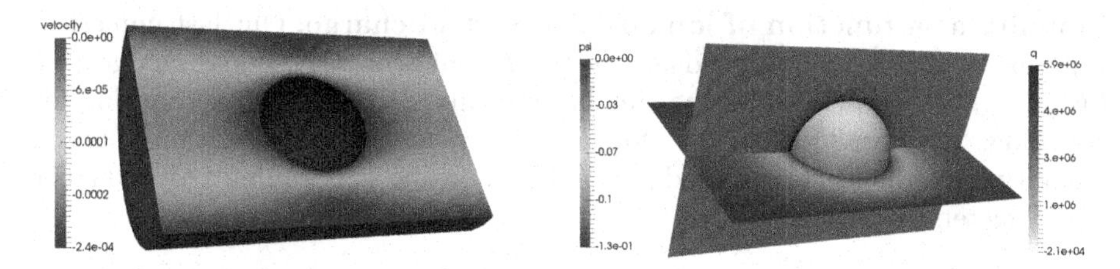

Fig. 1. Nano-channel geometry showing the PNP solution. Left: axial component of the velocity [m/s]. Right: Charge density [C] and electric potential [V] in two channel sections.

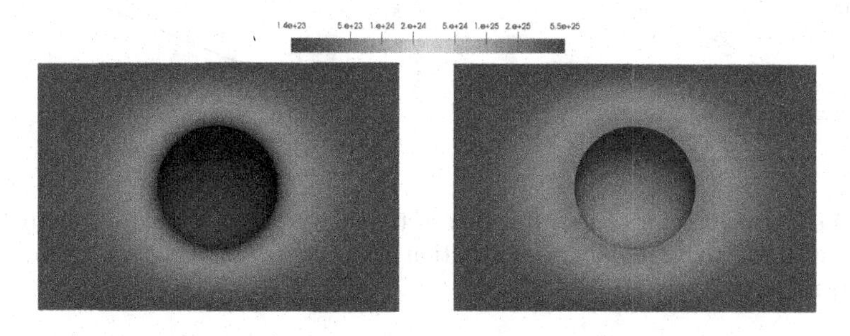

Fig. 2. Channel section showing the density of the positive ions computed with the PNP model (left) and PNP* model (right), with $a_\alpha = 5$ nm.

Mobility characterization. The colloid mobility μ_c is defined as $\mu_c = \frac{E_{app}}{v_c}$, where E_{app} is the magnitude of the applied electric field (oriented as the channel axis). The mobility can be assumed constant (i.e. the velocity is directly proportional to the applied electric field) in a wide range of applications.

In this study we verify that all models in our hierarchy, i.e., the PB and PNP equations and their enriched versions PB* and PNP*, respectively, admit this mobility characterization by computing the velocity of a charged colloid (with total charge 200e) for a varying applied electric field. Figure 3, left, shows the colloid velocity as a function of the applied electric field. In all four cases this functional dependence is linear, thereby confirming that the mobility, given by the slope of each line, is constant. We also note that the colloid velocities computed by PNP and its augmented version PNP* are very close to the velocities computed by PB and its augmented version PB*, respectively. Since computationally the cost of adding the crowding model is negligible and PB is twice as cheap as PNP, these results suggest that in many cases one can obtain reasonable results using the PB* model, or even the basic PB equations. However, inclusion of the crowding model does influence the mobility, as indicated by the difference in the slopes of the PNP and PB velocities on the one hand and the PNP* and PB* velocities on the other hand. Moreover, our experiments reveal that, for high electric fields or high colloid charge, the use of the crowding model makes the problem better conditioned and easier to solve by the linear solvers.

Mobility as a function of ion size and colloid charge. Our last numerical experiment studies the dependence of the mobility on the colloid charge (see Fig. 3, right). We performed the simulation using PNP* model, accounting for crowding. The results show that for large ion diameters, the relation between mobility and colloid charge tends to be linear, while it is sublinear for smaller ion diameters.

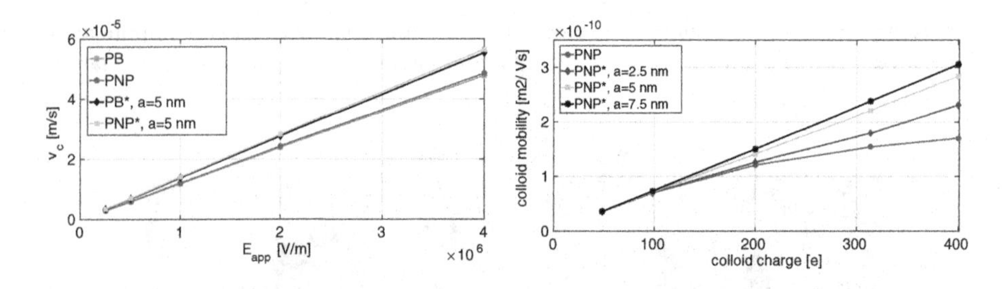

Fig. 3. Left: colloid velocity as a function of the applied electric field, computed with different models. Right: mobility as a function of the colloid total charge, for increasing size of the ions.

4 Conclusions

We have presented a multifidelity capability for the solution of electrokinetic equations comprising the PB and PNP models and their enriched versions PB* and PNP*, respectively. We found that the computational cost of the enriched models is essentially the same as for the original PB and PNP equations. Thus, PB* and PNP* could provide a cost-effective alternative to more sophisticated nonlocal models in situations where PB and PNP are inadequate, yet the full physical fidelity of cDFT is not necessary to accurately represent the crowding effects. Finally, we note that the cost of PNP and PNP* is roughly twice the cost of the simpler PB and PB* models.

We have examined various aspects of the multifidelity capability by computing the colloid mobility in a nano-channel. For this particular problem, the enriched PB* model proved to be the best trade-off between accuracy and computational costs. There are many applications where the PB model cannot be used because the thermal-equilibrium assumption is strongly violated. In such cases one can resort to more complex models such as PNP.

We plan to continue the development of the multifidelity capability by improving the accuracy of the discretization using high-order compatible finite elements and enhancing the efficiency and robustness of the solvers. Future plans include a thorough comparison with the cDFT model, and/or experimental data available in the literature, to asses the validity of the crowding approximation in the context of electrokinetic applications.

Acknowledgments. This work was supported by the U.S. Department of Energy Office of Science as part of the Collaboratory on Mathematics for Mesoscopic Modeling of Materials (CM4), under Award Number DE-SC0009247. The work of B. Hong was performed during a Computer Science Research Institute (CSRI) summer internship at Sandia National Laboratories.

References

1. Borukhov, I., Andelman, D., Orland, H.: Adsorption of large ions from an electrolyte solution: a modified Poisson-Boltzmann equation. Electrochimica Acta **46**, 221–229 (2000)
2. Evans, R.: The nature of the liquid-vapour interface and other topics in the statistical mechanics of nonuniform, classical fluids. Adv. Phys. **28**(2), 143–200 (1979)
3. Hainan, W., Thiele, A., Pilon, L.: Simulations of cyclic voltammetry for electric double layers in asymmetric electrolytes: a generalized modified Poisson-Nernst-Planck model. J. Phys. Chem. C **117**(36), 18286–18297 (2013)
4. Heroux, M., Bartlett, R., Hoekstra, V.H.R., Hu, J., Kolda, T., Lehoucq, R., Long, K., Pawlowski, R., Phipps, E., Salinger, A., Thornquist, H., Tuminaro, R., Willenbring, J., Williams, A.: An overview of trilinos, Technical report SAND2003-2927, Sandia National Laboratories (2003)
5. Hong, B.D., Perego, M.: A comparative study of simplified models for ion crowding and correlation effects in electrolyte solutions. In: Carleton, J.B., Parks, M.L. (eds.) Center for Computing Research Summer Proceedings 2016, Technical report SAND2017-1294R, Sandia National Laboratories, pp. 191–200 (2016)

6. Hsu, J.-P., Yeh, L.-H., Ku, M.-H.: Evaluation of the electric force in electrophoresis. J. Colloid Interface Sci. **305**, 324–329 (2007)
7. Tezduyar, T.E.: Stabilized finite element formulations for incompressible flow computations. Adv. Appl. Mech. **28**, 1–44 (1991)
8. Kornyshev, A.A.: Double-layer in ionic liquids: paradigm change? J. Phys. Chem. B **111**, 5545–5557 (2007)
9. Lin, P.T., Shadid, J.N., Tuminaro, R.S., Sala, M., Hennigan, G.L., Pawlowski, R.P.: A parallel fully coupled algebraic multilevel preconditioner applied to multiphysics PDE applications: drift-diffusion, flow/transport/reaction, resistive MHD. Int. J. Numer. Meth. Fluids **64**(10–12), 1148–1179 (2010)
10. Roth, R., Evans, R., Lang, A., Kahl, G.: Fundamental measure theory for hard-sphere mixtures revisited: the white bear version. J. Phys.: Condens. Matter **14**, 12063 (2002)
11. Shepherd, J., et al.: Cubit Mesh Generation Toolkit, SAND2000-2647. Sandia National Laboratories, Albuquerque (2000)
12. Wu, J.: Density functional theory for chemical engineering: from capillarity to soft materials. AIChE J. **52**, 1169–1193 (2006)

Advanced Numerical Methods for Nonlinear Elliptic Partial Differential Equations

On a Problem of Optimal Control of Convection-Diffusion Processes

Aigul Manapova(✉) and Fedor Lubyshev

Bashkir State University, Zaki Validi Street, 32, Republic of Bashkortostan, Russia
aygulrm@yahoo.com

Abstract. We study questions of the difference approximation of optimal control problems (OCPs) described by the Dirichlet problem for semilinear elliptic equations with non-self-adjoint operators and an imperfect contact matching condition. The coefficients of the convective transport of a state equation and in the matching boundary condition are used as a control function. Finite difference approximations for OCPs are constructed, the approximation error is estimated with respect to the state and the cost functional. We prove weak convergence of the approximations with respect to control and regularize them using Tikhonov regularization.

Keywords: Semi-linear elliptic equations · Optimal control problem
Difference solution method · Operator

1 Introduction and Setting of the Problem

The present work is devoted to construction and investigation of difference schemes approximating optimal control problems (OCPs) described by the Dirichlet problem for semilinear elliptic equations of second order with non-self-adjoint operators—convection-diffusion problems with an imperfect contact matching condition. The coefficients of the convective transport of the state equation and in the matching boundary condition are used as a control function. Note that currently, the most profound results in the theory of numerical solution to problem for PDEs and OCPs are obtained for processes with self-adjoint operators. But OCPs for second-order elliptic equations with minor terms are typical for mathematical models of liquid and gas mechanics, since heat and impurities transfer can occur not only due to diffusion, but also to the motion of the medium (see [1]). In particular, such problems arise in environmental problems associated with the description of the impurities distribution processes in the atmosphere and water reservoirs, and modeling of groundwater pollution.

This work provides a substantial complement to the results of [2–5]. Difference approximations of OCPs for semilinear elliptic equations with controls involved in the coefficients of the state equation were studied in [2]. Unlike the work [2] results of the present work are established for optimization problems described by semilinear elliptic equations with discontinuous data and states

I. Lirkov and S. Margenov (Eds.): LSSC 2017, LNCS 10665, pp. 167–174, 2018.
https://doi.org/10.1007/978-3-319-73441-5_17

subject to the boundary interface conditions of imperfect type (i.e., problems with a jump of the coefficients and the solution on the interface; the jump of the solution is proportional to the normal component of the flux, see [7]). In [3–5] the convergence and regularization of the approximations are analyzed for controls in the discontinuous coefficients of the right-hand-side of the state equation, in the matching boundary condition, and in the coefficients multiplying the highest derivatives, correspondingly. The principal difference from the current paper is that the main object of study is OCPs for a semilinear elliptic equation with non-self-adjoint operators, with controls multiplying lower order solution derivatives. Due to the non-self-adjoint operator, certain difficulties arise in the study of the approximations for differential equations describing discontinuous states of control processes, in particular, in proving the well-posedness of the difference approximations, and in the study the relation between the original OCP and the approximating one (see, e.g., Theorems 1 and 3). Note that the subject of this work is also related to [6], where we prove differentiability of the corresponding functional of an OCP related to a semi-linear elliptic equation with discontinuous coefficients, but with self-adjoint operators, define an iterative method for the solution of the discrete problem, and discuss its convergence property.

Let $\Omega = \{r = (r_1, r_2) \in \mathbf{R}^2 : 0 \le r_\alpha \le l_\alpha, \alpha = 1, 2\}$ with a boundary $\partial\Omega = \Gamma$. Suppose that the domain Ω is splitted by an "internal contact boundary" $\overline{S} = \{r_1 = \xi,\ 0 \le r_2 \le l_2\}$, where $0 < \xi < l_1$, into subdomains $\Omega_1 = \{0 < r_1 < \xi, 0 < r_2 < l_2\}$ and $\Omega_2 = \{\xi < r_1 < l_1, 0 < r_2 < l_2\}$ with boundaries $\partial\Omega_1$ and $\partial\Omega_2$. Thus, $\Omega = \Omega_1 \cup \Omega_2 \cup \overline{S}$, while $\partial\Omega$ is the outer boundary of Ω. Let $\overline{\Gamma}_k$ denote the boundaries of Ω_k without S, $k = 1, 2$. Therefore $\partial\Omega_k = \overline{\Gamma}_k \cup S$, and $\overline{\Gamma}_1 \cup \overline{\Gamma}_2 = \partial\Omega = \Gamma$. We assume that S is a straight line along which the coefficients and solutions of the boundary value problems (BVPs) are discontinuous, while in domains Ω_1 and Ω_2 they possess certain smoothness. Consider the problem of minimizing the functional $J : U \to \mathbf{R}^1$ of the form

$$g \to J(g) = \int\limits_{\Omega_1} |u(r_1, r_2; g) - u_0^{(1)}(r)|^2 d\Omega_1 = I(u(r; g)), \tag{1}$$

where $u_0^{(1)} \in W_2^1(\Omega_1)$ is a given function, on the solutions $u(g)$ to problem

$$-\sum_{\alpha=1}^{2} \frac{\partial}{\partial r_\alpha}\left(k(r)\frac{\partial u}{\partial r_\alpha}\right) + \sum_{\alpha=1}^{2} \vartheta^{(\alpha)} \frac{\partial u}{\partial r_\alpha} + d(r)q(u) = f(r),\ r \in \Omega_1 \cup \Omega_2, \tag{2}$$

and the conditions

$$\begin{gathered} u(r) = 0, \quad r \in \partial\Omega = \overline{\Gamma}_1 \cup \overline{\Gamma}_2, \\ \left[k(r)\frac{\partial u}{\partial r_1}\right] = 0, \quad \left(k_1(r)\frac{\partial u_1}{\partial r_1}\right) = \theta(r_2)[u], \quad x \in S, \end{gathered} \tag{3}$$

associated with all admissible controls $g = (g_1, g_2, g_3, g_4, g_5) = \Big(\vartheta_1^{(1)}, \vartheta_1^{(2)}, \vartheta_2^{(1)},$ $\vartheta_2^{(2)}, \theta\Big) \in U = \prod\limits_{k=1}^{5} U_k \subset (L_\infty(\Omega_1))^2 \times (L_\infty(\Omega_2))^2 \times L_2(S) = B,$

$$\begin{aligned} U_\beta &= \left\{ g_\beta = \vartheta_1^{(\beta)} \in L_\infty(\Omega_1) : \zeta_\beta \le g_\beta(r) \le \overline{\zeta}_\beta, \text{ a.e. on } \Omega_1 \right\}, \beta = 1,2, \\ U_\alpha &= \left\{ g_\alpha = \vartheta_2^{(\alpha)} \in L_\infty(\Omega_2) : \zeta_\alpha \le g_\alpha(r) \le \overline{\zeta}_\alpha, \text{ a.e. on } \Omega_2 \right\}, \alpha = 3,4, \\ U_5 &= \left\{ g_5 = \theta \in L_\infty(S) : 0 < \zeta_5 \le g_5(r) \le \overline{\zeta}_5, \text{ a.e. on } S \right\}. \end{aligned} \tag{4}$$

Here, in the formulation of the problem

$$u(r) = \begin{cases} u_1(r), \, r \in \Omega_1; \\ u_2(r), \, r \in \Omega_2, \end{cases} \quad q(\xi) = \begin{cases} q_1(\xi_1), \, \xi_1 \in \mathbf{R}; \\ q_2(\xi_2), \, \xi_2 \in \mathbf{R}, \end{cases}$$

$$k(r), d(r), f(r), \vartheta^{(\alpha)}(r) = \begin{cases} k_1(r), q_1(r), f_1(x), \vartheta_1^{(\alpha)}(r), \, r \in \Omega_1; \\ k_2(r), q_2(r), f_2(r), \vartheta_2^{(\alpha)}(r), \, r \in \Omega_2, \end{cases} \quad \alpha = 1,2,$$

$[u] = u_2(r) - u_1(r) = u^+(r) - u^-(r)$ is the jump of the function $u(r)$ on S; $k_\alpha(r)$, $\alpha = 1,2$, $d(r)$, $f(r)$ are given functions defined independently in Ω_1 and Ω_2, and having a first kind jump at S; $q_\alpha(\xi_\alpha)$, $\alpha = 1,2$ are given functions defined for $\xi_\alpha \in \mathbf{R}$, $\alpha = 1,2$. The given functions are assumed to satisfy the following conditions: $k_\alpha(r) \in W_\infty^1(\Omega_1) \times W_\infty^1(\Omega_2)$, $\alpha = 1,2$, $d(r) \in L_\infty(\Omega_1) \times L_\infty(\Omega_2)$, $f(r) \in L_2(\Omega_1) \times L_2(\Omega_2)$; $0 < \nu \le k_\alpha(r) \le \overline{\nu}$, $\alpha = 1,2$, $d_0 \le d(r) \le \overline{d}_0$, $r \in \Omega_1 \cup \Omega_2$; ν, $\overline{\nu}$, d_0, $\overline{d}_0$, - are given constants; and the functions $q_\alpha(\xi_\alpha)$, $\alpha = 1,2$ defined on $\mathbf{R}$ with values on $\mathbf{R}$ satisfy the conditions $q_\alpha(0) = 0$, $0 \le q_0 \le (q_\alpha(\xi_1) - q_\alpha(\xi_2))/(\xi_1 - \xi_2) \le L_q < \infty$ for all $\xi_1, \xi_2 \in \mathbf{R}$, $\xi_1 \ne \xi_2$, $L_q = Const.$ We also assume that: $-m_1 \le \zeta_1 \le \overline{\zeta}_1 \le m_1$, $-p_1 \le \zeta_2 \le \overline{\zeta}_2 \le p_1$, $-m_2 \le \zeta_3 \le \overline{\zeta}_3 \le m_2$, $-p_2 \le \zeta_4 \le \overline{\zeta}_4 \le p_2$, $m_\alpha, p_\alpha = Const > 0$, $\alpha = 1,2$,

$$\begin{aligned} \delta_\alpha &= \max_{\substack{\epsilon_1, \epsilon_2 > 0 \\ \epsilon_1 + \epsilon_2 \le \nu_\alpha}} \left\{ \frac{\nu_\alpha - (\epsilon_1 + \epsilon_2)}{C_{\Omega_\alpha}^2} + \lambda - \frac{m_\alpha^2}{4\epsilon_1} - \frac{p_\alpha^2}{4\epsilon_2} \right\} > 0, \ \alpha = 1,2, \\ C_{\Omega_1}^2 &= \left(8/\xi_1^2 + 8/l_2^2\right)^{-1}, \ C_{\Omega_2}^2 = \left(8/(l_1 - \xi_1)^2 + 8/l_2^2\right)^{-1}; \end{aligned} \tag{5}$$

here λ is any of the following constants: (1) $\lambda = q_0 \, d_0$, $\quad d_0 \ge 0$; (2) $\lambda = d_0$ is an arbitrary constant as $q_\alpha(u_\alpha) = u_\alpha$; (3) $\lambda = -L_q \zeta_0$, where $\zeta_0 = \max\left\{|d_0|, |\overline{d}_0|\right\}$.

We say that a function $u(g) \in \overset{\circ}{V}_{\Gamma_1,\Gamma_2}(\Omega^{(1,2)})$ is a generalized solution to the problem (2) with a fixed control $g \in U$, satisfying the identity:

$$\int\limits_\Omega \left[\sum_{\alpha=1}^{2} k \, \frac{\partial u}{\partial r_\alpha} \frac{\partial v}{\partial r_\alpha} + \sum_{\alpha=1}^{2} \vartheta^{(\alpha)} \, \frac{\partial u}{\partial r_\alpha} v + d \, q(u) v \right] d\Omega_0 + \int\limits_S \theta \, [u][v] \, dS = \int\limits_\Omega f \, v \, d\Omega_0.$$

Remark 1. We use here some spaces introduced in work [4].

2 Difference Approximation of Control Problems

Based on the grid method [7], we associate OCPs (1)–(5) with the following difference approximations: minimize the grid functional

$$J_h(\Phi_h) = \sum_{\overline{\omega}^{(1)}} \left| y(\Phi_h) - u_{0h}^{(1)} \right|^2 \hbar_1 \hbar_2 = \| y(\Phi_h) - u_{0h}^{(1)} \|_{L_2(\overline{\omega}^{(1)})}^2, \tag{6}$$

provided grid function $y \in \overset{\circ}{V}_{\gamma^{(1)}\gamma^{(2)}}(\overline{\omega}^{(1,2)})$ satisfies the summation identity

$$\sum_{\alpha=1}^{2}\Big\{\sum_{\omega_1^{(\alpha)+}}\sum_{\omega_2} a_{1h}^{(\alpha)} y_{\alpha\bar{x}_1} v_{\alpha\bar{x}_1} h_1 h_2 + \sum_{\omega_1^{(\alpha)}}\sum_{\omega_2^{+}} a_{2h}^{(\alpha)} y_{\alpha\bar{x}_2} v_{\alpha\bar{x}_2} h_1 h_2$$
$$+\frac{1}{2}\sum_{\omega_2^{+}} a_{2h}^{(\alpha)}(\xi,x_2) y_{\alpha\bar{x}_2}(\xi,x_2) v_{\alpha\bar{x}_2}(\xi,x_2) h_1 h_2\Big\} + \sum_{\omega_2} \Phi_{5h}(x_2)[y(\xi,x_2)]\,[v(\xi,x_2)]\,h_2$$
$$+\sum_{\alpha=1}^{2}\Big\{\sum_{\omega^{(1)}} \Phi_{\alpha,h}(x) y_{1\overset{\circ}{x}_\alpha}(x) v_1(x) h_1 h_2 + \frac{1}{2}\sum_{\omega_2} \Phi_{\alpha,h}(\xi,x_2) y_{1\overset{\circ}{x}_\alpha}(\xi,x_2) v_1(\xi,x_2) h_1 h_2$$
$$+\sum_{\omega^{(2)}} \Phi_{\alpha+2,h}(x) y_{2\overset{\circ}{x}_\alpha}(x) v_2(x) h_1 h_2 + \frac{1}{2}\sum_{\omega_2} \Phi_{\alpha+2,h}(\xi,x_2) y_{2\overset{\circ}{x}_\alpha}(\xi,x_2) v_2(\xi,x_2) h_1 h_2$$
$$+\sum_{\omega^{(\alpha)}} d_{\alpha h}(x) q_\alpha(y_\alpha(x)) v_\alpha(x) h_1 h_2 + \frac{1}{2}\sum_{\omega_2} d_{\alpha h}(\xi,x_2) q_\alpha(y_\alpha(\xi,x_2)) v_\alpha(\xi,x_2) h_1 h_2\Big\}$$
$$=\sum_{\alpha=1}^{2}\Big(\sum_{\omega^{(\alpha)}} f_{\alpha h}\, v_\alpha\, h_1 h_2 + \frac{1}{2}\sum_{\omega_2} f_{\alpha h}(\xi,x_2) v_\alpha(\xi,x_2) h_1 h_2\Big),\ \forall v \in \overset{\circ}{V}_{\gamma^{(1)}\gamma^{(2)}}(\overline{\omega}^{(1,2)}),$$
(7)

while the grid controls $\Phi_h = (\Phi_{1h}, \Phi_{2h}, \Phi_{3h}, \Phi_{4h}, \Phi_{5h}) \in \prod_{\alpha=1}^{5} U_{\alpha h} = U_h \subset B_h$,

$$\begin{aligned} U_{ph} &= \{\Phi_{ph} \in L_\infty(\bar{\omega}^{(1)}) : \zeta_p \le \Phi_{ph}(x) \le \bar{\zeta}_p, x \in \bar{\omega}^{(1)}\},\ p = 1,2,\\ U_{\beta h} &= \{\Phi_{\beta h} \in L_\infty(\bar{\omega}^{(2)}) : \zeta_\beta \le \Phi_{\beta h}(x) \le \bar{\zeta}_\beta, x \in \bar{\omega}^{(2)}\},\ \beta = 3,4,\\ \Phi_{5h} \in U_{ph} &= \{\Phi_{5h} \in L_2(\gamma_S) : 0 < \zeta_5 \le \Phi_{5h}(x) \le \bar{\zeta}_5, x \in \gamma_S\}. \end{aligned} \quad (8)$$

Here, $a_{\alpha h}^{(1)}(x)$, $a_{\alpha h}^{(2)}(x)$, $d_{\alpha h}(x)$, $\alpha = 1,2$, and $u_{0h}^{(1)}(x)$ are grid approximations of the functions $k_\alpha^{(1)}(r)$, $k_\alpha^{(2)}(r)$, $d_\alpha(r)$, $\alpha = 1,2$, and $u_0^{(1)}(r)$ defined via Steklov averages. For the definition of the grids $\overline{\omega}^{(1,2)}$, $\omega^{(\alpha)} \cup \gamma_S$, $\alpha = 1,2$ and the grid spaces $\overset{\circ}{V}_{\gamma^{(1)}\gamma^{(2)}}(\overline{\omega}^{(1,2)})$ see [4]. Problem (7) is a grid analogue of the original BVP (2).

Theorem 1. *The problem of finding a solution to difference scheme (7) with any fixed control $\Phi_h \in U_h$ is uniquely solvable, moreover, $\forall \Phi_h \in U_h$ we have*

$$\|y(\Phi_h)\|_{\overset{\circ}{V}_{\gamma^{(1)},\gamma^{(2)}}(\bar{\omega}^{(1,2)})} \le M \sum_{k=1}^{2} \|f_{kh}\|_{L_2(\omega^{(k)} \cup \gamma_S)} = \hat{M},\ M = Const > 0. \quad (9)$$

Proof. Using the constraints on the input data in problem (2)–(5), Cauchy-Schwarz and Hölder inequalities, and difference analogues of embedding theorems, one can show that $Q_h(y,v)$ with a fixed $y \in \overset{\circ}{V}_{\gamma^{(1)}\gamma^{(2)}}(\overline{\omega}^{(1,2)})$ and $\forall \Phi_h \in U_h$ defines a linear continuous functional $A_h y$ over $\overset{\circ}{V}_{\gamma^{(1)}\gamma^{(2)}}(\overline{\omega}^{(1,2)})$ so that $\langle A_h y, v\rangle = Q_h(y,v), \forall y, v \in \overset{\circ}{V}_{\gamma^{(1)}\gamma^{(2)}}$. Indeed, we have for $\forall y, v \in \overset{\circ}{V}_{\gamma^{(1)}\gamma^{(2)}}(\overline{\omega}^{(1,2)})$

$$|Q_h(y,v)| \leq \sum_{\alpha=1}^{2}\Big\{\overline{\nu}_\alpha \sum_{\omega_1^{(\alpha)+}\times\omega_2} |y_{\alpha\bar{x}_1} v_{\alpha\bar{x}_1}| h_1 h_2 + \overline{\nu}_\alpha \sum_{\omega_1^{(\alpha)}\times\omega_2^{+}} |y_{\alpha\bar{x}_2} v_{\alpha\bar{x}_2}| h_1 h_2$$
$$+\frac{1}{2}\overline{\nu}_\alpha \sum_{\omega_2^{+}} |y_{\alpha\bar{x}_2}(\xi,x_2) v_{\alpha\bar{x}_2}(\xi,x_2)| h_1 h_2 + m_\alpha \sum_{\omega^{(\alpha)}} |y_{\alpha \overset{\circ}{x}_1}(x) v_\alpha(x)| h_1 h_2$$
$$+p_\alpha \sum_{\omega^{(\alpha)}} |y_{\alpha \overset{\circ}{x}_2}(x) v_\alpha(x)| h_1 h_2 + \frac{1}{2} p_\alpha \sum_{\omega_2} |y_{\alpha \overset{\circ}{x}_2}(\xi,x_2) v_\alpha(\xi,x_2)| h_1 h_2\Big\}$$
$$+\overline{\theta}_0 \sum_{\omega_2} \big|[y(\xi,x_2)]\,[v(\xi,x_2)]\big|\, h_2 \leq C_0(\|y\|_{\overset{\circ}{V}_{\gamma^{(1)}\gamma^{(2)}}(\overline{\omega}^{(1,2)})}) \|v\|_{\overset{\circ}{V}_{\gamma^{(1)}\gamma^{(2)}}(\overline{\omega}^{(1,2)})},$$

i.e., $Q_h(y,v)$ is a continuous function. Besides, $Q_h(y,v)$ is linear in the argument v. On the other hand, the right-hand side of (7) also generates a linear bounded functional F_h over $\overset{\circ}{V}_{\gamma^{(1)}\gamma^{(2)}}(\overline{\omega}^{(1,2)})$: $\langle F_h, v\rangle = l_h(v)$, $\forall v \in \overset{\circ}{V}_{\gamma^{(1)}\gamma^{(2)}}$. Hence, the identity (7) can be written in the form: $\langle A_h y, v\rangle = \langle F_h, v\rangle$, from which, since $v \in \overset{\circ}{V}_{\gamma^{(1)}\gamma^{(2)}}(\overline{\omega}^{(1,2)})$ is arbitrary, we derive the equation $A_h y = F_h$.

Now we show that there exists a unique solution $y \in \overset{\circ}{V}_{\gamma^{(1)}\gamma^{(2)}}(\overline{\omega}^{(1,2)})$ satisfying identity (7). By virtue of Browder's result [8], it is sufficient to prove that the difference operator A_h is continuous and strongly monotone. Let us show that A_h is strongly monotone. Using the ε_1 and ε_2-Cauchy inequalities, we obtain

$$\begin{aligned}\langle A_h y - A_h v, y - v\rangle &\geq \sum_{\alpha=1}^{2}\Big\{\nu_\alpha \sum_{\omega_1^{(\alpha)+}\times\omega_2} \big[(y_\alpha - v_\alpha)_{\bar{x}_1}\big]^2 h_1 h_2\\
&+\nu_\alpha \sum_{\omega_1^{(\alpha)}\times\omega_2^{+}} \big[(y_\alpha - v_\alpha)_{\bar{x}_2}\big]^2 h_1 h_2 + \frac{1}{2}\nu_\alpha \sum_{\omega_2^{+}} \big[(y_\alpha - v_\alpha)_{\bar{x}_2}(\xi,x_2)\big]^2 h_1 h_2\\
&+\sum_{\omega^{(1)}} \Phi_{\alpha h}(y_1 - v_1)_{\overset{\circ}{x}_\alpha}(y_1 - v_1)\hbar_1 h_2 + \sum_{\omega^{(2)}} \Phi_{\alpha+2,h}(y_2 - v_2)_{\overset{\circ}{x}_\alpha}(y_2 - v_2)\hbar_1 h_2\\
&+\sum_{\omega^{(\alpha)}\cup\gamma_S} d_0\big[q_\alpha(y_\alpha) - q_\alpha(v_\alpha)\big](y_\alpha - v_\alpha)\hbar_1 h_2\Big\}\\
&+\theta_0 \sum_{\omega_2} \big[y - v\big]^2 h_2 \geq \sum_{\alpha=1}^{2}\big[\nu_\alpha - (\varepsilon_1 + \varepsilon_2)\big]\|y_\alpha - v_\alpha\|^2_{W_2^1(\overline{\omega}^{(\alpha)})}\\
&+\sum_{\alpha=1}^{2}\big[\lambda - \frac{m_\alpha^2}{4\varepsilon_1} - \frac{p_\alpha^2}{4\varepsilon_2}\big]\|y_\alpha - v_\alpha\|^2_{L_2(\omega^{(\alpha)})} + \theta_0 \sum_{\gamma_S} [y - v]^2 h_2,\ \varepsilon_1, \varepsilon_2 > 0.\end{aligned} \tag{10}$$

The right-hand side of (10), due to the Friedreich inequality $\|y_\alpha\|^2_{L_2(\omega^{(\alpha)})} \leq C^2_{\Omega_\alpha} |y_\alpha|^2_{W_2^1(\overline{\omega}^{(\alpha)})}$, $\alpha = 1, 2$, $C^2_{\Omega_\alpha}$ defined in (5), is no less than the expression $\sum_{\alpha=1}^{2}\Big((\nu_\alpha - (\varepsilon_1 + \varepsilon_1))/C^2_{\Omega_\alpha} + \lambda - m_\alpha^2/(4\varepsilon_1) - p_\alpha^2/(4\varepsilon_2)\Big)\|y_\alpha - v_\alpha\|^2_{L_2(\omega^{(\alpha)})}$, for all $\varepsilon_1 + \varepsilon_2 > 0$, $\varepsilon_1 + \varepsilon_2 \leq \nu_\alpha$, $\alpha = 1, 2$. Therefore,

$$\langle A_h y - A_h v, y - v\rangle \geq \sum_{\alpha=1}^{2} \delta_\alpha \|y_\alpha - v_\alpha\|^2_{L_2(\omega^{(\alpha)})} \geq \delta_\alpha \|y_\alpha - v_\alpha\|^2_{L_2(\omega^{(\alpha)})}, \tag{11}$$

$\alpha = 1, 2$, where δ_α and λ are defined by (5). If $\delta_\alpha > 0$, then we have $\|y_\alpha - v_\alpha\|^2_{L_2(\omega^{(\alpha)})} \le \frac{1}{\delta_\alpha}\langle A_h y - A_h v, y - v\rangle$, $\alpha = 1, 2$. Setting $\varepsilon_k = \nu_k/4$, $k = 1, 2$, in (10) and taking into account (11), we obtain

$$\sum_{\alpha=1}^{2} \frac{\nu_\alpha}{2} |y_\alpha - v_\alpha|^2_{W_2^1(\overline{\omega}^{(\alpha)})} + \theta_0 \|[y - v]\|^2_{L_2(\gamma_S)} \le$$
$$\le \langle A_h y - A_h v, y - v\rangle + \sum_{\alpha=1}^{2} \Big(\frac{m_\alpha^2}{4\varepsilon_1} - \frac{p_\alpha^2}{4\varepsilon_2} - \lambda\Big) \|y_\alpha - v_\alpha\|^2_{L_2(\omega^{(\alpha)})} \le$$
$$= \Big[1 + \sum_{\alpha=1}^{2} \frac{1}{\delta_\alpha} \max\{0; \frac{m_\alpha^2}{4\varepsilon_1} + \frac{p_\alpha^2}{4\varepsilon_2} - \lambda\}\Big] \langle A_h y - A_h v, y - v\rangle.$$

Then, $\langle A_h y - A_h v, y - v\rangle \big[1 + \sum_{\alpha=1}^{2} \frac{1}{\delta_\alpha} \max\{0; \frac{m_\alpha^2}{4\varepsilon_1} + \frac{p_\alpha^2}{4\varepsilon_2} - \lambda_\alpha\}\big] \ge \nu \|y - v\|^2_{\overset{\circ}{V}_{\gamma^{(1)}\gamma^{(2)}}}$, where $\nu = \min\{\nu_1/2, \nu_2/2, \theta_0\}$. This estimate yields

$$\langle A_h y - A_h v, y - v\rangle \ge \delta_* \|y - v\|^2_{\overset{\circ}{V}_{\gamma^{(1)}\gamma^{(2)}}}, \quad \forall y, v \in \overset{\circ}{V}_{\gamma^{(1)}\gamma^{(2)}} (\overline{\omega}^{(1,2)}),$$

where $\delta_* = \nu \big[1 + \sum_{\alpha=1}^{2} \frac{1}{\delta_\alpha} \max\{0; m_\alpha^2/(4\varepsilon_1) - p_\alpha^2/(4\varepsilon_2) - \lambda_\alpha\}\big]^{-1} > 0$, i.e., the operator $A_h : \overset{\circ}{V}_{\gamma^{(1)}\gamma^{(2)}} (\overline{\omega}^{(1,2)}) \to \overset{\circ}{V}_{\gamma^{(1)}\gamma^{(2)}} (\overline{\omega}^{(1,2)})$ is strongly monotone. Let us prove that the operator A_h is Lipschitz continuous. Obviously, we have the estimate $|\langle A_h y - A_h v, \eta\rangle| \le M \|y - v\|_{\overset{\circ}{V}_{\gamma^{(1)}\gamma^{(2)}}} \|\eta\|_{\overset{\circ}{V}_{\gamma^{(1)}\gamma^{(2)}}}$, $\forall y, v, \eta \in \overset{\circ}{V}_{\gamma^{(1)}\gamma^{(2)}} (\overline{\omega}^{(1,2)})$. Therefore, $\|A_h y - A_h v\|_{\overset{\circ}{V}_{\gamma^{(1)}\gamma^{(2)}}} \le M \|y - v\|_{\overset{\circ}{V}_{\gamma^{(1)}\gamma^{(2)}}}$, $\forall y, v \in \overset{\circ}{V}_{\gamma^{(1)}\gamma^{(2)}} (\overline{\omega}^{(1,2)})$, i.e., A_h is Lipschitz continuous. Thus, the equation $A_h y = F_h$ is uniquely solvable. Further, since A_h is strongly monotone, it is coercive: $\langle A_h y, y\rangle_{\overset{\circ}{V}_{\gamma^{(1)}\gamma^{(2)}}} \ge \delta_* \|y\|^2_{\overset{\circ}{V}_{\gamma^{(1)}\gamma^{(2)}}}$. This inequality and the chain of easily established inequalities

$$\delta_* \|y\|^2_{\overset{\circ}{V}_{\gamma^{(1)}\gamma^{(2)}}} \le \langle A_h y, y\rangle_{\overset{\circ}{V}_{\gamma^{(1)}\gamma^{(2)}}} \le M_1 \left(\sum_{k=1}^{2} \|f_{kh}\|_{L_2(\omega^{(k)} \cup \gamma_S)} \right) \|y\|_{\overset{\circ}{V}_{\gamma^{(1)}\gamma^{(2)}}}$$

yield to (9). The theorem is proved.

Theorem 2. *For every $h > 0$, there exists at least one optimal control $\Phi_{h*} \in U_h$ in the sequence of the grid (difference) optimization problems (6)–(8); i.e. $J_{h*} = \inf\{J_h(\Phi_h) : \Phi_h \in U_h\} > -\infty$, and $U_{h*} = \{\Phi_{h*} \in U_h : J_h(\Phi_{h*}) = J_{h*}\} \ne \emptyset$.*

Proof. The theorem is proved by showing that $J_h(\Phi_h)$ is continuous on U_h in the weak topology of $H_h = \big(L_2(\overline{\omega}^{(1)})\big)^2 \times \big(L_2(\overline{\omega}^{(2)})\big)^2 \times L_2(\gamma_S)$ with the use of the weak compactness of U_h in H_h and by applying the result of [9], p. 49, Theorem 2.

3 Convergence Properties of the Approximating OCPs

Let us estimate the convergence rate of the approximations (6)–(8) to the OCP (1)–(5) with respect to the state and the cost functional, prove weak convergence of the approximations with respect to control.

Theorem 3. *Let $g \in U$ and $\Phi_h \in U_h$ be arbitrary controls, and let $u(g)$ and $y(\Phi_h)$ be the corresponding solutions of state problems in (1)–(5) and (6)–(8). Then, (1) for any $h > 0$, we have the convergence rate estimate for the grid method with respect to the state:*

$$
\begin{aligned}
\|y(\Phi_h) - u(g)\|_{\overset{\circ}{V}_{\gamma^{(1)},\gamma^{(2)}}} &\leq C\Bigg\{|h|\Bigg[\sum_{\alpha=1}^{2}\Bigg(\|k_\alpha\|_{L_\infty(\Omega_\alpha)} + L_q\|d_\alpha\|_{L_\infty(\Omega_\alpha)} \\
&+ \sum_{\beta=1}^{2}\|\vartheta_\alpha^{(\beta)}\|_{L_\infty(\Omega_\alpha)}\Bigg)\|u_\alpha\|_{W_2^2(\Omega_\alpha)} + \|\theta\|_{L_\infty(0,l_\alpha)}\sum_{\alpha=1}^{2}\|u_\alpha\|_{W_2^2(\Omega_\alpha)}\Bigg] \\
&+ \sum_{\alpha=1}^{2}\Bigg(\Bigg\|\Phi_{\alpha,h}(r) - \frac{1}{\hbar_1 h_2}\int_{e^1(x)}\vartheta_1^{(\alpha)}(r)\,dr\Bigg\|_{L_\infty(\omega^{(1)}\cup\gamma_S)}\|u_1\|_{W_2^2(\Omega_1)} \\
&+ \Bigg\|\Phi_{\alpha+2,h}(r) - \frac{1}{\hbar_1 h_2}\int_{e^2(x)}\vartheta_2^{(\alpha)}(r)\,dr\Bigg\|_{L_\infty(\omega^{(2)}\cup\gamma_S)}\|u_2\|_{W_2^2(\Omega_2)}\Bigg) \\
&+ \Bigg\|S^{x_2}\theta(x_2) - \Phi_{5h}\Bigg\|_{L_\infty(\omega_2)}\sum_{\alpha=1}^{2}\|u_\alpha\|_{W_2^2(\Omega_\alpha)}\Bigg\}.
\end{aligned}
\tag{12}
$$

(2) $\lim J_{h*} = J_*$ as $|h| \to 0$, and we have the convergence rate estimate

$$|J_{h*} - J_*| \leq M|h|,$$

where J_* and J_{h*} are the infima of the functionals $J(g)$ and $J_h(\Phi_h)$, respectively.

(3) the approximations (6)–(8) converge weakly to original OCP (1)–(5) with respect to control, namely, if $\{\Phi_{h\epsilon_h}\} \subset U_h$ is a sequence of grid controls giving an approximate solution to problem (6)–(8) in the sense that $J_{h*} \leq J_h(\Phi_{h\epsilon_h}) \leq J_{h*} + \epsilon_h$, $\Phi_{h\epsilon_h} \in U_h$, where $\epsilon_h \geq 0$ and $\epsilon_h \to 0$ as $|h| \to 0$, then $\lim J(F_h\Phi_{h\epsilon_h}) = J_*$ as $|h| \to 0$ and we have the convergence rate estimate $0 \leq J(F_h\Phi_{h\epsilon_h}) - J_* \leq C|h| + \epsilon_h$, where $\{F_h\Phi_{h\epsilon_h}\} = \{F_{\alpha h}\Phi_{\alpha h\epsilon_h}\}_{\alpha=1}^{5}$, is a sequence of piecewise constant extensions of grid controls $\Phi_{\alpha h\epsilon_h}$, $\alpha = \overline{1,5}$. The sequence $\{F_h\Phi_{h\epsilon_h}\}$ converges weakly in H to the set $U_* = \{g_* \in U : J(g_*) = J_*\} \neq \emptyset$.

Proof. To prove the statement 1, first, we reduce the approximation error $\langle\psi_h, v\rangle = \langle F_h - A_h u, v\rangle = l_h(v) - Q_h(u,v)$, $\forall v \in \overset{\circ}{V}_{\gamma^{(1)},\gamma^{(2)}}(\bar{\omega}^{(1,2)})$ of scheme (7) with respect to the state to special form after rather cumbersome algebra by using the difference summation-by-parts formulas, Green's identity. To estimate the convergence rate of the approximations with respect to the state, it suffices to estimate quantities in the obtained representation for the truncation error ψ_h. Taking into account the strong monotonicity of the operator A_h and applying difference analogues of the Sobolev embedding theorems, equivalent norms of $\overset{\circ}{V}_{\gamma^{(1)},\gamma^{(2)}}(\bar{\omega}^{(1,2)})$ (see [3–5]), and the Cauchy–Schwarz and Hölder inequalities we derive the estimates (the proofs of estimates are omitted in view of the limited volume of the paper). The other two statements are proved by using the above results and applying the ideas of [2,9]. Note that the main difficulty here is to construct mappings $R_h : H \to H_h$ and $N_h : H_h \to H$, and study their properties. In particularly, we define the mappings as $R_h g = \Phi_h$, where $g = \left(\vartheta_1^{(1)}, \vartheta_1^{(2)}, \vartheta_2^{(1)}, \vartheta_2^{(2)}, \theta\right)$, and $N_h \Phi_h = g$, where $g = \left(F_h \Phi_{1h}, F_{2h}\Phi_{2h}, F_{3h}\Phi_{3h}, F_{4h}\Phi_{4h}, F_{5h}\Phi_{5h}\right)$, where $R_h^{(\alpha)}$, $\alpha = \overline{1,5}$, - some discretizations of the functions of continuous argument, and $F_{\alpha h}$, $\alpha = \overline{1,5}$, - some piecewise constant extensions of the grid functions.

Note that OCP (1)–(5) is not a well posed minimization problem in the sense of Tikhonov in the strong topology of H. Similarly to paper [3, pp. 1112–1113] (see also a related paper [4, pp. 1722–1723]) a Tikhonov regularization algorithm is applied to regularize the family of grid OCPs (6)–(8). It generates a minimizing sequence (based on the difference approximations) that converges strongly to the set of Ω-normal solutions to the original OCP (1)–(5). The set of Ω-normal solutions of the OCP (1)–(5) is defined as $U_{**} = \{g_{**} \in U_* : \Omega(g_{**}) = \inf\{\Omega(g_*) : g_* \in U_*\} = \Omega_*\}$, where a stabilizing functional $\Omega(g) = \|g\|_H^2$, $g \in U$.

References

1. Samarskii, A.A., Vabishchevich, P.N.: Computational Heat Transfer. Wiley, New York (1996). Librokom, Moscow (2009)
2. Lubyshev, F.V., Manapova, A.R.: On some optimal conrol problems and their finite difference approximations and regularization for quasilinear elliptic equations with controls in the coefficients. Comput. Math. Math. Phys. **47**(3), 361–380 (2007)
3. Lubyshev, F.V.: Finite difference approximations of optimal control problems for semilinear elliptic equations with discontinuous coefficients and solutions. Comput. Math. Math. Phys. **52**(8), 1094–1114 (2012)
4. Lubyshev, F.V., Manapova, A.R., Fairuzov, M.E.: Approximations of optimal control problems for semilinear elliptic equations with discontinuous coefficients and solutions and with control in matching boundary conditions. Comput. Math. Math. Phys. **54**(11), 1700–1724 (2014)
5. Lubyshev, F.V., Fairuzov, M.E.: Approximations of optimal control problems for semilinear elliptic equations with discontinuous coefficients and states and with controls in the coefficients multiplying the highest derivatives. Comput. Math. Math. Phys. **56**(7), 1238–1263 (2016)
6. Manapova, A.R., Lubyshev, F.V.: Numerical solution of optimization problems for semi-linear elliptic equations with discontinuous coefficients and solutions. Appl. Numer. Math. **104**, 182–203 (2016)
7. Samarskii, A.A., Andreev, V.B.: Difference Methods for Elliptic Equations. Nauka, Moscow (1976). (in Russian)
8. Browder, F.E.: Transactions of the symposium on partial differential equations. Sib. Otd. Akad. Nauk SSSR, Novosibirsk (1963)
9. Vasil'ev, F.P.: Optimization Methods. Faktorial, Moscow (2002). (in Russian)

Verifications of Primal Energy Identities for Variational Problems with Obstacles

Sergey Repin[1,2] and Jan Valdman[3,4(✉)]

[1] V.A. Steklov Institute of Mathematics in St.-Petersburg, 191011 Fontanka 27, Saint Petersburg, Russia
[2] University of Jyväskylä, P.O. Box 35, 40014 Jyväskylä, Finland
[3] Faculty of Science, Institute of Mathematics and Biomathematics, University of South Bohemia, Branišovská 31, 37005 České Budějovice, Czech Republic
jvaldman@prf.jcu.cz
[4] Institute of Information Theory and Automation, Academy of Sciences, Pod vodárenskou věží 4, 18208 Praha 8, Czech Republic

Abstract. We discuss error identities for two classes of free boundary problems generated by obstacles. The identities suggest true forms of the respective error measures which consist of two parts: standard energy norm and a certain nonlinear measure. The latter measure controls (in a weak sense) approximation of free boundaries. Numerical tests confirm sharpness of error identities and show that in different examples one or another part of the error measure may be dominant.

Keywords: Variational problems with obstacles · Coincidence set
Error identities

1 Introduction

New types of error identities were recently derived [8] for two types of inequalities generated by obstacle type conditions: a classical obstacle problem and a two-phase obstacle problem. Both problems belong to the class of variational problems

$$\inf_{v \in V} J(v), \qquad J(v) = G(\Lambda v) + F(v), \tag{1}$$

where $\Lambda : V \to Y$ is a bounded linear operator, $G : Y \to \mathbb{R}$ is a convex, coercive, and lower semicontinuous functional, $F : V \to \mathbb{R}$ is another convex lower semicontinuous functional, and Y and V are reflexive Banach spaces. Henceforth, we use results of [6] related to derivation of a posteriori error estimates for this class of problems.

1.1 The Classical Obstacle Problem

The classical obstacle problem (see, e.g. [2,3]) is characterized by

$$G(\Lambda v) = \frac{1}{2} \int\limits_{\Omega} A \nabla v \cdot \nabla v \, dx, \qquad F(v) = - \int\limits_{\Omega} f v \, dx + \chi_K(v),$$

I. Lirkov and S. Margenov (Eds.): LSSC 2017, LNCS 10665, pp. 175–182, 2018.
https://doi.org/10.1007/978-3-319-73441-5_18

where the characteristic functional is defined as

$$\chi_K(v) := \begin{cases} 0 & \text{if } \phi \le v \le \psi, \\ +\infty & \text{else} \end{cases}$$

and the admissible set reads

$$K := \{v \in V_0 := H_0^1(\Omega) \mid \phi(x) \le v(x) \le \psi(x) \text{ a.e. in } \Omega\}.$$

Here, $H_0^1(\Omega)$ denotes the Sobolev space of functions vanishing on $\partial\Omega$ (hence we consider the case $u_D = 0$), $\Omega \subset \mathbb{R}^d$ ($d \in \{1,2,3\}$) is a bounded domain with a Lipschitz continuous boundary $\partial\Omega$ and $\phi, \psi \in H^2(\Omega)$ are two given functions (lower and upper obstacles) such that

$$\phi(x) \le 0 \text{ on } \partial\Omega, \quad \psi(x) \ge 0 \text{ on } \partial\Omega, \quad \phi(x) \le \psi(x), \quad \forall x \in \Omega.$$

It is assumed that A is a symmetric matrix subject to the condition

$$A(x)\xi \cdot \xi \ge c_1 |\xi|^2 \qquad c_1 > 0, \qquad \forall \xi \in \mathbb{R}^d \tag{2}$$

almost everywhere in Ω. Under the assumptions made, the unique solution $u \in K$ exists. The mechanical motivation of the obstacle problem is to find the equilibrium position of an elastic membrane whose boundary is held fixed, and which is constrained to lie between given lower and upper obstacles ϕ and ψ.

1.2 The Two-Phase Obstacle Problem

The functional $J(v)$ of the two-phase-obstacle problem (see, e.g. [9]) is defined by the relation

$$J(v) := \int_\Omega \left(\frac{1}{2} A\nabla v \cdot \nabla v - fv + \alpha_+(v)_+ + \alpha_-(v)_- \right) dx. \tag{3}$$

The functional $J(v)$ is minimized on the set

$$V_0 + u_D := \{v = v_0 + u_D \,:\, v_0 \in V_0,\ u_D \in H^1(\Omega)\}.$$

Here u_D is a given bounded function that defines the boundary condition (u_D may attain both positive and negative values on different parts of the boundary $\partial\Omega$). It is assumed that the coefficients $\alpha_+, \alpha_- : \Omega \to \mathbb{R}$ are positive constants (without essential difficulties the consideration and main results can be extended to the case where they are positive Lipschitz continuous functions). Also, it is assumed that $f \in L^\infty(\Omega)$, $A \in L^\infty(\Omega, \mathbb{R}^{d\times d})$, and the condition (2) holds. Since the functional $J(v)$ is strictly convex and continuous on V, existence and uniqueness of a minimizer $u \in V_0 + u_D$ is guaranteed by well known results of the calculus of variations. The mechanical motivation of the two-phase obstacle problem is to find the equilibrium position of an elastic membrane in the two-phase matter with different gravitation densities related to α_- and α_+.

2 Error Identities

The solution u of the classical obstacle problem divides Ω into three sets:

$$\begin{aligned}\Omega_-^u &:= \{x \in \Omega \mid u(x) = \phi(x)\}, \\ \Omega_+^u &:= \{x \in \Omega \mid u(x) = \psi(x)\}, \\ \Omega_0^u &:= \{x \in \Omega \mid \phi(x) < u(x) < \psi(x)\}.\end{aligned} \tag{4}$$

The sets Ω_-^u and Ω_+^u are the *lower* and *upper coincidence sets* and Ω_0^u is an open set, where u satisfies the Poisson equation $\mathrm{div}(A\nabla u) + f = 0$. Thus, the problem involves *free boundaries*, which are unknown a priori. Let v be an approximation of u. It defines approximate sets

$$\begin{aligned}\Omega_-^v &:= \{x \in \Omega \mid v(x) = \phi(x)\}, \\ \Omega_+^v &:= \{x \in \Omega \mid v(x) = \psi(x)\}, \\ \Omega_0^v &:= \{x \in \Omega \mid \phi(x) < v(x) < \psi(x)\}.\end{aligned} \tag{5}$$

Notice that unlike the sets in (4), the sets (5) are known.

Theorem 1 [8]. *Let $v \in K$ be any approximation of the exact solution $u \in K$ of the classical obstacle problem. Then it holds*

$$\tfrac{1}{2}||\nabla(u-v)||_A^2 + \mu_{\phi\psi}(v) = J(v) - J(u), \tag{6}$$

where

$$\mu_{\phi\psi}(v) := \int_{\Omega_-^u} \mathsf{W}_\phi(v-\phi)\,dx + \int_{\Omega_+^u} \mathsf{W}_\psi(\psi - v)\,dx, \tag{7}$$

and $\mathsf{W}_\phi := -(\mathrm{div}A\nabla\phi + f), \mathsf{W}_\psi := \mathrm{div}A\nabla\psi + f$ are two nonnegative weight functions generated by the source term f, the obstacles ψ, ϕ and the diffusion A.

Here, $\mu_{\phi\psi}(v)$ represents a certain (non-negative) measure, which controls (in a weak integral sense) whether or not the function v coincides with obstacles ψ, ϕ on true coincidence sets Ω_-^u and Ω_+^u.

Remark 1. The error identity (6) was derived for the homogeneous boundary condition $u = 0$ on $\partial\Omega$, but it is possible to extend it in the same form to for the nonhomogeneous boundary condition $u \neq 0$ on $\partial\Omega$.

For the two-phase obstacle problem, we introduce two decompositions (different from the classical obstacle problem) of Ω associated with the minimizer u and an approximation v:

$$\begin{aligned}\Omega_-^u &:= \{x \in \Omega \mid u(x) < 0\}, \\ \Omega_+^u &:= \{x \in \Omega \mid u(x) > 0\}, \\ \Omega_0^u &:= \{x \in \Omega \mid u(x) = 0\},\end{aligned} \tag{8}$$

and

$$\begin{aligned} \Omega_-^v &:= \{x \in \Omega \mid v(x) < 0\}, \\ \Omega_+^v &:= \{x \in \Omega \mid v(x) > 0\}, \\ \Omega_0^v &:= \{x \in \Omega \mid v(x) = 0\}. \end{aligned} \tag{9}$$

These decompositions generate exact and approximate free boundaries. If we introduce new sets

$$\omega_+ := \Omega_+^v \cap \Omega_0^u, \quad \omega_- := \Omega_-^v \cap \Omega_0^u, \quad \omega_\pm := \{\Omega_+^v \cap \Omega_-^u\} \cup \{\Omega_-^v \cap \Omega_+^u\},$$

we can formulate an error identity for the two-phase obstacle problem.

Theorem 2 [7,8]. *Let $v \in V_0 + u_D$ be any approximation of the exact solution $u \in V_0 + u_D$ of the two-phase obstacle problem. Then it holds*

$$\frac{1}{2}||\nabla(u-v)||_A^2 + \mu_\omega(v) = J(v) - J(u), \tag{10}$$

where

$$\mu_\omega(v) := \int_\omega \alpha(x)|v|\,dx, \quad \omega := \omega_+ \cup \omega_- \cup \omega_\pm \tag{11}$$

and

$$\alpha(x) := \begin{cases} \alpha(x) = \alpha_+ & \text{if } x \in \omega_+, \\ \alpha(x) = \alpha_- & \text{if } x \in \omega_-, \\ \alpha(x) = \alpha_+ + \alpha_- & \text{if } x \in \omega_\pm. \end{cases} \tag{12}$$

Here, $\mu_\omega(v)$ represents another nonlinear measure (which differs from $\mu_{\phi\psi}$).

3 Numerical Verifications

We verify a posteriori error identities (6) and (10) for both obstacle problems and focus on interpretation of their nonlinear measures $\mu_{\phi\psi}(\cdot)$ and $\mu_\omega(\cdot)$. Another goal is to present examples with different balance between two components of the overall error measure.

3.1 The Classical Obstacle Problem in 2D

We assume a 2D example taken from [5]. In this example, $\Omega = (-1,1)^2, A = \mathbb{I}, \phi = 0, \psi = +\infty$. It is known that for

$$f(x,y) = \begin{cases} -16(x^2+y^2) + 8R^2 & \text{if } \sqrt{x^2+y^2} > R \\ -8(R^4+R^2) + 8R^2(x^2+y^2) & \text{if } \sqrt{x^2+y^2} \le R \end{cases},$$

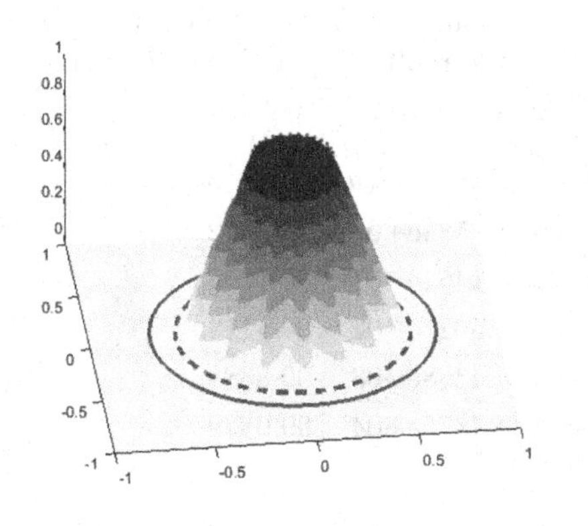
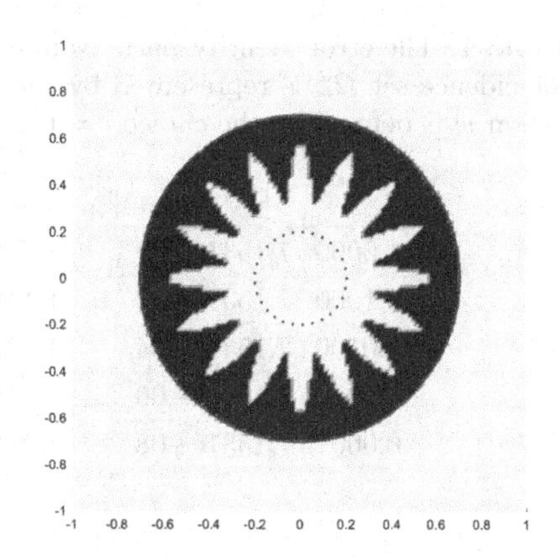

Fig. 1. A perturbation function w (left) generated by parameters $R = 0.7, r = 0.2, k = 16$ and the same corresponding coincidence set Ω_-^v (right) for all approximative solutions $v = u + \epsilon\, w$, where $\epsilon > 0$. The boundary of Ω_-^u is indicated by the full circle, the inner radius r by the dotted circle and the intermediate radius $\frac{r+3R}{4}$ by the dashed circle.

where $R \in [0, 1)$ is given, the exact solution to the obstacle problem reads

$$u(x,y) = \begin{cases} \left(\max\{x^2 + y^2 - R^2, 0\}\right)^2 & \text{if } (x,y) \in \Omega \\ \left(x^2 + y^2 - R^2\right)^2 & \text{if } (x,y) \in \partial\Omega \end{cases}.$$

The corresponding energy can be computed (see [4]) and it reads

$$J(u) = 192\left(\frac{12}{35} - \frac{28R^2}{45} + \frac{R^4}{3}\right) - 32R^2\left(\frac{28}{45} - \frac{4R^2}{3} + R^4\right) + \frac{2}{3}\pi R^8.$$

We consider approximations v in the form

$$v_\epsilon := u + \epsilon\, w, \tag{13}$$

where $\epsilon > 0$ is a given amplitude and w is a solution perturbation defined in polar coordinates (ρ, θ) as

$$w(\rho,\theta) := \begin{cases} 1, & \text{if } \rho \le r \\ 1 - \frac{\rho - r}{\tilde{r}(\theta) - r}, & \text{if } r \le \rho \le \tilde{r}(\theta) \\ 0, & \text{if } \rho \ge \tilde{r}(\theta) \end{cases} \tag{14}$$

Here, $0 < r < R$ is given internal radius and a variable radius $\tilde{r}(\theta)$ is defined as

$$\tilde{r}(\theta) := r + (R - r)\left(\frac{2 + \cos(k\theta)}{4}\right) \tag{15}$$

Table 1. The error identity parts computed for various $v_\epsilon = u + \epsilon w$, where the exact coincidence set Ω_-^u is represented by the circle of the radius $R = 0.7$ and the perturbation w is defined by the choice $r = 0.2, k = 16$.

ϵ	$\frac{1}{2}\|\nabla(u-v_\epsilon)\|_A^2$	$\mu_{\phi\psi}(v_\epsilon)$	$J(v_\epsilon) - J(u)$	$\kappa(v_\epsilon)$ [%]
1.0000	7.1531e+00	4.4311e+00	1.1584e+01	38.2512
0.1000	7.1531e−02	4.4311e−01	5.1464e−01	86.1008
0.0100	7.1531e−04	4.4311e−02	4.5027e−02	98.4113
0.0010	7.1531e−06	4.4311e−03	4.4388e−03	99.8388
0.0001	7.1531e−08	4.4311e−04	4.4375e−04	99.9839

for some $k \in \mathbb{Z}$. This construction ensures that

$$r < \frac{3r+R}{4} \le \tilde{r}(\theta) \le \frac{r+3R}{4} < R \tag{16}$$

and consequently ∇w is bounded. An examples of perturbations w is visualized in Fig. 1 together with corresponding coincidence sets Ω_-^v. For given k and r, there is always a convergence in the energy error

$$v_\epsilon \to u \qquad \text{(in } K) \qquad \text{as } \epsilon \to 0 \tag{17}$$

and consequently the nonlinear measure must also converge

$$\mu_{\phi\psi}(v_\epsilon) \to \mu_{\phi\psi}(u) = 0 \qquad \text{as } \epsilon \to 0. \tag{18}$$

It should be noted the shape of $\Omega_-^{v_\epsilon}$ depends on k and r only and it is completely independent of ϵ. Therefore, $\Omega_-^{v_\epsilon}$ never approximates $\Omega_-^u = \{x \in \Omega : ||x|| \le R\}$ for any choice of ϵ!

Table 1 reports on values of terms in the energy identity (6) for few approximations v_ϵ, where ϵ decreases to 0 and u and w are given by the choice of R and r, k. If ϵ tends to zero, the term $\frac{1}{2}\|\nabla(u-v_\epsilon)\|_A^2$ converges quadratically to 0 and the nonlinear measure $\mu_{\phi\psi}(v_\epsilon)$ only linearly to 0. The contribution of the nonlinear measure to the energy identity is measured by the quantity

$$\kappa(v_\epsilon) := 100 \, \frac{\mu_{\phi\psi}(v_\epsilon)}{J(v_\epsilon) - J(u)} \quad [\%]. \tag{19}$$

We see in this example, the contribution of $\mu_{\phi\psi}(v_\epsilon)$ dominates over the contribution of $\frac{1}{2}\|\nabla(u-v_\epsilon)\|_A^2$.

3.2 The Two-Phase Obstacle Problem in 1D

This subsection extends results of [7]. We consider the two-phase obstacle problem in 1D from [1]. Here, $\Omega = (-1,1), f = 0, A = \mathbb{I}, \alpha_\oplus = \alpha_\ominus = 8$ and the

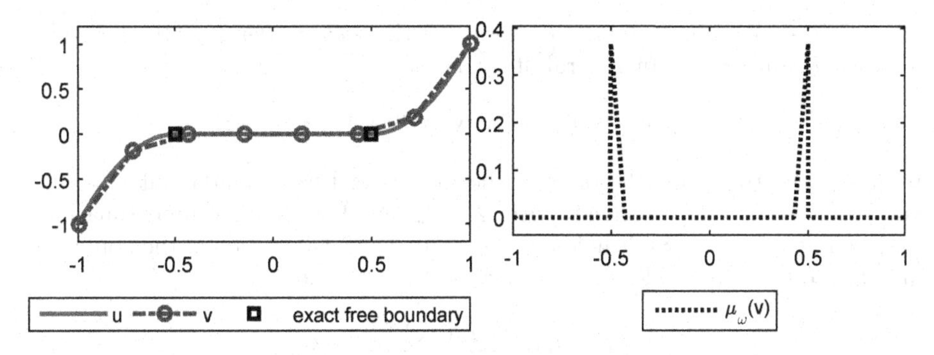

Fig. 2. Exact solution u of the two-phase obstacle problem and its approximations v_N (left) for $N = 8$ and the distributions of $\mu_\omega(v_N)$ (right).

Dirichlet boundary conditions $u(-1) = -1, u(1) = 1$. The exact solution is given by

$$u(x) = \begin{cases} -4x^2 - 4x - 1, & x \in [-1, -0.5], \\ 0, & x \in [-0.5, 0.5], \\ 4x^2 - 4x + 1, & x \in [0.5, 1] \end{cases}$$

and $J(u) = 5\frac{1}{3}$. We consider a sequence of approximations

$$v_N(x) = I_N(u)(x), \quad x \in [-1, 1],$$

where I_N (for $N = 2, 3, \ldots$) denotes a piecewise linear nodal interpolant of the function u in N uniformly distributed nodes $\{-1, -1+h, \ldots, 1-h, 1\}$, where

Table 2. The error identity terms computed for various approximation v_N.

N	$\frac{1}{2}\|\nabla(u - v_N)\|_A^2$	$\mu_\omega(v_N)$	$J(v_N) - J(u)$	$\kappa(v_N)$ [%]
2	1.67e+00	2.00e+00	3.67e+00	54.55
5	6.67e−01	0	6.67e−01	0.00
6	3.59e−01	7.20e−02	4.31e−01	16.72
7	2.59e−01	7.41e−02	3.33e−01	22.22
8	2.16e−01	2.62e−02	2.42e−01	10.82
9	1.67e−01	0	1.67e−01	0.00
10	1.20e−01	1.23e−02	1.32e−01	9.33
30	1.23e−02	3.69e−04	1.27e−02	2.91
60	3.06e−03	4.38e−05	3.10e−03	1.41
120	7.53e−04	5.34e−06	7.58e−04	0.70

$h = 2/(N-1)$. Table 2 reports on terms in the energy identity (10) for some increasing values of N. In general, it holds

$$\mu_\omega(v_N) = 0 \qquad \text{for } N = 4k+1, k \in \mathbb{N}.$$

In these cases, two interpolation nodes lie on the exact free boundary at $x = \pm 0.5$ and sets $\Omega^-_{v_N}, \Omega^0_{v_N}, \Omega^+_{v_N}$ coincide with $\Omega^-_u, \Omega^0_u, \Omega^+_u$. For all other approximations v_N (see Fig. 2 for $N = 8$), it holds $\mu_\omega(v_N) > 0$. The contribution of the nonlinear measure to the energy identity is measured by the quantity

$$\kappa(v_N) := 100 \, \frac{\mu_\omega(v_N)}{J(v_N) - J(u)} \quad [\%]. \tag{20}$$

We see in this benchmark, the contribution of $\frac{1}{2}\|\nabla(u - v_N)\|^2_A$ dominates over the contribution of the nonlinear measure term $\mu_\omega(v_N)$.

Acknowledgments. The first author acknowledges the support of RICAM during Special Semester on Computational Methods in Science and Engineering 2016, Linz, Austria. The second author has been supported by GA CR through the projects GF16-34894L and 17-04301S.

References

1. Bozorgnia, F.: Numerical solutions of a two-phase membrane problem. Appl. Numer. Math. **61**(1), 92–107 (2011)
2. Duvaut, G., Lions, G.-L.: Inequalities in Mechanics and Physics. Springer, Berlin (1976). https://doi.org/10.1007/978-3-642-66165-5
3. Ekeland, I., Temam, R.: Convex Analysis and Variational Problems. SIAM, Amsterdam (1976)
4. Harasim, P., Valdman, J.: Verification of functional a posteriori error estimates for obstacle problem in 2D. Kybernetika **50**(6), 978–1002 (2014)
5. Nochetto, R.H., Seibert, K.G., Veeser, A.: Pointwise a posteriori error control for elliptic obstacle problems. Numer. Math. **95**, 631–658 (2003)
6. Repin, S.: A posteriori error estimation for variational problems with uniformly convex functionals. Math. Comput. **69**(230), 481–500 (2000)
7. Repin, S., Valdman, J.: A posteriori error estimates for two-phase obstacle problem. J. Math. Sci. **20**(2), 324–336 (2015)
8. Repin, S., Valdman, J.: Error identities for variational problems with obstacles. ZAMM (in print)
9. Shahgholian, H., Uraltseva, N.N., Weiss, G.S.: The two-phase membrane problem regularity of the free boundaries in higher dimensions. Int. Math. Res. Not. **2007**(8), ID rnm026 (2007)

Control and Optimization of Dynamical Systems

An Optimal Control Problem with a Risk Zone

Sergey M. Aseev(✉)

Steklov Mathematical Institute of Russian Academy of Sciences, Moscow, Russia
aseev@mi.ras.ru
http://www.mathnet.ru/eng/person/8838

Abstract. We consider an optimal control problem for an autonomous differential inclusion with free terminal time in the situation when there is a set M ("risk zone") in the state space $\mathbb{R}^n$ which is unfavorable due to reasons of safety or instability of the system. Necessary optimality conditions in the form of Clarke's Hamiltonian inclusion are developed when the risk zone M is an open set. The result involves a nonstandard stationarity condition for the Hamiltonian. As in the case of problems with state constraints, this allows one to get conditions guaranteeing nondegeneracy of the developed necessary optimality conditions.

Keywords: Risk zone · State constraints · Optimal control
Differential inclusion · Hamiltonian inclusion · Stationarity condition

1 Statement of the Problem and Preliminaries

Consider the following problem (P):

$$J(T, x(\cdot)) = \varphi(T, x(0), x(T)) + \lambda \int_0^T \delta_M(x(t))\, dt \to \min\,, \tag{1}$$

$$\dot{x}(t) \in F(x(t)), \tag{2}$$

$$x(0) \in M_0, \qquad x(T) \in M_1. \tag{3}$$

Here $x \in \mathbb{R}^n$ is a state vector, M_0, M_1 are nonempty closed sets in $\mathbb{R}^n$, λ is a positive real, $F\colon \mathbb{R}^n \rightrightarrows \mathbb{R}^n$ is a locally Lipschitz multivalued mapping with nonempty convex compact values, $\varphi\colon [0,\infty) \times \mathbb{R}^n \times \mathbb{R}^n \mapsto \mathbb{R}^1$ is a locally Lipschitz function; $\delta_M(\cdot)$ is the characteristic function of a set M ("risk zone") in $\mathbb{R}^n$, i.e.

$$\delta_M(x) = \begin{cases} 1, & x \in M, \\ 0, & x \notin M. \end{cases} \tag{4}$$

We assume that M is a nonempty open set, $G = \mathbb{R}^n \setminus M \neq \emptyset$, and for any $x \in G$ the Clarke tangent cone $T_G(x)$ (see [9]) has nonempty interior, i.e. int $T_G(x) \neq \emptyset$. The terminal time $T > 0$ in problem (P) is assumed to be free; accordingly, the class of admissible trajectories in (P) consists of all absolutely continuous

I. Lirkov and S. Margenov (Eds.): LSSC 2017, LNCS 10665, pp. 185–192, 2018.
https://doi.org/10.1007/978-3-319-73441-5_19

solutions $x(\cdot)$ of differential inclusion (2) defined on corresponding time intervals $[0,T]$, $T > 0$, and satisfying boundary conditions (3). An admissible trajectory $x_*(\cdot)$ defined on a time interval $[0,T_*]$, $T_* > 0$, is optimal in problem (P) if the functional $J(\cdot,\cdot)$ (see (1)) reaches the minimal possible value at $(T_*, x_*(\cdot))$.

Notice, that the peculiarity of problem (P) consists of the presence of discontinuous integrand $\delta_M(\cdot)$ in the integral term in the functional $J(\cdot,\cdot)$. Substantially, the integral term penalizes the states in the risk zone M. Such risk zones could appear in statements of different applied problems when there is an admissible but unfavorable set M in the state space $\mathbb{R}^n$. In economics the set M can correspond to the states with high probability of bankruptcy; in ecology the set M can correspond to the states with high probability of the system degradation; in engineering such sets can correspond to the states of overloading or instability of the system.

In classical optimal control theory the presence of such unfavorable set M is modeled usually via introducing an additional state constraint (see [15, Chapt. 6])

$$x(t) \in G = \mathbb{R}^n \setminus M, \qquad t \in [0,T].$$

Substantially, this means that presence of the state variable $x(\cdot)$ in the set M is prohibited. The set G ("safety zone") is assumed to be closed in this case (i.e. the set M is open).

An optimal control problem with a closed convex risk zone M was initially considered in [16] in the case of linear control system, and under some a priori regularity assumptions on behavior of an optimal trajectory $x_*(\cdot)$. In particular, it was assumed in [16] that the optimal trajectory $x_*(\cdot)$ had a finite number of intersection points with the boundary of the set M. In [17] under the same linearity and regularity assumptions the case of time dependent closed convex set $M = M(t)$, $t \in [0,T]$, was considered. In [7,8] the problem of optimal crossing a given closed risk zone M was studied and necessary optimality conditions for affine in control system were developed without any a priori assumptions on the behavior of the optimal trajectory. In [18] this result (in the case of closed set M) was generalized to the case of more general integral utility functional. The main novelty of the present work is that the risk zone M is assumed to be open. In this case introducing of the risk zone M in the statement of problem (P) can be considered as a weakening of the classical concept of the state constraint in optimal control. Notice also, that the approach developed in [7,8,18] for the case of the closed set M does not work if the set M is open.

In that follows $N_A(a) = T_A^*(a)$ and $\hat{N}_A(a)$ are the Clarke normal cone [9] and the cone of generalized normals [13] to the closed set $A \subset \mathbb{R}^n$ at a point $a \in A$, respectively; ∂A is the boundary of the set A; $H(F(x), \psi) = \max_{f \in F(x)} \langle f, \psi \rangle$ is the value of the Hamiltonian $H(F(\cdot),\cdot)$ of differential inclusion (2) at a point $(x,\psi) \in \mathbb{R}^n \times \mathbb{R}^n$; $\partial H(F(x),\psi)$ is the Clarke subdifferential of the locally Lipschitz function $H(F(\cdot),\cdot)$ at a point $(x,\psi) \in \mathbb{R}^n \times \mathbb{R}^n$ [9], and $\hat{\partial}\varphi(T, x_1, x_2)$ is the generalized gradient of locally Lipschitz function $\varphi(\cdot,\cdot,\cdot)$ at a point $(T, x_1, x_2) \in [0,\infty) \times \mathbb{R}^n \times \mathbb{R}^n$ [13].

For $i \in \mathbb{N}$ and an arbitrary $x \in \mathbb{R}^n$ set $\tilde{\delta}_i(x) = \min\{i\rho(x, G), \delta_M(x)\}$ where $\rho(x, G) = \min\{\|x - \xi\| : \xi \in G\}$ is the distance from a point x to the nonempty closed set $G = \mathbb{R}^n \setminus M$ and the function $\delta_M(\cdot)$ is defined by equality (4).

Further, for $i \in \mathbb{N}$ let us define the function $\delta_i : \mathbb{R}^n \mapsto \mathbb{R}^1$ by equality

$$\delta_i(x) = \int_{\mathbb{R}^n} \tilde{\delta}_i(x + y)\omega_i(y)\, dy. \tag{5}$$

Here $\omega_i(\cdot)$ is a smooth ($C^\infty(\mathbb{R}^n)$) probabilistic density such that $\operatorname{supp} \omega_i(\cdot) \subset 1/2^i B$ where B is the closed unit ball in $\mathbb{R}^n$ with the center in 0. Then for any $i \in \mathbb{N}$ the function $\delta_i(\cdot)$ is smooth as a convolution with $\omega_i(\cdot)$.

The following auxiliary statements hold.

Lemma 1. *For any $x \in \mathbb{R}^n$ we have*

$$\delta_i(x) \le \delta_M(x) + \frac{i}{2^i}, \qquad i \in \mathbb{N}. \tag{6}$$

Proof. Indeed, if $x \in M$ then $\delta_M(x) = 1$. Since $\delta_i(x) \le 1$, $i \in \mathbb{N}$, inequality (6) is obviously satisfied. Now assume $x \notin M$. Then $\delta_M(x) = 0$, and for any $y \in \operatorname{supp} \omega_i(\cdot)$, $i \in \mathbb{N}$, we have $\tilde{\delta}_i(x + y) \le i\rho(x + y, G) \le iy \le i/2^i$. Due to the definition of the function $\delta_i(\cdot)$ (see (5)) we get

$$\delta_i(x) = \int_{\mathbb{R}^n} \tilde{\delta}_i(x + y)\omega_i(y)\, dy \le \frac{i}{2^i}, \qquad i \in \mathbb{N}.$$

Since $\delta_M(x) = 0$ inequality (6) also holds in this case. □

Lemma 2. *Let a sequence $\{x_i(\cdot)\}_{i=1}^\infty$ of continuous functions $x_i : [0, T] \mapsto \mathbb{R}^n$ defined on some time interval $[0, T]$, $T > 0$, converges uniformly to a continuous function $\tilde{x} : [0, T] \mapsto \mathbb{R}^n$. Then*

$$\liminf_{i \to \infty} \int_0^T \delta_i(x_i(t))\, dt \ge \int_0^T \delta_M(\tilde{x}(t))\, dt. \tag{7}$$

Proof. Assume that for some $t \in [0, T]$ we have $\tilde{x}(t) \in M$. Then $\delta_M(\tilde{x}(t)) = 1$, and since the set M is open and the sequence $\{x_i(\cdot)\}_{k=1}^\infty$ converges uniformly to $\tilde{x}(\cdot)$ there are $\varepsilon_0 > 0$ and $i_0 \ge 1/\varepsilon_0$ such that for all $i \ge i_0$ we have $x_i(t) + \varepsilon_0 B \subset M$. Then for all $i \ge i_0$ due to definition of function $\delta_i(\cdot)$ (see (5)) we get equality $\delta_i(x_i(t)) = 1$. Hence, $\lim_{i\to\infty} \delta_i(x_i(t)) = \delta_M(\tilde{x}(t)) = 1$ in this case. Now, assume that $t \in [0, T]$ is such that $\tilde{x}(t) \notin M$. Then $\delta_M(\tilde{x}(t)) = 0$. As far as $\delta_i(x_i(t)) \ge 0$ for any $t \in [0, T]$ and all $i \in \mathbb{N}$ (see (5)) we have $\liminf_{i\to\infty} \delta_i(x_i(t)) \ge \delta_M(\tilde{x}(t))$ in this case.

Thus, for any $t \in [0, T]$ the following inequality holds:

$$\liminf_{i \to \infty} \delta_i(x_i(t)) \ge \delta_M(\tilde{x}(t)).$$

From this inequality due to Fatou's lemma (see [10, Lemma 8.7.i.]) we get (7). □

As an immediate corollary of the lemmas above we get the following result.

Theorem 1. *The integral functional* $J_M\colon C([0,T],\mathbb{R}^n) \mapsto \mathbb{R}^1$, $T > 0$, *defined by the equality*

$$J_M(x(\cdot)) = \int_0^T \delta_M(x(t))\,dt$$

is lower semicontinuous.

Proof. Indeed, let $T > 0$ and a sequence $\{x_i(\cdot)\}_{i=1}^{\infty}$ of continuous functions $x_i\colon [0,T] \mapsto \mathbb{R}^n$ converges to a continuous function $\tilde{x}(\cdot)$ in $C([0,T],\mathbb{R}^n)$. Then due to Lemma 1 we have

$$J_M(x_i(\cdot)) = \int_0^T \delta_M(x_i(t))\,dt \geq \int_0^T \delta_i(x_i(t))\,dt - \frac{iT}{2^i}, \qquad i \in \mathbb{N}.$$

Hence, due to Lemma 2 passing to a limit as $i \to \infty$ we get

$$\liminf_{i\to\infty} J_M(x_i(\cdot)) \geq \liminf_{i\to\infty} \int_0^T \delta_i(x_i(t))\,dt \geq \int_0^T \delta_M(\tilde{x}(t))\,dt = J_M(\tilde{x}(\cdot)).$$

□

2 Main Result

Let $x_*(\cdot)$ be an optimal admissible trajectory in (P), and let $T_* > 0$ be the corresponding optimal terminal time. In that follows we always assume that $x_*(\cdot)$ is defined on the time interval $[T_*,\infty)$ as a constant: $x_*(t) \equiv x_*(T_*)$, $t \geq 0$. Define also the sets $\tilde{M}_0$ and $\tilde{M}_1$ by the equalities

$$\tilde{M}_0 = \begin{cases} M_0, & x_*(0) \in M, \\ M_0 \bigcap G, & x_*(0) \in G \end{cases} \quad \text{and} \quad \tilde{M}_1 = \begin{cases} M_1, & x_*(T_*) \in M, \\ M_1 \bigcap G, & x_*(T_*) \in G. \end{cases} \tag{8}$$

Next theorem is the main result of the present paper.

Theorem 2. *Let* $x_*(\cdot)$ *be an optimal admissible trajectory in problem* (P), *and let* $T_* > 0$ *be the corresponding optimal terminal time. Then there are a constant* $\psi^0 \geq 0$, *an absolutely continuous function* $\psi\colon [0,T_*] \mapsto \mathbb{R}^n$ *and a bounded regular Borel vector measure* η *on* $[0,T_*]$ *such that the following conditions hold:*

(1) the measure η *is concentrated on the set* $\mathfrak{M} = \{t \in [0,T_*]\colon x_*(t) \in \partial G\}$, *and it is nonpositive on the set of continuous functions* $y\colon \mathfrak{M} \mapsto \mathbb{R}^n$ *with values* $y(t) \in T_G(x_*(t))$, $t \in \mathfrak{M}$, *i.e.*

$$\int_{\mathfrak{M}} y(t)\,d\eta \leq 0;$$

(2) for a.e. $t \in [0, T_*]$ the Hamiltonian inclusion holds:

$$(-\dot{\psi}(t), \dot{x}_*(t)) \in \partial H(x_*(t), \psi(t) + \lambda \int_0^t d\eta);$$

(3) for $t = T_*$ and for any $t \in [0, T_*)$ which is a point of right approximate continuity[1] *of the function $\delta_M(x_*(\cdot))$ the following stationarity condition holds:*

$$H(x_*(t), \psi(t) + \lambda \int_0^t d\eta) - \psi^0 \lambda \delta_M(x_*(t)) = H(x_*(0), \psi(0)) - \psi^0 \lambda \delta_M(x_*(0));$$

(4) the transversality condition holds:

$$(H(x_*(T_*), \psi(T_*) + \lambda \int_0^{T_*} d\eta), \psi(0), -\psi(T_*) - \lambda \int_0^{T_*} d\eta) \in \psi^0 \hat{\partial} \phi(T_*, x_*(0), x_*(T_*)) + \{0\} \times \hat{N}_{\tilde{M}_0} \times \hat{N}_{\tilde{M}_1};$$

(5) the nontriviality condition holds:

$$\psi^0 + \|\psi(0)\| + \|\eta\| \neq 0.$$

The proof of Theorem 2 is based on approximation of problem (P) by a sequence of approximating problems with Lipschitz data for which the corresponding necessary optimality conditions are known (see [9, Theorem 5.2.1]).

Let $x_*(\cdot)$ be an optimal admissible trajectory in problem (P), and let $T_* > 0$ be the corresponding optimal terminal time. For $i \in \mathbb{N}$ consider the following optimal control problem (P_i):

$$J_i(T, x(\cdot)) = \varphi(T, x(0), x(T)) + (T - T_*)^2 + \int_0^T \left[\lambda \delta_i(x(t)) + \|x(t) - x_*(t)\|^2\right] dt \to \min, \quad (9)$$

$$\dot{x}(t) \in F(x(t)), \quad (10)$$

$$|T - T_*| \leq 1, \qquad \|x(t) - x_*(t)\| \leq 1, \quad t \in [0, T], \quad (11)$$

$$x(0) \in \tilde{M}_0, \qquad x(T) \in \tilde{M}_1. \quad (12)$$

Here the function $\varphi(\cdot, \cdot, \cdot)$, the multivalued mapping $F(\cdot)$ and the number $\lambda > 0$ are the same as in (P). The sets $\tilde{M}_0$ and $\tilde{M}_1$ are defined in (8). As in the problem (P), the set of admissible trajectories in (P_i), $i \in \mathbb{N}$, consists of all absolutely continuous solutions $x(\cdot)$ of differential inclusion (10) defined on their own time

[1] Recall, that $t \in [0, T)$, $T > 0$, is a point of right approximate continuity of a real function $\xi(\cdot)$ defined on $[0, T]$ if there is a Lebesgue measurable set $E \subset [t, T]$ such that t is its density point, and the function $\xi(\cdot)$ is continuous from the right at t along E (see [14, Chapt. 9, Sect. 5]).

intervals $[0,T], T>0$, and satisfying constraints in (11) and boundary conditions in (12).

For any $i \in \mathbb{N}$ the problem (P_i) is a standard optimal control problem for the differential inclusion with Lipschitz data, state and terminal constraints (see [9, Sect. 3.6]). Since $x_*(\cdot)$ is an admissible trajectory in (P_i), $i \in \mathbb{N}$, due to Filippov's existence theorem (see, [10, Theorem 9.3.i]) for any $i \in \mathbb{N}$ there is an optimal admissible trajectory $x_i(\cdot)$ in (P_i) which is defined on the corresponding time interval $[0,T_i]$, $T_i>0$. We will assume bellow that for any $i \in \mathbb{N}$ the trajectory $x_i(\cdot)$ is extended to the infinite time interval $[T_i,\infty)$ as a constant: $x_i(t) \equiv x_i(T_i))$, $t \geq T_i$.

We will call $\{(P_i)\}_{k=1}^{\infty}$ a sequence of approximating problems corresponding to the optimal trajectory $x_*(\cdot)$.

Theorem 3. *Let $x_*(\cdot)$ be an optimal admissible trajectory in problem (P), and let T_* be the corresponding optimal terminal time. Let $\{(P_i)\}_{i=1}^{\infty}$ be the sequence of approximating problems corresponding to $x_*(\cdot)$, and let $x_i(\cdot)$, $T_i>0$, be an optimal admissible trajectory and the corresponding optimal time, respectively in (P_i), $i \in \mathbb{N}$. Then*

$$\lim_{i\to\infty} T_i = T_*, \tag{13}$$

$$\lim_{i\to\infty} x_i(\cdot) = x_*(\cdot) \qquad in \quad C([0,T_*],\mathbb{R}^n), \tag{14}$$

$$\lim_{i\to\infty} \dot{x}_i(\cdot) = \dot{x}_*(\cdot) \qquad weakly\,in \quad L^1([0,T_*],\mathbb{R}^n), \tag{15}$$

$$\lim_{i\to\infty} \int_0^{T_i} \delta_i(x_i(t))\,dt = \int_0^{T_*} \delta_M(x_*(t))\,dt. \tag{16}$$

Proof. Since $x_i(\cdot)$ is an optimal admissible trajectory in (P_i), $i \in \mathbb{N}$, and $x_*(\cdot)$ is an admissible trajectory in (P_i), due to Lemma 1 we have (see (9) and (6)):

$$\begin{aligned}
\varphi(T_i, x_i(0), x_i(T_i)) + (T_i - T_*)^2 + \int_0^{T_i} \left[\lambda\delta_i(x_i(t)) + \|x_i(t)-x_*(t)\|^2\right] dt \\
\leq \varphi(T_*, x_*(0), x_*(T_*)) + \lambda \int_0^{T_*} \delta_i(x_*(t))\,dt \\
\leq \varphi(T_*, x_*(0), x_*(T_*)) + \lambda \int_0^{T_*} \delta_M(x_*(t))\,dt + \frac{i\lambda T_*}{2^i}.
\end{aligned} \tag{17}$$

Since $|T_i - T_*| \leq 1$, $i \in \mathbb{N}$, without loss of generality we can assume that $\lim_{i\to\infty} T_i = \tilde{T} \leq T_* + 1$. Further, the set of all admissible trajectories of (10) satisfying the state constraint (11) is a compactum in $C([0,\tilde{T}],\mathbb{R}^n)$. Let $\tilde{x}(\cdot)$ be a limit point of $\{x_i(\cdot)\}_{i=1}^{\infty}$ in $C([0,\tilde{T}],\mathbb{R}^n)$. Then $\tilde{x}(\cdot)$ is an admissible trajectory in (P), and passing to a subsequence we can assume that $\lim_{i\to\infty} x_i(\cdot) = \tilde{x}(\cdot)$ in $C([0,\tilde{T}],\mathbb{R}^n)$. Further, $x_*(\cdot)$ is an optimal trajectory in (P), while $\tilde{x}(\cdot)$ is an admissible one in this problem. Hence,

$$\varphi(T_*, x_*(0), x_*(T_*)) + \lambda \int_0^{T_*} \delta_M(x_*(t))\,dt \leq \varphi(\tilde{x}(0), \tilde{x}(\tilde{T})) + \lambda \int_0^{\tilde{T}} \delta_M(\tilde{x}(t))\,dt.$$

Hence, for $i \in \mathbb{N}$ due to (17) we get

$$\varphi(T_i, x_i(0), x_i(T_i)) - \varphi(\tilde{T}, \tilde{x}(0), \tilde{x}(\tilde{T})) + \lambda \int_0^{T_i} \delta_i(x_i(t))\, dt - \lambda \int_0^{\tilde{T}} \delta_M(\tilde{x}(t))\, dt$$
$$+ (T_i - T_*)^2 + \int_0^{T_i} \|x_i(t) - x_*(t)\|^2\, dt \le \frac{i\lambda T_*}{2^i}. \quad (18)$$

Since $\lim_{i\to\infty} T_i = \tilde{T}$ and $\lim_{i\to\infty} x_i(\cdot) = \tilde{x}(\cdot)$ in $C([0,\tilde{T}], \mathbb{R}^n)$ due to Lemma 2 for any $\varepsilon > 0$ there is a natural i_0 such that for all $i \ge i_0$ we have

$$\varphi(T_i, x_i(0), x_i(T_i)) - \varphi(\tilde{T}, \tilde{x}(0), \tilde{x}(\tilde{T})) \ge -\varepsilon,$$
$$\int_0^{T_i} \delta_i(x_i(t))\, dt - \int_0^{\tilde{T}} \delta_M(\tilde{x}(t))\, dt \ge -\varepsilon.$$

From these inequalities due to (18) for any $i \ge i_0$ we get

$$(T_i - T_*)^2 + \int_0^{T_i} \|x_i(t) - x_*(t)\|^2\, dt \le \varepsilon(1 + \lambda) + \frac{i\lambda T_*}{2^i}.$$

Passing to a limit as $i \to \infty$ in the inequality above we get

$$\limsup_{i\to\infty} \left[(T_i - T_*)^2 + \int_0^{T_i} \|x_i(t) - x_*(t)\|^2\, dt \right] \le \varepsilon(1 + \lambda).$$

Since $\varepsilon > 0$ is an arbitrary positive number this implies

$$\lim_{i\to\infty} T_i = T_*, \qquad \lim_{i\to\infty} \int_0^{T_*} \|x_i(t) - x_*(t)\|^2\, dt = 0.$$

Thus, equality (13) is proved. Since $\lim_{i\to\infty} T_i = \tilde{T} = T_*$ and $\tilde{x}(\cdot)$ is an arbitrary limit point of the sequence $\{x(\cdot)\}_{i=1}^{\infty}$ in $C([0,\tilde{T}], \mathbb{R}^n)$ we get (14). Equality (15) is followed by (14) and the fact that the sequence $\{\dot{x}_i(\cdot)\}_{i=1}^{\infty}$ is bounded in $L_\infty([0,T_*], \mathbb{R}^n)$. Finally, due to Lemma 2 equality (16) follows from (13), (14) and (18). □

Due to condition (14) of Theorem 3 for all sufficiently large numbers i the terminal time and state constraints in (11) hold as strict ones. Hence, the Clarke necessary conditions (see [9, Theorem 5.2.1])) hold for optimal trajectories $x_i(\cdot)$ in problems (P_i) for all sufficiently large numbers i. The subsequent proof of Theorem 2 is based on the limiting procedure in these necessary optimality conditions applied to problems (P_i), $i \in \mathbb{N}$, as $i \to \infty$. It is similar to the proof of analogous results for problems with state constraints (see [3,5, Theorem 1]). The detailed proof of a similar result for problem (P) in the case of a fixed time interval $[0, T]$, $T > 0$, is presented in [6].

Notice, that Theorem 2 is similar to the necessary conditions for optimality for an optimal control problem for the differential inclusion with state constraints proved in [3]. As in [3], the stationarity condition (3) allows one to get sufficient conditions for nondegeneracy of the developed necessary optimality conditions (Theorem 2). Other results on nondegeneracy of different versions of the maximum principle for problems with state constraints and further references can be found in [1–5,11,12].

Acknowledgements. This work is supported by the Russian Science Foundation under grant 14-50-00005.

References

1. Arutyunov, A.V.: Perturbations of extremal problems with constraints and necessary optimality conditions. J. Sov. Math. **54**(6), 1342–1400 (1991)
2. Arutyunov, A.V.: Optimality Conditions: Abnormal and Degenerate Problems. Kluwer, Dordrecht (2000)
3. Arutyunov, A.V., Aseev, S.M.: Investigation of the degeneracy phenomenon of the maximum principle for optimal control problems with state constraints. SIAM J. Control Optim. **35**(3), 930–952 (1997)
4. Arutyunov, A.V., Karamzin, D.Y., Pereira, F.L.: The maximum principle for optimal control problems with state constraints by R.V. Gamkrelidze: revisited. J. Optim. Theor. Appl. **149**(3), 474–493 (2011)
5. Aseev, S.M.: Methods of regularization in nonsmooth problems of dynamic optimization. J. Math. Sci. **94**(3), 1366–1393 (1999)
6. Aseev, S.M.: Optimization of dynamics of a control system in the presence of risk factors. Trudy Inst. Mat. i Mekh UrO RAN **23**(1), 27–42 (2017). (in Russian)
7. Aseev, S.M., Smirnov, A.I.: The Pontryagin maximum principle for the problem of optimal crossing of a given domain. Dokl. Math. **69**(2), 243–245 (2004)
8. Aseev, S.M., Smirnov, A.I.: Necessary first-order conditions for optimal crossing of a given region. Comput. Math. Model. **18**(4), 397–419 (2007)
9. Clarke, F.: Optimization and Nonsmooth Analysis. Wiley, New York (1983)
10. Cesari, L.: Optimization - Theory and Applications: Problems with Ordinary Differential Equations. Springer, New York (1983). https://doi.org/10.1007/978-1-4613-8165-5
11. Ferreira, M.M.A., Vinter, R.B.: When is the maximum principle for state constrained problems nondegenerate? J. Math. Anal. Appl. **187**(2), 438–467 (1994)
12. Fontes, F.A.C.C., Frankowska, H.: Normality and nondegeneracy for optimal control problems with state constraints. J. Optim. Theor. Appl. **166**(1), 115–136 (2015)
13. Mordukhovich, B.S.: Approximation Methods in Problems of Optimization and Control. Nauka, Moscow (1988). (in Russian)
14. Natanson, I.P.: Theory of Functions of a Real Variable. Frederick Ungar, New York (1961)
15. Pontryagin, L.S., Boltyanskii, V.G., Gamkrelidze, R.V., Mishchenko, E.F.: The mathematical theory of optimal processes. Pergamon, Oxford (1964)
16. Pshenichnyi, B.N., Ochilov, S.: On the problem of optimal passage through a given domain. Kibern. Vychisl. Tekh. **99**, 3–8 (1993)
17. Pshenichnyi, B.N., Ochilov, S.: A special problem of time-optimal control. Kibern. Vychisl. Tekhn. **101**, 11–15 (1994)
18. Smirnov, A.I.: Necessary optimality conditions for a class of optimal control problems with discontinuous integrand. Proc. Steklov Inst. Math. **262**, 213–230 (2008)

Spreading Rumors and External Actions

Séverine Bernard, Ténissia César, and Alain Piétrus(✉)

LAboratoire de Mathématiques Informatique et Applications, EA 4540,
Université des Antilles, Campus de Fouillole, BP 250,
97159 Pointe à Pitre Cedex, Guadeloupe
{Severine.Bernard,Tenissia.Cesar,Alain.Pietrus}@univ-antilles.fr

Abstract. In this paper, we consider a population of a social network in which a fake news propagates and divides it into four categories: ignorants, spreaders, stiflers who accept the rumor, and stiflers who oppose the rumor. Starting from a SIR type model describing the propagation of e-rumor, we modify it by adding some external actions and control them in order to reduce the spread of a bad information. To carry out this investigation, we use known facts from optimal control theory. Numerical simulations illustrate the efficiency of the obtained control strategy.

Keywords: Fake news spreading · External actions
Optimal control · Pontryagin's maximum principle

1 Introduction

In the age of media, social networks became excellent and easy tools for propagation of rumors, no matter whether true or not, since they allow rumors to occur and spread broadly and quickly. But rumors have an important impact on people's life which can cause a lot of damages in a society, since they can circulate around politics, financial markets and private life domains for instance. Consequently, it is important to know the dynamics of information spreading on social networks in order to counter its effects and to apply good strategies to control it.

The propagation of information has been early assimilated to an epidemic problem, by using SIR type models which are more accurate in describing the dynamic of social network [10]. We can cite in particular the standard model proposed by Daley and Kendall in [5,6] in the 1960's and the MK model in [11]. The construction of their model is based on the consideration of a network of individuals that they divide into three groups: the one of ignorants, the one of spreaders and the one of stiflers, that is those who know the rumor but do not spread it. In the litterature, the reader could find many studies focussed on this subject in order to firstly describe the dynamic of this phenomenon and secondly to better prevent its propagation [2–6,8,10,11,14].

In particular in [8], Huang and Jin presented a model for which a random and targeted immunization strategy has been introduced in order to prevent the spread of rumor when the size of the social network is not too large.

I. Lirkov and S. Margenov (Eds.): LSSC 2017, LNCS 10665, pp. 193–200, 2018.
https://doi.org/10.1007/978-3-319-73441-5_20

Later, Bernard et al. modified their model by putting all the transmission parameters depending on the time, which seems more realistic, and proposed an optimal control approach to restrict the rumor's propagation in [2,3]. The model that they used is the following:

$$\begin{cases} \dot{I}(t) = -(\lambda(t) + \alpha(t) + \beta(t))I(t)S(t), \\ \dot{S}(t) = \lambda(t)I(t)S(t) - \theta(t)S(t)(S(t) + RA(t) + RU(t)), \\ \dot{RA}(t) = \theta(t)S(t)(S(t) + RA(t) + RU(t)) + \alpha(t)I(t)S(t), \\ \dot{RU}(t) = \beta(t)I(t)S(t), \end{cases} \tag{1}$$

with $I(0) = \frac{N-1}{N}$, $S(0) = \frac{1}{N}$ and $RA(0) = RU(0) = 0$, where N is the number of persons of the network and I(.), S(.), RA(.), RU(.) the density of ignorants, spreaders, stiflers who accept the rumor and stiflers who oppose the rumor respectively. Moreover, $\lambda(.)$ is the rate at which an ignorant becomes a spreader after having met the last one, $\alpha(.)$ is the rate at which an ignorant becomes a stifler who accept the rumor after having met a spreader, $\beta(.)$ is the rate at which an ignorant becomes a stifler who oppose the rumor after having met a spreader and $\theta(.)$ is the rate at which a spreader becomes a stifler who accept the rumor after having met a spreader or a stifler. Consequently, for all $t \in \mathbb{R}_+$, the coefficients $\alpha(t)$, $\beta(t)$, $\lambda(t)$ and $\theta(t)$ are non negative real numbers and $\alpha(t)+\beta(t)+\lambda(t) \leq 1$. By noting that, at all time t, $I(t) + S(t) + RA(t) + RU(t) = 1$, we can simplify (1) as

$$\begin{cases} \dot{I}(t) = -(\lambda(t) + \alpha(t) + \beta(t))I(t)S(t), \\ \dot{S}(t) = \lambda(t)I(t)S(t) - \theta(t)S(t)(1 - I(t)). \end{cases} \tag{2}$$

In their first (respectively second) approach [2] (respectively [3]), Bernard et al. control the rate at which an ignorant becomes a spreader after having met the last one (respectively the rate for which a spreader becomes a stifler who accepts the rumor after having met a spreader or a stifler) in order to minimize the propagation of the rumor. In each cases, these authors showed existence and gave a characterization of the optimal control parameter.

In the following, as it has been done for epidemic models in [7,9] for example, some realistic external actions have been added in (2) and controlled separately in [4] and simultaneously in this work, with the same objective of fake news spreading minimization. In this state of minds, we present the modified model and show the existence of an optimal control in Sect. 2, in order to characterize it in Sect. 3. The last part will be devoted to numerical experiments in order to highlight our theoritical results. Note that this kind of strategy has been investigated for epidemic models in [7,9] but in these last works, each action takes place only on one of the two densities of the population and the control is bang-bang due to the choice of a linear objective function. In the present work, we consider that a fake news propagates on a social network and we act by introducing a counter-information in this network and this last one acts both on the density of ignorants and the one of spreaders. The second action consists of isolating the spreaders, for example the more active ones. The other difference with what has been done for epidemics in [7,9] is that we choose a nonlinear objective function in order to minimize the propagation of the fake news.

2 Control of both Counter-Information and Isolated Spreaders

In this work, we consider a social network on which circulates a fake news and add two external actions, one may be considered as the probability of persons who learn a counter-information of the fake news and a second as the isolation of some of the spreaders. These considerations imply a modification of model (2) as

$$\begin{cases} \dot{I}(t) = -(\lambda(t) + \alpha(t) + \beta(t))I(t)S(t) - u_v(t)I(t), \\ \dot{S}(t) = \lambda(t)I(t)S(t) - \theta(t)S(t)(1 - I(t)) - (u_v(t) + u_i(t))S(t), \end{cases} \tag{3}$$

with $I(0) = \frac{N-N_S-N_{RA}-N_{RU}}{N}$, $S(0) = \frac{N_S}{N}$, $RA(0) = \frac{N_{RA}}{N}$, $RU(0) = \frac{N_{RU}}{N}$, where N_S, N_{RA}, N_{RU} are the number of spreaders, stiflers who accept the rumor and stiflers who oppose the rumor respectively and u_v (respectively u_i) represents the probability of people who learned the counter-information (respectively the probability of isolated spreaders). We assume here that the bad information has already begun to spread in order to investigate the actions strategy, which explains our choice of initial conditions. By choosing (u_v, u_i) as control, the problem can be written as

$$\begin{cases} \dot{X}(t) = F(t, X(t), U(t)), \ \forall t > 0 \\ X(0) = X_0, \end{cases} \tag{4}$$

where $X(t) = (I(t), S(t))$, $X_0 = (I(0), S(0))$, $U(t) = (u_v(t), u_i(t))$, for all $t > 0$, and

$$F(t, X(t), U(t)) = \begin{pmatrix} -(\lambda(t) + \alpha(t) + \beta(t))I(t)S(t) - u_i(t)I(t) \\ \lambda(t)I(t)S(t) - \theta(t)S(t)(1 - I(t)) - (u_v(t) + u_i(t))S(t) \end{pmatrix}.$$

Theorem 1. *Let J be an interval of $\mathbb{R}$, V and W open sets of $\mathbb{R}^2$ and $F : J \times V \times W \to V$ the function defined as previously. For all fixed control (u_v, u_i), there exists one and only one maximal solution $([0, t_m(u_v, u_i)], X_{(u_v,u_i)}(.))$ of the Cauchy problem (4), with $t_m(u_v, u_i) \in \mathbb{R}_+ \cup \{+\infty\}$.*

Proof. Since u_v and u_i represent probabilities then, for all $t \geq 0$, we have $0 \leq u_v(t), u_i(t) \leq 1$, which implies that $u_v, u_i \in L^\infty(J, \mathbb{R})$. The fact that F satisfies the assumptions of Cauchy-Lipschitz's theorem completes the proof. □

Let us set $T > 0$, $T \in J$ and $\mathcal{U}_T = \{(u_v, u_i) \ / \ X_{(u_v,u_i)}(.) \ exists\}$ that is $T < t_m(u_v, u_i)$ and let us define the cost by

$$C(u_v, u_i) = \int_0^T (aS(t) - b[\ln(1 + u_v(t)) + \ln(1 + u_i(t))])\, dt, \ a, b > 0.$$

By minimizing $C(u_v, u_i)$, we minimize the density of spreaders and maximize the density of isolated spreaders and the number of persons who learn the counter-information.

Theorem 2. *Let $X_0 \in \mathbb{R}^2$ such that there is a control (u_v, u_i) satisfying (4). There exists an optimal control (u_v^*, u_i^*) on $[0,T]$ such that the associated trajectory $X_{(u_v^*,u_i^*)}$ satisfies (4) and which minimizes the cost $C(.)$.*

Proof. For all $(u_v, u_i) \in \mathcal{U}_T$, t $\in [0,T]$, $||X_{(u_v,u_i)}(t)|| = ||\left(I_{(u_v,u_i)}(t), S_{(u_v,u_i)}(t)\right)||$ is bounded by construction of the model (1). Moreover, the right hand side of the equations of (3) and the integrand of the cost function are convex with respect to u_v and u_i. These two arguments give the existence of the optimal control. □

3 Characterization of the Optimal Control

It is well-known that Pontryagin's maximum principle (see [12,13]) is a good tool in order to characterize the optimal control. The application of this principle gives the existence of an adjoint vector $P(.) = (P_I(.), P_S(.)) : [0,T] \to \mathbb{R}^2$ absolutly continuous and $P^0 \leq 0$ such that (P, P^0) is non trivial and for almost all $t \in [0,T]$,

$$\begin{cases} \dot{P_I} = -\frac{\partial H}{\partial I} = [(\lambda+\alpha+\beta)S + u_v]\, P_I - [\lambda+\theta]\, SP_S, \\ \dot{P_S} = -\frac{\partial H}{\partial S} = [\lambda+\alpha+\beta]\, IP_I + [-\lambda I + \theta(1-I) + u_v + u_i]\, P_S - aP^0, \end{cases}$$

where H is the associated Hamiltonian defined by

$$\begin{aligned} H(t,I,S,u_v,u_i,P_I,P_S,P^0) &= P_I\left[-(\lambda+\alpha+\beta)IS - u_v I\right] \\ + \; P_S\left[\lambda IS - \theta S(1-I) - (u_v+u_i)S\right] &+ \; P^0\left[aS - b[\ln(1+u_v) + \ln(1+u_i)]\right] \end{aligned} \tag{5}$$

and where we omit some parameters for convenience of writting. Remark that we did not fix a final state so the transversality condition is $P_I(T) = P_S(T) = 0$ and $P^0 < 0$. Moreover, we have the well-known maximization condition, almost everywhere on [0, T],

$$H(t,I,S,u_v^*,u_i^*,P_I,P_S) = \max_{(w,z)\in[0,1]^2} H(t,I,S,w,z,P_I,P_S).$$

Theorem 3. *The optimal control (u_v^*, u_i^*), whose existence has been proved previously, is*

$$\begin{cases} \left(\min\left(1, \max\left(0, -1 - \frac{bP^0}{IP_I+SP_S}\right)\right),\ \min\left(1, \max\left(0, -1 - \frac{bP^0}{SP_S}\right)\right)\right) & \text{if } IP_I+SP_S \neq 0, SP_S \neq 0, \\ \left(\min\left(1, \max\left(0, -1 - \frac{bP^0}{IP_I+SP_S}\right)\right),\ 1\right) & \text{if } IP_I+SP_S \neq 0, SP_S = 0 \\ \left(1,\ \min\left(1, \max\left(0, -1 - \frac{bP^0}{SP_S}\right)\right)\right) & \text{if } IP_I+SP_S = 0, SP_S \neq 0 \\ (1,1) & \text{if } IP_I+SP_S = SP_S = 0. \end{cases}$$

Proof. The proof is divided into two parts, one which consists of characterizing the maximum of the Hamiltonian on the edges and inside the domain $[0,1]^2$ and another which consists of comparing the different obtained values for the Hamiltonian. We first consider the case $(u_v, u_i) = (u_v, 0)$, where the Hamiltonian is

$$H(t, I, S, u_v, 0, P_I, P_S, P^0) = P_I [-(\lambda + \alpha + \beta)IS - u_v I]$$
$$+ P_S [\lambda IS - \theta S(1 - I) - u_v S] + P^0 [aS - b\ln(1 + u_v)].$$

It is easy to verify that, if $IP_I + SP_S \neq 0$, then

$$(u_v^*, 0) = \left(\min\left(1, \max\left(0, -1 - \frac{bP^0}{IP_I + SP_S}\right)\right), 0\right),$$

and if $IP_I + SP_S = 0$, by using the maximization condition defined previously, the fact that $P^0 < 0$ and u_v is a probability, then we obtain $(u_v^*, 0) = (1, 0)$. Thus

$$(u_v^*, 0) = \begin{cases} \left(\min\left(1, \max\left(0, -1 - \frac{bP^0}{IP_I + SP_S}\right)\right), 0\right) & if\ IP_I + SP_S \neq 0, \\ (1, 0) & if\ IP_I + SP_S = 0. \end{cases}$$

The case $(u_v, u_i) = (0, u_i)$ is developed as previously and we find

$$(0, u_i^*) = \begin{cases} \left(0, \ \min\left(1, \max\left(0, -1 - \frac{bP^0}{SP_S}\right)\right)\right) & if\ SP_S \neq 0, \\ (0, 1) & if\ SP_S = 0. \end{cases}$$

In the case where $(u_v, u_i) = (u_v, 1)$, the Hamiltonian is

$$H(t, I, S, u_v, 1, P_I, P_S, P^0) = P_I [-(\lambda + \alpha + \beta)IS - u_v I]$$
$$+ P_S [\lambda IS - \theta S(1 - I) - u_v S - S] + P^0 [aS - b\ln(1 + u_v) - b\ln 2]$$

and we obtain

$$(u_v^*, 1) = \begin{cases} \left(\min\left(1, \max\left(0, -1 - \frac{bP^0}{IP_I + SP_S}\right)\right), 1\right) & if\ IP_I + SP_S \neq 0, \\ (1, 1) & if\ IP_I + SP_S = 0. \end{cases}$$

The case $(u_v, u_i) = (1, u_i)$ is treating in the same way to obtain

$$(1, u_i^*) = \begin{cases} \left(1, \ \min\left(1, \max\left(0, -1 - \frac{bP^0}{SP_S}\right)\right)\right) & if\ SP_S \neq 0, \\ (1, 1) & if\ SP_S = 0. \end{cases}$$

It remains the case $(u_v, u_i) \in]0, 1[^2$ where the Hamiltonian is defined by (5). By using the fact that its partial derivatives are equal to zero in the optimal case and

u_i and u_v are between 0 and 1, we obtain the optimal value specified in the theorem. In the following, we write $H(u_v^*, u_i^*)$ instead of $H(t, I, S, u_v^*, u_i^*, P_I, P_S, P^0)$ for convenience of writing. The second part of the proof consists of comparing the Hamiltonian calculated with these previous optimal values. We have for example that

$$H(u_v^*, u_i^*) - H(u_v^*, 1) = SP_S[1 - u_i^*] - bP^0\left[\ln(1 + u_i^*) - \ln 2\right].$$

After applying the mean value theorem which gives the existence of a $\tilde{u_i} \in]u_i^*, 1[$, we obtain

$$H(u_v^*, u_i^*) - H(u_v^*, 1) = (1 - u_i^*)\left[SP_S + \frac{bP^0}{1 + \tilde{u_i}}\right],$$

whose sign depends on the one of the quantity $SP_S + \dfrac{bP^0}{1 + \tilde{u_i}}$. Moreover,

$$H(u_v^*, u_i^*) - H(1, u_i^*) = (1 - u_v^*)\left[IP_I + SP_S + \frac{bP^0}{1 + \tilde{u_v}}\right], \text{ with } \tilde{u_v} \in]u_v^*, 1[,$$

$$H(u_v^*, u_i^*) - H(0, u_i^*) = -u_v^*\left[IP_I + SP_S + \frac{bP^0}{1 + \tilde{u_v}}\right], \text{ with } \tilde{u_v} \in]0, u_v^*[,$$

and

$$H(u_v^*, u_i^*) - H(u_v^*, 0) = -u_i^*\left[SP_S + \frac{bP^0}{1 + \tilde{u_i}}\right], \text{ with } \tilde{u_i} \in]0, u_i^*[.$$

In order to find the maximum of the Hamiltonian on $[0, 1]^2$, it is sufficient to take all possible combinations of signs for the quantities between brackets $SP_S + \dfrac{bP^0}{1 + \tilde{u_i}}$ and $IP_I + SP_S + \dfrac{bP^0}{1 + \tilde{u_v}}$. □

4 Numerical Simulations

In order to solve numerically our problem, we use a direct method which is possible by the use of OCP (Optimal Control Problem) software. This method consists of discretizing the state and the control, by changing the problem to one of non linear optimization, which gives us an estimate of the optimal value. Then the software Maple is used to plot the different curves.

For these numerical experiments, we fix $\alpha = \frac{1}{16}$, $\beta = \frac{3}{4}$, $\lambda = \frac{1}{8}$, $\theta = \frac{3}{8}$, we choose $P^0 = -1$ and $a = b = 1$. Remark that other choices of parameters would be possible and do not really change the conclusion. Moreover, we assume that the fake news has already begun to spread so we choose the initial conditions as $N_S = 0.1N$ and $N_{RA} = N_{RU} = 0.05N$. Our aim is to illustrate the behavior of the density of spreaders S (in blue) and ignorants I (in red) as functions of time, without action, with only one optimal action, with two simultaneous optimal actions and for different sizes of social network. For example, for $N = 10^3$, we obtain Figs. 1, 2, 3, and 4 for different actions values.

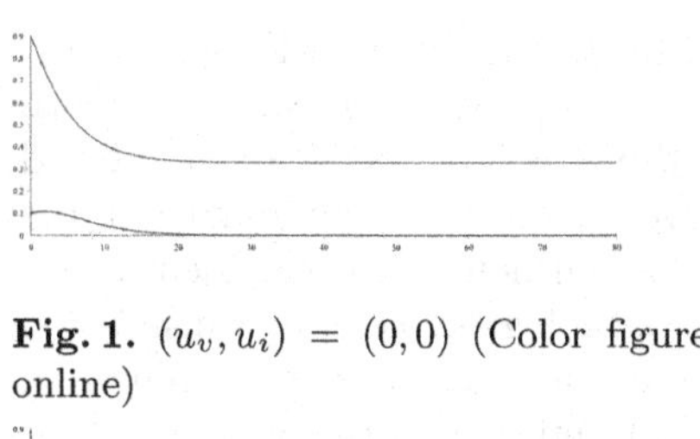

Fig. 1. $(u_v, u_i) = (0, 0)$ (Color figure online)

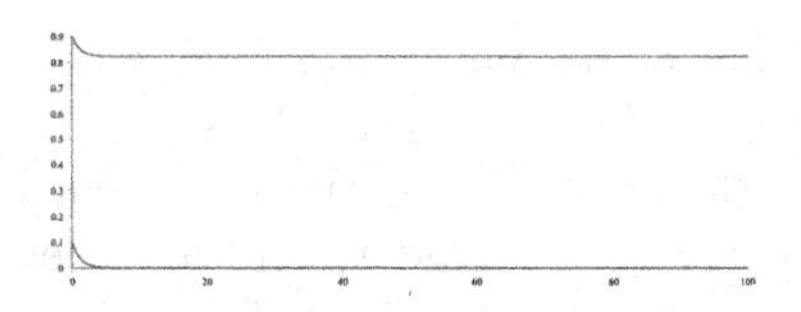

Fig. 2. $(u_v, u_i^*) = (0, 1)$ (Color figure online)

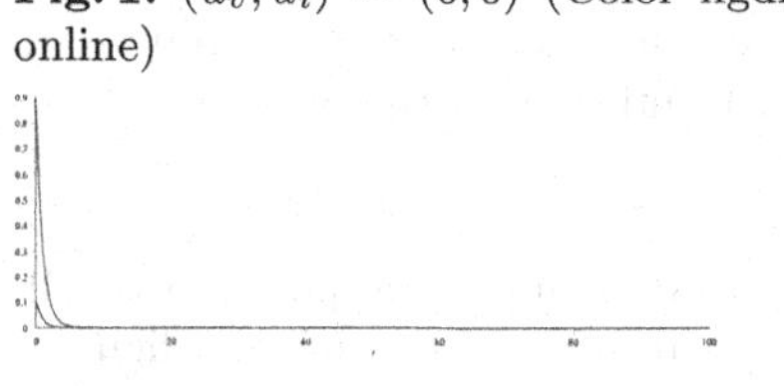

Fig. 3. $(u_v^*, u_i) = (1, 0)$ (Color figure online)

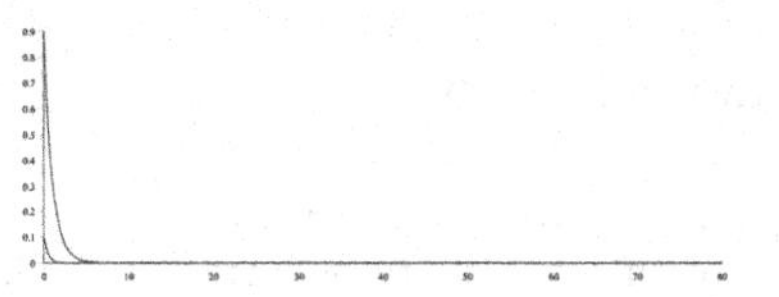

Fig. 4. $(u_v^*, u_i^*) = (1, 1)$ (Color figure online)

Then, for the optimal simultaneous action, we obtain the same behavior however the size of the social network in Figs. 5, 6 and 7.

We can notice that the density of ignorants decreases rapidly until cancelled and the density of spreaders decreases until cancelled, in the separated and simultaneous optimal cases. The difference of behaviors without actions, with only one optimal action and with two simultaneous optimal actions is clear. The best case is this last one and this however the size of the social network.

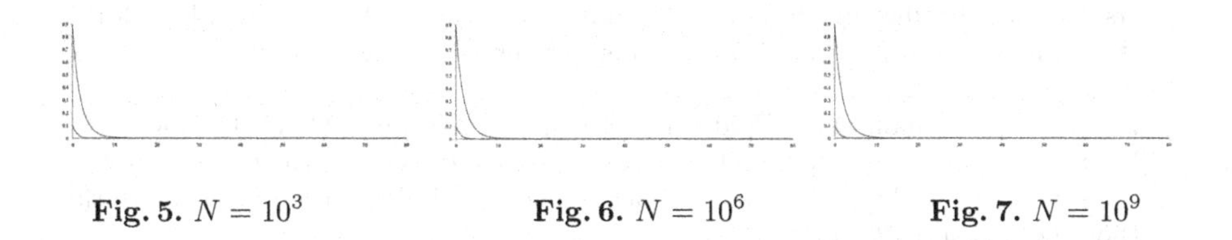

Fig. 5. $N = 10^3$ **Fig. 6.** $N = 10^6$ **Fig. 7.** $N = 10^9$

5 Conclusion

In this paper, we showed existence and gave a characterization of the optimal external actions, namely the probability of people who learned the counter-information and the probability of isolated spreaders, in order to reduce the propagation of a fake news. Numerical results are given in order to illustrate the behavior of the density of spreaders and ignorants, for different size of networks and for different values of actions and seem to be conform to the reality and strengthen our theoretical study. In the following, it will be interesting to investigate different models taking into account the crossing of an individual of a category to another one on the network and the socio-psychological dimension of the phenomenon. Indeed, the transmission of an information is done as the transmission of a disease but the parameters involved in the decision for a person to become a spreader are not the same than the ones leading a susceptible to become infected, as it has been recently pointed out in [1]. In both cases, the transmission can only occur after a direct contact between an infected (respectively a spreader) and a susceptible (respectively an ignorant) but the transmission depends on different parameters as virulence of the epidemy, sensibility of the individual, subject or not to pathologies, genetic transmission possibility, ... (respectively caracteristics of the information, demographic and socio-psychological factors, external sources, ...) and it is in this direction that we have to look for.

Acknowledgment. The authors would like to thank Mr. Raphael Pasquier, of the C3I center of the University of "Antilles", who helped us to perform the different figures. We also thank the anonymous referees for their valuable remarks and comments which improved the presentation of this manuscript.

References

1. Albi, G., Pareschi, L., Toscani, G., Zanella, M.: Recent advances in opinion modeling: control and social influence. In: Bellomo, N., Degond, P., Tadmor, E. (eds.) Active Particles, Volume 1. MSSET, pp. 49–98. Springer, Cham (2017). https://doi.org/10.1007/978-3-319-49996-3_2
2. Bernard, S., Bouza, G., Piétrus, A.: An optimal control approach for e-rumor. Revista Investigación Operacional **36**(2), 108–114 (2015)
3. Bernard, S., Bouza, G., Piétrus, A.: An e-rumour model with control on the spreaders. Comptes rendus de l'Académie Bulgare des Sciences **69**(11), 1407–1414 (2016)
4. Bernard, S., Cesar, A.T.: Piétrus, Some actions to control e-rumor. e-J. Carib. Acad. Sci. **9**(1) (2017)
5. Daley, D.J., Kendall, D.G.: Epidemics and rumors. Nature **204**, 11–18 (1964)
6. Daley, D.J., Kendall, D.G.: Stochastic rumors. IMA J. Appl. Math. **1**, 42–55 (1965)
7. Hansen, E., Day, T.: Optimal control of epidemics with limited resources. J. Math. Biol. **62**, 423–451 (2011)
8. Huang, J., Jin, X.: Preventing rumor spreading on small-world networks. J. Syst. Sci. Complex **24**, 449–456 (2011)
9. Ledzewicz, U., Schattler, H.: On optimal singular controls for a general SIR-model with vaccination and treatment. Discrete Continuous Dyn. Syst. Ser. B, (Suppl.), 981–990 (2011)
10. Liu, J., Niu, K., He, Z., Lin, J.: Analysis of rumor spreading in communities based on modified SIR model in microblog. In: Agre, G., Hitzler, P., Krisnadhi, A.A., Kuznetsov, S.O. (eds.) AIMSA 2014. LNCS (LNAI), vol. 8722, pp. 69–79. Springer, Cham (2014). https://doi.org/10.1007/978-3-319-10554-3_7
11. Maki, D.: Mathematical Models and Applications, with Emphasis on Social, Life and Management Sciences. Prentice Hall College Div, Englewood Cliffs (1973)
12. Pontryaguin, L., Boltyanski, V., Gamkrelidze, R., Michtchenko, E.: Théorie mathématique des processus optimaux. Editions Mir, Moscou (1974)
13. Trélat, E.: Contrôle optimal: théorie et applications. Vuibert, Paris (2008)
14. Zanette, D.H.: Dynamics of rumor propagation on small-world networks. Phys. Rev. E **65**, 041908 (2002)

Superposition Principle for Differential Inclusions

Giulia Cavagnari[1], Antonio Marigonda[2(✉)], and Benedetto Piccoli[1]

[1] Department of Mathematical Sciences, Rutgers University - Camden, 311 N. 5th Street, Camden, NJ 08102, USA
giulia.cavagnari@rutgers.edu, piccoli@camden.rutgers.edu
[2] Department of Computer Sciences, University of Verona, Strada Le Grazie 15, 37134 Verona, Italy
antonio.marigonda@univr.it

Abstract. We prove an extension of the Superposition Principle by Ambrosio-Gigli-Savaré in the context of a control problem. In particular, we link the solutions of a finite-dimensional control system, with dynamics given by a differential inclusion, to a solution of a continuity equation in the space of probability measures with admissible vector field. We prove also a compactness and an approximation result for admissible trajectories in the space of probability measures.

Keywords: Continuity equation · Differential inclusions
Optimal transport · Superposition principle

1 Introduction

This paper aims to provide a relation between the *macroscopic* and the *microscopic* approaches describing the evolution of a mass of particles/agents in a controlled context. The microscopic dynamics of the particles/agents is governed by a control system given in the form of a differential inclusion $\dot{x}(t) \in F(x(t))$, where $F(\cdot)$ is a given set-valued function stating the set of admissible velocities for each point in $\mathbb{R}^d$. This makes not trivial the construction of the corresponding macroscopic evolution and of its driving vector field in the space of probability measures. Indeed, from a macroscopic point of view, the evolving mass is described by a time-dependent family of probability measures $\boldsymbol{\mu} = \{\mu_t\}_{t\in[0,T]}$, solving in the distributional sense a (*controlled*) homogeneous continuity equation (thus a PDE), and driven by an admissible vector field that has to be chosen among the $L^1_{\mu_t}$-selections of F.

In a non-controlled framework, if the finite-dimensional dynamics is given by an ODE driven by a Lipschitz vector field v_t (locally Lipschitz continuous in the space variable uniformly w.r.t. t), then we have existence and uniqueness of the solution of the PDE. The solution μ_t at time t of the continuity equation is characterized by the push forward of the initial state μ_0 w.r.t. a map T_t called *transport map*, i.e. $\mu_t = T_t \sharp \mu_0$ for a.e. t, where $\dot{T}_t(x) = v_t(T_t(x))$, $T_0(x) = x$ is the

I. Lirkov and S. Margenov (Eds.): LSSC 2017, LNCS 10665, pp. 201–209, 2018.
https://doi.org/10.1007/978-3-319-73441-5_21

characteristic system. However, a relation between μ_t and the (integral) solutions of the characteristic system is possible even for nonsmooth vector fields, where uniqueness of solutions is no longer granted. This powerful result, called the *Superposition Principle*, appeared for the first time in the appendix of [17] and has been studied by different authors in [1,2,4], and in [15] in a non-homogeneous context. The idea is to take into account the possible non-uniqueness of the solution of the characteristic system by introducing a measure $\boldsymbol{\eta} \in \mathscr{P}(\mathbb{R}^d \times \Gamma_T)$, where $\Gamma_T = C^0([0,T];\mathbb{R}^d)$, concentrated on the set of $(\gamma(0),\gamma)$ where γ is any integral solution of the characteristic system. Indeed, under general assumptions, for any solution $\boldsymbol{\mu} := \{\mu_t\}_{t\in[0,T]} \subseteq \mathscr{P}(\mathbb{R}^d)$ of an homogeneous continuity equation, there exists such a (possibly not-unique) *representation* $\boldsymbol{\eta} \in \mathscr{P}(\mathbb{R} \times \Gamma_T)$ satisfying $\mu_t = e_t \sharp \boldsymbol{\eta}$ where $e_t : \mathbb{R}^d \times \Gamma_T \to \mathbb{R}^d, (x,\gamma) \mapsto \gamma(t)$, is the *evaluation operator*. Conversely, any $\boldsymbol{\eta} \in \mathscr{P}(\mathbb{R}^d \times \Gamma_T)$ concentrated on the characteristics yields a solution of the correspondent continuity equation by setting $\mu_t = e_t \sharp \boldsymbol{\eta}$.

We stress the fact that, given $\boldsymbol{\mu}$, its probabilistic representation $\boldsymbol{\eta}$ may be not unique, in particular different weights to the characteristics could lead to the same macroscopic evolution $\boldsymbol{\mu}$ (some examples are sketched in the forthcoming [8]). In the present paper we exploit the non-uniqueness of a probabilistic representative by extending the reverse implication of the Superposition Principle (see Theorem 8.2.1 in [2]) in a controlled setting. In particular, we replace the underlying characteristics' ODE with a differential inclusion. In Theorem 1 we prove that, under some natural assumptions of the set valued map F, a measure $\boldsymbol{\eta}$ concentrated on the Carathéodory solutions of the differential inclusion $\dot{\gamma}(t) \in F(\gamma(t))$, $\gamma(0) = x$, induces a macroscopic admissible trajectory $\boldsymbol{\mu} = \{\mu_t\}_{t\in[0,T]} \subseteq \mathscr{P}(\mathbb{R}^d)$, where μ_t is a solution of a continuity equation driven by a *mean vector field* v_t. More precisely, $v_t(y)$ turns out to be the integral average w.r.t. $\boldsymbol{\eta}$ of the underlying admissible vector fields crossing position $y \in \mathbb{R}^d$ at time t. In other words, the macroscopic evolution μ_t of our mass looses the information about the velocity field chosen by each single particle, providing only their average behaviour.

The results of this paper could be used to investigate further properties of control problems in $\mathscr{P}(\mathbb{R}^d)$, possibly requiring extremality conditions (e.g. time minimality to reach a target). For instance, one may improve the analysis made in [5–12] where the authors studied time-optimal control problems in the space of measures making large use of the Superposition Principle of [2]. Another potential application could be in the field of crowd dynamics, where the importance of a multiscale approach has been underlined for instance in [13,14]. Indeed, we can now collect together the microscopic behaviour of the single agents, even when they are subject to different vector fields, into a unique macroscopic mean description.

The paper is structured as follows: in Sect. 2 we state the notation and define the objects used, Sect. 3 contains the statement and proof of the extended Superposition Principle together with a compactness and approximation result.

2 Preliminaries

Let X be a separable metric space. We denote with $\mathscr{P}(X)$ the space of Borel probability measures on X endowed with the narrow topology induced by $(C_b^0(X))'$. When $p \geq 1$, $\mathscr{P}_p(X)$ denotes the space of Borel probability measures with finite *p-moment*, i.e., the measures $\mu \in \mathscr{P}(\mathbb{R}^d)$ satisfying $\mathrm{m}_p(\mu) := \int_{\mathbb{R}^d} |x|^p \, d\mu < +\infty$, endowed with the topology induced by the *p-Wasserstein distance* $W_p(\cdot,\cdot)$. We call $\mathscr{M}(\mathbb{R}^d;\mathbb{R}^d)$ the space of vector-valued Radon measures on $\mathbb{R}^d$ endowed with the w^*-topology. When $\nu \in \mathscr{M}(\mathbb{R}^d;\mathbb{R}^d)$, $|\nu|$ denotes its *total variation*, and we write $\sigma \ll \mu$ to say that σ is *absolutely continuous* w.r.t. μ, for a pair of measures σ, μ on $\mathbb{R}^d$. Preliminaries on measure theory can be found in Chap. 5 in [2].

We recall now the definition of *admissible trajectory* in $\mathscr{P}(\mathbb{R}^d)$ that, together with its probabilistic representation, is the central object of the present paper and it was introduced in [5,6,8–11] for the study of time-optimal control problems in the space of measures.

Definition 1. *Let $F : \mathbb{R}^d \rightrightarrows \mathbb{R}^d$ be a set-valued map, $\bar{\mu} \in \mathscr{P}(\mathbb{R}^d)$.*

1. *Let $T > 0$. We say that $\boldsymbol{\mu} = \{\mu_t\}_{t\in[0,T]} \subseteq \mathscr{P}(\mathbb{R}^d)$ is an* admissible trajectory *defined on $[0,T]$ and starting from $\bar{\mu}$ if there exists $\boldsymbol{\nu} = \{\nu_t\}_{t\in[0,T]} \subseteq \mathscr{M}(\mathbb{R}^d;\mathbb{R}^d)$ such that $|\nu_t| \ll \mu_t$ for a.e. $t \in [0,T]$, $\mu_0 = \bar{\mu}$, $\partial_t \mu_t + \operatorname{div} \nu_t = 0$ in the sense of distributions and $v_t(x) := \dfrac{\nu_t}{\mu_t}(x) \in F(x)$ for a.e. $t \in [0,T]$ and μ_t-a.e. $x \in \mathbb{R}^d$. In this case, we will say also that $\boldsymbol{\mu}$ is* driven *by $\boldsymbol{\nu}$.*
2. *Let $T > 0$, $\boldsymbol{\mu}$ be an admissible trajectory defined on $[0,T]$ starting from $\bar{\mu}$ and driven by $\boldsymbol{\nu} = \{\nu_t\}_{t\in[0,T]}$. We will say that $\boldsymbol{\mu}$* is represented *by $\boldsymbol{\eta} \in \mathscr{P}(\mathbb{R}^d \times \Gamma_T)$ if we have $e_t \sharp \boldsymbol{\eta} = \mu_t$ for all $t \in [0,T]$, where $e_t : \mathbb{R}^d \times \Gamma_T \to \mathbb{R}^d, (x,\gamma) \mapsto \gamma(t)$, and $\boldsymbol{\eta}$ is concentrated on the pairs $(x,\gamma) \in \mathbb{R}^d \times \Gamma_T$ where γ is an absolutely continuous solution of the underlying characteristic system*

$$\begin{cases} \dot{\gamma}(t) \in F(\gamma(t)), & \text{for a.e. } 0 < t \leq T \\ \gamma(0) = x. \end{cases} \tag{1}$$

Note that to have the existence of a probabilistic representation $\boldsymbol{\eta} \in \mathscr{P}(\mathbb{R}^d \times \Gamma_T)$ it is sufficient that the driving vector field associated with $\boldsymbol{\mu}$ satisfies the integrability hypothesis of the Superposition Principle (see Theorem 8.2.1 in [2]).

Finally, let X be a set, $A \subseteq X$. The *indicator function of* A is $I_A : X \to \{0, +\infty\}$ defined as $I_A(x) = 0$ for all $x \in A$ and $I_A(x) = +\infty$ for all $x \notin A$. The *characteristic function of* A is the function $\chi_A : X \to \{0,1\}$ defined as $\chi_A(x) = 1$ for all $x \in A$ and $\chi_A(x) = 0$ for all $x \notin A$.

3 Results

Throughout the paper we will require the following assumptions on the set-valued function $F : \mathbb{R}^d \rightrightarrows \mathbb{R}^d$ governing the finite-dimensional differential inclusion:

(F_0) $F(x) \neq \emptyset$ is compact and convex for every $x \in \mathbb{R}^d$, moreover $F(\cdot)$ is continuous with respect to the Hausdorff metric.
(F_1) $F(\cdot)$ has linear growth, i.e. there exists a constant $C > 0$ such that $F(x) \subseteq B(0, C(|x|+1))$ for every $x \in \mathbb{R}^d$.

The following simple Lemma states the possibility to approximate in the W_p-distance every measure $\mu \in \mathscr{P}_p(X)$, where X is a complete separable Banach space, by a sequence $\{\mu^k\}_{k\in\mathbb{N}}$ of empirical measures (i.e. convex combinations of Dirac deltas) concentrated on its support. A proof can be found for instance in [16] (see Lemma 6.1) for the case $X = \mathbb{R}^d$, but it can be easily extended to the general setting of complete separable Banach spaces.

Lemma 1 (Empirical approximation in Wasserstein). *Let X be a separable Banach space. For all $p \geq 1$ we have $\mathscr{P}_p(X) = \mathrm{cl}_{W_p}(\mathrm{co}\{\delta_x : x \in X\})$.*

We now consider the following problem: taking any probability measure $\boldsymbol{\eta}$ on the set of the admissible trajectories for (1), it is possible to construct a global vector field $v_t(\cdot)$, time-depending selection of $F(\cdot)$, such that $e_t\sharp\boldsymbol{\eta}$ yields a family of time-depending probability measures on $\mathbb{R}^d$ solving the continuity equation driven by v_t. This can be viewed as a partial extension to the Superposition Principle (see Theorem 8.2.1 in [2]) to the case of differential inclusions.

Theorem 1 (SP for differential inclusions). *Assume (F_0), (F_1), $p \geq 1$. Let $\boldsymbol{\eta} \in \mathscr{P}(\mathbb{R}^d \times \Gamma_T)$ be concentrated on the set of pairs $(\gamma(0), \gamma) \in \mathbb{R}^d \times \Gamma_T$ such that $\gamma \in AC([0,T];\mathbb{R}^d)$ is a Carathéodory solution of the differential inclusion $\dot{\gamma}(t) \in F(\gamma(t))$. For all $t \in [0,T]$, set $\mu_t := e_t\sharp\boldsymbol{\eta}$, and let $\{\eta_{t,y}\}_{y\in\mathbb{R}^d} \subseteq \mathscr{P}(\mathbb{R}^d\times\Gamma_T)$ be the disintegration of $\boldsymbol{\eta}$ w.r.t. the evaluation operator $e_t : \mathbb{R}^d \times \Gamma_T \to \mathbb{R}^d$, i.e. for all $\varphi \in C_b^0(\mathbb{R}^d \times \Gamma_T)$*

$$\iint_{\mathbb{R}^d\times\Gamma_T} \varphi(x,\gamma)\, d\boldsymbol{\eta}(x,\gamma) = \int_{\mathbb{R}^d}\int_{e_t^{-1}(y)} \varphi(x,\gamma)\, d\eta_{t,y}(x,\gamma)\, d\mu_t(y).$$

Then if $\mu_0 \in \mathscr{P}_p(\mathbb{R}^d)$, the curve $\boldsymbol{\mu} := \{\mu_t\}_{t\in[0,T]} \subseteq \mathscr{P}_p(\mathbb{R}^d)$, is an admissible trajectory driven by $\boldsymbol{\nu} = \{\nu_t\}_{t\in[0,T]}$, where $\nu_t = v_t\mu_t$ and the vector field

$$v_t(y) = \int_{e_t^{-1}(y)} \dot{\gamma}(t)\, d\eta_{t,y}(x,\gamma). \tag{2}$$

is well-defined for a.e. $t \in [0,T]$ and μ_t-a.e. $y \in \mathbb{R}^d$.

Proof. We define

$$\mathscr{N} := \{(t,x,\gamma) \in [0,T] \times \mathbb{R}^d \times \Gamma_T : \text{ either } \dot{\gamma}(t) \text{ does not exists or } \dot{\gamma}(t) \notin F(\gamma(t))\}.$$

Since $\mathscr{L}^1_{|[0,T]} \otimes \boldsymbol{\eta}(\mathscr{N}) = 0$, we have $\dot{\gamma}(t) \in F(\gamma(t))$ for $\boldsymbol{\eta}$-a.e. $(x,\gamma) \in \mathbb{R}^d \times \Gamma_T$ and a.e. $t \in [0,T]$, and so $v_t(y)$ is well-defined for a.e. $t \in [0,T]$ and μ_t-a.e. $y \in \mathbb{R}^d$.

We prove first that the map $t \mapsto \mu_t$ is Lipschitz continuous from $[0,T]$ to $\left(C^1_c(\mathbb{R}^d)\right)'$. For all $\tau \in [0,T]$ and $\boldsymbol{\eta}$-a.e. $(x,\gamma) \in \mathbb{R}^d \times \Gamma_T$ it holds

$$\begin{aligned}|\gamma(\tau)-\gamma(0)| &\le \int_0^\tau |\dot\gamma(s)|\,ds \le C\int_0^\tau (|\gamma(s)|+1)\,ds\\ &\le C\tau(1+|\gamma(0)|)+C\int_0^\tau |\gamma(s)-\gamma(0)|\,ds,\end{aligned}$$

thus, by Gronwall's inequality,

$$|\gamma(\tau)-\gamma(0)| \le C\tau(1+|\gamma(0)|)e^{C\tau} \le CTe^{CT}(1+|\gamma(0)|),$$

Since for any $\varphi \in C^1_c(\mathbb{R}^d)$ we have

$$\begin{aligned}\left|\int_{\mathbb{R}^d}\varphi(x)\,d\mu_s(x) - \int_{\mathbb{R}^d}\varphi(x)\,d\mu_t(x)\right| &\le \int_s^t \iint_{\mathbb{R}^d\times\Gamma_T} |\langle\nabla\varphi(\gamma(\tau)),\dot\gamma(\tau)\rangle|\,d\boldsymbol{\eta}(x,\gamma)\,d\tau\\ &\le C\|\nabla\varphi\|_\infty \int_s^t \iint_{\mathbb{R}^d\times\Gamma_T} (|\gamma(\tau)|+1)\,d\boldsymbol{\eta}(x,\gamma)\,d\tau\\ &\le C(CTe^{CT}+1)\|\nabla\varphi\|_\infty \int_s^t \iint_{\mathbb{R}^d\times\Gamma_T} (|\gamma(0)|+1)\,d\boldsymbol{\eta}(x,\gamma)\,d\tau\\ &\le C(CTe^{CT}+1)\left(\mathrm{m}_p^{1/p}(\mu_0)+1\right)\|\nabla\varphi\|_\infty|t-s|,\end{aligned}$$

we have $\|\mu_s-\mu_t\|_{(C^1_c(\mathbb{R}^d))'} \le C(CTe^{CT}+1)\left(\mathrm{m}_p^{1/p}(\mu_0)+1\right)|t-s|$.

According to Theorem 3.5 in [3], we have that for a.e. $t \in [0,T]$ the map $t \mapsto \mu_t$ is differentiable, and for all $\varphi \in C^1_c(\mathbb{R}^d)$

$$\frac{d}{dt}\int_{\mathbb{R}^d}\varphi(x)\,d\mu_t(x) = \iint_{\mathbb{R}^d\times\Gamma_T}\nabla\varphi(\gamma(t))\cdot\dot\gamma(t)\,d\boldsymbol{\eta}(x,\gamma) = \int_{\mathbb{R}^d}\nabla\varphi(y)\cdot v_t(y)\,d\mu_t(y),$$

which implies $\partial_t\mu_t + \operatorname{div}\nu_t = 0$ with $\nu_t = v_t\mu_t$. Finally, thanks to the convexity of $F(y)$, we can use Jensen's inequality to get that $v_t(y) \in F(y)$ for μ_t-a.e. $y \in \mathbb{R}^d$ and a.e. $t \in [0,T]$. To conclude the proof, it is enough to show the estimates on the p-moments of μ_t. Indeed, by Gronwall's inequality we have

$$\mathrm{m}_p^{1/p}(\mu_t) \le (CTe^{CT}+1)(1+\mathrm{m}_p^{1/p}(\mu_0)).$$

Moreover, by (F_1) we have that every Borel selection of $F(\cdot)$ is in L^p_μ for any $\mu \in \mathscr{P}_p(\mathbb{R}^d)$, hence $v_t \in L^p_{\mu_t}$ for a.e. $t \in [0,T]$. □

A possible interpretation of $v_t(y)$ is provided by the following remark.

Remark 1. By definition, we have $e_t^{-1}(y) = \{(x,\gamma) \in \mathbb{R}^d \times \Gamma_T : \gamma(t) = y\}$, so, by (2), for a.e. $t \in [0,T]$ and μ_t-a.e. $y \in \mathbb{R}^d$, we have that $v_t(y)$ corresponds to a weighted average of the velocity of the trajectories $\gamma \in AC([0,T];\mathbb{R}^d)$ of the differential inclusions $\dot\gamma(t) \in F(\gamma(t))$ satisfying $\gamma(t) = y$.

The next example provides a situation where the velocities of a nonnegligible set of curves differs from the mean field for a nonnegligible amount of time.

Example 1. The ambient space is $\mathbb{R}^2$. Define

- $\mathscr{A} = \{\gamma_{x,y}(\cdot)\}_{(x,y)\in\mathbb{R}^2} \subseteq AC([0,2])$ where $\gamma_{x,y}(t) = (x+t, y - t\operatorname{sgn} y)$ for any $(x,y)\in\mathbb{R}^2$, $t\in[0,2]$, and we set $\operatorname{sgn}(0)=0$;
- $F:\mathbb{R}^2 \rightrightarrows \mathbb{R}^2$ by $F(x,y)\equiv[-1,1]\times[-1,1]$ for all $(x,y)\in\mathbb{R}^2$;
- $\mu_0 = \dfrac{1}{2}\delta_0\otimes\mathscr{L}^1_{|[-1,1]} \in \mathscr{P}(\mathbb{R}^2)$, $\boldsymbol{\eta} = \mu_0\otimes\delta_{\gamma_{x,y}} \in \mathscr{P}(\mathbb{R}^2\times\Gamma_2)$, $\boldsymbol{\mu} = \{\mu_t\}_{t\in[0,2]}$ with $\mu_t = e_t\sharp\boldsymbol{\eta}$;
- Q be the open square of vertice $\{(0,0),(1,0),(1/2,\pm1/2)\}$.

We notice that

- F satisfies (F_0) and (F_1) and $\dot\gamma(t)\in F(\gamma(t))$ for all $\gamma\in\mathscr{A}$ and $t\in]0,2[$.
- The product measure $\boldsymbol{\eta}$ is well-defined since $(x,y)\mapsto\gamma_{x,y}(\cdot)$ is a Borel map, thus $\boldsymbol{\mu}$ is an admissible trajectory and we denote with $\boldsymbol{\nu} = \{\nu_t\}_{t\in[0,2]}$ its driving family of Borel vector-valued measures.
- For any $P=(p_x,p_y)\in Q$ with $p_y\neq0$ there are exactly two elements $\gamma\in\mathscr{A}$ satisfying $\gamma(0)\in\{0\}\times]-1,1[$ and crossing at P. These elements are $\gamma_{0,p_y\pm p_x}(\cdot)$ and we notice that $P=\gamma_{0,p_y+p_x}(t)=\gamma_{0,p_y-p_x}(t)$ if and only if $t=p_x$.

Denoted by $v_t = \dfrac{\nu_t}{\mu_t}$ the mean vector field, this implies $v_t(x,y)=(1,0)$ for all $(x,y)\in Q\setminus(\mathbb{R}\times\{0\})$ and $t=x$. For every $\gamma\in\mathscr{A}$ satisfying $\gamma(0)=\{0\}\times]-1,1[$ and $\gamma(0)\neq(0,0)$, there exists an interval $I_\gamma\subseteq[0,1]$ of Lebesgue measure $1/2$ such that $\gamma(t)\in Q$ if and only if $t\in I_\gamma$, thus

$$\mathscr{L}^1_{|[0,2]}\otimes\boldsymbol{\eta}\left(\{(t,x,\gamma)\in[0,2]\times\mathbb{R}^2\times\Gamma_2 : \dot\gamma(t)\neq v_t(\gamma(t))\}\right)=\frac{1}{2}.$$

With techniques similar to Theorem 1, it is possible to prove a result of relative compactness of the admissible trajectories even in the critical case $p=1$.

Proposition 1 (Relative compactness of admissible trajectories). *Assume (F_0), (F_1), $p\geq1$. Let $\{\boldsymbol{\eta}^N\}_{N\in\mathbb{N}}\subseteq\mathscr{P}(\mathbb{R}^d\times\Gamma_T)$ be a sequence of measures concentrated on the set of pairs $(\gamma(0),\gamma)\in\mathbb{R}^d\times\Gamma_T$ where $\gamma\in AC([0,T];\mathbb{R}^d)$ is a Carathéodory solution of the differential inclusion $\dot\gamma(t)\in F(\gamma(t))$ and such that $\{\mathrm{m}_p(e_0\sharp\boldsymbol{\eta}^N)\}_{N\in\mathbb{N}}$ is uniformly bounded. Denote with $\{\boldsymbol{\mu}^N\}_{N\in\mathbb{N}}$ the sequence of admissible trajectories represented by $\{\boldsymbol{\eta}^N\}_{N\in\mathbb{N}}$, and with $\{\boldsymbol{\nu}^N\}_{N\in\mathbb{N}}\subseteq\mathscr{M}(\mathbb{R}^d;\mathbb{R}^d)$ the sequence of their driving families of Borel vector-valued measures.*

Then, up to a non relabeled subsequence, we have that there exists $\boldsymbol{\eta}\in\mathscr{P}(\mathbb{R}^d\times\Gamma_T)$ such that $\boldsymbol{\eta}^N\rightharpoonup^*\boldsymbol{\eta}$, and $\boldsymbol{\mu}:=\{\mu_t\}_{t\in[0,T]}\subseteq\mathscr{P}_p(\mathbb{R}^d)$ defined by $\mu_t=e_t\sharp\boldsymbol{\eta}$ is an admissible curve driven by $\boldsymbol{\nu}=\{\nu_t\}_{t\in[0,T]}$, with $\nu_t^N\rightharpoonup^*\nu_t$ for a.e. $t\in[0,T]$.

Proof. We prove that $\{\boldsymbol{\eta}^N\}_{N\in\mathbb{N}}$ is relatively compact in $\mathscr{P}(\mathbb{R}^d \times \Gamma_T)$. Indeed, by exploiting the estimates of Theorem 1, and the uniformly boundedness of $\mathrm{m}_p(e_0\sharp\boldsymbol{\eta}^N)$, we have

$$\iint_{\mathbb{R}^d\times\Gamma_T} (|x| + |\gamma(t)|)\, d\boldsymbol{\eta}^N(x,\gamma) \le (CTe^{CT}+2)(1+\mathrm{m}_p^{1/p}(e_0\sharp\boldsymbol{\eta}^N)) \le K < +\infty,$$

by Remark 5.1.5 in [2] we have that $\{\boldsymbol{\eta}^N\}_{N\in\mathbb{N}}$ is tight, and so, up to a non relabeled subsequence, we have that there exists $\boldsymbol{\eta} \in \mathscr{P}(\mathbb{R}^d \times \Gamma_T)$, concentrated on the set of pairs $(\gamma(0),\gamma) \in \mathbb{R}^d \times \Gamma_T$ where $\gamma \in AC([0,T];\mathbb{R}^d)$ is a Carathéodory solution of the differential inclusion $\dot\gamma(t) \in F(\gamma(t))$, such that $\boldsymbol{\eta}^N \rightharpoonup^* \boldsymbol{\eta}$.

In particular, for all $t \in [0,T]$ we have $\mu_t^N = e_t\sharp\boldsymbol{\eta}^N \rightharpoonup^* e_t\sharp\boldsymbol{\eta} = \mu_t$. By Theorem 1 we have that $\boldsymbol{\mu} = \{\mu_t\}_{t\in[0,T]}$ is an admissible trajectory and it is driven by $\boldsymbol{\nu} = \{\nu_t\}_{t\in[0,T]}$, with $\nu_t = v_t\mu_t$ and v_t is a suitable $L^p_{\mu_t}$-selection of $F(\cdot)$ for a.e. $t \in [0,T]$.

Let us now conclude by proving that $\nu_t^N \rightharpoonup^* \nu_t$ for all $N \in \mathbb{N}$ and for a.e. $t \in [0,T]$. By Theorem 1, by w^*-convergence of μ_t^N to μ_t and by admissibility of $\boldsymbol{\mu}^N$, we have that for every $\varphi \in C_c^1([0,T]\times\mathbb{R}^d)$

$$\begin{aligned}
&-\iint_{[0,T]\times\mathbb{R}^d} \nabla\varphi(t,x)\cdot d\nu_t\, dt = \iint_{[0,T]\times\mathbb{R}^d} \partial_t\varphi(t,x)\, d\mu_t\, dt \\
&= \lim_{N\to+\infty} \iint_{[0,T]\times\mathbb{R}^d} \partial_t\varphi(t,x)\, d\mu_t^N\, dt = \lim_{N\to+\infty} -\iint_{[0,T]\times\mathbb{R}^d} \nabla\varphi(t,x)\, d\nu_t^N\, dt,
\end{aligned}$$

hence the statement follows. □

Finally, combining Lemma 1, Theorem 1, and Proposition 1, we have convergence of a suitable discrete approximation.

Corollary 1. *Assume (F_0), (F_1), $p \ge 1$. Let $\boldsymbol{\eta} \in \mathscr{P}(\mathbb{R}^d \times \Gamma_T)$ be concentrated on the set of pairs $(\gamma(0),\gamma) \in \mathbb{R}^d \times \Gamma_T$ such that $\gamma \in AC([0,T];\mathbb{R}^d)$ is a Carathéodory solution of the differential inclusion $\dot\gamma(t) \in F(\gamma(t))$ with $\mathrm{m}_p(e_0\sharp\boldsymbol{\eta}) < +\infty$. Then there exists a sequence $\{\boldsymbol{\eta}^N\}_{N\in\mathbb{N}} \subseteq \mathrm{co}\{\delta_x \otimes \delta_{\gamma_x}\} \subseteq \mathscr{P}(\mathbb{R}^d \times \Gamma_T)$ with $\gamma_x \in AC([0,T];\mathbb{R}^d)$, $\gamma(0) = x$ and $\dot\gamma(t) \in F(\gamma(t))$ for a.e. $t \in [0,T]$, such that $\boldsymbol{\eta}^N$ converges to $\boldsymbol{\eta}$ in W_p and for all $t \in [0,T]$*

$$\lim_{N\to+\infty} W_p(e_t\sharp\boldsymbol{\eta}^N, e_t\sharp\boldsymbol{\eta}) = 0.$$

Proof. We take $X = \mathbb{R}^d \times \Gamma_T$, endowed with norm $\|(x,\gamma)\|_X = |x| + \|\gamma\|_\infty$. We prove that $\mathrm{m}_p(\boldsymbol{\eta}) < +\infty$. Indeed, for all $t \in [0,T]$ and $\boldsymbol{\eta}$-a.e. $(x,\gamma) \in \mathbb{R}^d \times \Gamma_T$ we have $|\gamma(t)| \le (CTe^{CT}+1)(1+|\gamma(0)|)$, so $\|\gamma\|_\infty \le (CTe^{CT}+1)(1+|\gamma(0)|)$, hence

$$\begin{aligned}
\iint_{\mathbb{R}^d\times\Gamma_T} (|x| + \|\gamma\|_\infty)^p\, d\boldsymbol{\eta}(x,\gamma) &\le 2^p(CTe^{CT}+1)^p \iint_{\mathbb{R}^d\times\Gamma_T} (1+|\gamma(0)|)^p\, d\boldsymbol{\eta}(x,\gamma) \\
&= 2^p(CTe^{CT}+1)^p \int_{\mathbb{R}^d} (1+|x|)^p\, d(e_0\sharp\boldsymbol{\eta})(x) < +\infty.
\end{aligned}$$

By Lemma 1, we can construct a sequence $\{\boldsymbol{\eta}^N\}_{N\in\mathbb{N}} \subseteq \mathrm{co}\{\delta_x \otimes \delta_{\gamma_x}\} \subseteq \mathscr{P}(\mathbb{R}^d \times \Gamma_T)$ W_p-converging to $\boldsymbol{\eta}$. Moreover, we have $\mathrm{supp}\,\boldsymbol{\eta}^N \subseteq \mathrm{supp}\,\boldsymbol{\eta}$, which, by Theorem 1, implies that $\boldsymbol{\mu}^N = \{\mu_t^N = e_t \sharp \boldsymbol{\eta}^N\}_{t\in[0,T]}$ is an admissible trajectory. □

Acknowledgments. The authors acknowledge the endowment fund of the Joseph and Loretta Lopez Chair and the support of the INdAM-GNAMPA Project 2016 *Stochastic Partial Differential Equations and Stochastic Optimal Transport with Applications to Mathematical Finance.*

References

1. Ambrosio, L.: Transport equation and Cauchy problem for non-smooth vector fields. In: Dacorogna, B., Marcellini, P. (eds.) Contribute in Calculus of Variations and Nonlinear Partial Differential Equations. Lecture Notes in Mathematics, vol. 1927, pp. 1–41. Springer, Heidelberg (2008). https://doi.org/10.1007/978-3-540-75914-0_1
2. Ambrosio, L., Gigli, N., Savaré, G.: Gradient Flows in Metric Spaces and in the Space of Probability Measures. Lectures in Mathematics ETH Zürich, 2nd edn. Birkhäuser Verlag, Basel (2008)
3. Ambrosio, L., Kirchheim, B.: Rectifiable sets in metric and Banach spaces. Math. Ann. **318**(3), 527–555 (2000)
4. Bernard, P.: Young measures, superpositions and transport. Indiana Univ. Math. J. **57**(1), 247–276 (2008)
5. Cavagnari, G.: Regularity results for a time-optimal control problem in the space of probability measures. Math. Control Relat. Fields **7**(2), 213–233 (2017)
6. Cavagnari, G., Marigonda, A.: Time-optimal control problem in the space of probability measures. In: Lirkov, I., Margenov, S.D., Waśniewski, J. (eds.) LSSC 2015. LNCS, vol. 9374, pp. 109–116. Springer, Cham (2015). https://doi.org/10.1007/978-3-319-26520-9_11
7. Cavagnari, G., Marigonda, A.: Measure-theoretic Lie brackets for nonsmooth vector fields (Submitted)
8. Cavagnari, G., Marigonda, A., Nguyen, K.T., Priuli, F.S.: Generalized control systems in the space of probability measures. Set-Valued Var. Anal. (2017). https://doi.org/10.1007/s11228-017-0414-y
9. Cavagnari, G., Marigonda, A., Orlandi, G.: Hamilton-Jacobi-Bellman equation for a time-optimal control problem in the space of probability measures. In: Bociu, L., Désidéri, J.-A., Habbal, A. (eds.) CSMO 2015. IAICT, vol. 494, pp. 200–208. Springer, Cham (2016). https://doi.org/10.1007/978-3-319-55795-3_18
10. Cavagnari, G., Marigonda, A., Piccoli, B.: Averaged time-optimal control problem in the space of positive Borel measures. ESAIM: Control Optim. Calc. Var. (2017). https://doi.org/10.1051/cocv/2017060
11. Cavagnari, G., Marigonda, A., Piccoli, B.: Optimal synchronization problem for a multi-agent system. Netw. Heterogen. Media **12**(2), 277–295 (2017). https://doi.org/10.3934/nhm.2017012
12. Cavagnari, G., Marigonda, A., Priuli, F.S.: Attainability property for a probabilistic target in Wasserstein spaces (preprint)
13. Cristiani, E., Piccoli, B., Tosin, A.: Multiscale modeling of granular flows with application to crowd dynamics. Multiscale Model. Simul. **9**(1), 155–182 (2011)

14. Cristiani, E., Piccoli, B., Tosin, A.: Multiscale Modeling of Pedestrian Dynamics. MS&A Modeling, Simulation and Applications, vol. 12. Springer, Cham (2014). https://doi.org/10.1007/978-3-319-06620-2
15. Maniglia, S.: Probabilistic representation and uniqueness results for measure-valued solutions of transport equations. J. Math. Pures Appl. **87**(6), 601–626 (2007)
16. Siegfried, G., Harald, L.: Foundations of Quantization for Probability Distributions. Lecture Notes in Mathematics, vol. 1730. Springer, Berlin (2000). https://doi.org/10.1007/BFb0103945
17. Young, L.C.: Lectures on the calculus of variations and optimal control theory. American Mathematical Society, vol. 304. AMS Chelsea Publishing, Providence (1980)

Estimation of Star-Shaped Reachable Sets of Nonlinear Control Systems

Tatiana F. Filippova(✉)

Krasovskii Institute of Mathematics and Mechanics of RAS,
Ekaterinburg, Russian Federation
ftf@imm.uran.ru

Abstract. The problem of estimating reachable sets of nonlinear dynamical control systems with uncertainty in initial states is studied when it is assumed that only the bounding set for initial system positions is known and any additional statistical information is not available. We study the case when the system nonlinearity is generated from one side by bilinear terms in the matrix elements included in the state velocities and from the other side by quadratic functions in the right-hand part of system differential equations. Using results of the theory of trajectory tubes of control systems and techniques of differential inclusions theory and also results of ellipsoidal calculus we find set-valued estimates of reachable sets of such nonlinear uncertain control system.

Keywords: Control systems · Nonlinearity · Uncertainty
Ellipsoidal estimates

1 Introduction

Among important problems in the optimal control theory [19–21, 23, 29] is the construction and the study of reachable (attainable) sets for a dynamical system, i.e., the sets of all the states of the space to which the state vector can be moved from the initial position in a prescribed time with the help of admissible controls. Unfortunately very often it is difficult to find reachable sets exactly and well-developed technique of ellipsoidal calculus [6, 22, 23] may be helpful in constructing external and internal estimates of reachable sets for control systems with linear dynamics and for some special classes of nonlinear control systems [10, 13, 18, 25, 27]. The motivations for these studies come from applied areas ranged from engineering problems in physics to economics as well as to ecological and biomedical modeling [1, 2, 4, 5]. Some approaches to the nonlinear estimation problems and discrete approximation techniques for differential inclusions through a set-valued analogy of well-known Euler's method were developed in [3, 20, 21, 24, 30].

In this paper the modified state estimation approaches which use the special structure of nonlinearity of studied control system are presented. We assume here that the system nonlinearity is generated by the combination of two types

I. Lirkov and S. Margenov (Eds.): LSSC 2017, LNCS 10665, pp. 210–218, 2018.
https://doi.org/10.1007/978-3-319-73441-5_22

of functions in related differential equations, one of which is bilinear with ellipsoidal constraints on matrix parameters and the other one is quadratic. We find here the set-valued estimates of related reachable sets of such nonlinear uncertain control system. The algorithm of constructing the ellipsoidal estimates for studied nonlinear systems is given. Possibility of numerical simulations related to the proposed techniques are discussed with addressing also to impulsive control problems with nonlinearity and uncertainty.

2 Problem Formulation

Let us introduce the following basic notations. Let R^n denote the n-dimensional Euclidean space, $\operatorname{comp} R^n$ be the set of all compact subsets of R^n, $R^{n\times m}$ stands for the set of all real $n\times m$-matrices, $x'y = (x,y) = \sum_{i=1}^n x_i y_i$ be the usual inner product of $x, y \in R^n$ with prime as a transpose,

$$\|x\| = \|x\|_2 = (x'x)^{1/2}, \quad \|x\|_\infty = \max_{1\le i\le n} |x_i|$$

be vector norms for $x \in R^n$, $I \in R^{n\times n}$ be the identity matrix, $\operatorname{tr}(A)$ be the trace of $n \times n$-matrix A (the sum of its diagonal elements) and $|A|$ will denote its determinant. We denote by $B(a,r) = \{x \in R^n : \|x-a\| \le r\}$ the ball in R^n with a center $a \in R^n$ and a radius $r > 0$ and by

$$E(a,Q) = \{x \in R^n : \ (Q^{-1}(x-a),(x-a)) \le 1\}$$

the ellipsoid in R^n with a center $a \in R^n$ and with a symmetric positive definite $n \times n$-matrix Q.

Consider the following system

$$\dot{x} = A(t)x + f(x)d + u(t), \quad x_0 \in \mathcal{X}_0, \quad t \in [t_0, T], \tag{1}$$

where $x, d \in R^n$, $f(x)$ is the nonlinear function, which is quadratic in x, $f(x) = x'Bx$, with a given symmetric and positive definite $n \times n$-matrix B.

Control functions $u(t)$ in (1) are assumed to be Lebesgue measurable on $[t_0, T]$ and satisfying the constraint $u(t) \in \mathcal{U}$ for a.e. $t \in [t_0, T]$ (here $\mathcal{U}$ is a given set, $\mathcal{U} \in \operatorname{comp} R^n$).

We assume here that the $n \times n$-matrix function $A(t)$ in (1) has the form

$$A(t) = A^0 + A^1(t), \tag{2}$$

where the $n \times n$-matrix A^0 is given and the measurable $n \times n$-matrix $A^1(t)$ is unknown but bounded, $A^1(t) \in \mathcal{A}^1$ $(t \in [t_0, T])$,

$$A(t) \in \mathcal{A} = A^0 + \mathcal{A}^1. \tag{3}$$

Here

$$\begin{aligned} \mathcal{A}^1 = \big\{A=\{a_{ij}\}\in R^{n\times n} : a_{ij} = 0 \ \text{ for } \ i \ne j, \ \text{ and} \\ a_{ii} = a_i, \ \ i = 1,\ldots,n, \ \ a = (a_1,\ldots,a_n), \ \ a'Da \le 1\big\}, \end{aligned} \tag{4}$$

where $D \in R^{n\times n}$ is a symmetric and positive definite matrix.

We will assume that $\mathcal{X}_0$ in (1) is an ellipsoid, $\mathcal{X}_0 = E(a_0, Q_0)$, with a symmetric and positive definite matrix $Q_0 \in R^{n\times n}$ and with a center a_0.

Let the absolutely continuous function $x(t) = x\big(t; u(\cdot), A(\cdot), x_0\big)$ be a solution to dynamical system (1)–(3) with initial state $x_0 \in \mathcal{X}_0$, with admissible control $u(\cdot)$ and with a matrix $A(\cdot)$ satisfying (2)–(4). We assume that all these solutions $\{x(t)\}$ are extendable up to the instant T and are bounded $\|x(t)\| \le K$ (with some $K > 0$) (see e.g. [14] for detailed discussion). The reachable set $\mathcal{X}(t)$ at time t ($t_0 < t \le T$) of system (1)–(3) is defined as the following set

$$\mathcal{X}(t) = \{x \in R^n : \exists\, x_0{\in}\mathcal{X}_0,\ \exists\, u(\cdot){\in}\mathcal{U},\ \exists\, A(\cdot){\in}\mathcal{A},\ x = x(t) = x\big(t; u(\cdot), A(\cdot), x_0\big)\}.$$

Using the analysis of a special type of nonlinear control systems with uncertain initial data we consider here the techniques which allow us to find the external ellipsoidal estimate $E(a^+(t), Q^+(t))$ (with respect to the inclusion of sets) of the reachable set $\mathcal{X}(t)$ ($t_0 < t \le T$).

3 Auxiliary Results

3.1 Systems with Bilinear Dynamics

Bilinear dynamic systems constitute a special class of nonlinear systems representing a variety of important physical processes. A great number of results related to control problems for such systems has been developed over past decades, among them we mention here [7,15,27,28]. Reachable sets of bilinear systems in general are not convex, but have special properties (for example, may be star-shaped [15,21]). We, however, consider here the guaranteed state estimation problem and use ellipsoidal calculus for the construction of external estimates of reachable sets of such systems.

Consider the following control system of bilinear type

$$\dot{x} = A(t)\,x + u(t), \quad t_0 \le t \le T, \;\; x_0 \in \mathcal{X}_0 = E(a_0, Q_0), \tag{5}$$

where $x, a_0 \in R^n$, with a matrix Q_0 being symmetric and positive definite. We will assume that

$$u(t) \in \mathcal{U} = E(\hat{a}, \hat{Q}). \tag{6}$$

The bilinearity of the system (5) is due to the fact that the measurable matrix function $A(t) \in R^{n\times n}$ is not known but satisfies the constraint

$$A(t) \in \mathcal{A}, \;\; t_0 \le t \le T, \tag{7}$$

where

$$\mathcal{A} = \{A \in R^{n\times n} : A = \mathrm{diag}\,\{a\}, \;\; a \in A_0\}, \tag{8}$$

$$A_0 = \{a \in R^n : \; \sum_{i=1}^{n} |a_i|^2 \le 1\}. \tag{9}$$

Assumption 1. *We will assume further that* $0 \in \mathcal{X}_0$ *and* $0 \in \mathcal{U}$.

With this assumption the reachable sets $\mathcal{X}(t)$ of the system (5) are compact and star-shaped [15] and the following equality is true for Minkowski (gauge) functional [8,15],

$$h_M(z) = \inf\{t > 0 : z \in tM, x \in R^n\},$$

namely the next theorem is true.

Theorem 1 [25]. *The following external estimate is true*

$$\mathcal{X}(t_0+\sigma) \subseteq E(a^+(\sigma), Q^+(\sigma)) + o(\sigma)B(0,1), \quad \lim_{\sigma \to +0} \sigma^{-1} o(\sigma) = 0, \tag{10}$$

where $a^+(\sigma) = a_0 + \sigma\hat{a}$, $Q^+(\sigma) = (p^{-1}+1)Q_1(\sigma) + (p+1)\sigma^2\hat{Q}$,

$$Q_1(\sigma) = \operatorname{diag}\{(p_*^{-1}+1)\sigma^2 a_{0i}^2 + (p_*+1)r^2(\sigma) \mid i = 1,\ldots,n\},$$

$$r(\sigma) = \max_z \|z\| \cdot (h_{(I+\sigma\mathcal{A})*\mathcal{X}_0}(z))^{-1}, \quad (I+\sigma\mathcal{A}) * \mathcal{X}_0 = \bigcup_{A \in \mathcal{A}} (I+\sigma A)\mathcal{X}_0,$$

with $p_* > 0$ *and* $p > 0$ *being unique positive roots respectively of the related equations* $\sum_{i=1}^{n} \frac{1}{p+\alpha_i} = \frac{n}{p(p+1)}$ *where* $\alpha_i = \alpha_i(\sigma) \geq 0$ $(i = 1,...,n)$ *are roots of the equations having respectively in the case with* p_*, *the form* $\prod_{i=1}^{n}(\sigma^2(a_i^0)^2 - \alpha r^2(\sigma)) = 0$ *and in the case with* p, *the form* $|Q_1(\sigma) - \alpha\sigma^2\hat{Q}| = 0$.

Notes and Comments. Note that the estimate (10) may be easily modified for the case of constraints (2)–(4). The result presented in Theorem 1 is more convenient for computational implementation than the estimate given in [7,13] for systems with separate constraints on elements of unknown matrix.

3.2 Quadratic Nonlinearity

Problems of state estimation for nonlinear systems of type

$$\dot{x} = A(t)x + f(x)d + u(t), \quad x_0 \in \mathcal{X}_0, \; u(t) \in \mathcal{U}, \quad t \in [t_0, T], \tag{11}$$

with a quadratic function $f(x) = x'Bx$, with a given symmetric and positive definite $n \times n$-matrix B and a known matrix $A(t)$ were studied in [11,14], later these results have been extended to the case of quadratic functions $f(x) = x'Bx$ (11) without the assumption of a positive definiteness of B [12]. In [17] we examined the more complicated case when the system contains both types of nonlinearity, of quadratic type and also we assumed that matrix $A(t)$ was unknown but with bounded elements

$$A(t) \in \mathcal{A} = A^0 + \mathcal{A}^1, \quad \mathcal{A}^1 = \{A = \{a_{ij}\} \in R^{n\times n} : |a_{ij}| \leq c_{ij}, \; i,j = 1,\ldots n\}, \tag{12}$$

where $A^0 \in R^{n\times n}$, $c_{ij} \geq 0$ $(i,j = 1,\ldots n)$ are given.

In this paper we extend and modify the approach to the case when the uncertain system contains nonlinearities defined by a quadratic form in the right-hand sides of differential equations and also nonlinearities defined by uncertainty in coefficients of linear terms of the systems with quadratic constraints.

4 Main Results

Consider the system

$$\begin{aligned} \dot{x} = A(t)x + f(x)d + u(t), \ x_0 \in \mathcal{X}_0 = E(a_0, Q_0), \\ u(t) \in \mathcal{U} = E(\hat{a}, \hat{Q}), \ t \in [t_0, T], \end{aligned} \tag{13}$$

where $f(x) = x'Bx$, with a given symmetric and positive definite $n \times n$-matrix B. Here the $n \times n$-matrix function $A(t)$ satisfies the equality

$$A(t) \in \mathcal{A} = A^0 + \mathcal{A}^1, \tag{14}$$

where the $n \times n$-matrix A^0 is given and the measurable $n \times n$-matrix $A^1(t)$ is unknown but bounded, $A^1(t) \in \mathcal{A}^1$ ($t \in [t_0, T]$), with a constraint

$$\begin{aligned} \mathcal{A}^1 = \{A{=}\{a_{ij}\}{\in}R^{n\times n} : a_{ij} = 0 \ \text{ for } \ i \neq j, \ \text{ and} \\ a_{ii} = a_i, \ \ i = 1, \ldots, n, \ \ a = (a_1, \ldots, a_n), \ \ a'Da \leq 1\}, \end{aligned} \tag{15}$$

here $D \in R^{n \times n}$ is a symmetric and positive definite.

We need first the following auxiliary result.

Lemma 1. *For $\mathcal{X}_0 = E(0, Q_0)$ and $\mathcal{A}^1$ with the constraint (15) the Minkowski function of the set $(I + \sigma\mathcal{A}^1) * \mathcal{X}_0 = \bigcup\limits_{A \in \mathcal{A}^1} (I + \sigma A)\mathcal{X}_0$ has the form*

$$\begin{aligned} h_{(I+\sigma\mathcal{A}^1)*\mathcal{X}_0}(z) = \Big(||Q_0^{-1/2}z||^2 - 2\sigma\big(\sum_{i,j=1}^{n} w_i^2(z)(D^{-1/2})_{ij} \cdot w_j^2(z)\big)^{1/2}\Big)^{1/2} \\ + o(\sigma)||Q_0^{-1/2}z||, \ \ w(z) = Q_0^{-1/2}z, \ \ \lim_{\sigma \to +0} \sigma^{-1}o(\sigma) = 0. \end{aligned} \tag{16}$$

Proof. The formula follows from the definition of Minkowski functional and is proved by the schemes of papers [15,25] with minor modifications due to the presence of the matrix D in (15). □

The following result presents the external estimate of reachable sets of system (as in Theorem 1 we consider here the Assumption 1 fulfilled).

Theorem 2. *Let $\mathcal{X}_0 = E(a_0, k^2B^{-1})$, $k \neq 0$. Then for all $\sigma > 0$ the following external estimate is true*

$$\mathcal{X}(t_0 + \sigma) \subseteq E(a^+(\sigma), Q^+(\sigma)) + o(\sigma)B(0, 1), \ \ \lim_{\sigma \to +0} \sigma^{-1}o(\sigma) = 0, \tag{17}$$

where

$$\begin{aligned} a^+(\sigma) = a_0 + \sigma(A^0a_0 + \hat{a} + k^2d + a_0'Ba_0 \cdot d), \\ Q^+(\sigma) = (p^{-1} + 1)Q_1(\sigma) + (p + 1)\sigma^2\hat{Q}^*, \end{aligned}$$

$$Q_1(\sigma) = \text{diag}\,\{(p_*^{-1}+1)\sigma^2 a_{0i}^2 + (p_*+1)r^2(\sigma) \mid i = 1,\dots,n\},$$

$$r(\sigma) = \max_z ||z|| \cdot (h_{(I+\sigma\mathcal{A}^1)*\mathcal{X}_0}(z))^{-1}, \quad (I+\sigma\mathcal{A}^1) * \mathcal{X}_0 = \bigcup_{A\in\mathcal{A}^1} (I+\sigma A)\mathcal{X}_0\ ,$$

here numbers p_*, p *are the unique positive roots respectively of the related equations* $\sum_{i=1}^{n} \frac{1}{p+\alpha_i} = \frac{n}{p(p+1)}$ *with* $\alpha_i = \alpha_i(\sigma) \geq 0$ $(i = 1,...,n)$ *satisfing for* p_* *the equations* $\prod_{i=1}^{n}(\sigma^2(a_i^0)^2 - \alpha r^2(\sigma)) = 0$ *and for* p*, the equation* $|Q_1(\sigma) - \alpha\sigma^2\hat{Q}| = 0$*, here also* $E(0,\hat{Q}^*)$ *denotes the ellipsoid with minimal volume such that*

$$E(0,\hat{Q}) + (2d \cdot a_0' B \cdot + A^0)E(0, k^2B^{-1}) \subseteq E(0,\hat{Q}^*).$$

Proof. The funnel equation [22,23] which describes the time evolution of the reachable set $X(t) = X(t,t_0,\mathcal{X}_0)$ of the system (13)–(15) has the form here

$$\lim_{\sigma\to+0} \sigma^{-1}h(X(t+\sigma,t_0,\mathcal{X}_0), \bigcup_{x\in X(t,t_0,\mathcal{X}_0)} \{x + \sigma(\mathcal{A}x + f(x)d + E(\hat{a},\hat{Q}))\} = 0,$$

$$X(t_0,t_0,\mathcal{X}_0) = \mathcal{X}_0, \quad t_0 \leq t \leq T. \tag{18}$$

If $x_0 \in \partial\mathcal{X}_0$ where $\partial\mathcal{X}_0$ means the boundary of $\mathcal{X}_0$, we have

$$f(x_0) = k^2 + 2a'Bx - a'Ba$$

and from (4) for any $A \in \mathcal{A}$ we have also

$$\bigcup_{x_0\in\partial\mathcal{X}_0} \{(I+\sigma A)x_0 + \sigma f(x_0)d\} = \bigcup_{x_0\in\partial\mathcal{X}_0} \{(I+\sigma R)x_0 + \sigma(k^2 - a'Ba)d\}, \tag{19}$$

where $R = A + 2d \cdot a_0' B$. Note that if the ellipsoid in (17) gives the tube estimate for the system with $\partial\mathcal{X}_0$ as starting set, then also for the system with $\mathcal{X}_0$ as starting set (this idea follows the scheme of [14]). Applying Lemma 1 and Theorem 1 and taking into account the equality (19) and the above remark we come to the estimate (17). □

The following algorithm is based on Theorem 2 and may be used to produce the external ellipsoidal estimates for the reachable sets of the system (13)–(15).

Algorithm. Subdivide the time segment $[t_0, T]$ into subsegments $[t_i, t_{i+1}]$, where $t_i = t_0 + ih$ $(i = 1,\dots,m)$, $h = (T - t_0)/m$.

1. Take $\sigma = h$ and for given $\mathcal{X}_0 = E(a_0, Q_0)$ define the smallest $k_0 > 0$ such that $E(a_0,Q_0) \subseteq E(a_0, k_0^2B^{-1})$ (k_0^2 is the maximal eigenvalue of the matrix $B^{1/2}Q_0B^{1/2}$ ([11,12]).
2. For $\mathcal{X}_0 = E(a_0, k_0^2B^{-1})$ as an initial set define by Theorem 2 the upper estimate $\mathcal{X}_1 = E(a^+(\sigma), Q^+(\sigma))$ of the set $X(t_0+\sigma, t_0, \mathcal{X}_0)$.

3. Consider the system on the next subsegment $[t_1, t_2]$ with the initial ellipsoid $E(a_1, k_1^2 B^{-1})$ found as in step 1.
4. The next step repeats the previous iteration beginning with new initial data.

At the end of the process we will get the external estimate tube $E(a^+(t), Q^+(t))$ of the reachable sets $\mathcal{X}(t)$ $(t_0 \le t \le T)$ of the system (13)–(15). Discussions and numerical examples may be found also in [11–13,26].

Notes and Comments. It could be underlined here that the diagonal elements of A^1 in the differential equations of the studied control system (1)–(3) may be considered also as essentially additional controls with ellipsoidal control region. Therefore the studied system presents one more interesting class of control systems with nonlinear dynamics and uncertainty in initial system states.

Some approaches related to the case of systems with impulse control and uncertainty in initial data and parameters and also examples of model systems that illustrate the ideas may be found in [9,13,16,25].

Acknowledgments. The research was supported by Russian Science Foundation (RSF Project No.16-11-10146).

References

1. Apreutesei, N.C.: An optimal control problem for a prey-predator system with a general functional response. Appl. Math. Lett. **22**(7), 1062–1065 (2009)
2. August, E., Lu, J., Koeppl, H.: Trajectory enclosures for nonlinear systems with uncertain initial conditions and parameters. In: Proceedings of the 2012 American Control Conference, Fairmont Queen Elizabeth, Montréal, Canada, pp. 1488–1493. June 2012
3. Baier, R., Büskens, C., Chahma, I.A., Gerdts, M.: Approximation of reachable sets by direct solution methods of optimal control problems. Optim. Methods Softw. **22**, 433–452 (2007)
4. Boscain, U., Chambrion, T., Sigalotti, M.: On some open questions in bilinear quantum control. In: European Control Conference (ECC), Zurich, Switzerland, pp. 2080–2085. July 2013
5. Ceccarelli, N., Di Marco, M., Garulli, A., Giannitrapani, A.: A set theoretic approach to path planning for mobile robots. In: Proceedings of the 43rd IEEE Conference on Decision and Control, Atlantis, Bahamas, pp. 147–152. December 2004
6. Chernousko, F.L.: State Estimation for Dynamic Systems. Nauka, Moscow (1988)
7. Chernousko, F.L., Ovseevich, A.I.: Properties of the optimal ellipsoids approximating the reachable sets of uncertain systems. J. Optim. Theory Appl. **120**(2), 223–246 (2004)
8. Demyanov, V.F., Rubinov, A.M.: Quasidifferential Calculus. Optimization Software, New York (1986)
9. Filippova, T.F.: Set-valued solutions to impulsive differential inclusions. Math. Comput. Modell. Dyn. Syst. **11**(2), 149–158 (2005)
10. Filippova, T.F.: Construction of set-valued estimates of reachable sets for some nonlinear dynamical systems with impulsive control. Proc. Steklov Inst. Math. **269**(Suppl. 2), 95–102 (2010)

11. Filippova, T.F.: Differential equations of ellipsoidal state estimates in nonlinear control problems under uncertainty. In: Discrete and Continuous Dynamical Systems, Supplement 2011, Dynamical Systems, Differential Equations and Applications, vol. 1. pp. 410–419, Springfield, American Institute of Mathematical Sciences (2011)
12. Filippova, T.F.: State estimation for uncertain systems with arbitrary quadratic nonlinearity. In: Proceedings of the PHYSCON 2015, Istanbul, Turkey, August 1922, pp. 1–6 (2015)
13. Filippova, T.F.: Estimates of reachable sets of impulsive control problems with special nonlinearity. In: Proceedings of the AIP Conference, vol. 1773, p. 100004, 1–8 (2016)
14. Filippova, T.F., Berezina, E.V.: On state estimation approaches for uncertain dynamical systems with quadratic nonlinearity: theory and computer simulations. In: Lirkov, I., Margenov, S., Waśniewski, J. (eds.) LSSC 2007. LNCS, vol. 4818, pp. 326–333. Springer, Heidelberg (2008). https://doi.org/10.1007/978-3-540-78827-0_36
15. Filippova, T.F., Lisin, D.V.: On the estimation of trajectory tubes of differential inclusions. Proc. Steklov Inst. Math. Probl. Control Dyn. Syst. Suppl. 2, pp. S28–S37 (2000)
16. Filippova, T.F., Matviychuk, O.G.: Reachable sets of impulsive control system with cone constraint on the control and their estimates. In: Lirkov, I., Margenov, S., Waśniewski, J. (eds.) LSSC 2011. LNCS, vol. 7116, pp. 123–130. Springer, Heidelberg (2012). https://doi.org/10.1007/978-3-642-29843-1_13
17. Filippova, T.F., Matviychuk, O.G.: Estimates of reachable sets of control systems with bilinearquadratic nonlinearities. Ural Math. J. **1**(1), 45–54 (2015)
18. Gusev, M.I.: Application of penalty function method to computation of reachable sets for control systems with state constraints. In: Proceedings of the AIP Conference, vol. 1773 p. 050003, 1–8 (2016)
19. Kurzhanski, A.B.: Control and Observation Under Conditions of Uncertainty. Nauka, Moscow (1977)
20. Kurzhanski, A.B., Veliov, V.M. (eds.): Set-valued Analysis and Differential Inclusions: Progress in Systems and Control Theory, vol. 16. Birkhäuser, Boston (1990)
21. Kurzhanski, A.B., Filippova, T.F.: On the theory of trajectory tubes – a mathematical formalism for uncertain dynamics, viability and control. In: Kurzhanski, A.B. (ed.) Advances in Nonlinear Dynamics and Control: A Report from Russia. Progress in Systems and Control Theory, vol. 17, pp. 122–188. Birkhäuser, Boston (1993). https://doi.org/10.1007/978-1-4612-0349-0_4
22. Kurzhanski, A.B., Valyi, I.: Ellipsoidal Calculus for Estimation and Control. Birkhäuser, Boston (1997)
23. Kurzhanski, A.B., Varaiya, P.: Dynamics and Control of Trajectory Tubes: Theory and Computation. Systems & Control, Foundations & Applications, vol. 85. Basel, Birkhäuser (2014)
24. Krastanov, M.I., Veliov, V.M.: High-order approximations to nonholonomic affine control systems. In: Lirkov, I., Margenov, S., Waśniewski, J. (eds.) LSSC 2009. LNCS, vol. 5910, pp. 294–301. Springer, Heidelberg (2010). https://doi.org/10.1007/978-3-642-12535-5_34
25. Matviychuk, O.G.: Ellipsoidal estimates of reachable sets of impulsive control systems with bilinear uncertainty. Cybern. Phys. **5**(3), 96–104 (2016)
26. Matviychuk, O.G.: Internal ellipsoidal estimates for bilinear systems under uncertainty. In: Proceedings of the AIP Conference, vol. 1789, p. 060008, 1–8 (2016)

27. Mazurenko, S.S.: A differential equation for the gauge function of the star-shaped attainability set of a differential inclusion. Doklady Math. **86**(1), 476–479 (2012)
28. Polyak, B.T., Nazin, S.A., Durieu, C., Walter, E.: Ellipsoidal parameter or state estimation under model uncertainty. Automatica **40**, 1171–1179 (2004)
29. Schweppe, F.C.: Uncertain Dynamical Systems. Prentice-Hall, Englewood Cliffs (1973)
30. Veliov, V.: Second-order discrete approximation to linear differential inclusions. SIAM J. Numer. Anal. **29**(2), 439–451 (1992)

On Reachability Analysis of Nonlinear Systems with Joint Integral Constraints

Mikhail Gusev(✉)

N.N. Krasovskii Institute of Mathematics and Mechanics,
S. Kovalevskaya str., 16, 620990 Ekaterinburg, Russia
gmi@imm.uran.ru

Abstract. The problems of reachability for linear control systems with joint integral constraints on the state and input functions have been studied in the literature on the theory of set-valued state estimation. In this paper we consider a reachability problem for a nonlinear affine-control system on a finite time interval. The constraints on the state and control variables are given by the joint integral inequality, which assumed to be quadratic in the control variables. Assuming the controllability of the linearized system, we prove that any admissible control, that steers the control system to the boundary of its reachable set, is a local solution to an optimal control problem with integral cost functional.

Keywords: Optimal control · Integral constraints · Reachable set
Boundary points · Maximum principle

1 Introduction

We consider a reachability problem for a nonlinear affine-control system with joint integral constraints on the state and control. The reachability properties of nonlinear systems with integral constraints were investigated in the papers [10,14]. The algorithms for construction of reachable sets based on discrete approximations were proposed in [6,7]. The problems of control and estimation under integral constraints were studied in many papers (see, for example, [1,3,5,8]).

For the systems with pointwise constraints on the control it is known (see, for example, [13]) that the control, which steers the trajectory to the boundary of the reachable set, satisfies the Pontryagin maximum principle. In this paper we consider the reachability problem for a nonlinear affine-control system. The constraints on the state and control variables are given by the joint integral inequality which supposed to be quadratic in the control. Assuming the controllability property of the linearized system, we prove that any admissible control that steers the control system to the boundary of the projection of its reachable set onto given subspace is a local solution to some optimal control problem with an integral cost functional. This leads to the numerical method, based on Pontryagin maximum principle, for calculation of the reachable set boundary.

I. Lirkov and S. Margenov (Eds.): LSSC 2017, LNCS 10665, pp. 219–227, 2018.
https://doi.org/10.1007/978-3-319-73441-5_23

Algorithms for the approximation of reachable sets using solutions of optimal control problems were considered in [2,12,14].

Further we use the following notation. By $A^\top$ we denote the transpose of a real matrix A, I is an identity matrix, 0 stands for a zero vector of appropriate dimension. For $x, y \in \mathbb{R}^n$ let $(x, y) = x^\top y$ denotes the inner product, $x^\top = (x_1, \ldots, x_n)$, $\|x\| = (x, x)^{\frac{1}{2}}$ be the Euclidean norm, and $B_r(\bar{x})$: $B_r(\bar{x}) = \{x \in \mathbb{R}^n : \|x - \bar{x}\| \leq r\}$ be a ball of radius $r > 0$ centered at $\bar{x}$. For a set $S \subset \mathbb{R}^n$ let ∂S be the boundary of S; $\frac{\partial f}{\partial x}(x)$ is the Jacobi matrix of a vector-valued function $f(x)$. For a real $k \times m$ matrix A a matrix norm is denoted as $\| A \|$. The symbols $\mathbb{L}_1, \mathbb{L}_2$ and $\mathbb{C}$ stand for the spaces of summable, square summable and continuous functions respectively. The norms in these spaces are denoted as $\| \cdot \|_{\mathbb{L}_1}$, $\| \cdot \|_{\mathbb{L}_2}$, $\| \cdot \|_{\mathbb{C}}$.

Consider the linear control system

$$\dot{x}(t) = A(t)x(t) + B(t)u(t), \; t \in [t_0, t_1], \; x(t_0) = x^0, \tag{1}$$

$x \in \mathbb{R}^n$, $u \in \mathbb{R}^r$, on the fixed time interval $[t_0, t_1]$. Let $m \leq n$ and let P be an $m \times n$ full rank real matrix. Define the output of system (1) by the equality $y = Px$. Recall the following definition: the system (1) is said to be controllable on $[t_0, t_1]$ w.r.t. the output $y = Px$ if for any $y^1 \in \mathbb{R}^m$ there exists a control $u(\cdot) \in \mathbb{L}_2$ that transfers (1) from the zero initial state $x(t_0) = 0$ to the final state $x(t_1)$ such that $Px(t_1) = y^1$.

Define the symmetric matrix $W(t)$ (the controllability Gramian) by the equality $W(t) = \int_{t_0}^{t} X(t, \tau)B(\tau)B^\top(\tau)X^\top(t, \tau)d\tau$, where $X(t, \tau)$ is the fundamental Cauchy matrix of (1) ($X_t(t, \tau) = A(t)X(t, \tau)$, $X(\tau, \tau) = I$). The system (1) is controllable on $[t_0, t_1]$ w.r.t. the output $y = Px$ if and only if the matrix $V = PW(t_1)P^\top$ is positive definite.

Consider the following integral quadratic functional

$$I(u(\cdot)) = \| u(\cdot) \|^2_{\mathbb{L}_2} = \int_{t_0}^{t_1} u^\top(t)u(t)dt,$$

and define the set $U(\mu) = \{u(\cdot) \in \mathbb{L}_2 : J(u(\cdot)) \leq \mu^2\}$, $\mu > 0$ is a given number.

Denote by $G(\mu)$ the (output) reachable set of the system (1) at the time t_1 for the fixed x^0 and the integral constraints:

$$G(\mu) = \{y \in \mathbb{R}^m : \exists u(\cdot) \in U(\mu), y = Px(t_1, u(\cdot))\},$$

here $x(t, u(\cdot))$ is a trajectory of system (1) corresponding to $u(\cdot)$. If, for example, $P = [I, 0]$, then $G(\mu)$ is the projection of the reachable set in the state space on the space of first m coordinates. If system (1) is controllable on $[t_0, t_1]$ w.r.t. the output $y = Px$, then $G(\mu)$ is an ellipsoid in $\mathbb{R}^m$:

$$G(\mu) = \{y \in \mathbb{R}^m : (y - \hat{y})^\top V^{-1}(y - \hat{y}) \leq \mu^2\}, \; \hat{y} = PX(t_1, t_0)x^0.$$

Standard arguments, using the convexity of reachable sets, lead to the following assertion.

Proposition 1. *Let system (1) be controllable w.r.t.* $y = Px$ *on* $[t_0, t_1]$. *The control* $u(\cdot) \in U(\mu)$ *steers the trajectory of system (1) to the point* y^1 *belonging to the boundary of the reachable set* $G(\mu)$, *if and only if it solves the following optimal control problem for system (1)*

$$I(u(\cdot)) \to \min, \; u(\cdot) \in \mathbb{L}_2, \; x(t_0) = x^0, \; Px(t_1) = y^1$$

and the minimum of J *equals to* μ^2.

Note that if system (1) if not controllable, the reachable set $G(\mu)$ is a degenerate ellipsoid which belongs to the subspace $\mathbb{R}^q$, $q < m$. In this case all points of $G(\mu)$ are boundary points, and the assertion of proposition 1 is, obviously, not true.

The aim of this paper is to prove the necessary conditions of the above type for nonlinear control systems with joint integral constraints.

2 Reachable Sets of Affine-Control Systems

2.1 Definitions and Auxiliary Results

We consider the control system

$$\dot{x}(t) = f_1(t, x(t)) + f_2(t, x(t))u(t), \; x(t_0) = x^0, \tag{2}$$

where $t_0 \leq t \leq t_1$, $x \in \mathbb{R}^n$, $u \in \mathbb{R}^r$, $f_1 : \mathbb{R}^{n+1} \to \mathbb{R}^n$, $f_2 : \mathbb{R}^{n+1} \to \mathbb{R}^{n \times r}$ are continuous mappings.

The functions f_1 and f_2 are assumed to be continuously differentiable in x and satisfying the following conditions:

$$\|f_1(t, x)\| \leq l_1(t)(1 + \| x\|), \; \|f_2(t, x)\|_{n \times r} \leq l_2(t), \; t_0 \leq t \leq t_1, \; x \in \mathbb{R}^n, \tag{3}$$

where $l_1(\cdot) \in \mathbb{L}_1$, $l_2(\cdot) \in \mathbb{L}_2$. Under these assumptions for any $u(\cdot) \in \mathbb{L}_2$ there exists a unique absolutely continuous solution $x(t)$ of system (2) which satisfies the initial condition $x(t_0) = x_0$ and is defined on the interval $[t_0, t_1]$.

Further we denote as $J(u(\cdot))$ the following integral functional

$$J(u(\cdot)) = \int_{t_0}^{t_1} Q(t, x(t)) + u^\top(t)R(t, x(t))u(t)dt.$$

Here $x(t)$ is a solution of system (2) corresponding to the control $u(t)$ and the initial vector x^0, the function $Q(t, x)$ and symmetric matrix $R(t, x)$ are assumed to be continuous on $[t_0, t_1] \times \mathbb{R}^n$. Define the set $U(\mu) = \{u(\cdot) \in \mathbb{L}_2 : J(u(\cdot)) \leq \mu^2\}$, and let P be a given $m \times n$ full rank real matrix, $m \leq n$. Denote as earlier by $G(\mu)$ the (output) reachable set of the system (2) at the time t_1 for the fixed x^0 and the integral constraints:

$$G(\mu) = \{y \in \mathbb{R}^m : \exists u(\cdot) \in U(\mu), y = Px(t_1, u(\cdot))\},$$

$x(t, u(\cdot))$ is a trajectory of system (2) corresponding to $u(\cdot)$.

Assumption 1. *The inequalities* $Q(t,x) \geq 0$, $u^\top R(t,x)u \geq \alpha\|u\|^2$ *hold for some* $\alpha > 0$ *and any* $(t,x,u) \in [t_0,t_1] \times \mathbb{R}^n \times \mathbb{R}^r$.

Proposition 2 *(See, for example, [9]).* Let $u_p(\cdot) \in U(\mu)$ be a sequence of controls and $x_p(\cdot)$ the corresponding sequence of trajectories. If $u_p(\cdot) \to \hat{u}(\cdot)$ with respect to weak topology in $\mathbb{L}_2$, then $x_p(\cdot) \to \hat{x}(\cdot)$ in $C[t_0,t_1]$, where $\hat{x}(\cdot)$ is the trajectory corresponding to $\hat{u}(\cdot)$.

Corollary 1. *The functional* $J(u(\cdot))$ *is continuous in* $\mathbb{L}_2$.

Proof. The proof is straightforward.

Proposition 3. *Let functions* $f_1(t,x), f_2(t,x)$ *be continuous, continuously differentiable in* x*, and satisfy conditions (3). Let Assumption 1 be fulfilled. Then the set of trajectories of system (2), corresponding to controls* $u(\cdot) \in U(\mu)$*, is a compact set in* $C[t_0,t_1]$.

Proof. Without loss of generality it can be assumed that $U(\mu)$ is nonempty. By Assumption 1 for any $u(\cdot) \in U(\mu)$ we have $\| u(\cdot) \|_{\mathbb{L}_2} \leq \mu^2/\alpha$. Let $x_p(\cdot)$ be a sequence of trajectories, corresponding to $u_p(\cdot) \in U(\mu)$. Since $\| u_p(\cdot) \|_{\mathbb{L}_2} \leq \mu^2/\alpha$, $u_p(\cdot)$ contains a weakly convergent in $\mathbb{L}_2$ subsequence. By Proposition 2 we may assume that $u_p(\cdot)$ weakly converges to $\hat{u}(\cdot)$ and $x_p(t)$ uniformly converges to $\hat{x}(t)$ as $p \to \infty$, here $\hat{x}(t)$ is the trajectory corresponding to $\hat{u}(\cdot)$. Due to convexity of the functional

$$J_1(u(\cdot) = \int_{t_0}^{t_1} (Q(t,\hat{x}(t)) + u^\top(t)R(t,\hat{x}(t))u(t)dt$$

we have

$$J_1(\hat{u}(\cdot)) \leq \liminf_{p\to\infty} J_1(u_p(\cdot)). \tag{4}$$

From the inequality

$$|\int_{t_0}^{t_1} u^{p\top}(t)(R(t,\hat{x}(t)) - R(t,x_p(t)))u^\top(t)dt| \leq \psi(u_p(\cdot)),$$

where $\psi(u_p(\cdot)) = \max_t \|R(t,\hat{x}(t)) - R(t,x_p(t))\| \|u_p(\cdot)\|^2_{\mathbb{L}_2}$, and the uniform convergence of $x_p(t)$ to $\hat{x}(t)$ it follows that $|J_1(u_p(\cdot)) - J(u_p(\cdot))|$ tends to zero as $p \to \infty$. Taking into account (4) we obtain $J(\hat{u}(\cdot)) \leq \mu^2$. This completes the proof.

Definition 1. *Let* $u(\cdot) \in \mathbb{L}_2$ *be a control,* $x(t)$ *be the corresponding trajectory. A linear control system with matrices*

$$A(t) = \frac{\partial f_1}{\partial x}(t,x(t)) + \frac{\partial}{\partial x}[f_2(t,x(t))u(t)],\ B(t) = f_2(t,x(t))$$

is said to be a linearization of (2) along the pair $(x(t),u(t))$.

Consider the sequences of controls $u_p(\cdot)$ and trajectories $x_p(\cdot)$. Denote as $A_p(t)$, $B_p(t)$ the matrices of the linearization of system (2), corresponding to $(x_p(\cdot), u_p(\cdot))$.

Lemma 1. *If $u_p(\cdot) \to u(\cdot)$ in $\mathbb{L}_2$ and the pair $(A(t), B(t))$, corresponding to $u(\cdot)$, is controllable with respect to output $y = Px$, then, for sufficiently large p, the pair $(A_p(t), B_p(t))$ is also controllable.*

Proof. The proof follows to the proof of Lemma 1 in [9].

2.2 Extremal Properties of the Boundary Points

Define the map $F : \mathbb{L}_2 \to \mathbb{R}^m$ by the equality $Fu(\cdot) = Px(t_1)$, where $x(t)$ is the trajectory of system (2), corresponding to a control $u(\cdot)$. The following assertion is true (see also [9,14]).

Lemma 2. *Let the functions $f_1(t, x), f_2(t, x)$ be continuous, continuously differentiable in x, and satisfy inequalities (3). Then F has a continuous Fréchet derivative $F' : \mathbb{L}_2 \to \mathbb{R}^n$ which is defined $\forall u(\cdot) \in \mathbb{L}_2[t_0, t_1]$ by the equality*

$$F'(u(\cdot))\delta u(\cdot) = P\delta x(t_1). \tag{5}$$

Here $\delta x(t)$ is a solution of the linearization along $(u(t), x(t))$ of system (2), corresponding to a zero initial state and the control $\delta u(t)$. If the linearized system is controllable on $[t_0, t_1]$ w.r.t. output $y = Px$, then $ImF'(u(\cdot)) = \mathbb{R}^m$.

Consider the following auxiliary optimal control problem

$$J(u(\cdot)) \to \min,\ u(\cdot) \in \mathbb{L}_2,\ x(t_0) = x^0,\ Px(t_1) = y^1. \tag{6}$$

The only constraint in this problem is the terminal constraint $Px(t_1) = y^1$.

To prove the next result we need the Graves theorem (see, for example, [4,11]). This theorem states the following: if the mapping F from a Banach space X to a Banach space Y is continuously differentiable at a point $\bar{x}$ and satisfy the condition $ImF'(\bar{x}) = Y$, then there exists $s > 0$ such that for all sufficiently small $r > 0$ $F-$ image of the ball $B(\bar{x}, r)$ contains the ball of radius sr with center $F(\bar{x}) : B(F(\bar{x}), sr) \subset F(B(\bar{x}, r))$. Using Lemmas 1 and 2 and the Graves theorem we prove the following.

Theorem 1. *Suppose that:*

(1) $y^1 \in \partial G(\mu)$;
(2) $\hat{u}(\cdot) \in U(\mu)$ is a control that steers the system from the state $x(t_0) = x^0$ to $\hat{x}(t_1)$, $P\hat{x}(t_1) = y^1$, $\hat{x}(t)$ is the corresponding trajectory;
(3) the linearization along $(\hat{x}(t), \hat{u}(t))$ of system (2) is controllable on $[t_0, t_1]$ w.r.t. output $y = Px$;

Then there exists $\sigma > 0$ such that $J(u(\cdot)) \geq \mu^2$ for any $u \in B(\hat{u}(\cdot), \sigma)$ satisfying the condition $Px(t_1) = y^1$. Since $J(\hat{u}(\cdot)) \leq \mu^2$, this implies that $J(\hat{u}(\cdot)) = \mu^2$ and the control $\hat{u}(\cdot)$ provides a local minimum in (6).

Proof. Assume the opposite. Then for every $p \in \mathbb{N}$ there exists $u_p(\cdot)$, which satisfies the terminal constraints and the inequalities $\| \hat{u}(\cdot) - u_p(\cdot) \|_{\mathbb{L}_2} < \frac{1}{p}$, $J(u_p(\cdot)) < \mu^2$. This implies the existence of a sequence $u_p(\cdot) \to \hat{u}(\cdot)$ in $\mathbb{L}_2$, $p \to \infty$ such that $Px_p(t_1) = y^1$, and $J(u_p(\cdot)) < \mu^2$. Take $\bar{p}$ so large that the pair $(A_{\bar{p}}(t), B_{\bar{p}}(t))$, corresponding to $(u_{\bar{p}}(t), x_{\bar{p}}(t))$, is controllable on $[t_0, t_1]$ (see Lemma 1). Denote $\delta = \mu^2 - J(u_{\bar{p}}(\cdot)) > 0$. From Lemma 1 it follows that there exist $\varepsilon > 0$ such that

$$v(\cdot) \in \mathbb{L}_2, \ \| v(\cdot) - u_{\bar{p}}(\cdot) \|_{\mathbb{L}_2} < \varepsilon \Rightarrow |J(v(\cdot)) - J(u_{\bar{p}}(\cdot))| < \delta/2,$$

the last implies that $J(v(\cdot)) < J(u_{\bar{p}}(\cdot)) + \delta/2 < \mu^2 - \delta/2$. Applying Lemma 2, we get that the linearization along $(u_{\bar{p}}(\cdot), x_{\bar{p}}(\cdot))$ of system (2) is controllable, hence $ImF'(u_{\bar{p}}(\cdot)) = \mathbb{R}^m$. Then, by the Graves theorem, there exists $s > 0$ such that for all sufficiently small $0 < r < \varepsilon$ the F-image of the ball $B(u_{\bar{p}}(\cdot), r)$ contains the ball of radius sr with center $y^1 = F(u_{\bar{p}}(\cdot)) : B(y^1, sr) \subset F(B(u_{\bar{p}}(\cdot), r))$. Thus, $B(u_{\bar{p}}(\cdot), r) \subset U(\mu) \Rightarrow F(B(u_{\bar{p}}(\cdot), r)) \subset F(U(\mu)) \subset G(\mu)$, the last implies $B(y^1, sr) \subset G(\mu)$, this contradicts the assumption that $y^1 \in \partial G(\mu)$.

2.3 The Maximum Principle for Boundary Trajectories

Since the local minimum in $\mathbb{L}_2$ admits the needle variations of the control, the local $\mathbb{L}_2$-minimizer satisfies Pontryagin's maximum principle. To formulate it, let us introduce the Pontryagin function (Hamiltonian) associated with (6)

$$H(p, t, x, u) = -p_0 f_0(t, x, u) + p^\top (f_1(t, x) + f_2(t, x)u),$$

$p_0 \geq 0$, $f_0(t, x, u) = Q(t, x) + u^\top R(t, x)u$. Assume that $Q(t, x)$, $R(t, x)$ are continuously differentiable in x. A locally optimal control $u(t)$ for (6) satisfies the maximum principle: there exist $p_0 \geq 0$, $l \in \mathbb{R}^m$, $(p_0, l) \neq 0$, and a function $p(t)$ such that

$$H(p(t), t, x(t), u(t)) = \max_{v \in \mathbb{R}^r} H(p(t), t, x(t), v),$$

$$\dot{p(t)} = -\frac{\partial f}{\partial x} H(p(t), x(t), u(t)) = -A^\top(t)p(t) + p_0 \frac{\partial}{\partial x} f_0(t, x, u), \ p(t_1) = P^\top l.$$

The maximum principle implies that $H_u(p(t), t, x(t), u) = 0$, hence $p^\top(t)B(t) = p_0 R(t, x(t))u(t)$, $B(t) = f_2(t, x(t)$. Since the terminal constraints are regular ($rankP = m$), we have $p_0 + \|p(t)\| \neq 0$, $t \in [t_0, t_1]$. As previously, we denote here by $(A(t), B(t))$ the matrices of the linearization along $(x(t), u(t))$ of system (2). If this system is controllable w.r.t. $y = Px$, then $p_0 > 0$.

Really, if $p_0 = 0$, then $p(\cdot)$ is a non zero solution of the equation $\dot{p}(t) = -A^\top(t)p(t)$, $p(t_1) = P^\top l$, and $p^\top(t)B(t) \equiv 0$. Represent $p(t)$ in the form $p(t) = X^\top(t_1, t)P^\top l$, then $\|l^\top PX(t_1, t)B(t)\|^2 = 0$, $t \in [t_0, t_1]$. Integrating both sides of the last equality over $[t_0, t_1]$, we get $l^\top Vl = 0$. This contradicts to the controllability of $(A(t), B(t))$ w.r.t. $y = Px$, since $l \neq 0$. Thus we can take $p_0 = \frac{1}{2}$, that implies that $u(t) = u(t, x(t), p(t))$, where $u(t, x, p) = R^{-1}(t, x)f_2^\top(t, x)p$.

Let us describe shortly the following algorithm for calculating boundary points of reachable sets based on the maximum principle. Further we assume that $P = [I, 0]$ if $m < n$ or $P = I$ if $m = n$. In this case the transversality conditions take the form $p_i(t_1) = 0,\ i = m+1, \ldots, n$. Letting $\dot{x}_0(t) = f_0(t, x(t), u(t)),\ x_0(t_0) = 0$, we get $J(u(\cdot)) = x_0(t_1)$. Substituting $u(t, x, p)$ into differential equations, we obtain the following system

$$\begin{aligned}
\dot{x}(t) &= f_1(t, x(t)) + f_2(t, x(t))u(t, x(t), p(t)), x(t_0) = x^0 \\
\dot{p}(t) &= -\frac{\partial f}{\partial x} H(p(t), x(t), u(t, x(t), p(t))), p(t_0) = q \\
\dot{x}_0(t) &= f_0(t, x(t), u(t, x(t), p(t))), x_0(t_0) = 0.
\end{aligned} \tag{7}$$

Since x^0 is fixed, the solution of (7) depends only on the vector $q \in \mathbb{R}^n$, denote this solution as $x(t, q), p(t, q), x_0(t, q)$. These functions have continuous derivatives with respect to q, which can be found by integrating the linearization of (7) along the trajectory. Consider the following functions $\phi_0(q) = x_0(t_1, q) - \mu^2$, $\phi_i(q) = p_{i+m}(t_1, q)$, $i = 1, \ldots, n-m$. These functions are continuously differentiable; it's derivatives can be obtained by numerical integration of corresponding differential equations. The calculations of boundary points requires the solving a nonlinear system $\phi_i(q) = 0,\ i = 0, \ldots, n-m$ and the integrating (7) for the solutions of this system as the initial points.

2.4 Example

To illustrate the procedure consider the Duffing equation

$$\dot{x}_1 = x_2, \quad \dot{x}_2 = -x_1 - 10x_1^3 + u,\ t \in [0, t_1],\ x_1(0) = 0,\ x_2(0) = 0, \tag{8}$$

which describes the motion of nonlinear stiff spring on impact of an external force u. Consider the integral constraint

$$\int_0^{t_1} (ax_1^2(t) + bx_2^2 + u^2(t))dt \leq 2,$$

where a, b are positive parameters, and take $P = I$. It is easy to verify that the controllability assumption of Theorem 1 is satisfied here. To calculate the boundary of $G(2)$ here we need to solve the equation $\phi_0(q) = 0$.

Represent $q \in \mathbb{R}^2$ in polar coordinates: $q_1(\theta) = r(\theta)\cos(\theta + \theta_0) + q_1^0$, $q_2(\theta) = r(\theta)\sin(\theta + \theta_0) + q_2^0$. Here $r(\theta)$ is a distance from a reference point q^0 and θ is an angle between $q - q^0$ and the reference direction $\bar{q} = (\cos\theta_0, \sin\theta_0)$. Differentiating the identity $\phi_0(q(\theta)) = 0$, we get a differential equation for $r(\theta)$

$$\dot{r}(\theta) = r(\theta)\frac{\phi_{0q_1}(q(\theta))\sin(\theta + \theta_0) - \phi_{0q_2}(q(\theta))\cos(\theta + \theta_0)}{\phi_{0q_1}(q(\theta))\cos(\theta + \theta_0) + \phi_{0q_2}(q(\theta))\sin(\theta + \theta_0)}, \quad 0 \leq \theta \leq 2\pi.$$

To start the solution we use one-dimensional search for the root of equation $\phi_0(q^0 + r\bar{q}) = 0$ and take this root as the initial state for differential equation.

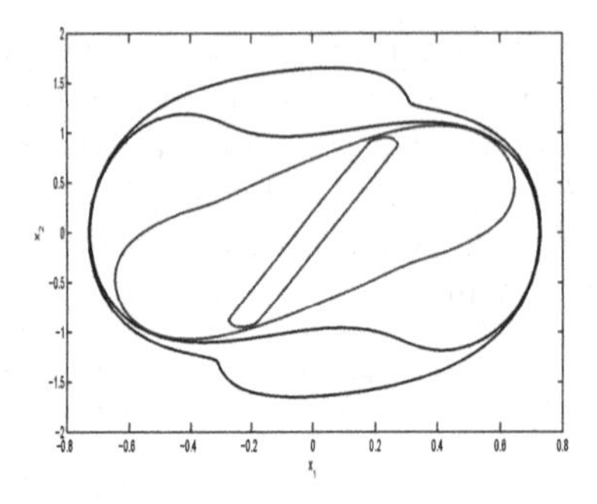

Fig. 1. Reachable sets for $t_1 = 0.5, 1.0, 1.5, 2.0; a, b = 0$

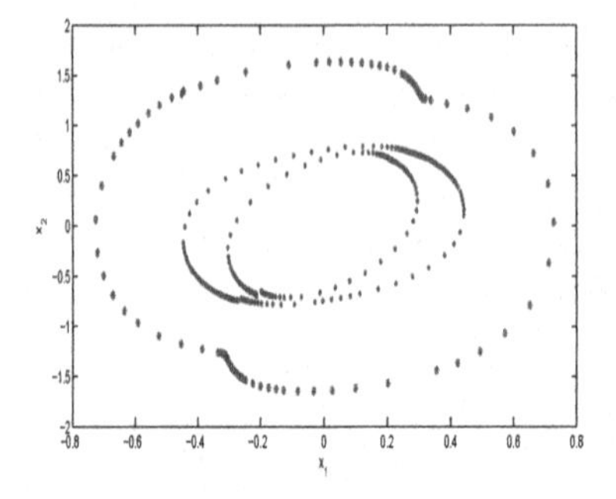

Fig. 2. Reachable sets for different values of a, b

The Fig. 1 shows the plot of the reachable sets boundaries for $t_1 = 0.5, 1, 1.5$, and 2 respectively, and for $a = 0,\ b = 0$. The reachable sets boundaries for the values $a = 0, b = 0; a = 5, b = 10; a = 30, b = 15$ and $t_1 = 2$ are presented in the Fig. 2.

Acknowledgments. The research is supported by Russian Science Foundation, project № 16-11-10146.

References

1. Anan'ev, B.I.: Motion correction of a statistically uncertain systemunder communication constraints. Autom. Remote Control **71**(3), 367–378 (2010)
2. Baier, R., Gerdts, M., Xausa, I.: Approximation of reachable sets using optimal control algorithms. Numer. Algebra Control Optim. **3**(3), 519–548 (2013)
3. Dar'in, A.N., Kurzhanskii, A.B.: Control under indeterminacy and double constraints. Differ. Equ. **39**(11), 1554–1567 (2003)
4. Donchev, A.: The Graves theorem revisited. J. Convex Anal. **3**(1), 45–53 (1996)
5. Filippova, T.F.: Estimates of reachable sets of impulsive control problems with special nonlinearity. In: AIP Conference Proceedings, vol. 1773, pp. 1-10 (2016). Article number 100004
6. Guseinov, K.G., Ozer, O., Akyar, E., Ushakov, V.N.: The approximation of reachable sets of control systems with integral constraint on controls. Nonlinear Differ. Equ. Appl. **14**(1–2), 57–73 (2007)
7. Guseinov, K.G., Nazlipinar, A.S.: Attainable sets of the control system with limited resources. Trudy Inst. Mat. i Mekh. Uro RAN **16**(5), 261–268 (2010)
8. Gusev, M.I.: On optimal control problem for the bundle of trajectories of uncertain system. In: Lirkov, I., Margenov, S., Waśniewski, J. (eds.) LSSC 2009. LNCS, vol. 5910, pp. 286–293. Springer, Heidelberg (2010). https://doi.org/10.1007/978-3-642-12535-5_33
9. Gusev, M.I., Zykov, V.I.: On extremal properties for boundary points of reachable sets under integral constraints on the control. Trudy Inst. Mat. Mekh. UrO RAN **23**(1), 103–115 (2017). (in Russian)

10. Huseyin, N., Huseyin, A.: Compactness of the set of trajectories of the controllable system described by an affine integral equation. Appl. Math. Comput. **219**, 8416–8424 (2013)
11. Ioffe, A.D.: Metric regularity and subdifferential calculus. Russ. Math. Surv. **55**(3), 501–558 (2000)
12. Kurzhanski, A.B., Varaiya, P.: Dynamic optimization for reachability problems. J. Optim. Theor. Appl. **108**(2), 227–251 (2001)
13. Lee, E.B., Marcus, L.: Foundations of Optimal Control Theory. Willey, Hoboken (1967)
14. Polyak, B.T.: Convexity of the reachable set of nonlinear systems under L2 bounded controls. Dyn. Contin. Discret. Impuls. Syst. Ser. A: Math. Anal. **11**, 255–267 (2004)

Existence Theorem for Infinite Horizon Optimal Control Problems with Mixed Control-State Isoperimetrical Constraint

Valeriya Lykina(✉)

Research Unit ORCOS (E105-4), Vienna University of Technology,
Wiedner Hauptstraße 8, 1040 Vienna, Austria
valeriya.lykina@tuwien.ac.at

Abstract. In this paper a class of infinite horizon optimal control problems with a mixed control-state isoperimetrical constraint, also interpreted as a budget constraint, is considered. Herein a linear both in the state and in the control dynamics is allowed. The problem setting includes a weighted Sobolev space as the state space. For this class of problems, we establish an existence theorem. The proved theoretical result is applied to a mixed control-state budget constrained advertisement model.

Keywords: Mixed control-state isoperimetrical constraint
Infinite horizon · Optimal control · Existence theorem
Weighted sobolev spaces · Advertisement model

1 Introduction

Infinite horizon optimal control problems represent a class of problems having very broad applications e.g. in the economics, biology and stabilization problems as well, cf. [7,8]. Since several decades, many authors have contributed to investigations of this class of problems in proving necessary and sufficient optimality conditions, c.f. [1,2,4], in developing numerical results [7,17], in deriving existence results [3,5] among many others. It turned out that in many models it is not possible to assure the existence of an optimal solution, cf. [15]. In [13] it was investigated how the introduction of an additional isoperimetrical state constraint into a problem statement influences the existence of optimal solutions and whether it forces it. The specifical interest of introducing an isoperimetric constraint in an optimal control problem is also motivated by many other reasons, e.g. from the point of view of multicriterial control problems, see [19, p. 176], or by evident economical arguments, cf. [8, p. 122], or by therapeutic evidences, cf. [16, p. 3] and [11, p. 307]. To the best of author's knowledge, there are no results

V. Lykina—This research was supported by the German Research Foundation (DFG), grant number LY 149/1-1.

I. Lirkov and S. Margenov (Eds.): LSSC 2017, LNCS 10665, pp. 228–236, 2018.
https://doi.org/10.1007/978-3-319-73441-5_24

in the broad literature which concern the existence of optimal solutions for the considered class of mixed budget-constrained control problems, formulated in weighted functional spaces. Due to variety of optimality notions for infinite horizon control problems, the comparability of the results is hardly possible. The main idea here is to choose a weighted Sobolev space as the state space, whereas the weight is specified by exploiting the underlying dynamics. The used optimality notion is the "classical" one, applied only to the Lebesgue interpretation of the integral in the objective.

The present paper generalizes the result obtained in [13] with respect to the more general isoperimetrical constraint which is allowed to depend linearly on both the state and the control variables. This class of problems has numerous reasonable applications such as medical or epidemiological, where the overall dosages of some drug or vaccine indicated to apply over the whole treatment period is bounded from above by a constant or even prescribed exactly by this constant. In this paper we consider an economic application of the presented class of problems.

The paper is organized as follows. The Sect. 2 contains important definitions. Section 3 introduces the problem statement and the optimality criterion. The derivation of the existence result is contained in Sect. 4. In the next section we analyse the mixed control-state budget-constrained advertisement model in order to demonstrate the applicability of the theoretical result proved before. Finally a summary of the results follows.

2 Main Definitions

Let us write $[0,\infty) = \mathbb{R}_+$. We denote by $M(\mathbb{R}_+), L_p(\mathbb{R}_+)$ and $C^0(\mathbb{R}_+)$ the spaces of all functions $x : \mathbb{R}_+ \to \mathbb{R}$ which are Lebesgue measurable, in the pth power Lebesgue integrable or continuous, respectively, see [6]. The Sobolev space $W^1_p(\mathbb{R}_+)$ is then defined as the space of all functions $x : \mathbb{R}_+ \to \mathbb{R}$ which belong to $L_p(\mathbb{R}_+)$ and admit distributional derivative $\dot{x}$ [21, p. 49] belonging to $L_p(\mathbb{R}_+)$ as well.

Definition 1. (a) *A continuous function $\nu : \mathbb{R}_+ \to \mathbb{R}$ with positive values is called a weight function.*

(b) *A weight function ν will be called a density function iff it is Lebesgue integrable over $\mathbb{R}_+$, i.e. $\int_0^\infty \nu(t)\,dt < \infty$, (cf. [10, p. 18]).*

(c) *By means of a weight function $\nu \in C^0(\mathbb{R}_+)$ we define for any $1 \le p < \infty$ the weighted Lebesgue space*

$$L_p(\mathbb{R}_+,\nu) = \left\{ x \in M(\mathbb{R}_+) \;\middle|\; \|x\|_{L_p(\mathbb{R}_+,\nu)} = \Big(\int_0^\infty |x(t)|^p \,\nu(t)\,dt \Big)^{1/p} < \infty \right\}.$$

(d) *For $x \in L_p(\mathbb{R}_+,\nu)$ let the distributional derivative $\dot{x}$ be defined according to [21, p. 46]. We introduce the weighted Sobolev space of all $L_p(\mathbb{R}_+,\nu)$ functions* *having their distributional derivative in* $L_p(\mathbb{R}_+, \nu)$*:*

$$W_p^1(\mathbb{R}_+, \nu) = \Big\{ x \in M(\mathbb{R}_+) \;\big|\; x \in L_p(\mathbb{R}_+, \nu),\, \dot{x} \in L_p(\mathbb{R}_+, \nu) \Big\}$$

(see [10, p. 11]).

Equipped with the norm

$$\|x\|_{W_p^1(\mathbb{R}_+,\nu)} = \|x\|_{L_p(\mathbb{R}_+,\nu)} + \|\dot{x}\|_{L_p(\mathbb{R}_+,\nu)},$$

$W_p^1(\mathbb{R}_+, \nu)$ becomes a Banach space (this can be confirmed analogously to [10, p. 19]).

3 Problem Formulation

The main control problem being considered in the present paper is:

$$(\mathrm{P})_\infty^B : \qquad J_\infty(x,u) = \int_0^\infty r_0(t, x(t), u(t))\nu_0(t)dt \longrightarrow \text{Min}! \tag{1}$$

$$(x,u) \in W_2^{1,n}(\mathbb{R}_+, \nu_1) \times L_2^m(\mathbb{R}_+, \nu_1), \tag{2}$$

$$\dot{x}(t) = A(t)x(t) + B(t)u(t) + C(t),\; x(0) = x_0 > 0, \tag{3}$$

$$d = \int_0^\infty \left\{ D_1^0(t)^T x(t) + D_2^0(t)^T u(t) \right\} \nu_2(t)\, dt \tag{4}$$

$$u(t) \in U \text{ a. e. on } \mathbb{R}_+, \tag{5}$$

where the state Eq. (3) has to be satisfied almost everywhere on $\mathbb{R}_+$. Furthermore, U denotes a compact convex subset of $\mathbb{R}^m$. Further, let functions ν_0, ν_1 and ν_2 be weight functions as defined in Sect. 2. The functions x and u are called the state and the control function respectively. The integral in (1) is understood in Lebesgue sense. The fact that we have to distinguish between different integral types in infinite horizon optimal control problems was discussed in detail in [14,18]. Our considerations are based on the following

Definition 2. **(a)** The set of all *admissible pairs*, denoted by $\mathcal{A}$, consists of all processes satisfying (2)–(5) and making the Lebesgue integral in (1) finite.
(b) Let processes (x,u), $(x^*, u^*) \in \mathcal{A}$ be given. Then the pair $(x^*, u^*) \in \mathcal{A}$ is called *global optimal for* $(P)_\infty^B$, if for any pair $(x,u) \in \mathcal{A}$ holds

$$\int_0^\infty r_0(t, x(t), u(t))\nu_0(t)\, dt - \int_0^\infty r_0(t, x^*(t), u^*(t))\nu_0(t)\, dt \geq 0.$$

Remark 1. As weight or even density functions it is imaginable to consider the following well known densities: the exponential density function $\nu(t) := e^{-\rho t}$ often used in economic applications, the Weibull density function $\nu(t) := t^k e^{-\rho t}$ or the density of the normal distribution $\nu(t) := e^{-\rho t^2}, \rho > 0, k > 0$.

4 Existence Theorem for $(\mathrm{P})_\infty^B$

Remark 2. With the denotations $r(t,x(t),u(t)) := r_0(t,x(t),u(t))$ and $D_i(t) := D_i^0(t)\nu_2(t)\nu_1^{-1}(t)$, $i = 1,2$, an equivalent formulation of $(\mathrm{P})_\infty^B$ can be given, wherein only two density functions ν_0, ν_1 appear.

We use this equivalent formulation of $(P)_\infty^B$, namely

$$J_\infty(x,u) = \int_0^\infty r(t,x(t),u(t))\nu_0(t)dt \longrightarrow \min! \tag{6}$$

$$(x,u) \in W_2^{1,n}(\mathbb{R}_+,\nu_1) \times L_2^m(\mathbb{R}_+,\nu_1), \tag{7}$$

$$\dot{x}(t) = A(t)x(t) + B(t)u(t) + C(t) \text{ a. e. on } \mathbb{R}_+, x(0) = x_0 > 0, \tag{8}$$

$$d = \int_0^\infty \left\{D_1^T(t)x(t) + D_2^T(t)u(t)\right\}\nu_1(t)\,dt \tag{9}$$

$$u(t) \in U \text{ a. e. on } \mathbb{R}_+ \tag{10}$$

Before proving the existence result, we cite an auxiliary lemma, proved in [13]:

Lemma 1. *Any admissible trajectory $x(\cdot)$ of the problem $(P)_\infty^B$ satisfies the inequality*

$$|x(t)| \leqslant \beta(t), \tag{11}$$

where

$$\beta(t) := \alpha(t) + \int_0^t \alpha(\tau)\cdot|A(\tau)|e^{\int_\tau^t |A(\sigma)|d\sigma}\,d\tau \tag{12}$$

with $\alpha(t) = |x_0| + \int_0^t \left(|B(\tau)|\cdot\max_{v\in U_1}|v| + |C(\tau)|\right)d\tau.$ □

As far as it is known to the author, there are no existence results to the considered class of problems in the literature. The existence result in [13] captures only control problems with a budget constraint depending solely on the state variable. Therefore, we now derive an existence theorem for our main optimal control problem which represents a generalization of the theorem in the cited source. Let us assume:

Assumption 1. The function $r(t,\xi,v)$ is continuous in t, continuously differentiable in ξ and v, and convex on U for all $(t,\xi) \in \mathbb{R}_+ \times \mathbb{R}^n$ and the functions $A : \mathbb{R}_+ \to \mathbb{R}^n \times \mathbb{R}^n, B : \mathbb{R}_+ \to \mathbb{R}^n \times \mathbb{R}^m$; $C, D_i : \mathbb{R}_+ \to \mathbb{R}^n$ $(i = 1,2)$ satisfy $A, B \in L_\infty(\mathbb{R}_+), C, (D_1,D_2) \in L_2^{n+m}(\mathbb{R}_+,\nu_1)$.

Assumption 2. The integrand $r(t,\xi,v)$ satisfies the growth condition

$$\left|r(t,\xi,v)\right| \leq \frac{A_1(t)}{\nu_0(t)} + B_1\cdot\sum_{k=1}^n \frac{|\xi_k|^2}{\nu_0(t)}\cdot\nu_1(t) + B_1\cdot\sum_{k=1}^m \frac{|v_k|^2}{\nu_0(t)}\cdot\nu_1(t) \tag{13}$$

$\forall\,(t,\xi,v) \in \mathbb{R}_+ \times \mathbb{R}^n \times U$ with a function $A_1 \in L_1(\mathbb{R}_+)$ and a constant $B_1 > 0$.

Assumption 3. The gradient $\nabla_v r(t, \xi, v)$ satisfies the growth condition

$$\left|\nabla_v r\big(t,\, \xi,\, v\big) \cdot \tfrac{\nu_0(t)}{\nu_1(t)}\right| \leq A_2(t)\, \nu_1(t)^{-1/2} + B_2 \cdot \sum_{k=1}^{n} |\xi_k| + B_2 \cdot \sum_{k=1}^{m} |v_k| \quad (14)$$

for all $(t, \xi, v) \in \mathbb{R}_+ \times \mathbb{R}^n \times U$ with a function $A_2 \in L_2(\mathbb{R}_+)$ and a constant $B_2 > 0$.

Theorem 1. *Let the weight function ν_1 be a density function chosen in such a manner that the function $\beta : \mathbb{R}_+ \to \mathbb{R}$ defined by*

$$\beta(t) := \alpha(t) + \int_0^t \alpha(\tau) \cdot |A(\tau)| e^{\int_\tau^t |A(\sigma)| d\sigma} d\tau \quad (15)$$

with $\alpha(t) = |x_0| + \int_0^t \Big(|B(\tau)| \cdot \max_{v \in U_1} |v| + |C(\tau)| \Big) d\tau$ *belongs to the space* $L_2(\mathbb{R}_+, \nu_1)$*. Additionally, assume that Assumptions 1–3 are satisfied for a problem of class* $(P)_\infty^B$ *and the admissible set is not empty. Then the control problem* $(P)_\infty^B$ *possesses an optimal solution.*

Proof. Due to Lemma 1 and to the assumption of the theorem, the so called *natural state constraint* of the form

$$|x(t)| \leq \beta(t), \quad \beta \in L_2(\mathbb{R}_+, \nu_1) \quad (16)$$

holds for any admissible state trajectory $x(\cdot)$ for any time $t > 0$ and with $\beta(t)$ as in (15).

The proof of the weak lower semicontinuity of the functional (1) can now be completely taken from [12], Theorem 3.1, p. 56 ff. In order to get the weak compactness of the admissible set of $(P)_\infty^B$ we can use the proof of the weak compactness of the admissible set for a state constrained control problem with the state constraint of the form (16) provided in [12], Theorem 4.1, p. 63 ff. However, we still need to show that the isoperimetrical constraint (4) remains valid, if one passes to a weak limit $x_N \rightharpoonup x_0$ in $W_2^{1,n}(\mathbb{R}_+, \nu_1)$ and weak limit $u_N \rightharpoonup u_0$ in $L_2^m(\mathbb{R}_+, \nu_1)$ as $N \to \infty$ for an arbitrary sequence $\{(x_N, u_N)\}_{N=1}^\infty$ of admissible processes. Indeed, the continuity of the embedding $W_p^{1,n}(\mathbb{R}_+, \nu_1) \subset L_p^n(\mathbb{R}_+, \nu_1)$ implies the weak convergence $x_n \rightharpoonup x_0$ also in the space $L_2^n(\mathbb{R}_+, \nu_1)$ as $N \to \infty$. Further, since $(D_1, D_2) \in L_2^{n+m}(\mathbb{R}_+, \nu_1)$ holds, the left hand side of (4) can be interpreted as a linear continuous functional on the space $L_2^{n+m}(\mathbb{R}_+, \nu_1)$, i.e. for the elements of the sequence $\{x_N, u_N\}_{N=1}^\infty$ we have

$$\int_0^\infty \left(D_1^T(t) x_N(t) + D_2^T(t) u_N(t)\right) \nu_1(t)\, dt = f(x_N, u_N) \quad (17)$$

for some $f \in \left[L_2^{n+m}(\mathbb{R}_+, \nu_1)\right]^*$. Due to the weak convergence of (x_N, u_N) one obtains

$$d = f(x_N, u_N) \to f(x_0, u_0), \quad N \to \infty. \tag{18}$$

Therefore, $f(x_0, u_0) = d$. The same would hold, if (=d) in (4) is replaced by (⩽d). Thus, the admissible set of $(P)^B_\infty$ is weakly compact and the generalized Weierstraß theorem can be applied to assure the existence of an optimal solution.

Remark 3. Theorem 1 remains valid, if the growth conditions posed in Assumptions 2 and 3 as well as the differentiability assumptions on the integrand r are satisfied only on the set

$$R := \left\{(t, \xi, v) \in \mathbb{R}_+ \times \mathbb{R}^n \times \mathbb{R}^m \,\middle|\, t \in \mathbb{R}_+, \quad |\xi| \leq \beta(t),\, v \in U\right\}. \tag{19}$$

5 Application to an Advertisement Model

We consider the following linear-quadratic advertisement model with a mixed control-state isoperimetrical constraint:

$$J_\infty(x, u) = \int_0^\infty \left\{u^2(t) - \pi x(t)\right\} e^{-\rho t} dt \longrightarrow \min! \tag{20}$$

$$(x, u) \in W_2^1(\mathbb{R}_+, e^{-\rho t}) \times L_2(\mathbb{R}_+, e^{-\rho t}), \tag{21}$$

$$\dot{x}(t) = u(t) - \delta x(t) \text{ a. e. on } \mathbb{R}_+,\; x(0) = 2, \tag{22}$$

$$u(t) \in [-\bar{u}, \bar{u}] \text{ a. e. on } \mathbb{R}_+, \tag{23}$$

$$d = \int_0^\infty ((1 + \alpha)u(t) - \alpha\pi x(t))\, e^{-\rho t} dt. \tag{24}$$

This model is a slightly changed version of the advertisement model in [8, p. 126]. The problem (20)–(24) is a problem of class $(\mathrm{P})^B_\infty$ with weight functions $\nu_1(\cdot), \nu_2(\cdot)$ defined by $\nu_0(t) = \nu_1(t) = e^{-\rho t}$, with functions D_1, $D_2 : \mathbb{R}_+ \to \mathbb{R}$ defined by $D_1(t) \equiv -\alpha\pi, D_2(t) \equiv (1 + \alpha), \alpha \in [0, 1], \rho \in (0, 1)$. The isoperimetrical constraint (24) means that some back into advertisement budget.

Proposition 1. *For $\rho > 2\delta$ there exists an optimal solution for the advertisement model with isoperimetrical constraint, i.e. for (20)–(24), and is given through*

$$x^*(t) = \left(2 - \frac{K}{\delta}\right) e^{-\delta t} + \frac{K}{\delta}; \tag{25}$$

$$u^*(t) \equiv K \tag{26}$$

with

$$K = -\frac{\pi(1 + \mu_0\alpha)}{2(\rho + \delta)} + \frac{\mu_0}{2}(1 + \alpha); \tag{27}$$

$$\mu_0 = \frac{2d\delta\rho^2(\rho+\delta)^2 + \pi(1+\alpha)(\rho+\delta)\delta + 2\pi\alpha\rho^2\delta(\rho+\delta) + \alpha\pi^2\rho^2}{(\rho+\delta)^2\delta(1+\alpha)^2 - \pi\alpha\delta + \alpha^2\pi^2\rho^2 - \alpha\pi(1+\alpha)\rho\delta(\rho+\delta)}. \tag{28}$$

Remark 4. Let

$$\bar{u} := \left| -\frac{\pi(1+\mu_0\alpha)}{2(\rho+\delta)} + \frac{\mu_0}{2}(1+\alpha) \right| + 1$$

with μ_0 from (28). This setting of the constant $\bar{u}$ assures that the optimal control $u^*(\cdot)$ lies in the interior of the control set at any point of time $t > 0$.

Proof. The existence of an optimal solution to the considered mixed budget-constrained control problem is guaranteed due to Theorem 1 for the following reasons. Firstly, the growth conditions from Assumptions (2) and (3) are satisfied with $A_1(t) := e^{-\rho t}, B_1 := \pi$ and $A_2(t) \equiv 0, B_2 := 2$ respectively and the majorant function $\beta(t) := \left(2 + \frac{\bar{u}}{\delta}\right) e^{\delta t} - \frac{\bar{u}}{\delta}$, which was computed according to formula (15) belongs to the weighted Lebesgue space $L_2^1(\mathbb{R}^+, \nu_1)$ for $\rho > 2\delta$. Secondly, the admissible set is not empty, as the pair (x^*, u^*) from (25), (26) is included in it. Besides, the constant functions $D_1(\cdot)$ and $D_2(\cdot)$ are in the space $L_\infty(\mathbb{R}^+) \subset L_2^1(\mathbb{R}^+, \nu_1)$. Therefore, all the assumptions of the existence Theorem 1 are satisfied.

The solution of the considered problem can be computed by means of Pontryagin's Type Maximum Principle, cf. Fig. 1. However, we do not provide here the proof of this result.

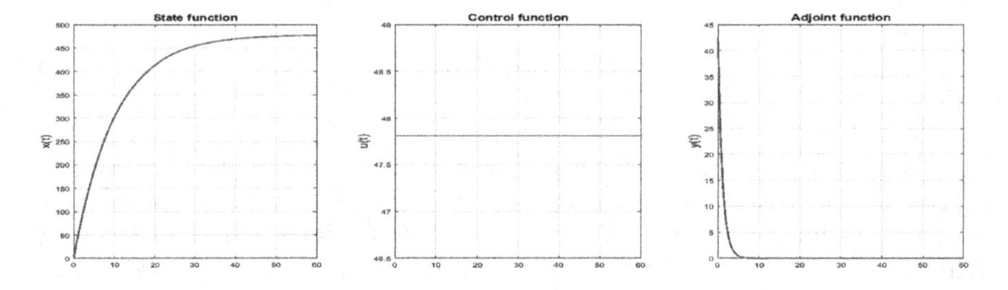

Fig. 1. Optimal solution for parameter values: $\alpha = \delta = 0.1, \rho = 0.9, d = 100$.

Remark 5. It turns out that constant advertisement efforts are optimal, but their height depends on the overall budget d, the starting capital stock and its relation to the parameters α, ρ and δ. If the parameters are so that $K/\delta < 2$, then the overall advertisement budget d is not sufficient for gaining new capital and the capital stock would decrease with growing time.

6 Conclusions

A class of infinite horizon optimal control problems with mixed control-state isoperimetrical constraint formulated in weighted functional spaces was investigated. We obtained an existence theorem which has been successfully applied to a budget-constrained advertisement model.

References

1. Aseev, S.M., Kryazhimskii, A.V.: The Pontryagin maximum principle and optimal economic growth problems. Proc. Steklov Inst. Math. **257**, 1–255 (2007)
2. Aseev, S.M., Veliov, V.M.: Maximum principle for problems with dominating discount. Dyn. Contin. Discret. Impuls. Syst. Seri. B **19**(1–2b), 43–63 (2012)
3. Bogucz, D.: On the existence of a classical optimal solution and of an almost strongly optimal solution for an infinite-horizon control problem. J. Optim. Theory Appl. **156**(3), 650–682 (2013)
4. Carlson, D.A., Haurie, A.B., Leizarowitz, A.: Infinite Horizon Optimal Control. Springer, Heidelberg (1991). https://doi.org/10.1007/978-3-642-76755-5
5. Dmitruk, A.V., Kuz'kina, N.V.: Existence theorem in the optimal control problem on an infinite time interval. Math. Notes **78**(4), 466–480 (2005)
6. Elstrodt, J.: Maß und Integrationstheorie. Springer, Berlin (1996). https://doi.org/10.1007/978-3-662-08527-1
7. Grass, D., Caulkins, J.P., Feichtinger, G., Tragler, G., Behrens, D.A.: Optimal Control of Nonlinear Processes. Springer, Berlin (2008). https://doi.org/10.1007/978-3-540-77647-5
8. Feichtinger, G., Hartl, R.F.: Optimale Kontrolle ökonomischer Prozesse. de Gruyter, Berlin (1986)
9. Hamdache, A., Elmouki, I., Saadi, S.: Optimal control with an isoperimetric constraint applied to cancer immunotherapy. Int. J. Comput. Appl. **94**(15), 31–37 (2014)
10. Kufner, A.: Weighted Sobolev Spaces. Wiley, Chichester (1985)
11. Ledzewicz, U., Maurer, H., Schaettler, H.: Optimal and suboptimal protocols for a mathematical model for tumor anti-angiogenesis in combination with chemotherapy. Math. Biosci. Eng. **8**(2), 307–323 (2011)
12. Lykina, V.: An existence theorem for a class of infinite horizon optimal control problems. J. Optim. Theory Appl. **69**(1), 50–73 (2016)
13. Lykina, V.; Pickenhain, S.: Budget-constrained infinite horizon control problems with linear dynamics. In: Proceedings of 55th Conference on Decision and Control (2016, accepted for publication)
14. Lykina, V., Pickenhain, S., Wagner, M.: Different interpretations of the improper integral objective in an infinite horizon control problem. Math. Anal. Appl. **340**, 498–510 (2008)
15. Lykina, V., Pickenhain, S., Wagner, M.: On a resource allocation model with infinite horizon. Appl. Math. Comput. **204**, 595–601 (2008)
16. Maurer, H., de Pihno, M.D.R.: Optimal control of epidemiological SEIR models with L^1 - objectives and control-state constraints. AIMS J. (2015, submitted)
17. Pickenhain, S., Burtchen, A., Kolo, K., Lykina, V.: An indirect pseudospectral method for linear-quadratic infinite horizon optimal control problems. Optimization **65**(3), 609–633 (2016)

18. Pickenhain, S., Lykina, V., Wagner, M.: On the lower semicontinuity of functionals involving Lebesgue or improper Riemann integrals in infinite horizon optimal control problems. Control Cybern. **37**(2), 451–468 (2008)
19. Torres, D.F.M.: A noether theorem on unimprovable conservation laws for vector-valued optimization problems in control theory. Georgian Math. J. **13**(1), 173–182 (2006)
20. Walter, W.: Gewöhnliche Differentialgleichungen. Springer, New York (2000). https://doi.org/10.1007/978-3-642-57240-1
21. Yosida, K.: Functional Analysis. Springer, New York (1974). https://doi.org/10.1007/978-3-642-96208-0

On the Regularity of Linear-Quadratic Optimal Control Problems with Bang-Bang Solutions

J. Preininger[1], T. Scarinci[2], and V. M. Veliov[1(✉)]

[1] Institute of Statistics and Mathematical Methods in Economics, Vienna University of Technology, Vienna, Austria
{jakob.preininger,vladimir.veliov}@tuwien.ac.at
[2] Department of Statistics and Operations Research, University of Vienna, Vienna, Austria
teresa.scarinci@univie.ac.at
http://orcos.tuwien.ac.at/people/veliov/

Abstract. The paper investigates the stability of the solutions of linear-quadratic optimal control problems with bang-bang controls in terms of metric sub-regularity and bi-metric regularity. New sufficient conditions for these properties are obtained, which strengthen the known conditions for sub-regularity and extend the known conditions for bi-metric regularity to Bolza-type problems.

Keywords: Optimal control · Regularity
Linear-quadratic problems · Bang-bang controls

1 Introduction

In this paper we investigate the stability with respect to perturbations of the solutions of the following optimal control problem:

$$\begin{aligned} &\text{minimize } J(x,u) && \text{(P)}\\ &\text{subject to } \dot{x}(t) = A(t)x(t) + B(t)u(t) + d(t), \quad t \in [0,T],\\ &\qquad u(t) \in U := [-1,1]^m,\\ &\qquad x(0) = x_0, \end{aligned}$$

where

$$J(x,u) := g(x(T)) + \int_0^T \left(\frac{1}{2}x(t)^\top W(t)x(t) + x(t)^\top S(t)u(t)\right) dt. \tag{1}$$

This research is supported by the Austrian Science Foundation (FWF) under grant No. P26640-N25. The second author is also supported by the Doctoral Programme "Vienna Graduate School on Computational Optimization" funded by the Austrian Science Fund (FWF), project No. W1260-N35.

I. Lirkov and S. Margenov (Eds.): LSSC 2017, LNCS 10665, pp. 237–245, 2018.
https://doi.org/10.1007/978-3-319-73441-5_25

Here, admissible controls are all measurable functions $u : [0,T] \to [-1,1]^m$, while $x(t) \in \mathbb{R}^n$ denotes the state of the system at time $t \in [0,T]$. The initial state x_0, the final time T and the terminal function $g : \mathbb{R}^n \to \mathbb{R}$ are given, as well as the matrices $A(t), W(t) \in \mathbb{R}^{n\times n}$, $B(t), S(t) \in \mathbb{R}^{n\times m}$ and $d(t) \in \mathbb{R}^n$, $t \in [0,T]$.

The stability of the solutions of this problem is investigated within the general framework of *metric regularity* (see e.g. [2, Sect. 3E]) of the associated Pontryagin system of necessary optimality conditions.

The issue is challenging due to the linearity of the problem with respect to the control, which may result in bang-bang solutions. Only a few results are known in the literature that deal with the regularity of this problem, among which we mention [1,4,6]. The paper [6] introduces the notion of bi-metric regularity as an appropriate extension of the established notion of metric regularity, which is more relevant to problems with discontinuous optimal controls. However, the result in [6] applies to Mayer-type problems only, where the integral term in the objective functional (1) is missing. The integral term brings a substantial difference, due the presence of the state and the control in the adjoint equation.

In this paper we obtain a strengthened version of the Hölder sub-regularity result obtained in [1, Theorem 8], which provides a basis for further investigations, including error analysis of approximation schemes. We also announce a result about strong bi-metric regularity of the Pontryagin system of necessary conditions associated with problem (P), extending [6] to Bolza problems with bang-bang solutions.

2 Preliminaries

We begin with formulation of assumptions.

Assumption (A1). The matrix-functions A, B, W, S and d are Lipschitz continuous. The matrix $W(t)$ is symmetric for every $t \in [0,T]$. The function g is differentiable with locally Lipschitz derivative.

Let $(\hat{x}, \hat{u})$ be a solution of problem (P), from now on fixed; a standard compactness argument implies existence.

Assumption (A2). For every admissible pair (x,u) of (P) it holds that

$$\langle \nabla g(x(T)) - \nabla g(\hat{x}(T)), \Delta x(T)\rangle + \int_0^T (\langle W(t)\Delta x, \Delta x\rangle + 2\langle S(t)\Delta u, \Delta x\rangle)dt \geq 0,$$

where $\Delta x := x(t) - \hat{x}(t)$ and $\Delta u := u(t) - \hat{u}(t)$, and $\langle \cdot,\cdot\rangle$ is the scalar product.

By the Pontryagin maximum (here minimum) principle, there exists an absolutely continuous function $\hat{p}$ such that the triple $(\hat{x}, \hat{p}, \hat{u})$ solves for a.e. $t \in [0,T]$ the system

$$\begin{aligned}
0 &= \dot{x}(t) - A(t)x(t) - B(t)u(t) - d(t), \\
0 &= \dot{p}(t) + A(t)^\top p(t) + W(t)x(t) + S(t)u(t), \\
0 &\in B(t)^\top p(t) + S(t)^\top x(t) + N_U(u(t)), \\
0 &= p(T) - \nabla g(x(T)),
\end{aligned} \qquad \text{(PMP)}$$

where $N_U(u)$ is the normal cone to U at $u \in \mathbb{R}^m$:

$$N_U(u) := \begin{cases} \emptyset & \text{if } u \notin U \\ \{l \in \mathbb{R}^m : \langle l, v - u \rangle \le 0 \ \forall v \in U\} & \text{if } u \in U. \end{cases}$$

We recall that $\hat{\sigma} := B^\top \hat{p} + S^\top \hat{x}$ is the so-called *switching function* corresponding to the triple $(\hat{x}, \hat{p}, \hat{u})$. For every $j \in \{1, \ldots, m\}$, denote by $\hat{\sigma}_j$ its j-th component.

The following assumption requires that the optimal control $\hat{u}$ is *strictly bang-bang*, with a finite number of switching times, and that the switching function exhibits a certain growth in a neighborhood of any zero. A similar assumption is introduced in [4] in the case $\kappa = 1$ and in [7] for $\kappa > 1$.

Assumption (A3). There exist real numbers $\kappa \ge 1$ and $\alpha, \tau > 0$ such that for each $j \in \{1, \ldots, m\}$ and $s \in [0, T]$ with $\hat{\sigma}_j(s) = 0$ it holds that

$$|\hat{\sigma}_j(t)| \ge \alpha |t - s|^\kappa \quad \forall t \in [s - \tau, s + \tau] \cap [0, T].$$

The Pontryagin minimum principle (PMP) can be recast as

$$0 \in F(x, p, u), \tag{2}$$

where $F : \mathcal{X} \rightrightarrows \mathcal{Y}$ is a set-valued map defined as

$$F(x, p, u) := \begin{pmatrix} \dot{x} - Ax - Bu - d \\ \dot{p} + A^\top p + Wx + Su \\ B^\top p + S^\top x + N_U(u) \\ p(T) - \nabla g(x(T)) \end{pmatrix}. \tag{3}$$

We will investigate the stability under perturbations of the solution of problem (P) by studying the stability of the generalized equation $y \in F(x, p, u)$ with respect to a perturbation y. The mapping F is considered as acting in the space

$$\mathcal{X} := W^{1,1}_{x_0}([0, T], \mathbb{R}^n) \times W^{1,1}([0, T], \mathbb{R}^n) \times L^1([0, T], \mathbb{R}^m)$$

with values in the space

$$\mathcal{Y} := L^1([0, T], \mathbb{R}^n) \times L^1([0, T], \mathbb{R}^n) \times L^\infty([0, T], \mathbb{R}^m) \times \mathbb{R}^n,$$

which restricts the set of considered selections of the mapping $t \mapsto N_U(u(t))$ to essentially bounded ones. Here $W^{1,1}_{x_0}([0, T], \mathbb{R}^n) := \{x \in W^{1,1}([0, T], \mathbb{R}^n) : x(0) = x_0\}$. The spaces $\mathcal{X}$ and $\mathcal{Y}$ are endowed with the usual norms for $(x, p, u) \in \mathcal{X}$ and $(\xi, \pi, \rho, \nu) \in \mathcal{Y}$:

$$\|(x, p, u)\|_{\mathcal{X}} := \|x\|_{1,1} + \|p\|_{1,1} + \|u\|_1, \quad \|(\xi, \pi, \rho, \nu)\| := \|\xi\|_1 + \|\pi\|_1 + \|\rho\|_\infty + |\nu|.$$

3 Metric Sub-regularity

We begin with an auxiliary result that is similar in spirit to [7, Lemma 1.3] (also cf. [8, Theorem 2.1]) but is proved on slightly less restrictive assumptions.

Lemma 1. *Let $l : [0,T] \to \mathbb{R}^m$ be a continuous function satisfying assumption (A3) (with l at the place of $\hat{\sigma}$). Then there exists a constant $c > 0$ such that for any $v \in L^\infty([0,T],\mathbb{R}^m)$ the following inequality holds:*

$$\|v\|_\infty^k \int_0^T \sum_{j=1}^m |l_j(t)v_j(t)|\,dt \geq c\|v\|_1^{\kappa+1}. \tag{4}$$

Proof. The claim of this lemma is trivial when $v = 0$. If $v \neq 0$ then due to the homogeneity with respect to v of order $\kappa + 1$ of the two sides of (4), it is enough to prove the lemma in the case $\|v\|_\infty = 1$, which will be assumed in the remaining part of the proof. For any $0 < \delta \leq \tau$, we set

$$I_j(\delta) := \bigcup_{s\in[0,T]:\, l_j(s)=0} (s-\delta, s+\delta) \cap [0,T], \qquad I(\delta) := \bigcup_{1\leq j\leq m} I_j(\delta).$$

Since l is continuous and Assumption (A3) holds for l_j, we have that

$$l_{\min} := \min_{1\leq j\leq m} \min_{t\in[0,T]\setminus I_j(\tau)} |l_j(t)| > 0.$$

Now we choose $\bar{\delta} \in (0,\tau)$ such that $\alpha\bar{\delta}^\kappa < l_{\min}$. Then for all $\delta \in (0,\bar{\delta})$ and $j \in \{1,\ldots,m\}$ we have

$$|l_j(t)| \geq \alpha\delta^\kappa \quad \forall t \in [0,T] \setminus I(\delta). \tag{5}$$

Indeed, if $t \notin I_j(\tau)$ then $|l_j(t)| \geq l_{\min} > \alpha\bar{\delta}^\kappa \geq \alpha\delta^\kappa$. If $t \in I_j(\tau) \setminus I(\delta)$, then $t \in I_j(\tau) \setminus I_j(\delta)$. Thus there exists a zero s of l_j such that $\delta \leq |t-s| < \tau$. According to Assumption (A3), $|l_j(t)| \geq \alpha|t-s|^\kappa \geq \alpha\delta^\kappa$. Hence,

$$\begin{aligned}
\phi(v) := \int_0^T \sum_{j=1}^m |l_j(t)v_j(t)|\,dt &\geq \int_{[0,T]\setminus I(\delta)} \sum_{j=1}^m |l_j(t)v_j(t)|\,dt \\
&\geq \alpha\delta^\kappa \sum_{j=1}^m \int_{[0,T]\setminus I(\delta)} |v_j(t)|\,dt \geq \alpha\delta^\kappa \left(\|v\|_1 - \sum_{j=1}^m \int_{I(\delta)} |v_j(t)|\,dt \right) \\
&\geq \alpha\delta^\kappa(\|v\|_1 - 2\lambda\delta),
\end{aligned}$$

where λ is sum of the maximum of the number of zeros of l_j over all $j \in \{1,\ldots m\}$ (notice that Assumption (A3) implies $\lambda \leq mT/2\tau + m$). If $\|v\|_1 \geq 4\lambda\bar{\delta}$ then we choose $\delta := \bar{\delta}$ to get

$$\phi(v) \geq \frac{\alpha\bar{\delta}^\kappa}{2}\|v\|_1$$

and since $\|v\|_1 \le T\|v\|_\infty = T$ we have that $\phi(u) \ge \frac{\alpha\bar{\delta}^\kappa}{2T^\kappa}\|v\|_1^{\kappa+1}$. If, on the other hand, $\|v\|_1 \le 4\lambda\bar{\delta}$ then we choose $\delta := \frac{\|v\|_1}{4\lambda} \le \bar{\delta}$ to get

$$\phi(v) \ge \frac{\alpha}{2^{2\kappa+1}\lambda^\kappa}\|v\|_1^{\kappa+1}.$$

Hence choosing $c := \min\{\frac{\alpha\bar{\delta}^\kappa}{2T^\kappa}, \frac{\alpha}{2^{2\kappa+1}\lambda^\kappa}\}$ we obtain that

$$\phi(v) \ge c\|v\|_1^{\kappa+1}.$$

Q.E.D.

The following theorem establishes a property of the mapping F associated with system (PMP), which is a somewhat stronger form of the well known property of *metric sub-regularity*, [2, Sect. 3H]. It extends [1, Theorem 8] in several directions: Assumption (A3) is weaker than the corresponding assumption there, the norms are different, and the function g is not necessarily quadratic and convex.

Theorem 1. *Let $(\hat{x}, \hat{p}, \hat{u})$ be a solution of (PMP) such that (A1)–(A3) are fulfilled. Then for any $b > 0$ there exists $c > 0$ such that for any $y \in \mathcal{Y}$ with $\|y\| \le b$, there exists a triple $(x, p, u) \in \mathcal{X}$ solving $y \in F(x, p, u)$, and any such triple satisfies*

$$\|(x, p, u) - (\hat{x}, \hat{p}, \hat{u})\|_{\mathcal{X}} \le c\|y\|^{\frac{1}{\kappa}}.$$

Proof. Since the inclusion $y \in F(x, p, u)$ represents a system of necessary optimality conditions of a problem of the form of (P) with appropriate, bounded in L^1, perturbations defined by y (a simple and well known fact), the evident existence of an optimal solution of this perturbed version of (P) implies existence of a solution (x, p, u) of the inclusion $y \in F(x, p, u)$.

Now let $b > 0$ be arbitrarily chosen and let (x, p, u) be a solution of $y \in F(x, p, u)$, where $y = (\xi, \pi, \rho, \nu) \in \mathcal{Y}$ and $\|y\| \le b$. The following notations will be used. As before, $\hat{\sigma}(t) := B(t)^\top \hat{p}(t) + S(t)^\top \hat{x}(t)$, while $\sigma(t) := B(t)^\top p(t) + S(t)^\top x(t) - \rho(t)$. Furthermore, we denote $\Delta x := x(t) - \hat{x}(t)$, $\Delta p = p(t) - \hat{p}(t)$, $\Delta u := u(t) - \hat{u}(t)$, $\Delta\sigma := \sigma(t) - \hat{\sigma}(t)$, and skip the argument t whenever clear.

Integrating by parts, we have

$$\int_0^T \langle \Delta\dot{p}, \Delta x\rangle\, dt = \langle \Delta p(T), \Delta x(T)\rangle - \int_0^T \langle \Delta p, \Delta\dot{x}\rangle\, dt.$$

Substituting here the expressions for Δx and Δp resulting from the inclusions $y \in F(x, p, u)$ and $0 \in F(\hat{x}, \hat{p}, \hat{u})$ in view of (3), we obtain that

$$\int_0^T \langle -A^\top \Delta p - W\Delta x - S\Delta u + \pi, \Delta x\rangle\, dt$$
$$= \langle \nabla g(x(T)) - \nabla g(\hat{x}(T)) + \nu, \Delta x(T)\rangle - \int_0^T \langle \Delta p, A\Delta x + B\Delta u + \xi\rangle\, dt.$$

Rearranging the terms in this equality and using (A2) we get

$$\int_0^T (\langle \Delta p, B\Delta u\rangle + \langle S\Delta u, \Delta x\rangle)\, dt + \int_0^T (\langle \pi, \Delta x\rangle + \langle \xi, \Delta p\rangle)\, dt - \langle \nu, \Delta x(T)\rangle$$
$$= \langle \nabla g(x(T)) - \nabla g(\hat{x}(T)), \Delta x(T)\rangle + \int_0^T (\langle W\Delta x, \Delta x\rangle + 2\langle S\Delta u, \Delta x\rangle)\, dt \geq 0.$$

Using this inequality and the definitions of the functions σ and $\hat{\sigma}$ we obtain

$$\int_0^T \langle \Delta\sigma, \Delta u\rangle\, dt = \int_0^T \langle B^\top \Delta p + S^\top \Delta x - \rho, \Delta u\rangle\, dt \geq$$
$$\geq \int_0^T (-\langle \pi, \Delta x\rangle - \langle \xi, \Delta p\rangle - \langle \rho, \Delta u\rangle)\, dt + \langle \nu, \Delta x(T)\rangle. \quad (6)$$

The third component of the inclusion $y \in F(x, p, u)$ reads as $-\sigma(t) \in N_U(u(t))$, which implies $\langle -\sigma(t), \hat{u}(t) - u(t)\rangle \leq 0$. Then

$$-\int_0^T \langle \Delta\sigma, \Delta u\rangle\, dt = \int_0^T [-\langle \sigma, \Delta u\rangle + \langle \hat{\sigma}, \Delta u\rangle]\, dt \geq \int_0^T \langle \hat{\sigma}, \Delta u\rangle\, dt.$$

From here, using that $-\hat{\sigma}_j(t) \in N_{[-1,1]}(\hat{u}_j(t))$, hence $\hat{\sigma}_j(t)\, \Delta u_j(t) \geq 0$ for each j, Lemma 1 implies that

$$-\int_0^T \langle \Delta\sigma, \Delta u\rangle\, dt \geq \int_0^T \sum_{j=1}^m |\hat{\sigma}_j\, \Delta u_j|\, dt \geq c_1 \|\Delta u\|_1^{\kappa+1},$$

where the constant c_1 is independent of y and (x, p, u). Then using (6) and the Hölder inequality we obtain

$$\|\pi\|_1\, \|\Delta x\|_\infty + \|\xi\|_1\, \|\Delta p\|_\infty + |\nu|\, |\Delta x(T)| + \|\rho\|_\infty\, \|\Delta u\|_1 \geq c_1 \|\Delta u\|_1^{\kappa+1}. \quad (7)$$

Using Assumption (A1) and the Cauchy formula for Δx and Δp we get

$$\|\Delta x\|_\infty \leq c_2(\|\xi\|_1 + \|\Delta u\|_1) \quad (8)$$

and

$$\|\Delta p\|_\infty \leq c_3(\|\xi\|_1 + \|\pi\|_1 + \|\Delta u\|_1 + |\nu|) \quad (9)$$

for some constants c_2 and c_3 that are independent of y and (x, p, u). Therefore, using (7) we obtain that

$$(\|y\|^2 + \|y\| \|\Delta u\|_1) \geq c_4 \|\Delta u\|_1^{\kappa+1} \quad (10)$$

for some constant c_4, also independent of y and (x, p, u).

Now we distinguish two cases. First, if $\|\Delta y\| \leq \|u\|_1$ then

$$2\|y\| \|\Delta u\|_1 \geq c_4 \|\Delta u\|_1^{\kappa+1},$$

which implies

$$\|\Delta u\|_1 \le \left(\frac{2}{c_4}\|y\|\right)^{1/\kappa}. \tag{11}$$

Otherwise, if $\|\Delta u\|_1 \le \|y\| \le b$ then

$$\|\Delta u\|_1 \le \|y\|^{1/\kappa}\|y\|^{(\kappa-1)/\kappa} \le b^{(\kappa-1)/\kappa}\|y\|^{1/\kappa}. \tag{12}$$

Inequality (11) and (12) imply that for any $b > 0$ there exists $c_5 > 0$ such that for any and $\|y\| \le b$,

$$\|\Delta u\|_1 \le c_5\|y\|^{1/\kappa}.$$

Then the claim of the theorem follows with a suitable constant c from the above estimate, (8) and (9). Q.E.D.

We mention that the property established in Theorem 1 is stronger than metric sub-regularity (as defined e.g. in [2, Sect. 3H]) in that it is global with respect to the solution $(x, p, u) \in \mathcal{X}$, and also with respect to the size b of the "disturbance" y, although the constant c in the theorem may depend on b.

4 Bi-metric Regularity

We begin this section by introducing appropriate modifications of the spaces $\mathcal{X}$ and $\mathcal{Y}$ defined in Sect. 2. First, we consider the set $\mathcal{U} \subset L^\infty([0,T],\mathbb{R}^m)$ of admissible controls (that is, the set of all measurable functions $u : [0,T] \to U$) as a metric space with the metric

$$d^{\#}(u_1, u_2) = \text{meas}\,\{t \in [0,T] :\, u_1(t) \ne u_2(t)\},$$

in $L^\infty([0,T],\mathbb{R}^m)$, where "meas" stands for the Lebesgue measure in $[0,T]$. This metric is shift-invariant and we shall shorten $d^{\#}(u_1,u_2) = d^{\#}(u_1 - u_2, 0) =: d^{\#}(u_1 - u_2)$. Moreover, $\mathcal{U}$ is a complete metric space with respect to $d^{\#}$ (see [3, Lemma 7.2]). Then the triple (x,p,u) is considered as an element of the space

$$\widetilde{\mathcal{X}} = W^{1,1}_{x_0}([0,T],\mathbb{R}^n) \times W^{1,1}([0,T],\mathbb{R}^n) \times \mathcal{U},$$

endowed with the (shift-invariant) metric

$$d_{\sim}(x,p,u) = \|x\|_{1,1} + \|p\|_{1,1} + d^{\#}(u). \tag{13}$$

Clearly $\widetilde{\mathcal{X}}$ is a complete metric space. We also define the space $\widetilde{\mathcal{Y}} \subset \mathcal{Y}$ as

$$\widetilde{\mathcal{Y}} := L^\infty([0,T],\mathbb{R}^n) \times L^\infty([0,T],\mathbb{R}^n) \times W^{1,\infty}([0,T],\mathbb{R}^m) \times \mathbb{R}^n$$

with the usual norm of $y = (\xi,\pi,\rho,\nu) \in \widetilde{\mathcal{Y}}$:

$$\|(\xi,\pi,\rho,\nu)\|_{\sim} := \|\xi\|_\infty + \|\pi\|_\infty + \|\rho\|_{1,\infty} + |\nu|. \tag{14}$$

The paper [6] introduces the notion of bi-metric regularity as a concept of regularity that is relevant to problems with bang-bang optimal controls. In the particular context of the present paper the definition of bi-metric regularity of the set-valued mapping $F : \widetilde{\mathcal{X}} \rightrightarrows \mathcal{Y}$ (see (3)) reads, in a somewhat more general form, as follows.

Definition 1. *The map* $F : \widetilde{\mathcal{X}} \rightrightarrows \mathcal{Y}$ *is strongly bi-metrically regular relative to (disturbance space)* $\widetilde{\mathcal{Y}} \subset \mathcal{Y}$ *at* $\hat{z} \in \widetilde{\mathcal{X}}$ *for* $0 \in \widetilde{\mathcal{Y}}$ *if* $(\hat{z}, 0) \in \operatorname{graph}(F)$ *and there exist numbers* $\varsigma \geq 0$, $\beta > 0$ *and* $a > 0$ *such that the map* $B_{\widetilde{\mathcal{Y}}}(0;\beta) \ni y \mapsto F^{-1}(y) \cap B_{\widetilde{\mathcal{X}}}(\hat{z};a)$ *is single-valued and*

$$d_{\sim}(F^{-1}(y') \cap B_{\widetilde{\mathcal{X}}}(\hat{z};a),\, F^{-1}(y) \cap B_{\widetilde{\mathcal{X}}}(\hat{z};a)) \leq \varsigma \|y' - y\| \tag{15}$$

for all $y, y' \in B_{\widetilde{\mathcal{Y}}}(0;\beta)$. *Here* $B_{\widetilde{\mathcal{X}}}(\hat{z};a)$ *is the ball of radius* a *centered at* $\hat{z}$ *in the space* $\widetilde{\mathcal{X}}$, *and* $B_{\widetilde{\mathcal{Y}}}(0;\beta)$ *is the ball of radius* β *(in the norm* $\|\cdot\|_{\sim}$*) centered at* $0 \in \widetilde{\mathcal{Y}}$.

The following theorem extends the result for bi-metric regularity of F obtained in [6] for Mayer's problems for linear systems to the present Bolza problem. For that we need the following strengthened forms of assumptions (A1) and (A2).

Assumption (A1'). The functions A, W and d are Lipschitz continuous, B and S have first order Lipschitz derivatives. The matrices $W(t)$ and $S^{\top}(t)B(t)$ are symmetric for every $t \in [0,T]$. The function g is differentiable with locally Lipschitz derivative.

Assumption (A2'). The function J is convex on the set of admissible pairs (x, u).

Theorem 2 (Bi-metric regularity). *Let Assumptions (A1') and (A2') be fulfilled. Let* $(\hat{x}, \hat{p}, \hat{u})$ *be a solution to (PMP) such that (A3) is fulfilled with* $\kappa = 1$. *Then the mapping* $F : \widetilde{\mathcal{X}} \rightrightarrows \mathcal{Y}$ *introduced in* (3) *is strongly bi-metrically regular (relative to* $\widetilde{\mathcal{Y}} \subset \mathcal{Y}$*) at* $(\hat{x}, \hat{p}, \hat{u}) \in \widetilde{\mathcal{X}}$ *for* $0 \in \widetilde{\mathcal{Y}}$.

The proof of this theorem is too long to be placed here, therefore it will be presented as a part of a full size paper. This also applies to applications of Theorems 1 and 2 in qualitative analysis and error analysis of numerical approximations in the spirit of [5].

We mention, that the strong bi-metric regularity for Mayer's problems is proved in [6] for a general polyhedral set U and also in the case $\kappa > 1$. Extension of Theorem 2 to a general compact polyhedral U set is a matter of modification of Assumption (A3) and technicalities that we avoid in this paper, while the case $\kappa > 1$ is still open and challenging for the Bolza problem.

References

1. Alt, W., Schneider, C., Seydenschwanz, M.: Regularization and implicit Euler discretization of linear-quadratic optimal control problems with bang-bang solutions. Appl. Math. Comput. **287–288**, 104–105 (2016)
2. Dontchev, A.L., Rockafellar, R.T.: Implicit Functions and Solution Mappings: A View from Variational Analysis. SSORFE. Springer, New York (2014). https://doi.org/10.1007/978-1-4939-1037-3
3. Ekeland, I.: On the variational principle. J. Math. Anal. Appl. **47**, 324–353 (1974)

4. Felgenhauer, U.: On stability of bang-bang type controls. SIAM J. Control Optim. **41**(6), 1843–1867 (2003)
5. Pietrus, A., Scarinci, T., Veliov, V.M.: High order discrete approximations to Mayer's problems for linear systems. Research report 2016–04, ORCOS, TU Wien (2016, to appear)
6. Quincampoix, M., Veliov, V.: Metric regularity and stability of optimal control problems for linear systems. SIAM J. Control Optim. **51**(5), 4118–4137 (2013)
7. Seydenschwanz, M.: Convergence results for the discrete regularization of linear-quadratic control problems with bang-bang solutions. Comput. Optim. Appl. **61**(3), 731–760 (2015)
8. Veliov, V.M.: On the convexity of integrals of multivalued mappings: applications in control theory. J. Optim. Theory Appl. **54**(3), 541–563 (1987)

HPC and Big Data: Algorithms and Applications

On Monte Carlo and Quasi-Monte Carlo for Matrix Computations

Vassil Alexandrov[1,2,4], Diego Davila[2], Oscar Esquivel-Flores[4], Aneta Karaivanova[3(✉)], Todor Gurov[3], and Emanouil Atanassov[3]

[1] ICREA - Catalan Institution for Advanced Research Studies, Barcelona, Spain
[2] Barcelona Supercomputing Center, Barcelona, Spain
[3] IICT, Bulgarian Academy of Sciences, Sofia, Bulgaria
anet@parallel.bas.bg
[4] Inst. Tech. y de Estudios Superiores de Monterrey, Monterrey, Nuevo León, Mexico

Abstract. This paper focuses on minimizing further the communications in Monte Carlo methods for Linear Algebra and thus improving the overall performance. The focus is on producing set of small number of covering Markov chains which are much longer that the usually produced ones. This approach allows a very efficient communication pattern that enables to transmit the sampled portion of the matrix in parallel case. The approach is further applied to quasi-Monte Carlo. A comparison of the efficiency of the new approach in case of Sparse Approximate Matrix Inversion and hybrid Monte Carlo and quasi-Monte Carlo methods for solving Systems of Linear Algebraic Equations is carried out. Experimental results showing the efficiency of our approach on a set of test matrices are presented. The numerical experiments have been executed on the MareNostrum III supercomputer at the Barcelona Supercomputing Center (BSC) and on the Avitohol supercomputer at the Institute of Information and Communication Technologies (IICT).

Keywords: Monte Carlo for linear algebra
Quasi-Monte Carlo for linear algebra · Hybrid methods

1 Introduction

Solving systems of linear algebraic equations (SLAE) in the form of $Ax = b$ or inverting a real matrix A is of unquestionable importance in many scientific fields. Iterative solvers are used widely to compute the solutions of these systems and such approaches are often the method of choice due to their predictability and reliability when considering accuracy and speed. They, however, may become prohibitive for large-scale problems [1].

Monte Carlo methods (MCMs) complexity is linear of the matrix size [2,3] on the other hand and can quickly yield a rough estimate of the solution. For some problems an estimate is sufficient or even favorable, due to the accuracy of the underlying data. Therefore, it should be pointed out, that MCMs may be efficiently used as preconditioners.

I. Lirkov and S. Margenov (Eds.): LSSC 2017, LNCS 10665, pp. 249–257, 2018.
https://doi.org/10.1007/978-3-319-73441-5_26

Depending on the method used to compute the preconditioner, the savings and end-results vary. A very sparse preconditioner may be computed quickly, but it is unlikely to improve the quality of the solution. On the other hand, computing a rather dense preconditioner is computationally expensive and might be time or cost prohibitive. Therefore, finding a good preconditioner that is computationally efficient, while still providing substantial improvement to the iterative solution process, is a worthwhile research topic.

A variety of parallel MCMs have been developed within the past 20 years. A comprehensive compendium of the Monte Carlo (MC) functions and strategies of parallelization can be found for example, in [2,3,12,13].

In this work we present an enhanced version of a SPAI preconditioner that is based on parallel MCMs presented in [2,3]. This new optimized version is compared against the previous one taken as a baseline, as well as against the *state-of-the-art* MSPAI, which is the main accepted deterministic algorithm for SPAI preconditioning. Further research, to solve this issues, is proposed within the context of quasi-MCMs. The next section gives and overview of related work. MC and quasi-MC algorithms as a SPAI preconditioner, parallelization and methodology are presented in Sect. 3. Sections 4 and 5 present individual results for each improvement and the overall result respectively. The last section consists of the conclusion and outlines the future work.

2 Related Work

Research efforts in the past have been directed towards optimizing the approach of sparse approximate inverse preconditioners. Improvements to the Frobenius norm have been proposed for example by concentrating on sparse pattern selection strategies [4], or building a symmetric preconditioner by averaging off-diagonal entries [5]. Further, it has been shown that the sparse approximate inverse preconditioning approach is also a viable course of action on large-scale dense linear systems [6]. This is of special interest to us, as the MC method we are proposing in this paper is part of a bigger family. It includes serial and parallel MC algorithms for the inversion of sparse, as well as dense matrices, and the solution of systems of linear algebraic equations. The proposed MC algorithm has been developed and enhanced upon in the last decades, and several key advances in serial and parallel MCMs for solving such problems have been made [2,3,12,13,17,18]. There is an increased research interest in parallel MCMs for Linear Algebra in the past few years, and recent example is the Monte Carlo Synthetic Acceleration (MCSA) developed through MCREX project at ORNL [7]. Future work that deals with a parallel implementation of the presented algorithm is being considered further in this Section and in Sect. 3.

In the past there have been differing approaches and advances towards a parallelization of the SPAI preconditioner. In recent years the class of Frobenius norm minimizations that has been used in the original SPAI implementation [8] was modified and is provided in a parallel SPAI software package. One implementation of it, by the original authors of SPAI, is the Modified SParse Approximate Inverse (MSPAI [9]).

This version provides a class of modified preconditioners such as MILU (modified ILU), interface probing techniques and probing constraints to the original SPAI, apart from a more efficient, parallel Frobenius norm minimization. Further, this package also provides two novel optimization techniques. One option is using a dictionary in order to avoid redundant calculations, and to serve as a lookup table. The second option is an option to switch to a less computational intensive, sparse QR decomposition whenever possible. This optimized code runs in parallel, together with a dynamic load balancing.

2.1 Using SParse Approximate Inverse as Preconditioner (SPAI)

The SPAI algorithm [10] is used to compute a sparse approximate inverse matrix M for a given sparse input matrix B. This is done by minimizing $||BM - I||$ in the Frobenius norm. The algorithm explicitly computes the approximate inverse, which is intended to be applied as a preconditioner of an iterative method. The SPAI application provides the option to fix the sparsity pattern of the approximate inverse a priori or capture it automatically. Since the introduction of the original SPAI in 1996, several advances, building upon the initial implementation, have been made. Two newer implementations are provided by the original authors, the before mentioned MSPAI, and the highly scalable Factorized SParse Approximate Inverse (FSPAI [11]). The intended use of both differs depending on the problem at hand. Whereas MSPAI is used as a preconditioner for large sparse and ill-conditioned systems of linear equations, FSPAI is applicable only to symmetric positive definite systems of this kind. FSPAI is based around an inherently parallel implementation, generating the approximate inverse of the Cholesky factorization for the input matrix. MSPAI on the other hand is using an extension of the well-known Frobenius norm minimization that has been introduced in the original SPAI.

The algorithm attempts to solve a system of linear equations of the form $Bx = b$. Its input is a sparse, square coefficient matrix B. The right hand side vector b can either be provided by the user, or is arbitrarily defined by the software implementation. In the case of the SPAI application suite, if no right hand side vector is handed to the algorithm, it constructs one by multiplying matrix B with a vector consisting of all ones. In a general case, an input matrix B is passed to SPAI as a file. The program then computes a preconditioner using the Frobenius norm, afterwards it uses this intermediate result as an input to an appropriate solver.

3 Monte Carlo and Quasi-Monte Carlo Approach

The MCMs are probabilistic methods, that use random numbers to either simulate a stochastic behavior or to estimate the solution of a problem. Quasi-MC on the other hand employ quasi-random sequences such as Sobol or Halton, for example, in the computation process. They are both good candidates for parallelization because of the fact that many independent samples are used to estimate the solution. These samples can be calculated in parallel, thereby speeding up the solution finding process. The so designed and developed parallel MCMs possess the following main generic properties [2,3]: efficient distribution of the compute data, minimum communication during the computation and increased precision being achieved by adding extra refinement computations.

The MC algorithm for matrix inversion can be roughly explained within the following 5 phases (Notice that phases 1 and 5 are only necessary when the initial matrix is not *diagonally dominant (ddm)*): (1) Initial matrix is transformed into a ddm; (2) Transformation of ddm for suitable *Neumann series expansion*; (3) The MCM is applied to calculate sparse approximation of the inverse matrix; (4) Given 2, calculate the inverse of the ddm from 3; (5) Recovery process is applied to calculate the inverse of the original matrix due to the transformation in 1.

This algorithm was originally designed for a HPC cluster composed of single-core compute nodes. It is written in C and uses the MPI library. It also makes use of the BeBOP sparse matrix converter [14] to translate the input matrix format into a CSR format.

The quasi-MC algorithm for matrix inversion is presented in [17]. Theoretical estimates can be found in [18]. Here the algorithm is based on the new version of the low discrepancy sequences, optimized for parallel computations on the Avitohol supercomputer [19].

4 Used Algorithms

All the changes and improvements are carried out on the algorithm for matrix inversion in an incremental fashion [12,13]. Experiments are carried out on Marenostrum III supercomputer at BSC. It currently consists of 3056 compute nodes equipped with 2 Intel Xeon 8-core processors and 64 GB of RAM interconnected via InfiniBand FDR-10. The Monte Carlo-based algorithm is written in C and it uses the open MPI-1.8.0[1] implementation of MPI and Intel-13.0.1 OpenMP. The solver used to measure the time needed to calculate the solution for the preconditioned system, is the paralution-1.1.0[2] implementation of GMRES. We selected a matrix set from 3 different sources: The Matrix Market [15], The University of Florida Sparse Matrix Collection [16] and some real-life problems from our collaborators.

5 Overall Results and Evaluation

The optimized MC algorithm was compared with the latest version of the MSPAI application. Parameters of the MC implementation have been adjusted accordingly ($\epsilon = 7 \times 10^{-1}$ and $\epsilon = 1 \times 10^{-1}, delta = 1 \times 10^{-1}, \alpha = 5$) to produce comparable results with those obtained with MSPAI's default configuration.

[1] www.open-mpi.org/.
[2] www.paralution.com/.

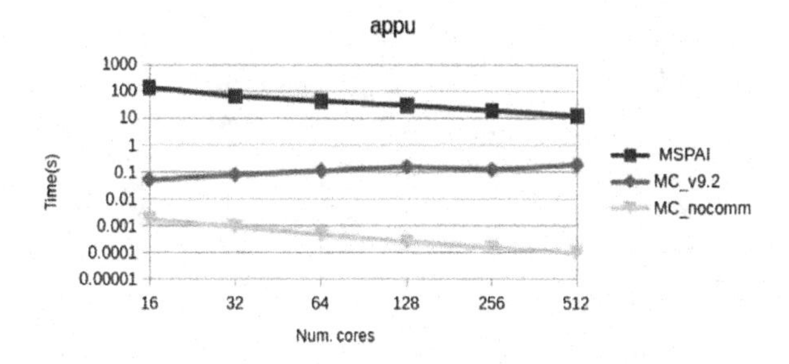

Fig. 1. Scalability comparison MSPAI and MC for matrix appu.

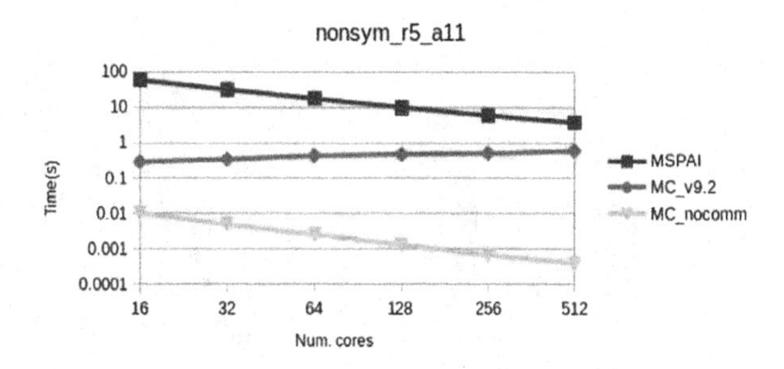

Fig. 2. Scalability comparison MSPAI and MC for matrix nonsym_r5_a11.

5.1 Scalability Comparison

Scalability plots (Figs. 1, 2 and 3) are presented to show the behavior of both algorithms when the number of cores is scaled. In all these plots a metric called *MC_nocomm* is shown to provide an insight of the scalability of the MC process itself.

It is quite obvious that the MC algorithm performs several times faster than MSPAI, despite the communication hampering the scalability, which is mainly affected by the 3 following aspects: the communication overhead, the optimizations applied and the reduction in the number or iterations.

In the case of MSPAI, scalability issues regarding the small matrices (Fig. 3) also persist.

Figure 4 summarizes the scalability plots by showing the *fastest time* achieved, for the preconditioner calculation, by each of the algorithms.

5.2 Monte Carlo vs Quasi-Monte Carlo

In this section we consider the performance on MC vs quasi-MC for matrix inversion. Initially we have employed both Sobol and modified Halton sequences, and experiments have shown better behavior of the algorithm using Halton ones. The experiments were run on the Bulgarian supercomputer Avitohol deployed at the IICT-BAS. It consists of 150 computational servers HP SL250s Gen8, equipped with two Intel Xeon E5-2650v2 CPUs and two Intel Xeon Phi 7120P

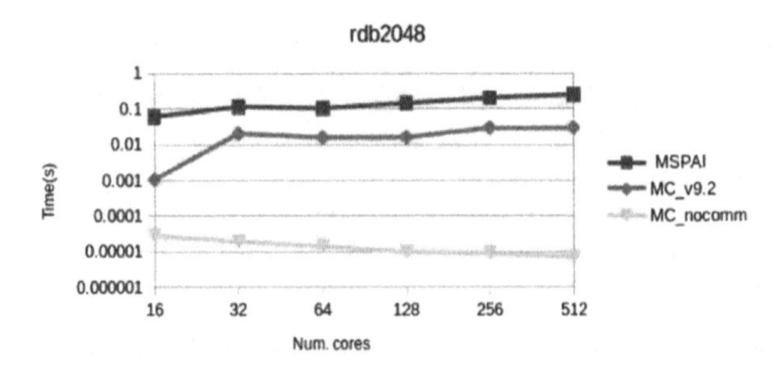

Fig. 3. Scalability comparison MSPAI and MC for matrix rdb2048.

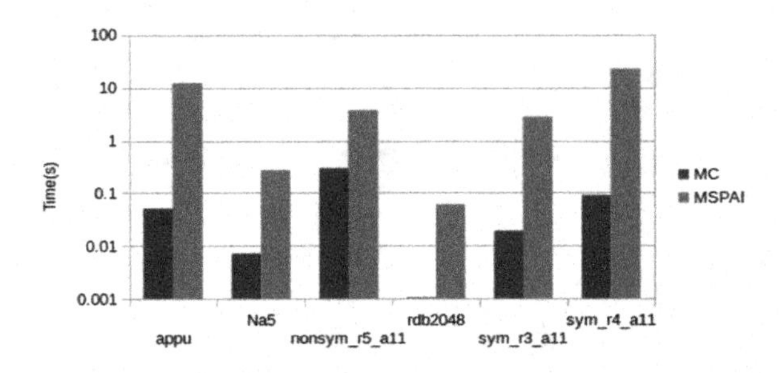

Fig. 4. Fastest execution time achieved during the preconditioner calculation.

coprocessors, 64 GB RAM, two 500 GB hard drives, interconnected with non-blocking FDR InfiniBand running at 56 Gbp/s line speed. The total number of cores is 20700 and the total RAM is 9600 GB, respectively.

The experiments show that the approximate inverse based on the MC algorithm using random sequences is slightly faster than the approximate inverse based on Halton quasirandom ones. The possible gain comes from the quality of the inverse obtained. For certain matrices the quasi-MC yields faster convergence of GMRES (see Table 1). But this is not the general case since there are instance where the MC based inverse leads to faster GMRES runs for certain problems. It should be noted that increasing the precision in generating the inverse leads to denser inverse matrices and after certain threshold is prohibitive.

5.3 Quality and Efficiency Comparison

We have already measured the time of construction, now we evaluate the quality of the preconditioner. A metric must be selected to measure the quality of the preconditioner but also its efficiency. A good choice of such a metric is the time needed by the solver to find the solution for the preconditioned system. Other metrics like the number of iterations required by the solver to calculate the solution of the preconditioned system, reflects the quality but not the efficiency. In that sense it has been observed that in many cases the number of iterations required by MC preconditioned systems is smaller than in the case of

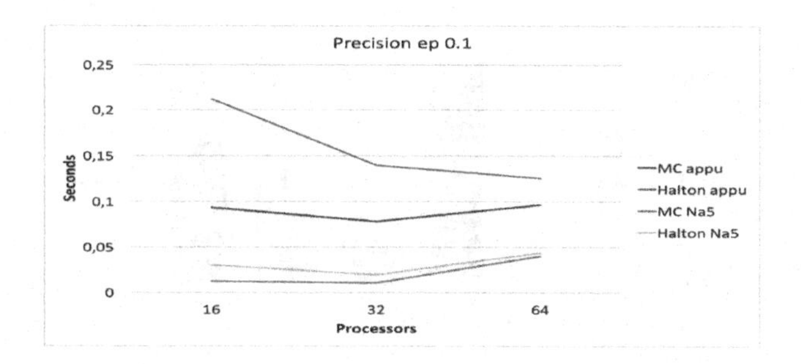

Fig. 5. MC vs QMC - epsillon = 0.1

Table 1. GMRES execution times for MC and QMC

Matrix	QMC-Halton	MC	QMC-Halton	MC
	eps = 0.5	eps = 0.5	eps = 0.1	eps = 0.1
Appu	0.135902	0.139417	1.62762	2.78041
bcsstm13	106.616	107.129	119.986	120.103
Na5	0.012513	0.013384	0.027767	0.025393
Rdb2048	0.158109	0.197112	0.17864	0.170368
Si10H16	0.0312071	0.0299697	0.436809	0.494146

those preconditioned by MSPAI. Similar is the situation in the case of quasi-MC preconditioning. On the other hand comparing the quality of MC and quasi-MC preconditioners shows that for some instances the solver converges faster after the quasi-MC preconditioner and for some matrices it converges faster after the MC preconditioner have been employed (Fig. 5).

To view the overall time, we provide Fig. 6 in which the times of the preconditioner construction and the time needed by the solver (Fig. 4) is added. Here we can see that sometimes the time invested in the quality of the preconditioner can be compensated with a reduction in the solver execution time. One example, which contradicts this is the case of matrix *nonsym_r5_a11* where MC approach produces a preconditioner faster that MSPAI but the solver needs more iterations to find the solution of the resulting system and the combined execution time of the MC preconditioner plus solver time is worse than the combined MSPAI and the solver execution time.

Observe that the MC performs much better than MSPAI for both the symmetric matrices as well as in most cases for non-symmetric matrices too. Similar is the situation with quasi-MC. We have been testing new communication strategies employing covering chains and preliminary results show some minimization of the communication time being between the MC timing and the yellow graph on the pictures above. Further more detailed experiments are required to validate fully these strategies.

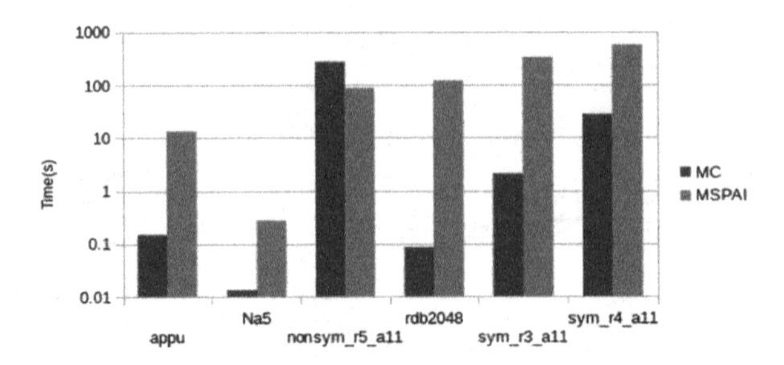

Fig. 6. Total time = Preconditioner construction time + Solver execution time.

6 Conclusions and Future Work

An efficient and enhanced MC and quasi-MC algorithms that produces a SParse Approximate Inverse-based preconditioners has been presented. The implementation of this method has been *optimized* by a factor of 25x in average, and 70x in the best case. These gains in performance were obtained directly by improvements within the algorithm. A mixed MPI + OpenMP version was developed in order to adapt the code to contemporary architectures. Results of the evaluation of the enhanced version made evident that the code runs efficiently in case of sufficiently large matrices and also pointed to the need for further research in quasi-MCMs like [17] which could be used to boost the performance by improving the memory access patterns. The usability of this method was demonstrated for cases of very large matrices. A detailed comparison was carried out between the MC algorithm and the *state-of-the-art* MSPAI. Results have shown that in general the parallel MC algorithm outperforms the MSPAI. Comparison between different approaches was carried out with precision given the methodology proposed to quantify the error for a given solution.

Acknowledgments. The work of the authors (V.A., D.D., and O.E-F.) is supported by Severo Ochoa program of excellence, Spain. The work of the authors (A.K. and T.G.) is supported by the NSF of Bulgaria under Grant DFNI-I02/8.

References

1. Golub, G., Loan, C.: Matrix Computations. Johns Hopkins Studies in the Mathematical Sciences. Johns Hopkins University Press, Baltimore (1996)
2. Straßburg, J., Alexandrov, V.N.: Enhancing Monte Carlo preconditioning methods for matrix computations. In: Proceedings ICCS 2014, pp. 1580–1589 (2014)
3. Alexandrov, V.N., Esquivel-Flores, O.A.: Towards Monte Carlo preconditioning approach and hybrid Monte Carlo algorithms for matrix computations. CMA **70**(11), 2709–2718 (2015)
4. Carpentieri, B., Duff, I., Giraud, L.: Some sparse pattern selection strategies for robust Frobenius norm minimization preconditioners in electromagnetism. Numer. Linear Algebra Appl. **7**, 667–685 (2000)

5. Carpentieri, B., Duff, I., Giraud, L.: Experiments with sparse preconditioning of dense problems from electromagnetic applications, CERFACS, Toulouse, France. Technical report (2000)
6. Alléon, G., Benzi, M., Giraud, L.: Sparse approximate inverse preconditioning for dense linear systems arising in computational electromagnetics. Num. Algorithms **16**(1), 1–15 (1997)
7. Evans, T., Hamilton, S., Joubert, W., Engelmann, C.: MCREX - Monte Carlo Resilient Exascale Project. http://www.csm.ornl.gov/newsite/documents
8. Benzi, M., Meyer, C., Tůma, M.: A sparse approximate inverse preconditioner for the conjugate gradient method. SIAM J. Sci. Comput. **17**(5), 1135–1149 (1996)
9. Huckle, T., Kallischko, A., Roy, A., Sedlacek, M., Weinzierl, T.: An efficient parallel implementation of the MSPAI preconditioner. Parallel Comput. **36**(5–6), 273–284 (2010)
10. Grote, M., Hagemann, M.: SPAI: SParse Approximate Inverse Preconditioner. Spaidoc. pdf paper in the SPAI, vol. 3, p. 1 (2006)
11. Huckle, T.: Factorized sparse approximate inverses for preconditioning. J. Supercomput. **25**(2), 109–117 (2003)
12. Strassburg, J., Alexandrov, V.: On scalability behaviour of Monte Carlo sparse approximate inverse for matrix computations. In: Proceedings of the ScalA 2013 Workshop, Article no. 6. ACM (2013)
13. Vajargah, B.F.: A new algorithm with maximal rate convergence to obtain inverse matrix. Appl. Math. Comput. **191**(1), 280–286 (2007)
14. Hoemmen, M., Vuduc, R., Nishtala, R.: BeBOP sparse matrix converter. University of California at Berkeley (2011)
15. Boisvert, R.F., Pozo, R., Remington, K., Barrett, R.F., Dongarra, J.J.: Matrix market: a web resource for test matrix collections. In: Boisvert, R.F. (ed.) QNS 1997. IFIPAICT, pp. 125–137. Springer, Boston (1997). https://doi.org/10.1007/978-1-5041-2940-4_9
16. Davis, T.A., Hu, Y.: The University of Florida sparse matrix collection. ACM Trans. Math. Softw. (TOMS) **38**(1), 1 (2011)
17. Alexandrov, V., Esquivel-Flores, O., Ivanovska, S., Karaivanova, A.: On the preconditioned Quasi-Monte Carlo algorithm for matrix computations. In: Lirkov, I., Margenov, S.D., Waśniewski, J. (eds.) LSSC 2015. LNCS, vol. 9374, pp. 163–171. Springer, Cham (2015). https://doi.org/10.1007/978-3-319-26520-9_17
18. Karaivanova, A.: Quasi-Monte Carlo methods for some linear algebra problems. Convergence and complexity. Serdica J. Comput. **4**, 57–72 (2010)
19. Atanassov, E., Gurov, T., Karaivanova, A., Ivanovska, S., Durchova, M., Dimitrov, D.: On the parallelization approaches for Intel MIC architecture. In: AIP Conference Proceedings, vol. 1773, p. 070001 (2016). https://doi.org/10.1063/1.4964983

On the Parallel Implementation of Quasi-Monte Carlo Algorithms

E. Atanassov, T. Gurov(✉), S. Ivanovska, A. Karaivanova, and T. Simchev

Institute of Information and Communication Technologies,
Bulgarian Academy of Sciences,
Acad. G. Bonchev str., bl. 25A, 1113 Sofia, Bulgaria
gurov@bas.bg

Abstract. The quasi-Monte Carlo algorithms utilize deterministic low-discrepancy sequences in order to increase the rate of convergence of stochastic simulation algorithms. Such kinds of algorithms are widely applicable and consume large share of the computational time on advanced HPC systems. The recent advances in HPC are increasingly rely on the use of accelerators and other similar devices that improve the energy efficiency and offer better performance for certain type of computations. The Xeon Phi coprocessors combine efficient vector floating point computations with familiar operational and development environment. One potentially difficult part of the conversion of a Monte Carlo algorithm into a quasi-Monte Carlo one is the generation of the low-discrepancy sequences. On such specialized equipment as the Xeon Phi, the value of memory increases due to the presence of a large number of computational cores. In order to allow quasi-Monte Carlo algorithms to make use of hybrid OpenMP+MPI programming, we implemented generation routines that save both memory space and memory bandwidth, with the aim to widen the applicability of quasi-Monte Carlo algorithms in environments with an extremely large number of computational elements. We present our implementation and compare it with regular Monte Carlo using a popular pseudorandom number generator, demonstrating the applicability and advantages of our approach.

Keywords: Low-discrepancy sequences
Quasi-Monte Carlo algorithms · High performance computing

1 Introduction

The quasi-Monte Carlo algorithms attempt to improve the rate of convergence of Monte Carlo methods, which usually reach accuracy of $O(N^{-1/2})$, where N is the number of samples. Instead of pseudo-random numbers the quasi-Monte Carlo methods use specially constructed sequences, so that for certain classes of problems better convergence rates may be obtained. Many of the Monte Carlo and consequently quasi-Monte Carlo methods can be considered as methods for

I. Lirkov and S. Margenov (Eds.): LSSC 2017, LNCS 10665, pp. 258–265, 2018.
https://doi.org/10.1007/978-3-319-73441-5_27

solving numerical integration problems. In such cases the dimension of the function space is called "constructive dimensionality" of the problem (see, e.g., [8]). The integration error of quasi-Monte Carlo methods is frequently connected with some measure of the irregularity of distribution of the sequences used. Perhaps the most popular such measure is the discrepancy, which is related to the integration error for functions with bounded variation (see, e.g., [7]). Sequences that can be proved to attain rate of convergence of their discrepancy to zero in the order of $O(N^{-1} \log^s N)$, where s is the dimension, are called low-discrepancy sequences. Many constructions of such sequences have been devised, but the Sobol sequences (see, e.g., [1]) remain one of the most popular in theoretical and practical sense. Because of their relation to the binary number system they are easier to implement on digital computers and are widely used in practice. More complex constructions of low-discrepancy sequences may require substantial computational resources for generating the terms of the sequence and thus the increased accuracy of the quasi-Monte Carlo method may be outweighed by the increased computational time. This trade-off becomes more important to consider for algorithms with high constructive dimension or where many terms of the sequence are required. Such kinds of problems have to be resolved taking into account the particular hardware system where the algorithm is going to run. The computational accelerators become increasingly popular due to their high energy and cost-efficiency and take significant share of the installed computational power, available to researchers. Even a single workstation nowadays may be equipped with one or more computational accelerators, whether GPGPUs or Intel Xeon Phi cards, [2,3].

In this paper we concentrate on the usage of Intel Xeon Phi accelerators, which are built using the Intel's Multiple Integrated Core (MIC) technology. Essentially the Intel Xeon Phi cards behave like a regular CPU with the important distinction that powerful vector instructions enable much higher rates of floating point operations per second. Although Intel's compiler has the capability to generate vector instructions automatically, in many cases manually developed codes outperform significantly. In our previous works we have shown how one can use the vector instructions in order to speed-up the generation of low-discrepancy sequences. However, it is important to consider the modes of parallel execution that are available on the Xeon Phi in order to match them with corresponding generation routines. The previously developed codes for Intel Xeon Phi were only suitable for pure MPI parallel execution. The problem with this is that since the MIC architecture provides multiple cores and allows for hyperthreading (using up to four times more logical threads of execution than the number of physical cores), it is important to decrease memory utilization. Consequently algorithms must strive to share as much as possible of their state and conserve memory. We also point out that the memory bandwidth is a well known weak point for such architectures with large number of computing elements. That is why in the next two sections we explain how we adapted our algorithms to the different modes of parallel execution available for the Xeon Phi and how we were able to save memory bandwidth. Then we present and discuss the results we obtained in practical application of the devised algorithms.

2 Adapting Generators for Low-Discrepancy Sequences to Different Modes of Parallel Execution

The initial editions of Intel Xeon Phi accelerators required regular CPU to control the system, while the newer versions are able to function stand-alone. That is why the so-called "offload" mode, where computations are offloaded from the main processor to the Xeon Phi co-processor, can be used on the first generations of Intel Xeon Phis, (see e.g. [6]). Because this type of execution is being dropped, we are not going to discuss it, although our generation routines can be used in such mode. In the native execution mode the applications are run directly on the Xeon Phi coprocessor, while the symmetric mode combines processors and coprocessors into one MPI job. In both of these modes it is important to use in the most efficient way the resources of the corresponding computing elements - regular CPUs or accelerators. There are several ways to achieve that. One can use pure MPI, pure OpenMP or hybrid MPI+OpenMP parallelization. We decided to develop an implementation that is suitable for the hybrid mode, so that the other modes result as special cases.

It is obvious that the memory model of OpenMP enables for savings in the use of memory for the generation routines. In order to make use of this property, we put in the common memory the "twisted direction numbers" (see, e.g., [9] for a definition). The substantial savings in memory requirements result in capability to achieve much higher constructive dimensions without overflowing the rather limited memory available at the accelerators. The remaining part of the state that is private to each thread of execution is proportional to the constructive dimension—one floating point number per coordinate, plus one long integer for the position in the sequence.

While parallel pseudo-random number generators only need to ensure independence between the streams generated at different processing elements, the accuracy of quasi-Monte Carlo methods depends on using precisely those terms of the low-discrepancy sequence that are prescribed. That is why two strategies for their parallelization are used—blocking and leap-frogging. For the Sobol sequence blocking is achieved through suitable modification of the private part of the generator's state, while leap-frogging (only by power-of-two) also requires modification of the shared part of the generator's state. These modifications are not specific to accelerators and can be seen in [4].

Thus we developed a thread-safe version of the generation routines, optimized for the vector instructions of the Xeon Phi, with small memory footprint. In the next section we discuss further improvements to it.

3 Saving Memory Bandwidth While Implementing Quasi-Monte Carlo Algorithms

Once we optimized the memory requirements of our generation routines, we consider the memory bandwidth that they use. In general the speed of access to memory, measured with bandwidth and latency, improves at much slower rate compared to the speed of processing. That is why many computations are actually memory-bound. There are different ways to improve the execution through optimization of memory access. One can decrease the total amount of data being transferred or tune the patterns of access in order to improve the use of the various caches. For the Sobol sequence, there is one low-hanging fruit to be had in this direction. It is the observation that if we are generating consecutive terms of the sequence, half of the time the direction number being used is actually the same over all dimensions—it corresponds to changing the most significant binary digit after the binary point. Thus we do not need to load this number from memory. The resulting improvement is substantial in any kind of benchmark and may be even more important in real usage, since space in the caches is saved. Taking this approach one step further, we can make use of the fact that the matrices of binary numbers in the Sobol sequences are triangular. This means that for the first 8 positions 8 bits or one byte is enough to hold all the necessary information. Since we generate in double precision we usually need to load 64 bits or 8 bytes. Thus the savings in memory bandwidth are substantial, when we compress the corresponding "twisted direction numbers". The expansion happens with vector operations, by shifting appropriately and adding the omitted zeroes. Unfortunately, the Xeon Phi seems not to be efficient in such kinds of integer operations and thus this approach does not outperform in benchmarks. However, our benchmarks do not strain the use of caches and therefore can not capture the advantage of this approach. On the CPUs a similar approach has been winning in previous tests. Since the special handling of the first direction number had clear advantage, we leave the choice of using the compression for the next 7 direction numbers to the user.

The generation codes are provided under the GNU Public license and are available at http://parallel.bas.bg/~emanouil/sequences/micmemory.tgz.

4 Numerical and Timing Results

With the standard methods of parallelization of quasi-Monte Carlo algorithms the final numerical result is always the same (up to rounding errors), independent on the number of processing elements. Since this paper focuses on the computer implementation of these methods, we selected several multidimensional functions that were used in other benchmarks as follows:

$$F_1 = \prod_{i=1}^{s} \left({x_i}^3 + \frac{3}{4} \right), \quad F_2 = \prod_{i=1}^{s} |4x_i - 2|, \quad F_3 = \prod_{i=1}^{s} (-1)^{i+1} x_i$$

$$F_4 = \prod_{i=1}^{s} \frac{\pi}{2} \sin \pi x_i \quad \text{and} \quad F_5 = (1 + 1/s)^s \prod_{i=1}^{s} (x_i)^{1/s}$$

where the integration domain is the $s-$ dimensional unite cube $[0, 1]^s$. We denote by I_k the corresponding integral of the function F_k. In the Figs. 1 and 2 we can see a comparison of the computational time and the integration error for the

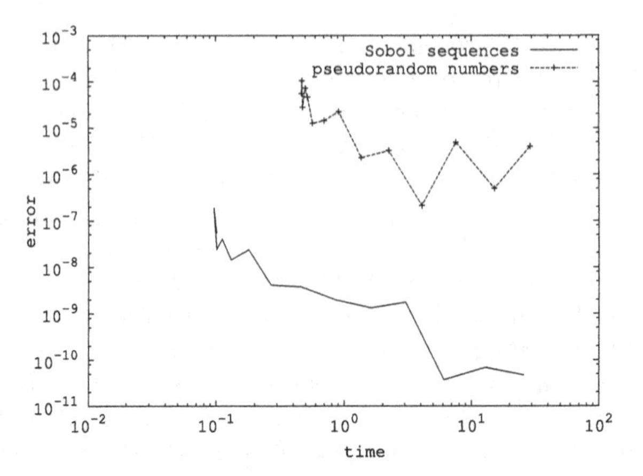

Fig. 1. Scalability properties, I_3

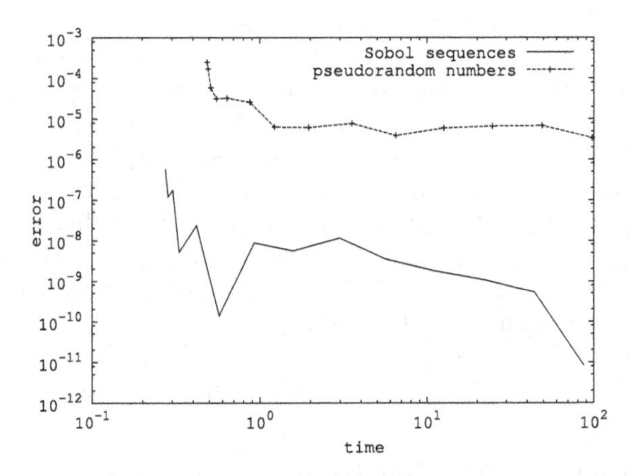

Fig. 2. Scalability properties, I_5

quasi-Monte Carlo method using the Sobol sequence and the crude Monte Carlo method, when the dimension is $s = 64$ and 8 Xeon Phi cards were used with a hybrid MPI-OpenMP execution (Table 1).

One can observe that for these two integrals the integration error of the Sobol sequences decreases much faster. The pseudorandom number generator that we use for comparison in this paper is the Mersenne Twister [5] and its particular implementation is from the Intel MKL [10]. For these two integrals the Sobol sequences achieve vastly superior rate of convergence of the numerical integration. For the other integrals it is difficult to say which method is better because of the relatively high dimension. When we want to use optimally a single Xeon Phi accelerator card we can vary several parameters of the parallel execution. Because of the possibility for hyperthreading on Xeon Phi, it is sensible to launch parallel threads of execution up to 4 times the number of

Table 1. Timing results for different number of threads using 10^9 terms of the Sobol sequences

Threads	Processes	Cards	I_1	I_2	I_3	I_4	I_5
244	1	1	6.15	6.46	5.86	9.68	20.73
122	2	1	6.08	6.27	5.80	9.07	19.24
61	4	1	5.14	5.11	4.62	7.16	15.53
244	1	2	3.20	3.38	2.97	4.78	10.33
122	2	2	3.03	3.39	2.99	4.44	10.30

physical cores. These parallel threads of execution may be parallel processes in one pure MPI application, parallel threads in an OpenMP application or some combination of threads and processes in a hybrid OpenMP+MPI application. The same division between threads and processes can be made for the Monte Carlo application, since the Intel implementation of the Mersenne twister allows for parallel execution. In the next table we can see comparison of timing results obtained for different number of threads, starting with 61 (equal to the number of the physical cores available) and reaching 244, which is the maximum sensible number of threads to be tested. We also try different combinations of number of MPI processes and number of OpenMP threads, since the same total number of independent execution threads can be achieved in different ways. One can conclude that it is beneficial for this type of problem to use the maximum number of hardware threads (244), since this provides significant speedup with respect to the use of only physical cores. We also observe that using 61 threads with 4 MPI processes seems to be better than using only OpenMP. In practical applications the user should take into account the increased memory requirements resulting from using more MPI processes (Table 2).

Table 3 shows comparison of execution times and speedup when increasing the total number of threads and using 1 or 2 physical cards. The speedup is computed with respect to the use of only the physical cores of one card. We can see that the speedup is not ideal, but is acceptable and one may expect that more computationally intensive problems will provide better speedup, since the

Table 2. Timing results for different number of threads using 10^9 pseudorandom numbers generated using Mersenne Twister (MT2203 from MKL)

Threads	Processes	Cards	I_1	I_2	I_3	I_4	I_5
244	1	1	5.68	5.68	7.26	11.81	23.15
122	2	1	5.51	5.67	7.08	11.69	23.04
61	4	1	5.40	5.48	7.04	11.77	23.26
244	1	2	3.14	3.18	3.96	6.36	12.08
122	2	2	2.99	3.07	3.80	6.12	11.85

Table 3. Comparison of execution times and speedup using different number of threads for I_1 with 10^9 points

Cards		1					2			
Generator	Processors	1	61	122	183	244	61	122	183	244
Sobol	Time (s)	923	14.57	9.59	7.73	6.47	8.46	5.07	3.87	3.12
	Speedup		63.35	96.25	119.40	142.65	109.06	181.89	237.95	295.61
	Speedup vs 61			1.52	1.88	2.25		1.67	2.19	2.71
MT2203	Time (s)	571	10.34	6.70	6.03	5.61	5.52	3.87	3.58	3.16
	Speedup		55.22	85.22	94.69	101.78	103.44	147.54	159.50	180.70
	Speedup vs 61			1.54	1.71	1.84		1.43	1.54	1.75

initialization times will have smaller impact. The Sobol sequence shows comparable speedup and execution times with the popular psedorandom number generator MT2203. When more than one card is used it is impossible to use pure OpenMP, so the competing implementations remain MPI only vs OpenMP+MPI. Theoretically the parallel efficiency should be close to the maximum possible, but we must take into account that the quasi-Monte Carlo algorithm requires pre-processing, when the generator's state is computed. Similar results that are not shown here were obtained with larger number of Xeon Phi cards. However, the computational power of the accelerators is so large that the computational times become too small to obtain good speedups.

5 Conclusions and Future Work

We developed a flexible implementation of generation routines for the Sobol low-discrepancy sequence, which can be used in pure MPI or OpenMP or hybrid MPI+OpenMP parallel quasi-Monte Carlo applications on the Intel Xeon Phi coprocessor. Significant savings have been achieved in the use of memory space and bandwidth. The comparisons with a widely used pseudo-random number generator show that the quasi-Monte Carlo methods can compete not only on accuracy but also on speed. By decreasing the memory footprint we widen the practical area of applicability towards higher constructive dimensions and bigger number of parallel processing elements, allowing for flexible combination of MPI and OpenMPI suitable for the respective application. In the future we plan to test the developed algorithms in quasi-Monte Carlo applications that do require large dimensions and number of terms of the sequence for obtaining good accuracy.

Acknowledgments. This work was supported by the National Science Fund of Bulgaria under Grant #DFNI-I02/8 and by the European Commission under H2020 project VI-SEEM (Contract Number 675121).

References

1. Atanassov, E.I.: A new efficient algorithm for generating the scrambled Sobol' sequence. In: Dimov, I., Lirkov, I., Margenov, S., Zlatev, Z. (eds.) NMA 2002. LNCS, vol. 2542, pp. 83–90. Springer, Heidelberg (2003). https://doi.org/10.1007/3-540-36487-0_8
2. Atanassov, E., Dimitrov, D., Ivanovska, S.: Efficient implementation of the Heston model using GPGPU. Monte Carlo Methods and Applications, De Gruyter, pp. 21–28 (2012). ISBN: 978-3-11-029358-6, ISSN: 0929-9629
3. Atanassov, E., Gurov, T., Karaivanova, A., Ivanovska, S., Durchova, M., Georgiev, D., Dimitrov, D.: Tuning for Scalability on Hybrid HPC Cluster. Mathematics in Industry, pp. 64–77. Cambridge Scholar Publishing, Cambridge (2014)
4. Atanassov, E., Karaivanova, A., Ivanovska, S.: Tuning the generation of Sobol sequence with owen scrambling. In: Lirkov, I., Margenov, S., Waśniewski, J. (eds.) LSSC 2009. LNCS, vol. 5910, pp. 459–466. Springer, Heidelberg (2010). https://doi.org/10.1007/978-3-642-12535-5_54
5. Matsumoto, M., Nishimura, T.: Mersenne twister: a 623-dimensionally equidistributed uniform pseudo-random number generator. ACM Trans. Model. Comput. Simul. **8**(1), 3–30 (1998)
6. Meswani, M., Carrington, L., Unat, D., Snavely, A., Baden, S., Poole, S.: Modeling and predicting application performance on hardware accelerators. Int. J. High Perform. Comput. (2012)
7. Niederreiter, H.: Random Number Generation and Quasi-Monte Carlo Methods. Society for Industrial and Applied Mathematics, Philadelphia (1992)
8. Sobol, I.M.: Uniformly distributed sequences with an additional uniform property. Zh. Vych. Mat. Mat. Fiz. **16**, 1332–1337 (1976, in Russian). U.S.S.R Comput. Maths. Math. Phys. **16**, 236–242 (1976, in English)
9. Sobol, I., Asotsky, D., Kreinin, A., Kucherenko, S.: Construction and comparison of high-dimensional Sobol generators. Wilmott J. **56**, 64–79 (2011)
10. Intel Math Kernel Library (MKL). http://software.intel.com/en-us/articles/intel-math-kernel-library-documentation

TVRegCM Numerical Simulations - Preliminary Results

Georgi Gadzhev[1(✉)], Vladimir Ivanov[1], Kostadin Ganev[1], and Hristo Chervenkov[2]

[1] National Institute of Geophysics, Geodesy and Geography, Bulgarian Academy of Sciences, Acad. G. Bonchev str., bl. 3, 1113 Sofia, Bulgaria
ggadjev@geophys.bas.bg

[2] National Institute of Meteorology and Hydrology, Bulgarian Academy of Sciences, Tsarigradsko Shose blvd. 66, 1784 Sofia, Bulgaria

Abstract. The oncoming climate changes at the moment are the biggest challenge the mankind is faced with. They will exert influence on the ecosystems, on the all branches of the national economy, and on the quality of life. The climate changes and their consequences have a great number of regional features, which the global models cannot predict. That is why an operation plan for adaptation to climate changes has to be based on scientifically well-grounded assessments, giving an account of regional features in the climate changes and their consequences.

The purpose of the current research is to develop a method that permits a set of validated models, tuned to the physical geographic and climate conditions of the region will be able reliably to predict the regional climate changes for different global climate scenarios. The comprehensive and detail computer simulations will be done for the present climate. Here an evaluation of the ERA-Interim-driven regional climate model RegCM v4.4 over Southeastern Europe is presented. The study documents the performance of 20 different model configurations in representing the basic spatial and temporal patterns of the SE European climate for the period 1999–2009. Model evaluation focuses on near-surface air temperature and precipitation, and uses the EOBS data set as observational reference.

The study reveals that no particular model configuration can be judged as the best one, nevertheless seven ones indicate better performance for the precipitation during the summer.

Keywords: Regional climate simulation · RegCM4.4 · EOBS
High performance computing

1 Introduction

Regional climate models (RCMs) have been developed and extensively applied for dynamically downscaling coarse resolution information from different sources, such as general circulation models (GCMs) and reanalysis, for different purposes

I. Lirkov and S. Margenov (Eds.): LSSC 2017, LNCS 10665, pp. 266–274, 2018.
https://doi.org/10.1007/978-3-319-73441-5_28

including past climate simulations and future climate projection. The RCMs are tools that greatly enhance the usability of climate simulations made by the GCMs for studying climate and its change and impacts on a regional scale. The outputs of GCMs can be used as driving fields for the nested RCMs running with higher resolution, allowing capturing the local features of the climate [7]. Main manifestation of the flexibility of the modern RCMs is the possibility for selection among different initial and boundary conditions (ICBC) data-sets, parameterization schemes/modules within the model, various constants and closure assumptions, etc., combining them in practically countless model setups. Obviously the simulation output from such model setups will differ from one another, and, more or less, from the "reality". Thus a necessary prerequisite before any model implementation is to select the optimal RCM-configuration. The main aim of the present work is to perform extensive investigation of all 20 possible combinations between the main modules of the RCM RegCM4.4, which is our simulation tool. The EOBS one is used as a reference data-set and the comparisons are performed on seasonal basis.

The article is structured as follows. The different model setups and the used computational resources are described in the second section. The performed computations and the obtained results are presented in the third. Some general remarks of the main outcome of the study are placed in the conclusion.

2 Data and Methods

The simulations with the RCM RegCM version 4.4 [3,4] were made for the Southeastern Europe. The simulation domain covers the Balkan Peninsula, a minor part of Italy and a part of Asia Minor Peninsula. The model grid is in Lambert Conformal Conic projection, and is characterized by spatial resolution equal to 10 km, time step equal to 25 s, and 27 vertical levels. All simulations are for ten years period from 01.12.1999 to 30.11.2009. We made 20 control simulations with relaxation exponential technique lateral boundary condition scheme, BATS land-surface model, and combining boundary layer parameterization schemes, moisture/large-scale precipitation (M) schemes and cloud convection (CC) parameterization schemes. In such a way, we have big enough set of model configurations, and we are able to choose as good as possible combination(s) of them for the considered domain. The ICBC are taken from ERA-Interim (noted here EIN15) [2]. The planetary boundary layer (PBL) schemes used in these model configurations are the proposed by Holstlag et al., and the University of Washington (UW) PBL parameterization schemes. The large-scale precipitation schemes which have been used in the control simulations are Subgrid Explicit Moisture Scheme (SUBEX) and a new cloud microphysics scheme, proposed by Nogherotto and Tompkins (NT). The cumulus convections parameterization include Grell scheme with Arakawa-Schubert (AS) and Fritsch-Chappell (FC) closure assumption, Emanuel scheme, Tiedtke scheme and Kain-Fritsch scheme. The simulations with Kuo convective parameterization scheme have shown instability and interruptions of the model simulations at some periods, so we do not use it in our research.

The considered model configurations are listed in Table 1.

Table 1. Considered model configurations and their notations

Index	Notation	ICBC	PBL-scheme	M-scheme	CC-scheme
1	r11111	EIN15	Holtslag	SUBEX	Grell/FC
2	r11112	EIN15	Holtslag	SUBEX	Grell/AS
3	r11133	EIN15	Holtslag	SUBEX	Emanuel
4	r11144	EIN15	Holtslag	SUBEX	Tiedtke
5	r11155	EIN15	Holtslag	SUBEX	Kain-Fritsch
6	r11221	EIN15	Holtslag	Nogherotto/Tompkins	Grell/AS
7	r11222	EIN15	Holtslag	Nogherotto/Tompkins	Grell/FC
8	r11233	EIN15	Holtslag	Nogherotto/Tompkins	Emanuel
9	r11244	EIN15	Holtslag	Nogherotto/Tompkins	Tiedtke
10	r11255	EIN15	Holtslag	Nogherotto/Tompkins	Kain-Fritsch
11	r12121	EIN15	UW	SUBEX	Grell/AS
12	r12122	EIN15	UW	SUBEX	Grell/FC
13	r12133	EIN15	UW	SUBEX	Emanuel
14	r12144	EIN15	UW	SUBEX	Tiedtke
15	r12155	EIN15	UW	SUBEX	Kain-Fritsch
16	r12221	EIN15	UW	Nogherotto/Tompkins	Grell/AS
17	r12222	EIN15	UW	Nogherotto/Tompkins	Grell/FC
18	r12233	EIN15	UW	Nogherotto/Tompkins	Emanuel
19	r12244	EIN15	UW	Nogherotto/Tompkins	Tiedtke
20	r12255	EIN15	UW	Nogherotto/Tompkins	Kain-Fritsch

In Table 2 it can be seen that the computer resource requirements for the performed RegCM simulations are big enough. On the other hand the planned numerical experiments were organized in the effective HPC environment. The calculations were performed on the Supercomputer System "Avitohol" at the Institute of Information and Communication Technologies at the Bulgarian Academy of Sciences (IICT-BAS). It is built with HP Cluster Platform SL250S GEN8 (150 servers), Intel Xeon E5-2650 v2 8C 2.6 GHz CPUs (300 CPUs). Storage is provided by a storage system with 96 TB of raw disk storage capacity. The simulations for the selected domain were organized in separate jobs, which again make the jobs run time for 3 months real time fairly reasonable [1].

Table 2. Computer resource requirements on 16CPU-s for the application TVRegCM

	1 Month	1 Year	20 Cases
Time (h)	6	720	14400
HDD (GB)	6	720	14400

The well-known and freely available daily data from the EOBS dataset version 12.0 with $0.25° \times 0.25°$ regular grid spacing is used [5]. It is used as reference in the model validation.

3 Results

The reference for the mean seasonal temperature in winter (traditionally accepted as December, January and February) is shown on the upper leftmost subplot of Fig. 1 and the other subplots depicts the bias of the considered model configurations, according the notations in Table 1.

Some important features of bias distribution in the different cases are noticeable. The cases with schemes (r12221, r12222, r12233, r12244, r12255) are with significantly colder biases than other ones. Generally for them, the biases are smaller than 0.5 °C and mostly negative, i.e. the model output is colder than the reference. This is most remarkable in East Bulgaria, Hungary, and Greece. The biases in ±0.5 °C, are observed in the most areas of Carpathian Mountains in Romania. The other feature in the bias distribution is typical for the cases with schemes (r11121, r11122, r11133, r11144, r11155). They are characterized by bigger areas with temperature bias in ±0.5 °C, and warm biases bigger than 0.5 °C mainly in Romania. The cold biases are observed over significant areas of Bulgaria, Greece, FYR Macedonia, South Serbia, and Kosovo. Warm biases are distributed in south parts of Romania and Romanian part of the Carpathian

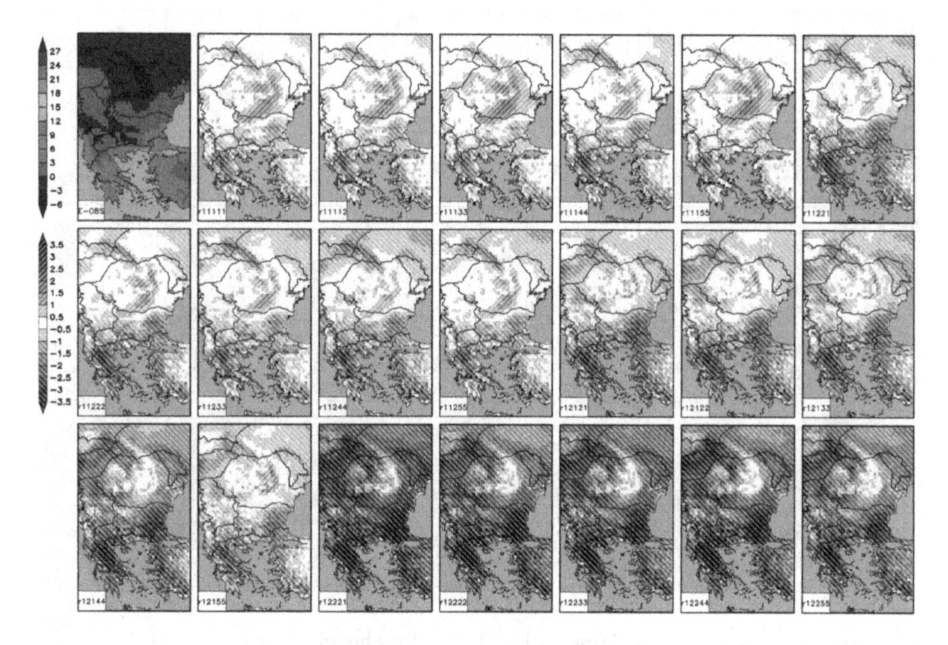

Fig. 1. Mean winter temperature (upper leftmost subplot, units: °C) and biases (units: °C) of the considered model configurations

mountains. The cases with schemes (r11121, r11122, r11133, r11144, r11155) have more areas with warm bias, in many parts of Romania, in the most northern parts of Northwestern Bulgaria, and parts of Croatia. Common for these cases is that they also have more areas with temperature bias in ±0.5 °C, than other ones. The case with scheme (r12144) can be also considered as an intermediate case. Its temperature bias is not as smaller as in the cases with the former bias feature, and there are some grid-cells with warm biases in the Carpathian mountains.

The results for the bias of the mean summer (June, July, August) temperature are shown in Fig. 2 similar to Fig. 1 manner. The bias distribution for the summer is more diverse than the winter. The cases with schemes (r11111, r11112, r11133, r11144, r11155), as well as with the schemes (r11233, r11255) are characterized with relatively big areas with prevailing warm biases, typically below 2 °C. The warmest biases in these configurations are distributed in Hungary, Croatia, and Albania. The case with scheme (r11155) differs from other ones with warm biases distributed on the biggest area of the domain, and with the biggest values, reaching 3–3.5 °C, especially over Hungary, Croatia, and Danube lowlands. The model set-ups with other combinations of parameterization schemes results in cold biases in most of the domain area, and especially in Bulgaria and Greece. The case with scheme (r12155) has got a bias in ±0.5 °C. The model configurations with schemes (r12144, r12244) are noticeable with the coldest bias, especially over Bulgaria and Greece.

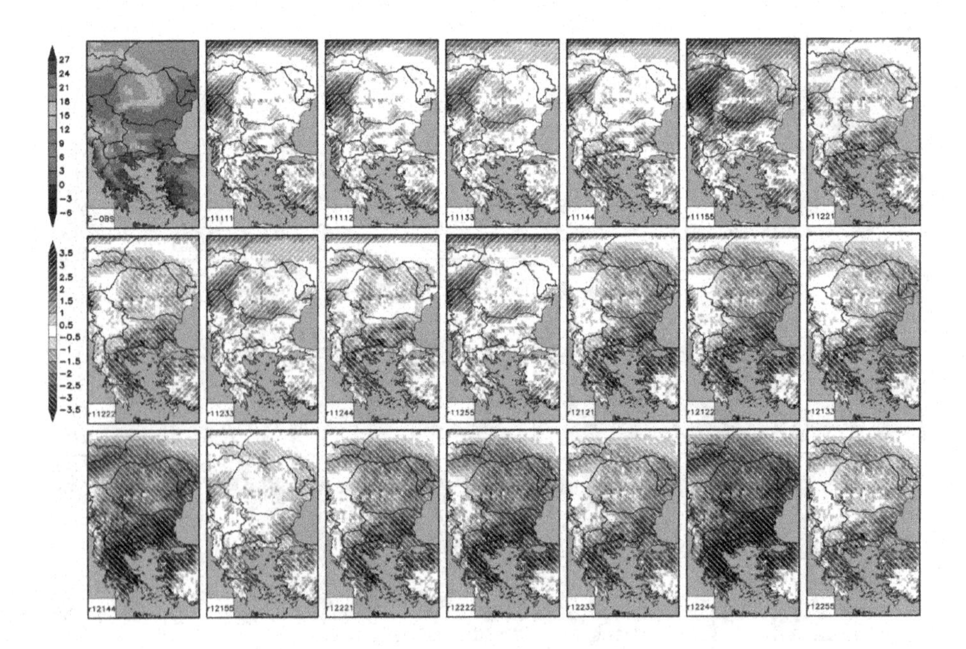

Fig. 2. Same as Fig. 1, but for the summer

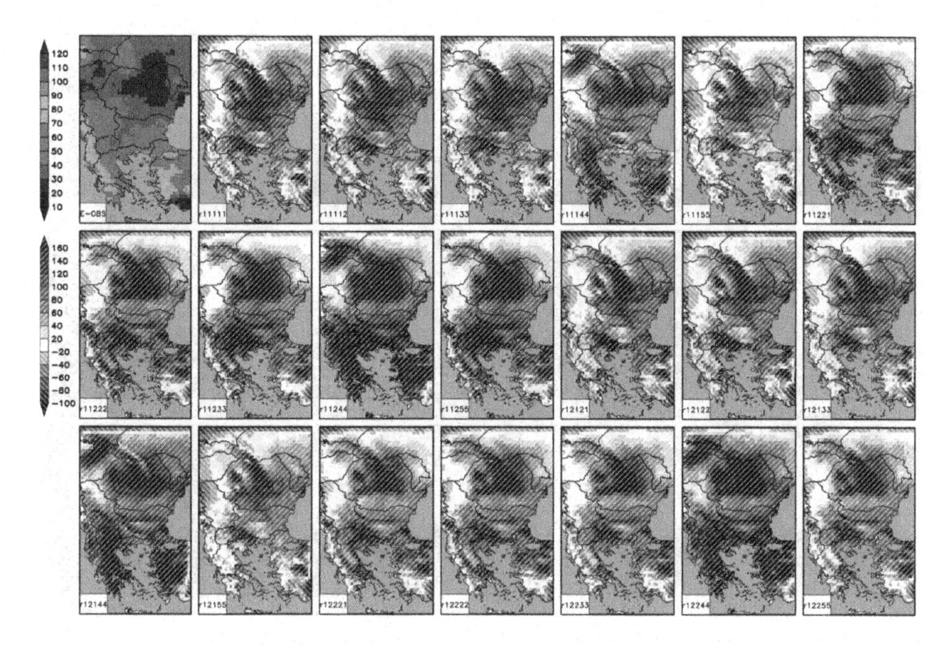

Fig. 3. Mean winter precipitation sum (upper leftmost subplot, units: mm/month) and relative biases of the considered model configurations

The reference for the mean seasonal precipitation sum in winter is shown on the upper leftmost subplot of Fig. 3 and the other subplots depicts the relative bias of the considered model configurations, according the notations in Table 1. Generally speaking, there are almost no differences between simulations with different combinations of parameterization schemes. The model simulations results in dominating positive biases reaching more than 100% notably in Romania without its most southern areas, in Greece, in the mountain areas of Bulgaria. There are some minor differences between models cases noticeable in most of the simulations with schemes (r12121, r12122, r12133, r12144, r12155, r12221, r12222, r12233, r12244, r12255), which have got more areas with smaller precipitation bias. The boundary conditions have some influence to the model results, noticeable as a spurious grids on the northern boundary of the model domain.

The relative bias for precipitation in summer shown in Fig. 4, suggest much more clear distinctions in behavior of some cases. Most significant is the presence of some ones with discernibly lower bias. Possible explanation is the bigger share of the convective precipitation in the total amount during the summer than the winter and, correspondingly, the role of the convective scheme. These are the cases (r11111, r11112, r11155, r12121, r12122, r12155, r12221). The other cases have worse representativeness of the precipitation in the domain area, although generally the biases in the Danube lowlands are the smallest ones. The worst case appears to be the schemes (r12244, r11244). We can see that the areas with the wettest biases up to 160% include Albania, parts of Ukraine and Moldova,

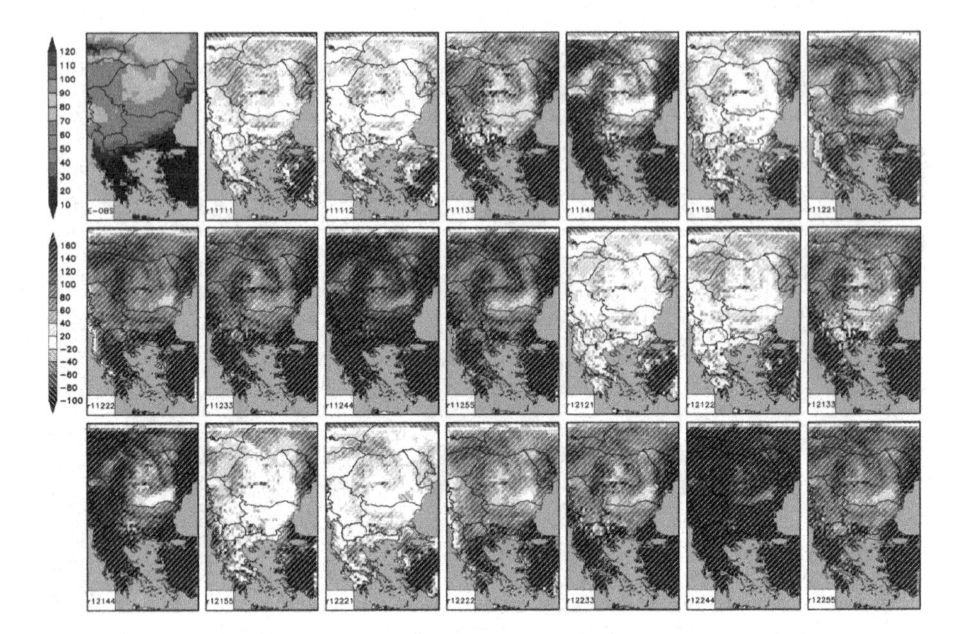

Fig. 4. Same as Fig. 3, but for the summer

and especially Greece and Turkey. It is also noticeable, that Bulgaria and Romania are with the wettest biases in the cases with the NT scheme and Tiedtke convective scheme. The northern part of the domain is the most influenced from the boundary conditions.

4 Conclusion

The differences in temperature biases are pronounced in winter, as well as in summer. The biases are negative in the southern part of the domain, and these with the biggest absolute value are obtained by simulations with UW PBL and the NT scheme in Eastern Bulgaria, Greece, and Hungary. The results for summer however, show bigger deviations from the model values. The warm summer biases are spread mostly in the northern part of the domain. The temperature biases are in the range from about $-3.5\,^{\circ}$C to $3.5\,^{\circ}$C.

The precipitation biases have almost equal distribution in the winter. They are positive with some minor exceptions, on that season. The summer biases however, show significant distinction in their distribution and magnitude. They are positive, with some minor exception in Greece. The simulations with the smallest biases are with Grell schemes for convection. The case with UM PBL scheme, NT scheme, and Kain-Fritsch convective scheme also has got significantly better results. The precipitation biases however, are very big. Their values vary from below -100% to above 160%.

The simulation outcomes for the both considered variables, the near surface temperature and precipitation shows principal agreement with most of the EURO-CORDEX results [6]. The last ones are done using seven different models (including WRF), and two additional configurations of WRF with different parameterization. The bias is in the same or almost the same range, although with some spatial differences. The relative bias for precipitation in some locations in our simulations are bigger than the results from CORDEX experiments with 50% and more.

The necessary next step is to investigate more deeply, including on subseasonal basis, these model configurations, which shows overall better simulation capabilities. The optimal configuration improves the model skill and increases the benefit of any model implementation.

Acknowledgment. Deep gratitude to the organizations and institutes (ICTP, ECMWF, NCEP-NCAR, ECA&D, Unidata, MPI-M and all others), which provides free of charge software and data. Without their innovative data services and tools this study would be not possible. Special thanks to: VRE for regional Interdisciplinary communities in Southeastern Europe and the Eastern Mediterranean (VI-SEEM), Horizon 2020 project 675121, the EC-FP7 grant PIRSES-GA-2013-612671 (project REQUA), the Bulgarian National Science Fund (grant DN-04/2/13.12.2016), and Program for career development of young scientists, BAS.

References

1. Atanassov, E., Gurov, T., Karaivanova, A., Ivanovska, S., Durchova, M., Dimitrov, D.: On the parallelization approaches for intel MIC architecture. In: AIP Conference Proceedings, vol. 1773, p. 070001 (2016). https://doi.org/10.1063/1.4964983
2. Dee, D.P., Uppala, S.M., Simmons, A.J., Berrisford, P., Poli, P., Kobayashi, S., Andrae, U., Balmaseda, M.A., Balsamo, G., Bauer, P., Bechtold, P., Beljaars, A.C.M., van de Berg, L., Bidlot, J., Bormann, N., Delsol, C., Dragani, R., Fuentes, M., Geer, A.J., Haimberger, L., Healy, S.B., Hersbach, H., Hèlm, E.V., Isaksen, L., Kållberg, P., Köhler, M., Matricardi, M., McNally, A.P., Monge-Sanz, B.M., Morcrette, J.-J., Park, B.-K., Peubey, C., de Rosnay, P., Tavolato, C., Thèpaut, J.-N., Vitart, F.: The ERA-Interim reanalysis: configuration and performance of the data assimilation system. Q. J. R. Meteorol. Soc. **137**, 553–597 (2011)
3. Elguindi, N., Bi, X., Giorgi, F., Nagarajan, B., Pal, J., Solmon, F., Rauscher, S., Zakey, A., O'Brien, T., Nogherotto, R., Giuliani, G.: Regional Climate Model RegCM User Manual Version 4.4 (2014). http://gforge.ictp.it/gf/download/docmanfileversion/71/1223/ReferenceMan.pdf. Accessed 20 Feb 2017
4. Giorgi, F., et al.: RegCM: model description and preliminary tests over multiple CORDEX domains. Clim. Res. **52**, 7–29 (2012)
5. Haylock, M.R., Hofstra, N., Klein Tank, A.M.G., Klok, E.J., Jones, P.D., New, M.: A European daily high-resolution gridded dataset of surface temperature and precipitation. J. Geophys. Res. (Atmos.) **113**, D20119 (2008). https://doi.org/10.1029/2008JD010201

6. Kotlarski, S., Keuler, K., Christensen, O.B., Colette, A., Dèquè, M., Gobiet, A., Goergen, K., Jacob, D., Lüthi, D., van Meijgaard, E., Nikulin, G., Schär, C., Teichmann, C., Vautard, R., Warrach-Sagi, K., Wulfmeyer, V.: Regional climate modeling on European scales: a joint standard evaluation of the EURO-CORDEX RCM ensemble. Geosci. Model Dev. **7**(1297–1333), 2014 (2014)
7. Xue, Y., Janjic, Z., Dudhia, J., Vasic, R., De Sales, F.: A review on regional dynamical downscaling in intraseasonal to seasonal simulation/prediction and major factors that affect downscaling ability. Atmos. Res. **147–148**, 68–85 (2014)

Territorial Design Optimization for Business Sales Plan

Laura Hervert-Escobar[1(✉)] and Vassil Alexandrov[1,2,3]

[1] Instituto Tecnológico y de Estudios Superiores de Monterrey, Monterrey, Nuevo León, Mexico
laura.hervert@itesm.mx
[2] ICREA - Catalan Institution for Research and Advanced Studies, Barcelona, Spain
[3] Barcelona Supercomputing Center, Barcelona, Spain

Abstract. A well designed territory enhances customer coverage, increases sales, fosters fair performance and rewards systems and lower travel cost. This paper considers a real life case study to design a sales territory for a business sales plan. The business plan consists in assigning the optimal quantity of sellers to a territory including the scheduling and routing plans for each seller. The problem is formulated as a combination of assignment, scheduling and routing optimization problems. The solution approach considers a meta-heuristic using stochastic iterative projection method for large systems. Several real life instances of different sizes were tested with stochastic data to represent raise/fall in the customers demand as well as the appearance/loss of customers.

Keywords: Territory design · Projection methods · Optimization

1 Introduction

Network models and integer programs are applicable to an enormous known variety of decision problems. In a real life case, the cost efficient management decision is defined by a combination of different models/problems. This paper considers a real life case study that determines the minimum number of sellers required to serve a set of customers located in a certain region together with the weekly schedule plan for visits and the optimal route. Therefore, the decision should consider the demand of the customers as well as the daily capacity of the sellers to fulfill the demand. Additionally, it is important to define the seller's route of visiting the customers per day.

The solution method for the problem in question combines objectives and constraints of three classical approaches, the clustering of customers, the scheduling of visits, and the routing plan. Consider the location of the customers (such as points in the area or nodes of a network) with a given distance between every pair of points. We wish to find a cluster of customers using the nearest neighbor approach. Therefore, each cluster will represent a seller in the solution. This objective can be interpreted as the tightest cluster of m points. This is similar

I. Lirkov and S. Margenov (Eds.): LSSC 2017, LNCS 10665, pp. 275–282, 2018.
https://doi.org/10.1007/978-3-319-73441-5_29

to the one facility version of the max-cover problem (for a network or discrete formulation [4], for planar models [3], for one facility [6], and for several facilities [9]) where we wish to find the location of several facilities which cover the maximum number of points within a given distance. After defining the clusters of customers, an important logistic problem to solve is the scheduling and routing problem.

The following step in the procedure is the scheduling of the visits to customers. Effective scheduling systems aim at matching demand with capacity so that resources are better utilized and waiting times are minimized. Tuga and Emre [8] provide a comprehensive survey on appointment scheduling in outpatient services. The underlying problem applies to a wide variety of environments of outpatient scheduling, and is modeled using queuing system representing the unique set of conditions for the design of the patient appointments. The authors present a complete survey of problem definitions and formulations considering the nature of Decision-Making and Modeling of Clinic Environments. In addition, they mention a variety of performance criteria used in the literature to evaluate appointment systems, which are grouped as: (a) Cost-Based Measures, (b) Time-Based Measures, (c) Congestion Measures, (d) Fairness Measures.

Finally, the goal of the problem described above is to optimize the distribution process from depots to customers (routing design) in such a way that customer's demand of goods is satisfied without violating any problem-specific constraints. In the literature, these kind of logistic problems are known as Vehicle Routing Problems (VRP) and the objective regularly is the minimization of the complete distance traveled by the vehicles while servicing all the customers. The VRP is an interesting problem in operations research due to its practical relevance and the difficulty to be solved exactly. Moreover, it is one of the most demanding NP-hard problems [5]. In reality, the task of finding the best set of vehicle tours by solving optimization problems has a high computational cost, prohibitive for medium and large real applications.

Caceres et al. [2] present a survey on VRP's applied to real life problems. The authors call these VRP's as Rich (realistic) VRP's (RVRP's) and classify their variants according to the company decision levels and the routing elements involved. A classification that applies for this case study is Multi-Period/Periodic VRP with Multiple Visits/Split deliveries. In this classification, the clients are visited several times as vehicles may deliver a fraction of the customer's demand. Moreover, optimization is made over a set of days, considering a different frequency of visits to each client.

In this work, a real business strategy for sales in different territories is modeled using the formulation of three classical problems: cluster, scheduling and VRP. Particularities of the modeling approach include scheduling constraints of visits spread over the week, service and traveling times; as well as time capacity to ensure the fulfillment of the clients demand.

The paper is organized as follows. A general mixed integer linear programming (MILP) formulation for the problem is presented in Sect. 2. Section 3 describes a projection method for large systems of linear equations and two optimization models to cut the problem size and to reduce the execution time. In Sect. 4, the algorithms are tested for different scenarios. Finally, conclusions are presented in Sect. 5.

2 Mathematical Formulation

Consider a set of customers $C = \{1, 2, ..., N\}$ dispersed in a given region with geographical coordinates $(long, lat)$. It is desired to design a business plan that assigns the customers to a set of sellers $S = \{1, 2, ..., S\}$. The sellers will attend the demand of customers during the weekdays $W = \{1, 2, 3, 4, 5, 6\}$ denoted by index t in the scheduling plan per week. Finally, it is desired to get the optimal daily routing. The distance between two location is computed using the Haversine formula given in the following equation:

$$d_{i,j} = 2r \arcsin\left(\sqrt{\sin^2\left(\frac{lat_j - lat_i}{2}\right) + \cos(lat_i)\cos(lat_j)\sin^2\left(\frac{lon_j - lon_i}{2}\right)}\right)$$

where latitude(lat) and longitude(lon) are given in radians, and R is earth's radius (mean radius = 6,371 km).

The mathematical formulation of the model is defined in the following equations:

The objective function (1) represents the sum of two goals, the minimization of the distance between customers assigned to the sellers $d_{s,i}$ as well as the traveling distance to visit each customer for each routing plan $d_{i,j}$. This equation has two binary decision variables: y_i^s with value of 1 if the customer i is assigned to seller s, and $x_{i,j}^{s,t}$ with value of 1 is customer j is visited after customer i by seller s on day t.

$$min \sum_{s \in S} \sum_{i \in \mathbb{N}} d_{s,i} y_i^s + \sum_i \sum_j \sum_s \sum_t d_{i,j} x_{i,j}^{s,t} \tag{1}$$

As for constraints, (2) ensures that a customer is attended by only one seller.

$$\sum_{s \in S} y_i^s = 1; \qquad \forall i \tag{2}$$

Part of the goal in the problem is to define the scheduling plan for visits during the week. Then a binary variable $v_i^{s,t}$ takes a value of 1 if the customer i assigned to seller s is visited on day t. Next Eq. (3) relates two binary variables y and v in order to guaranties that a customer assigned to the seller is actually visited.

$$v_i^{s,t} \leq y_i^s; \qquad \forall i, t, s \tag{3}$$

Equations (4) and (5) link the scheduling variables to the routing ones.

$$x_{i,j}^{s,t} \leq v_i^{s,t}; \qquad \forall i, j, t, s; \qquad i \neq j \in N \tag{4}$$

$$x_{i,j}^{s,t} \leq v_j^{s,t}; \qquad \forall i, j, t, s; \qquad i \neq j \in N \tag{5}$$

Equations (6) and (7) are used for connectivity purposes. Equation (6) establishes that the number of incoming links to a client node must be equal to the number of outgoing links, whereas Eq. (7) sets for each client one arrival and one departure at the time (degree of the node is 2).

$$\sum_i x_{i,j}^{s,t} = \sum_j x_{i,j}^{s,t}; \qquad \forall i, j, t, s; \qquad i \neq j \in N \tag{6}$$

$$\sum_i x_{i,j}^{s,t} + \sum_j x_{i,j}^{s,t} = 2v_i^{s,t}; \qquad \forall i, j, t, s; \qquad i \neq j \in N \tag{7}$$

Next Eq. (8) ensures that sum the service time per customer θ_i during the day does not exceed the available time of the seller τ^s.

$$\sum_i \theta_i v_i^{s,t} \leq \tau^s; \qquad \forall t, s \tag{8}$$

The total visits per week are given by the frequency σ. The frequency is computed by dividing the demand and capacity. In this way, Eq. (9) establishes the number of visits to be carried out per customer according to the given frequency. Equation (10) avoids consecutive visits to those customers whose frequency is less than 4 visits per week.

$$\sum_t \sum_s v_i^{s,t} = \sigma_i; \qquad \forall i \tag{9}$$

$$v_i^{s,t} + v_i^{s,t+1} \leq 1; \qquad \forall i, s, t; \qquad t \leq 5 \qquad \sigma_i \leq 3 \tag{10}$$

Finally, Eqs. (11) and (12) allow to assign the proper order of visits to customers during the routing plan to avoid sub-tours. Here, a continuous variable e is introduced. This variable denotes the order in which customer i is visited in the route plan of seller s during day t. Equation (11) ensures that difference between the order of visits to two consecutive customers is one, whereas Eq. (12) limits the maximum order of visit to the customer.

$$e_i^{s,t} - e_j^{s,t} + N x_{i,j}^{s,t} \leq N - 1; \qquad \forall i, j, t, s; \qquad i \neq j \in N \tag{11}$$

$$e_i^{s,t} \leq \sum_j v_j^{s,t}; \qquad \forall i, j, t, s; \qquad i \neq j \in N \tag{12}$$

3 Solution Methods

The proposed approach combines assignment, scheduling and routing problem formulations. The above three problems individually have been shown to be NP-complete [1]. Therefore, the time required to achieve an optimal solution of the whole model (problem) increases exponentially with the growth of the problem size. To overcome this difficulty, the problem is divided into three sub-problems (or phases). The objective is to work with smaller problems at each phase. The first phase is treated as an assignment optimization problem that minimizes the distance between sellers and customers. The second phase of the problem is focused on solving the scheduling problem for each seller. Finally, the third phase solves the routing problem for each active working day defined during the second phase.

The first phase tackles the largest part of the problem size, therefore, a projection method for large system of linear equations, adapted to deal with mixed equality an inequality constraints, is used in order to assign the sellers to the customers. Projection methods for solving systems of linear equations have been known for some time. Projection methods are so named because an $m - dimensional$ method will map m components of the approximate solution vector onto a subspace determined for example by $k \leq m$. The method used in this research was proposed by Sabelfeld [7]. The method is a randomized block projection method combined with the Johnson-Lindenstrauss dimension reduction. The steps of the method are as follows:

1. Take an $m \times n$ matrix A, and a column vector b, choose an integer parameter $s = 8log(m)/\varepsilon^2$, ε is the desired accuracy, which corresponds to the probability level $1 - m^{-2}$ in the J-L theorem, and the initial approximation $x0$.
2. We aim at an approximation of the solution x to $Ax = b$.
3. Set $k = 0$, generate an $n \times s$ random matrix R satisfying the independence and symmetry condition, say, a Gaussian matrix $R = \{g_{ij}\}$ where the entries g_{ij} are independent zero mean Gaussian random numbers with variance 1/s, and calculate the rows $hi = (AR)_i = a_i R$.
4. Sample a set of $N = m$ rows at random, say, according to $p_i = ||a_i||_2^2 = ||A||_F^2$. Uniform sampling is also possible. Note that N can be taken less than m which implies that the random search of rows is carried out only among a part of all the rows. For each row, calculate the distance $\Delta_i = |b_i - (h_i; R^T x_k)|/||h_i||_2$ and choose a_j as the row with the maximal Δ_j. This row is used to calculate the next projection step in our iteration process (13). An hedging step can be introduced here to prevent the case that the chosen j is worse than the random row actually used in the calculation of the projection. So along with the chosen a_j take also an arbitrary row a_p in the set of chosen rows, and calculate

$$\overline{\Delta}_j = \frac{|b_j - (a_j, x_k)|}{||a_j||_2}, \qquad \overline{\Delta}_p = \frac{|b_p - (a_p, x_k)|}{||a_p||_2}$$

If $\overline{\Delta}_p \geq \overline{\Delta}_j$ set $j = p$.

Calculate the projection

$$x_{k+1} = x_k + \frac{b_j - (a_j, x_k)}{||a_j||_2^2} a_j^T \tag{13}$$

5. $k + k + 1$, go to 4.

The projection methods deals with the assignment of the customers to sellers. The next step is the scheduling plan for visits. The sequence of visits for each seller is determined by solving the corresponding scheduling problem. Its objective function is given in Eq. (14).

$$min \sum_{i} \sum_{t} r_i \cdot v_i^t \tag{14}$$

The objective function (14) minimizes the radius of coverage (r_i) subject to constraints (8)–(10). This objective function helps to schedule the nearest customers per day. Finally, the routing is solved per day and per seller in phase 3. The model of the third phase is based on the routing problem formulation. The objective function is given in (15).

$$min \sum_{i} \sum_{j} d_{i,j} \cdot x_{i,j} \tag{15}$$

The objective function (14) minimizes the total distance of the routes subject to constraints given by Eqs. (4)–(7), (11) and (12).

4 Tests and Results

To test the performance of the proposed approach, several instances were tested. The data for each instance correspond to a real life case taken from a soft-drinks manufacturer. Table 1 shows the nomenclature used to identify instances, where the rows are used to describe the territory and the columns denote the kind of seller required for the type of products distributed. For each couple of territory and seller, the table provides the total number of customers to be assigned.

As shown in the Table 1, the variety of size is good enough to prove whether the proposed approach is efficient for a business plan. The execution time is an important issue for the company due to the deadline to generate the business plan each week. Therefore, the results are given in terms of both objective functions as well as execution times. The projection method algorithm was implemented in matlab 2016a v9.0.0.341360. The scheduling and routing methods algorithms were implemented using AMPL to call the optimizer gurobi v6.5. The results for the phase 1 are given in Table 2 and Fig. 1. For each combination of territory

Table 1. Customers assigned per territory and type of seller

Territory/type of seller	A	B	C	D	E	F	G
T1	48	37	163	15	970	145	10812
T2	33	12	463	4	1645	186	-
T3	26	18	405	9	1219	112	-
T4	59	40	448	17	1981	243	22475

Table 2. Results of phase 1-projection method

Instance	OF-OPTIMAL	OF-PM	GAP	Instance	OF-OPTIMAL	OF-PM	GAP
T1B	397.79	397.79	0.0%	**T1D**	126.81	128.66	1.5%
T2B	440.11	488.52	11.0%	**T4D**	271.34	336.46	24.0%
T3B	236.70	260.37	10.0%	**T3F**	1375.22	1498.99	9.0%
T4B	703.25	752.48	7.0%	**T1F**	1359.63	1645.15	21.0%
T3A	423.58	495.59	17.0%	**T2A**	763.45	977.22	28.0%
T3D	150.90	161.46	7.0%	**T4A**	1266.49	1418.47	12.0%
T2D	95.40	109.71	15.0%	**T2F**	1118.54	1398.17	25.0%
T1A	301.70	381.48	26.4%	**T4F**	4137.24	4675.08	13.0%

and seller, the table provides the optimal value of the objective function, the result from the projection method and the gap of the solution.

As Table 1 shows, the gap between two methods is not related to the size of the instance but to a combination of distance and demand of the customers. Figure 1 shows a comparison of the execution time provided by the optimal exact method and the projection method. As the figure shows, the time rapidly increases for the exact method. For this case, the size of the instance impacts greatly the execution time. In the initial experiments it was set to achieve the solution in seconds and aim for the approximate solution. This led to a large gap in the objective function values when applying the exact and projection methods in some cases (up to 28 per cent in the worst case). But for some problems the optimal solution was achieved in both cases. Additional experiments have shown that in case of the projection method, the convergence is improved when the selected block is set to the minimum size of 2.

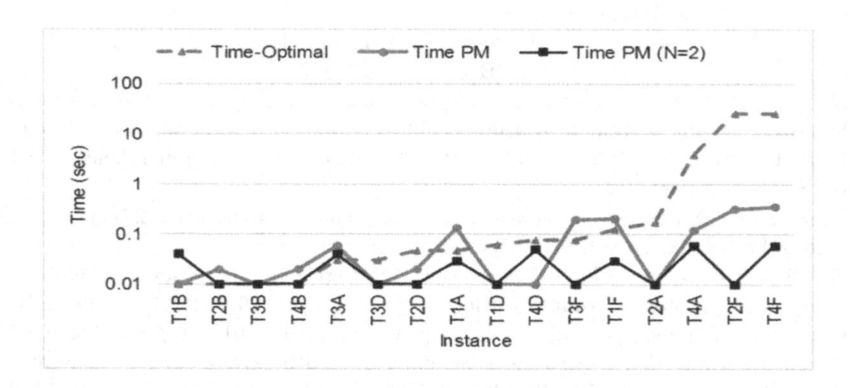

Fig. 1. Solving time results for the projection method

5 Conclusion

Assignment, scheduling and routing problems have attracted the attention of both researchers and practitioners for several decades due to the practical value of solving such problems efficiently in decision-making contexts, the ever-present need and desire to incorporate increasingly realistic constraints and objectives into the models, the challenges associated with solving the problems, and the ability of the basic formulations to represent important decision-making issues in business contexts. These four factors continue to be important to this day and are likely to be present and influence the algorithmic developments for years to come.

This work presents a three-phase approach to solve a real life problem with different problem sizes. The first phase uses a projection method to assign customers to sellers, the method proved an efficient solving time, with gaps from 0 to 28% in the objective function.

The described approach allows also to tackle the uncertainties stemming from practical problems such as different sizes of territory and particular features of the demand such as the distance and the service time.

References

1. Baase, S., Gelder, A.V.: Computer Algorithms: Introduction to Design and Analysis. Addison Wesley, Boston (1999)
2. Caceres-Cruz, J., Arias, P., Guimarans, D., Riera, D., Juan, A.A.: Rich vehicle routing problem: survey. ACM Comput. Surv. **47**(2), 32:1–32:28 (2014). http://doi.acm.org/10.1145/2666003
3. Current, J., Daskin, M., Schilling, D.: Discrete network location models. In: Drezner, Z., Hamacher, H.W. (eds.) Facility Location: Applications and Theory, chap. 3, pp. 81–118. Springer, Heidelberg (2002)
4. Daskin, M.S.: Network and Discrete Location: Models, Algorithms, and Applications. Wiley, New York (2013). ISBN: 978-0-470-90536-4
5. Dondo, R., Cerda, J.: A cluster-based optimization approach for the multi-depot heterogeneous fleet vehicle routing problem with time windows. Eur. J. Oper. Res. **176**(3), 1478–1507 (2007). http://www.sciencedirect.com/science/article/pii/S0377221705008672
6. Drezner, Z.: Note-on a modified one-center model. Manag. Sci. **27**(7), 848–851 (1981). http://dx.doi.org/10.1287/mnsc.27.7.848
7. Sabelfeld, K., Loshchina, N.: Stochastic iterative projection methods for large linear systems. Monte Carlo Methods Appl. **16**(3–4), 343–359 (2010). http://EconPapers.repec.org/RePEc:bpj:mcmeap:v:16:y:2010:i:3–4:p:343–359:n:13
8. Tugba, C., Emre, V.: Outpatient scheduling in health care: a review of literature. Prod. Ope. Manag. **12**(4), 519–549 (2003). http://dx.doi.org/10.1111/j.1937-5956.2003.tb00218.x
9. Watson-Gandy, C.: Heuristic procedures for the m-partial cover problem on a plane. Eur. J. Oper. Res. **11**(2), 149–157 (1982). http://www.sciencedirect.com/science/article/pii/0377221782901096

Monte Carlo Algorithms for Problems with Partially Reflecting Boundaries

Nikolai A. Simonov([✉])

Institute of Computational Mathematics and Mathematical Geophysics SB RAS, Lavrentjeva 6, Novosibirsk 630090, Russian Federation
nas@osmf.sscc.ru

Abstract. We consider diffusion problems with partially reflecting boundaries that can be formulated in terms of an elliptic equation. To solve boundary value problems with the Robin condition, we propose a Monte Carlo method based on a randomization of an integral representation. The algorithm behaviour is analysed in its application for solving a model problem.

Keywords: Monte Carlo · Random walk
Partially reflecting boundary condition · Laplace equation
Markov chain · Robin problem · Third boundary value problem

1 Introduction

Monte Carlo methods for elliptic equations are well known and have a history of successful practical use when solving boundary value problems in various applications [1–4]. The common feature of such algorithms is that they are based on a randomization of mean-value relations and Green's formulas written out for simple geometric structures. The resulting Markov chain makes possible an efficient simulation of the diffusion process' exit points on the boundary of the computational domain. For the absorbing boundary conditions, due to the probabilistic representation [5,6], a solution can then be computed as the mean of known boundary values.

In many practical applications, however, it is more natural to consider partially or fully reflecting boundary conditions. From a mathematical point of view, this leads to the need of solving problems with boundary conditions, which include flux or the solution's normal derivative [7]. To solve such problems, Monte Carlo methods can also be used. In particular, to compute a solution in a domain with a compact simply connected boundary, the random walk on boundary algorithm can be applied [3,8,9]. For the Green's function based Monte Carlo methods, the common approach is to use a finite-difference approximation to the normal derivative and simulate the diffusion reflection in accordance with the randomized treatment of the resulting relation [10,11]. Such algorithm introduces an additional bias into the computed value and lacks efficiency. The exception is

I. Lirkov and S. Margenov (Eds.): LSSC 2017, LNCS 10665, pp. 283–291, 2018.
https://doi.org/10.1007/978-3-319-73441-5_30

the problem with the Neumann boundary conditions and plane boundaries. For this case, an unbiased Monte Carlo estimate can be constructed [12].

To overcome the inconsistencies caused by the finite-difference approximation, in [13,14] we proposed a new approach, which is based on mean-value formulas for boundary points with the Neumann or the flux continuity conditions. A randomization of these relations makes possible use of the walk-on-spheres algorithm not only inside the domain but also after hitting the boundary.

Recently [15], we proposed to use an analogous approach to construct a Monte Carlo method for solving the third boundary value problem. In this paper, we clarify the properties of the developed random walk algorithm and apply it to solving a model problem.

2 Problem Statement and Standard Approach

Consider the Laplace equation

$$\Delta u(x) = 0 \tag{1}$$

in a (possibly infinite) domain $G \subset \mathbb{R}^m$, and let u satisfy the conditions of the third kind

$$\alpha(y)\frac{\partial u}{\partial n}(y) + \beta(y)u(y) = g(y),\ y \in \Gamma \tag{2}$$

on the boundary of the domain, $\Gamma = \partial G$. Here $\alpha \geq 0$, $\beta \geq 0$, $\alpha + \beta \geq c > 0$, and the normal vector, $n(y)$, is considered to be external with respect to G. Suppose the boundary is piece-wise smooth and the parameters of the problem given in (1), (2) ensure existence and uniqueness of its solution [16].

Suppose that the boundary value problem data (geometry, boundary conditions) ensure existence of a probabilistic representation of the solution to this problem. Therefore

$$u(x_0) = \mathbb{E}\left[u(x^*)\right], \tag{3}$$

where $x^* \in \Gamma$ is an exit point for the Brownian motion originated at $x_0 \in G$. A numerical computation of the exit point is a non-trivial algorithmic problem, and there are many papers devoted to its solution based on a simulation of various random walks (Markov chains of points). Therefore, here we concentrate our efforts on the boundary condition per se, and suppose that x^* is already known.

Denote $x = x^*$. Without loss of generality we can consider x either an elliptic point or an internal point of a plane part of the boundary. Suppose that $\alpha(x)$ is separated from zero. Also, for definiteness, we set $m = 3$. A generalization to multi-dimensional cases is obvious.

Consider first the standard approach, which is based on the first order finite-difference approximation to the normal derivative. For the external n,

$$u(x) = \frac{\alpha(x)}{\alpha(x) + h\beta(x)}u(x - nh) + \frac{h\beta(x)}{\alpha(x) + h\beta(x)}\frac{g(x)}{\beta(x)} + O(h^2). \tag{4}$$

In the framework of the statistical paradigm, this relation can be considered as the total probability formula. This means with probability $1-p_h = \alpha(x)/(\alpha(x)+h\beta(x))$ a random walk trajectory is 'reflected' in the direction opposite to the normal vector, and the next point is taken to be $x-nh$. Here, we suppose this point lies inside G. With the complementary probability, p_h, the simulation is stopped, and the solution estimate takes the value $g(x)/\beta(x)$. With this approach, a random walk simulation terminates only on the boundary. The mean number of reflections equals the inverse of the termination probability. Therefore, for small h this number is $O(h^{-1})$, and from (4) it follows that the total bias of the solution estimate is $O(h)$.

Commonly, for general geometries the walk-on-spheres algorithm is used for simulating the Brownian motion exit point. The position of such a point is approximated by the first hit into a strip near the boundary. To adjust the error induced by such approach to the finite-difference approximation error, the strip width, ε, should be taken equal to $O(h^2)$.

3 The Mean-Value Relation for a Boundary Point

As an alternative to the finite-difference approximation for the normal derivative we proposed [13] to use an exact integral relation written out at a point on the boundary.

Consider a ball $B(x,a)$ of radius a and having x as its center. The Green's function for the Dirichlet problem in this ball is known exactly. For a point $y_0 \neq x$, it equals

$$\Phi(y_0,y) = -\frac{1}{4\pi}\left(\frac{1}{|y-y_0|} - \frac{a}{|y_0-x|}\frac{1}{|y-\overline{y_0}|}\right), \tag{5}$$

where $\overline{y_0} = x + \dfrac{a^2}{|y_0-x|^2}(y_0-x)$ is the point conjugate to y_0 (inverse of y_0) relative to the spherical surface $S(x,a)$ (see, e.g. [17]). For the central point, x, this formula can be simplified.

By definition, $\Delta_y\Phi(y_0,y) = \delta(y_0-y)$, where δ is the standard Dirac delta-function concentrated at y_0, and $\Phi(y_0,y)=0$ for $y \in S(x,a)$.

Denote by $B_i(x,a) = B(x,a)\bigcap G$ the part of the ball that lies inside the computational domain, and let $S_i(x,a)$ be the part of its spherical surface that lies in G. Denote by $\Gamma S = S_i(x,a)\bigcup(\Gamma\bigcap B(x,a))$ the boundary of $B_i(x,a)$.

Consider the pair of functions, $u(y)$ and $\Phi(y_0,y)$, inside $B_i(x,a)$ and write down the Green's integral formula. To have a possibility of doing this, we take into account smoothness properties of these functions and the surface Γ, limiting properties of a double-layer potential [18] and properties of the normal derivative of a harmonic function. Thus we have

$$u(y_0) = \int_{\Gamma S} 2\frac{\partial\Phi}{\partial n(y)}\, u\, d\sigma(y) - \int_{\Gamma\bigcap B(x,a)} 2\Phi\frac{\partial u}{\partial n(y)}\, d\sigma(y). \tag{6}$$

Consider now the boundary condition (2) and substitute it into the second integral. Then the relation (6) can be rewritten as the following integral representation for the solution boundary values inside the ball, $B_i(x,a)$:

$$u = K_0 u - K_{1,\alpha}\beta u + f_\alpha. \tag{7}$$

Here

$$K_0 u(y_0) = \int_{\Gamma S} \left[\frac{1}{2\pi} \frac{\cos \varphi_{yy_0}}{|y - y_0|^2} - k_{0,1}(y_0, y) \right] u(y)\, d\sigma(y), \tag{8}$$

where $k_{0,1} = 0$ for $y_0 = x$, and $k_{0,1} = \frac{a}{|y_0 - x|} \frac{1}{2\pi} \frac{\cos \varphi_{y\overline{y_0}}}{|y - \overline{y_0}|^2}$ otherwise. Here, φ_{yy_0} is the angle between $n(y)$ and $y - y_0$. For points on the spherical surface, $y \in S(x,a)$, the kernel of this integral operator coincides with the doubled Poisson kernel:

$$k_0(y_0, y) = \frac{1}{2\pi a} \frac{a^2 - |y_0 - x|^2}{|y - y_0|^3}. \tag{9}$$

K_0 operates in the space of functions defined on ΓS and provides values at the points y_0 that lie on the boundary of the computational domain, Γ.

The rest of the terms in (7) are defined by the following integrals

$$K_{1,\alpha}[\beta u](y_0) = \int_{\Gamma \bigcap B(x,a)} [-2\Phi(y_0, y)] \frac{\beta(y)}{\alpha(y)} u(y)\, d\sigma(y), \tag{10}$$

$$f_\alpha(y_0) = \int_{\Gamma \bigcap B(x,a)} [-2\Phi(y_0, y)] \frac{g(y)}{\alpha(y)}\, d\sigma(y). \tag{11}$$

Further we will assume the coefficients in the boundary condition (2) are piece-wise constant and choose the radius of the auxiliary ball, $B(x,a)$, in such a way that both α and β are constant inside it. With these assumptions, the Eq. (7) can be rewritten in the following form

$$\alpha u = \alpha K_0 u - \beta K_{1,1} u + f_1, \tag{12}$$

where $f_1 = K_{1,1} g$. By the Green's function properties (see, e.g. [17]), the kernel of the integral operator $K_{1,1}$ is positive. Denote by $A(y_0) = K_{1,1}[1](y_0)$ and $K_1 = K_{1,1}/(a/2)$. Then (12) is equivalent to

$$\begin{aligned} u(y_0) = & \frac{\alpha}{\alpha + \beta a/2} K_0 u(y_0) + \frac{\beta a/2}{\alpha + \beta a/2} K_1[g/\beta](y_0) \\ & + \frac{\beta a/2}{\alpha + \beta a/2} (I - K_1) u(y_0). \end{aligned} \tag{13}$$

Here I is the identity operator. Note that in its essence this representation has the same form as (4), however, the additional term that involves an integral of u over the domain boundary complicates the construction of an algorithm.

4 A Monte Carlo Estimate for the Solution Boundary Value

To construct a Monte Carlo estimate, we randomize (13) using the same considerations that were utilized in the case of the finite-difference approximation to the normal derivative.

Denote by K_2 the integral operator $p_a(I - K_1)$, and assume the Neumann series for it converges. Here $p_a = \dfrac{\beta a/2}{\alpha + \beta a/2}$. This means we have

$$u = \sum_{n=0}^{\infty} K_2^n F. \tag{14}$$

Here $F = [(1 - p_a)K_0 u + p_a K_1[g/\beta]]$. To have a possibility of using standard Monte Carlo estimates, we also require that the Neumann series for the operator $K_2^+ = p_a(I + K_1)$ converges. With this assumptions, an absorption conjugate estimate (or terminal estimator) [4,19].

$$\eta^* = Q_N^* \frac{F(z_N)}{q(z_N)} \tag{15}$$

can be utilized. Here, $\{z_n, n = 0, 1, ...\}$ is a Markov chain of points in $\Gamma \bigcap B(x,a)$, $z_0 = y_0$, z_N is the absorption point at which the simulation chain is stopped, $Q_0^* = 1$, $Q_n^* = Q_{n-1}^* k_2(z_{n-1}, z_n)/p_2(z_{n-1}, z_n)$, where $p_2(z_{n-1}, z_n)$ is a transition density of the chain, and $q(z_n)$ is the absorption probability.

With some natural assumptions [4,20], and for p_2 consistent with the kernel k_2 of the integral operator K_2, this estimate is unbiased, i.e.

$$\mathbb{E}\eta^* = u(y_0).$$

Here, as usual, we can utilize the consecutive randomization technique and use an unbiased estimator in place of the exact value of $F(z_N)$.

An alternative method of constructing a Monte Carlo estimate for $u(y_0)$ can also be used. With that approach, we consider (13) as the total probability formula and sequentially randomize the sum and the integrals. The comparative efficiency of two methods depends on particular features of a boundary value problem.

For a strictly convex inside $B(x,a)$ domain boundary, a Monte Carlo estimate construction is performed in the same way. The main difference is that the randomized computation of $K_0 u(y_0)$ leads to simulation of a point on ΓS, which includes a part of Γ. This results in an additional probability for the random walk coming back to the boundary.

In the case Γ is strictly concave inside $B(x,a)$, the kernel of K_0 is alternating, but the standard approach can be used [8]. The next point of the Markov chain can be chosen equiprobably from two intersections of a random direction with the boundary surface.

5 Algorithm Construction

Properties of K_0 and K_1 (including the convergence of the Neumann series for K_2 and K_2^+) heavily depend on the geometric properties of the domain boundary. To take the geometry into account, let, first, ΓS be convex. In this case the kernel of K_0 is positive and the norm of K_0 as an operator in L_∞ equals one. K_1 operates in the space of functions defined on the part of the domain boundary contained in a ball, $B(x, a)$.

First we construct an estimate, $\xi[F]$, for F. Clearly, this function has the form of the total probability formula. Hence, it is possible to randomize directly this sum and proceed as follows:

(a) with probability $1-p_a$, we set $\xi[F] = K_0 u(y_0)$, and by further randomization, $\xi[F] = u(y_1)$. Here, y_1 is simulated in accordance with the Poisson density (9) on the half-sphere $S(x, a)$ (or distributed uniformly in the solid angle for $y_0 = x$). The random walk trajectory is 'reflected' to $S(x, a)$ and the Markov chain simulation is continued inside the solution domain;

(b) with probability p_a, the random walk simulation is stopped, and we set $\xi[F] = K_1[g/\beta](y_0)$. By further randomization, $\xi[F] = k_1(y_0, y')/p_1(y_0, y')\ g(y')/\beta$. Here, p_1 is some consistent probability density of a random point $y' \in \Gamma \bigcap B(x, a)$ and $k_1(y_0, y') = -2\Phi(y_0, y')/(a/2)$ is the kernel of K_1.

The same procedure can be utilized when choosing a transition density of the Markov chain $\{z_n, n = 0, 1, ...\}$. For p_2 to be consistent with the kernel of the integral operator K_2, it must be a mix of the delta-function $\delta(z_{n-1} - z_n)$ (the distribution density of the identity operator) and p_1. Simulation in accordance with p_2 results in an additional factor in the weight Q_n^* computation. It is essential, however, that the kernel k_2 depends on the parameter a in such a way that by adjusting its value we can control both the convergence of the Neumann series and the estimate's variance. Note that in contrast to the step in the finite-difference approximation, a is not required to have the same order as the desired accuracy of a computed solution. With the described algorithm, the mean number of reflections equals the inverse of p_a, the termination probability, i.e. it is equal to $1 + 2\alpha/(\beta a)$, which in fact is $O(1)$. It depends on the ratio of coefficients in the boundary condition and on the geometric complexity of Γ.

6 Plane Boundary

To clarify the algorithm construction, suppose the domain boundary within the auxiliary ball is plane. In this case, $\Gamma \bigcap B(x, a) = C(x, a)$, the circle of radius a centered at x, $A(y_0) = a/2$ for $y_0 = x$ and monotonically decreases with the increasing distance, $|y_0 - x|$. As a density consistent with the kernel of the integral operator K_1, we can choose $p_1(z, z') = 1/(2\pi a|z - z'|)$. If $z = x$ then in the polar coordinate system centered at this point the chosen probabilistic density corresponds to the isotropic distribution of the polar angle and the uniform distribution of the radial coordinate, $r = |z - z'|$. Let $d = |z - x|$. For $d > 0$, the range of r depends on the angle value. Thus, the normalizing constant changes, and we can take $p_1(z, z') = 1/(4\ Elliptic(d^2/a^2)\ a|z - z'|)$, where $Elliptic(x^2)$ is the complete elliptic integral of the second kind. From here it follows that

$$\frac{k_1(z,z')}{p_1(z,z')} = 2\ \frac{4}{2\pi} Elliptic(d^2/a^2) \left[1 - \frac{ar}{d\overline{r}}\right] \equiv 2\ q_1(z,z').$$

Here $\overline{r} = |\overline{z} - z'|$ where $\overline{z}$ is the point conjugate to z.

The transition density of Markov chain is chosen as $k_2(z_{n-1}, z_n) = p_a\delta(z_{n-1} - z_n) + (1 - p_a)p_1(z_{n-1}, z_n)$. By definition $k_2(z_{n-1}, z_n) = p_a\delta(z_{n-1} - z_n) - p_a k_1(z_{n-1}, z_n)$ and hence

(a) with probability p_a, $z_n = z_{n-1}$ and $q_2(z_{n-1}, z_n) = 1$;
(b) with probability $1 - p_a$, the Markov chain either terminates with the conditional probability $1 - a\ q_1(z_{n-1}, z_n)$, or with the complementary conditional probability the next point, z_n, is simulated in accordance with $p_1(z_{n-1}, z_n)$. In this case, $q_2(z_{n-1}, z_n) = -\beta/\alpha$. Here $q_2(z_{n-1}, z_n) = k_2(z_{n-1}, z_n)/p_2(z_{n-1}, z_n)$ is the weight factor.

Following the absorption, a point y_1 inside the computational domain G is simulated in accordance with the Poisson density (9) (where y_0 is set to z_N) on the half-sphere $S(x, a)$. Further we simulate an exit point for the Brownian motion starting at this point. To perform this task, we can use, e.g., the random-walk-on-spheres algorithm. Other Green's function based Monte Carlo methods can also be used. Finally, the boundary conditions of the problem have to guarantee that the random walk simulation terminates with probability one.

7 Computational Experiments

To clarify the computational features of the proposed approach, we consider the model problem of solving (1) in the unit cube, $\{x \in \mathbb{R}^3 : 0 < x_i < 1, i = 1, 2, 3\}$. We consider the Robin conditions (2) with $\alpha = 1, \beta = 1$ to be given only for $x_3 = 0$, whereas on all other sides of the cube we take $\alpha = 0, \beta = 1$, i.e. the Dirichlet conditions are given. To calculate the solution value at the center of the face $x_3 = 0$, we took it as the initial point of the Markov chain. The random walk parameters were set to $a = \min\{0.1, \text{distance to side face}\}$, $\varepsilon = 10^{-3}$. For the test exact solution $u = 1 + x_1 + x_2 + x_3$ averaging over 10^6 simulated trajectories provided the result $u = 1.9991$ with the statistical error (two sigmas) 0.0013. The exact value equals 2.0. Note that to get to the same accuracy with the finite-difference approximation would require taking $h = 10^{-3}$ and $\varepsilon = 10^{-6}$, and hence approximately 100 times more computational time.

8 Conclusion

The proposed Monte Carlo algorithm for taking into account partially reflecting boundary conditions provides a possibility to construct an unbiased estimator for the solution boundary value. Using this estimate leads to more efficient random walk methods for solving boundary value problems for the Laplace (and for other elliptic) equations.

Acknowledgements. This work was supported by the Bulgarian Science Fund Grant DFNI-I02/8.

References

1. Müller, M.E.: Some continuous Monte Carlo methods for the Dirichlet problem. Ann. Math. Stat. **27**(3), 569–589 (1956)
2. Elepov, B.S., Kronberg, A.A., Mikhailov, G.A., Sabelfeld, K.K.: Solution of Boundary Value Problems by the Monte Carlo Method. Nauka, Novosibirsk (1980). (in Russian)
3. Sabelfeld, K.K.: Monte Carlo Methods in Boundary Value Problems. Springer, Berlin/Heidelberg/New York (1991)
4. Ermakov, S.M., Nekrutkin, V.V., Sipin, A.S.: Random Processes for Classical Equations of Mathematical Physics. Kluwer Academic Publishers, Dodrecht (1989)
5. Kac, M.: On some connection between probability theory and differential and integral equations. In: Proceedings of the Second Berkeley Symposium on Mathematical Statistics and Probability, University of California Press, pp. 189–215 (1951)
6. Freidlin, M.: Functional Integration and Partial Differential Equations. Princeton University Press, Princeton (1985)
7. Hahn, D.W., Ozisik, M.N.: Heat Conduction, 3rd edn. Wiley, Hoboken (2012)
8. Sabelfeld, K.K., Simonov, N.A.: Random Walks on Boundary for Solving PDEs. VSP, Utrecht (1994)
9. Sabelfeld, K.K., Simonov, N.A.: Stochastic Methods for Boundary Value Problems: Numerics for High-Dimensional PDEs and Applications. De Gruyter, Berlin/Boston (2016)
10. Haji-Sheikh, A., Sparrow, E.M.: The floating random walk and its application to Monte Carlo solutions of heat equations. SIAM J. Appl. Math. **14**(2), 570–589 (1966)
11. Makarov, R.N.: Monte Carlo methods for solving boundary value problems of second and third kinds. Russ. J. Numer. Anal. Math. Model. **13**(2), 117–132 (1998)
12. Sipin, A.S.: On stochastic algorithms for solving boundary value problems for the Laplace operator. Trans. POMI Sci. Semin. **442**, 133–142 (2015). (in Russian)
13. Simonov, N.A.: Monte Carlo methods for solving elliptic equations with boundary conditions containing the normal derivative. Doklady Math. **74**, 656–659 (2006)
14. Simonov, N.A.: Random walk on spheres algorithms for solving mixed and Neumann boundary value problems. Siberian J. Numer. Math. **10**(2), 209–220 (2007). (in Russian)
15. Simonov, N.A.: Walk-on-spheres algorithm for solving third boundary value problem. Appl. Math. Lett. **64**, 156–161 (2017)
16. Miranda, C.: Partial Differential Equations of Elliptic Type. Springer, Berlin/Heidelberg/New York (1970). https://doi.org/10.1007/978-3-642-87773-5
17. Helms, L.L.: Introduction to Potential Theory. Wiley-Interscience, New York (1969)
18. Günter, N.M.: La theorie du potentiel et ses applications aux problemes fondamentaux de la physique mathematique. Gauthier-Villars, Paris (1934)

19. Spanier, J., Gelbard, E.M.: Monte Carlo Principles and Neutron Transport Problems. Addison-Wesley Publishing Company, Boston (1969)
20. Ermakov, S.M., Mikhailov, G.A.: Statistical Simulation. Nauka, Moscow (1982). (in Russian)

Toward Exascale Computation

Renormalization Based MLMC Method for Scalar Elliptic SPDE

Oleg Iliev[1,2(✉)], Jan Mohring[1], and Nikolay Shegunov[1]

[1] Fraunhofer ITWM, Kaiserslautern, Germany
{iliev,mohring,shegunov}@itwm.fraunhofer.de
[2] Institute of Mathematics and Informatics, Bulgarian Academy of Sciences, Sofia, Bulgaria

Abstract. Previously the authors have presented MLMC algorithms exploiting Multiscale Finite Elements and Reduced Bases as a basis for the coarser levels in the MLMC algorithm. In this paper a Renormalization based Multilevel Monte Carlo algorithm is discussed. The advantage of the renormalization as a basis for the coarse levels in MLMC is that it allows in a cheap way to create a reduced dimensional space with a variation which is very close to the variation at the finest level. This leads to especially efficient MLMC algorithms. Parallelization of the proposed algorithm is also considered and results from numerical experiments are presented.

1 Introduction

Stochastic PDEs have attracted a great attention due to their importance in modeling a variety of environmental and industrial processes, and thanks to the increased computational power solution of such problems can be facilitated. In this paper we consider a scalar elliptic SPDE, although the approach here is not limited to this problem. To name just a few applications, one can consider saturated flow in subsurface, or heat conduction in Metal Matrix Composites, or in other composite materials. In this paper we target at computing the mean flux through saturated porous media with prescribed pressure drop and known distribution of the random coefficients.

Multilevel Monte Carlo is one of the powerful methods for solving SPDE. As it is known, the standard Monte Carlo method converges very slowly (i.e., a large number of deterministic PDE problems have to be solved for different realizations of the permeability field). The idea of the MLMC is to combine in a proper way fewer expensive computations with a plenty of cheap computations, so that the targeted expected value is computed at significantly lower costs compared to the standard Monte Carlo algorithm. One of the key components of MLMC is the selection of the coarser levels. In [4] authors use the number of the terms in the Kahrunen-Loewe expansion in order to define coarser levels. In [8] the authors consider coarser grids approximations of the fine grid problem build with AMG. In [2], similarly to our approach, to define the coarser levels in MLMC, simple

I. Lirkov and S. Margenov (Eds.): LSSC 2017, LNCS 10665, pp. 295–303, 2018.
https://doi.org/10.1007/978-3-319-73441-5_31

arithmetic averaging for the coefficients is used. In our previous papers we had considered different approaches. The number of the basis functions in a Reduced Basis mixed Multiscale Finite Element Method algorithm (MsFEM) is used to build the coarser levels in [6]. MsFEM is used to build the coarser level of MLMC in [10].

Here we construct the coarser levels in MLMC by renormalization. This technique has been widely used in the past (and is still intensively used by many groups) for upscaling hydraulic conductivity in heterogeneous media. For details we refer to two review papers and the references there [11,12]. Note that the effective hydraulic conductivity obtained as a result of the renormalization can be used to calculate an effective flux, and in this sense the renormalization in our case has a similar target as the calculation of the mean flux formulated above. In [9] it is shown that the standard renormalization is not very good for upscaling, but in this paper we show that it can be efficiently used for significantly reducing the computational costs of the MLMC algorithms.

Due to the high complexity of the considered problem, HPC systems are usually necessary for solving stochastic PDEs. With the rapidly advancing area of computer science, methods based on Monte Carlo sampling are of great interest. Such methods are suitable for parallelization and can give reasonable predictions at reasonable computational costs, such as multilevel Monte Carlo method, which overcomes the slow convergence in the classical Monte Carlo.

2 Model Problem and Discretization

We consider a simple model problem in a unit cube domain, steady state single phase flow in random porous media. This problem illustrates well the challenges in solving stochastic PDE, the advantages of the MLMC algorithm compared to the standard Monte Carlo algorithm, and the advantage of using renormalization for building the coarse levels in MLMC.

$$\begin{aligned} -\nabla \cdot [k(x,\omega)\,\nabla p(x,\omega)] = 0 \quad & \text{for } x \in D = (0,1)^d\,, \quad \omega \in \Omega \\ p\big|_{x_1=0} = 1\,, \quad p\big|_{x_1=1} = 0\,, \quad & \tfrac{\partial p}{\partial n} = 0 \text{ on other boundaries}\,, \end{aligned} \tag{1}$$

with dimension $d \in \{2,3\}$, pressure p, scalar permeability k, and random vector ω. Let $K(x,\omega) = \log k(x,\omega)$ be the logarithm of the permeability. We assume that its expected value and spacial covariance are shift invariant and satisfy

$$E\left[K(x,\cdot)\right] = 0, \quad E\left[K(x,\cdot)\,K(y,\cdot)\right] = C(x-y) = C(y-x) \text{ for } x,y \in D. \tag{2}$$

The quantity of interest is the mean (expected) value of the total flux through the unit square, where the flux is given by (2D case is considered here):

$$Q(x,\omega) := \int_{x_1=0} k(x,\omega)\partial_n p(x,\omega)dx_2. \tag{3}$$

Here, we consider a practically relevant covariance of type (2) proposed in [3]:

$$C(h) = \sigma^2 \exp\left(- \|h\|_2 / \lambda\right), h \in [-1,1]^d \quad (4)$$

with variance $1 \leq \sigma^2 \leq 4$ and correlation length $0.05 \leq \lambda \leq 0.3$. The selected bounds include SPDE problems with larger range of computational effort, additionally they allow comparison with the results of other authors solving the same problem.

Generation of random permeability fields is an essential topic in solving SPDEs. Several ways of generating random permeability fields applicable to flow simulations are known in the literature. Here we use an algorithm based on forward and inverse Fourier transformation over a circulant covariance matrix, more details can be found in our previous paper [10].

A finite volume method on a cell centered grid is used as a discretization for the elliptic PDEs corresponding to each realization of the permeability field. This discretization naturally leads to harmonic averaging of the discontinuous coefficients, and by our opinion this is important not only for the accuracy on a single grid, but also for the performance of the MLMC algorithm presented here.

Let us subdivide $D = [0,1]^2$ uniformly into $m \times m$ square cells. Let $k_{i,j}$ denote the value of k at $x_{i,j}$ for a fixed realization of the permeability field, and let $p_{i,j}$ denote the approximation to p at $x_{i,j}$.

To approximate k on the edge, we use harmonic average $\overline{k}_{i+1/2,j}$ of $k_{i,j}$ and $k_{i+1,j}$.

The standard five point stencil discretization of (1), for each realization of the permeability field, can be written as:

$$\begin{aligned}
&- \overline{k}_{i-1/2,j} p_{i-1,j} - \overline{k}_{i,j-1/2} p_{i,j-1} + (\overline{k}_{i-1/2,j} + \overline{k}_{i,j-1/2} + \overline{k}_{i+1/2,j} + \overline{k}_{i,j+1/2}) p_{i,j} \\
&- \overline{k}_{i+1/2,j} p_{i+1,j} - \overline{k}_{i,j+1/2} p_{i,j+1} = 0, \quad i = 1, 2, ...m; \ j = 1, 2, ..., m
\end{aligned}$$

Neumann boundary condition, i.e. prescribed flux $-k \nabla p \cdot n$, is straightforward to approximate by simply substituting the corresponding boundary term by 0. For the Dirichlet boundaries we use midpoint rule and one sided difference for ∇p and take the permeability coefficient in the boundary cell.

3 Standard and Multilevel Monte Carlo Approaches

Standard Monte Carlo for solving the SPDE of interest

Due to the limited space, we will not discuss here the discretization error for the considered SPDE. For its balancing with the stochastic error, we refer, e.g., to [4,6] and to corresponding references therein for discussions on this issue.

Let $\omega_M : \Omega \longmapsto R^M$ be a random vector over some probability space (Ω, F, P) and consider quantity of interest Q_M, defined by some functional, depending on ω_M. Assume also that $E[Q_M]$ can be made arbitrary close to $E[Q]$ by choosing M sufficiently large. In solving Eq. (1), M is related to the number of grid cells (fine grid means large M). The quantity of interest in our case is an approximation to the mean flux $E[Q_M]$, and more precisely, an approximation obtained via Monte Carlo method

$$E[Q_{M,N}] = \frac{1}{N}\sum_{i}^{N} Q_M^i,$$

where N is the number of the samples and the flux is given by Eq. (3). It is known that for $N \to \infty$ one has $E[Q_{M,N}] \to E[Q_M]$, a.s. The rate of convergence of the error of the Monte Carlo approximation,

$$e_N^{MC} = |E[Q_{M,N}] - E[Q_M]|,$$

is proportional to the variance

$$V_N^{MC} = E\left[Q_{M,N} - E[Q_{M,N}]\right]^2$$

and inversely proportional to N. Thus to obtain accuracy ϵ^{MC}, we need to take sufficiently large N, such that

$$\epsilon^{MC} < N^{-\frac{1}{2}}\sqrt{V_N^{MC}}. \tag{5}$$

Two conclusions can be drawn from this. First, that the standard Monte Carlo method converges rather slow and a large number of samples is needed to achieve the desired convergence. Second, for the problems with large variation a practical way to improve the convergence rate would be to find a cheap way to reduce the variance, i.e., to apply variance reduction method. In general, Multilevel Monte Carlo method, can be classified as a variance reduction method.

Multilevel Monte Carlo

For larger M, i.e. solving Eq. (1) on fine grid, the computational costs are high for each sample (each realization of the permeability field), and the necessity for solving the PDE for many realizations can make the costs prohibitive. However, a significant reduction of the computational costs can be achieved by designing and using a multilevel variant of the Monte Carlo method. The idea was first suggested by Heinrich in 1999 in connection with computation of multi-dimensional integrals, for a recent overview on MLMC for different applications we refer to [7]. We shortly describe the MLMC idea.

Let $\{M_l : l = 0 \dots L\} \subset N$ be sequence of increasing numbers, with $M_L = M$. We will say that each l defines a level, and in our particular case the levels are linked to a fine and coarser grids. At each level we can calculate quantities $\{Q_{M_l}\}_{l=0}^L$. Defining $Y_l = Q_{M_l} - Q_{M_{l-1}}$ and setting $Y_0 = Q_{M_0}$, we can write the following telescopic sum for $E[Q_M]$

$$E[Q_M] = E[Q_{M_0}] + \sum_{l=1}^{L} E[Q_{M_l} - Q_{M_{l-1}}] = \sum_{l=0}^{L} E[Y_l] \tag{6}$$

The expectation on the finest level (in our case, for $M_L = M$) is equal to the expectation on the coarsest level plus sum of expectations of corrections. The terms in Eq. (6) can now be approximated using standard Monte Carlo method, taking N_l samples at each level. For each of the terms, the convergence of the error will be estimated as

$$\epsilon_l^{Y_l} < N_l^{-\frac{1}{2}} \sqrt{V_l[Y_{N_l}]}, l = 0, ..., L. \tag{7}$$

To have an efficient MLMC algorithm, one has to define the levels in such a way that the variance $V_l[Y_{N_l}]$ is small for larger l (i.e., on finer grids), while the variance is large for $l = 0$, i.e., on the coarsest grid. The optimal way would be to have $V[Y_0] \approx V[Q_M]$, and very small variances $V_l[Y_{N_l}]$ for $l = 1, ..., L$.

The goal of this paper is to show that a renormalization based MLMC is satisfying the above requirements.

An essential part of any MLMC algorithm is the procedure for determining the required number N_l at each level, so that the errors are balanced. We have already discussed possible approaches for determining optimal set of N_l in [10], see also [4,6–8]. Here we will not recall the details, only the final results will be listed. Denote $v_l = V[Y_l]$, and let t_l be the average time for computing once the pair of problems Y_l. Let $T = \sum_{l=0}^{L} N_l t_l$ be the total time for the computation. Minimizing T under the above constraint and turning it to integer value gives us:

$$N_l = ceil[\alpha\sqrt{(v_l/t_l)}] \text{ with Lagrangian multiplier } \alpha = \frac{1}{\epsilon^2} \sum_{l=0}^{L} \sqrt{(v_l/t_l)} \tag{8}$$

Recall, that initially a certain number of samples have to be solved to evaluate the variances. These are used in estimating required numbers of samples at each level. In fact, this procedure can be repeated several times to improve the estimate.

Recall that our definition of levels is based on the spatial discretization of sequence of coarser grids, combined with a renormalization procedure.

3.1 Renormalization

For details on the renormalization approach we refer to two review papers [11,12]. Here we have implemented simplified renormalization procedure listed in [9].

Consider $D = [0,1]^2$ and divided it to $M \times M$ square cells, corresponding to level L of our MLMC algorithm. The generation of one realization of the random permeability field means that a permeability value $K_L(x)$ is associated with each cell. Consider now a twice coarser grid, corresponding to level $L-1$. The permeability field at the coarser grid is calculated in renormalization manner by applying recursive composition of harmonic, arithmetic and geometric means. If two neighboring cells are in series with respect to the flow direction, then the equivalent permeability is estimated as a harmonic mean, if the two cells are in parallel with respect to the flow direction then equivalent permeability is calculated as arithmetic mean of the two, as illustrated on Fig. 1. So for each four by four cells from a 2D domain with $M \times M$ grid, we can get one equivalent permeability coefficient on twice coarser grid by taking geometric mean $K_{1234} = \sqrt{K_{1234}^{ah} K_{1234}^{ha}}$. The procedure is repeated recursively for more than two levels.

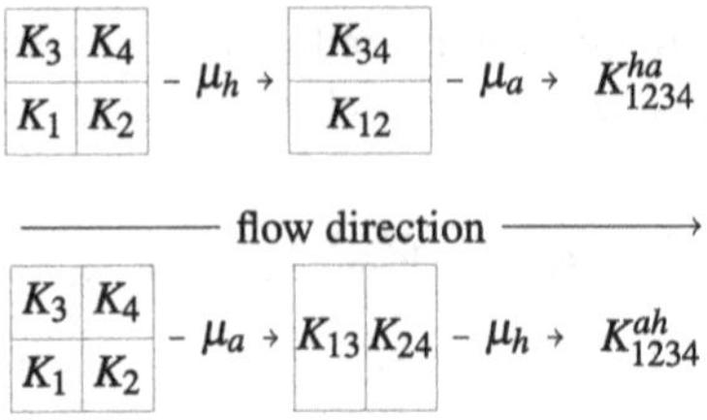

K_3	K_4		
K_1	K_2		

$\begin{bmatrix} K_3 & K_4 \\ K_1 & K_2 \end{bmatrix}$ - μ_h → $\begin{bmatrix} K_{34} \\ K_{12} \end{bmatrix}$ - μ_a → K^{ha}_{1234}

flow direction →

$\begin{bmatrix} K_3 & K_4 \\ K_1 & K_2 \end{bmatrix}$ - μ_a → $\begin{bmatrix} K_{13} & K_{24} \end{bmatrix}$ - μ_h → K^{ah}_{1234}

Fig. 1. Simplified renormalization, where μ_h denotes harmonic mean, and μ_a arithmetic mean

4 Numerical Results

Our numerical tests have been conducted on the Cluster at Fraunhofer ITWM, on 32 processors, without usage of accelerators. Parallelization strategy and obtained results will be explained elsewhere. The fine grid has $2^{10} \times 2^{10}$ cells. Conjugate Gradient iterative method with tolerance $1e{-}7$, preconditioned with AMG, is used to solve the PDEs for each realization (sample) of the random permeability field. We examine the performance of the standard Monte Carlo algorithm and of the multilevel Monte Carlo with two and three levels. The tolerance for MC and MLMC is $\epsilon = 5e{-}3$. The time for generating the permeability fields is negligible compared to the time for solving the PDE.

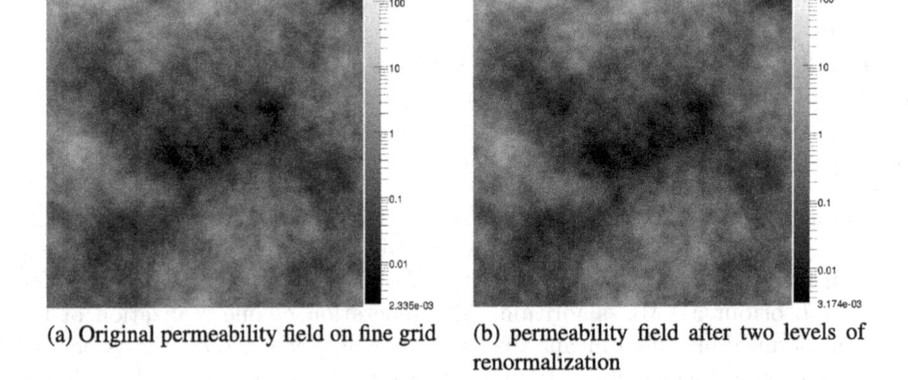

(a) Original permeability field on fine grid (b) permeability field after two levels of renormalization

Fig. 2. A realization of permeability field with $\sigma = 2$ and correlation length $\lambda = 0.3$

On Fig. 2, one can observe the effect of the renormalization on the permeability field. As any averaging technique, it has smoothing effect. However, the smoothing here is less pronounced, compared to, e.g., arithmetic averaging, and visually the variance of the renormalized field is not far from the variance of the original field. This statement can be quantitatively confirmed by the data

Table 1. Simulation with permeability generating parameters $\sigma = 2$, $\lambda = 0.2$, and with Monte Carlo method tolerance $\epsilon = 0.005$

	Mean flux	$V[Y_l]$	Grid size and Samples N_l
MC	1.3969	$V[Y_0]$: 1.9132	$2^{10} \times 2^{10}$ 76529
Two level MLMC	1.3964	$V[Y_0]$: 1.9066	$2^{9} \times 2^{9}$ 75310
		$V[Y_1]$: $1.284e-07$	$2^{10} \times 2^{10}$ 10
Three level MLMC	1.3960	$V[Y_0]$: 1.90186	$2^{8} \times 2^{8}$ 76416
		$V[Y_1]$: $4.284e-06$	$2^{9} \times 2^{9}$ 57
		$V[Y_2]$: $1.284e-07$	$2^{10} \times 2^{10}$ 5

presented in Table 1. Indeed, the variances presented in the third column confirm both: (i) after renormalization the variance at the coarsest level is close to the variance on the original fine grid, and (ii) the variances for the corrections in MLMC are decaying very fast. The results presented in the table are for $\sigma = 2$, $\lambda = 0.2$. On Fig. 3 we observe the fast decay of the empirical variance over different levels of three level MLMC for $\sigma = 3$, $\lambda = 0.3$.

Let us elaborate a little bit more on the results presented in Table 1. The second column shows that the mean flux computed with MC and MLMC satisfies

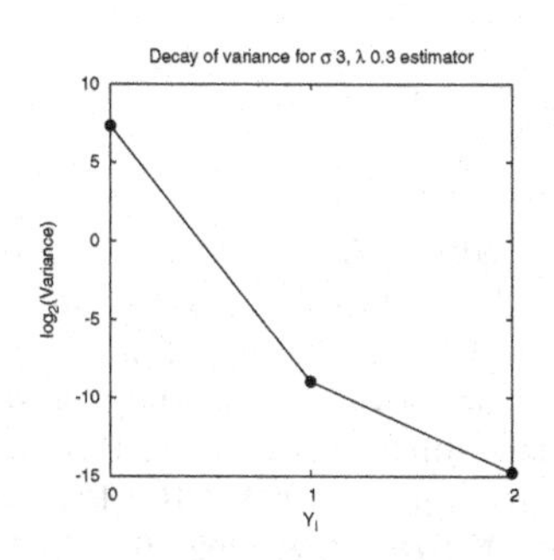

Fig. 3. Decay of variance

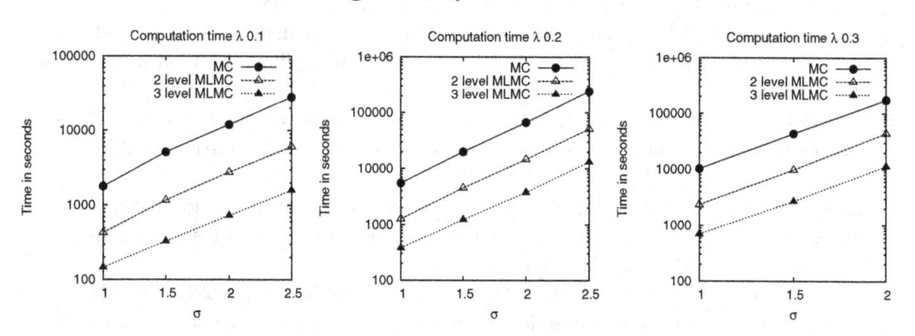

Fig. 4. Speedup of MLMC with respect to MC

the prescribed accuracy tolerance $\epsilon = 5e{-3}$. The fourth column shows that while in MC we solve for tens of thousands of realizations on the finest grid, in the three level MLMC almost the same number of realizations are needed on 16 times coarser grid, while only five realizations are needed on the finest grid.

Comparison of the computational times for MC and MLMC is presented on Fig. 4. One case see that a significant speedup is achieved, e.g., up to 17 times for the three level case.

5 Conclusions

The described MLMC method gives substantial speedup, compared to MC method. The usage of the renormalization provides a cheap way to build coarse levels in the MLMC. The variance at the coarser levels is very close to the variance at the fine level, what makes the presented particular MLMC method a very efficient variance reduction method. A more detailed comparison of the performance of the renormalization based MLMC in comparison with other MLMC approaches will be subject of forthcoming paper. Since the samples are independent, the method is very suitable for parallel computation on HPC systems. The parallel performance of the algorithm will be discussed in a forthcoming paper.

Acknowledgments. This research was funded by the DFG SPP 1648 'Software for Exascale Computing'.

References

1. Bastian, P., Blatt, M., Dedner, A., Engwer, C., Klöfkorn, R., Ohlberger, M., Sander, O.: A generic grid interface for parallel and adaptive scientific computing Part I abstract framework. Computing **82**(2–3), 103–119 (2008)
2. Blaheta, R., Béreš, M., Domesová, S.: A study of stochastic FEM method for porous media flow problem. In: Bris, R., Dao, P. (eds.) Applied Mathematics in Engineering and Reliability, pp. 281–289. CRC Press (2016). https://doi.org/10.1201/b21348-47, Print ISBN 978-1-138-02928-6, eBook ISBN 978-1-315-64165-2
3. Hoeksema, R.J., Kitanidis, P.K.: Analysis of the spatial structure of properties of selected aquifers. Water Resour. Res. **21**(4), 563–572 (1985)
4. Cliffe, K., Giles, M., Scheichl, R., Teckentrup, A.L.: Multilevel Monte Carlo methods and applications to elliptic PDEs with random coefficients. Comput. Vis. Sci. **14**(1), 3–15 (2011)
5. Dietrich, C., Newsam, G.N.: Fast and exact simulation of stationary Gaussian processes through circulant embedding of the covariance matrix. SIAM J. Sci. Comput. **18**(4), 1088–1107 (1997)
6. Efendiev, Y., Iliev, O., Kronsbein, C.: Multilevel Monte Carlo methods using ensemble level mixed MsFEM for two-phase flow and transport simulations. Comput. Geosci. **17**(5), 833–850 (2013)
7. Giles, M.B.: Multilevel Monte Carlo methods. Acta Numerica **24**, 259–328 (2015)
8. Kalchev, D., Ketelsen, C., Vassilevski, P.S.: Two-level adaptive algebraic multigrid for a sequence of problems with slowly varying random coefficients. SIAM J. Sci. Comput. **35**(6), B1215–B1234 (2013)

9. Lunati, I., Bernard, D., Giudici, M., Parravicini, G., Ponzini, G.: A numerical comparison between two upscaling techniques: non-local inverse based scaling and simplified renormalization. Adv. Water Resour. **24**(8), 913–929 (2001)
10. Mohring, J., Milk, R., Ngo, A., Klein, O., Iliev, O., Ohlberger, M., Bastian, P.: Uncertainty quantification for porous media flow using multilevel Monte Carlo. In: Lirkov, I., Margenov, S.D., Waśniewski, J. (eds.) LSSC 2015. LNCS, vol. 9374, pp. 145–152. Springer, Cham (2015). https://doi.org/10.1007/978-3-319-26520-9_15
11. Renard, P., De Marsily, G.: Calculating equivalent permeability: a review. Adv. Water Resour. **20**(5), 253–278 (1997)
12. Wen, X.H., Gómez-Hernández, J.J.: Upscaling hydraulic conductivities in heterogeneous media: an overview. J. Hydrol. **183**(1–2), ix–xxxii (1996)

Performance Analysis of MG Preconditioning on Intel Xeon Phi: Towards Scalability for Extreme Scale Problems with Fractional Laplacians

Nikola Kosturski, Svetozar Margenov(✉), and Yavor Vutov

Institute of Information and Communication Technologies, Bulgarian Academy of Sciences, Sofia, Bulgaria
{kosturski,margenov,vutov}@parallel.bas.bg

Abstract. The Intel Xeon Phi architecture is currently a popular choice for supercomputers, with many entries of the Top 500 list, using it either as main processors or as accelerators/coprocessors. In this paper, we explore the performance and scalability of the Intel Xeon Phi chips in the context of large sparse linear systems, commonly arising from the discretization of PDEs. At the first step, the PCG [1] is applied as a basic iterative solution method in the case of sparse SPD problems. The parallel multigrid (MG) implementation from Trilinos ML package is utilized as a preconditioner. A matrix free algebraic multilevel solver is used to reduce the memory requirements, thus allowing the cores to be more efficiently utilized. The second part of the paper is devoted to the fractional Laplacian, that is, we consider the equation $-\Delta^\alpha \mathbf{u} = \mathbf{f}, 0 < \alpha < 1$, $\Omega \subset \mathbb{R}^d$. The related elliptic boundary value problem describes anomalous diffusion phenomena also referred to as super-diffusion. The implemented method approximates the solution of the nonlocal problem by a series of local elliptic problems. The currently available numerical methods for fractional diffusion Laplacian have computational complexity, comparable e.g., to the complexity of solving local elliptic problem in $\tilde{\Omega} \subset \mathbb{R}^{d+1}$. The presented parallel results are for $\Omega = (0,1)^3$, including meshes of very large scale. The numerical experiments are run on the Avitohol computer at the Institute of Information and Communication Technologies, IICT-BAS. The presented results show very good scalability when the CPU-cores and MIC work together for a certain number of compute nodes.

1 Introduction

The paper is aimed at development of highly parallel algorithms for fractional diffusion problems with computational complexity of extreme scale. For this purpose, we investigate the parallel efficiency on the hybrid architecture of the supercomputer Avitohol (http://www.iict.bas.bg/avitohol/). The supercomputer consists of 150 compute nodes, each equipped with two 8 core (up to 16 threads)

I. Lirkov and S. Margenov (Eds.): LSSC 2017, LNCS 10665, pp. 304–312, 2018.
https://doi.org/10.1007/978-3-319-73441-5_32

Intel Xeon E5-2650 processors and two 61 core (up to 244 threads) Intel Xeon Phi 7120P coprocessors. Each node has 32 GB of RAM and the accelerators have 16 GB. The nodes are connected via InfiniBand FDR.

Two model problems leading to large linear systems with sparse symmetric positive definite (SPD) matrices are considered. The first one is the Laplace's equation $-\Delta u = f$, in the unit cube $\Omega = [0,1]^3$ with a seven point stencil and homogeneous Dirichlet boundary conditions. The implementation of the developed parallel solution method is based on the `ml_MatrixFree` example distributed along with the Trilinos libraries. The second part of the study is devoted to the case of elliptic problems with fractional Laplacians. The numerical solution of such nonlocal problems is very expensive. A straightforward approach to the related discrete problems lead to linear systems with dense matrices. Three techniques to avoid this difficulty were recently proposed. They are based on transformation of the problem

$$\mathcal{L}^\alpha u = f$$

to a local elliptic [6] or pseudo-parabolic [7,8] problem, or on a proper integral representation of the solution [2]. For all of them, the computational complexity is comparable to the complexity of solving local problems in $\tilde{\Omega} \subset R^{d+1}$. A comparative analysis of parallel properties of the related three algorithms for distributed memory computer architecture is presented in [3]. Some substantial advantages of the algorithm from [2] are observed there, which is the motivation to implement it in the present study.

More recently, an alternative approach aimed at reducing the computational complexity was proposed in [5], see also [4]. The linear algebraic system $\mathcal{A}^\alpha \mathbf{u} = \mathbf{f}$, $0 < \alpha < 1$ is considered, where $\mathcal{A}$ is a properly scaled sparse SPD matrix. The method is based on best uniform rational approximations (BURA) of the function $t^{\beta-\alpha}$ for $0 < t \leq 1$ and small natural β (e.g. $\beta = 1, 2$). It is important, that the algorithmic implementation of this method is practically identical to the method based on the integral representation of the solution.

The rest of the paper is organized as follows. The fractional diffusion elliptic problem and the method from [2] are presented in the next section. The developed parallel implementation approach is described in Sect. 3. The next Sect. 4 contains numerical tests of the parallel solvers for very large-scale problems. Short conclusions are given at the end.

2 Fractional Laplacian

Let us consider the elliptic boundary value problem: find $u \in V$ such that

$$a(u,v) := \int_\Omega (\mathbf{a}(x)\nabla u(x) \cdot \nabla v(x) + q(x))\, dx = \int_\Omega f(x)v(x)dx, \quad \forall v \in V, \quad (1)$$

where $V := \{v \in H^1(\Omega) : \ v(x) = 0 \ \text{on} \ \Gamma_D\}$, $\Gamma = \partial\Omega$, and $\Gamma = \bar{\Gamma}_D \cup \bar{\Gamma}_N$. The bilinear form $a(\cdot,\cdot)$ defines a linear operator $\mathcal{L} : V \to V^*$ with V^* being the dual of V. Namely, for all $u, v \in V$ $a(u,v) := \langle \mathcal{L}u, v\rangle$, where $\langle \cdot,\cdot \rangle$ is the pairing between V and V^*. One possible way to introduce $\mathcal{L}^\alpha$, $0 < \alpha < 1$, is through its spectral decomposition, i.e.

$$\mathcal{L}^\alpha u(x) = \sum_{i=1}^{\infty} \lambda_i^\alpha c_i \psi_i(x), \quad \text{where} \quad u(x) = \sum_{i=1}^{\infty} c_i \psi_i(x).$$

Here c_i are the Fourier coefficients of u in the L_2-orthogonal basis, $\{\psi_i(x)\}_{i=1}^\infty$ are the eigenfunctions of $\mathcal{L}$, orthonormal in L_2-inner product and $\{\lambda_i\}_{i=1}^\infty$ are the corresponding positive real eigenvalues.

As already noted, the numerical solution of the nonlocal problem $\mathcal{L}^\alpha u = f$ is computationally rather expensive. The following representation of the solution u is used in [2] in order to overcome the problem of non-locality:

$$\mathcal{L}^{-\alpha} = \frac{2\sin(\pi\alpha)}{\pi} \int_0^\infty t^{2\alpha-1} \left(\mathcal{I} + t^2\mathcal{L}\right)^{-1} dt.$$

Among others, an exponentially convergent quadrature scheme is introduced in [2]. Then, the approximation of u only involves evaluations of $(\mathcal{I} + t_i\mathcal{A})^{-1} f$, where $t_i \in (0, \infty)$ is related to the current quadrature node, and where $\mathcal{I}$ and $\mathcal{A}$ stand for the identity and the stiffness matrix corresponding to a certain approximation of the (local) diffusion Eq. (1). The computational complexity depends on the number of quadrature nodes. More precisely, the following quadrature formula is implemented in our parallel code:

$$L^{-\alpha} \approx \frac{2k\sin(\pi\alpha)}{\pi} \sum_{\ell=-m}^{M} e^{2\alpha y_\ell} \left(I + e^{2y_\ell} L\right)^{-1}, \tag{2}$$

where

$$k > 0, \qquad m = \left\lceil \frac{\pi^2}{4\alpha k^2} \right\rceil, \qquad M = \left\lceil \frac{\pi^2}{4(1-\alpha)k^2} \right\rceil, \qquad y_\ell = \ell k.$$

Let us assume that the utilized parallel AMG solver of the systems $(\mathcal{I} + t_i\mathcal{A})$ $u = f$ has optimal complexity of $O(N)$, where N is the number of unknowns. Then, the computational complexity of the fractional diffusion solver is $O((m + M)N)$.

3 Parallel Implementation Approach

The developed code uses MPI for parallelization and is run on the Avitohol supercomputer at IICT-BAS introduced above. In MPI terms, the communicators encapsulate communication context and represent groups of processes that are able to communicate. All processors within a communicator have an unique number (rank). This number is used as an address for communications. All processes within a communicator participate in collective operations. There is a predefined communicator `MPI_COMM_WORLD`, which consists of all started processes. We spawn one extra process, and use it as a master which distributes the tasks.

Smaller communicators for the Laplace subproblems are created according to several parameters: *pph* – processors per host, *hpp* – hosts per (sub)problem, *ppa* – processors per accelerator and *app* – accelerators per (sub)problem. We create smaller communicators for our subtasks using `MPI_Comm_split` function. It has a color input argument and a source communicator. New communicators are created after the call, containing processors from the source one, with the same color.

The first step we do is to create the communicator `ALL_COMM` which does not contain the master processor (see Fig. 1). Then using color obtained by hashing the host name, obtained from `MPI_Get_processor_name`, the communicators `HOST_COMM` are created (see Fig. 2). The ranks from `HOST_COMM` are used as colors to obtain `RANK_COMM`, (see Fig. 3). As a final step from the ranks of `RANK_COMM` and `HOST_COMM` along with the parameters *pph*, *hpp*, *ppa*, *app* communicators `COMP_COMM` are created (see Fig. 4). More precisely, the ranks of `HOST_COMM` are used to limit the number of active processes (*pph* and *ppa*) and the ones from `RANK_COMM` are used to group multiple hosts/accelerators into one communicator when required (by the *hpp* and *app* parameters). The examples in Figs. 1, 2, 3 and 4 are for distribution with following parameters: $pph = 8, hpp = 1, ppa = 16, app = 2$.

Since memory access speed is the bottleneck of sparse matrix operations for large matrices, we have applied a matrix-free multigrid algorithm. We use the

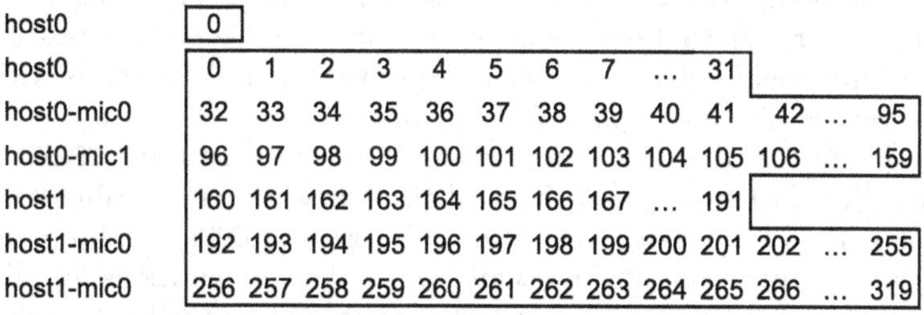

host0	0												
host0	0	1	2	3	4	5	6	7	...	31			
host0-mic0	32	33	34	35	36	37	38	39	40	41	42	...	95
host0-mic1	96	97	98	99	100	101	102	103	104	105	106	...	159
host1	160	161	162	163	164	165	166	167	...	191			
host1-mic0	192	193	194	195	196	197	198	199	200	201	202	...	255
host1-mic0	256	257	258	259	260	261	262	263	264	265	266	...	319

Fig. 1. Example of `ALL_COMM` communicator, where 32 processes are spawned on each host and 64 on each accelerator. An additional master process is started on `host0` to distribute the problems among groups of worker processes.

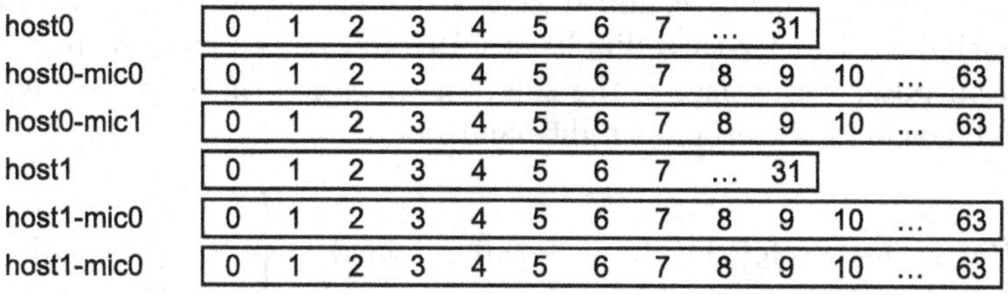

host0	0	1	2	3	4	5	6	7	...	31			
host0-mic0	0	1	2	3	4	5	6	7	8	9	10	...	63
host0-mic1	0	1	2	3	4	5	6	7	8	9	10	...	63
host1	0	1	2	3	4	5	6	7	...	31			
host1-mic0	0	1	2	3	4	5	6	7	8	9	10	...	63
host1-mic0	0	1	2	3	4	5	6	7	8	9	10	...	63

Fig. 2. Example of `HOST_COMM` communicators. These communicators represent all the processes on a particular host. It is used to construct `RANK_COMM` (see Fig. 3) and limit the number of active processes.

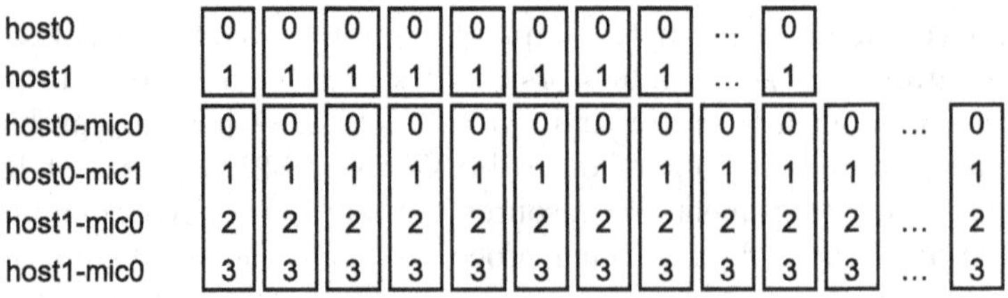

host0	0	0	0	0	0	0	0	0	...	0			
host1	1	1	1	1	1	1	1	1	...	1			
host0-mic0	0	0	0	0	0	0	0	0	0	0	0	...	0
host0-mic1	1	1	1	1	1	1	1	1	1	1	1		1
host1-mic0	2	2	2	2	2	2	2	2	2	2	2	...	2
host1-mic0	3	3	3	3	3	3	3	3	3	3	3	...	3

Fig. 3. Example of `RANK_COMM` communicators. They are used to group multiple hosts or accelerators into the same `COMP_COMM` (see Fig. 4).

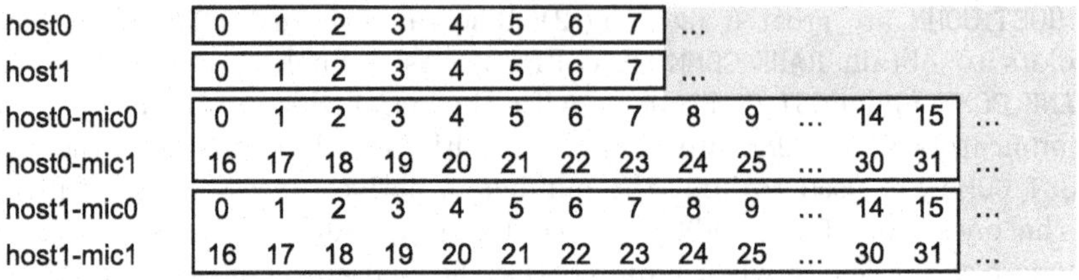

host0	0	1	2	3	4	5	6	7	...					
host1	0	1	2	3	4	5	6	7	...					
host0-mic0	0	1	2	3	4	5	6	7	8	9	...	14	15	...
host0-mic1	16	17	18	19	20	21	22	23	24	25	...	30	31	...
host1-mic0	0	1	2	3	4	5	6	7	8	9	...	14	15	...
host1-mic1	16	17	18	19	20	21	22	23	24	25	...	30	31	...

Fig. 4. Example of `COMP_COMM` communicators, where each diffusion problem is to be solved on either one node's CPUs or it's two accelerators. Gray areas correspond to idle processes, $pph = 8$, $ppa = 16$, $hpp = 1$, $app = 2$;

multigrid preconditioner from the Trilinos ML package. The domain is partitioned in the three spacial directions across the available parallel processes. In the case of fractional diffusion problem, we solve in parallel systems with diagonally perturbed discrete Laplacians in the form $(\frac{1}{t_i}\mathcal{I} + \mathcal{A})$, $t_i > 0$.

To compute (2), the processor with rank 0 from each `COMP_COMM` requests work from the master processor. Then if there is work left, a value fore index ℓ is supplied from the master. After that processors in `COMP_COMM` proceed with the solution, accumulating partial sums. When the computations for all values of ℓ are done, a global summation is performed to obtain the final result.

4 Parallel Experiments

The first set of experiments is aimed at establishing the optimal number of processes (including a hyper-threading) per CPU and per accelerator respectively. For this purpose, we measured the solution times for the Laplace's equation, with the following checkerboard right-hand side

$$f(x,y,z) = g(x)g(y)g(z), \qquad \text{where} \qquad g(x) = \begin{cases} -1, & \text{if } x < 0.5, \\ 0, & \text{if } x = 0.5, \\ 1, & \text{if } x > 0.5. \end{cases} \tag{3}$$

The PCG is applied as a basic iterative solution method in the case of linear systems with sparse SPD matrices where a parallel multigrid (MG) implementation from the Trilinos ML package is the preconditioner. A matrix free solver is used to reduce the memory requirements thus allowing the cores to be more efficiently utilized. The PCG tolerance is set to 10^{-10} in all reported experiments. The C++ language is used in both Trilinos and our code.

Table 1 shows results on the CPUs. Even for very large problems requiring a lot of communications, the solver is able to utilize all available cores with reasonable efficiency. Moreover, using hyper-threading, albeit less efficiently, provides a faster solution in all considered cases.

Table 2 shows similar results for the case when the solver is run on the Xeon Phi accelerators. The optimal number of cores to be used for solving the problem in the case of Laplacian is 32 for the smaller problems and even 16 for the largest one, i.e. considerably smaller than the number of physical cores – 61. Table 3 compares the weak scalability of the solver between running on the CPUs and on the accelerators respectively. The observed scalability on the CPUs is better. Some of the differences can be attributed to the fact that the accelerators cannot use the InfiniBand FDR interconnect directly, but instead rely on the host to perform the underlying communications, potentially increasing the latency and decreasing the bandwidth.

To achieve peak performance for the Xeon-Phi processor, one should run 4 threads per physical core. This allows for utilizing all of the available AVX2 vector units. As we see from Table 5 the performance degrades with the number of threads used. There are two reasons for this. First the performance of sparse matrix and large vector operations are memory speed bound – efficient caching is hard to achieve. The second reason is that the Pentium derived cores can issue maximum of 2 instructions per cycle. The performance of modern C++ code suffers from low instruction per clock ratio.

The next set of parallel experiments is focused on the solver for fractional Laplace's equation. Here we consider fractional powers of $\alpha = 0.25, 0.5, 0.75$. The parameter $k = \frac{1}{3}$ is used in the quadrature formula (2). The corresponding approximation error estimates and numbers of systems with diagonally perturbed discrete Laplacians to be solved are given in Table 4. The same right-hand side as in (3) is used. Tables 5, 6 and 7 show the parallel times and the scalability of the solver on up to 4 nodes. The following communication parameters were used: $pph = 32$, $hpp = 1$, $ppa = 32$, $app = 2$. The results demonstrate a rather impressive scalability of the proposed parallel implementation approach.

Table 1. Laplacian: parallel times and efficiency of the MG PCG solver on the CPUs.

DOFs	Nodes	Processes/Node	1	2	4	8	16	32
128^3	1	Time [s]	37.3	19.3	10.2	5.6	3.2	3.0
		Efficiency [%]		97	92	84	72	39
256^3	1	Time [s]	374	187	103	53	30	21
		Efficiency [%]		100	91	88	77	55
1024^3	32	Time [s]	878	512	242	126	76	58
		Efficiency [%]		86	91	87	72	48

Table 2. Laplacian: parallel times and efficiency of the MG PCG solver on the accelerators.

DOFs	Nodes	Processes/MIC	1	2	4	8	16	32	64	128
128^3	1	Time [s]	154	82	48	25	16	13	15	66
		Efficiency [%]		94	80	76	59	36	17	2
256^3	1	Time [s]	1390	720	361	191	98	70	110	
		Efficiency [%]		97	96	91	89	62	20	
512^3	8	Time [s]	1698	863	464	243	199	453	2757	
		Efficiency [%]		98	91	87	53	12	1	

Table 3. Laplacian: weak scalability of the MG PCG solver with respect to the number of nodes.

	CPUs		Accelerators	
Nodes	1	32	1	8
DOFs	256^3	1024^3	256^3	512^3
Time [s]	21.4	57.7	69.8	199.4
Efficiency [%]		74		35

Table 4. Numerical solution of the problem with fractional Laplacian: error estimates and corresponding number of systems to be solved for different values of α.

α	0.5	0.75	0.25
Error	1.80E−7	1.01E−7	3.06E−7
N_{SYS}	91	120	120

Table 5. Fractional Laplacian: parallel times and efficiency for $\alpha = 0.25$.

DOFs	128^3		256^3	
Nodes	Time [s]	Efficiency [%]	Time [s]	Efficiency [%]
1	147		971	
2	65	113	461	105
4	26	101	241	101

Table 6. Fractional Laplacian: parallel times and efficiency for $\alpha = 0.5$.

DOFs	128^3		256^3	
Nodes	Time [s]	Efficiency [%]	Time [s]	Efficiency [%]
1	146		989	
2	59	124	455	109
4	37	98	244	101

Table 7. Fractional Laplacian: parallel times and efficiency for $\alpha = 0.75$.

DOFs	128^3		256^3	
Nodes	Time [s]	Efficiency [%]	Time [s]	Efficiency [%]
1	235		1623	
2	103	115	729	111
4	52	113	360	113

5 Concluding Remarks

A parallel solver for the fractional Laplace's equation is implemented and tested on a heterogeneous system, demonstrating very promising parallel efficiency. The developed pioneering parallelization approach allows to use in parallel all levels of the heterogeneous architecture of the supercomputer Avitohol, including a number of nodes, each of them integrating CPUs and MIC accelerators. Results of extreme scale numerical tests are reported. We conclude also, that some further improvements of the parallel performance of the basic PCG MG solver on Xeon Phi accelerators is desired.

The major contribution of this study is the proposed general approach leading to efficient solution of fractional diffusion problems on a heterogeneous Intel Xeon Phi architecture, utilizing a given commonly available parallel AMG solver. We would emphasize that the results presented in the last three tables demonstrate the efficiency when the CPU-cores and MIC work together, that is they have been mixed in the related runs. In the particular case, `ml_MatrixFree` is used. Here, the important message is that the developed algorithm and software tools are directly portable if another faster parallel solver is available.

References

1. Axelsson, O.: Iterative Solution Methods. Cambridge University Press, New York (1996)
2. Bonito, A., Pasciak, J.: Numerical approximation of fractional powers of elliptic operators. Math. Comput. **84**, 2083–2110 (2015)
3. Ciegis, R., Starikovicius, V., Margenov, S., Kriauzienė, R.: Parallel solvers for fractional power diffusion problems. Concurrency Comput. Pract. Exper. **29**, e4216 (2017). https://doi.org/10.1002/cpe.4216
4. Harizanov, S., Margenov, S.: Positive Approximations of the Inverse of Fractional Powers of SPD M-Matrices, arXiv:1706.07620, June 2017, submitted
5. Harizanov, S., Lazarov, R., Margenov, S., Marinov, P., Vutov, Y.: Optimal Solvers for Linear Systems with Fractional Powers of Sparse SPD Matrices, arXiv:1612.04846v1, December 2016, submitted
6. Chen, L., Nochetto, R., Enrique, O., Salgado, A.J.: Multilevel methods for nonuniformly elliptic operators and fractional diffusion. Math. Comput. **85**, 2583–2607 (2016)

7. Vabishchevich, P.N.: Numerically solving an equation for fractional powers of elliptic operators. J. Comput. Phys. **282**, 289–302 (2015)
8. Lazarov, R., Vabishchevich, P.: A Numerical Study of the Homogeneous Elliptic Equation with Fractional Order Boundary Conditions, arXiv:1702.06477v1, February 2017, submitted

Application of Metaheuristics to Large-Scale Problems

Training Feed-Forward Neural Networks Employing Improved Bat Algorithm for Digital Image Compression

Adis Alihodzic(✉)

Department of Mathematics, University of Sarajevo, Zmaja od Bosne 33-35, 71000 Sarajevo, Bosnia and Herzegovina
adis.alihodzic@pmf.unsa.ba

Abstract. Training of feed-forward neural networks is a well-known and a vital optimization problem which is used to digital image lossy compression. Since the inter-pixel relationship in the picture is highly non-linear and unpredictive in the absence of a prior knowledge of the picture itself, it has shown that the neural networks combined with metaheuristics can be very efficient optimization method for image compression issues. In this paper, we propose an improved bat algorithm for training the input-output weights of the network which contains input-output layers of the equal sizes and a hidden layer of smaller size in-between. It has applied on five standard digital images. From the experimental analysis, it can be shown that the proposed method produces an acceptable quality of the compressed image as well as a good ratio of compression.

Keywords: Large scale problems · Lossy image compression
Metaheuristics · Swarm intelligence · Bat algorithm

1 Introduction

During last few decades, with the advent of more and more data to be transmitted through the network or to saved on servers, digital image compression is frequently employed to reduce the storage space as well as transmission costs that occur during data transfer through Internet, social networks, telecommunications systems, and so forth. Based on statistics, it has already shown that uncompressed digital images require significant storage capacity and wider transmission bandwidth for effective utilization of data. Image compression is aimed to transmit the pixels with minor bits. Recognition of redundancies in image, perfect and suitable encoding algorithms are the fruitful factors for digital image compression. In general, there are in the literature two popular categories of image compression techniques such as lossy and lossless image compressions. Lossless image compression techniques are based on the fact that the reconstructed data is a replica of the original one [1,2]. On the other hand, lossy image compression methods produce approximations with the perceptually good

I. Lirkov and S. Margenov (Eds.): LSSC 2017, LNCS 10665, pp. 315–323, 2018.
https://doi.org/10.1007/978-3-319-73441-5_33

quality of the original image. Although lossless image compression methods such as cosine transform, wavelet transform produce excellent results, they are time-consuming and produce lower compression ratio [3]. On the other hand, lossy compression methods such as JPEG generate a high compression ratio up to 40:1 or more while maintaining an acceptable perceptual quality of the reconstructed image [4–6]. In the last time, since the artificial neural network can process data in parallel and does require less time, several authors have applied them for lossy image compression [7–10]. Very popular algorithms such as back-propagation (BP) algorithm and the Levenberg-Marquardt (LM) algorithm are commonly exploited for optimizing the network performance by training feed-forward neural networks [11,12]. Although these algorithms have good performance in the process of training a single hidden-layer feedforward neural networks (SLNF) [15], they have their drawbacks such as slow convergence and trapping into local minima. To overcome these lacks, in this paper we will investigate the use of metaheuristic algorithms based on a random selection to improve the overall performance [13]. Several studies prove that the hybridization of the SLFN with swarm intelligence and nature-inspired algorithms such as genetic algorithm (GA), differential evolution (DE), simulated annealing, particle swarm optimization (PSO), artificial bee colony (ABC) has produced excellent results [14]. Despite these studies, further improvements are still possible, and there is still a strong need to develop even faster-learning methods. In this paper, a new method called IBA-SLFN combining the SLFN with an improved bat algorithm (IBA) is proposed as a novel learning method for tuning single hidden feed forward neural networks to obtain better image compression performance. In the proposed method, the improved bat algorithm is adjusted and applied to optimize input-output weights according to the mean squared error. From the experimental results, it will be shown that this approach will increase the performance and decrease the convergence time as well as it will provide a high compression ratio and acceptable quality of decompressed image. In subsequent sections, we foremost introduce ourselves with the training of SLFN to image compression. Section 3 describes some improvements of pure BA. In Sect. 4 the discussion and analysis of obtained experimental results are presented. Finally, some conclusions will be drawn briefly in Sect. 5.

2 Training of SLFN to Image Compression

This section of the paper briefly describes the process of training SLFN for lossy image compression. Suppose that we have a training set $\mathbf{T} = \{(\mathbf{x}_s, \mathbf{t}_s)\}_{s=1}^{\tilde{N}}$ of $\tilde{N}$ input-desired samples, where $\mathbf{x}_s = (x_{s,1}, x_{s,2}, \dots, x_{s,n})$ presents s-th input sample, while $\mathbf{t}_s = (t_{s,1}, t_{s,2}, \dots, t_{s,m})$ denotes s-th desired sample. SLFN with L hidden neurons and activation $g(x)$ function can be defined as follows:

$$\mathbf{o}_{s,k} = \sum_{j=1}^{L} \beta_{j,k} g(\mathbf{x}_s \cdot \mathbf{w}_j^T + b_j) \tag{1}$$

where $\mathbf{o}_{s,k}$ $(s = 1 \ldots N,\ k = 1 \ldots m)$ is the s-th output to be obtained from the SLFN, and $\mathbf{W} = \{w_{l,j}\}$ is $L \times n$ random input weight matrix connecting the hidden neurons and the input neurons. In addition, $\mathbf{w}_j^T$ is a vector column of matrix $\mathbf{W}$, while b_j are random biases assigned hidden layer and $\boldsymbol{\beta} = \{\beta_{j,k}\}$ is $L \times m$ output weight matrix connecting the hidden neurons and the output neurons. For image compression purposes, we will rewrite (1) in a matrix form:

$$\boldsymbol{H} \cdot \boldsymbol{\beta} = \boldsymbol{O} \tag{2}$$

where $\boldsymbol{H}$ is an $\tilde{N} \times L$ hidden matrix given by

$$\mathbf{H} = \begin{bmatrix} g_1(\mathbf{x}_1 \cdot \mathbf{w}_1^T + b_1), \ldots, & g_L(\mathbf{x}_1 \cdot \mathbf{w}_L^T + b_L) \\ g_1(\mathbf{x}_2 \cdot \mathbf{w}_1^T + b_1), \ldots, & g_L(\mathbf{x}_2 \cdot \mathbf{w}_L^T + b_L) \\ \vdots & \vdots \\ g_1(\mathbf{x}_{\tilde{N}} \cdot \mathbf{w}_1^T + b_1), \ldots, & g_L(\mathbf{x}_{\tilde{N}} \cdot \mathbf{w}_L^T + b_L) \end{bmatrix}$$

while the output weight matrix $\boldsymbol{\beta}$ and output matrix $\boldsymbol{O}$ (consisting of $\tilde{N}$ patterns) can be expressed as follows:

$$\boldsymbol{\beta} = \begin{bmatrix} \beta_{1,1} & \beta_{1,2}, & \ldots, & \beta_{1,m} \\ \beta_{2,1} & \beta_{2,2}, & \ldots, & \beta_{2,m} \\ \vdots & \vdots & & \vdots \\ \beta_{L,1} & \beta_{L,2}, & \ldots, & \beta_{L,m} \end{bmatrix}_{L \times m} \qquad \mathbf{O} = \begin{bmatrix} o_{1,1} & o_{1,2} & \ldots & o_{1,m} \\ o_{2,1} & o_{2,2} & \ldots & o_{2,m} \\ \vdots & \vdots & & \vdots \\ o_{\tilde{N},1} & o_{N,2} & \ldots & o_{\tilde{N},m} \end{bmatrix}_{\tilde{N} \times m}$$

Below we will explain how the process of training SLFN is connected to image compression. Namely, by using (2), it can be observed that if we have stored data in the matrices $\mathbf{H}$ and $\boldsymbol{\beta}$, the original image can be approximately reconstructed. More specifically, since the training of SLFN has finished, then quantized data from the matrices $\mathbf{H}$ and $\boldsymbol{\beta}$ are used for encoding of the original image. Before that, it is required to be made several samples by dividing the original image into small non-overlapping square blocks of pixels which are then used as the patterns for the training of the neural network. The number of neurons in the input-output layers are chosen due to the size of some block. The size of each block in the picture is usually taken to be of the square form 2×2, 4×4, 8×8, 16×16, and so forth. In this paper we have determined that the size of each subdivided block is 8×4 pixels, and it will be treated as an input-output training vector consisting of 32 dimensions. So, after division of the original training image to the blocks of sizes 8×4, there are $\frac{M}{8} \times \frac{N}{4}$ input-output samples available for neural network training, whereby $M \times N$ is a number of pixels in the original image. Therefore, for each sample in the input layer consisting of 32 neurons as well as for a fixed number of hidden neurons L, it is necessary to be learned both $32 \times L$ input weights and $L \times 32$ output weights by using some training algorithm. In this way, for a given image of size 256×256, for each of $\tilde{N} = 2048$ input-output patterns will be generated by one row in the matrix $\mathbf{H}$ by multiplying input neurons with input weights and passing the sum of that multiplication through the selected activation function. Similarly, the data from matrix **H** will be multiplied by the output weights to be obtained reconstructed (decoded) pixels for each of $\tilde{N}$ samples. Hence, for the purpose of image decoding, it is important to highlight here that packed quantized data in matrices **H** and $\boldsymbol{\beta}$ will be used to generate the reconstructed image. According to the large number of input patterns $\tilde{N}$, in this paper, we propose the improved bat algorithm (IBA) to increase the speed of training of SLFN and to get an acceptable quality of the decompressed image. Applying the IBA method on optimization of (2), deviation between matrices $\mathbf{H} \cdot \boldsymbol{\beta}$ and **O** expressed through the mean square error (MSE) will be minimized. The MSE is calculated by averaging of the squared difference of the intensity pixels between the original image I_O and decompressed image I_R as follows:

$$MSE = \frac{1}{M \times N} \sum_{i=1}^{M} \sum_{j=1}^{N} (I_O(i,j) - I_R(i,j))^2 \tag{3}$$

where $M \times N$ is the size of the original image.

As an alternative to MSE as indication of decompressed image quality, the peak signal to noise ratio (PSNR) is assessed:

$$PSNR = 10 \times \log \frac{MAX_I^2}{MSE} (dB) \tag{4}$$

where MAX_I is the maximum pixel intensity of an image, and MSE is given in (3). At this point, it is essential to say that all values of the pixels in the original image before training of the SLFN should be normalized on the closed interval $[0, 1]$ with respect to their maximum values. In this way, the overall efficiency of the SLFN is being achieved. Also, the performance of the neural network is connected to the initialization of the input-output weights. Namely, based on the experimental results, it has been proved that the best results are obtained when the first input-output weights are initialized to small random values uniformly drawn from the interval $(-1.0, 1.0)$. When the training of the SLFN is over, the data contained in the matrices $\boldsymbol{H}$ and $\boldsymbol{\beta}$ should be quantized. During the quantization process, we will reserve between 8 and 16 bits for matrix $\boldsymbol{\beta}$, while for the matrix $\boldsymbol{H}$, we will choose between 3 and 8 bits. In the case of the matrix $\boldsymbol{H}$, it should be rescaled to the interval $(0, 1)$, and after that each $h \in \boldsymbol{H}$ is saved in the file as an unsigned integer on the following way:

$$h_{int} = \lfloor h \cdot 2^{k_1} \rfloor \tag{5}$$

where k_1 denotes the number of bits for a component of the matrix $\boldsymbol{H}$. Now, we can compute the compression ratio (CR) as follows:

$$CR = \frac{I_O^{TNB}}{I_R^{TNB}} \quad (bpp) \tag{6}$$

where I_O^{TNB} is the total number of bits required to store the original image (in our case $256 \times 256 \times 8$), and I_R^{TNB} denotes the total number of bits to be recorded the compressed image. It is defined by:

$$I_R^{TNB} = \tilde{N} \times L \times k_1 + L \times 32 \times l_1 \tag{7}$$

where l_1 is the required number of bits to present an element of matrix $\boldsymbol{\beta}$ as an integer. It is known that at the quantization process, some significant bits after the decimal point will be lost. As a result, a mesh of black dots ("salt and pepper noise") will appear especially in the high-intensity image after reconstruction, and the quality of the reconstructed image will be weakened. In order to mitigate these drawbacks, it is required to perform 2D median filtering on each of 3×3 blocks until the whole decompressed image is covered.

3 The Improvements of Standard Bat Algorithm

In this section, we introduce two improvements of the original Bat algorithm (BA) proposed by Xin-She Yang [16]. Due to the shortage of room for writing, at this place, we will not describe the basic structure of BA, because we have already described it in detail in the paper [18]. To achieve more efficient search for synaptic weights during the training of SLFN for image compression purposes, and also to avoid drawbacks of the simple bat algorithm such as the chance to get stuck into local optima, we suggest the improved version of bat algorithm (IBA). IBA combines principles of the artificial bee colony algorithm (ABC) [19] and partly solution search equations of differential evolution (DE) algorithm [20]. The IBA consisting of two essential improvements as follows:

- (I1) **The first improvement** relates to the exploration capability of BA. To establish a right balance between intensification and diversification, IBA supported by the operator's mutation and crossover from DE algorithm perform more efficiently exploration and exploitation of a new search space, and it does not allow to be stuck into local optima [17].
- (I2) **The second improvement** is the refinement of the first improvement in terms that it does not allow some solutions to remain all time stuck in some local minima. Hence, this improvement is inspired by the launch of the scouts in the Scout phase of the ABC algorithm which helps that when some solution gets trapped into local optima after a certain number of iterations, it will eventually exceed the predetermined number of allowed trials called "limit." When a solution unchanged exceeds the limit trials, it is redirected to search a new space by using the random walk [18].

4 Experimental Analysis

In this section, our proposed method incorporated into the SLFN was compared against PSO and BA. All training algorithms have been implemented in C# programming language. All simulations were done on an Intel Core i7 3770 K, 3.5 GHz with 16 GB of RAM and Windows 10 ×64 Professional operating system. For evaluating the algorithms, the grayscale test images of sizes 256×256 with pixel amplitude resolution of 8 bits are used. The test images (a) and (b) shown in Fig. 1 are presented in BMP format, while the others are in TIFF format. For each image, the experiments were repeated 30 times, and the training process was stopped when the maximum number of 600 evaluations is reached.

Fig. 1. (a) Lena (b) Baboon (c) Cameraman (d) Moon surface (e) Aerial

For algorithms BA and IBA, the control parameters f_{min}, f_{max}, r_i^0 and A_i^0 were set to 0, 2, 0.5 and 0.99, respectively. The control parameters introduced from DE algorithms, such as differential weight F and crossover probability C_r respectively were set to 0.75 and 0.95. The parameter "limit" borrowed from the ABC algorithm related to the second improvement was set to 100. Parameters for PSO were set as follows: $w_{min} = 2.0$, $w_{max} = 2.0$. The population size for PSO, BA and IBA was set to 15, while the maximum number of optimization epochs was set to 20. In this paper, we have chosen the blocks of sizes 8×4 instead of square blocks, since they allow a wider opportunity being selected between CR and the quality of the decompressed image. In Table 1 the statistical results obtained by IBA in the case of use eight quantization bits for matrices $\mathbf{H}$ and $\boldsymbol{\beta}$ were shown. Based on these results, it can be observed that for a small number of hidden neurons IBA produces an excellent compression ratio but a low quality of the decompressed image and vice versa. This means that our novel model of SLFN allows a user to trade between required image quality and compression ratio by selecting various values of parameters such as L, k_1, and l_1. On another hand, the statistical parameters such as *Best of MSE* and *Std. Dev.* prove that IBA is a very stable and accurate algorithm, while the parameter *Mean time* says that IBA is capable of generating acceptable results in a reasonable time. Due to the statistical parameter *Best of MSE* presented in Table 2, it can be concluded that IBA performs the best prediction for the test image "Moon surface," since it has the simplest structure compared to the remaining images. Also, IBA outperforms both PSO and original BA algorithms for each test image related to the *Best of MSE* and PSNR. Moreover, it always produces better results considering both accuracy and quality of the decompressed image. Therefore, our method can be used as a fruitfull technique for lossy image compression.

Table 1. Statistical results obtained by the proposed IBA method for different number of hidden neurons over 30 runs for the standard test image **Cameraman**

H.N	Best of MSE	Mean of MSE	Std. Dev.	Mean time (s)	PSNR	CR
1	0.0091464903	0.0091549391	4.78E−08	2.98	20.387	31.51
2	0.0071143160	0.0073829699	1.53E−04	3.72	21.479	15.75
4	0.0051851414	0.0056150334	2.01E−04	4.74	22.852	7.88
8	0.0033777867	0.0035416620	1.03E−04	6.98	24.714	3.93
12	0.0020921733	0.0023304306	8.05E−05	9.79	26.794	2.63
16	0.0014110390	0.0015208147	4.92E−05	11.76	28.505	1.97
20	0.0008615488	0.0009573181	3.29E−05	14.59	30.647	1.57
24	0.0004858462	0.0005482598	2.49E−05	16.77	33.135	1.31
28	0.0002236706	0.0002455267	1.14E−05	20.34	36.504	1.16

Table 2. Results for 30 independent runs of the training SLFN produced by PSO, BA and IBA using linear activation function in all layers with 8 hidden neurons

Algorithms	PSO		BA		IBA	
Images	Best of MSE	PSNR	Best of MSE	PSNR	Best of MSE	PSNR
Moon surface	0.003113145	25.07	0.0019101408	27.19	**0.0009004251**	**30.45**
Lena	0.0039146061	24.07	0.0030146150	25.21	**0.0018946167**	**27.22**
Cameraman	0.0051010861	22.92	0.0042011685	23.77	**0.0033777867**	**24.71**
Aerial	0.0059816475	22.23	0.0049928748	23.02	**0.0038404657**	**24.17**
Baboon	0.0096511454	20.15	0.0085600680	20.67	**0.0072604542**	**21.39**

5 Conclusion

In this paper an improved bat algorithm (IBA) incorporated to SLFN has applied for direct lossy image compression. The proposed algorithm was used to optimize the input-output weights and minimize the norm least-square error. We compared the results with the results which were obtained by the PSO and BA. To validate and demonstrate the performance and effectiveness of the proposed IBA-SLFN method, it was tested on five test images. The performance of the proposed algorithm is better than the fulfillment of the PSO and BA. It has also been shown that our proposed algorithm is more accurate in the sense of MSE compared to the remaining algorithms. Due to the achieved results, it can be concluded that other swarm intelligence algorithms can be investigated as a future research for lossy image compression issues.

References

1. Bovik, A.: Handbook of Image and Video Processing, 2nd edn. Academic Press, University of Texas, Austin (2005)
2. Shukla, J., Alwani, M., Tiwari, A.K.: A survey on lossless image compression methods. In: Proceedings of 2nd International Conference on Computer Engineering and Technology (ICCET), vol. 6, pp. 136–141. IEEE, Chengdu (2010)
3. Miaou, S.-G., Lin, C.-L.: A quality-on-demand algorithm for wavelet-based compression of electrocardiogram signals. IEEE Trans. Biomed. Eng. **49**(3), 233–239 (2002)
4. Yang, J., Zhu, G., Shi, Y.-Q.: Analyzing the effect of JPEG compression on local variance of image intensity. IEEE Trans. Image Process. **25**(6), 2647–2656 (2006)
5. Tsolakis, D., Tsekouras, G.E., Niros, A.D., Rigos, A.: On the systematic development of fast fuzzy vector quantization for grayscale image compression. J. Neural Netw. **36**, 83–96 (2012)
6. Grailu, H., Lotfizad, M., Sadoghi-Yazdi, H.: A lossy/lossless compression method for printed typeset bi-level text images based on improved pattern matching. Int. J. Doc. Anal. Recogn. (IJDAR) **11**(4), 159–182 (2009)
7. Gaidhane, V., Singh, V., Kumar, M.: Image compression using PCA and improved technique with MLP neural network. In: Proceedings of IEEE International Conference on Advances in Recent Technologies in Communication and Computing, pp. 106–110. IEEE, Kottayam (2010)
8. Feng., H., Tang., M., Qi, J.: A back-propagation neural network based on a hybrid genetic algorithm and particle swarm optimization for image compression. In: 4th International Congress on Image and Signal Processing (CISP), pp. 1315–1318. IEEE, Shanghai (2011)
9. Gaidhane, V.H., Singh, V., Hote, Y.V., Kumar, M.: New approaches for image compression using neural network. J. Intell. Learn. Syst. Appl. **3**(4), 220–229 (2011)
10. Jiang, J.: Image compression with neural networks - a survey. Signal Process. Image Commun. **14**(9), 737–760 (1999)
11. Haykin, S.: Neural Networks and Learning Machines. Prentice Hall, New York (2008)
12. Huang, G.-B., Chen, L., Siew, C.-K.: Universal approximation using incremental constructive feedforward networks with random hidden nodes. IEEE Trans. Neural Netw. **17**(4), 879–892 (2006)
13. Yang, X.-S.: Efficiency analysis of swarm intelligence and randomization techniques. J. Comput. Theoret. Nanosci. **9**(2), 189–198 (2012)
14. Ojha, V.K., Abraham, A., Snasel, V.: Metaheuristic design of feedforward neural networks: A review of two decades of research. Eng. Appl. Artif. Intell. **60**(2017), 97–116 (2017)
15. Seifollahi, S., Yearwood, J., Ofoghi, B.: Novel weighting in single hidden layer feedforward neural networks for data classification. Comput. Math. Appl. **64**(2), 128–136 (2012)
16. Yang, X.-S.: A new metaheurisitic bat-inspired algorithm. Stud. Comput. Intell. **284**, 65–74 (2010)
17. Alihodzic, A., Tuba, M.: Improved bat algorithm applied to multilevel image thresholding. Sci. World J. **2014**, 1–16 (2014)
18. Tuba, M., Alihodzic, A., Bacanin, N.: Cuckoo search and bat algorithm applied to training feed-forward neural networks. In: Yang, X.-S. (ed.) Recent Advances in Swarm Intelligence and Evolutionary Computation. SCI, vol. 585, pp. 139–162. Springer, Cham (2015). https://doi.org/10.1007/978-3-319-13826-8_8

19. Bullinaria, J.A., AlYahya, K.: Artificial bee colony training of neural networks. In: Terrazas, G., Otero, F., Masegosa, A. (eds.) Nature Inspired Cooperative Strategies for Optimization (NICSO 2013). Studies in Computational Intelligence, vol. 512, pp. 191–201. Springer, Cham (2014). https://doi.org/10.1007/978-3-319-01692-4_15
20. Piotrowski, A.P.: Differential evolution algorithms applied to neural network training suffer from stagnation. Appl. Soft Comput. **21**, 382–406 (2014)

Modeling and Optimization of Pickup and Delivery Problem Using Constraint Logic Programming

Amelia Bădică[1], Costin Bădică[1](✉), Florin Leon[2], and Ion Buligiu[1]

[1] University of Craiova, Craiova, Romania
ameliabd@yahoo.com, cbadica@software.ucv.ro
[2] Technical University "Gheorghe Asachi" of Iaşi, Iaşi, Romania
fleon@cs.tuiasi.ro

Abstract. Our research was conducted in a project that aims to develop an intelligent freight broker agent for providing logistics brokerage services for the efficient allocation of transport resources (vehicles or trucks) to the transport applications. This agent coordinates transportation arrangements of customers (usually shippers and consignees) with resource providers or carriers, following the freight broker business model. The scheduling function of the freight broker agent was formulated as a special type of vehicle routing with pickup and delivery problem. This research is based on our recently proposed declarative model of the freight broker agent using constraint logic programming. This model allows the computation of the feasible transportation schedules. In this paper we augment this model with a declarative representation of optimal schedules and then we show how these optimal schedules can be computed using the ECLiPSe constraint logic programming system.

Keywords: Combinatorial optimization
Pickup and delivery problem · Constraint logic programming

1 Introduction

Our research is part of a project that aims to develop an intelligent freight broker agent for providing logistics brokerage services for the efficient allocation of transport resources (vehicles or trucks) to the transport applications. This agent coordinates transportation arrangements of customers (usually shippers and consignees) with resource providers or carriers, following the freight broker business model. This agent is part of an agent-based architecture of a freight brokering system consisting of several interacting agents [3]. The freight broker agent is using a mathematical optimization service to compute the optimal schedule of vehicles that fulfils the customer requirements [4].

The scheduling function of the freight broker agent is defined as a special type of vehicle routing with pickup and delivery problem [6]. This research is

I. Lirkov and S. Margenov (Eds.): LSSC 2017, LNCS 10665, pp. 324–332, 2018.
https://doi.org/10.1007/978-3-319-73441-5_34

based on our recently proposed model of the freight broker agent using constraint logic programming [1]. The model allows the computation of the set of feasible transportation schedules.

The contribution of this paper is a proposal for augmenting this model with a declarative representation of optimal transport schedules. We also show how these optimal schedules can be effectively computed using the ECLiPSe state-of-the-art constraint logic programming (CLP) system [8,9].

The paper is structured as follows. We present in Sect. 2 our proposed constraint-based model for the pickup and delivery problem. In Sect. 3 we discuss the CLP implementation using ECLiPSe CLP and present some experimental results. Section 4 presents our conclusions and planned future work.

2 Formal Model of Pickup and Delivery Problem

2.1 Problem Definition

Definition 1. *A* vehicle routing with pickup and delivery problem *is a tuple* $\langle \mathcal{L}, \mathcal{O}, \mathcal{T}, \Delta \rangle$ *such that:*

a. $\mathcal{L}$ *is the* set of locations of interest, *including the pickup points, the delivery points, and the truck home locations. We assume that* $\mathcal{L} = \{1, 2, \ldots, k+h\}$ *such that* $k > 0$ *and* $h \geq 0$. *The set* $\mathcal{P} = \{1, 2, \ldots, k\}$ *contains the pickup, as well as the delivery points, while the set* $\mathcal{H} = \{k+1, \ldots, k+h\}$ *contains the truck home locations not already included in the set of pickup and delivery points. The elements of* $\mathcal{L}$ *represent locations of a certain geographical region.*
b. $\mathcal{O}$ *is the* set of customer orders, $|\mathcal{O}| = n > 0$. *Each order is a triple* (OS_i, OD_i, C_i) *such that* $OS_i, OD_i \in \mathcal{P}$, $OS_i \neq OD_i$ *are the pickup, respectively the delivery points and* $C_i > 0$ *is the requested capacity of order* i, *for all* $1 \leq i \leq n$. *Note that* $2 \leq k \leq 2n$.
c. $\mathcal{T}$ *is the* set of trucks, $|\mathcal{T}| = t > 0$. *Each truck is a pair* (H_i, Γ_i) *such that* $H_i \in \mathcal{L}$ *and* $\Gamma_i > 0$ *are the home location or origin and respectively the maximum provided transportation capacity of truck* $i = 1, \ldots, t$.
d. Δ *is an* $(k+h) \times (k+h)$ *positive real matrix such that* $\Delta_{ij} > 0$ *is the distance between any two locations* $1 \leq i \neq j \leq k+h$.

Definition 2. *A* schedule *of the vehicle routing with pickup and delivery problem* $\langle \mathcal{L}, \mathcal{O}, \mathcal{T}, \Delta \rangle$ *can be represented as a tuple* $\langle X, M, S, D \rangle$ *such that:*

a. $X \in \{1, 2, \ldots, k\}^m$ *is a vector of size* m *that captures all the sequences of hops that determine the necessary truck routes to serve all the customer orders. Each of the* k *locations must be visited, so* $m \geq k$. *Moreover, each order requires two hops, one for pickup, the other for delivery, so maximum* $2n$ *hops are needed. It follows that* $k \leq m \leq 2n$.
b. $M \in \{0, 1, \ldots, m\}^t$ *is a vector of size* t *such that* M_l *defines the number of hops of each truck* $l = 1, 2, \ldots, t$. *The total number of hops is* m *so* $\sum_{l=1}^{t} M_l = m$. *Note that if* $M_l = 0$ *then truck* l *is not part of the solution. So, setting* $M_l \geq 0$ *allows solutions using "at most"* t *trucks, while setting* $M_l \geq 1$ *constraints solutions to use "exactly"* t *trucks.*

c. $S, D \in \{0,1\}^{m \times n}$ *are two Boolean matrices such that* $S_{ij} = 1$ *if and only if* X_i *is the pickup point of order* j, *otherwise* $S_{ij} = 0$ *and* $D_{ij} = 1$ *if and only if* X_i *is the delivery point of order* j, *otherwise* $D_{ij} = 0$.

A schedule can be intuitively interpreted as follows. We partition the interval $[1, m]$ into t (possibly empty) intervals $I_l = [A_l, B_l]$ defined as follows: $A_l = 1 + \sum_{\alpha=1}^{l-1} M_\alpha$ and $B_l = \sum_{\alpha=1}^{l} M_\alpha$, for $1 \leq l \leq t$. Note that if $M_l = 0$ then $I_l = \emptyset$, so truck l does not contribute to the schedule. On the other hand, each interval $I_l \neq \emptyset$ defines the hops of truck l. The route of truck l (adding also its home point, as well as the departure/return segments from/to the home point) is $H_l \rightarrow X_{A_l} \rightarrow X_{A_l+1} \rightarrow \cdots \rightarrow X_{B_l} \rightarrow H_l$. Note that the total number of hops of truck l, excluding its departure from and return to its home point, is equal to $B_l - A_l + 1$.

2.2 Constraints

Following [1], we can now formulate the set of constraints involving the pickup and delivery points and requested capacities of all the customer orders, as well as the available trucks and their capacities. Concerning the use of quantified variables i, j, k, and l in the specification of the constraints let us assume that i, k represent hops, i.e. $i, k \in \{1, 2, \ldots, m\}$, j represents orders, i.e. $j \in \{1, 2, \ldots, n\}$ and l represents trucks, i.e. $l \in \{1, 2, \ldots, t\}$.

For each hop i and order j the pickup point of order j is OS_j so:

$$(\forall i, j)((S_{ij} = 1) \Rightarrow (X_i = OS_j)) \tag{1}$$

Similarly, for each hop i and order j the delivery point of order j is OD_j so:

$$(\forall i, j)((D_{ij} = 1) \Rightarrow (X_i = OD_j)) \tag{2}$$

There exists at least one load or unload operation in each hop i, i.e.:

$$(\forall i) \textstyle\sum_{j=1}^{n} (S_{ij} + D_{ij}) \geq 1 \tag{3}$$

For all hops i, k if there exists an order j such that i is the pickup point of order j and k is the delivery point of order j then i must precede k, so:

$$(\forall i, k)((\exists j)((S_{ij} = 1) \wedge (D_{ij} = 1)) \Rightarrow (i < k)) \tag{4}$$

For each order j there is a unique load point and a unique unload point, so:

$$(\forall j)(\textstyle\sum_{i=1}^{m} S_{ij} = 1) \wedge (\sum_{i=1}^{m} D_{ij} = 1) \tag{5}$$

Each order is completely served by a unique truck. So, for each truck l, the orders with load point assigned to l have also the unload point assigned to l, i.e.:

$$(\forall l, j) \textstyle\sum_{i=A_l}^{B_l} S_{ij} = \sum_{i=A_l}^{B_l} D_{ij} \tag{6}$$

We must also specify the constraints stating that the capacities of the trucks are not overflowed along each route. Let T_i denote the transported capacity of truck l between hops $i-1$ and i, for all $i = A_l, \ldots, B_l$. Initially $T_0 = 0$. For each hop i the value of T_i is obtained from T_{i-1} by adding the capacities loaded and subtracting the capacities unloaded in hop i. The constraints on capacities usage can be formulated as follows:

$$\begin{array}{ll} T_0 = 0 & \\ T_i = T_{i-1} + \sum_{j=1}^{n}(S_{ij} - D_{ij})C_j & \text{for all } 1 \leq l \leq t \text{ and } A_l \leq i \leq B_l \\ T_i \leq \Gamma_i & \text{for all } 1 \leq l \leq t \end{array} \quad (7)$$

Note that for each truck $1 \leq l \leq t$ the remaining capacity to be transported along the last segment connecting the last hop to the truck home location is 0, i.e. $T_{B_l} = 0$ holds.

If there is a solution that specifies that a truck l can perform a load as well as an unload (for a different order) in hop j then, according to the constraints stating that there must be at least one pickup or delivery in each hop, the constraint solver can generate redundant solutions actually representing the same solution. For example, a solution could specify only a single hop j containing both the pickup and the delivery, as well as two successive, but identical hops, the first representing the pickup and the second representing the delivery. This redundancy can be eliminated by stating that for each truck l, each of its two consecutive hops i and $i+1$ with $j \in [A_l, B_l - 1]$ must be different, i.e.:

$$(\forall i \in [A_l, B_l - 1]) X_i \neq X_{i+1} \quad (8)$$

Definition 3. *A* feasible schedule *of the vehicle routing with pickup and delivery problem is a schedule that satisfies the constraints specified by Eqs. (1), (2), (3), (4), (5), (6), (7), and (8).*

2.3 Optimization Criterion

Many different optimization criteria can be set for the vehicle routing with pickup and delivery problem. Actually, each optimization criterion defines a different problem. Example criteria can take into account: total travelled distance of the trucks, profit of the truck company, delivery time, a.o. In this paper we are interested to optimize the total distance that is travelled by trucks. Nevertheless, the method can be generalized to other criteria, depending on the specifics of the problem.

The total distance D_l travelled by truck $l = 1, 2, \ldots, t$, as well as the total distance $DIST$ travelled by all the trucks, can be determined as follows:

$$\begin{array}{l} D_l = \begin{cases} 0 & \text{if } I_l = \emptyset \\ \Delta_{H_l X_{A_l}} + \Delta_{X_{B_l} H_l} + \sum_{i=A_l}^{B_l - 1} \Delta_{X_i X_{i+1}} & \text{otherwise} \end{cases} \\ DIST = \sum_{l=1}^{t} D_l \end{array} \quad (9)$$

Definition 4. *An* optimal schedule *of the vehicle routing with pickup and delivery problem is a feasible schedule that minimizes the cost function $DIST$ specified by Eq. (9).*

3 CLP Implementation

3.1 Model Development

We propose a CLP model based on the state-of-the-art ECLiPSe CLP system. This model is in fact an upgrade of our proposed model already introduced in [1]. The upgrade contains a declarative representation of the optimization criterion that defines the optimal transportation schedules. We consider the total distance criterion defined by Eq. (9), as example. Nevertheless, the same approach can be used to define other suitable optimization criteria, for specific problems.

A CLP program is a set of logic statements – facts and rules, composed of predicates. CLP distinguishes between normal Prolog predicates encountered in standard Prolog programming and constraints that are specific to CLP. Constraints are handled by specialized constraint satisfaction algorithms that provide more efficient problem solving methods than the standard Prolog's backtracking search algorithm.

Following the methodology proposed in [5], an ECLiPSe-based optimization model has three parts: (i) definition of variables and domains; (2) definition of constraints (3) definition of cost variable; (4) search for optimal solution.

Following [1], a solution is represented in ECLiPSe by a tuple of logic variables: M is the number of hops, N is the number of orders, T is the number of trucks, MM is an ECLiPSe array [8] with T elements such that for each I from 1 to T the value of $MM[I]$ is the number of hops on the route of truck I, S and D are $M \times N$ ECLiPSe Boolean arrays for capturing the load/unload points associated to each hop, X is an M-sized ECLiPSe array for capturing the hops, and $DIST$ the cost of the optimal solution (see Listing 1).

We assume for simplicity that the problem is given as a set of Prolog facts, as in [1]. Note that the values of variables N, T and $Delta$ are extracted by predicate `domains_and_variables` directly from the facts that define the problem.

Listing 1. Specification of the solution predicate

```
solution(M,MM,S,D,X,DIST) :-
    domains_and_variables(M,N,T,MM,S,D,X,Delta),
    constraints(M,N,T,MM,S,D,X),
    compute_distance_variable(M,T,MM,Delta,X,K,DIST),
    solve(M,MM,S,D,X,DIST).
```

The definitions of predicates `domains_and_variables` and `constraints` follow our model introduced in [1]. So we now focus on the definition of the distance variable that represents the optimization criterion. This is achieved using predicate `compute_distance_variable`.

The cost of a schedule is computed using Eq. (9). Although apparently simple, using this equation has a tricky aspect: the variables X_i that are part of the solution occur as subscripts of matrix Δ, so $DIST$ cannot be directly expressed by an algebraic expression of X, as required by the general format of ECLiPSe constraints. Nevertheless, the problematic values $\Delta_{H_l X_{A_l}}$, $\Delta_{X_{B_l} H_l}$, and $\Delta_{X_j X_{j+1}}$ occurring in Eq. (9) can be rewritten algebraically using Eq. (10). Here the

Boolean expression $x = y$ is evaluated to 1 if x equals y, otherwise it evaluates to 0.

$$\begin{aligned} \Delta_{H_l X_{A_l}} &= \sum_{x=1}^{k} \Delta_{H_l x}(x = X_{A_l}) \\ \Delta_{X_{B_l} H_l} &= \sum_{x=1}^{k} \Delta_{H_l x}(x = X_{B_l}) \\ \Delta_{X_i X_{i+1}} &= \sum_{x,y=1}^{k} \Delta_{xy}(x = X_i)(y = X_{i+1}) \end{aligned} \tag{10}$$

Note however that using Eq. (10) has the drawback of incurring an $O(k)$ computational overhead, as compared with the evaluation using Eq. (9).

Now, using Eq. (10), the declarative specification of the cost variable can be realized using the ECLiPSe code presented in Listing 2. Note that this specification is using logical loops, firstly proposed in [7].

Listing 2. Specification of the cost variable

```
compute_distance_variable(M,T,MM,Delta,X,K,Dist) :-
  (for(L,1,T), param(X,K,MM,Delta),
    fromto([0,0],[S0,Dist0],[BT,Dist1],[M,Dist]) do
    BT #= MM[L]+S0, AT #= 1+S0,
    (AT =< BT ->
      truck(L,H,_),
      var_d_h(X,K,AT,BT,H,Delta,DHAT,DBTH),
      DistToHome #= DHAT+DBTH,
      (for(I,AT,BT-1), param(X,K,Delta), fromto(DistToHome,D0,D1,D) do
        var_d_i(X,K,I,Delta,DI),
        D1 #= D0+DI
      );
      D #= 0
    ),
    Dist1 #= Dist0+D
  ).
var_d_h(X,K,I,J,H,Delta,DHI,DJH) :-
  (for(V,1,K), param(X,I,J,H,Delta),
    fromto([0,0],[DHI0,DJH0],[DHI1,DJH1],[DHI,DJH]) do
    DHI1 #= DHI0+Delta[H,V]*(V #= X[I]),
    DJH1 #= DJH0+Delta[V,H]*(V #= X[J])
  ).
var_d_i(X,K,I,Delta,DI) :-
  (multifor([V,W],[1,1],[K,K]), param(X,I,Delta),
    fromto(0,DI0,DI1,DI) do
    DI1 #= DI0+Delta[V,W]*(V #= X[I])*(W #= X[I+1])
  ).
```

3.2 Results and Discussion

We experimented using the 64-bit version of ECLiPSe 6.1_224 on an *x64*-based PC with Intel(R) Core(TM) i7-5500U CPU at 2.40 GHz running Windows 10. We considered the data set described in Table 1. Even these small parameter values determine a large search space. For $n = 5$ and $k = 6$ this size is $\sum_{m=k}^{2n} 2^{mn} \times 2^{mn} \times m^k > 2^{36}$ for each $m = 6 \ldots 10$ and $m_1, m_2 \geq 1, m_1 + m_2 = m$!

The experiment was focused on determining the optimal schedule that satisfies the constraints, for each data set (i.e. we considered a OS-type problem according to [5]). We considered the solutions consisting of exactly t trucks.

Table 1. Data set.

#	n = # orders	k = # cities	t = # trucks	$m \in k \ldots 2n$	Cost	# opt. solutions
1	5	6	2	$6 \ldots 10$	821	2

We used the built-in `bb_min/3` ECLiPSe predicate [9] for performing a branch-and-bound search. This means that whenever a new and better solution is found, it is remembered together with its (best so far solution) cost. Then the search continues using a supplementary constraint requiring that the future solutions must have a lower cost than the current solution.

The search process is divided in two stages. In first stage M and MM get instantiated, while incompletely instantiated template structures are defined for X, S, and D. This stage is achieved by predicate `domains_and_variables`. The subset of solutions defined by the given values of M and MM is then explored in the second stage using the `bb_min(Goal,Cost,Options)` query. Here *Goal* denotes a predicate that nondeterministically explores the set of feasible solutions, while *Cost* represents the cost variable. Usually *Goal* is based on the `search/6` built-in ECLiPSe predicate, although this is not mandatory. *Cost* gets instantiated as soon as a new feasible solution is determined. For example, the two sample queries from Listing 3 were used in our experiments

Listing 3. Sample optimization queries

```
solve1(M,MM,S,D,X,DIST) :-
  bb_min(search([](X,S,D),0,first_fail,indomain,complete,[]),DIST,_).
solve2(M,MM,S,D,X,DIST) :-
  bb_min(
    (solution(M,MM,S,D,X,Dist)
     bb_min(search([](X,S,D),0,first_fail,indomain,complete,[]),DIST,_)),
    DIST,_).
```

Note that query `solve1` returns an optimal schedule separately for each valid instantiation of the variables M and MM, while query `solve2` provides a unique solution for the whole search, so it is more convenient to use than `solve1`.

We experimented with various search parameters *Select*, *Choice*, and *Method* available for the built-in ECLiPSe predicate `search/6`. *Select* denotes the variable selection method, *Choice* denotes the value-to-variable assignment method during the search process, while *Method* denotes the search algorithm.

Table 2 presents the values used in our experiments, together with the user cpu time and cost value recorded for running the search. The value of the user cpu time was retrieved with the help of the `statistics/2` ECLiPSe built-in predicate, using the goal `statistics(runtime,[,T])`.

For use cases from 1 to 4 we performed a complete search. The optimal solution is obtained faster if we use the most constrained variable selection method.

The search space for our problem can be too large to search exhaustively. Therefore, for use cases from 5 to 10 we performed an incomplete search using limited discrepancy search [2] that is available in ECLiPSe. Firstly, we observed the value-to-variable assignment method had an impact on the execution time.

Table 2. Experimental results.

#	Query	Select	Choice	Method	Cost	Runtime [ms]
1	solve1	first_fail	indomain	complete	821	99078
2	solve1	most_constrained	indomain	complete	821	46406
3	solve2	first_fail	indomain	complete	821	96953
4	solve2	most_constrained	indomain	complete	821	43860
5	solve2	first_fail	indomain	lds(1)	–	6004
6	solve2	first_fail	indomain_median	lds(1)	829	4910
7	solve2	most_constrained	indomain_median	lds(2)	1001	5484
8	solve2	most_constrained	indomain_median	lds(2)	957	10638
9	solve2	most_constrained	indomain_median	lds(3)	821	18830
10	solve2	max_regret	indomain_median	lds(2)	821	16044

Secondly, we have used different values of the *Disc* parameter that limits the number of discrepancies allowed on each path. ECLiPSe documentation [9] recommends values from 1 to 3 for *Disc*. For our test problem, we were able to find the optimal solution in two situations: (i) using the most constrained variable selection method and $Disc = 3$ (use case 9) and (ii) using the maximum regret variable selection method and $Disc = 2$ (use case 10).

4 Conclusions and Future Works

In this paper we have presented a CLP model for computing optimal schedules of the vehicle routing with pickup and delivery problem. We provided experimental results obtained with various search options available in the ECLiPSe CLP system. As future work we plan to strengthen our results by performing more experiments with different problems and search options. Moreover, we are working on integrating the ECLiPSe-based scheduling into an agent-based application for freight brokering.

References

1. Bădică, C., Bădică, A., Leon, F., Luncean, L.: Declarative representation and solution of vehicle routing with pickup and delivery problem. In: Proceedings of the International Conference of Computational Science - ICCS 2017 (2017)
2. Harvey, W.D., Ginsberg, M.L.: Limited discrepancy search. In: Proceedings of the 14th International Joint Conference on Artificial Intelligence - IJCAI 1995, pp. 607–613 (1995)
3. Luncean, L., Bădică, C., Bădică, A.: Agent-based system for brokering of logistics services – initial report. In: Nguyen, N.T., Attachoo, B., Trawiński, B., Somboonviwat, K. (eds.) ACIIDS 2014. LNCS (LNAI), vol. 8398, pp. 485–494. Springer, Cham (2014). https://doi.org/10.1007/978-3-319-05458-2_50

4. Leon, F., Bădică, C.: A freight brokering system architecture based on web services and agents. In: Borangiu, T., Dragoicea, M., Nóvoa, H. (eds.) IESS 2016. LNBIP, vol. 247, pp. 537–546. Springer, Cham (2016). https://doi.org/10.1007/978-3-319-32689-4_41
5. Niederliński, A.: A Gentle Guide to Constraint Logic Programming via ECLiPSe, 3rd edn. Jacek Skalmierski Computer Studio, Gliwice (2014)
6. Parragh, S.N., Doerner, K.F., Hartl, R.F.: A survey on pickup and delivery problems. Part II: transportation between pickup and delivery locations. Journal für Betriebswirtschaft **58**(2), 81–117 (2008). https://doi.org/10.1007/s11301-008-0036-4
7. Schimpf, J.: Logical loops. In: Stuckey, P.J. (ed.) ICLP 2002. LNCS, vol. 2401, pp. 224–238. Springer, Heidelberg (2002). https://doi.org/10.1007/3-540-45619-8_16
8. Schimpf, J., Shen, K.: ECLiPSe - from LP to CLP. Theory Pract. Log. Program. **12**(1–2), 127–156 (2012). https://doi.org/10.1017/S1471068411000469. Cambridge University Press
9. The ECLiPSe Constraint Programming System. http://www.eclipseclp.org/. Accessed Jan 2017

Intercriteria Analysis over Intuitionistic Fuzzy Data

Veselina Bureva[1], Evdokia Sotirova[1], Vassia Atanassova[2], Nora Angelova[2,3], and Krassimir Atanassov[1,2(✉)]

[1] Asen Zlatarov University — Burgas 1 “Prof. Yakimov” Blvd, 8010 Burgas, Bulgaria
{vbureva,esotirova}@btu.bg, deyanmegara@gmail.com

[2] Department of Bioinformatics and Mathematical Modelling, Institute of Biophysics and Biomedical Engineering, Bulgarian Academy of Sciences, 105 Acad. G. Bonchev Str., 1113 Sofia, Bulgaria
vassia.atanassova@gmail.com, krat@bas.bg

[3] Faculty of Mathematics and Informatics, Sofia University “St. Climent Ochridski”, 5, James Bourchier Str., 1126 Sofia, Bulgaria
metida.su@gmail.com

Abstract. The possibility for application of Intercriteria Analysis over intuitionistic fuzzy data is discussed. An example in the area of mathematical logic is given as an illustration of the application of the Intercriteria Analysis.

Keywords: Data · Intercriteria analysis
Intuitionistic fuzzy index matrix · Intuitionistic fuzzy pair

AMS Classification: 03E72

1 Introduction

The concept of InterCriteria Analysis was introduced in [4,7]. The intercriteria analysis is based on the apparatus of the Index Matrices (IMs, see [4]) and of Intuitionistic Fuzzy Sets (IFSs, see, e.g., [3]). The paper is a continuation of [1,6,7,9–11,14,15]. Here, for the first time we discuss the possibility, the data, that will be processed by intercriteria analysis, to be Intuitionistic Fuzzy Pairs (IFP, see [8]), variables or formulas, or more general - intuitionistic fuzzy data (see [12]).

2 Short Notes on Intuitionistic Fuzzy Pairs

The Intuitionistic Fuzzy Pair (IFP) is an object in the form $\langle a, b\rangle$, where $a, b \in [0, 1]$ and $a + b \leq 1$, that is used as an evaluation of some object or process

I. Lirkov and S. Margenov (Eds.): LSSC 2017, LNCS 10665, pp. 333–340, 2018.
https://doi.org/10.1007/978-3-319-73441-5_35

and which components (a and b) are interpreted as degrees of membership and non-membership, or degrees of validity and non-validity, or degree of correctness and non-correctness, etc. One of the geometrical interpretations of the IFPs is shown on Fig. 1.

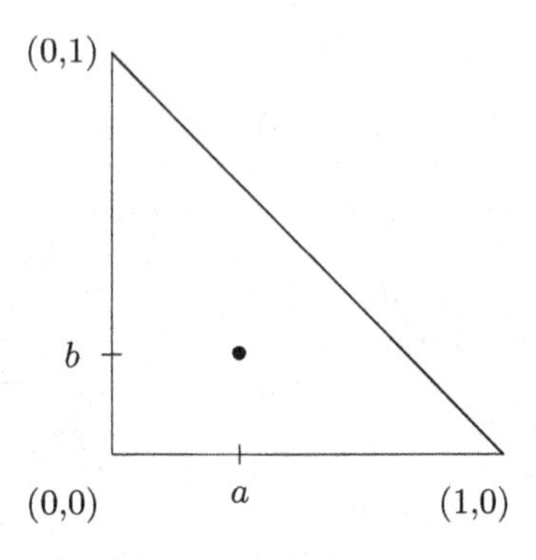

Fig. 1. Geometrical interpretation of an IFP.

Let us have two IFPs $x = \langle a, b\rangle$ and $y = \langle c, d\rangle$. We define the relations

$$\begin{array}{lll} x < y & \text{iff} & a < c \text{ and } b > d \\ x > y & \text{iff} & a > c \text{ and } b < d \\ x \geq y & \text{iff} & a \geq c \text{ and } b \leq d \\ x \leq y & \text{iff} & a \leq c \text{ and } b \geq d \\ x = y & \text{iff} & a = c \text{ and } b = d \end{array}$$

3 Short Remarks on Index Matrices

The concept of Index Matrix (IM) was discussed in a series of papers collected in [4].

Let I be a fixed set of indices and $\mathcal{R}$ be the set of the real numbers. By IM with index sets K and L ($K, L \subset I$), we denote the object:

$$[K, L, \{a_{k_i,l_j}\}] \equiv \begin{array}{c|cccc} & l_1 & l_2 & \ldots & l_n \\ \hline k_1 & a_{k_1,l_1} & a_{k_1,l_2} & \ldots & a_{k_1,l_n} \\ k_2 & a_{k_2,l_1} & a_{k_2,l_2} & \ldots & a_{k_2,l_n} \\ \vdots & \vdots & \vdots & \ddots & \vdots \\ k_m & a_{k_m,l_1} & a_{k_m,l_2} & \ldots & a_{k_m,l_n} \end{array},$$

where $K = \{k_1, k_2, ..., k_m\}$, $L = \{l_1, l_2, ..., l_n\}$, for $1 \leq i \leq m$, and $1 \leq j \leq n$: $a_{k_i,l_j} \in \mathcal{R}$.

In [2,4], different operations, relations and operators are defined over IMs. For the needs of the present research, we will introduce the definitions of some of them.

When elements a_{k_i,l_j} are some variables, propositions or formulas, we obtain an extended IM with elements from the respective type. Then, we can define the evaluation function V that juxtaposes to this IM a new one with elements – IFPs $\langle \mu, \nu \rangle$, where $\mu, \nu, \mu + \nu \in [0, 1]$. The new IM, called Intuitionistic Fuzzy IM (IFIM), contains the evaluations of the variables, propositions, etc., i.e., it has the form

$$V([K, L, \{a_{k_i,l_j}\}]) = [K, L, \{V(a_{k_i,l_j})\}] = [K, L, \{\langle \mu_{k_i,l_j}, \nu_{k_i,l_j} \rangle\}]$$

$$= \begin{array}{c|ccccc} & l_1 & \dots & l_j & \dots & l_n \\ \hline k_1 & \langle \mu_{k_1,l_1}, \nu_{k_1,l_1} \rangle & \dots & \langle \mu_{k_1,l_j}, \nu_{k_1,l_j} \rangle & \dots & \langle \mu_{k_1,l_n}, \nu_{k_1,l_n} \rangle \\ \vdots & \vdots & \ddots & \vdots & \ddots & \vdots \\ k_i & \langle \mu_{k_i,l_1}, \nu_{k_i,l_1} \rangle & \dots & \langle \mu_{k_i,l_j}, \nu_{k_i,l_j} \rangle & \dots & \langle \mu_{k_i,l_n}, \nu_{k_i,l_n} \rangle \\ \vdots & \vdots & \ddots & \vdots & \ddots & \vdots \\ k_m & \langle \mu_{k_m,l_1}, \nu_{k_m,l_1} \rangle & \dots & \langle \mu_{k_m,l_j}, \nu_{k_m,l_j} \rangle & \dots & \langle \mu_{k_m,l_n}, \nu_{k_m,l_n} \rangle \end{array},$$

where for every $1 \leq i \leq m, 1 \leq j \leq n$: $V(a_{k_i,l_j}) = \langle \mu_{k_i,l_j}, \nu_{k_i,l_j} \rangle$ and $0 \leq \mu_{k_i,l_j}, \nu_{k_i,l_j}, \mu_{k_i,l_j} + \nu_{k_i,l_j} \leq 1$.

4 Intercriteria Analysis Applied over Intuitionistic Fuzzy Data

Now, following [4,7], here we describe shortly the intercriteria analysis, but from intuitionistic fuzzy point of view.

Let us have the set of objects $O = \{O_1, O_2, ..., O_n\}$ that must be evaluated by criteria from the set $C = \{C_1, C_2, ..., C_m\}$.

Let us have an IM

$$A = \begin{array}{c|ccccccc} & O_1 & \cdots & O_i & \cdots & O_j & \cdots & O_n \\ \hline C_1 & a_{C_1,O_1} & \cdots & a_{C_1,O_i} & \cdots & a_{C_1,O_j} & \cdots & a_{C_1,O_n} \\ \vdots & \vdots & \ddots & \vdots & \ddots & \vdots & \ddots & \vdots \\ C_k & a_{C_k,O_1} & \cdots & a_{C_k,O_i} & \cdots & a_{C_k,O_j} & \cdots & a_{C_k,O_n} \\ \vdots & \vdots & \ddots & \vdots & \ddots & \vdots & \ddots & \vdots \\ C_l & a_{C_l,O_1} & \cdots & a_{C_l,O_i} & \cdots & a_{C_l,O_j} & \cdots & a_{C_l,O_n} \\ \vdots & \vdots & \ddots & \vdots & \ddots & \vdots & \ddots & \vdots \\ C_m & a_{C_m,O_1} & \cdots & a_{C_m,O_i} & \cdots & a_{C_m,O_j} & \cdots & a_{C_m,O_n} \end{array},$$

where for every p, q $(1 \leq p \leq m,\ 1 \leq q \leq n)$:

(1) C_p is a criterion, taking part in the evaluation,
(2) O_q is an object, being evaluated.
(3) a_{C_p,O_q} is a variable, formula or $a_{C_p,O_q} = \langle \alpha_{C_p,O_q}, \beta_{C_p,O_q} \rangle$ is an intuitionistic fuzzy pair, that is comparable about relation R with the other a-objects, so that for each i, j, k: $R(a_{C_k,O_i}, a_{C_k,O_j})$ is defined. Let $\overline{R}$ be the dual relation

of R in the sense that if R is satisfied, then $\overline{R}$ is not satisfied and vice versa. For example, if "R" is the relation "<", then $\overline{R}$ is the relation ">", and vice versa.

Let $S^{\mu}_{k,l}$ be the number of cases in which

$$\langle \alpha_{C_k,O_i}, \beta_{C_k,O_i} \rangle \leq \langle \alpha_{C_k,O_j}, \beta_{C_k,O_j} \rangle$$

and

$$\langle \alpha_{C_l,O_i}, \beta_{C_l,O_i} \rangle \leq \langle \alpha_{C_l,O_j}, \beta_{C_l,O_j} \rangle,$$

or

$$\langle \alpha_{C_k,O_i}, \beta_{C_k,O_i} \rangle \geq \langle \alpha_{C_k,O_j}, \beta_{C_k,O_j} \rangle$$

and

$$\langle \alpha_{C_l,O_i}, \beta_{C_l,O_i} \rangle \geq \langle \alpha_{C_l,O_j}, \beta_{C_l,O_j} \rangle$$

are simultaneously satisfied.

Let $S^{\nu}_{k,l}$ be the number of cases in which

$$\langle \alpha_{C_k,O_i}, \beta_{C_k,O_i} \rangle \geq \langle \alpha_{C_k,O_j}, \beta_{C_k,O_j} \rangle$$

and

$$\langle \alpha_{C_l,O_i}, \beta_{C_l,O_i} \rangle \leq \langle \alpha_{C_l,O_j}, \beta_{C_l,O_j} \rangle,$$

or

$$\langle \alpha_{C_k,O_i}, \beta_{C_k,O_i} \rangle \leq \langle \alpha_{C_k,O_j}, \beta_{C_k,O_j} \rangle$$

and

$$\langle \alpha_{C_l,O_i}, \beta_{C_l,O_i} \rangle \geq \langle \alpha_{C_l,O_j}, \beta_{C_l,O_j} \rangle$$

are simultaneously satisfied.

Obviously,

$$S^{\mu}_{k,l} + S^{\nu}_{k,l} \leq \frac{n(n-1)}{2}.$$

Now, for every k, l, such that $1 \leq k < l \leq m$ and for $n \geq 2$, we define

$$\mu_{C_k,C_l} = 2\frac{S^{\mu}_{k,l}}{n(n-1)}, \quad \nu_{C_k,C_l} = 2\frac{S^{\nu}_{k,l}}{n(n-1)}.$$

Hence,

$$\mu_{C_k,C_l} + \nu_{C_k,C_l} = 2\frac{S^{\mu}_{k,l}}{n(n-1)} + 2\frac{S^{\nu}_{k,l}}{n(n-1)} \leq 1.$$

Therefore, $\langle \mu_{C_k,C_l}, \nu_{C_k,C_l} \rangle$ is an IFP. Now, we can construct the IM

$$\begin{array}{c|ccc} & C_1 & \cdots & C_m \\ \hline C_1 & \langle \mu_{C_1,C_1}, \nu_{C_1,C_1} \rangle & \cdots & \langle \mu_{C_1,C_m}, \nu_{C_1,C_m} \rangle \\ \vdots & \vdots & \ddots & \vdots \\ C_m & \langle \mu_{C_m,C_1}, \nu_{C_m,C_1} \rangle & \cdots & \langle \mu_{C_m,C_m}, \nu_{C_m,C_m} \rangle \end{array},$$

that determines the degrees of correspondence between criteria $C_1, \ldots, C_m$.

5 An Example

Here, we discuss the results of application of the ICA over truth-values of some formulas, a part of which – axioms of intuitionistic logic IL [13]. Let us use the following formulas:

$$F1 : A \to A$$
$$F2 : A \to (B \to A)$$
$$F3 : A \to (B \to (A \wedge B))$$
$$F4 : (\neg A \vee B) \to (A \to B)$$
$$F5 : \neg(A \vee B) \to (\neg A \wedge \neg B)$$
$$F6 : (\neg A \wedge \neg B) \to \neg(A \vee B)$$

Let variables A and B can obtain values from set $\{\langle 0, 0\rangle, \langle 0, 0.5\rangle, \langle 0, 1\rangle, \langle 0.5, 0\rangle, \langle 0.5, 0.5\rangle, \langle 1, 0\rangle\}$. Then the truth-values of the six formulas for respective values of variables A and B are given in Table 1.

Now, interpreting the separate pairs of values of A and B as objects and the formulas as criteria, we can apply the ICA over above data and in a result, we will obtain the IM R of the relations between the separate axions, as follows

$$R = [\{F1, F2, F3, F4, F5, F6\}, \{F1, F2, F3, F4, F5, F6\}, \{\langle \mu_{i,j}, \nu_{i,j}\rangle\}]$$

=

	$F1$	$F2$	$F3$	$F4$	$F5$	$F6$
$F1$	⟨1.00,0.00⟩	⟨0.98,0.00⟩	⟨0.82,0.00⟩	⟨0.76,0.00⟩	⟨0.76,0.00⟩	⟨0.66,0.01⟩
$F2$	⟨0.98,0.00⟩	⟨1.00,0.00⟩	⟨0.84,0.00⟩	⟨0.79,0.00⟩	⟨0.78,0.00⟩	⟨0.68,0.01⟩
$F3$	⟨0.82,0.00⟩	⟨0.84,0.00⟩	⟨1.00,0.00⟩	⟨0.89,0.00⟩	⟨0.89,0.00⟩	⟨0.83,0.00⟩
$F4$	⟨0.75,0.00⟩	⟨0.78,0.00⟩	⟨0.89,0.00⟩	⟨1.00,0.00⟩	⟨1.00,0.00⟩	⟨0.89,0.00⟩
$F5$	⟨0.75,0.00⟩	⟨0.78,0.00⟩	⟨0.89,0.00⟩	⟨1.00,0.00⟩	⟨1.00,0.00⟩	⟨0.89,0.00⟩
$F6$	⟨0.66,0.01⟩	⟨0.68,0.01⟩	⟨0.83,0.00⟩	⟨0.89,0.00⟩	⟨0.89,0.00⟩	⟨1.00,0.00⟩

From IM R we see, e.g., that formulas F_4 and F_5 coincide, which really is true. Value of R that corresponds to formulas F_1 and F_2, which are tautologies in classical logic, are also in very near to point $\langle 1, 0\rangle$. In this case, following [7], we say that they are in strong consonance.

Now, following the idea from [5], we can show the geometrical interpretation of the elements of the IM R in the intuitionistic fuzzy interpretational triangle from Fig. 2.

Table 1. Intuitionistic fuzzy evaluations of formulas F_1 to F_6.

A	B	F1	F2	F3	F4	F5	F6
$\langle 0,0\rangle$	$\langle 0,0\rangle$	$\langle 0,0\rangle$	$\langle 0,0\rangle$	$\langle 0,0\rangle$	$\langle 0,0\rangle$	$\langle 0,0\rangle$	$\langle 0,0\rangle$
$\langle 0,0.5\rangle$	$\langle 0,0\rangle$	$\langle 0.5,0\rangle$	$\langle 0.5,0\rangle$	$\langle 0.5,0\rangle$	$\langle 0.5,0\rangle$	$\langle 0,0\rangle$	$\langle 0,0\rangle$
$\langle 0,1\rangle$	$\langle 0,0\rangle$	$\langle 1,0\rangle$	$\langle 1,0\rangle$	$\langle 1,0\rangle$	$\langle 1,0\rangle$	$\langle 0,0\rangle$	$\langle 0,0\rangle$
$\langle 0.5,0\rangle$	$\langle 0,0\rangle$	$\langle 0.5,0\rangle$	$\langle 0.5,0\rangle$	$\langle 0,0\rangle$	$\langle 0,0\rangle$	$\langle 0.5,0\rangle$	$\langle 0.5,0\rangle$
$\langle 0.5,0.5\rangle$	$\langle 0,0\rangle$	$\langle 0.5,0.5\rangle$	$\langle 0.5,0\rangle$	$\langle 0.5,0\rangle$	$\langle 0.5,0\rangle$	$\langle 0.5,0\rangle$	$\langle 0.5,0\rangle$
$\langle 1,0\rangle$	$\langle 0,0\rangle$	$\langle 1,0\rangle$	$\langle 1,0\rangle$	$\langle 0,0\rangle$	$\langle 0,0\rangle$	$\langle 1,0\rangle$	$\langle 1,0\rangle$
$\langle 0,0\rangle$	$\langle 0,0.5\rangle$	$\langle 0,0\rangle$	$\langle 0.5,0\rangle$	$\langle 0.5,0\rangle$	$\langle 0,0\rangle$	$\langle 0,0\rangle$	$\langle 0,0\rangle$
$\langle 0,0.5\rangle$	$\langle 0,0.5\rangle$	$\langle 0.5,0\rangle$	$\langle 0.5,0\rangle$	$\langle 0.5,0\rangle$	$\langle 0.5,0\rangle$	$\langle 0.5,0\rangle$	$\langle 0.5,0\rangle$
$\langle 0,1\rangle$	$\langle 0,0.5\rangle$	$\langle 1,0\rangle$	$\langle 1,0\rangle$	$\langle 1,0\rangle$	$\langle 1,0\rangle$	$\langle 0.5,0\rangle$	$\langle 0.5,0\rangle$
$\langle 0.5,0\rangle$	$\langle 0,0.5\rangle$	$\langle 0.5,0\rangle$	$\langle 0.5,0\rangle$	$\langle 0.5,0\rangle$	$\langle 0.5,0\rangle$	$\langle 0.5,0\rangle$	$\langle 0.5,0\rangle$
$\langle 0.5,0.5\rangle$	$\langle 0,0.5\rangle$	$\langle 0.5,0.5\rangle$	$\langle 0.5,0\rangle$	$\langle 0.5,0\rangle$	$\langle 0.5,0.5\rangle$	$\langle 0.5,0.5\rangle$	$\langle 0.5,0.5\rangle$
$\langle 1,0\rangle$	$\langle 0,0.5\rangle$	$\langle 1,0\rangle$	$\langle 1,0\rangle$	$\langle 0.5,0\rangle$	$\langle 0.5,0\rangle$	$\langle 1,0\rangle$	$\langle 1,0\rangle$
$\langle 0,0\rangle$	$\langle 0,1\rangle$	$\langle 0,0\rangle$	$\langle 1,0\rangle$	$\langle 1,0\rangle$	$\langle 0,0\rangle$	$\langle 0,0\rangle$	$\langle 0,0\rangle$
$\langle 0,0.5\rangle$	$\langle 0,1\rangle$	$\langle 0.5,0\rangle$	$\langle 1,0\rangle$	$\langle 1,0\rangle$	$\langle 0.5,0\rangle$	$\langle 0.5,0\rangle$	$\langle 0.5,0\rangle$
$\langle 0,1\rangle$	$\langle 0,1\rangle$	$\langle 0.1,0\rangle$	$\langle 1,0\rangle$	$\langle 1,0\rangle$	$\langle 0.1,0\rangle$	$\langle 0.1,0\rangle$	$\langle 0.1,0\rangle$
$\langle 0.5,0\rangle$	$\langle 0,1\rangle$	$\langle 0.5,0\rangle$	$\langle 1,0\rangle$	$\langle 1,0\rangle$	$\langle 0.5,0\rangle$	$\langle 0.5,0\rangle$	$\langle 0.5,0\rangle$
$\langle 0.5,0.5\rangle$	$\langle 0,1\rangle$	$\langle 0.5,0.5\rangle$	$\langle 1,0\rangle$	$\langle 1,0\rangle$	$\langle 0.5,0.5\rangle$	$\langle 0.5,0.5\rangle$	$\langle 0.5,0.5\rangle$
$\langle 1,0\rangle$	$\langle 0,1\rangle$	$\langle 1,0\rangle$	$\langle 1,0\rangle$	$\langle 1,0\rangle$	$\langle 1,0\rangle$	$\langle 1,0\rangle$	$\langle 1,0\rangle$
$\langle 0,0\rangle$	$\langle 0.5,0\rangle$	$\langle 0,0\rangle$	$\langle 0,0\rangle$	$\langle 0,0\rangle$	$\langle 0.5,0\rangle$	$\langle 0.5,0\rangle$	$\langle 0.5,0\rangle$
$\langle 0,0.5\rangle$	$\langle 0.5,0\rangle$	$\langle 0.5,0\rangle$	$\langle 0.5,0\rangle$	$\langle 0.5,0\rangle$	$\langle 0.5,0\rangle$	$\langle 0.5,0\rangle$	$\langle 0.5,0\rangle$
$\langle 0,1\rangle$	$\langle 0.5,0\rangle$	$\langle 1,0\rangle$	$\langle 1,0\rangle$	$\langle 1,0\rangle$	$\langle 1,0\rangle$	$\langle 0.5,0\rangle$	$\langle 0.5,0\rangle$
$\langle 0.5,0\rangle$	$\langle 0.5,0\rangle$	$\langle 0.5,0\rangle$	$\langle 0.5,0\rangle$	$\langle 0.5,0\rangle$	$\langle 0.5,0\rangle$	$\langle 0.5,0\rangle$	$\langle 0.5,0\rangle$
$\langle 0.5,0.5\rangle$	$\langle 0.5,0\rangle$	$\langle 0.5,0.5\rangle$	$\langle 0.5,0.5\rangle$	$\langle 0.5,0.5\rangle$	$\langle 0.5,0\rangle$	$\langle 0.5,0\rangle$	$\langle 0.5,0\rangle$
$\langle 1,0\rangle$	$\langle 0.5,0\rangle$	$\langle 1,0\rangle$	$\langle 1,0\rangle$	$\langle 0.5,0\rangle$	$\langle 0.5,0\rangle$	$\langle 1,0\rangle$	$\langle 1,0\rangle$
$\langle 0,0\rangle$	$\langle 0.5,0.5\rangle$	$\langle 0,0\rangle$	$\langle 0.5,0\rangle$	$\langle 0.5,0\rangle$	$\langle 0.5,0\rangle$	$\langle 0.5,0\rangle$	$\langle 0.5,0\rangle$
$\langle 0,0.5\rangle$	$\langle 0.5,0.5\rangle$	$\langle 0.5,0\rangle$	$\langle 0.5,0\rangle$	$\langle 0.5,0\rangle$	$\langle 0.5,0\rangle$	$\langle 0.5,0.5\rangle$	$\langle 0.5,0.5\rangle$
$\langle 0,1\rangle$	$\langle 0.5,0.5\rangle$	$\langle 1,0\rangle$	$\langle 1,0\rangle$	$\langle 1,0\rangle$	$\langle 1,0\rangle$	$\langle 0.5,0.5\rangle$	$\langle 0.5,0.5\rangle$
$\langle 0.5,0\rangle$	$\langle 0.5,0.5\rangle$	$\langle 0.5,0\rangle$	$\langle 0.5,0\rangle$	$\langle 0.5,0.5\rangle$	$\langle 0.5,0.5\rangle$	$\langle 0.5,0\rangle$	$\langle 0.5,0\rangle$
$\langle 0.5,0.5\rangle$	$\langle 0.5,0.5\rangle$	$\langle 0.5,0.5\rangle$	$\langle 0.5,0.5\rangle$	$\langle 0.5,0.5\rangle$	$\langle 0.5,0.5\rangle$	$\langle 0.5,0.5\rangle$	$\langle 0.5,0.5\rangle$
$\langle 1,0\rangle$	$\langle 0.5,0.5\rangle$	$\langle 1,0\rangle$	$\langle 1,0\rangle$	$\langle 0.5,0.5\rangle$	$\langle 0.5,0.5\rangle$	$\langle 1,0\rangle$	$\langle 1,0\rangle$
$\langle 0,0\rangle$	$\langle 1,0\rangle$	$\langle 0,0\rangle$	$\langle 0,0\rangle$	$\langle 0,0\rangle$	$\langle 1,0\rangle$	$\langle 1,0\rangle$	$\langle 1,0\rangle$
$\langle 0,0.5\rangle$	$\langle 1,0\rangle$	$\langle 0.5,0\rangle$	$\langle 0.5,0\rangle$	$\langle 0.5,0\rangle$	$\langle 1,0\rangle$	$\langle 1,0\rangle$	$\langle 1,0\rangle$
$\langle 0,1\rangle$	$\langle 1,0\rangle$	$\langle 1,0\rangle$	$\langle 1,0\rangle$	$\langle 1,0\rangle$	$\langle 1,0\rangle$	$\langle 1,0\rangle$	$\langle 1,0\rangle$
$\langle 0.5,0\rangle$	$\langle 1,0\rangle$	$\langle 0.5,0\rangle$	$\langle 0.5,0\rangle$	$\langle 0.5,0\rangle$	$\langle 1,0\rangle$	$\langle 1,0\rangle$	$\langle 1,0\rangle$
$\langle 0.5,0.5\rangle$	$\langle 1,0\rangle$	$\langle 0.5,0.5\rangle$	$\langle 0.5,0.5\rangle$	$\langle 0.5,0.5\rangle$	$\langle 1,0\rangle$	$\langle 1,0\rangle$	$\langle 1,0\rangle$
$\langle 1,0\rangle$	$\langle 1,0\rangle$	$\langle 1,0\rangle$	$\langle 1,0\rangle$	$\langle 1,0\rangle$	$\langle 1,0\rangle$	$\langle 1,0\rangle$	$\langle 1,0\rangle$

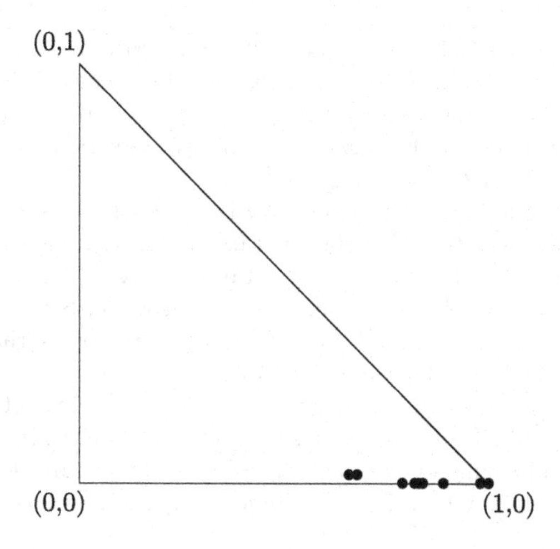

Fig. 2. Geometrical interpretation of the elements of the IM R.

6 Conclusion

In the presented research, an Intercriteria Analysis over intuitionistic fuzzy data is discussed. It extends the area of objects which the Intercriteria Analysis uses. Intercriteria Analysis can be applied over intuitionistic fuzzy data to determine possible correlations between the pairs of criteria.

In a next research of the authors, the above described constructions will be extend to the case of 3-dimensional IMs.

Acknowledgments. The authors are thankful for the support provided by the Bulgarian National Science Fund under Grant Ref. No. DFNI-I-02-5.

References

1. Angelova, N., Atanassov, K., Riecan, B.: Intercriteria analysis of the intuitionistic fuzzy implication properties. In: Proceedings of the 11th International Workshop on Intuitionistic Fuzzy Sets, 30 October 2015, Banská Bystrica, Slovakia (2015). Int. J. Notes Intuitionistic Fuzzy Sets **21**(5), 20–23 (2015)
2. Atanassov, K.: Generalized index matrices. Comptes rendus de l'Academie Bulgare des Sciences **40**(11), 15–18 (1987)
3. Atanassov, K.: On Intuitionistic Fuzzy Sets Theory. STUDFUZZ, vol. 283. Springer, Heidelberg (2012). https://doi.org/10.1007/978-3-642-29127-2
4. Atanassov, K.T.: Index Matrices: Towards an Augmented Matrix Calculus. SCI, vol. 573. Springer, Cham (2014). https://doi.org/10.1007/978-3-319-10945-9
5. Atanassov, K., Atanassova, V., Gluhchev, G.: InterCriteria analysis: ideas and problems. Notes Intuitionistic Fuzzy Sets **21**(1), 81–88 (2015)
6. Atanassov, K., Atanassova, V., Chountas, P., Mitkova, M., Sotirova, E., Sotirov, S., Stratiev, D.: Intercriteria analysis over normalized data. In: Proceedings of the 8th IEEE Conference Intelligent Systems, Sofia, 4–6 September 2016, pp. 136–138 (2016)

7. Atanassov, K., Mavrov, D., Atanassova, V.: Intercriteria decision making: a new approach for multicriteria decision making, based on index matrices and intuitionistic fuzzy sets. Issues Intuitionistic Fuzzy Sets Generalized Nets **11**, 1–8 (2014)
8. Atanassov, K., Szmidt, E., Kacprzyk, J.: On intuitionistic fuzzy pairs. Notes Intuitionistic Fuzzy Sets **19**(3), 1–13 (2013)
9. Atanassova, V., Doukovska, L., Mavrov, D., Atanassov, K.: InterCriteria decision making approach to EU member states competitiveness analysis: temporal and threshold analysis. In: Angelov, P., Atanassov, K.T., Doukovska, L., Hadjiski, M., Jotsov, V., Kacprzyk, J., Kasabov, N., Sotirov, S., Szmidt, E., Zadrożny, S. (eds.) Intelligent Systems'2014. AISC, vol. 322, pp. 95–106. Springer, Cham (2015). https://doi.org/10.1007/978-3-319-11313-5_9
10. Atanassova, V., Doukovska, L., Atanassov, K., Mavrov, D.: InterCriteria decision making approach to EU member states competitive analysis. In: Proceedings of 4th International Symposium on Business Modeling and Software Design, Luxembourg, Grand Duchy of Luxembourg, 24–26 June 2014, pp. 289–294 (2014). ISBN 978-989-758-032-1
11. Atanassova, V., Mavrov, D., Doukovska, L., Atanassov, K.: Discussion on the threshold values in the InterCriteria decision making approach. Notes Intuitionistic Fuzzy Sets **20**(2), 94–99 (2014)
12. Kolev, B., El-Darzi, E., Sotirova, E., Petronias, I., Atanassov, K., Chountas, P., Kodogianis, V.: Generalized Nets in Artificial Intelligence, vol. 3. Generalized Nets, Relational Data Bases and Expert Systems. "Prof. M. Drinov" Academic Publishing House, Sofia (2006)
13. Rasiowa, H., Sikorski, R.: The Mathematics of Metamathematics. Polish Academy of Sciences, Warszawa (1963)
14. Stratiev, D., Sotirov, S., Shishkova, I., Nedelchev, A., Sharafutdinov, I., Vely, A., Mitkova, M., Yordanov, D., Sotirova, E., Atanassova, V., Atanassov, K., Stratiev, D.D., Rudnev, N., Ribagin, S.: Investigation of relationships between bulk properties and fraction properties of crude oils by application of the intercriteria analysis. Petroleum Sci. Technol. **34**(13), 1113–1120 (2016)
15. Todinova, S., Mavrov, D., Krumova, S., Marinov, P., Atanassova, V., Atanassov, K., Taneva, S.: Blood plasma thermograms dataset analysis by means of intercriteria and correlation analyses for the case of colorectal cancer. Int. J. Bioautomation **20**(1), 115–124 (2016)

Genetic Algorithm with Optimal Recombination for the Asymmetric Travelling Salesman Problem

Anton V. Eremeev and Yulia V. Kovalenko(✉)

Sobolev Institute of Mathematics, 4, Akad. Koptyug avenue,
630090 Novosibirsk, Russia
eremeev@ofim.oscsbras.ru, julia.kovalenko.ya@yandex.ru

Abstract. We propose a new genetic algorithm with optimal recombination for the asymmetric instances of travelling salesman problem. The algorithm incorporates several new features that contribute to its effectiveness: 1. Optimal recombination problem is solved within crossover operator. 2. A new mutation operator performs a random jump within 3-opt or 4-opt neighborhood. 3. Greedy constructive heuristic of Zhang and 3-opt local search heuristic are used to generate the initial population. A computational experiment on TSPLIB instances shows that the proposed algorithm yields competitive results to other well-known memetic algorithms for asymmetric travelling salesman problem.

Keywords: Genetic algorithm · Optimal recombination · Local search

1 Introduction

Travelling Salesman Problem (TSP) is a well-known NP-hard combinatorial optimization problem [8]. Given a complete digraph G with the set of vertices $V = \{v_1, \ldots, v_n\}$, the set of arcs $A = \{(v_i, v_j) : v_i, v_j \in V, i \neq j\}$ and arc weights (lengths) $c_{ij} \geq 0$ of each arc $(v_i, v_j) \in A$, the TSP asks for a Hamiltonian circuit of minimum length. If $c_{ij} \neq c_{ji}$ for at least one pair (v_i, v_j) then the TSP is called the Asymmetric Travelling Salesman Problem (ATSP). Numerous metaheuristics and heuristics have been proposed for the TSP and the genetic algorithms (GAs) are among them (see e.g. [2,3,7,16,18]).

The performance of GAs depends significantly upon the choice of the *crossover* operator, where the components of parent solutions are combined to build the offspring. A supplementary problem that emerges in some versions of crossover operator is called *Optimal Recombination Problem* (ORP). Given two feasible parent solutions, ORP consists in finding the best possible offspring in view of the basic principles of crossover [13]. Experimental results [3,16,18] indicate that ORP may be used successfully in genetic algorithms.

Y. V. Kovalenko—This research is supported by the Russian Science Foundation grant 15-11-10009.

I. Lirkov and S. Margenov (Eds.): LSSC 2017, LNCS 10665, pp. 341–349, 2018.
https://doi.org/10.1007/978-3-319-73441-5_36

In this paper, we propose a new GA using the ORP with adjacency-based representation to solve the ATSP. Two simple crossover-based GAs for ATSP using ORPs were investigated in [7] but no problem-specific local search procedures or fine-tuning of parameters were used. In comparison to the GAs from [7], the GA proposed in this paper uses a 3-opt local search heuristic and a problem-specific heuristic of Zhang [19] to generate the initial population. In addition, this GA applies a new mutation operator, which performs a random jump within 3-opt or 4-opt neighborhood. The current GA is based on the steady state replacement [14], while the GAs in [7] were based on the elitist recombination (see e.g. [9]). The experimental evaluation on instances from TSPLIB library shows that the proposed GA yields results competitive to those obtained by some other well-known evolutionary algorithms for the ATSP.

2 Genetic Algorithm

The genetic algorithm is a random search method that models a process of evolution of a population of *individuals* [14]. Each individual is a sample solution to the optimization problem being solved. The components of an individual are called *genes*. Individuals of a new population are built by means of reproduction operators (crossover and/or mutation). The crossover operator produces the offspring from two parent individuals by combining and exchanging their genes. The mutation adds small random changes to an individual.

The formal scheme of the GA with steady state replacement is as follows:

Steady State Genetic Algorithm

STEP 1. Construct the initial population and assign $t := 1$.

STEP 2. Repeat steps 2.1–2.4 until some stopping criterion is satisfied:

2.1. Choose two parent individuals $\mathbf{p}_1, \mathbf{p}_2$ from the population.

2.2. Apply mutation to $\mathbf{p}_1$ and $\mathbf{p}_2$ and obtain individuals $\mathbf{p}'_1, \mathbf{p}'_2$.

2.3. Create an offspring $\mathbf{p}'$, applying a crossover to $\mathbf{p}'_1$ and $\mathbf{p}'_2$.

2.4. Choose a least fit individual in population and replace it by $\mathbf{p}'$.

2.5. Set $t := t + 1$.

STEP 3. The result is the best found individual w.r.t. objective function.

Our implementation of the GA is initiated by generating N initial solutions, and the population size N remains constant during the execution of the GA. Two individuals of the initial population are constructed by means of the problem-specific heuristic of Zhang [19]. The heuristic first solves the Assignment Problem, and then patches the cycles of the optimum assignment together to form a feasible tour. Karp [12] proposed two variants of the patching. In the first one, some cycle of maximum length is selected and the remaining cycles are patched into it. In the second one, cycles are patched one by one in a special sequence, starting with a shortest cycle. All other $N-2$ individuals of the initial population are generated using the *arbitrary insertion* method [18], followed by a local search heuristic with a 3-opt neighborhood (see Subsect. 2.2).

Each parent on Step 2.1 is chosen by *s-tournament selection*: sample randomly s individuals from the current population and select a fittest among them.
Operators of crossover and mutation are described in Subsects. 2.1 and 2.3.

2.1 Recombination Operators

Suppose that feasible solutions to the ATSP are encoded as vectors of adjacencies, where the immediate predecessor is indicated for each vertex. Then the *optimal recombination problem* with adjacency-based representation [6] consists in finding a shortest travelling salesman's tour which coincides with two given feasible parent solutions in arcs belonging to both solutions and does not contain the arcs absent in both solutions. These constraints are equivalent to a requirement that the recombination should be respectful and gene transmitting as defined in [13]. The ORP with adjacency-based representation for the ATSP is shown to be NP-hard but it can be reduced to the TSP on graphs with bounded vertex degrees [6]. The resulting TSP may be solved in $O(n2^{\frac{n}{2}})$ time by means of an adaptation of the algorithm proposed by Eppstein [5]. A detailed description of the reduction can be found in [6]. An experimental evaluation of the ORP with adjacency-based representation in a crossover-based GA was carried out in [7]. The experiments showed that the CPU cost of solving the ORPs in this GA is acceptable and decreases with iterations count, due to decreasing population diversity. In what follows, the optimized crossover operator, which solves the ORP with adjacency-based representation will be called Optimized Directed Edge Crossover (ODEC). This operator may be considered as a deterministic "direct descendant" of Directed Edge Crossover (DEC) [17]. Unlike DEC, Optimized Directed Edge Crossover guarantees genes transmission.

An alternative way for solution encoding to the ATSP is the position-based representation, where a feasible solution is encoded as a sequence of the vertices of the TSP tour. The computational experiment performed in [7] indicates that the ORP for the adjacency-based representation has an advantage over the ORP for the position-based representation on ATSP instances from TSPLIB library. Therefore, in this paper we consider only the adjacency-based representation.

Note that most of the known GAs for the TSP (see e.g. [2,10,16]) apply a local search on GA iterations. However an optimal recombination may be considered as a best-improving move in a neighborhood defined by two parent solutions. So we use a local search only at the initialization stage.

2.2 Local Search Heuristic

In general, k-opt neighborhood for TSP is defined as the set of tours that can be obtained from a given tour by replacing k arcs. Our Local Search Heuristic is a typical local search heuristic that explores a subset of 3-opt neighborhood.

We try to improve the current tour by changing three of its arcs (see Fig. 1). To this end, we consider all possible arcs of the current tour as candidates for arc (v_{i_1}, v_{i_2}) to be deleted in the order of decreasing length. Observe that, in our search, the possibilities for choosing v_{i_3} (arc (v_{i_1}, v_{i_3}) is added) may be limited

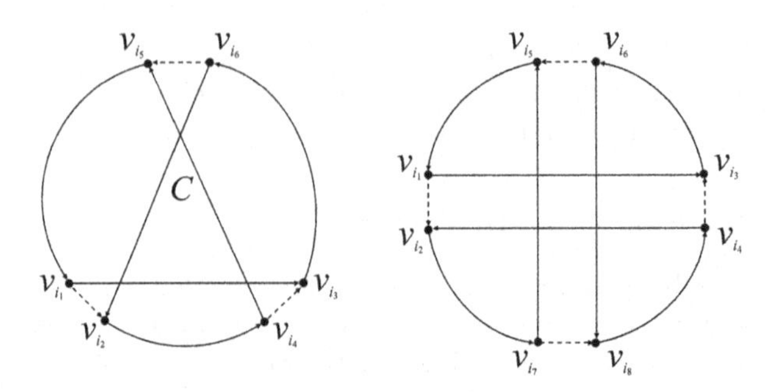

Fig. 1. 3-change and quad change.

to those vertices that are closer to v_{i_1} than v_{i_2}. To use this property, for each vertex v we store a list of the remaining vertices in the order of increasing length from v. Considering candidates for v_{i_3}, we start at the beginning of v_{i_1}'s list and proceed down the list until a vertex u with $c_{v_{i_1},u} \geq c_{v_{i_1},v_{i_2}}$ is reached. Moreover, only the $\lceil 0.2n \rceil$ nearest vertices are stored in the sorted list of each vertex, which allows to reduce the running time and the memory usage as observed in [10]. Finally, among all vertices belonging to the closed cycle C created by (v_{i_1}, v_{i_3}), we choose a vertex v_{i_5} that would produce the most favorable 3-change. Local Search Heuristic stops if no favorable 3-change is possible, otherwise it proceeds to the next step with a new tour obtained.

In order to reduce the running time of the presented local search heuristic, we use the well-known "don't look bits" and "first improving move" strategies presented in [10] for local search based on 3-opt neighborhood.

2.3 Mutation Operators

The mutation is applied to each parent solution on Step 2.2 with probability p_{mut}, which is a tunable parameter of the GA. We implement two mutation operators that perform a random jump within 3-opt or 4-opt neighborhood. Each time one of the operators is used for mutation with equal probability.

The first mutation operator makes a 3-change (see Sect. 2.2). First of all, an arc (v_{i_1}, v_{i_2}) is chosen at random among all arcs of the given tour. After that, an arc (v_{i_1}, v_{i_3}) is chosen using the following idea from [11]. For each possible arc (v_{i_1}, u), we calculate $F(u) = c_{v,u} + |C(u)| \cdot c_{aver}$, where v is the immediate predecessor of u in the given tour, $|C(u)|$ is the length of the cycle created by u and c_{aver} is the average weight of arcs in graph G. Then an arc (v_{i_1}, v_{i_3}) is chosen uniformly at random among the top 50% of arcs (v_{i_1}, u) w.r.t. $F(u)$ value. The reason for taking into account the value of $|C(u)|$ when (v_{i_1}, v_{i_3}) is being chosen is that the bigger the value of $C(u)$ the more options for (v_{i_4}, v_{i_5}) will become available subsequently. Finally, we choose the arc (v_{i_6}, v_{i_5}) among the arcs of $C(v_{i_3})$ so that the most favorable 3-change is produced.

The second mutation operator is based on 4-opt neighborhood and implements a *quad change* [11] (see Fig. 1). Here two arcs (v_{i_1}, v_{i_2}) and (v_{i_7}, v_{i_8}) are selected randomly and removed, while the other two arcs for deletion are chosen so that the most favorable quad change is obtained.

3 Computational Experiments

This section presents the results of computational experiments on the ATSP instances from TSPLIB library [15]. The GA was programmed in Java (NetBeans IDE 7.2.1) and tested on a computer with Intel Core 2 Duo CPU E7200 2.53 GHz processor, 2 Gb RAM. In the experiments, we set the population size $N = 100$, the tournament size $s = 10$ and the mutation probability $p_{mut} = 0.1$.

Our GA is restarted every time as soon as the current iteration number becomes twice the iteration number when the best incumbent was found, until the overall execution time reached the limit. Moreover, if the greedy heuristic of Zhang generates only one subcycle, this indicates that the ATSP instance was solved to optimality, and the algorithm stops. The best solution found over all restarts was returned as the result. We have also tested an alternative approach, where the GA runs for the whole given period of time without restarts but it turned to be inferior to the GA with the restart rule.

The first computational experiment is aimed at comparison of the performance of our GA based on ODEC (GA_{ODEC}) with SAX/RAI memetic algorithm ($MA_{SAX/RAI}$) from [2], which has one of the best results in the literature on metaheuristics for the ATSP. In order to put the considered algorithms into equal positions, GA_{ODEC} was given the CPU-time limit (denoted as T) by a factor 3 less than the CPU resource used by $MA_{SAX/RAI}$ in [2]. This scaling factor chosen on the basis a rough comparison of computers by means of performance table [4]. For a statistical comparison, on each instance we executed GA_{ODEC} 1000 times. In each execution GA_{ODEC} was given the same CPU-time limit indicated above. In [2], $MA_{SAX/RAI}$ was run 20 times on each instance. Table 1 shows the obtained results, where F_{opt} represents the frequency of finding an optimum, Δ_{err} is the average percentage deviation of the length of a resulting solution from the optimum, Δ_{init} denotes the average percentage deviation of the length of the best initial solution from the optimum. As seen from Table 1, GA_{ODEC} achieved 100% success rate on 17 out of 26 instances. On each instance, GA_{ODEC} found optima in not less than 91% of runs.

The statistical analysis of experimental data was carried out using a significance test of the null hypothesis from [1], Chap. 8, Sect. 2. Suppose that two algorithms are compared in terms of probability of "success", where "success" corresponds to finding an optimal solution. Let P_1 and P_2 denote the probabilities of success for the considered algorithms. The null hypothesis is expressed by $P_1 = P_2$.

The test procedure is as follows. Under the null hypothesis, the estimate of common success rate is $\hat{P} = \frac{\hat{P}_1 N_1 + \hat{P}_2 N_2}{N_1 + N_2}$, where $\hat{P}_1$ denotes the frequency of success in N_1 runs for the first algorithm and $\hat{P}_2$ is the frequency of success in N_2 runs for the second algorithm. Then the difference $\hat{P}_1 - \hat{P}_2$ is expressed in units of the standard deviation by calculating the statistic $A = \frac{|\hat{P}_1 - \hat{P}_2|}{\hat{SD}}$, where $\hat{SD} = \sqrt{\frac{\hat{P}(1-\hat{P})}{N_1} + \frac{\hat{P}(1-\hat{P})}{N_2}}$ is the estimation of the standard deviation.

It is supposed that statistic A is normally distributed. To test the null hypothesis versus the alternative one at a confidence level α, we compare the computed

Table 1. Computational results for the ATSP instances

Instance	Genetic algorithms								$MA_{SAX/RAI}$			
	GA_{ODEC}			GA_{ER}				T,				
	Δ_{init}	F_{opt}	Δ_{err}	Δ_{init}	F_{opt}	Δ_{err}	A	sec.	Δ_{init}	F_{opt}	Δ_{err}	A
ftv33	0.00	1	0.00	0.00	1	0.00	0	0.097	12.83	1	0.00	0
ftv35	0.00	1	0.00	0.00	1	0.00	0	0.11	0.14	1	0.00	0
ftv38	0.00	1*	0.00	0.131	1	0.00	0	0.103	0.13	0.25	0.10	27.6
p43	0.00	1*	0.00	0.00	1	0.00	0	0.16	0.05	0.55	0.01	21.3
ftv44	0.098	1*	0.00	0.167	0.874	0.078	11.6	0.137	7.01	0.35	0.44	25.7
ftv47	0.199	1	0.00	0.338	1	0.00	0	0.157	2.70	1	0.00	0
ry48p	0.978	**0.997***	0.0001	3.511	0.520	0.092	24.9	0.187	5.42	0.85	0.03	8.8
ft53	0.438	1	0.00	5.073	0.668	0.035	19.9	0.187	18.20	1	0.00	0
ftv55	0.002	1	0.00	0.002	1	0.00	0	0.167	3.61	1	0.00	0
ftv64	0.032	1	0.00	0.376	0.989	0.002	3.3	0.22	3.81	1	0.00	0
ft70	0.367	1*	0.00	0.321	0.583	0.013	22.9	0.32	1.88	0.4	0.03	24.6
ftv70	1.025	1*	0.00	1.525	0.660	0.098	20.2	0.277	3.33	0.95	0.01	7.1
ftv90	0.063	0.976	0.003	0.318	0.516	0.007	23.6	0.317	3.67	1	0.00	0.7
ftv100	0.386	0.92	0.013	1.092	0.784	0.016	8.6	0.4	3.24	1	0.00	1.3
kro124p	0.164	**0.996***	0.0001	0.288	0.322	0.033	31.8	0.457	6.46	0.90	0.01	5.6
ftv110	0.287	**0.972**	0.003	0.305	0.854	0.025	9.4	0.57	4.7	0.90	0.02	1.9
ftv120	0.156	**0.912***	0.008	2.463	0.430	0.506	22.9	0.73	8.31	0.35	0.14	8.3
ftv130	0.342	**0.934**	0.008	1.841	0.361	0.068	26.8	0.727	3.12	0.90	0.01	0.6
ftv140	0.111	**0.947***	0.004	0.601	0.463	0.065	23.7	0.887	2.23	0.70	0.08	4.7
ftv150	0.739	**0.982***	0.002	1.358	0.532	0.068	23.5	0.897	2.3	0.90	0.01	2.6
ftv160	0.026	1*	0.00	0.958	0.491	0.099	26.1	1.093	1.71	0.80	0.02	14.2
ftv170	0.108	1*	0.00	0.334	0.222	0.141	35.7	1.307	1.38	0.75	0.05	15.9
rbg323	0.00	1	0.00	0.00	1	0.00	0	0.03	0.00	1	0.00	0
rbg358	0.00	1	0.00	0.00	1	0.00	0	0.03	0.00	1	0.00	0
rbg403	0.00	1	0.00	0.00	1	0.00	0	0.032	0.00	1	0.00	0
rbg443	0.00	1	0.00	0.00	1	0.00	0	0.033	0.00	1	0.00	0
Average	0.212	0.986	0.0016	0.808	0.741	0.0518	12.9	0.371	3.701	0.829	0.0369	6.6

A to the quantile of standard normal distribution $z_{\alpha/2}$. If A is larger than $z_{\alpha/2}$, the null hypothesis is rejected. Otherwise the null hypothesis is accepted. At $\alpha = 0.05$ we have $z_{0.025} = 1.96$. The values of statistic A for algorithms GA_{ODEC} and $MA_{SAX/RAI}$ are found and presented in the last column of Table 1 ('*' indicates the statistical significance difference between GA_{ODEC} and $MA_{SAX/RAI}$ at level $\alpha = 0.05$).

In 14 out of 26 instances, GA_{ODEC} finds an optimum more frequently than $MA_{SAX/RAI}$ (in 12 cases among these, the difference between the frequencies of finding an optimum is statistically significant). Both algorithms demonstrate 100% frequency of obtaining an optimum on 10 problems. Note that the heuristic of W. Zhang is very efficient on series rbg and the optimal solutions to all rbg instances were found in the considered algorithms at the initialization stage. $MA_{SAX/RAI}$ slightly outperforms GA_{ODEC} only on two instances ftv90 and ftv100, but the differences are not statistically significant. Moreover, the average quality of the resulting solutions for GA_{ODEC} is approximately in 23 times better than the average quality for $MA_{SAX/RAI}$. The quality of initial solutions is better in our algorithm. (Note that we use the local search at the initialization stage, while $MA_{SAX/RAI}$ applies a local search only on GA iterations.)

Recently, Tinós et al. [16] proposed a GA with new crossover operator GAPX, which presents very competitive results in terms of frequencies of finding an optimum, but its CPU resource usage is significantly higher than that of GA_{ODEC}. On all of 16 TSPLIB-instances tested in [16] GA with GAPX demonstrated 100% success, while GA_{ODEC} displayed 99.96% success on average. However, the average CPU-time T of our GA was 0.22 s on these instances, and the overall CPU-time of GA with GAPX was 98.38 s on a similar computer.

In the second experiment, we compare our steady state GA to the similar GA with the population management strategy known as *elitist recombination* [9] (GA_{ER}) under the same CPU-time limit. The results are also listed in Table 1. The eighth column represents the values of statistic A for comparison of GA_{ER} against GA_{ODEC} on all ATSP instances. We estimate the average frequency of finding optimal solutions for GA_{ER} as approximately 60% of the average frequency for GA_{ODEC} (the difference between the frequencies is statistically significant), except for 10 of 26 instances where both algorithms have 100% success. Note that the GA with elitist recombination maintains the population diversity better. Due to this reason, the GA with elitist recombination outperformed the steady state GA in our preliminary experiments with no restarts, which were organized analogously to the experiments in [7]. The restarts performed in GA_{ODEC} allow to avoid localization of the search and restore the population diversity, leading to better results.

We carried out the third experiment in order to compare the optimized crossover ODEC to its randomized prototype DEC. This experiment clearly showed an advantage of ODEC over DEC. The modification of GA_{ODEC}, where operator DEC substitutes ODEC, on average gave only 45% frequency of obtaining an optimum within the same CPU time limit. Moreover, for the large-scale problems such as ftv120, ftv130, ftv140, ftv150, and ftv170 the GA with DEC found optimal solution no more than once out of 1000 runs.

We also estimate that the average frequency of success of the GAs with optimal recombination reported in [7] is twice as small compared to such frequency for GA_{ODEC}, even though the GAs in [7] were given more CPU time.

4 Conclusions

We proposed a steady-state GA with adjacency-based representation using an optimal recombination and a local search to solve the ATSP. An experimental evaluation on instances from TSPLIB library shows that the proposed GA yields results competitive to those of some other state-of-the-art genetic algorithms. The experiments also indicate that the proposed GA dominates a similar GA based on the population management strategy, known as elitist recombination.

The restarts performed in the proposed GA allow to avoid localization of search and restore the population diversity, leading to better results when the steady-state population management is used. The experiments also show an advantage of the deterministic optimized crossover over its randomized prototype.

References

1. Brown, B.W., Hollander, M.: Statistics: A Biomedical Introduction. Wiley Inc., New York (1977)
2. Buriol, L.S., Franca, P.M., Moscato, P.: A new memetic algorithm for the asymmetric traveling salesman problem. J. Heuristics **10**, 483–506 (2004)
3. Cook, W., Seymour, P.: Tour merging via branch-decomposition. INFORMS J. Comput. **15**(2), 233–248 (2003)
4. Dongarra, J.J.: Performance of various computers using standard linear equations software. Technical Report CS-89-85, 110 p. University of Manchester (2014)
5. Eppstein, D.: The traveling salesman problem for cubic graphs. J. Graph Algorithms Appl. **11**(1), 61–81 (2007)
6. Eremeev, A.V., Kovalenko, J.V.: Optimal recombination in genetic algorithms for combinatorial optimization problems: Part II. Yugoslav J. Oper. Res. **24**(2), 165–186 (2014)
7. Eremeev, A.V., Kovalenko, J.V.: Experimental evaluation of two approaches to optimal recombination for permutation problems. In: Chicano, F., Hu, B., García-Sánchez, P. (eds.) EvoCOP 2016. LNCS, vol. 9595, pp. 138–153. Springer, Cham (2016). https://doi.org/10.1007/978-3-319-30698-8_10
8. Garey, M.R., Johnson, D.S.: Computers and Intractability. A Guide to the Theory of NP-completeness. W. H. Freeman and Company, San Francisco (1979)
9. Goldberg, D., Thierens, D.: Elitist recombination: An integrated selection recombination GA. In: First IEEE World Congress on Computational Intelligence, vol. 1, pp. 508–512. IEEE Service Center, Piscataway, New Jersey (1994)
10. Johnson, D.S., McGeorch, L.A.: The traveling salesman problem: a case study. In: Aarts, E., Lenstra, J.K. (eds.) Local Search in Combinatorial Optimization, pp. 215–336. Wiley Ltd. (1997)
11. Kanellakis, P.C., Papadimitriou, C.H.: Local search for the asymmetric traveling salesman problem. Oper. Res. **28**, 1086–1099 (1980)
12. Karp, R.M.: A patching algorithm for the nonsymmetric traveling-salesman problem. SIAM J. Comput. **8**, 561–573 (1979)
13. Radcliffe, N.J.: The algebra of genetic algorithms. Ann. Math. Artif. Intell. **10**(4), 339–384 (1994)
14. Reeves, C.R.: Genetic algorithms for the operations researcher. INFORMS J. Comput. **9**(3), 231–250 (1997)
15. Reinelt, G.: TSPLIB - a traveling salesman problem library. ORSA J. Comput. **3**(4), 376–384 (1991)
16. Tinós, R., Whitley, D., Ochoa, G.: Generalized asymmetric partition crossover (GAPX) for the asymmetric TSP. In: The 2014 Annual Conference on Genetic and Evolutionary Computation, pp. 501–508. ACM, New York (2014)
17. Whitley, D., Starkweather, T., Shaner, D.: The traveling salesman and sequence scheduling: Quality solutions using genetic edge recombination. In: Davis, L. (ed.) Handbook of Genetic Algorithms, pp. 350–372. Van Nostrand Reinhold (1991)

18. Yagiura, M., Ibaraki, T.: The use of dynamic programming in genetic algorithms for permutation problems. Eur. J. Oper. Res. **92**, 387–401 (1996)
19. Zhang, W.: Depth-first branch-and-bound versus local search: A case study. In: 17th National Conference on Artificial Intelligence, Austin, pp. 930–935 (2000)

Heuristic Algorithm for 2D Cutting Stock Problem

Georgi Evtimov and Stefka Fidanova(✉)

Institute of Information and Communication Technologies,
Bulgarian Academy of Sciences, Acad. G. Bonchev Str., bl. 25A,
1113 Sofia, Bulgaria
gevtimov@abv.bg, stefka@parallel.bas.bg

Abstract. Every day optimization problems arise in our life and industry. Many of them require huge amount of calculations and need special type of algorithms to be solved. An important industrial problem is cutting stock problem (CSP). Cutting with less possible waste is significant in some industries. The aim of this work is to cut 2D items from rectangular stock, minimizing the waste. Even the simplified version of the problem, when the items are rectangular is NP hard. When the number of items increases, the computational time increases exponentially. It is impossible to find the optimal solution for a reasonable time. Only for very small problems the exact algorithms and traditional numerical methods can be applied. We propose a stochastic algorithm which solves the problem, when the items are irregular polygons.

1 Introduction

The 2D CSP is a significant problem coming from the industry. Most of the authors solve the simplified model when the items are rectangular. CSP with rectangular items appear in paper and glass industries [6], container loading, Very-large-scale integration (VLSI) design, and various scheduling tasks [8]. When the items are not rectangular the problem becomes much more complicate. This problem arises in building constructions in fasteners production, clothes production, shoes production and so on. In some applications the rotation is not possible, for example in some types of clothes, while in other it is possible and can be used for minimizing the waste.

In [6] the main topic is a two-dimensional orthogonal packing problem, where a fixed group of small rectangles must be fitted into a large rectangle so that, most of the material is used, and the unused area of the large rectangle is minimized. The algorithm combines a replacement method with a genetic algorithm. In [1] a number of heuristic algorithms for two-dimensional cutting problems (on large scales) are developed. In this study, there is a large primary stock that has to be cut into smaller pieces, so as to maximize the value of the pieces. They developed a greedy randomized adaptive search procedure. Cintra et al. [3] propose an exact algorithm based on dynamic programming. This kind of algorithms are appropriate for small problems, because the problem is NP-hard. For these problems

I. Lirkov and S. Margenov (Eds.): LSSC 2017, LNCS 10665, pp. 350–357, 2018.
https://doi.org/10.1007/978-3-319-73441-5_37

some method based on stochastic search is more appropriate to apply. Stochastic search do not guarantee finding optimal solution, but for the practitioners, solution which is about 5% from the optimal solution is acceptable. Dusberger and Raidl [4,5] propose two metaheuristic algorithms based on variable neighborhood search.

Above mentioned works solve the simplified problem with rectangular items. In this paper investigation on more general problem is presented, when the items are arbitrary polygons including convex and concave. Proposed algorithm is tested on real data. The performance is compared with commercial software.

The rest of the paper is organized as follows. In Sect. 2 the CSP problem is described. In Sect. 3 we propose a stochastic algorithm for solving CSP. In Sect. 4 experimental results and comparison with other algorithms are shown. In Sect. 5 some concluding remarks and directions for future work are done.

2 Problem Formulation

The CSP where the items are arbitrary polygons is very difficult. Therefore most of the authors prefer to concentrate their effort on simplified variant of the problem, where the items are rectangles. The rectangular variant of the problem arises in paper and glass industries. Some authors try to solve the problem with arbitrary polygons by completing the items to rectangle, but in most of the cases it is not effective [2]. Completing some polygons to rectangle, the received rectangles can have surface more than two times larger, than the original polygon. In some of the existing softwares, the ordered polygons are sorted by their surface. After, only the first half of the polygons with larger surface are completed to rectangles. It is verified if some of the polygons with smaller surface, can be positioned in some of the rectangles, without overlapping with the other polygon. This variant of the algorithm improve the achieved solutions, but is not effective too.

In our variant of CSP rectangular sheet with fixed width and unlimited length is done. The set $E = \{i_1, i_2, \ldots, i_n\}$ of ordered items are polygons, which can be convex and concave. On Fig. 1 examples of items, which are plates from metal building construction, are shown.

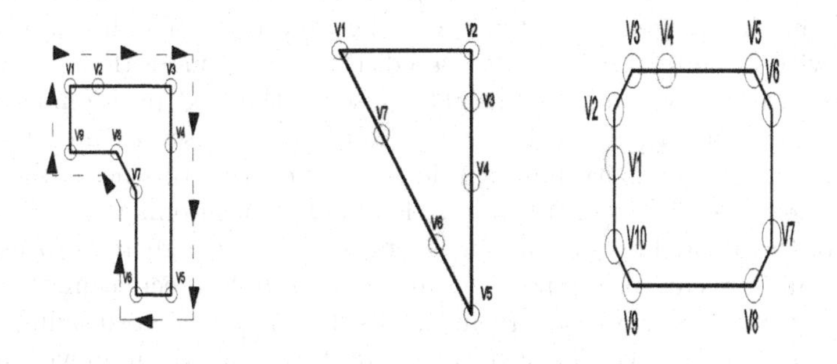

Fig. 1. The polygons which will be cut.

Every item is specified with the coordinates of his nodes and number of orders d_i, for $i = 1, \ldots, n$. When positioning, the items can be rotated.

The objective is to find a cutting pattern P with a minimal waste. The solution is the arrangement of the items from E on the stock sheet, without overlapping. Let x is the width of the sheet and y to be cutting height of P. The cutting pattern with a minimal waste is equivalent to the cutting pattern with a minimal cutting height, because the area of all ordered items is fixed and the width sheet is fixed. So the objective function $C(P)$ is:

$$C(P) = min(y). \tag{1}$$

The solution can be represented by cutting sequence and coordinates of the nodes of the cutting items.

3 Algorithm Description

In many industries a task of cutting stock arises. Cutting with minimal waste is important for practitioners. The problem needs exponential number of calculations, but even the simplified problem, where the cutting items are rectangular continue to be NP-hard [2,7]. Before starting the main optimization algorithm we verify if the input data are correct: if there is self-crossing; if all the points, describing the polygon, are the line; are there more than two points on the line, redundant points.

The cutting items are not necessary to be convex, they need to be correct. Each edge is linear and the nodes of the polygon describe it. The sheet, from which we will cut the item is rectangular with fixed width and infinity length. A random item is chosen from the set E of items and one of its nodes is fixed on the point with minimal height in the sheet. Let m is the number of the nodes of the current item. We will translate the item m times and will rotate it as it is shown on Fig. 2. Feasible positioning is the one for which all points inside the polygon are in the sheet of cutting. The best positioning between all feasible is one with a minimal cutting height. In our example from Fig. 2 only two of possible positioning are feasible and the cutting with minimal cutting height is shown on Fig. 3.

When the next item is positioned, again it is chosen in a random way from the polygons in the set E of items and the procedure of fixing it with in a position with a minimal height is repeated. If there are more than one possible positioning with a same minimal height, we select the next positioning between them in a random way. If it is impossible to position chosen item on this point we select the next one with minimal height. We do this till there are no more items in the set E. Figure 4 illustrates this kind of positioning. The stochastic element of the algorithm is a way for diversification of the search process.

When all items are positioned we run a procedure for decreasing the cutting height. In above described algorithm a possible nodes of positioning are the nodes of positioned items. The next procedure starts from the item with minimal positioning height, which is more than 0. We try to decrease this height with a

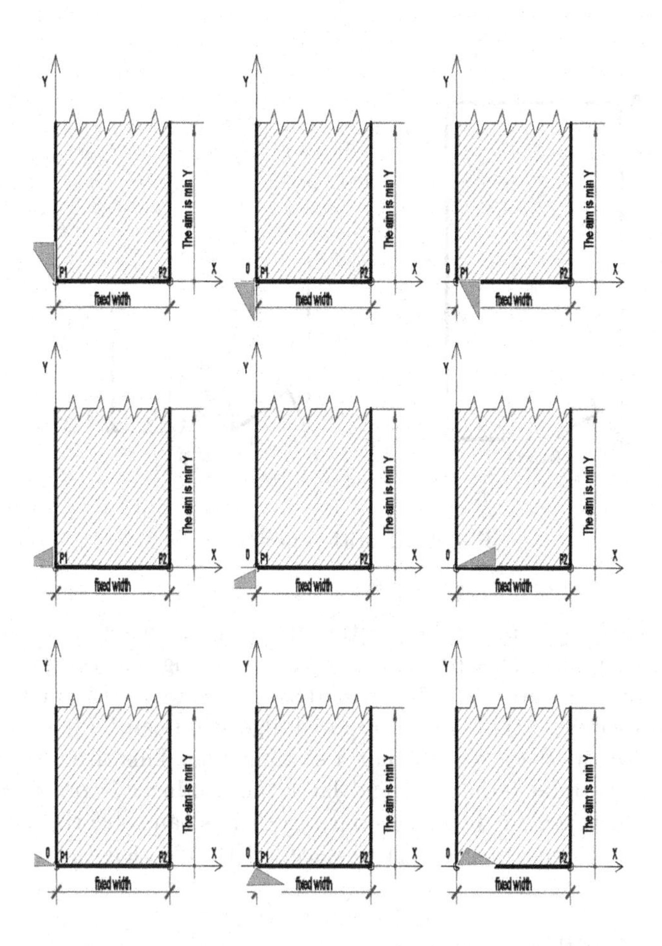

Fig. 2. Translations and rotations of a polygon

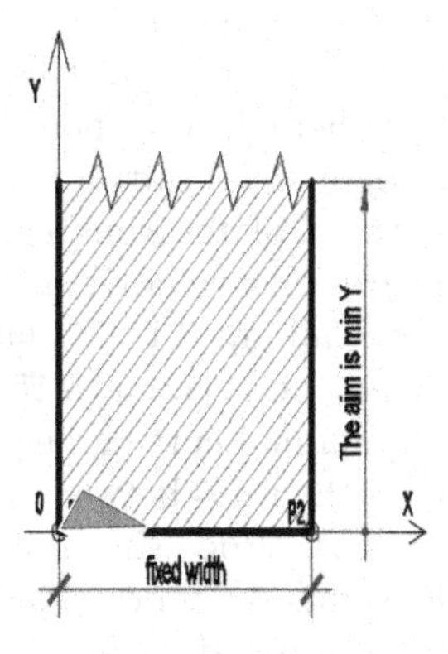

Fig. 3. Best positioning of the cutting item

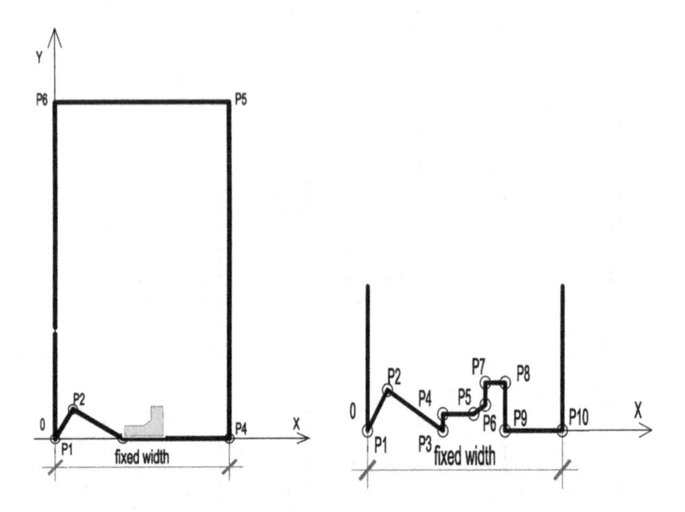

Fig. 4. Algorithm performance

fixed step. Thus an item can be positioned in a point which is on some of the edges of other items. This procedure is like adjusting, "shaking" the solution with aim to improve it. With this procedure we decrease the cutting height of all items, respectively the waste. Proposed algorithm can be applied when the bottom part of the stock sheet is not a straight line. This situation arise when cutting of some order is finished and later the producer is prepared to cut a new order. With very small changes our algorithms can solve the variant of the problem, where there are several stock sheets with fixed length.

4 Test Results

The CSP is a real industrial problem and is very important if we can test it on real data. Assessment of the results require comparison with other algorithms. Although many scientists are working hard in this direction, published results are not sufficient and require further efforts. Therefore our results are compared with the results, achieved by commercial software.

Our test example consists of steel plates from real steel structure, Fig. 5. Some of them are convex and others are concave polygons. The plates are 242 different shapes and 1958 targets. The overall area of the plates is 129,053,789 mm^2. The width of the cutting sheet is fixed and is equal to 1500 mm. The minimum waste is equal to find a solution with minimal cutting height and respectively minimal filling factor. The filling factor is the ratio between the sum of the area of the all plates and the cutting area ($x \times y$). Processing the input data, it is taken in to account that the cut width is 5 mm. We process the input data, before running the optimization algorithm. The input data are polygons, described by points and edges between them. We verify the input data and we remove redundant

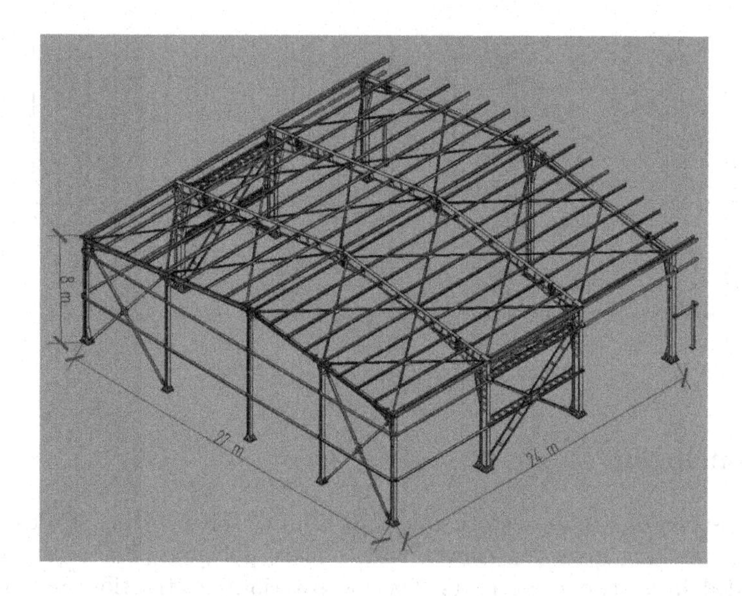

Fig. 5. Steel structure. Composite from steel profiles and steel plates.

points. After the processing the plates are described only with their nodes. The algorithm is run on desktop computer with 2.8 GHz CPU.

We run the proposed algorithm 30 times and we compare achieved best result with result achieved by one commercial software used by professionals. We chose a professional software, which do not complete the items to rectangular. In Table 1 we report the value of cutting height and filling factor of the commercial algorithm, algorithm without improvement procedure and the algorithm with improvement procedure. Regarding the table we observe that proposed algorithm achieves better result than the commercial one and improvement procedure improves the achieved results. Our algorithm achieves solution with smaller height and respectively less filling factor, which is equal to solution with less waste. We can conclude that proposed algorithm performs better and receives very encouraging results.

Table 1. Results comparison

Properties	Commercial algorithm	Proposed algorithm	Algorithm with "shaking"
Cutted height	200,014 mm	170,273 mm	136,133
Filling factor	0.43	0.505	0.632

The Fig. 6 shows part of the solution, found by the algorithm with improvement procedure.

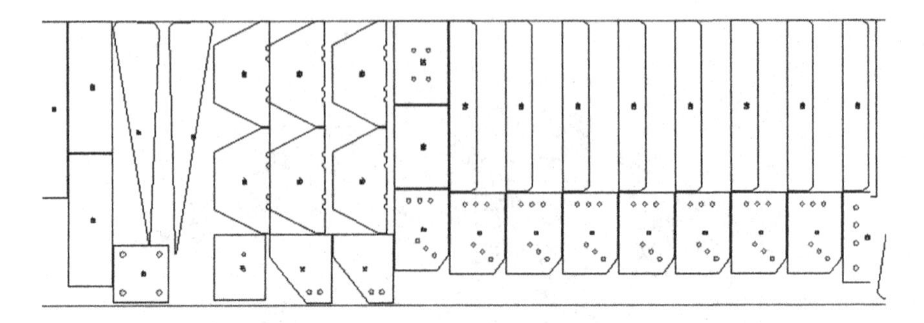

Fig. 6. Arrangement of the plates.

5 Conclusion

In this paper we propose a stochastic algorithm to solve cutting stock problem where the cutting items are convex and concave polygons. The algorithm can be divided in to two main parts, feasible solution construction and solution improvement. The algorithm is applied on real data from steal structure. We compare the results achieved by our algorithm with the results achieved by one commercial software. We show that our algorithm performs better than other. For a future work we plan to propose an heuristic algorithm for 2D cutting stock problem.

Acknowledgment. Work presented here is partially supported by the Bulgarian National Scientific Fund under Grants DFNI I02/20 "Efficient Parallel Algorithms for Large Scale Computational Problems" and DFNI DN 02/10 "New Instruments for Data Mining and their Modeling".

References

1. Alvarez-Valdes, R., Parajon, A., Tamarit, J.M.: A computational study of heuristic algorithms for two-dimensional cutting stock problems. In: 4th Metaheuristics International Conference (MIC 2001), pp. 16–20 (2001)
2. Alvarez-Valdes, R., Parreno, F., Tamarit, J.M.: A Tabu Search algorithm for two dimensional non-guillotine cutting problems. Eur. J. Oper. Res. **183**(3), 1167–1182 (2007)
3. Cintra, G., Miyazawa, F., Wakabayashi, Y., Xavier, E.: Algorithms for two-dimensional cutting stock and strip packing problems using dynamic programming and column generation. Eur. J. Oper. Res. **191**, 61–85 (2008)
4. Dusberger, F., Raidl, G.R.: A variable neighborhood search using very large neighborhood structures for the 3-staged 2-dimensional cutting stock problem. In: Blesa, M.J., Blum, C., Voß, S. (eds.) HM 2014. LNCS, vol. 8457, pp. 85–99. Springer, Cham (2014). https://doi.org/10.1007/978-3-319-07644-7_7
5. Dusberger, F., Raidl, G.R.: Solving the 3-staged 2-dimensional cutting stock problem by dynamic programming and variable neighborhood search. Electron. Notes Discret. Math. **47**, 133–140 (2015)

6. Gonçalves, J.F.: A hybrid genetic algorithm-heuristic for a two-dimensional orthogonal packing problem. Eur. J. Oper. Res. **183**(3), 1212–1229 (2007)
7. Parmar, K., Prajapati, H., Dabhi, V.: Cutting stock problem: a survey of evolutionary computing based solution. In: Proceedings of Green Computing Communication and Electrical Engineering (2014). https://doi.org/10.1109/ICGCCEE.2014.6921411
8. Lodi, A., Martello, S., Vigo, D.: Recent advances on two-dimensional bin packing problems. Discret. Appl. Math. **123**, 379–396 (2002)

Influence of Ant Colony Optimization Parameters on the Algorithm Performance

Stefka Fidanova[1(✉)] and Olympia Roeva[2]

[1] Institute of Information and Communication Technologies – BAS, Acad. G. Bonchev Str., bl. 25A, 1113 Sofia, Bulgaria
stefka@parallel.bas.bg

[2] Institute of Biophysics and Biomedical Engineering – BAS, Acad. G. Bonchev Str., bl. 105, 1113 Sofia, Bulgaria
olympia@biomed.bas.bg

Abstract. In this paper an Ant Colony Optimization (ACO) algorithm for parameter identification of cultivation process models is proposed. In computational point of view it is a hard problem. To be solved problem with a high accuracy in reasonable time, metaheuristic techniques are used. The influence of ACO algorithm parameters, namely number of agents (ants) and number of iterations, to the quality of achieved solution is investigated. As a case study an *E. coli* fed-batch cultivation process is explored. Based on the parameter identification of *E. coli* MC4110 cultivation process model some conclusions for the optimal ACO parameter settings are done.

Keywords: Ant Colony Optimization · *E. coli* cultivation
Model parameter identification

1 Introduction

Ant Colony Optimization (ACO) is a population-based metaheuristics that can be used to find approximate solutions to optimization problems with big computational complexity [5]. ACO is implemented as a team of intelligent agents which simulate the ants behaviour, walking around the graph representing the problem to solve using mechanisms of cooperation and adaptation [6]. ACO is applicable for a broad range of optimization problems and it can be easily adapted to dynamic changes of the problem [2,7,8,16]. ACO can compete with other metaheuristic techniques like genetic algorithms, simulated annealing, etc. [9,13,15].

In metaheuristic algorithms, values of several algorithm components and parameters have to be set, due to their significant impact on the algorithm's efficacy and performance [3,10,17,18]. For this reason, it is important to study how the algorithm parameters affect the metaheuristic algorithms performance, and thus find which values of the parameters prove optimal for a particular optimization problem.

I. Lirkov and S. Margenov (Eds.): LSSC 2017, LNCS 10665, pp. 358–365, 2018.
https://doi.org/10.1007/978-3-319-73441-5_38

One of the important topics that require careful consideration in the field of swarm optimization is population size. Various results about the appropriate population size have been obtained [1,11,14,19]. In many cases, the choice of population size determines the solutions quality. It is generally considered that "small" population size yields poor solutions, while "large" population size yields solutions with high accuracy, but requires more computational time. Taking into consideration that the population size has impact on quality of the solution and on the computational time, this algorithm parameter will be more thoroughly investigated. Another parameter that have impacted on solutions quality and computational time is the number of iterations (generations) [4,11].

The main purpose of the study presented in this paper is to investigate the influence of ACO parameters's, namely population size (number of ants) and number of iterations, on the algorithm performance. As a case study an *E. coli* fed-batch cultivation process is considered. Bacteria *E. coli* is one of the most used host organisms in the cultivation processes.

A parameter identification problem of a non-linear mathematical model of *E. coli* cultivation is investigated. Due to considered hard combinatorial optimization problem, for which exact algorithms or traditional numerical methods are not efficient enough, the model parameter identification is performed applying ACO.

The paper is organized as follows. The problem formulation – parameter identification of an *E. coli* cultivation process model – is given in Sect. 2. The ACO algorithm is described in Sect. 3. The numerical results and a discussion are presented in Sect. 4. Concluding remarks are given in Sect. 5.

2 Problem Formulation

Application of the general state space dynamical model to the *E. coli* fed-batch cultivation process leads to the following non-linear differential equation system [12]:

$$\frac{dX}{dt} = \mu X - \frac{F_{in}}{V} X, \tag{1}$$

$$\frac{dS}{dt} = -\frac{1}{Y_{S/X}} \mu X + \frac{F_{in}}{V}(S_{in} - S), \tag{2}$$

$$\frac{dV}{dt} = F_{in}, \tag{3}$$

$$\mu = \mu_{max} \frac{S}{k_S + S}, \tag{4}$$

where: X is the biomass concentration, [g/l]; S is the substrate concentration, [g/l]; F_{in} is the feeding rate, [l/h]; V is the bioreactor volume, [l]; S_{in} is the substrate concentration in the feeding solution, [g/l]; μ is the specific growth rate, [h^{-1}]; μ_{max} is the maximum value of the specific growth rate, [h^{-1}]; k_S is the saturation constant, [g/l]; $Y_{S/X}$ is the yield coefficient, [g/g].

The mathematical formulation of the non-linear model (Eqs. (1)–(4)) of an *E. coli* fed-batch cultivation process is described according to the mass balance. The model is based on the following a priori assumptions [12]: (i) the bioreactor is completely mixed; (ii) the main product is biomass; (iii) the substrate glucose mainly is consumed oxidatively and its consumption can be described by Monod kinetics; (iv) variation in the growth rate and substrate consumption do not significantly change the elemental composition of biomass, thus balanced growth conditions are assumed; (v) parameters, e.g. temperature, pH, pO_2 are controlled at their individual constant set points.

Real experimental data of the *E. coli* MC4110 fed-batch cultivation process are used for the parameter identification problem. The cultivation condition and the experimental data are presented in [12]. The initial process conditions are as follows:

- $t_0 = 6.68$ h, $X(t_0) = 1.25$ g/l, $S(t_0) = 0.8$ g/l and $S_{in} = 100$ g/l.

For the considered here non-linear mathematical model Eqs. (1)–(4) three parameters should be identified:

- μ_{max} – maximum specific growth rate,
- k_S – saturation constant, and
- $Y_{S/X}$ – yield coefficient.

As an optimization criterion, mean square deviation between the modelled data and experimental data is used. Thus, the objective function is presented as follows:

$$J = \sum_{i=1}^{m} \sum_{j=1}^{n} \left(\mathbf{y}_{\exp_j}(i) - \mathbf{y}_{\mathrm{mod}_j}(i) \right)^2 \rightarrow \min \qquad (5)$$

where m is the number of state variables (in this case $m = 2$ – biomass (X) and substrate (S) concentrations (see Eqs. (1)–(4)); n is the number of the experimental data; $\mathbf{y}_{\exp}$ is the known experimental data for X and S; $\mathbf{y}_{\mathrm{mod}}$ is the model predictions of X and S with a given set of the model parameters – μ_{max}, k_S, and $Y_{S/X}$.

3 Ant Colony Optimization

This section gives a brief overview of the ACO algorithms which comes from real ant behavior [5]. When looking for a food the ant lays down a chemical substance called pheromone to mark its way back. An isolated ant moves essentially at random but an ant will detect previously laid pheromone and decide to follow it with high probability and will reinforce it with an other quantity of pheromone. In ACO algorithms the problem is represented by a graph. An ant start to create a solution from random node with a help of function called transition probability. At the end of every iteration the pheromone is updated according the quality of constructed solutions.

In our work the parameters μ_{max}, k_S, and $Y_{S/X}$ have to be estimated. First, the problem is represented by a graph. The aim is to find the optimal values of three parameters which are interrelated. Thus, the problem is represented with three-partitive graph. The graph is divided of three levels. Every level corresponds to a search area of one of the parameters that will be optimized. Every area is thus discretized, to consists of minimum 1000 points (nodes), which are uniformly distributed in the search interval of every parameter. The first level of the graph represents the parameter μ_{max}, the second level represents the parameter k_S and the third level represents the parameter $Y_{S/X}$. There are arcs between nodes from consecutive levels of the graph and there are no arcs between nodes from the same level. The pheromone is deposited on the arcs, to indicate how good this parameter combination is.

The process is iterative. At the beginning of every iteration the ants choose a node from the first level in a random way. Next, for nodes from the second and the third level, they apply the probabilistic rule. The transition probability depends only on the pheromone level. The heuristic information is not used. Thus the transition probability is as follows:

$$p_{i,j} = \frac{\tau_{i,j}}{\sum\limits_{k \in Unused} \tau_{i,k}}. \tag{6}$$

The ants choose the node with maximal quantity of pheromone on the arc (starting from the current node). If there is more than one candidate for next node, the ant chooses randomly between the candidates. At the end of every iteration the pheromone on the arcs is updated. The quality of the solutions is represented by the value of the objective function. The aim of the process is to minimize it, therefore the newly added pheromone by ant i is:

$$\Delta\tau = (1 - \rho)/J(i), \tag{7}$$

where $J(i)$ is the value of the objective function according the solution constructed by ant i. Thus the arcs corresponding to solutions with lower value of the objective function will receive more pheromone and will be more attractive in the next iteration. More details about ACO algorithm are published in the literature [5,6,9].

4 Numerical Results and Discussion

Applying ACO algorithm a model parameter identification of an *E. coli* fed-batch cultivation process is performed. To find the best solution for reasonable computational time it is necessary to find appropriate ACO parameters values. The main ACO parameters that essential influence on algorithm performance are the number of ants and the number of iterations. Too many ants will quickly accumulate too many pheromone on the elements of suboptimal solutions and the algorithm will converge to the optimal solution very slowly. Using several ants only will not produce enough pheromone accumulation and the algorithm convergence will be very slow too.

In order to investigate the influence of the number of ants, we propose a scheme where different number of ants and different number of iterations are combined to ensure constant algorithm running time. In our research we fixed the product of the number of ants and the number of iterations to be 2000. We run eight ACO algorithms with 5, 10, 20, 25, 40, 50, 100, and 200 ants for 400, 200, 100, 80, 50, 40, 20 and 10 iterations, respectively. The running time is the same for every combination of ants $\times$ iterations. The aim is to find the minimal number of ants we need to achieve good solution.

The initial pheromone and evaporation rate are another ACO parameters that should be defined according to the particular problem. The pheromone value lies between the ranges 0 to 1. Normally initial pheromone has a small value. The evaporation rate shows the importance of the last found solution as related to the previous ones. Usually the evaporation rate has a value between 0 and 1, too. These two ACO parameters were determined based on several pre-tests according to the considered here identification problem, as follows:

- initial pheromone = 0.5,
- evaporation rate = 0.1.

Due to the stochastic nature of the ACO we run each algorithm 30 times and calculate the average value of the objective function J (Eq. (5)). We apply ANOVA test to check whether the obtained results are statistically different. The results achieved by 5 ants and 400 iterations and by 100 ants and 20 iterations are statistically the same.

The resulting average, best, and worst values of the objective function J for the considered eight ACO algorithms are reported in Table 1.

Table 1. ACO algorithms performance

ACO algorithm	Ants $\times$ iterations	5 $\times$ 400	10 $\times$ 200	20 $\times$ 100	25 $\times$ 80	40 $\times$ 50	50 $\times$ 40	100 $\times$ 20	200 $\times$ 10
Value of J	Average	**5.86**	6.25	6.67	6.30	5.91	6.27	**5.78**	6.60
	Best	3.49	3.43	3.70	3.52	3.43	3.45	3.43	3.68
	Worst	13.83	23.26	18.79	12.10	16.32	13.28	10.78	20.42

We observe that the best found results for all variants (number of ants) are very similar (statistically equivalent). Regarding the average results, the best performance is achieved of ACO with 5 ants $\times$ 400 iterations and ACO with 100 ants $\times$ 20 iterations.

The numerical results for objective function value J, over 30 runs of each ACO algorithm, are visualized on Fig. 1. We observe that the obtained values of J of ACO with 5 ants has less peaks compared to the ACO with 100 ants. The standard deviation of ACO with 5 ants $\times$ 400 iterations is less than of ACO with

100 ants $\times$ 20 iterations. Moreover more ants means more memory is needed. Therefore we can conclude that the best algorithm performance is when 5 ants and 400 iteration are used. Thus the minimal number of ants, necessary to achieve good solutions, is 5.

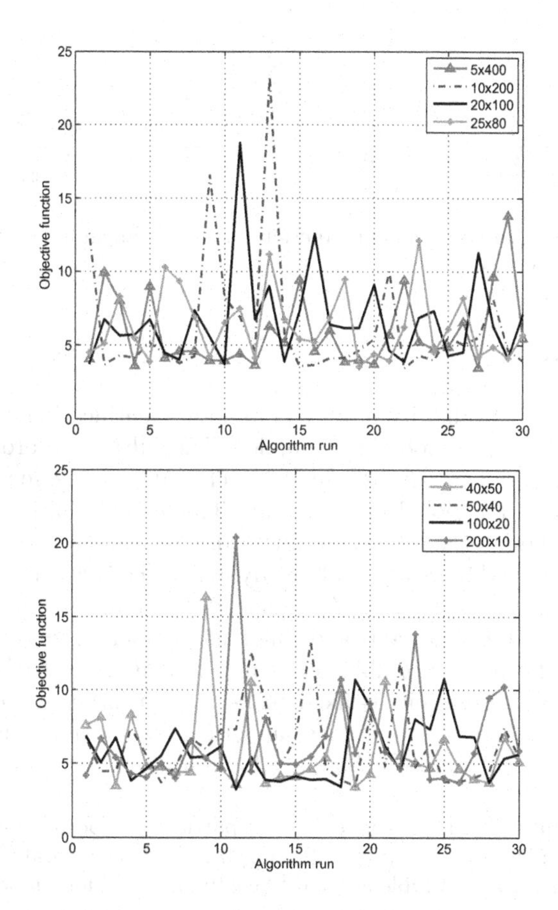

Fig. 1. Obtained objective function value J with the eight ACO algorithms

On Fig. 2 the best achieved biomass X and substrate S model predictions are shown compared to the real experimental data of an *E. coli* fed-batch cultivation process.

The presented Fig. 2 illustrate a very good correlation between the experimental and predicted data of the considered process variables. The obtained results show the proposed ACO algorithm performs very well for model parameter identification of an *E. coli* fed-batch cultivation process.

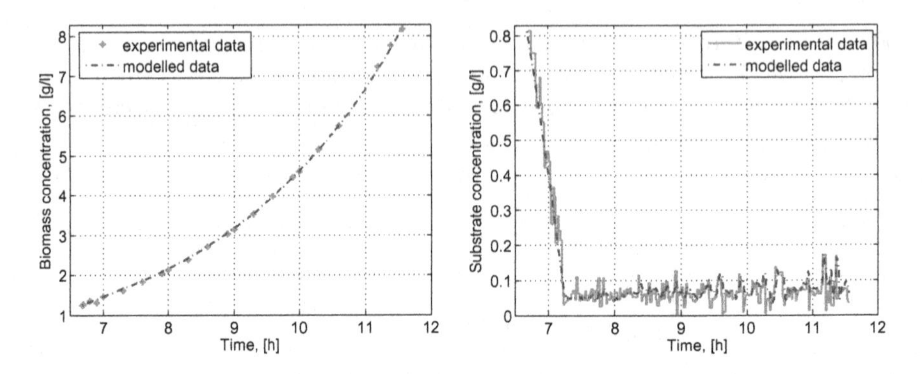

Fig. 2. Comparison between model predictions and experimental data for biomass and substrate concentrations

5 Conclusion

In this work eight ACO algorithms for a parameter identification of an *E. coli* fed-batch cultivation process model were applied. The cultivation process modelling is a problem of parameter-setting models. The ACO is chosen as one of the best methods used for global optimization. The *E. coli* cultivation process is modelled by a system of non-linear ordinary differential equation describing the biomass, substrate and bioreactor volume dynamics. For identification procedure real experimental data were used. A scheme where different number of ants (5, 10, 20, 25, 40, 50, 100, and 200) and different number of iterations (400, 200, 100, 80, 50, 40, 20, 10) are combined is applied. The product of the number of ants and number of iterations is fixed to be 2000 in order to ensure constant algorithm running time. The obtained numerical results show that the optimal ACO parameters combination is 5 ants $\times$ 400 iterations.

Acknowledgments. Work presented here is partially supported by the Bulgarian National Scientific Fund under Grants DFNI I02/20 "Efficient Parallel Algorithms for Large Scale Computational Problems" and "New Instruments for Knowledge Discovery from Data, and their Modelling" DN 02/10.

References

1. Alajmi, A., Wright, J.: Selecting the most efficient genetic algorithm sets in solving unconstrained building optimization problem. Int. J. Sustain. Built Environ. **3**(1), 18–26 (2014)
2. Barbosa, H.J.C.: Ant Colony Optimization – Techniques and Applications. InTech, Rijeka (2013)
3. de Moraes Barbosa, E.B., Senne, E.L.F., Silva, M.B.: Improving the performance of metaheuristics: an approach combining response surface methodology and racing algorithms. Int. J. Eng. Math. **2015**, 1–9 (2015). Article ID 167031

4. Cooray, P.L.N.U., Rupasinghe, T.D.: Machine learning-based parameter tuned genetic algorithm for energy minimizing vehicle routing problem. J. Ind. Eng. **2017**, 1–13 (2017). Article ID 3019523
5. Dorigo, M., Stützle, T.: Ant Colony Optimization. MIT Press, Cambridge (2004)
6. Dorigo, M., Stützle, T.: Ant colony optimization: overview and recent advances. In: Gendreau, M., Potvin, Y. (eds.) Handbook of Metaheuristics. International Series in Operations Research & Management Science, vol. 146, 2nd edn., pp. 227–263. Springer, New York (2010)
7. Fidanova, S., Lirkov, I.: 3D protein structure prediction. J. An. Univ. de Vest Timis. Ser. Mat. Inform. **XLVII**(2), 33–46 (2009)
8. Fidanova, S.: An improvement of the grid-based hydrophobic-hydrophilic model. Int. J. Bioautomation **14**(2), 147–156 (2010)
9. Haroun, S.A., Jamal, B., Hicham, E.H.: A performance comparison of GA and ACO applied to TSP. Int. J. Comput. Appl. **117**(19), 28–35 (2015)
10. Nowotniak, R., Kucharski, J.: GPU-based tuning of quantum-inspired genetic algorithm for a combinatorial optimization problem. Bull. Pol. Acad. Sci. **60**(2), 323–330 (2012)
11. Rexhepi, A., Maxhuni, A., Dika, A.: Analysis of the impact of parameters values on the Genetic Algorithm for TSP. IJCSI Int. J. Comput. Sci. **10**(1/3), 158–164 (2013)
12. Roeva, O., Pencheva, T., Tzonkov, S., Hitzmann, B.: Functional state modelling of cultivation processes: dissolved oxygen limitation state. Int. J. Bioautomation **19**(1), Suppl. 1, S93–S112 (2015)
13. Roeva, O., Fidanova, S., Paprzycki, M.: InterCriteria analysis of ACO and GA hybrid algorithms. In: Fidanova, S. (ed.) Recent Advances in Computational Optimization. SCI, vol. 610, pp. 107–126. Springer, Cham (2016). https://doi.org/10.1007/978-3-319-21133-6_7
14. Roeva, O., Fidanova, S., Paprzycki, M.: Influence of the population size on the genetic algorithm performance in case of cultivation process modelling. In: IEEE Proceedings of the Federated Conference on Computer Science and Information Systems (FedCSIS), pp. 371–376 (2013)
15. Saleem, W., Kharal, A., Ahmad, R., Saleem, A.: Comparison of ACO and GA techniques to generate neural network based Bezier-PARSEC parameterized airfoil. In: Proceedings of the 11th International Conference on Natural Computation (ICNC) (2015). https://doi.org/10.1109/ICNC.2015.7378152
16. Sharvani, G.S., Ananth, A.G., Rangaswamy, T.M.: Ant colony optimization based modified termite algorithm (MTA) with efficient stagnation avoidance strategy for manets. Int. J. Appl. Graph Theor. Wirel. Ad Hoc Netw. Sens. Netw. **4**(2/3), 39–50 (2012)
17. Veček, N., Mernika, M., Filipičb, B., Črepinšek, M.: Parameter tuning with chess rating system (CRS-Tuning) for meta-heuristic algorithms. Inf. Sci. **372**, 446–469 (2016)
18. Yang, X.S., Deb, S., Loomes, M., Karamanoglu, M.: A framework for self-tuning optimization algorithms. Neural Comput. Appl. **23**(7–8), 2051–2057 (2013)
19. Zhang, Y., Ma, Q., Sakamoto, M., Furutani, H.: Effects of population size on the performance of genetic algorithms and the role of crossover. Artif. Life Robot. **15**, 239–243 (2010). https://doi.org/10.1007/s10015-010-0836-1

2D Optimal Packing with Population Based Algorithms

Desislava Koleva[1], Maria Barova[2], and Petar Tomov[2(✉)]

[1] Department of Computer Science, University College London, Gower Street, London WC1E 6BT, UK
desislava.koleva.15@ucl.ac.uk

[2] Institute of Information and Communication Technologies, Bulgarian Academy of Sciences, Acad. G. Bonchev Str., Block 2, 1113 Sofia, Bulgaria
p.tomov@iit.bas.bg

Abstract. This study addresses application of population based optimization heuristics to the solution of packing problems as part of optimal cutting tasks in the field of operations research. Such problems are very common in the industrial material cutting. The problem has one, two or three dimensional variations. The focus of this paper is the two dimensional case of steel sheet cutting. A description of two dimensional plates is supplied as input for the algorithm. The output is in the form of coordinates of the plates in the steel sheet and angle of rotation for each plate. Population based global optimization heuristics are used for optimal packing. All experiments are done with open source libraries for 2D geometry and population based heuristics.

Keywords: Optimal cutting problem · Optimal packing
Evolutionary algorithms · Optimization

1 Introduction

Optimal packing problem is an optimization problem in mathematics that involves attempting to pack objects together into one or many containers. A typical goal is to fill a single container as densely as possible. This problem is related to real life packaging (in fact optimal cutting) problem as described in [1,2]. This study is related to the problem presented at ESGI120 [2] and ESGI113 [3]. Mathematical definition of the problem is well defined in the ESGI problem proposals. Papers are public and the interested readers can refer to them. The goal is to cut optimally into pieces a sheet of steel without overlapping them. The cutting result shapes are irregular not self-intersecting polygons. The problem presented at ESGI113 was less complicated than the problem presented at ESGI120, because both the pieces and the sheet were of rectangular shape. A similar problem is well presented in [4]. With irregular shapes and when the

I. Lirkov and S. Margenov (Eds.): LSSC 2017, LNCS 10665, pp. 366–373, 2018.
https://doi.org/10.1007/978-3-319-73441-5_39

angle of orientation of the given shape is unconstrained, the general nesting approaches are not particularly successful [5].

This section starts with description of the problem and refers to related work in a brief review. Section 2 points out the geometric considerations necessary to understand this work. Section 3 describes the underlying rules and criteria used to build the GA based heuristic approach, which is proposed in the same section. Section 4 describes the evaluation of the proposed approach and presents some of the obtained results. Finally, the last section draws conclusions and includes some comments about future work.

1.1 Problem Description

The placement of shapes can be found in literature under the keywords of "optimal packing". When packing on a limited stock sheet, the goal is to maximize the percentage of stock sheet utilization, which is equivalent to maximizing the number of shapes placed inside the plate. The problem proposed to be solved does not consider orientation constraints, so, any rotation of the given shape can be allowed. When stock sheet borders need to be taken into account, the orientation of the shapes may be extremely important as it will be shown in the experimental part of this study.

In this work, we propose to solve the problem of Packing of Irregular Shapes (PIS) on a limited stock sheet (a rectangle) by heuristic methods.

1.2 Related Work

There are references in the literature to the Packing Problem since the 10th century. Persia, Abul Wefa produced a square dissection problem which often reappears today. Henry Ernest Dudeney's dissection puzzles were famous in the early part of last century, and in three dimensions Piet Hein's Soma Cube (Van Delft and Botermans 1978) in which pieces have not only have to be packed into a cube, but must also be sufficiently stable to balance on a central point, is perhaps the most interesting packing problem to date [6]. However it is only recently that industrial packing problems have been approached from a scientific viewpoint. It should be noted that the rectangular packing problem is known to be NP-complete and therefore it is often not possible to provide exact solutions within a reasonable time limit. Practical problems are not restricted to those involving rectangles, but because of the increased complexity of non-rectangular problems a large proportion of the published work to date has been limited to the packing of rectangles or cuboids [6].

1.3 Genetic Algorithms

Genetic algorithms (GAs) are search heuristic inspired by the process of natural selection [7,8]. GAs are routinely used to generate points (candidate solutions) into solutions space. By application of techniques for inheritance (crossover), mutation and selection generated points can get closer to the optimums. GAs are classified also as population based algorithms, because each point into solution space represents an individual inside GAs population. Each individual has a set of properties which are subject of mutation and modification (usually crossover). Traditional representation of the properties is a binary sequence of 0s and 1s, but other encodings are also possible (binary tree for example) [9].

Optimization usually starts from random generated population of individuals, but this is also subject of implementation. The optimization process is iterative and the population in each iteration is called generation. For each individual of the generation fitness value is calculated. Fitness value usually represents the objective function which is a subject of optimization. The most fit individuals into the population are selected (according selection rule) and recombined (crossover and/or mutation) to form a new generation. This new generation is used in the next iteration of the algorithm. Algorithm termination is usually achieved by reaching maximum number of generations or by reaching the desired level of the fitness value [9].

In order to run GAs it is necessary to provide: (1) Genetic representation of the solution space (solution domain); (2) an appropriate fitness function to evaluate the solution domain. Once these two conditions are met GAs can proceed with population initialization and iterative population improvement by repetitive application of selection, crossover, mutation and individuals' evaluation [9].

2 Geometric Considerations

All shapes are represented as polygons. Arcs are approximated with small lines. Holes inside the shapes are not considered by definition [2]. Each polygon is represented as set of vertices. It is not allowed the polygons to overlap. All polygons must be entirely placed inside the stock sheet. To guarantee that two polygons do not overlap and are positioned as close as possible, a concept similar to no-fit-polygon (NFP) is used. The first polygon is positioned on the plane and is considered as the fixed piece. The second polygon is tracing sheet the edge of the fixed polygon. To ensure that any shape is entirely placed inside the stock sheet, a concept similar to inner-fit-polygon (IFP) is used. The IFP is the geometric place of all the points where the reference point of the piece to place can be positioned, so that the piece can be completely placed inside the stock sheet [5]. The computation of NFPs is a very time consuming operation for nonconvex polygons. As a new NFP needs to be recomputed whenever a new relative orientation between two polygons is considered, the complexity of this operation must be taken into account when developing heuristics.

The objective of the PIS problem is to place the largest number of irregular shapes inside a limited rectangular stock sheet, where all the shapes have different orientation. In its most general formulation, there are no constraints on the selection of the shapes orientation used to build the layout and different orientations should be tested in order to find the one that takes the most out of the stock sheet's borders.

To build a layout for the PIS problem, it is necessary to define some parameters:

- the orientation of each shape, i.e. the rotation angle relative to the original orientation;
- the placement point of each shape positioned;
- the order in which shapes are placed on the sheet;

One of the problems arising in all types of nesting problems is the intrinsic difficulty to deal with geometry, as usually shapes are not regular and not even convex. Besides this difficulty, the necessity of considering all the non-overlapping constraints naturally leads to the consideration of heuristic procedures [5].

3 Genetic Algorithm for Optimal Packing Order

Reading of the input data is done from a text file with listed coordinates of the polygons. During GA's population initialization, different individuals are initialized either with all plates oriented horizontally, vertically or in random angle of rotation. This way a better population diversity is achieved. Eight different plate orderings are applied to GA's individuals: unchained input order, randomly shuffled, sorted by plates width, sorted by plates height and bounding rectangle [10]. The initial population is then evaluated by packing with the length of the bounding rectangles.

As GA selection rule parent individuals are randomly chosen. For plates ordering in the GA's individuals permutation crossover is applied. The resulting child is kept on the place of the worst individual in the population (indirect elitism rule). Mutation is applied over the newly created individual in the form of rotation in a small random angle and change in the order of two randomly selected plates. For newly created individual to be evaluated, a procedure for plates' packing is applied. The individual fitness value is the length of the steel sheet used after packing is applied.

The basic algorithm is as follows:

1. Load plates information and store them as polygon objects;
2. Initialize random GA population;
3. Optimization:
   3.1 Select parents;
   3.2 Crossover;
   3.3 Mutation;
   3.4 Pack polygons;
   3.5 Measure used sheet length as fitness value;
   3.6 Keep newly generated chromosome;
   3.7 Stop if predefined number of generations is reached;
   3.8 Repeat from 3.1.

4 Experiments and Results

All experiments are done with open source software [10] developed by the authors with the support of Velbazhd Software LLC. The software solution is using Java AWT and JST library polygon functionality. Custom implementation of GA is proposed and capabilities for Apache GA Framework usage are provided. GA is applied with the parameters listed in Table 1. Three independent experiments were executed (Figs. 1, 2 and 3) with the same input data for the maximum execution time of 1000 min. The input data are taken from ESGI 120 case study [2], as real industrial task presented by STOBET Ltd. In the original case study the width of the steel sheet is 1500 mm and the height is 12000 mm. The best achieved solution during experiments in this study is 8373 mm of the length of the steel sheet with fixed width of 1500 mm. The useful steel sheet area in 1500×8373 mm is about 85%.

Table 1. Genetic algorithm parameters.

Parameter	Value
Generation gap	0.98
Crossover rate	0.90
Mutation rate	0.01
Maximum generations	100
Number of individuals	137
Number of variables	318
Inserted rate	100%

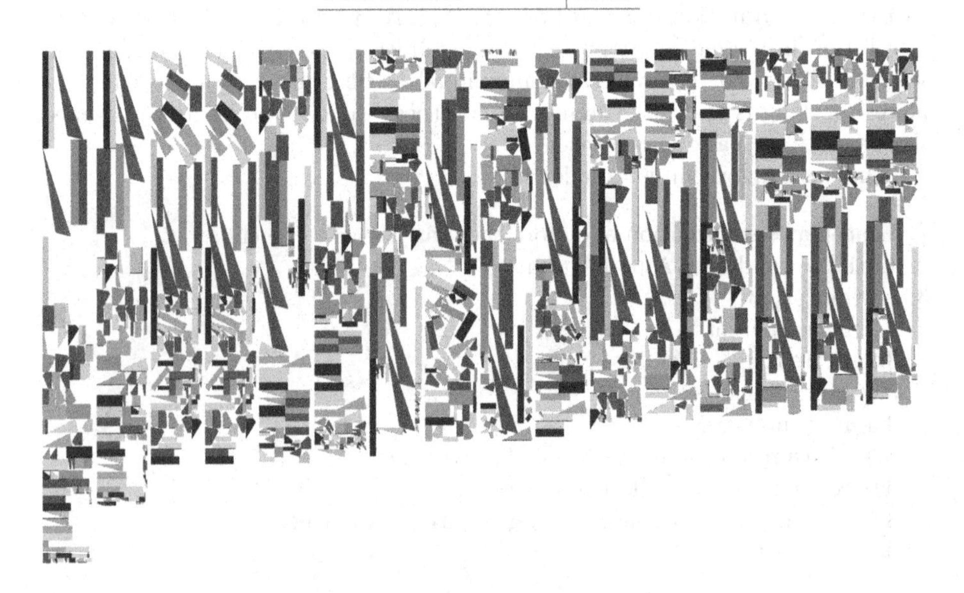

Fig. 1. Experiment 1 - final packing solution.

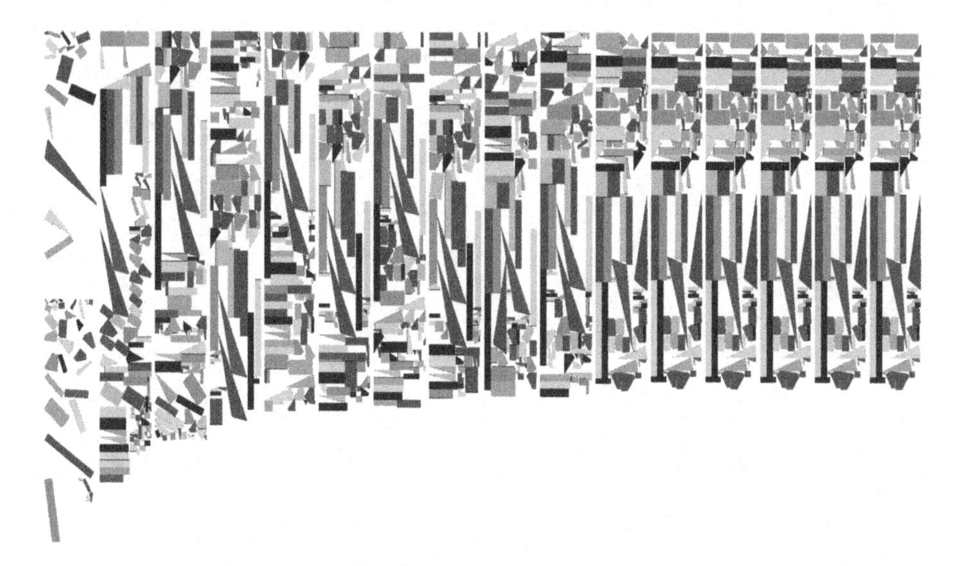

Fig. 2. Experiment 2 - final packing solution.

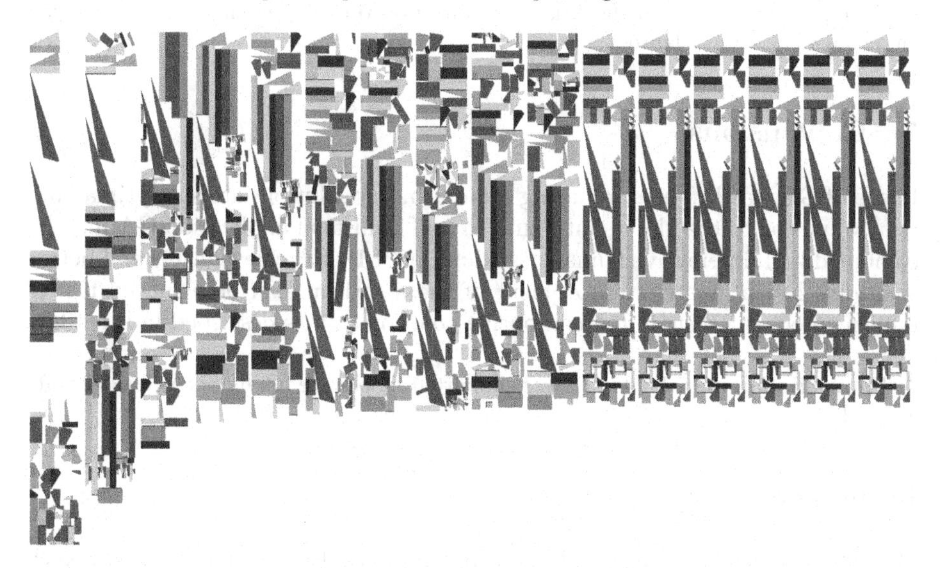

Fig. 3. Experiment 3 - final packing solution.

The convergence of the experiments is similar as it is shown on Fig. 4. Convergence is stairs like because GA is working on a discrete basis and elitism rule was applied.

It is obvious that the two biggest triangles are problematic in the optimal packing and GA is not very capable to deal with this local optima problem.

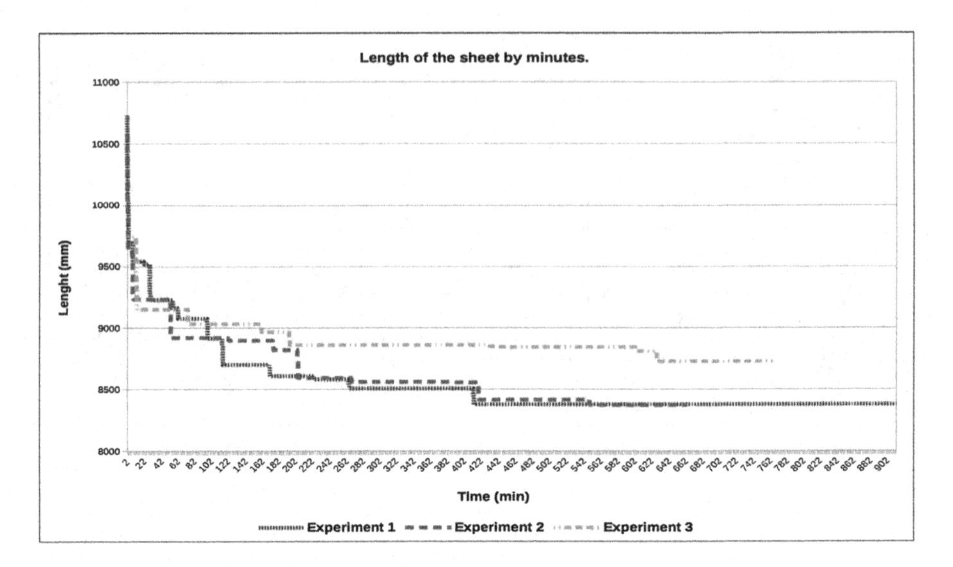

Fig. 4. Optimization convergence comparison. On the Y-axis the length of the used steel sheet is presented. On the X-axis optimization time is presented (in minutes) on an average performing laptop.

5 Conclusions

Usage of GAs for PIS is a promising approach, but solutions are suboptimal and relatively near to the global optimum. In real industrial application GA may be combined with interactive human assistance in order to achieve faster and better solutions. Much greater improvement of the algorithm can be achieved if it is implemented as a distributed computing solution as it is described in [11–13].

Acknowledgements. This work was supported by private funding of Velbazhd Software LLC.

References

1. Evtimov, G., Fidanova, S.: Ant colony optimization algorithm for 1D cutting stock problem. In: Proceedings of 11th Annual Meeting of the Bulgarian Section of SIAM, Sofia, Bulgaria, pp. 24–25. Fastumprint (2016)
2. Evtimov, G.: Project 2: optimal cutting problem. In: 120th European Study Group with Industry, Sofia, Bulgaria. Stobet Ltd. (2016)
3. Avdzhieva, A., Balabanov, T., Evtimov, G., Kirova, D., Kostadinov, H., Tsachev, T., Zhelezova, S., Zlateva N.: Optimal cutting problem. In: Problems and Final Reports of 113th European Study Group with Industry, Sofia, Bulgaria, pp. 49–61. Fastumprint (2015)
4. Martelloa, S., Monacib, M.: Models and algorithms for packing rectangles into the smallest square. Comput. Oper. Res. **63**, 161–171 (2015)

5. Teresa Costa, M., Miguel Gomes, A., Oliveira, J.: Heuristic approaches to large-scale periodic packing of irregular shapes on a rectangular sheet. Eur. J. Oper. Res. **192**, 29–40 (2009)
6. Dowsland, K., Dowsland, W.: Packing problems. Eur. J. Oper. Res. **56**, 2–14 (1992)
7. Eiben, A.E., Raué, P.-E., Ruttkay, Z.: Genetic algorithms with multi-parent recombination. In: Davidor, Y., Schwefel, H.-P., Männer, R. (eds.) PPSN 1994. LNCS, vol. 866, pp. 78–87. Springer, Heidelberg (1994). https://doi.org/10.1007/3-540-58484-6_252
8. Ting, C.-K.: On the mean convergence time of multi-parent genetic algorithms without selection. In: Capcarrère, M.S., Freitas, A.A., Bentley, P.J., Johnson, C.G., Timmis, J. (eds.) ECAL 2005. LNCS (LNAI), vol. 3630, pp. 403–412. Springer, Heidelberg (2005). https://doi.org/10.1007/11553090_41
9. Balabanov, T., Zankinski, I., Shumanov, B.: Slot machines RTP optimization with genetic algorithms. In: Dimov, I., Fidanova, S., Lirkov, I. (eds.) NMA 2014. LNCS, vol. 8962, pp. 55–61. Springer, Cham (2015). https://doi.org/10.1007/978-3-319-15585-2_6
10. Balabanov, T., Evtimov, G., Koleva, D.: ESGI 120 - Problem 2 - Genetic Algorithm Solver, Sofia, Bulgaria (2016). github.com/VelbazhdSoftwareLLC/ESGI120Problem2GeneticAlgorithmSolver
11. Balabanov, T.: Distributed evolutional model for music composition by human-computer interaction. In: Proceedings of International Scientific Conference UniTech 2015, Gabrovo, Bulgaria, vol. 2, 389–392. University Publishing House V. Aprilov (2015)
12. Balabanov, T.: Avoiding local optimums in distributed population based heuristic algorithms. In: Proceedings of XXIII International Symposium Management of Energy, Industrial and Environmental Systems, John Atanasoff Union of Automation and Informatics, Sofia, Bulgaria, pp. 83–86 (2015). (in Bulgarian)
13. Balabanov, T.: Heuristic forecasting approaches in distributed environment. In: Proceedings of Anniversary Scientific Conference 40 Years Department of Industrial Automation, Sofia, Bulgaria, pp. 163–166. UCTM (2011). (in Bulgarian)

A Non-dominated Sorting Approach to Bi-objective Optimisation of Mixed-Model Two-Sided Assembly Lines

Ibrahim Kucukkoc(✉)

Department of Industrial Engineering, Faculty of Engineering, Balikesir University, Cagis Campus, 10145 Balikesir, Turkey
ikucukkoc@balikesir.edu.tr
http://ikucukkoc.baun.edu.tr

Abstract. Assembly lines are of widely utilized mass production techniques emerged after the industrial revolution started in 18th century in England. Ever since, the changes in the global market and increasing interest in customized products forced companies to change their production systems in such a way that customer demands can be met in a more flexible environment. Assembly line balancing problem is an NP-hard class of combinatorial optimization problem for which exact solution techniques fail to solve large-scaled instances. This paper addresses to the problem of balancing mixed-model two-sided assembly lines, on which large-sized products (such as automobiles, trucks and buses) are assembled in an intermixed-sequence, with the aim of minimising two conflicting objectives (cycle time and number of workstations). A new ant colony optimization approach, called non-dominated sorting ant colony optimization (NSACO shortly), is proposed. Thus, the NSACO algorithm is used for the first time to solve an assembly line balancing problem. NSACO is described in details and a numerical example is solved to demonstrate its solution building mechanism. The results indicate that NSACO has a promising performance.

Keywords: Assembly line balancing
Mixed-model two-sided assembly · Non-dominated sorting
Ant colony optimisation · NSACO

1 Introduction

An assembly line is constituted from a sequential order of workstations linked to each other through a conveyor or moving belt. Each workstation is responsible from completing a set of tasks assigned to this particular workstation within a designated amount of time, called cycle time. Assembly lines are of mass production techniques utilised widely in various sectors ranging from electronics and automotive to home appliances and furniture. High efficiency is achieved through successful implementations of assembly lines, as they enable production of high-quality homogeneous products in mass quantities.

I. Lirkov and S. Margenov (Eds.): LSSC 2017, LNCS 10665, pp. 374–381, 2018.
https://doi.org/10.1007/978-3-319-73441-5_40

The advances in technology and global market forced companies to change their strategies. In parallel to these advancements, there have been major changes in the configuration and problem types of assembly lines since their first implementation by Ford and his colleagues [1]. More flexible and efficient assembly lines have gained more importance to meet customised demands at a reasonable cost. This yielded more complicated managerial decisions to make, such as line balancing and model sequencing among others, and more sophisticated solution methods to help make those. Battaia and Dolgui [2] presented a comprehensive taxonomy of assembly line balancing problems.

Mixed-model production strategy has emerged as a response to the effort for meeting the customised high-quality product demand at the right time. This concept was introduced to the academia by Thomopoulos [3] and studied widely thereafter. Many heuristic and meta-heuristic approaches have been proposed for solving various mixed-model lines; see, for example, Gokcen and Erel [4], Vilarinho and Simaria [5], Kara et al. [6], Yagmahan [7] and Simaria and Vilarinho [8].

Assembly lines are classified as one-sided lines and two-sided lines depending on the configuration of workstations across the line. In one-sided lines, tasks are performed in workstations located on only one side of the line (left or right side). However, in two-sided lines, workstations are located on both sides (left and right) of the line and tasks are performed in those workstations. Two-sided lines are mainly used in producing large-sized products (such as cars and buses) as it is not convenient to access one specific side (e.g. left) of the product from a workstation located on the other side (e.g. right). Two-sided line balancing problems have widely been researched recently. Please refer to Li et al. [9] and Make et al. [10] for a comprehensive review on two-sided assembly lines.

The minimisation of number of workstations and cycle time conflict with each other and these two aims correspond to two different problems (type-I and type-II, respectively) in the literature. Some studies aimed to optimise both objectives (number of workstations and cycle time) at the same time, e.g. see, Wei and Chao [11], Garcia-Villoria and Pastor [12], Manavizadeh et al. [13] and Kucukkoc and Zhang [14]. However, to the best of the authors' knowledge, mixed-model two-sided lines have not been balanced in this manner. Simaria and Vilarinho [15], proposed an ant colony optimisation approach for minimising the number of workstations considering zoning and synchronism constraints. Ozcan and Toklu [16] presented a new mathematical model and a simulated annealing algorithm for mixed-model two-sided lines. The primary aim of the proposed simulated annealing algorithm was to maximise the weighted line efficiency while the secondary aim was to minimise the smoothness index. Chutima and Chimklai [17] dealt with mixed-model two-sided assembly line balancing problem and developed a particle swarm optimization approach. Kucukkoc and Zhang [18] addressed to balancing and sequencing mixed-model parallel two-sided lines and proposed a mathematical model for solving the problem using an ant colony optimisation approach [19] and a hybrid genetic algorithm - ant colony algorithm optimisation approach [20]. The aim in those works was to minimise the number of mated-stations as the primary goal and the number of stations as the secondary goal.

As seen from this survey, there is no research which aims at simultaneously minimising cycle time and the number of workstations in mixed-model two-sided lines. Therefore, the current work contributes to literature as it addresses to type-E mixed-model two-sided assembly line balancing problem and proposes a new non-dominated sorting ant colony optimisation (NSACO) approach for solving it. The remainder of this paper is organised as follows. The problem is defined briefly in Sect. 2 and the proposed model is described in Sect. 3. The detailed balancing solution of a numerical example is also given in the same section and the paper is concluded in Sect. 4.

2 Problem Statement

The problem studied in this research regards to balancing a mixed-model two-sided assembly line. Two-sided assembly line balancing problem has an NP-hard complexity to solve, as other assembly line balancing problems. As the model variation adds further complexity, the problem becomes even harder to solve using conventional techniques when the problem size increases [16].

Tasks ($i = 1, 2, ..., ni$) are performed in workstations ($k = 1, 2, ..., K$) located in the left and right sides of the line. On the line, more than one model ($j = 1, 2, ..., nj$) of a product is assembled in an intermixed sequence. The aim is to find the optimal allocation of tasks to the workstations in such a way that the two conflicting objectives, namely cycle time and the number of workstations, are minimised. Each task needs to be assigned to exactly one workstation and the total workload (the sum of processing times of tasks) assigned to a workstation for a particular model cannot exceed the cycle time (C). There are precedence relationships between tasks which need to be satisfied during the assignment process. P_i represents the set of predecessors of task i. For example, if we assume $P_8 = (3, 5)$, tasks 3 and 5 must be completed before initialising task 8. Figure 1 shows the schematic representation of a mixed-model two-sided assembly line.

As seen from Fig. 1, two different models (A and B) are assembled on the line. The numbers given inside the bars represent task numbers (i) while their lengths correspond to task processing times (t_{ij}). There are a total of 9 tasks performed, some require different execution time while some are not needed for a specific model. For example, tasks 2 and 5 require different processing times

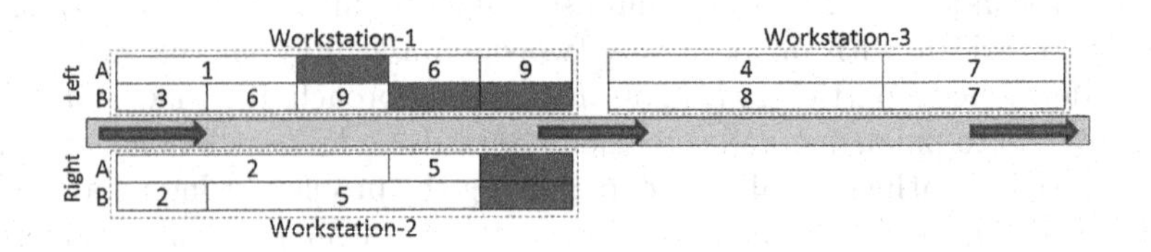

Fig. 1. Schematic representation of a mixed-model two-sided assembly line [16].

for models ($t_{2A} = 3, t_{5A} = 1, t_{2B} = 1$ and $t_{5B} = 3$ time-units). On the other hand task 3 is not needed for model A ($t_{3A} = 0$). In two-sided lines, precedence relationships make the problem even more complex to solve. This is because the relationships between tasks assigned to different sides of the line need to be considered carefully. This phenomenon is called interference and the violation of this rule may cause infeasibility. The following section describes the NSACO method proposed for solving mixed-model two-sided assembly line balancing problem aiming to minimise the cycle time and the number of workstations.

3 Proposed Method

Many optimisation problems in real world are multi-faced, requiring the optimization of several conflicting objectives concurrently. Between the conflicting objectives, there will be a set of optimal trade-offs, called Pareto-front [21]. Thus, the aim is to determine the optimal solution which maximises or minimises conflicting objectives considering problem specific constraints.

3.1 Non-dominated Sorting Algorithm

The NSACO method proposed in this research employs the non-dominated sorting mechanism [22] to find non-dominated solutions in terms of their cycle time and number of workstations. Figure 2 shows the general outline of the proposed algorithm. As seen from Fig. 2, the algorithm starts with initialising all sets and parameters. C is set to C_{min} (a user determined parameter) and K^* (best K) is set to a big number. A new colony, consisting of *colSize* number of ants, is released and each ant in the colony builds a balancing solution considering precedence relationships, capacity and assignment constraints. Pheromone is released

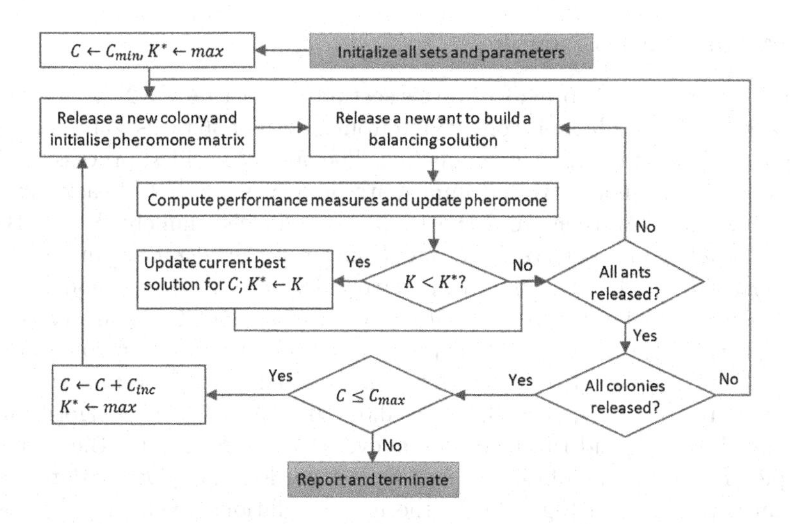

Fig. 2. Flowchart of the proposed algorithm.

between tasks and workstations in which they are assigned based on the performance criteria (K value) and this cycle continues until all ants complete their tour. If a solution with $K < K^*$ is obtained at any stage, the current best solution and K^* for current C value are updated. The solution parameters are reset after the colony completes its tour and a new colony is released again. All colonies are released one-by-one and this is continued until $maxIter$ (a user determined parameter) is achieved. C is increased by C_{inc}, sets are reset to default values (including the pheromone matrix) and $maxIter$ number of colonies are released again to find balancing solutions for current C value. This cycle continues until C_{max} is exceeded, the solutions obtained are reported and the algorithm is terminated thereafter.

The selection of a task to a workstation is determined using the formulation $p_{ik} = \frac{[\tau_{ik}]^{\alpha}[\eta_i]^{\beta}}{\sum_{h \in Z_i}[\tau_{hk}]^{\alpha}[\eta_h]^{\beta}}$; where α and β are weighting parameters determined by the user to signify the relative importance of pheromone and heuristic information in the task selection process. Z_i represents the list of candidate tasks when selecting task i and τ_{ik} is the pheromone amount existing between task i and workstation k. η_i is the heuristic information of task i comes from the well-known line balancing heuristic, ranked positional weight method [23]. In this method, the importance of a task is determined by the sum of its processing time and those of its successors. Eventually, the higher the p_{ik} value, the higher chance the task i to be selected [24].

The pheromone is deposited using the well-known expression $\tau_{ik} \leftarrow (1 - \rho)\tau_{ik} + \Delta\tau_{ik}$; where $\Delta\tau_{ik} = Q/K$, ρ is the evaporation rate and Q is a user determined parameter. K denotes the number of workstations required by the solution obtained. Hence, the lower the number of workstations, the more the pheromone amount deposited. Please note that the detailed procedure for building a balancing solution will not be provided due to page limit.

3.2 A Numerical Example

A numerical example is provided in this section to describe the NSACO method illustratively. The problem is constituted from a total of 36 tasks to be conducted on a mixed-model two-sided assembly line. The operation sides, processing times (for model A and model B) and immediate predecessors (*IP*) of tasks are provided in Table 1. As seen from the table, two models, namely A and B, are subject to balancing. Letters L, R, and E given in the *Side* column denote *left* side, *right* side, and *either* side, respectively. Demand rates for models A and B are assumed to be equal ($d_A = d_B = 0.5$). C_{min} and C_{max} are considered 12 ($= \frac{4}{3}max\{t_{ij}\}$) and 18 ($= 2max\{t_{ij}\}$), respectively. Based on some preliminary tests, the values of parameters are determined as follows: $maxIter = 50$, $colSize = 15$, $\alpha = 0.5$, $\beta = 0.3$, $\rho = 0.01$, and $Q = 50$. The algorithm has been coded in Java and run using the above parameters to solve the numerical example. The solutions obtained have been recorded and plotted Pareto-front diagram is provided in Fig. 3(a). In the figure, solutions given in yellow represent nondominant individuals. Among those, the best solution (given in blue) is

Table 1. Processing times and precedence relationships of the tasks for the numerical example.

Task	Side	Model A	Model B	IP	Task	Side	Model A	Model B	IP
1	E	9	6	–	19	E	8	5	15
2	L	5	5	–	20	R	2	2	16
3	E	7	4	–	21	E	8	9	19
4	E	2	2	1	22	L	1	1	–
5	E	6	6	1,2	23	R	1	3	18
6	E	4	4	–	24	R	7	7	20
7	L	4	4	3,4	25	E	5	5	19
8	E	1	4	5	26	E	5	5	22
9	R	0	3	3	27	E	9	5	23
10	R	8	8	6	28	L	1	3	21
11	E	4	2	7,9	29	E	2	2	17,24
12	E	6	6	7	30	E	5	5	25
13	L	7	7	8	31	L	5	5	26
14	E	1	3	10	32	E	0	4	26,27
15	R	5	5	9	33	E	7	7	28,29
16	E	6	6	12	34	E	3	3	32
17	E	2	2	13,14	35	E	8	6	31,33
18	R	7	7	–	36	E	8	8	34

taken in accordance with the efficiency of the line calculated using the equation $LE = \frac{\sum_{j \in J} \sum_{i \in I} d_j t_{ij}}{CxK}$; where C and K are cycle time and the number of workstations required by the individual. I and J are the sets of tasks and models, respectively, and d_j is the proportional demand of model j. Line efficiency is a well-respected measurement which indicates the quality of a balancing solution obtained. As seen from Fig. 3(a), the best balancing solution is obtained when $C = 17$ and $K = 11$. The balancing configuration of tasks to workstations for the best solution is provided in Fig. 4. The workload of workstations for this solution is also plotted in Fig. 3(b). Efficiency of the line is calculated as 90.37%, which corresponds to a successful rate based on common tendency in the literature. The smooth distribution of workloads across workstations also verifies the promising solution capacity of the proposed NSACO approach.

For the purpose of further validation, the lower bound [16] for K is also calculated under the assumption of $C = 17$ and it was observed that the solution found by NSACO requires one more workstation than the lower bound. Keeping in mind that the heterogeneous processing times of tasks and precedence relationships among them are not considered when calculating the lower bound, it can be concluded that the solution found by NSACO is optimal or near-optimal.

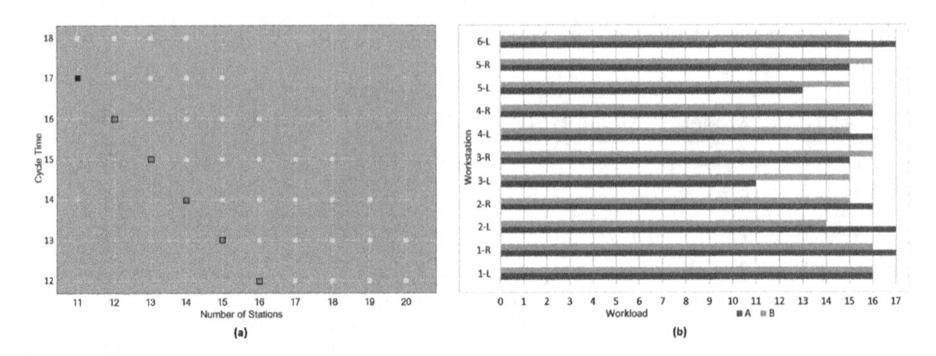

Fig. 3. (a) Pareto-front of the solutions obtained and (b) workloads of the workstations for the best solution ($C = 17$ and $K = 11$). (Color figure online)

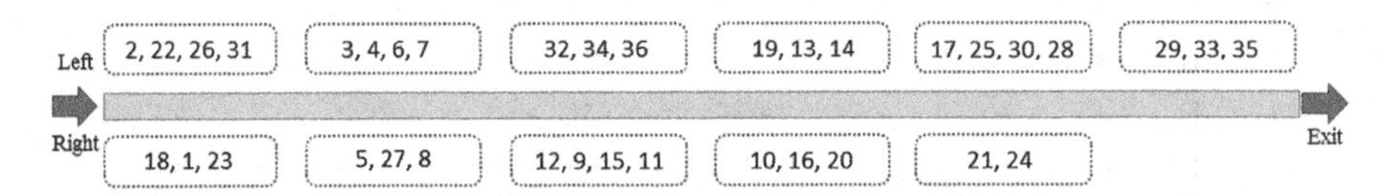

Fig. 4. The balancing configuration of tasks for the best solution ($C = 17$ and $K = 11$).

4 Conclusion

Multi-objective optimisation is a significant part of many engineering design problems. Assembly line balancing problem is a well-known and widely studied industrial engineering problem. A mixed-model two-sided assembly line balancing problem is addressed in this paper aiming to minimise two conflicting objectives, namely cycle time and number of workstations. Such problems are so called type-E in assembly line balancing domain and very limited effort have been put into practice for solving those problem types. In this study, the mixed-model two-sided assembly line balancing problem is explained with an example and a nondominated sorting approach, NSACO, is proposed as a solution method. NSACO is described using a flowchart and the solution of a numerical example is also provided. The result of the numerical example indicated that NSACO has a promising performance. Therefore, the author's future study focuses on testing the algorithm's ability for solving test problems and comparing its performance to other methods, like genetic algorithm, simulated annealing algorithm and tabu search.

References

1. Ford, H.: My Life and Work - An Autobiography of Henry Ford. Classic House Books, New York (2009)
2. Battaia, O., Dolgui, A.: A taxonomy of line balancing problems and their solution approaches. Int. J. Prod. Econ. **142**(2), 259–277 (2013)
3. Thomopoulos, N.T.: Line balancing-sequencing for mixed-model assembly. Manage. Sci. **14**(2), 59–75 (1967)

4. Gokcen, H., Erel, E.: A goal programming approach to mixed-model assembly line balancing problem. Int. J. Prod. Econ. **48**(2), 177–185 (1997)
5. Vilarinho, P.M., Simaria, A.S.: A two-stage heuristic method for balancing mixed-model assembly lines with parallel workstations. Int. J. Prod. Res. **40**(6), 1405–1420 (2002)
6. Kara, Y., Ozcan, U., Peker, A.: An approach for balancing and sequencing mixed-model JIT U-lines. Int. J. Adv. Manuf. Tech. **32**(11–12), 1218–1231 (2007)
7. Yagmahan, B.: Mixed-model assembly line balancing using a multi-objective ant colony optimization approach. Expert Syst. Appl. **38**(10), 12453–12461 (2011)
8. Simaria, A.S., Vilarinho, P.M.: A genetic algorithm based approach to the mixed-model assembly line balancing problem of type II. Comput. Ind. Eng. **47**(4), 391–407 (2004)
9. Li, Z., Kucukkoc, I., Nilakantan, J.M.: Comprehensive review and evaluation of heuristics and meta-heuristics for two-sided assembly line balancing problem. Comput. Oper. Res. **84**, 146–161 (2017). https://doi.org/10.1016/j.cor.2017.03.002
10. Make, M.R.A., Rashid, M.F.F., Razali, M.M.: A review of two-sided assembly line balancing problem. Int. J. Adv. Manuf. Tech. **89**(5), 1743–1763 (2017). https://doi.org/10.1007/s00170-016-9158-3
11. Wei, N., Chao, M.: A solution procedure for type E simple assembly line balancing problem. Comput. Ind. Eng. **61**(3), 824–830 (2011)
12. Garcia-Villoria, A., Pastor, R.: Erratum to "A solution procedure for type E simple assembly line balancing problem". Comput. Ind. Eng. **66**(1), 201–202 (2013)
13. Manavizadeh, N., Rabbani, M., Moshtaghi, D., Jolai, F.: Mixed-model assembly line balancing in the make-to-order and stochastic environment using multi-objective evolutionary algorithms. Expert Syst. Appl. **39**(15), 12026–12031 (2012)
14. Kucukkoc, I., Zhang, D.Z.: Type-E parallel two-sided assembly line balancing problem: mathematical model and ant colony optimisation based approach with optimised parameters. Comput. Ind. Eng. **84**, 56–69 (2015)
15. Simaria, A.S., Vilarinho, P.M.: 2-ANTBAL: an ant colony optimisation algorithm for balancing two-sided assembly lines. Comput. Ind. Eng. **56**(2), 489–506 (2009)
16. Ozcan, U., Toklu, B.: Balancing of mixed-model two-sided assembly lines. Comput. Ind. Eng. **57**(1), 217–227 (2009)
17. Chutima, P., Chimklai, P.: Multi-objective two-sided mixed-model assembly line balancing using particle swarm optimisation with negative knowledge. Comput. Ind. Eng. **62**(1), 39–55 (2012)
18. Kucukkoc, I., Zhang, D.Z.: Simultaneous balancing and sequencing of mixed-model parallel two-sided assembly lines. Int. J. Prod. Res. **52**(12), 3665–3687 (2014)
19. Kucukkoc, I., Zhang, D.Z.: Mathematical model and agent based solution approach for the simultaneous balancing and sequencing of mixed-model parallel two-sided assembly lines. Int. J. Prod. Econ. **158**, 314–333 (2014)
20. Kucukkoc, I., Zhang, D.Z.: Integrating ant colony and genetic algorithms in the balancing and scheduling of complex assembly lines. Int. J. Adv. Manuf. Tech. **82**(1), 265–285 (2016)
21. Bagherinejad, J., Dehghanib, M.: A non-dominated sorting ant colony optimization algorithm approach to the bi-objective multi-vehicle allocation of customers to distribution centers. J. Opt. Ind. Eng. **9**(19), 61–73 (2016)
22. Srinivas, N., Deb, K.: Muilti-objective optimization using non-dominated sorting in genetic algorithms. Evol. Comput. **2**, 221–248 (1994). MIT Press
23. Helgeson, W.B., Birnie, D.P.: Assembly line balancing using the ranked positional weight technique. J. Ind. Eng. **12**(6), 394–398 (1961)
24. Dorigo, M., Stutzle, T.: Ant Colony Optimization. The MIT Press, Cambridge (2004)

Development of Threshold Algorithms for a Location Problem with Elastic Demand

Tatyana Levanova[1,2](✉) and Alexander Gnusarev[1]

[1] Omsk Branch, Sobolev Institute of Mathematics,
Pevtsova str. 13, 644043 Omsk, Russia
alexander.gnussarev@gmail.com
[2] Dostoevsky Omsk State University, Prospekt Mira 55A, 644077 Omsk, Russia
levanova@ofim.oscsbras.ru

Abstract. This work is devoted to development of threshold algorithms for one static probabilistic competitive facility location and design problem in the following formulation. New Company plans to enter the market and to locate new facilities with different design scenarios. Clients of each point choose to use the facilities of Company or its competitors depending on their attractiveness and distance. The aim of the new Company is to capture the greatest number of customers thus serving the largest share of the demand. This share for the Company is elastic and depends on clients' decisions. We offer three types of threshold algorithms: Simulated annealing, Threshold improvement and Iterative improvement. Experimental tuning of parameters of algorithms was carried out. A comparative analysis of the algorithms, depending on the nature and value of the threshold on special test examples up to 300 locations is carried out. The results of numerical experiments are discussed.

Keywords: Discrete optimization · Integer programming
Location problems · Threshold algorithms · Simulated annealing

1 Introduction

Nowadays a lot of economic situations are described by the mathematical model of discrete optimal location discrete problems. The situations, where the decision is made by the monopolist only, have been mostly studied. Such situations are common for the normal economic behaviour. Thus there appear the problems aimed to minimise the expenses such as: simple plant location problem [7], p-median problem [13], capacitated plant location problem [16], which have already become classical. In the modern economic situation, it is often demanded to take into consideration the current rivalry at the market. From this point of view there is a great interest to the competitive facility location models, as they describe most complicated situations and require the special methods of solution. The models differ particularly in the competitors' behaviour [10].

T. Levanova—This research was supported by the Program of Fundamental Scientific Research of the State Academies of Sciences, task I.5.1.6.

I. Lirkov and S. Margenov (Eds.): LSSC 2017, LNCS 10665, pp. 382–389, 2018.
https://doi.org/10.1007/978-3-319-73441-5_41

The static competitive facility location problems imply that the competitor's decision is known and will not be changed. Let us take into consideration the situation when a Company plans to enter the market of existing products and services, so that its p shops served as big share of the market as it is possible. The facilities produce goods of the same kind; the prices are similar; the customers' choice of the shops depends only on the distance from the located ones [14]. If the customers take into account the size of the shop, its assortment, the quality of the service, then such circumstances are described by the static probabilistic models. The different kinds of which are described e.g. in [3,6,8].

Even more complicated is the situation when the competitor's decision is not known beforehand. The deterministic multilevel sequential problems can be considered among such formulations. The bi-level models distinguish the Leader (the company) and the Follower (the competitor). At first the Leader enters the market and establishes a set of facilities $S1$ which consists of p facilities. Then the Follower, being aware of that decision, establishes the set of facilities $S2$ which consists of r facilities. Each customer chooses the place to be served among the facilities from the sets $S1$ or $S2$ according to his/her own preferences. Each served customer brings definite profit, therefore all the market is divided between the Leader and the Follower. Those problems are called $(r|p)$-centroid problems. The interesting results for that kind of problem have been gained e.g. in [4,9,15].

Besides the probabilistic bi-level sequential problems, competitive facility location with competition of customers and others should be taken into consideration. Some solutions for them can be found in [5,10].

This paper deals with the static competitive facility location problems with the elastic demand. The three types of threshold algorithms [1] have been developed for that task. The computational experiments have been carried out in order to compare them.

The comparative analysis of the algorithms, according to the nature and the value of the threshold, has been carried out for the special test examples with up to 300 locations. The results of the computational experiments are given below.

2 Problem Formulation

Berman and Krass [6] and Aboolian et al. [3] develop a spatial interaction model with variable expenditure by introducing non-constant expenditure functions into spatial interaction location models. Aboolian et al. [2] proposed a new model where optimal location and design decisions for a set of new facilities are seeking. Here we develop approximate algorithms for the location and design problem described in [2]. In this problem, Company plans to locate its facilities which differ from one another in design: size, range, etc. Let R be the set of facility designs, $r \in R$. There are w_i customers at the point i of discrete set $N = \{1, 2, \dots, n\}$. All customers have the same demand, so each item can be considered as one client with weight w_i. Clients of each point choose to use the facilities of Company or its competitors depending on their attractiveness and distance. The distance d_{ij} between the points i and j is measured, for example, in Euclidean metric or equals to the shortest distance in the corresponding graph. It is assumed that $C \subset P$ is the set of pre-existing competitor facilities and it will not change. The Company may open its markets in $S = P \setminus C$ taking into account the budget B and the cost of opening c_{jr} facility $j \in S$ with design $r \in R$. The Company's goal is to attract a maximum number of customers, i.e. to serve the largest share of total demand.

Let us write out the mathematical model according to [2]. Variables $x_{jr} = 1$, if facility j is opened with design variant r and $x_{jr} = 0$ otherwise, $j \in S, r \in R$. Utility $u_{ij} = \sum_{r=1}^{R} k_{ijr}x_{jr}$, where the supplementary coefficients

$$k_{ijr} = a_{jr}(d_{ij} + 1)^{-\beta}.$$

They depend on the sensitivity β of customers to distance to facility and attractiveness a_{jr}. The total utility for the customers in point $i \in N$ from the facilities controlled by the competitors is $U_i(C) = \sum_{j \in C} u_{ij}$.

The demand function has an exponential form:

$$g(U_i) = 1 - \exp\Big(- \lambda_i U_i \Big),$$

where λ_i is the characteristic of elastic demand in point i, $\lambda_i > 0$; U_i is the total utility for a customer at $i \in N$ from all open facilities:

$$U_i = \sum_{j \in S} \sum_{r=1}^{R} k_{ijr}x_{jr} + U_i(C) = U_i(S) + U_i(C).$$

It is assumed that the demand of the customers at $g(U_i)$ is a concave increasing function.

The company's total share of facility $i \in N$ is measured by:

$$MS_i = \frac{U_i(S)}{U_i(S) + U_i(C)} = \frac{\sum_{j \in S} \sum_{r=1}^{R} k_{ijr}x_{jr}}{\sum_{j \in S} \sum_{r=1}^{R} k_{ijr}x_{jr} + \sum_{j \in C} u_{ij}}.$$

Then the mathematical model looks like:

$$\max \sum_{i \in N} w_i \cdot g(U_i) \cdot MS_i, \tag{1}$$

$$\sum_{j \in S} \sum_{r \in R} c_{jr}x_{jr} \le B, \tag{2}$$

$$\sum_{r \in R} x_{jr} \le 1, j \in S, \tag{3}$$

$$x_{jr} \in \{0, 1\}, \quad r \in R, j \in S. \tag{4}$$

Based on above notation, the objective function (1) looks as follows:

$$\max \sum_{i \in N} w_i \cdot \left(1 - \exp\left(-\lambda_i\left(\sum_{j \in S}\sum_{r=1}^{R} k_{ijr}x_{jr} + U_i(C)\right)\right)\right)$$
$$\cdot\left(\frac{\sum_{j \in S}\sum_{r=1}^{R} k_{ijr}x_{jr}}{\sum_{j \in S}\sum_{r=1}^{R} k_{ijr}x_{jr} + \sum_{j \in C} u_{ij}}\right). \tag{5}$$

The objective function (5) reflects the Company's goal to maximize the share of customers demand. Inequality (2) takes into account the available budget. Condition (3) shows that only one variant of the design can be selected.

3 Threshold Algorithms

The number of algorithms known for solution of the competitive facility and design problem is little. In [2], an adapted weighted greedy heuristic algorithm is proposed. Earlier we developed Variable Neighborhood Search algorithms [12]. In this paper, we offer Threshold Algorithms [1] for this problem.

Let s be a feasible solution of a combinatorial minimization problem. The general scheme of threshold algorithms is the following:

1. Select an initial solution s_0, compute the initial value of the objective function $f(s_0)$, define the record value as $f^* := f(s_0)$. Set the iteration counter $k = 0$, set the threshold value t_k and the type of neighborhood $N(s_k)$.
2. Until the stopping criterion is not satisfied, do the following:
   2.1 Select the new solution randomly in the neighborhood of the current one: $s_j \in N(s_k)$.
   2.2 If the difference does not exceed the threshold $f(s_j) - f(s_k) < t_k$, then $s_{(k+1)} := s_j$.
   2.3 If $f^* > f(s_k)$, then update the best found solution value $f^* := f(s_k)$.
   2.4 Set $k := k + 1$.

There are three variants of this algorithm depending on the setting of the sequence of the threshold value $\{t_k\}$:

(1) *Iterative improvement*: the sequence $t_k = 0, k = 0, 1, 2, ...$, is a variant of local descent with a monotonous improvement of the objective function;
(2) *Threshold accepting*: $t_k = c_k$, $k = 0, 1, 2, ..., c_k \geq 0$, $c_k \geq c_{k+1}$ and $\lim_{k\to\infty} c_k \to 0$ are the local search variants when the objective function deterioration is assumed until some fixed threshold, and the threshold continually goes down to zero;
(3) *Simulated annealing*: the sequence $t_k \geq 0, k = 0, 1, 2, ...$, is a random variable with the expectation $E(t_k) = c_k \geq 0$ which is the local search variant, when the objective function arbitrary deterioration is assumed but the transition

probability is inversely related to the deterioration value. For any two feasible solutions s_i and s_j the probability of accepting s_j from s_i at the iteration k is:

$$P_{s_i s_j} = \begin{cases} 1 & if \quad f(s_j) \le f(s_i), \\ \exp \frac{f(s_i)-f(s_j)}{c_k} & if \quad f(s_j) > f(s_i), \end{cases} s_j \in N(s_i).$$

At the present time the simulated annealing algorithm shows good results for a wide range of optimization problems. The randomized nature of this algorithm allows asymptotic convergence to optimal solutions under special conditions [1].

We adapted threshold algorithms for the maximization problem. These algorithms belong to the class of local search methods. The neighborhood selection plays an important role in their development for individual tasks. The Lin-Kernighan neighborhood have been used in the proposed variants of algorithms. In addition, a special neighborhood is constructed as follows. Let the vector $z = (z_i)$ be such that z_i corresponds to facility i: $z_i = r$ iff $x_{ir} = 1$. Then feasible solution z' is called neighboring for z if it can be obtained with the following moves: (a) choose one of the open facilities p with design variant z_p and reduce the number of design variant up to 0 (close the facility); (b) select the facility q which is closed; then open the facility q with the design variant z_p.

4 Computatinal Experiments

To study the algorithms a series of testing instances similar to the real data of the applied problem [2] has been constructed in [12]. The testing instances consist of two sets with uniform distribution of distances in the interval [0;30] (Series 1) and with Euclidean distances (Series 2). They contain 96 instances for location of 60, 80, 100, 150, 200 and 300 facilities; 3 types of design variants are used, the budget limited is 3, 5, 7 and 9; the demand parameter is $\lambda_i = 1, i \in N$; the customer sensitivity to the distance is high ($\beta = 2$).

It must be mentioned that there is a problem of the choice of the parameter values so that the algorithm could produce good results for a wide range of instances. After the series of preliminary experiments the following parameter values for the simulated annealing algorithm have been found: the temperature interval length $l = 10$, the initial temperature $t_0 = 150$, the cooling (minimal) temperature value $t_{cool} = 5$, the cooling coefficient $r = 0.99$, the number of points in the Lin-Kernighan neighborhood $K = 3$. The threshold value equal to 5 was chosen for the threshold accepting algorithm.

Table 1 shows the values of deviations from the upper bounds (UB) [12] for test instances with uniform distribution of distances in a single run of the algorithms. For example, the average deviations for the uniform distribution of distances instances of 300 locations are: 6.627% for the iterative improvement; 8.308% for the threshold accepting; 1.523% for the simulated annealing algorithm. The test instances with Euclidean distances proved to be difficult for all considered algorithms. The deviations in this case are: 15.325% for the iterative improvement; 18.296% for the threshold accepting; 14.231% for the simulated

Table 1. Deviations from upper bounds in case of uniform distribution of distances

N	Iterative improvement			Threshold accepting			Simulated annealing		
	min	max	av	min	max	av	min	max	av
60	0.010	5.196	1.562	0.066	11.043	3.511	0.008	4.727	0.818
80	0.019	5.698	1.017	0.027	9.030	3.452	0.014	1.810	0.342
100	0.000	6.596	1.043	0.050	22.014	3.737	0.000	3.463	0.648
150	0.000	12.279	3.434	0.013	13.492	6.638	0.000	4.804	0.668
200	0.000	10.176	3.402	0.000	24.627	6.796	0.000	9.589	1.796
300	0.000	19.050	6.627	0.000	21.510	8.308	0.000	11.340	1.523

annealing algorithm. In 8 test instances the best found goal function values of simulated annealing was within 0.001% from UB (in 6 and in 2 test instances for iterative improvement and threshold accepting respectively). It was noted that in some instances, vector solutions of the algorithms coincide with the vectors of UB: in 42 cases for iterative improvement, in 23 test instances for threshold accepting and in 75 cases for simulated annealing algorithm. For instances with Euclidean distances this values are 23, 11 and 27 respectively.

Since algorithms are of a probabilistic nature, they are tested repeatedly. We ran 1000 each of the algorithms for each of the test instances. For this computational experiment, the following results are obtained. The average deviations for the instances with uniform distribution of distances of 300 locations are: 6.771% for the iterative improvement; 8.36% for the threshold accepting algorithm; 1.044% for the simulated annealing algorithm. The deviations in case of the Euclidean distances are: 16.017% for the iterative improvement; 18.71% for the threshold accepting algorithm; 14.734% for the simulated annealing algorithm. This indicates that Series 2 is more complicated for the proposed algorithms. On the other hand, such deviations may display that the upper bounds for the second series are inaccurate. For Series 1, the 95% confidence interval for the probability of obtaining the deviations less than 0.001% for simulated annealing is between [0.111;0.115], for iterative improvement is [0.044;0.047] and for threshold accepting is [0.030;0.032]. For Series 2, the 95% confidence interval for the probability of obtaining the deviations less than 12% for simulated annealing is [0.112;0.116], for iterative improvement is [0.075;0.078] and for threshold accepting is [0.030;0.032]. The Wilcoxon test [11] showed statistically significant differences between the values of the objective functions of the investigated algorithms with significance level 0.05. The simulated annealing algorithm in both series of test instances gives the best value of the objective function. At the same time, an iterative improvement algorithm has an advantage over the threshold accepting algorithm, which is confirmed by statistically significant differences in both series.

Table 2 contains the information about minimal (min), average (av) and maximal (max) CPU time (in seconds) for the proposed algorithms until the stopping criterion was met. The experiments have been carried out using a PC Intel

Table 2. CPU time (sec)

N	Iterative improvement			Threshold accepting			Simulated annealing		
	min	av	max	min	av	max	min	av	max
60	0.085	0.109	0.142	0.078	0.104	0.123	0.905	1.190	1.462
80	0.132	0.161	0.203	0.130	0.162	0.226	1.544	1.866	2.281
100	0.177	0.210	0.235	0.176	0.217	0.260	2.228	2.228	2.853
150	0.332	0.411	0.504	0.342	0.416	0.523	4.120	4.876	5.783
200	0.525	0.603	0.660	0.554	0.618	0.668	6.367	7.213	7.824
300	1.242	1.330	1.581	1.227	1.325	1.529	15.197	16.176	18.004

i5-2450M, 2.50 GHz, memory 4 GB. The time for the instances with Euclidean distances and with the uniform distribution of distances is approximately the same. Note that well-known commercial software is rather time-consuming. For instance, for one of the instances of 60 locations the CPU time of CoinBonmin (GAMS) was 63 h and the objective function deviation from the upper bounds was 12.919%. All proposed algorithms yielded equal objective function values with deviation 11.796%, running time was less than 1.638 s for it.

5 Conclusion

This paper is devoted to the development of approximate algorithms for one variant of the static probabilistic competitive facility location and design problem. Its mathematical model is based on a nonlinear objective function, and the share of the served demand is elastic. That complicates the task of finding an optimal solution.

The threshold algorithms for the search of approximate solutions have been built, their parameter setting has been carried out. It should be noticed that the iterative improvement and the simulated annealing algorithms are comparable in the objective function for the instances up to 100 locations. The maximum counting duration of the built algorithms does not exceed 20 s. and the minimal deviations from upper bounds was less then 0.001%. The analysis of the algorithms for the instances with a large number of locations has proved the advantage of the simulated annealing algorithm over the threshold algorithms and its applicability to complex problems.

References

1. Aarts, E., Korst, J., Laarhoven, P.: Simulated annealing. In: Aarts, E., Lenstra, J.K. (eds.) Local Search in Combinatorial Optimization, pp. 91–120. Wiley, New York (1997)
2. Aboolian, R., Berman, O., Krass, D.: Competitive facility location and design problem. Eur. J. Oper. Res. **182**(1), 40–62 (2007)

3. Aboolian, R., Berman, O., Krass, D.: Competitive facility location model with concave demand. Eur. J. Oper. Res. **181**, 598–619 (2007)
4. Alekseeva, E., Kochetova, N., Kochetov, Y., Plyasunov, A.: Heuristic and exact methods for the discrete $(r|p)$-centroid problem. In: Cowling, P., Merz, P. (eds.) EvoCOP 2010. LNCS, vol. 6022, pp. 11–22. Springer, Heidelberg (2010). https://doi.org/10.1007/978-3-642-12139-5_2
5. Beresnev, V.L.: Upper bounds for objective function of discrete competitive facility location problems. J. Appl. Ind. Math. **3**(4), 3–24 (2009)
6. Berman, O., Krass, D.: Locating multiple competitive facilities: spatial interaction models with variable expenditures. Ann. Oper. Res. **111**, 197–225 (2002)
7. Cornuejols, G., Nemhauser, G.L., Wolsey, L.A.: The uncapacitated facility location problem discrete location problems. In: Mirchamdani, P.B., Franscis, R.L. (eds.) Discrete Location Problems, pp. 119–171. Wiley, New York (1990)
8. Drezner, T.: Competitive facility location in plane. In: Drezner, Z. (ed.) Facility Location. A Survey of Applications and Methods, pp. 285–300. Springer, Berlin (1995)
9. Davydov, I.A., Kochetov, Y.A., Mladenovic, N., Urosevic, D.: Fast metaheuristics for the discrete $r|p$-centroid problem. Autom. Remote Control **75**(4), 677–687 (2014)
10. Karakitsiou, A.: Modeling Discrete Competitive Facility Location. Springer, Heidelberg (2015)
11. Kerby, D.S.: The simple difference formula: an approach to teaching nonparametric correlation. Compr. Psychol. **3**(1) (2014). http://journals.sagepub.com/doi/full/10.2466/11.IT.3.1. Accessed 20 Dec 2017
12. Levanova, T., Gnusarev, A.: Variable neighborhood search approach for the location and design problem. In: Kochetov, Y., Khachay, M., Beresnev, V., Nurminski, E., Pardalos, P. (eds.) DOOR 2016. LNCS, vol. 9869, pp. 570–577. Springer, Cham (2016). https://doi.org/10.1007/978-3-319-44914-2_45
13. Mirchamdani, P.B.: The p-median problem and generalizations. In: Mirchamdani, P.B., Franscis, R.L. (eds.) Discrete Location Problems, pp. 55–118. Wiley, New York (1990)
14. Serra, D., ReVelle, C.: Competitive location in discrete space. In: Drezner, Z. (ed.) Facility Location. A Survey of Applications and Methods, pp. 367–386. Springer, Berlin (1995)
15. Spoerhose, J., Wirth, H.C.: (r, p)-centroid problems on paths and trees. Technical report No. 441, Institute of Computer Science, University of Wurzburg (2008)
16. Sridharan, R.: The capacitated plant location problem. Eur. J. Oper. Res. **87**, 203–213 (1995)

Investigation of Genetic Algorithm Performance Based on Different Algorithms for InterCriteria Relations Calculation

Tania Pencheva(✉), Olympia Roeva, and Maria Angelova

Institute of Biophysics and Biomedical Engineering, Bulgarian Academy of Sciences, 105 Acad. G. Bonchev Str., 1113 Sofia, Bulgaria
{tania.pencheva,olympia,maria.angelova}@biomed.bas.bg

Abstract. InterCriteria Analysis is a recently developed approach for the evaluation of the correlation between multiple objects against multiple criteria. As such, it is expected to prove any existing correlations between the criteria themselves or even to discover any new. In this investigation different algorithms for InterCriteria relations calculation are explored to render the influence of the genetic algorithm (GA) parameters on the algorithm performance. GA is chosen as an optimization technique as they are among the most widely used out of the biologically inspired approaches for global search. GA is here applied to parameter identification of a *S. cerevisiae* fed-batch fermentation process model.

Keywords: Genetic algorithm · InterCriteria Analysis
Parameter identification · Fermentation process · *S. cerevisiae*

1 Introduction

InterCriteria Analysis (ICrA) is an approach going beyond the nature of the criteria involved in a process of evaluation of multiple objects against multiple criteria [4]. Up to now ICrA has been successfully applied in different areas of science and practice. Some of the last ICrA applications are in the fields of e-learning [6], algorithms performance [7], medicine [12], etc. The ICrA approach is also applied to fermentation processes (FP) modelling. In [11] the approach has been implemented to establish the relations between parameters of the genetic algorithm (GA) and an *E. coli* FP model. As such, the approach discovers some existing correlations between the model parameters and/or optimization algorithm parameters, both defined as ICrA "criteria".

Meanwhile, the ICrA approach theory is continuously developed. In [10] four different algorithms for InterCriteria relations calculation, namely μ-biased, balanced, ν-biased and unbiased, are proposed. In order to gain some more generalized conclusions about the considered algorithms performance, different case studies should be investigated. Therefore, a further investigation of the four algorithms for InterCriteria relations calculation is performed in this research. As a case study, a model parameter identification of *S. cerevisiae* FP based on

I. Lirkov and S. Margenov (Eds.): LSSC 2017, LNCS 10665, pp. 390–398, 2018.
https://doi.org/10.1007/978-3-319-73441-5_42

GA is used. μ-biased, balanced, ν-biased and unbiased algorithms are applied to explore the correlations between *S. cerevisiae* FP model, from one side, and GA parameters – the number of individuals (*ind*) and the number of generations (*gen*), from the other side. The obtained results are compared with those discussed in [10], where an *E. coli* FP is considered as a case study.

2 Problem Formulation

2.1 Case Study: *S. cerevisiae* Fed-Batch Fermentation Process

Experimental data of *S. cerevisiae* fed-batch FP is obtained in the Institute of Technical Chemistry – University of Hanover, Germany, and described in details in [9]. Mathematical model of *S. cerevisiae* fed-batch FP is commonly described as follows, according to the mass balance and considering mixed oxidative functional state [9]:

$$\frac{dX}{dt} = \left(\mu_{2S}\frac{S}{S+k_S} + \mu_{2E}\frac{E}{E+k_E}\right)X - \frac{F}{V}X \quad (1)$$

$$\frac{dS}{dt} = -\frac{\mu_{2S}}{Y_{SX}}\frac{S}{S+k_S}X + \frac{F}{V}\left(S_{in} - S\right) \quad (2)$$

$$\frac{dV}{dt} = F, \quad (3)$$

where X is the concentration of biomass, [g/l]; S – concentration of substrate (glucose), [g/l]; E – concentration of ethanol, [g/l]; F – feeding rate, [l/h]; V – volume of bioreactor, [l]; S_{in} – glucose concentration in the feeding solution, [g/l]; μ_{2S}, μ_{2E} – maximum growth rates of substrate and ethanol, [1/h]; k_S, k_E – saturation constants of substrate and ethanol, [g/l]; Y_{SX} – yield coefficient, [g/g]. All functions are continuous and differentiable, and all model parameters fulfil the non-zero division requirement.

2.2 Model Identification Based on Genetic Algorithm

GAs are a stochastic global optimization technique [5] for hard problems solving. There are many operators, functions, parameters and settings in the GA that can be tuned when they are implemented to different tasks [5,7,11]. Among the most important GA parameters that have a significant influence on the algorithm effectiveness, are the number of individuals (*ind*) and the number of generations (*gen*). The number of individuals determines how many chromosomes are included in the population. If there are few chromosomes, GA will have fewer opportunities to perform crossover and only a small part of the search space will be investigated. On the other hand, if there are too many chromosomes, the algorithm convergence time will logically increase. Concerning the number of generations, different authors offer different solutions depending on the solved problem. The number of generations can significantly affect the accuracy of the solution and the convergence time of the algorithm.

In this paper the impact of *ind* and *gen* is going to be evaluated examining different values of both GA parameters. GA is applied with the following values of both parameters: $ind = \{20, 40, 60, 80, 100\}$ and $gen = \{100, 200, 500, 1000\}$. The choice of the rest GA parameters and operators is based on the recommendations in [5] and authors expertise [7–9,11]: crossover operator – double point; mutation operator – bit inversion; selection operator – roulette wheel selection; encoding – binary; crossover rate – 0.95; mutation rate – 0.05; generation gap – 0.8.

Aiming at the best fit to an experimental data set the model parameter vector $p = [\mu_{2S}, \mu_{2E}, k_S, k_E, Y_{SX}]$ of the non-linear mathematical model (Eqs. (1)–(3)) is going to be identified. As an optimization criterion, mean square deviation between the model output and the FP experimental data has been used:

$$J = \sum \left((S - S^*)^2 + (X - X^*)^2 \right) \rightarrow min, \tag{4}$$

where X, S are the experimental data and X^*, S^* – model predicted data.

2.3 InterCriteria Analysis

The detailed description of the ICrA approach, based on the apparatuses of Index Matrices (IM) [1] and Intuitionistic Fuzzy Sets (IFS) [2,3], might be found in [4]. Here, for completeness, the proposed idea is briefly presented.

The initial IM A is presented in the form of Eq. (5), where C_p is a criterion, O_q – an object, $C_p(O_q)$ – a real number, for every p, q $(1 \leq p \leq m, 1 \leq q \leq n)$.

$$A = \begin{array}{c|ccccc} & O_1 & \dots & O_q & \dots & O_n \\ \hline C_1 & C_1(O_1) & \dots & C_1(O_q) & \dots & C_1(O_n) \\ \vdots & \vdots & \ddots & \vdots & \ddots & \vdots \\ C_p & C_p(O_1) & \dots & C_p(O_q) & \dots & C_p(O_n) \\ \vdots & \vdots & \ddots & \vdots & \ddots & \vdots \\ C_m & C_m(O_1) & \dots & C_m(O_q) & \dots & C_m(O_n) \end{array} \tag{5}$$

Let $x_i = C(O_i)$. Then $C^*(O) \stackrel{\text{def}}{=} \{\langle x_i, x_j \rangle | i \neq j \,\&\, \langle x_i, x_j \rangle \in C(O) \times C(O)\}$.

Further, if $x = C(O_i)$ and $y = C(O_j)$, $x \prec y$ will be written iff $i < j$. In order to find the agreement between two criteria, the vectors of all internal comparisons for each criterion are constructed, which elements fulfil one of the three relations R, $\overline{R}$ and $\tilde{R}$. The nature of the relations is chosen such that for a fixed criterion C and any ordered pair $\langle x, y \rangle \in C^*(O)$:

$$\langle x, y \rangle \in R \Leftrightarrow \langle y, x \rangle \in \overline{R}, \tag{6}$$

$$\langle x, y \rangle \in \tilde{R} \Leftrightarrow \langle x, y \rangle \notin (R \cup \overline{R}), \tag{7}$$

$$R \cup \overline{R} \cup \tilde{R} = C^*(O). \tag{8}$$

Due to Eqs. (6)–(8), the subset $C^{\prec}(O) \stackrel{\text{def}}{=} \{\langle x, y\rangle | \, x \prec y \,\&\, \langle x, y\rangle \in C(O)\times C(O)\}$ is considered. For brevity, $c^{i,j} = \langle C(O_i), C(O_j)\rangle$. Then, for a fixed criterion C, the vector of lexicographically ordered pair elements is constructed:

$$V(C) = \{c^{1,2}, c^{1,3}, \ldots, c^{1,n}, c^{2,3}, c^{2,4}, \ldots, c^{2,n}, c^{3,4}, \ldots, c^{3,n}, \ldots, c^{n-1,n}\}. \quad (9)$$

$V(C)$ is replaced by $\hat{V}(C)$, where its k-th component $(1 \le k \le \frac{n(n-1)}{2})$ is as:

$$\hat{V}_k(C) = \begin{cases} 1, & \text{iff } V_k(C) \in R, \\ -1, & \text{iff } V_k(C) \in \overline{R}, \\ 0, & \text{otherwise.} \end{cases}$$

Here the following algorithms for calculation of the degree of "agreement" $(\mu_{C,C'})$ and the degree of "disagreement" $(\nu_{C,C'})$ are used [10]:

- **Algorithm 1, μ-biased**: Here the rule for $=, =$ for two criteria C and C' is assigned to $\mu_{C,C'}$.
- **Algorithm 2, balanced**: Here the rule for $=, =$ for two criteria C and C' is assigned a half to both $\mu_{C,C'}$ and $\nu_{C,C'}$. It should be noted that in such case a criteria compared to itself does not necessarily yield $\langle 1, 0\rangle$.
- **Algorithm 3, ν-biased**: In this case the rule for $=, =$ for two criteria C and C' is assigned to $\nu_{C,C'}$.
- **Algorithm 4, unbiased**: In this case the rule for $=, =$ for two criteria C and C' is assigned neither to $\mu_{C,C'}$, nor to $\nu_{C,C'}$. As such, the degrees of "uncertainty" $\pi_{C,C'}$ $(\pi_{C,C'} = 1 - \mu_{C,C'} - \nu_{C,C'})$ is increased.

3 Numerical Results and Discussion

GA with different values of *ind* and *gen* are consequently applied to a model parameter identification of *S. cerevisiae* fed-batch FP. Due to the algorithms' stochastic nature 30 independent runs of each GA have been performed. Based on the obtained average results for the model parameters, GA convergence time (T) and objective function (J), the two IMs A_{ind} and A_{gen} are constructed:

$A_{ind} =$

	$GA^{1,1}_{ind}$	$GA^{1,2}_{ind}$	$GA^{1,3}_{ind}$	$GA^{1,4}_{ind}$	$GA^{1,5}_{ind}$
J	0.022	0.022	0.022	0.022	0.022
T	79.2	145.5	212.8	295.5	375.1
ind	20	40	60	80	100
μ_{2S}	0.95	0.97	0.95	0.97	0.98
μ_{2E}	0.12	0.12	0.12	0.14	0.14
k_S	0.13	0.13	0.12	0.13	0.13
k_E	0.8	0.8	0.8	0.8	0.8
Y_{SX}	0.41	0.41	0.41	0.4	0.4

$A_{gen} =$

	$GA^{1,1}_{gen}$	$GA^{1,2}_{gen}$	$GA^{1,3}_{gen}$	$GA^{1,4}_{gen}$
J	0.022	0.022	0.022	0.022
T	76.8	148.1	368.3	771.3
gen	100	200	500	1000
μ_{2S}	0.96	0.95	0.95	0.97
μ_{2E}	0.13	0.12	0.12	0.14
k_S	0.13	0.12	0.12	0.13
k_E	0.8	0.8	0.8	0.8
Y_{SX}	0.41	0.42	0.41	0.4

(10)

IM A_{ind} consists of average estimations of the five model parameters μ_{2S}, μ_{2E}, k_S, k_E, Y_{SX}, as well as of T and J, respectively in the case of $ind = \{20, 40, 60,$

80, 100}, corresponding to $\mathrm{GA}_{ind}^{1,1} \div \mathrm{GA}_{ind}^{1,5}$. Average estimations of the model parameters and objective function values are rounded up to the second digit after the decimal point, while the convergence time is rounded up to the first digit. IM A_{gen} is constructed in a similar manner. It consists of the average and rounded results for μ_{2S}, μ_{2E}, k_S, k_E, Y_{SX}, T, and J, respectively in the case of $gen = \{100, 200, 500, 1000\}$, corresponding to $\mathrm{GA}_{gen}^{1,1} \div \mathrm{GA}_{gen}^{1,4}$.

In the presented IMs (Eq. (10)), there are many equal estimates for J and model parameters. As such, they are good examples to test the proposed four algorithms for InterCriteria relations calculations of $\mu_{C,C'}$, $\nu_{C,C'}$ and $\pi_{C,C'}$.

Four algorithms for InterCriteria relations calculation (μ-biased, balanced, ν-biased and unbiased) have been applied to the constructed above two IMs (Eq. (10)). Thus, the IF pairs $\langle \mu_{C,C'}, \nu_{C,C'} \rangle$ and $\pi_{C,C'}$ for every two pairs of considered criteria are calculated. The results are listed in the Tables 1 and 2, respectively for different values of *ind* and *gen*.

The results from all four investigated algorithms show, that the strongest correlation ($\mu = 1$) (i.e. positive consonance) has been observed between convergence time $T - ind$ and $T - gen$. Such correlation had been expected, since the convergence time logically increases with the increase of number of individuals or number of generations. The results achieved in [10] are similar, although GA parameter *ind* renders slightly bigger influence to T than to *gen*.

Very strong correlation ($\mu = 0$), i.e. negative consonance, has been also observed for $J - ind$ and $J - gen$, respectively. These results have been also expected since the objective function value logically decreases when the number of individuals or the number of generations increase. In other words, a solution with a higher accuracy is achieved either with investigation of a bigger part of the search space, or with performance of more algorithm generations, respectively. In comparison to the results in [10] in case of *E. coli* FP, such correlation is observed only for GA parameter *gen*. It is worth to note that in this case the calculated values of the degrees of "uncertainty" are $\pi_{C,C'} = 1$ due to the obtained equal values of J during all performed parameter identification procedures. From the other point of view, this is the evidence of excellent GA performance reaching the lowest J-value ($J = 0.022$) for all considered GA parameters *ind* and *gen*.

In case of μ-biased algorithm (**Algorithm 1**) another strong correlation has been observed between J and the model parameter k_E for both investigated GA parameters *ind* and *gen*. Such a result shows a high sensitivity of k_E in comparison to the other model parameters in vector $p = [\mu_{2S}, \mu_{2E}, k_S, k_E, Y_{SX}]$ of the non-linear mathematical model (Eqs. (1)–(3)).

In the case of *ind* variation a strong correlation between μ_{2S} and μ_{2E} is observed. **Algorithm 1** assigns the highest value to the $\mu_{C,C'} = 1$, i.e. strong positive consonance, **Algorithm 2** – the value of $\mu_{C,C'} = 0.92$, i.e. positive consonance, while **Algorithm 3** and **Algorithm 4** – the value of $\mu_{C,C'} = 0.83$, i.e. weak positive consonance. The pairs $\mu_{2S} - k_S$ and $\mu_{2E} - k_S$ show also $\mu_{C,C'} = 0.83$, i.e. weak positive consonance. These high $\mu_{C,C'}$-values are also expected due to the physical meaning of FP models parameters [9]. For the case of GA parameter *gen*, the obtained results are not so convinced.

Table 1. Results from the ICrA in case of *S. cerevisiae* FP – *ind*

Criteria pairs	μ-biased			Balanced			ν-biased			Unbiased		
	μ	ν	π	μ	ν	π	μ	ν	π	μ	ν	π
$J-T$	0	0	1	0	0	1	0	0	1	0	0	1
$J-ind$	0	0	1	0	0	1	0	0	1	0	0	1
$J-\mu_{2S}$	0.17	0	0.83	0.08	0.08	0.83	0	0.17	0.83	0	0	1
$J-\mu_{2E}$	0.17	0	0.83	0.08	0.08	0.83	0	0.17	0.83	0	0	1
$J-k_S$	0.33	0	0.67	0.17	0.17	0.67	0	0.33	0.67	0	0	1
$J-k_E$	1	0	0	0.50	0.50	0	0	1	0	0	0	1
$J-Y_{SX}$	0.17	0	0.83	0.08	0.08	0.83	0	0.17	0.83	0	0	1
$T-ind$	1	0	0	1	0	0	1	0	0	1	0	0
$T-\mu_{2S}$	0.50	0.33	0.17	0.50	0.33	0.17	0.50	0.33	0.17	0.50	0.33	0.17
$T-\mu_{2E}$	0.50	0.33	0.17	0.50	0.33	0.17	0.50	0.33	0.17	0.50	0.33	0.17
$T-k_S$	0.33	0.33	0.33	0.33	0.33	0.33	0.33	0.33	0.33	0.33	0.33	0.33
$T-k_E$	0	0	1	0	0	1	0	0	1	0	0	1
$T-Y_{SX}$	0.17	0.67	0.17	0.17	0.67	0.17	0.17	0.67	0.17	0.17	0.67	0.17
$ind-\mu_{2S}$	0.50	0.33	0.17	0.50	0.33	0.17	0.50	0.33	0.17	0.50	0.33	0.17
$ind-\mu_{2E}$	0.50	0.33	0.17	0.50	0.33	0.17	0.50	0.33	0.17	0.50	0.33	0.17
$ind-k_S$	0.33	0.33	0.33	0.33	0.33	0.33	0.33	0.33	0.33	0.33	0.33	0.33
$ind-k_E$	0	0	1	0	0	1	0	0	1	0	0	1
$ind-Y_{SX}$	0.17	0.67	0.17	0.17	0.67	0.17	0.17	0.67	0.17	0.17	0.67	0.17
$\mu_{2S}-\mu_{2E}$	1	0	0	0.92	0.08	0	0.83	0.17	0	0.83	0	0.17
$\mu_{2S}-k_S$	0.83	0	0.17	0.75	0.08	0.17	0.67	0.17	0.17	0.67	0	0.33
$\mu_{2S}-k_E$	0.17	0	0.83	0.08	0.08	0.83	0	0.17	0.83	0	0	1
$\mu_{2S}-Y_{SX}$	0	0.67	0.33	0	0.67	0.33	0	0.67	0.33	0	0.67	0.33
$\mu_{2E}-k_S$	0.83	0	0.17	0.75	0.08	0.17	0.67	0.17	0.17	0.67	0	0.33
$\mu_{2E}-k_E$	0.17	0	0.83	0.08	0.08	0.83	0	0.17	0.83	0	0	1
$\mu_{2E}-Y_{SX}$	0	0.67	0.33	0	0.67	0.33	0	0.67	0.33	0	0.67	0.33
k_S-k_E	0.33	0	0.67	0.17	0.17	0.67	0	0.33	0.67	0	0	1
k_S-Y_{SX}	0	0.50	0.50	0	0.50	0.50	0	0.50	0.50	0	0.50	0.50
k_E-Y_{SX}	0.17	0	0.83	0.08	0.08	0.83	0	0.17	0.83	0	0	1

The results show that four algorithms for InterCriteria relations calculation uniquely define $\mu_{C,C'}$-values for 13 out of 28 criteria pairs, when the influence of both GA parameter *ind* and *gen* has been investigated. For the rest 15 criteria pairs, the value of $\mu_{C,C'}$ varies depending on the applied ICrA algorithms, due to different assignment of the rule for =,=. Investigated in this paper different ICrA algorithms show discrepancies only for those criteria that have equal estimates by pairs, concerning input IMs (Eq. (10)). As can be seen from Tables 1 and 2,

Table 2. Results from the ICrA in case of *S. cerevisiae* FP – *gen*

Criteria pairs	μ-biased			Balanced			ν-biased			Unbiased		
	μ	ν	π	μ	ν	π	μ	ν	π	μ	ν	π
$J-T$	0	0	1	0	0	1	0	0	1	0	0	1
$J-gen$	0	0	1	0	0	1	0	0	1	0	0	1
$J-\mu_{2S}$	0.20	0	0.80	0.10	0.10	0.80	0	0.20	0.80	0	0	1
$J-\mu_{2E}$	0.40	0	0.60	0.20	0.20	0.60	0	0.40	0.60	0	0	1
$J-k_S$	0.60	0	0.40	0.30	0.30	0.40	0	0.60	0.40	0	0	1
$J-k_E$	1	0	0	0.50	0.50	0	0	1	0	0	0	1
$J-Y_{SX}$	0.40	0	0.60	0.20	0.20	0.60	0	0.40	0.60	0	0	1
$T-gen$	1	0	0	1	0	0	1	0	0	1	0	0
$T-\mu_{2S}$	0.70	0.10	0.20	0.70	0.10	0.20	0.70	0.10	0.20	0.70	0.10	0.20
$T-\mu_{2E}$	0.60	0	0.40	0.60	0	0.40	0.60	0	0.40	0.60	0	0.40
$T-k_S$	0.20	0.20	0.60	0.20	0.20	0.60	0.20	0.20	0.60	0.20	0.20	0.60
$T-k_E$	0	0	1	0	0	1	0	0	1	0	0	1
$T-Y_{SX}$	0	0.60	0.40	0	0.60	0.40	0	0.60	0.40	0	0.60	0.40
$gen-\mu_{2S}$	0.70	0.10	0.20	0.70	0.10	0.20	0.70	0.10	0.20	0.70	0.10	0.20
$gen-\mu_{2E}$	0.60	0	0.40	0.60	0	0.40	0.60	0	0.40	0.60	0	0.40
$gen-k_S$	0.20	0.20	0.60	0.20	0.20	0.60	0.20	0.20	0.60	0.20	0.20	0.60
$gen-k_E$	0	0	1	0	0	1	0	0	1	0	0	1
$gen-Y_{SX}$	0	0.60	0.40	0	0.60	0.40	0	0.60	0.40	0	0.60	0.40
$\mu_{2S}-\mu_{2E}$	0.60	0	0.40	0.55	0.05	0.40	0.50	0.10	0.40	0.50	0	0.50
$\mu_{2S}-k_S$	0.40	0	0.60	0.35	0.05	0.60	0.30	0.10	0.60	0.30	0	0.70
$\mu_{2S}-k_E$	0.20	0	0.80	0.10	0.10	0.80	0	0.20	0.80	0	0	1
$\mu_{2S}-Y_{SX}$	0.10	0.50	0.40	0.05	0.55	0.40	0	0.60	0.40	0	0.50	0.50
$\mu_{2E}-k_S$	0.40	0	0.60	0.30	0.10	0.60	0.20	0.20	0.60	0.20	0	0.80
$\mu_{2E}-k_E$	0.40	0	0.60	0.20	0.20	0.60	0	0.40	0.60	0	0	1
$\mu_{2E}-Y_{SX}$	0.40	0.60	0	0.20	0.80	0	0	1	0	0	0.60	0.40
k_S-k_E	0.60	0	0.40	0.30	0.30	0.40	0	0.60	0.40	0	0	1
k_S-Y_{SX}	0.20	0.20	0.60	0.10	0.30	0.60	0	0.40	0.60	0	0.20	0.80
k_E-Y_{SX}	0.40	0	0.60	0.20	0.20	0.60	0	0.40	0.60	0	0	1

different degrees of "agreement" and "disagreement" are obtained between J, T and model parameters $\mu_{2S}, \mu_{2E}, k_S, k_E$ and Y_{SX}, as well as between model parameters themselves.

As per authors expertise, comparing all considered here four algorithms for InterCriteria relations calculation, the most reliable results which do not contradict to the model parameters physical meaning (Eqs. (1)–(3)), are those obtained by **Algorithm 1**.

4 Conclusions

In this paper four different algorithms for InterCriteria relations calculation, namely μ-biased, balanced, ν-biased and unbiased are applied to *S. cerevisiae* fed-batch FP model identification procedure. As an optimization technique nine GA with different values for *ind* and *gen* are consequently performed. The ICrA approach is applied to establish correlations and dependencies of GA parameters *ind* and *gen* over the results of an FP model identification procedure, namely convergence time, model accuracy and model parameters estimates.

The ICrA application defines some strong correlations between the following parameter pairs $\mu_{2S} - \mu_{2E}$, $\mu_{2S} - k_S$ and $\mu_{2E} - k_S$, proving the FP model parameters physical meaning. From the point of view of the GA performance, strong correlations have been observed between $T - ind$ and $T - gen$, as well as between $J - ind$ and $J - gen$, which is logically expected due to the GA nature.

All obtained results show that for the considered here case study of *S. cerevisiae* fed-batch FP **Algorithm 1** is the most reliable one. As these results verify the obtained ones in [10] (a case study of an *E. coli* fed-batch FP), the μ-biased algorithm is the favourable one for the purposes of the analysis of parameter identification results considering fermentation processes.

Acknowledgements. This work is partially supported by the National Science Fund of Bulgaria under the Grants DFNI-I-02-5 "InterCriteria Analysis – A New Approach to Decision Making" and DM-07/1 "Development of New Modified and Hybrid Metaheuristic Algorithms".

References

1. Atanassov, K.T.: Index Matrices: Towards an Augmented Matrix Calculus. SCI, vol. 573. Springer, Cham (2014). https://doi.org/10.1007/978-3-319-10945-9
2. Atanassov, K.: Intuitionistic fuzzy sets, VII ITKR Session, Sofia (1983). Reprinted: Int. J. Bioautom. **20**(S1), S1–S6 (2016)
3. Atanassov, K.: On Intuitionistic Fuzzy Sets Theory. Springer, Berlin (2012). https://doi.org/10.1007/978-3-642-29127-2
4. Atanassov, K., Mavrov, D., Atanassova, V.: InterCriteria decision making: a new approach for multicriteria decision making, based on index matrices and intuitionistic fuzzy sets. Issues IFSs GNs **11**, 1–8 (2014)
5. Goldberg, D.: Genetic Algorithms in Search, Optimization and Machine Learning. Addison-Wiley Publishing Company, Massachusetts (1989)
6. Krawczak, M., Bureva, V., Sotirova, E., Szmidt, E.: Application of the intercriteria decision making method to universities ranking. In: Atanassov, K.T., et al. (eds.) Novel Developments in Uncertainty Representation and Processing. AISC, vol. 401, pp. 365–372. Springer, Cham (2016). https://doi.org/10.1007/978-3-319-26211-6_31
7. Pencheva, T., Angelova, M.: InterCriteria analysis of simple genetic algorithms performance. In: Georgiev, K., Todorov, M., Georgiev, I. (eds.) Advanced Computing in Industrial Mathematics. SCI, vol. 681, pp. 147–159. Springer, Cham (2017). https://doi.org/10.1007/978-3-319-49544-6_13

8. Pencheva, T., Angelova, M., Vassilev, P., Roeva, O.: InterCriteria analysis approach to parameter identification of a fermentation process model. In: Atanassov, K.T., et al. (eds.) Novel Developments in Uncertainty Representation and Processing. AISC, vol. 401, pp. 385–397. Springer, Cham (2016). https://doi.org/10.1007/978-3-319-26211-6_33
9. Pencheva, T., Roeva, O., Hristozov, I.: Functional State Approach to Fermentation Processes Modelling. Prof. Marin Drinov Academic Publishing House, Sofia (2006)
10. Roeva, O., Vassilev, P., Angelova, M., Su, J., Pencheva, T.: Comparison of different algorithms for InterCriteria relations calculation. In: Proceedings of the 8th International Conference on Intelligent Systems, pp. 567–572 (2016)
11. Roeva, O., Vassilev, P., Fidanova, S., Paprzycki, M.: InterCriteria analysis of genetic algorithms performance. In: Fidanova, S. (ed.) Recent Advances in Computational Optimization. SCI, vol. 655, pp. 235–260. Springer, Cham (2016). https://doi.org/10.1007/978-3-319-40132-4_14
12. Todinova, S., Mavrov, D., Krumova, S., Marinov, P., Atanassova, V., Atanassov, K., Taneva, S.G.: Blood plasma thermograms dataset analysis by means of intercriteria and correlation analyses for the case of colorectal cancer. Int. J. Bioautom. **20**(1), 115–124 (2016)

Free Search in Multidimensional Space M

Kalin Penev(✉)

School of Media, Art and Technology, Southampton Solent University,
East Park Terrace, Southampton SO14 0YN, UK
Kalin.Penev@solent.ac.uk

Abstract. In the modern world of billions connected things and exponentially growing data, search in multidimensional spaces and optimisation of multidimensional tasks will become a daily need for variety of technologies and scientific fields. Resolving multidimensional tasks with thousands parameters and more require time, energy and other resources and seems to be an embarrassing challenge for modern computational systems in terms of software abilities and hardware capacity. Presented study focuses on evaluation and comparison of thousands dimensional heterogeneous real-value numerical optimisation tests on two enhanced performance computer systems. The aim is to extend the knowledge on multidimensional search and identification of acceptable solutions with non-zero probability on heterogeneous tasks. It aims also to study computational limitations, energy consumptions and time. Use of energy and time are measured and analysed. Experimental results are presented and can be used for further research and evaluation of other methods.

Keywords: Free Search · 1000 dimensional Keane's Bump test
1000 dimensional Ackley test · 1000 dimensional Griewank test
1000 dimensional Rosenbrock test · 1000 dimensional Michalewicz test
1000 dimensional Rastrigin test · Overclocking XEON E5

1 Introduction

In the Zettabyte Era of data [11] and ubiquitous computing with billions of connected things [10] search in multidimensional spaces and optimisation of multidimensional tasks become a daily need for variety of technologies and scientific fields, such as traffic of information and goods, wireless connections, economics, business, finance and energy [2]. This study continues research efforts on essential, for multidimensional optimisation, research questions such as: What time, energy and other resources are required to resolve thousands dimensional task with acceptable level of precision? How dimensionality reflects on the search space complexity? The article presents an investigation on optimisation of 1000 dimensional scalable heterogeneous real-value numerical tests. Due to a specific

In Roman numeral system, letter M equals 1000, in decimal numeral system.

I. Lirkov and S. Margenov (Eds.): LSSC 2017, LNCS 10665, pp. 399–407, 2018.
https://doi.org/10.1007/978-3-319-73441-5_43

performance identified in earlier publications [7,16–22], optimisation method explored in this study is Free Search (FS) [6] only. For this purpose eight scalable numerical tests are used—Ackley [1], Griewank [4], Keane's Bump [7], Michalewicz [5], Rastrigin [13], Shwefel [12], Step [3] and Rosenbrock [9], test functions.

2 Test Problems

Criteria for tests selection are:—must be scalable for multidimensional format;—must be with heterogeneous landscape. Chosen numerical test are scalable and form different search spaces. All tests are transformed for maximisation.

- Ackley test is global [1]. Optimal value is 0.
- Griewank test is global [4]. Optimal value is 0.
- Keane's Bump test is global and constrained [7]. Optimal value depends on dimensions number and for variety of dimension is unknown.
- Michalewicz test is global test with unknown optimum [5]. Optimal value depends on dimensions number.
- Rastrigin test is global [13]. Optimal value is 0.
- Shwefel test is global [12]. Optimal value is 0.
- Step test introduces plateaus to the topology and the search process cannot rely on local correlation [3]. Optimal value depends on dimensions number and for variety of dimension is unknown.
- Rosenbrock test is smooth flat test with single optimal solution [9]. Optimal value is 0.

3 Optimization Method

Due to the abilities to produce acceptable results within feasible period of time identified in earlier publications [7,8,16,20], optimisation method selected for this study is Free Search. Detailed description of this method is published [6–8].

4 Experimental Methodology

Experimental Methodology aims to investigate method's performance, energy use and time for 1000 dimensional tests limited to 10^9 objective functions evaluations. All numerical tests are evaluated simultaneously in series of experiments with start from random initial locations, different for each experiment, for limited period of time on two computer systems with enhanced performance. Rosenbrock test only is evaluated on experiments limited to 10^{10} objective functions evaluations.

System I—processor XEON E5 1660 V2 overclocked at 4.748 GHz, 6 cores – 12 threads, Max TDP – 130 W, CPU water cooler and RAM at 2333 MHz, motherboard ASUS P9X79-E WS and solid state disk—SanDisk Extreme SSD

SATA III. All experiments on System I are completed simultaneously running in parallel 10 tests (two of these test are not presented in this study) reaching 83% of 12 threads CPU performance. The whole exercise is limited to 12 days equal to 17280 min.

System II—processor XEON E5 1620 V2 overclocked at 4.220 GHz, 4 cores – 8 threads, Max TDP – 130 W, CPU water cooler and RAM at 1825 MHz, motherboard ASUS P9X79-E WS and solid state disk—SanDisk Extreme SSD SATA III. All experiments on System II are completed simultaneously running in parallel 8 tests reaching 100% of 8 threads CPU performance. The whole exercise is limited to 36 days equal to 51840 min.

5 Experimental Results

Achieved results are analysed for maximal values, time for completion of one experiment limited to 10^9 objective function evaluations on System I and on System II.

Table 1. Maximal results for 1000 dimensions.

Test	FE	Maximal results
Ackley	10^9	−0.00070707454
Griewank	10^9	−0.00000089215
Keane's Bump	10^9	0.85127716000
Michalewicz	10^9	999.45900000000
Rastrigin	10^9	−0.00000145938
Schwefel	10^9	−0.12842600000
Step	10^9	2000.00000000000
Rosenbrock	10^{10}	−0.00245150932

Table 1 presents maximal achieved results. These results are identical for both System I and System II. However periods of time required for completion of the experiments on System I and System II are different. It should be noted also that 1000 dimensional Step test is resolved for much less than 10^9 function evaluations. For this test function evaluations required for completion of the test are measured. Average number is 266000 also identical for both System I and System II and periods of time for completion of these tests on System I and System II are also different. Energy consumption for completion of the experiments on System I and System II is measured, compared and analysed.

Table 2 presents average time in minutes for completion of one experiment limited to 10^9 functions evaluation on explored tests on System I. Low number of objective function evaluations for Step test is noted. Experiments on Rosenbrock test are limited to 10^{10} functions evaluation.

Table 2. Average time in minutes for one experiment on System I.

Test	FE	Time
Ackley	10^9	388
Griewank	10^9	463
Keane's Bump	10^9	5998
Michalewicz	10^9	2470
Rastrigin	10^9	395
Schwefel	266000 average	994
Step	10^9	0.062
Rosenbrock	10^{10}	2087

Table 3 presents average time in minutes for completion of one experiment limited to 10^9 functions evaluation on explored tests on System II. Low number of objective function evaluations for Step test is noted. Experiments on Rosenbrock test are limited to 10^{10} functions evaluation.

Table 3. Average time in minutes for one experiment on System II.

Test	FE	Time
Ackley	10^9	481
Griewank	10^9	585
Keane's Bump	10^9	8185
Michalewicz	10^9	3253
Rastrigin	10^9	493
Schwefel	266000 average	1294
Step	10^9	0.084
Rosenbrock	10^{10}	2858

Table 4 presents the time difference in percentage for Systems I and II. It is calculated as $T\% = (T_{SystemII}/T_{SystemI}) * 100$. Periods of time for calculation for System I are lower in average of 30% and vary within the interval between 23% and 37%.

Table 5 presents technical parameters of System I and System II which are compared in this study. Line 1 presents CPU (Central Processor Unit) frequencies in MHz. CPU for System I is 12.51% faster. Line 2 presents RAM (Random Access Memory) frequencies in MHz. RAM for System I is 127.84% faster. Line 3 presents CPU energy consumption for 0% CPU workload in Watts. CPU energy consumption for System II for 0% CPU workload is 153.92% lower. Lines 4 and 8 present number of tests performed simultaneously. System I performs 125% more tests. Lines 5 and 7 present CPU energy consumption during the tests

Table 4. Fast performance System I versus System II.

Test	FE	Fast %
Ackley	10^9	123.97%
Griewank	10^9	126.35%
Keane's Bump	10^9	136.46%
Michalewicz	10^9	131.70%
Rastrigin	10^9	124.81%
Schwefel	266000 average	128.17%
Step	10^9	135.48%
Rosenbrock	10^{10}	136.94%

Table 5. Parameters of System I and System II.

№	Parameter	System I	System I	Difference
1	CPU	4748 MHz	4220 MHz	112.51%
2	RAM	2333 MHz	1825 MHz	127.84%
3	CPU 0	48.5 W	19.1 W	253.92%
4	Tests	10–83% of 12	8	125.00%
5	CPU	109.6 W 83%	59.4 W 100%	184.51%
6	ΔW/Test	6.11	5.03	121.47%
7	CPU	109.6 W 83%	59.4 W 100%	184.51%
8	Tests	10–83% of 12	8	125.00%
9	W/Test	10.96	7.42	147.43%
10	PSU	650 W	650 W	100.00%
11	W/Test	65 W	81.25	80.00%

performance for System I – 83% CPU workload and for System II – 100% CPU workload in Watts. CPU energy consumption for System II for the tests performance is 184.51% lower. Line 6 presents energy consumption per test in Watts. It is calculated as difference between energy consumption for test performance minus energy consumption for 0% CPU workload then this difference is divided to the number of tests performed simultaneously. Energy consumption per test for System II is 121.47% lower. Line 9 presents CPU energy consumption per test in Watts. It is calculated as CPU energy consumption divided to the number of tests performed simultaneously. Energy consumption per CPU for System II is 147.43% lower. Line 10 presents PSU (Power Supply Unit) energy consumption based on PSU technical characteristics. It is the same for both systems. Line 11 presents estimated system energy consumption per test. It is calculated on estimated system energy consumption divided to the number of tests performed simultaneously. System I performs 80% more effective per test. Presented on

Tables 4, 5 and 6 results suggest that System I consumes less energy per test and performs faster. Assessment and conclusion on system effectiveness need additional investigation. Measured time and energy consumption per test, per CPU and per system are compared and estimated by cost based on £0.1602 per kWh within Tables 7 and 8.

6 Discussion

Analysis of experimental results suggests that 1000 dimensional Step tests can be resolved within 266000 function evaluations. 1000 dimensional Ackley, Griewank, Michalewicz and Rastrigin tests can be resolved with precision 0.001 for 10^9 function evaluations. 1000 dimensional Schwefel test can be resolved with precision 0.1 for 10^9 function evaluations. 1000 dimensional Rosenbrock test can be resolved with precision 0.01 for 10^{10} function evaluations. 1000 dimensional Keane's Bump test can be resolved with precision 0.001 for 10^9 function evaluations compared to the results published earlier [7].

Comparison of the periods of search for 1000 dimensional and 300 dimensional tests publishes earlier [8] confirms earlier conclusion that time increases higher than linearly and hardware and software speed appear as potential constraints. To improve capabilities of modern search methods time consuming events should be identified and optimised. For further investigation on high dimensional problems hardware speed should be improved. Regarding the time required to resolve multidimensional task with acceptable level of precision, presented on Tables 2 and 3 results suggest that, on used hardware configurations, selected 1000 dimensional tests could be resolved with high probability within the range of 6 to 137 h. For more general conclusion additional experiments with 1000 dimensional tests should be done.

Experimental results confirm conclusions published earlier [8] that complexity of task specific landscapes varies among the tests and for same dimensionality different number of functions evaluations could guarantee successful results. This is illustrated with Tables 1, 2, 3, 7 and 8.

Table 6. Time and energy consumption per test.

Test	Time I	Time II	Fast %	Energy I	Energy II	Low %
Ackley	388	481	123.97%	0.039511	0.040324	97.98%
Griewank	463	585	126.35%	0.047149	0.049043	96.14%
Keane's Bump	5998	8185	136.46%	0.610796	0.686176	89.01%
Michalewicz	2470	3253	131.70%	0.251528	0.27271	92.23%
Rastrigin	395	493	124.81%	0.040224	0.04133	97.32%
Schwefel	994	1274	128.17%	0.101222	0.106804	94.77%
Step	0.062	0.084	135.48%	6.31E-06	7.04E-06	89.63%
Rosenbrock	2087	2858	136.94%	0.212526	0.239596	88.70%

Analysis on time, energy use and cost suggests that time and energy will become an essential constraint for future growth of information technologies, in particular minimisation of computer systems overall time delays and energy use require further research.

Table 7. Time, energy consumption and cost per test, CPU and PSU for System I.

Test	Time	Energy test/CPU/PSU kWh			Cost* Test/CPU/PSU		
Michalewicz	2470	0.251528	0.450363	2.675833	£0.04	£0.07	£0.43
Rastrigin	395	0.040224	0.072022	0.427917	£0.01	£0.01	£0.07
Griewank	463	0.047149	0.08442	0.501583	£0.01	£0.01	£0.08
Step	0.062	6.31E−06	1.13E−05	6.72E−05	£0.00	£0.00	£0.00
Ackley	388	0.039511	0.070745	0.420333	£0.01	£0.01	£0.07
Schwefel	994	0.101222	0.181239	1.076833	£0.02	£0.03	£0.17
Keane's Bump	5998	0.610796	1.093635	6.497833	£0.10	£0.18	£1.04
Rosenbrock	2087	0.212526	0.38053	2.260917	£0.03	£0.06	£0.36

Table 8. Time, energy consumption and cost per test, CPU and PSU for System II.

Test	Time	Energy test/CPU/PSU kWh			Cost* Test/CPU/PSU		
Michalewicz	3253	0.27271	0.402288	4.405104	£0.04	£0.06	£0.71
Rastrigin	493	0.04133	0.060968	0.667604	£0.01	£0.01	£0.11
Griewank	585	0.049043	0.072345	0.792188	£0.01	£0.01	£0.13
Step	0.084	7.04E−06	1.04E−05	0.000114	£0.00	£0.00	£0.00
Ackley	481	0.040324	0.059484	0.651354	£0.01	£0.01	£0.10
Schwefel	1274	0.106804	0.157551	1.725208	£0.02	£0.03	£0.28
Keane's Bump	8185	0.686176	1.012212	11.08385	£0.11	£0.16	£1.78
Rosenbrock	2858	0.239596	0.353439	3.870208	£0.04	£0.06	£0.62

7 Conclusion

This article presents experimental evaluation of Free Search on 1000 dimensional tests. Identified are specific issues related with multidimensional optimisation. Experimental results are summarized and analysed. Further investigation could focus on evaluation and measure of time and computational resources sufficient for completion of other multidimensional tasks using parallel processing systems and accelerated processing. Algorithms analysis and improvement could be also subject of future research.

Acknowledgements. I would like to thank to my students Ashley Raven [14] and Petar Tashkov [15] for the design, implementation and overclocking of computer systems used for the experiments presented in this article.

References

1. Ackley, D.H.: A Connectionist Machine for Genetic Hillclimbing. Kluwer, Boston (1987)
2. Censor, Y.: Optimisation methods. In: Ralston, A., Reilly, E.D., Hemmendinger, D. (eds.) Encyclopedia of Computer Science, pp. 1339–1341. Nature Publishing Group, London (2000). ISBN: 0-333-77879-0
3. De Jung, K.A.: An analysis of the behaviour of a class of genetic adaptive systems. Ph.D. thesis, University of Michigan, USA, August 1975
4. Griewank, A.O.: Generalized decent for global optimization. J. Optim. Theory Appl. **34**, 11–31 (1981)
5. Michalewicz, Z.: Genetic Algorithms + Data Structures = Evolution Programs. Springer, New York (1992). https://doi.org/10.1007/978-3-662-03315-9
6. Penev, K.: Free Search of Real Value or How to Make Computers Think. St. Qu, UK (2008). ISBN 978-0-9558948-0-0
7. Penev, K.: Free search – comparative analysis 100. Int. J. Metaheuristics **3**(2), 118–132 (2014)
8. Penev, K.: Free search in multidimensional space II. In: Dimov, I., Fidanova, S., Lirkov, I. (eds.) NMA 2014. LNCS, vol. 8962, pp. 103–111. Springer, Cham (2015). https://doi.org/10.1007/978-3-319-15585-2_12
9. Rosenbrock, H.H.: An automate method for finding the greatest or least value of a function. Comput. J. **3**, 175–184 (1960)
10. van der Meulen, R.: Gartner Says 8.4 Billion Connected "Things" Will Be in Use in 2017, Up 31 Percent From 2016, Gartner. http://www.gartner.com/newsroom/id/3598917. Accessed 11 July 2017
11. Cisco: The Zettabyte Era—Trends and Analysis—Cisco, White Papers. http://www.cisco.com/c/en/us/solutions/collateral/service-provider/visual-networking-index-vni/vni-hyperconnectivity-wp.html. Accessed 11 July 2017
12. Schwefel, H.P.: Numerical Optimization of Computer Models. Wiley, New York (1977). English translation of Numerische Optimierung von Computer-Modellen mittels der Evolutionsstrategie
13. Mühlenbein, H., Schomisch, D., Born, J.: The parallel genetic algorithm as function optimizer. Parallel Comput. **17**, 619–632 (1991)
14. Raven, A.: Overclocking XEON E5 1660 V2 validation. http://valid.x86.fr/top-cpu/496e74656c2852292058656f6e28522920435055204535352d313636302076322040203332e373047487a. Accessed 11 July 2017
15. Tashkov, P.: Overclocking XEON E5 1620 V2 validation. http://valid.x86.fr/7q0py1. Accessed 11 July 2017
16. Nesmachnow, S.: An overview of metaheuristics: accurate and efficient methods for optimisation. Int. J. Metaheuristics **3**(4), 320–347 (2014)
17. Zhongda, T., Shujiang, L., Yanhong, W., Yi, S.: A prediction method based on wavelet transform and multiple models fusion for chaotic time series. Chaos Solitons Fractals **98**, 158–172 (2017). Elsevier
18. Sun, G., Zhao, R., Lan, Y.: Joint operations algorithm for large-scale global optimization. Appl. Soft Comput. **38**, 1025–1039 (2016)
19. Hultmann Ayala, H.V., Keller, P., De Fátima Morais, M., Mariani, V.C., Dos Santos Coelho, L., Venkata Rao, R.: Design of heat exchangers using a novel multiobjective free search differential evolution paradigm. Appl. Therm. Eng. **94**, 170–177 (2016). Elsevier

20. Hultmann Ayala, H.V., Dos Santos Coelho, L., Mariani, V.C., Askarzadeh, A.: An improved free search differential evolution algorithm: a case study on parameters identification of one diode equivalent circuit of a solar cell module. Energy **93**, 1515–1522 (2015)
21. Marinakis, Y., Marinaki, M.: A bumble bees mating optimization algorithm for the open vehicle routing problem. Swarm Evol. Comput. **15**, 80–94 (2014). Elsevier
22. Xu, W., Wang, R., Yang, J.: An improved league championship algorithm with free search and its application on production scheduling. J. Intell. Manuf., 1–10 (2015). Springer. Journal no. 10845. https://doi.org/10.1007/s10845-015-1099-4

Generalized Net Model of Adhesive Capsulitis Diagnosing

Simeon Ribagin[1(✉)], Evdokia Sotirova[2], and Tania Pencheva[1]

[1] Institute of Biophysics and Biomedical Engineering, Bulgarian Academy of Sciences, Sofia, Bulgaria
sim_ribagin@mail.bg, tania.pencheva@biomed.bas.bg

[2] Laboratory of Intelligent Systems, "Prof. Asen Zlatarov" University, 1 "Prof. Yakimov" blvd., 8010 Bourgas, Bulgaria
esotirova@btu.bg

Abstract. Adhesive capsulitis is a musculoskeletal condition of the shoulder characterized by pain and gradual loss of the global shoulder motion. Proper diagnosis of adhesive capsulitis is extremely important for designing a coordinated exercise program and reliable monitoring progress during treatment. In this investigation we present a successful example of Generalized Nets (GN) application in orthopedics and propose a novel approach to timely detection of adhesive capsulitis. The developed GN-model provides a framework that can be used by primary care practitioners to guide diagnostic processes for patients suspected to have adhesive capsulitis and might assist in optimizing patient outcomes and more effective treatment. The method proposed in this investigation accurately identifies the various steps during the diagnosing processes and significantly improve the health care level. Obtained so far results could be used to assist in the decision making in the diagnostic processes.

Keywords: Adhesive capsulitis · Generalized nets

1 Introduction

The diagnosis of adhesive capsulitis is made on the basis of the medical history, clinical and radiological examination and the exclusion of other shoulder pathologies. The basic issue associated with adhesive capsulitis diagnosing is whether it's a primary or secondary. Then as part of the assessment, an attempt should be made to define in which stage the disease is at present. Although the underlying pathology of stiff and painful shoulder is wide ranging, the shoulder physical examination should include inspection of range of motion, both passive and active. The purpose of the present study is to give an example how the apparatus of generalized nets might be successfully applied in orthopedics and as such—to be proposed as a novel mathematical approach for diagnosing the adhesive capsulitis. Generalized nets (GNs) (see [1,2]) are an apparatus for modeling of parallel and concurrent processes, developed as an extension of the concept of Petri nets and some of their modifications. In general, the GNs may

I. Lirkov and S. Margenov (Eds.): LSSC 2017, LNCS 10665, pp. 408–415, 2018.
https://doi.org/10.1007/978-3-319-73441-5_44

or may not have some of the components in their definition. GNs without some of their components form special classes called reduced GNs [2]. The presented GN-model shows similar features with previous models for medical diagnosing [3–5] but this is the first one highlighting the diagnostic algorithm for patients with stiff and painful shoulder, suspected to have adhesive capsulitis and, as such, representing an application of GNs in orthopedics and traumatology.

2 Generalized Net Model of Adhesive Capsulitis Diagnosing

A reduced GN-model which represents the diagnosing plan for adhesive capsulitis is developed here. The proposed GN-model is shown in Fig. 1.

The GN-model has 8 transitions and 26 places with the following meanings:

- Z_1 represents the personal data of the patient;
- Z_2 – the results from the history and risk factors;
- Z_3 – the current functional status of the patient and duration of symptoms;
- Z_4 – the set of physical examination techniques;
- Z_5 – the results from the physical examination;
- Z_6 – the results from the X-ray;
- Z_7 – the results from the special tests;
- Z_8 – the final diagnosis.

The GN-model contains 8 types of tokens: α, β, μ, ν, π, σ, ϕ and ψ. Some of the model transitions contain the so called "special place" where a token stays and collects information about the specific parts of the diagnosing process which are represented as follows:

- place l_3 collects the overall information obtained from the diagnostics steps in the personal record (personal data);
- place l_6 – information obtained from the history of the patient;

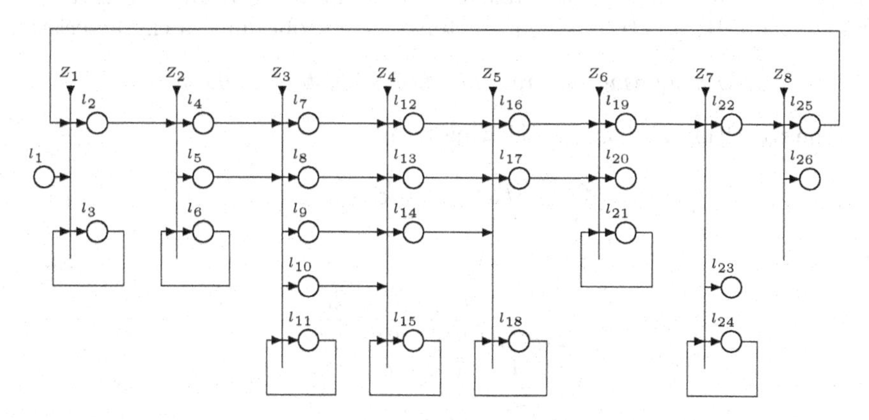

Fig. 1. GN-model of adhesive capsulitis diagnosing

- place l_{11} – information about the current functional status of the patient;
- place l_{15} – information about the possible physical examination techniques;
- place l_{18} – information about the results from physical examinations;
- place l_{21} – information about the results from the X-ray;
- place l_{24} – information about the results from the special tests.

During the GM-model functioning, the α-tokens will unite with the tokens from the other types: β, μ, ν, π, σ, ϕ and ψ. After that, some of these tokens can split in order to generate new α-tokens obtaining corresponding characteristics. When there are some α-tokens (α_1, α_2 and, eventually, α_3), on the next time-moment, all they will unite with a token from another type.

The token α enters the net in place l_1 with an initial characteristic:

"*patient with stiff and painful shoulder, who is sespected to have adhesive capsulitis*".

The transition Z_1 has the following form:

$$Z_1 = \langle \{l_1, l_3, l_{25}\}, \{l_2, l_3\}, r_1 \rangle$$

where

$$r_1 = \begin{array}{c|cc} & l_2 & l_3 \\ \hline l_1 & false & true \\ l_3 & W_{3,2} & true \\ l_{25} & false & true \end{array}$$

and

- $W_{3,2}$ = "*past medical history and physical examination are necessary*".

The tokens from the three input places of transition Z_1 enter place l_3 and unite with token β with the above mentioned characteristic. On the next time-moment, token β splits to two tokens – the same token β and token α. When the predicate $W_{3,2}$ is true, token α enters place l_2, obtaining a characteristic:

"*obtain information from the medical history of the patient*".

The transition Z_2 has the following form:

$$Z_2 = \langle \{l_2, l_6\}, \{l_4, l_5, l_6\}, r_2 \rangle$$

where

$$r_2 = \begin{array}{c|ccc} & l_4 & l_5 & l_6 \\ \hline l_2 & false & false & true \\ l_6 & W_{6,4} & W_{6,5} & true \end{array}$$

and

- $W_{6,4}$ = "*there is history of trauma or surgery of the affected shoulder or past history of adhesive capsulitis or underlying diabetes mellitus*";
- $W_{6,5} = \neg W_{6,4}$.

The tokens from all input places of transition Z_2 enter place l_6 and unite with token μ with the characteristic, as mentioned above. On the next time-moment, token μ splits to two tokens – the same token μ that stays permanently in the place l_6, and token α_1. When the predicate $W_{6,4}$ is true, token α_1 enters place l_4 and there it obtains a characteristic:

"*consider secondary adhesive capsulitis*".

When the predicate $W_{6,5}$ is true, token α_1 enters place l_5 and there it obtains a characteristic:

"*consider primery adhesive capsulitis*".

The transition Z_3 has the following form:

$$Z_3 = \langle \{l_4, l_5, l_{11}\}, \{l_7, l_8, l_9, l_{10}, l_{11}\}, r_3 \rangle$$

where

$$r_3 = \begin{array}{c|ccccc} & l_7 & l_8 & l_9 & l_{10} & l_{11} \\ \hline l_4 & false & false & false & false & true \\ l_5 & false & false & false & false & true \\ l_{11} & W_{11,7} & W_{11,8} & W_{11,9} & W_{11,10} & true \end{array}$$

and

- $W_{11,7}$ = "*symptoms have been present for less than 3 months, patient reports an aching pain and moderate limitations of the shoulder*";
- $W_{11,8}$ = "*symptoms have been present for 3 to 9 months, patient reports significant limitations of the shoulder mobility*";
- $W_{11,9}$ = "*symptoms have been present from 9 to 14 months, patient reports relatively painless but stiffened shoulder*";
- $W_{11,10}$ = "*the patient reports a prior history of similar symptoms, but this time the pain is negligible and there is a gradual returning of shoulder mobility*".

The tokens from all input places of transition Z_3 enter place l_{11} and unite with token ν with the characteristic, as mentioned above. On the next time-moment, token ν splits to two tokens – the same token ν that stays permanently in the place l_{11} and token α_1. When the predicate $W_{11,7}$ is true, token α_1 enters place l_7 and there it obtains a characteristic:

"*consider that the patient is present in stage 1 of adhesive capsulitis*".

When the predicate $W_{11,8}$ is true, token α_1 enters place l_8 and there it obtains a characteristic:

"*consider that the patient is present in stage 2 of adhesive capsulitis*".

When the predicate $W_{11,9}$ is true, token α_1 enters place l_9 and there it obtains a characteristic:

"*consider that the patient is present in stage 3 of adhesive capsulitis*".

When the predicate $W_{11,10}$ is true, token α_1 enters place l_{10} and there it obtains a characteristic:

"*consider that the patient is present in the recovery stage of adhesive capsulitis*".

The transition Z_4 has the following form:

$$Z_4 = \langle \{l_7, l_8, l_9, l_{10}, l_{15}\}, \{l_{12}, l_{13}, l_{14}, l_{15}\}, r_4 \rangle$$

where

$$r_4 = \begin{array}{c|cccc} & l_{12} & l_{13} & l_{14} & l_{15} \\ \hline l_7 & false & false & false & true \\ l_8 & false & false & false & true \\ l_9 & false & false & false & true \\ l_{10} & false & false & false & true \\ l_{15} & true & true & true & true \end{array}$$

The tokens from all input places of transition Z_4 enter place l_{15} and unite with token π with the characteristic, as mentioned above. On the next time-moment, token π splits to four tokens – the same token π that stays permanently in the place l_{18} and tokens α_1, α_2 and α_3.

The token α_1 enters place l_{12} and there it obtains a characteristic "*perform a coracoid pain test*".

The token α_2 enters place l_{13} and there it obtains the characteristics "*perform an examination of: passive and active range of motion of the shoulder, joint accessory mobility, resisted muscle tests*".

The token α_3 enters place l_{14} and there it obtains a characteristic "*perform an upper quarter neurological exam*".

The transition Z_5 has the following form:

$$Z_5 = \langle \{l_{12}, l_{13}, l_{14}, l_{18}\}, \{l_{16}, l_{17}, l_{18}\}, r_5 \rangle$$

where

$$r_5 = \begin{array}{c|ccc} & l_{16} & l_{17} & l_{18} \\ \hline l_{12} & false & false & true \\ l_{13} & false & false & true \\ l_{14} & false & false & true \\ l_{18} & W_{18,16} & W_{18,17} & true \end{array}$$

and

- $W_{18,16}$ = "*the results from the physical examination are: possitive coracoid pain test or reduced passive and active range of motion or the shoulder joint has a capsular pattern or weakness in shoulder abduction external and internal rotation or normal neurological findings*";
- $W_{18,17} = \neg W_{18,16}$.

The tokens from all input places of transition Z_5 enter place l_{18} and unite with token σ with the characteristic, as mentioned above. On the next time-moment, token σ splits to two or three tokens – the same token σ that stays permanently in the place l_{18} and tokens α_1 and α_2. When the predicate $W_{18,16}$ is true, token α_1 enters place l_{16} and there it obtains a characteristic:

"*perform a X-ray to rule out other potential shoulder pathologies*".

When the predicate $W_{18,17}$ is true, token α_2 enters place l_{17} and there it obtains a characteristic:

"*exclude adhesive capsulitis*".

The transition Z_6 has the following form:

$$Z_6 = \langle \{l_{16}, l_{17}, l_{21}\}, \{l_{19}, l_{20}, l_{21}, \}, r_6 \rangle$$

where

$$r_6 = \begin{array}{c|ccc} & l_{19} & l_{20} & l_{21} \\ \hline l_{16} & false & false & true \\ l_{17} & false & false & true \\ l_{21} & W_{21,19} & W_{21,20} & true \end{array}$$

and

- $W_{21,19}$ = "*anteroposterior, lateral and axillary X-ray views of the shoulder are normal*";
- $W_{21,20} = \neg W_{21,19}$.

The tokens from all input places of transition Z_6 enter place l_{21} and unite with token ϕ with the characteristic, as mentioned above. On the next time-moment, token ϕ splits to two tokens – the same token ϕ that stays permanently in the place l_{21} and token α_1. When the predicate $W_{21,19}$ is true, token α_1 enters place l_{19} and there it obtains a characteristic:

"*perform a set of special tests (i.e. external rotation test, shoulder "shrug" test, Neer test, Howkin's test) to make the final diagnosis*".

When the predicate $W_{21,20}$ is true, token α_1 enters place l_{20} and there it obtains a characteristic:

"*consider osteoarthritis, chronic dislocations of the shoulder, osteolysis or other osseus abnormalities of the shoulder*".

The transition Z_7 has the following form:

$$Z_7 = \langle \{l_{19}, l_{24}\}, \{l_{22}, l_{23}, l_{24}\}, r_7 \rangle$$

where

$$r_7 = \begin{array}{c|ccc} & l_{22} & l_{23} & l_{24} \\ \hline l_{19} & false & false & true \\ l_{24} & W_{24,22} & W_{24,23} & true \end{array}$$

and

- $W_{24,22}$ = "*the external rotation test is positive or the shoulder "shrug" test is positive or the Neer test and Howkin's test are negative*";
- $W_{24,23} = \neg W_{24,22}$.

The tokens from all input places of transition Z_7 enter place l_{26} and unite with token ψ with the characteristic, as mentioned above. On the next time-moment, token ψ splits to two or three tokens – the same token ψ that stays permanently in the place l_{26} and tokens α_1 and α_2. When the predicate $W_{24,22}$ is true, token α_1 enters place l_{22} and there it obtains a characteristic:

"*consider adhesive capsulitis as a final diagnosis*".

When the predicate $W_{24,23}$ is true, token α_2 enters place l_{23} and there it obtains a characteristic:

"*consider rotator cuff impingement or shoulder impingement as a differential diagnosis, send patient for MRI or MR arthrogram*".

The transition Z_8 has the following form:

$$Z_8 = \langle \{l_{22}\}, \{l_{25}, l_{26}\}, r_8 \rangle$$

where

$$r_8 = \begin{array}{c|cc} & l_{25} & l_{26} \\ \hline l_{22} & true & true \end{array}$$

The token α_1 splits to two tokens: token α that obtains the characteristic: "*the diagnosis of the patient is adhesive capulitis*" in place l_{26} and token β_1 with the same characteristic for the personal record of the patient, in place l_{25}. The token β_1 returns to place l_3 to extend the personal record of the current patient.

3 Conclusions

The so described GN-model may provide a framework that can be used by primary care practitioners to guide diagnostic processes for patient suspected to have adhesive capsulitis, enabling more accurate and efficient identification of that condition and would assist in optimizing patient outcomes and more effective treatment. The presented in this paper GN-model of diagnostic algorithm for patient with possible diagnosis of adhesive capsulitis is a part of a series of studies for diagnosing through GN-modeling assistance and can be improved in multiple ways to yield improvements in results. This model can be complicated and detailed, which will significantly improve the accuracy of the primary diagnosis and the reliability of the proposed algorithm.

Acknowledgements. Work presented here is partially supported by the Grant DFNP-142/2016 "Program for career development of young scientists, BAS".

References

1. Atanassov, K.: Generalized Nets. World Scientific, Singapore (1991)
2. Atanassov, K.: On Generalized Nets Theory. Prof. M. Drinov Academic Publishing House, Sofia (2007)
3. Ribagin, S., Atanassov, K., Shannon, A.: Generalized net model of shoulder pain diagnosis. Issues Intuitionistic Fuzzy Sets Gen. Nets **11**, 55–62 (2014)
4. Ribagin, S., Roeva, O., Pencheva, T.: Generalized net model of asymptomatic osteoporosis diagnosing. In: Proceedings of 2016 IEEE 8th International Conference on Intelligent Systems (IS), Sofia, pp. 604–608 (2016)
5. Shannon, A., Sorsich, J., Atanassov, K., Nikolov, N., Georgiev, P.: Generalized Nets in General and Internal Medicine, vol. 1. Prof. M. Drinov Academic Publishing House, Sofia (1998)

Adaptive Multi-agent System Based on Wasp-Like Behaviour for the Virtual Learning Game Sotirios

Dana Simian[1(✉)] and Florentin Bota[2]

[1] Faculty of Sciences, University "Lucian Blaga" of Sibiu, 5-7 dr. I. Raţiu str, Sibiu, Romania
dana.simian@ulbsibiu.ro

[2] Centre for the Study of Complexity, University "Babeş-Bolyai" Cluj-Napoca, 30 Fantanele str, Cluj-Napoca, Romania

Abstract. The aim of this paper is to propose a model for an adaptive multi-agent system based on wasp-like behaviour for dynamic allocation of puzzles and quests in the virtual learning game SOTIRIOS. This is a digital learning game integrated inside a First Person Shooter designed by the second author of this paper. The learning process is based on many puzzles hidden in the game flow. The multi-agent system is necessary to integrate a multiplayer mode into the game. The agents use wasp task allocation behaviour, combined with a model of wasp dominance hierarchy in order to create a unique multiplayer learning system, where each user has a different learning curve, based on his results. The wasp behaviour is required to create a balanced multiplayer mode and to optimize the results of teams within the game.

1 Introduction

The aim of this paper is to propose a model for an adaptive e-learning system based on the virtual learning game SOTIRIOS. The game is based on educational puzzles, integrated inside a First Person Shooter (FPS), where you progress by properly resolving informatics problems inserted into the game flow [2]. The game can be used in single or multiplayer mode. In all the cases an important problem to be solved is the distribution of the puzzles such that the learning results to be maximized. Therefore each player must have different problems according to the previous results and choices. SOTIRIOS was designed in a simpler form in 2012 (only single-player mode and without the multi-agent system for puzzle distribution) by the second author of this paper. The motivation of the project came after an analysis of most students' behavior against the learning process. Most of them get bored or tired after a short time, but can play a computer game for weeks. The video game follows the story of a future robot, Sotirios, who dared to question the axioms that ruled his world. The players follow the main character as he escapes from the brainwashing camp where he was being kept and explores the world outside the walls. In order to escape, he has to force his way out, using more and more difficult algorithms to solve

I. Lirkov and S. Margenov (Eds.): LSSC 2017, LNCS 10665, pp. 416–424, 2018.
https://doi.org/10.1007/978-3-319-73441-5_45

simple and complex problems. A door won't open for example if the cracking code is not semantically correct, and a wrong choice can alert the nearby guards. There are several rules that govern the game play, based on the player's answers. In essence, one cannot progress in the game until he answers correctly on specific quizzes (connected on doors, elevators, platforms etc.) or until a certain experience level is achieved (next level or map). A player can gain experience points by providing a correct answer or lose experience points otherwise. The amount of points is conforming to the attributes of a specific puzzle. In this way a player is encouraged to answer correctly and give him the chance to retry a puzzle until he answers a quiz correctly. The minimum and maximum possible points are capped by the total number of puzzles on a specific level. Note that in essence theoretical puzzles are items (questions) with one or many correct answers. The quests are different levels of the game (or maps) that have a specific goal or trajectory in order to move the players forward in the story. We realize the dynamic allocation of educational puzzles by creating a multi-agent model inspired by wasp colonies behaviour. The model is necessary for integrating a multiplayer mode into the game, allowing many players to perform as a team. This feature gives a social dimension to the game. The agents use wasp task allocation behaviour, combined with a model of wasp dominance hierarchy [3] in order to create a unique multiplayer learning system, where each user has a different learning curve, based on his results. The wasp behaviour is required to create a balanced multiplayer mode and to optimize the results of teams within the game.

The paper is organized as follows. In Sect. 2 we present the related work regarding the design of educational games in general and focus on several computer programming educational platforms. In Sect. 3 we describe the basic principle of the wasp based algorithms for multi agent systems. The proposed model for optimization of the results in SOTIRIOS team multiplayer mode is explained in Sect. 4. Conclusions and further directions of study are presented in Sect. 5.

2 Related Work

There are several educational frameworks proposed for the design of educational games and we analyzed two of the most interesting and similar to our approach.

Four-dimensional framework [5] is a model based on several principles: *context, representation, learner and pedagogy.* These concepts create a framework where the context will guide the scenarios (infrastructure, technical requirements, type of game, activities etc.) and each player will be represented by avatars with characteristics based on the context. The model focuses on the learner immersed in the game and should take into consideration the learner's preferences, previous knowledge and even collaborative learning.

EFM: Model for Educational Game Design is a multi-dimensional model, based on an effective learning environment and featuring flow experience and motivation [11]. In this model, the authors suggest four main strategy components that can be used to stimulate motivation for students: *Interest, Goal, Feedback and Challenge.* The designers should create an immersible challenging world, focused on learning and not on other distractions.

Because our game focuses on Computer Science knowledge, we also analyzed games teaching computer programming.

The Scratch environment was developed by the MIT Media Lab to assist creative people to easily accomplish their goals. It features steps like *imagine, create, play, share, reflect, imagine etc.* and the students can create programs, stories and games by using drag and drop tiles instead of writing code. The environment has multiple assets that motivate the students to create their own animations and games, play them and share them with their classmates.

CMX educational game [6] focuses on the education of the computer programming domain. It is a Massive Multiplayer online Role-Playing Game (MMORPG) and aims to familiarize students with core programming concepts (variables, loops, statements etc.). The authors followed a user-centric approach based on the user target, the teacher's aspirations and other several metrics (*CMX Design Strategy Puzzle*).

CodeCombat (codecombat.com) is a browser-based role-playing game that teaches programming and fundamentals of computer science. The player can only advance through the game by writing code and use it to move the characters or solve puzzles.

In [6], the authors identified several recommended features for educational games' design and development. A brief overview consists of: *Multi-player/Role-playing, Interaction/Experimentation, Collaboration, Scaffolding, Drag & drop lines of code, programming editor, multiple choice questions, scenarios, compiler, familiar metaphor, visualization of concepts and simplicity.*

These features are all compatible with our model and implementation. We could not find however any framework or implementation that presents dynamically allocated puzzles in a game-based learning environment. The progress in the studied games is linear and expects all the players to play on a predefined learning path.

3 Models Based on Wasp Behaviour

The model of self-organization in a colony of wasps was presented by Theraulaz et al. in [12]. The main characteristic of the wasp behaviour is the interaction between the individuals and the environment using a stimulus-response mechanism which governs the distributed task allocation. An individual wasp has a response threshold for each zone of the nest. The wasp's threshold for a given zone and the amount of stimulus from brood located in this zone are responsible for the decision of engaging or not the wasp in the task of foraging for this zone. A lowest response threshold for a given zone amounts to a higher likelihood of engaging in activity if given a stimulus. Algorithms simulating the wasp behaviour were used for building multi-agent systems for solving problems with dynamic character as: self-organizing groups of robots [12], dynamic distribution of resources and tasks control and coordination in a factory [3], distribution of tasks and projects in virtual learning environments [8], optimization problems [9], dynamic scheduling problems [7,10], knowledge mining [4], etc. All these problems can be connected with the classical problem of dynamic tasks allocation in a distributed manufacturing system with specialized machines considered by Cicirelo in [3]. In this problem each machine has associate an artificial wasp agent. Each artificial wasp will have a response threshold for every new task (command) that could appear in the manufacturing system; in turn each task will have associate a stimulus. Each artificial wasp, will bid for each unassigned task with a probability dependent on the corresponding threshold and stimulus: as lower the response thresholds is, as bigger the probability of binding a task is, but a wasp can bid for a task if a high enough stimulus is broadcasted. If two or more wasps bid for a task with the same probability a dominance contest will occur. The dominance hierarchy is based on a "force" function that is inverse proportional with the dominance probability. The wasp having the lowest force will gain [1,3]. Different models for updating the response thresholds were developed. In [1] the response threshold remain fixed while in [13] the idea of reinforcement of response threshold is presented.

4 The Proposed Model

In this section we present our approach to the problem of allocating dynamically the puzzles and quests from the database to corresponding players in multiplayer mode of the game. The quizzes are distributed at player level, but quests can be performed by the whole team if they decide to play together.

The originality of the multiplayer mode consists in the approach by the team point of view. Many players organized as a team fight against the bots in SOTIRIOS's world. The number of players once chosen remains constant. The purpose is to attend the higher levels of the game and implicitly to optimize the learning results of the team giving the possibility that all team members to participate almost equal in the game. This implies a specialization of the players within the team. Multiplayer mode uses puzzles with higher difficulty than the single-player mode and requires that any member of the team had already passed through the game in single-player mode and acquired at least two levels. Thus there is enough information for computing the specialization of every member of the team. Then, in the multiplayer mode, the puzzle distribution is realized using a multi-agents system based on wasp behavior such that the learning results of the team to be maximized and the game to be balanced. Any player has associated a wasp agent that we named learning player wasp. We define next the problem's terms.

(a) *Specialization of the player i for solving the puzzle k* - $PS_{i,k}$

$$PS_{i,k} = \max\left\{0, \frac{ns_{i,k} \cdot l_{i,k} - nus_{i,k} \cdot l_{i,k}}{ns_{i,k} + nus_{i,k}}\right\} \tag{1}$$

The terms in (1) denote:
$ns_{i,k}$ - the number of puzzles having one of the types of the puzzle k, successfully solved before by the player i;
$nus_{i,k}$ - the number of puzzles having one of the types of the puzzle k, incorrectly solved before by the player i;
$l_{i,k}$ - the level of the puzzle counted in $ns_{i,k}$ and $nus_{i,k}$.
Note that $PS_{i,k}$ is a dynamic property. First it is given by the specialization acquired in the single-player mode and then changes with any mission (or equivalent puzzle) received.

(b) *Maximal mission threshold* - $MAXMT$
$MAXMT$ represents the maximal allowable difference between the number of puzzles assigned to different players at any moment of the game.

(c) *Mission difference of player* i - MD_i

$$MD_i = \max_{j=1,\ldots,i;j\neq i} |n_i - n_j| \quad (2)$$

where
n_i - the number of puzzles already solved by the player i, within the multi-player mode of the game;
N - the number of the players in the team.

(d) *Complexity of the puzzle* k - C_k

$$C_k = lp_k \cdot cs_k \quad (3)$$

with
lp_k - the game level corresponding to puzzle k;
cs_k - the complexity score.
In the puzzle data base there is a field "complexity score" with values 1 if the puzzle is with unique answer and 2 if it is with multiple answers. Other values of complexity score could be introduced in the future for other different kind of puzzles.

Each player in our system has an associated learning player wasp. Each wasp is in charge of choosing which puzzle to bid for possible assignment to its associate player. Like in the wasp behavior model, each learning player wasp has a response threshold for any possible puzzle. The response thresholds vary in the interval $[w_{min}, w_{max}]$. We denote by $w_{i,k}$ the response threshold of wasp associated to player i for the puzzle k. Initially all the response thresholds are set on the average value $w_{med} = (w_{min} + w_{max})/2$. The puzzle k broadcasts to all of the learning player wasps a stimulus S_k which is proportional to the puzzle complexity C_k. The learning routing wasp i will bid for the puzzle k only if

$$MD_i \leq MAXMT \quad (4)$$

In this case the learning routing wasp i will bid for the puzzle k with the probability

$$P_{i,k} = \frac{(S_k)^\gamma}{(S_k)^\gamma + (w_{i,j})^\gamma} \tag{5}$$

The exponent γ is a system parameter.

If $\gamma \geq 0$ then as lower the response thresholds are, as bigger the probability of binding the puzzle is. Using this rule, a wasp can bid for a puzzle if a high enough stimulus is emitted. Each learning player wasp knows at all times all the information associated with its player (specialization, mission difference). This is necessary in order to adjust the response thresholds for the various puzzles. The update of the response thresholds occurs at each time step and is given by:

$$w_{i,k} = \begin{cases} w_{i,k} - \delta_1, & \mathit{if}\ PS_{i,k} \geq \max_{j=1,\dots N}(PS_{j,k}) - \varepsilon \\ w_{i,k} + \delta_2, & \mathit{if}\ PS_{i,k} < \max_{j=1,\dots N}(PS_{j,k}) - \varepsilon \\ w_{i,k} - \delta_3, & \mathit{if}\ PS_{i,k} = 0 \end{cases} \tag{6}$$

The δ_1, δ_2 and δ_3 are positive model parameters. The response thresholds for the current puzzle are reinforced as to encourage the learning player wasp of the players with a specialization closed to the maximum for this puzzle to bid. The last equation from (6) encourages a wasp associated to a player with a zero specialization for a puzzle to take whatever puzzle it can get, rather than remaining idle. This equation allows a balanced game within the components of the team and avoids the blockage of the game when a puzzle with a type different from all the types of the previous puzzles is introduced in the game.

If two or more learning player wasps respond positively to the same stimulus, i.e. they bid for the same puzzle, they enter in a dominance contest. In order to decide the winner, we introduce for any wasp from the competition the force for solving the puzzle k. The force $F_{i,k}$ of the wasp associated to the player i for solving the puzzle k is given by:

$$F_{i,k} = 1 + \frac{\alpha_{1,l}}{1 + SPS_i} + \frac{\alpha_{2,l}}{1 + MPS_i} + \frac{\alpha_{3,l}}{1 + LS_i} + \frac{\alpha_{4,l}}{1 + PCS_{i,k}} + \frac{\alpha_{5,l}}{1 + PS_{i,k}} + \alpha_{6,l} \cdot n_i \tag{7}$$

where we denoted by

SPS_i - the number of puzzles successfully solved by the player i in single-player mode (single player specialization);
MPS_i - the number of puzzles successfully solved by the player i in multi-player mode (multiplayer specialization);
LS_i - the number of puzzles successfully solved by the player i in the current level (level specialization);
$PCS_{i,k}$ - the number of puzzles with the same complexity score as current puzzle k successfully solved by the player i in multiplayer mode (player complexity score specialization);

$PS_{i,k}$ - the specialization given in (1);
n_i - the number of puzzles already solved by the player i within the multiplayer mode of the game;
$\alpha_{j,l} \geq 0$ (with $j \in \{1, \ldots, 6\}$ and l - the current level of the game) are system parameters. These parameters define the strategy of the contest model. Different strategies can be implemented for different levels of the game by modifying these parameters.

Let i and s be two learning player wasps in a dominance contest. Learning player wasp i will get the puzzle with the probability:

$$P_c(i, s) = P(i\ win|F_i, F_s) = \frac{F_s^2}{F_s^2 + F_i^2} \tag{8}$$

In this way, learning player wasps associated with players with equivalent specializations and equal number of puzzles already solved within the multiplayer mode of the game have equal probabilities of getting the puzzle. For the same specializations, the wasp with the smaller number of puzzles already solved has a higher probability of taking the new puzzle. Thus we obtain a balance of the game. For the same number of puzzles already solved the wasp with the higher specializations has a higher probability of taking the new puzzle. If the specializations of a wasp are lower but the number of puzzles already solved is also lower, the probability for this wasp may increase. The presence of the parameters $\alpha_{i,l}$ in (7) allows modifying the wasp contest strategy: e.g. if $\alpha_{1,l} = 0$ for a level l, the results obtained by the player in single-player mode are neglected. Strategies may vary depending on the data bases of puzzles used.

5 Conclusions and Further Directions of Study

The virtual game SOTIRIOS presented in this article creates the bases of an educational tool that can be used to improve the teaching experience and to develop team working skills. The novelty consists in the team multiplayer mode idea and in the puzzle allocation model. The proposed model aims to optimize the learning results while maintaining a balanced participation of the players. We chose a wasp behaviour based model for designing and implementing an adaptive multi-agent system for dynamic allocation of puzzles in team multiplayer mode taking into account the player specialization and the puzzle complexity. An adaptive reinforcement of the thresholds in the wasp model keeps a balance between the components of the team and avoids the blockage of the game caused by the introduction of a new type of puzzle in the game. Our model depends on many parameters allowing different strategies of the proposed multi-agent system. The parameters can be tuned by hand or other approaches could be developed for this purpose. Anyway the analysis of different models is required in order to obtain the better strategy.

One of the future developments of our work consists in model validation and analysis and comparison of results obtained using different sets of parameters and strategies in the model. Another direction of study is oriented towards a meta-level optimization of the control parameters.

You can follow the development of Sotirios here: http://botashop.ro/sotirios/.

Acknowledgement. First author, Dana Simian, was supported by the research grant LBUS-IRG-2015-01, project financed by the from Lucian Blaga University of Sibiu.

Second author, Florentin Bota, was supported by a grant of the Romanian National Authority for Scientific Research and Innovation, CNCS - UEFISCDI, project number PN-II-RU-TE-2014-4-2560.

References

1. Bonabeau, E., Theraulaz, G., Demeubourg, J.I.: Fixed response thresholds and the regulationof division of labor in insect societies. Bull. Math. Biol. **60**, 753–807 (1998)
2. Bota, F.: Game based learning. Project Sotirios. In: Proceedings of the second International Students Conference on Informatics, ICDD 2012, Sibiu, Romania, pp. 44–50 (2012)
3. Cicirelo, V.A., Smith, S.F.: Wasp-like agents for distributed factory coordination. Auton. Agents Multi Agent Syst. **8**(3), 237–267 (2004)
4. Dehuri, S., Cho, S.-B.: Knowledge Minning Using Intelligent Agents. Advanced in Computer Science and Engineering: Texts, vol. 6. Imperial College Press, London (2011)
5. Freitas S., Jarvis S.: A framework for developing serious games to meet learner needs. In: Interservice/Industry Training, Simulation and Education Conference, Orlando, Florida (2006)
6. Malliarakis, C., Satratzemi, M., Xinogalos, S.: Designing educational games for computer programming: a holistic framework. Electron. J. e-Learn. **12**(3), 281–298 (2014)
7. Santos, M., Martinho, C.: Wasp-like scheduling for unit training in real-time strategy games. In: Proceedings of the Seventh AAAI Conference on Artificial Intelligence and Interactive Digital Entertainment, pp. 195–200 (2011)
8. Simian, D., Simian, C., Moisil, I., Pah, I.: Computer mediated communication and collaboration in a virtual learning environment based on a multi-agent system with wasp-like behavior. In: Lirkov, I., Margenov, S., Waśniewski, J. (eds.) LSSC 2007. LNCS, vol. 4818, pp. 618–625. Springer, Heidelberg (2008). https://doi.org/10.1007/978-3-540-78827-0_71
9. Simian, D., Stoica, F., Simian, C.: Optimization of complex SVM kernels using a hybrid algorithm based on wasp behaviour. In: Lirkov, I., Margenov, S., Waśniewski, J. (eds.) LSSC 2009. LNCS, vol. 5910, pp. 361–368. Springer, Heidelberg (2010). https://doi.org/10.1007/978-3-642-12535-5_42
10. Simian, D.: Wasp based algorithms and applications. In: Proceedings of International Conference on Modelling and Development of Intelligent Systems, Sibiu, pp. 229–235 (2009)

11. Song, M., Zhang, S.: EFM: a model for educational game design. In: Pan, Z., Zhang, X., El Rhalibi, A., Woo, W., Li, Y. (eds.) Edutainment 2008. LNCS, vol. 5093, pp. 509–517. Springer, Heidelberg (2008). https://doi.org/10.1007/978-3-540-69736-7_54
12. Theraulaz, G., Bonabeau, E., Gervet, J., Demeubourg, J.I.: Task differention in policies wasp colonies. A model for self-organizing groups of robots. From animals to Animats. In: Proceedings of the First International Conference on Simulation of Adaptive behavior, pp. 346–355 (1991)
13. Theraulaz, G., Bonabeau, E., Demeubourg, J.I.: Response threshold reinforcements and division of labour in insect societies. Proc. R. Soc. Lond. B Biol. Sci. **265**(1393), 327–335 (1998)

Hybrid Approach Based on Combination of Backpropagation and Evolutionary Algorithms for Artificial Neural Networks Training by Using Mobile Devices in Distributed Computing Environment

Iliyan Zankinski(✉), Maria Barova, and Petar Tomov

Institute of Information and Communication Technologies, Bulgarian Academy of Sciences, Acad. G. Bonchev Str., Block 2, 1113 Sofia, Bulgaria
iliyan@hsi.iccs.bas.bg

Abstract. When Evolutionary Algorithms (EAs) are used for Artificial Neural Networks (ANNs) training, the most valuable advantage is the potential for this training to be done in parallel or even using distributed computing. With the capabilities of modern mobile devices, for example their use for distributed computations, they can be used much more extensively for scientific calculations. It is well known that distributed computing systems are limited by their communication bandwidth, because of network latency. In such environment some EAs are pretty suitable for distributed implementation. This is because of their high level of parallelism and relatively less intensive network communication needs. Subset of distributed computing is volunteer computing where users donate some of the computing power provided by devices under their control. This research proposes Android Live Wallpaper volunteer computing implementation of a system used for financial time series prediction. The forecasting module is organized as ANN, which is trained by hybrid combination of Backpropagation and EAs.

Keywords: Artificial Neural Networks · Evolutionary Algorithms
Distributed computing

1 Introduction

ANNs are very common in the field of machine learning. In its nature ANN training is an optimization problem. When the searching space is too big, global optimization EAs as Genetic Algorithms (GAs) can be very suitable. In the last two decades there are numerous attempts to combine EAs based optimization with ANNs training. This combination shows to be much more promising when it is implemented as distributed computing system [1–3]. With the rising popularity of mobile devices a lot of new possibilities for distributed computing can be investigated. Successful desktop distributed computing projects, as described

I. Lirkov and S. Margenov (Eds.): LSSC 2017, LNCS 10665, pp. 425–432, 2018.
https://doi.org/10.1007/978-3-319-73441-5_46

in [4,5] can be efficiently implemented on mobile devices. This idea can be applied to EA based ANN training into a mobile distributed computing environment [6]. In this paper a genetic algorithm in combination with backpropagation artificial neural network training is described. The training is done on a mobile devices as active Android wallpaper application. In addition, the traditional ANN's sigmoid function is replaced by fading sinusoidal function. The successful application of this hybrid approach is documented by series of experiments.

This paper is organized as follows. Section 1 gives an overview of the problem with a special emphasis on ANNs/GAs and their strengths/weaknesses when applied for the problem's solution. Section 2 introduces a distributed computing system based on mobile devices. Experiments and results are presented in Sect. 3. The final Sect. 4 concludes and some further work suggestions are provided.

1.1 Financial Time Series Forecasting

A time series is a collection of measurements made in a sequence through time. There are many examples for time series like the sales of the specific stock in successive months, the average daily temperature at a particular location, the electricity consumption in a particular areal for successive seasons and many other. In the field of the finance, decision makers are in the position of taking very responsible steps during formation of investment strategy. To invest it means to take acceptable risk with expectation of certain amount of profit. The most important aspect of investment is the balance between risk taken and expected profit. On the currency market (FOREX) the main trading is done by exchanging currencies. Currency is the most volatile in price changing object of trading. During the process of trading on a market as FOREX decision makers needs to take three important decisions: 1. Price will go up or down; 2. What volume to buy or sell; 3. How long to keep the opened position. Even if it sounds simple in fact it is very difficult to estimate price changing direction, because of the huge number of factors influencing it. The order volume is directly related to the amount of risk taken. High volume order can lead to high profit if price changing direction is well estimated, but it can lead to high loss in the other case. How long to keep the opened position is related to making the profit even bigger or making the loss as small as possible. Financial forecasting is most important for the traders on the currency market, because of the high price dynamics [2].

1.2 Artificial Neural Networks

Artificial Neural Network is a mathematical model inspired by research into the nature of the human brain. ANNs consists of five components: 1. Network topology which is represented by a directed graph where arcs are referred as links; 2. A variable which represents the state of each node; 3. For each link a real value variable represents its weight; 4. A real value bias which is supplied to each node; 5. Node transfer function which determines the state of the node. The transfer function consists of an activation function (in most cases sigmoid or hyperbolic tangent). The activation function accepts the sum of the inputs multiplied by the weights of the links between the node and the input nodes.

The input nodes in feed-forward network do not have input arcs. The input nodes should be supplied with values. After that the input information can be spread across the network. The other nodes are changing their state variable according to the propagation rules. In multilayer perceptron (MLP) any path from an input node to an output node traverses the same number of arcs. A hidden layer is a group of nodes which are neither input nor output and are on the same number of arcs distance from the input layer and the output layer. MLP is fully connected if each node in a particular layer is connected to each node in the previous layer.

Research shows that such MLP generalize well in practical problems. When trained on a relatively sparse set of data points, they often provide the right output for an input not presented in the training set. Secondly, gradient based training algorithms can be successfully applied in order to find a good set of weights in acceptable amount of time. The advantage is in calculating the gradient of the error according to the weights for a given input by back propagating the error through the network. Gradient based training works well on simple training problems, but when the problem complexity increases (most commonly because of the increased dimensionality and/or greater data complexity) the performance of the gradient based training falls off rapidly [7].

The performance slow down seems to come from the fact that complex spaces have nearly global optimum around the local optimum. Gradient search techniques tend to get trapped at the local optimum. With a high enough gain (or momentum), backpropagation can escape these local optima. However, it leaves them without knowing whether the next one it finds will be better or worse. When the nearly global optima are well hidden among the local optimum, backpropagation can end up bouncing between local optima without much overall improvement, thus making for very slow training. Another drawback of the gradient based training is the requirement for the activation function differentiability. Backpropagation can not handle discontinuous functions or discontinuous node activation functions.

1.3 Genetic Algorithms

Genetic Algorithms are meta heuristics for global optimization inspired from the ideas of biological evolution. They have five components: 1. Approach for encoding the problem in terms of chromosomes (population of individuals); 2. An evaluation function which is used to determine survival capabilities of each chromosome; 3. A strategy for initial population initialization; 4. Set of operator applied over parent chromosomes in order reproduction to appear (in most cases—selection, crossover, mutation and/or domain specific genetic material recombination); 5. Set of parameters applied over the population and the operators.

GA applies these components and operates in the following steps:

1. Initialization of the population (random or some prior information).
2. Chromosomes evaluation (relative ranking as result in this step).
3. Epochs of recombinations are executed until stopping criteria is met.
   3.1 Stochastic parents selection (parents with better fitness are preferred).
   3.2 Children are produced by recombination operators over the parents.
   3.3 Evaluation of the children is done to keep some of them in the population. Some times the elitism rule is applied by keeping the best found individuals to survive until the end of the evolution.

If appropriate problem encoding and recombination operators are selected the algorithm is capable of producing better and better solutions, converging finally on results close to a global optimum. In most of the real life problems (as the problem presented in this research) standard operators as crossover and mutation are sufficient. In this case GA can be used as black-box optimizer. It means that no specific knowledge of the problem domain is needed. As it is shown in this research by involving problem specific operators (crossover and mutation), the optimization efficiency can be improved.

GA does not have a scaling problem as it is the case with ANN [8]. Even more, GAs are very suitable for parallel computing and even for distributed computing, because GA improves the current best candidate monotonically. It is done by keeping the best found candidates as part of the population while searching for better candidates continues. GAs are not threatened by getting stuck in a local optimum. The mutation and crossover operators can step from a valley across a hill to an even lower valley with no more difficulty than descending directly into a valley.

1.4 Neuron Activation Function Proposal

The most often used activation functions in MLPs are the sigmoid function and the hyperbolic tangent function [9]. In this research sin function with exponent fading effect is proposed as neurons activation function (Fig. 1). Because of its fading effect neurons are more active in a specific input range. If the input is highly positive or highly negative the neuron stops acting as effect of over saturation.

$$f(x) = \frac{\pi \sin(x)}{e^{|x|}} \tag{1}$$

The proposed function (Eq. 1) is differentiable (Eqs. 2 and 3) which is one of the common requirements in gradient based training of ANNs.

$$\frac{d}{dx} f(x) = \frac{\pi(|x| \cos x - x \sin x)}{|x| e^{|x|}} \tag{2}$$

Function derivative can be expressed by the function itself as it is shown in Eq. 3.

$$\frac{d}{dx} f(x) = \begin{cases} f(x+\pi) - f(x), \, x > 0 \\ f(x+\pi) + f(x), \, x < 0 \\ +\infty, \qquad\qquad\quad x = 0 \end{cases} \tag{3}$$

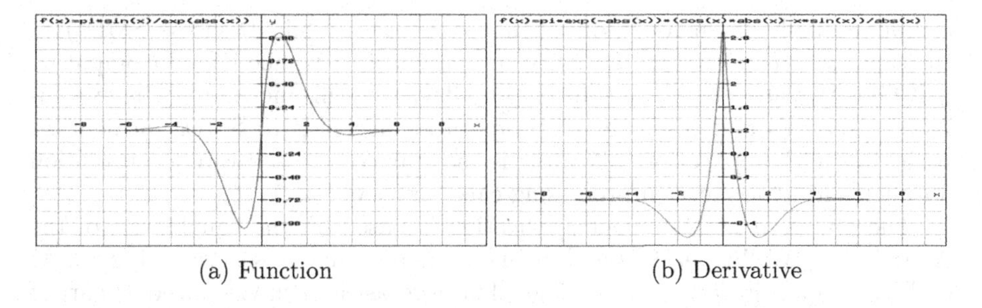

(a) Function (b) Derivative

Fig. 1. Exponent regulated sin activation function and its derivative.

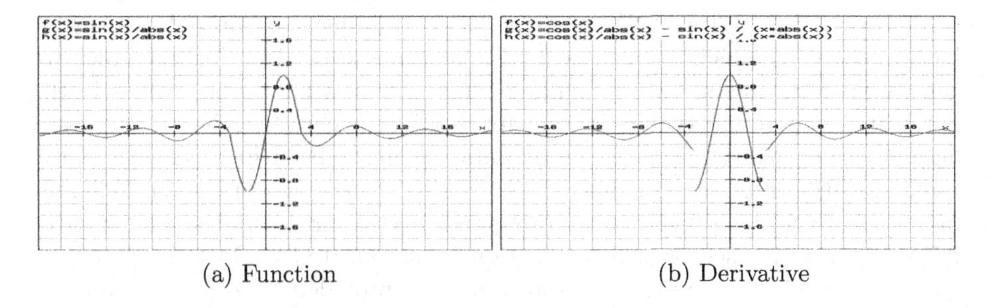

(a) Function (b) Derivative

Fig. 2. Fading sin activation function and its derivative.

The advantage of such activation function is that each neuron has loading level which it is capable to handle.

If periodic processes should be modeled with ANN it is possible to apply sin fading function (Fig. 2).

$$f(x) = \begin{cases} \sin(x), & -\pi <= x <= +\pi \\ \frac{\sin(x)}{|x|}, & +\pi < x < -\pi \end{cases} \tag{4}$$

The problem with the sin fading function is that it has break points at $-\pi$ and $+\pi$.

$$\frac{d}{dx} f(x) = \begin{cases} \cos(x), & -\pi <= x <= +\pi \\ \frac{\cos(x)}{|x|} - \frac{\sin(x)}{x|x|}, & +\pi < x < -\pi \end{cases} \tag{5}$$

The components of the function (Eq. 4) are differentiable functions (Eq. 5). This small inconvenience is not a problem when calculations are done with discrete number calculating computers. Representing real numbers in the computer memory is less accurate than the problems of analytic differentiability of the sin fading function.

2 Mobile Devices Distributed Computing

The latest developments in mobile devices technology have made smartphones as the future computing and service access devices. Users expect to run computationally intensive applications on Smart Mobile Devices (SMDs) in the same way as powerful stationary computers [10]. Mobile Cloud Computing (MCC) is the latest practical solution for alleviating this incapability by extending the services and resources of computational clouds to SMDs on demand basis. In MCC, application offloading is ascertained as a software level solution for augmenting application processing capabilities of SMDs [10].

The proposed mobile client application is based on the capabilities of the Android Live Wallpaper [6]. On a regular basis a wallpaper service wakes up. At each wake up a single forecast is done. This forecast is then visualized as part of a wallpaper image. The wake up finishes a single ANN training epoche. The forecasting model is organized as a three-layers MLP instantiated as Encog framework Java object. Local ANN gradient based training is organized as resilient backpropagation which is provided by Encog framework. Training examples are generated from time series by dividing the values in a lag frame (values in the past) and lead frame (values in the future). In this way ANN tries to learn lag and lead patterns [11]. All input data are normalized in the range of the used activation function (in this case -1 and $+1$, as shown in Fig. 1). Different instances of the ANN weights are presented as GA chromosomes. The error of ANN forecasts is used as fitness function in the local GA. On the side of the mobile client GA is represented as Java objects provided by Apache Commons Math Genetic Algorithms framework. In order for ANN training to be continuous and to be independent from Internet network connection, all ANN training examples are stored in a local SQLite database. The communication with the remote server is done only when there is better local solution. The information exchange between the client and the server is organized as HTTP sessions. All data exchanges are packaged as JSON messages. On the server side there is a PHP/MySQL based application which is described in [2].

As summary the weights of an ANN (as Encog Java objects) are optimized with a GA (as Apache Framework Java objects) and this optimization is done on the Android mobile device. Optimization results are collected on the remote PHP/MySQL server.

3 Experiments and Results

All experiments are done with Encog framework by using Java programming language [12]. MLP neural network (256-64-10 topology) is trained with examples of handwritten digits. Data set consists of 1593 handwritten digits from around 80 persons were scanned, stretched in a rectangular box 16×16 in a gray scale of 256 values [13]. Experiments are done on a single CPU core with the following parameters of the machine: Intel Core i7-4790 - 3.6 GHz - 4 cores - 8 threads, 8 GB RAM, Microsoft Windows 10, Encog Core v3.3.0 - Java Version, Java 8 Update 112 (64 bit).

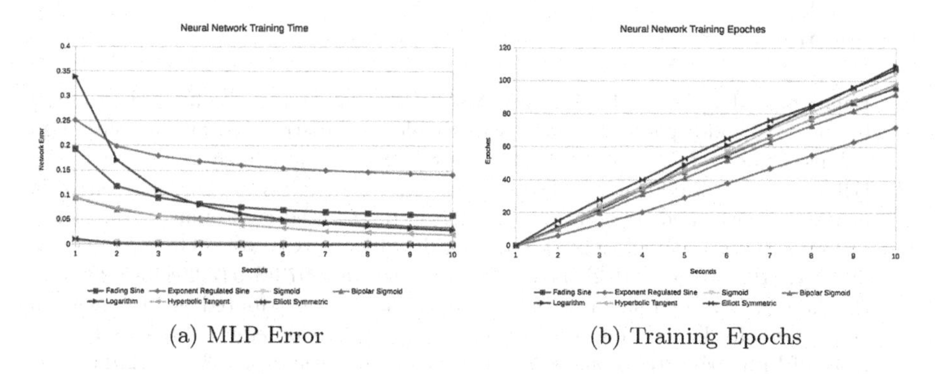

(a) MLP Error (b) Training Epochs

Fig. 3. Activation functions efficiency.

Comparison is done between seven activation functions - Fading Sin, Exponent Regulated Sin, Sigmoid, Bipolar Sigmoid, Logarithm, Hyperbolic Tangent, Elliott Symmetric. As it is shown in Fig. 3, between 70 and 110 training epoche are done for 10 s of training. The performance of exponent regulated sine is slower than the performance of fading sine. The network error for both functions decreases slower than the others, which is very logical because each neuron is limited in the activation signal which can be handled. This limitation gives better responsibilities separation between the neurons in each layer.

4 Conclusions

Even though GA based ANN training is slower than back propagation, when it is implemented as simultaneous computations in a distributed environment it can be efficient enough for practical use. Because modern mobile devices are used on a 24/7 basis to use them as distributed computing network is very cost effective if the approach is based on donated computational power. As further development it will be interesting if mobile devices distributed computing is combined with supercomputers. The central node will be able to provide access to a super computer. Such hybrid infrastructure can provide interesting research and industrial challenges. Another very interesting direction of further development is the capabilities of Android widgets. The client side calculations are possible to be done in a widget instead of a wallpaper. Android widgets are interactive components in the graphical user interface. Such computational widget can be used for user voting or even user's guess for the future forecast. The collection of human opinion will transfer this software solution into the human-computer based distributed computing.

Acknowledgements. This work was supported by private funding of Velbazhd Software LLC.

References

1. Balabanov, T., Zankinski, I., Barova, M.: Strategy for individuals distribution by incident nodes participation in star topology of distributed evolutionary algorithms. Cybern. Inf. Technol. **16**(1), 80–88 (2016). Sofia, Bulgaria
2. Balabanov, T., Zankinski, I., Dobrinkova, N.: Time series prediction by artificial neural networks and differential evolution in distributed environment. In: Lirkov, I., Margenov, S., Waśniewski, J. (eds.) LSSC 2011. LNCS, vol. 7116, pp. 198–205. Springer, Heidelberg (2012). https://doi.org/10.1007/978-3-642-29843-1_22
3. Balabanov, T.: Heuristic forecasting approaches in distributed environment (in Bulgarian). In: Proceedings of Anniversary Scientific Conference 40 Years Department of Industrial Automation, UCTM, Sofia, Bulgaria, pp. 163–166 (2011)
4. Balabanov, T.: Distributed evolutional model for music composition by human-computer interaction. In: Proceedings of International Scientific Conference UniTech15. University publishing house "V. Aprilov", Gabrovo, Bulgaria, vol. 2, pp. 389–392 (2015)
5. Balabanov, T.: Avoiding local optimums in distributed population based heuristic algorithms (in Bulgarian). In: Proceedings of XXIII International Symposium Management of Energy, Industrial and Environmental Systems, pp. 83–86. John Atanasoff Union of Automation and Informatics, Sofia, Bulgaria (2015)
6. Balabanov, T., Zankinski, I., Barova, M.: VitoshaTrade Distributed Computing Android Wallpaper, Sofia, Bulgaria (2017). https://github.com/TodorBalabanov/VitDisComp
7. Bas, E.: The training of multiplicative neuron model based artificial neural networks with differential evolution algorithm for forecasting. J. Artif. Intell. Soft Comput. Res. **6**(1), 5–11 (2016)
8. Magnusson, K., Olsson, T.: Training artificial neural networks with genetic algorithms for stock forecasting. A comparative study between genetic algorithms and the backpropagation of errors algorithms for predicting stock prices. Dissertation (2016). http://urn.kb.se/resolve?urn=urn:nbn:se:kth:diva-186447
9. Karlik, B., Vehbi, A.: Performance analysis of various activation functions in generalized MLP architectures of neural networks. Int. J. Artif. Intell. Expert Syst. (IJAE) **1**(4), 111–122 (2011)
10. Shiraz, M., Gani, A., Khokhar, R., Buyya, R.: A review on distributed application processing frameworks in smart mobile devices for mobile cloud computing. IEEE Commun. Surv. Tutorials **15**(3), 1294–1313 (2013)
11. Qiu, M., Song, Y.: Predicting the direction of stock market index movement using an optimized artificial neural network model. PLoS ONE **11**(5), e0155133 (2016). https://doi.org/10.1371/journal.pone.0155133
12. Zankinski, I.: Encog digits classification resilient training example with sin activation functions, Sofia, Bulgaria (2017). https://github.com/iliyanzan/DigitsResilient
13. Buscema, M., Terzi, S.: Semeion Handwritten Digit Data Set. Center for Machine Learning and Intelligent Systems, California, USA (2009). http://archive.ics.uci.edu/ml/datasets/Semeion+Handwritten+Digit

Large-Scale Models: Numerical Methods, Parallel Computations and Applications

Solution of the 3D Neutron Diffusion Benchmark by FEM

A. V. Avvakumov[1], P. N. Vabishchevich[2], A. O. Vasilev[3(✉)], and V. F. Strizhov[2]

[1] National Research Center Kurchatov Institute, Moscow, Russia
[2] Nuclear Safety Institute of RAS, Moscow, Russia
[3] North-Eastern Federal University, Yakutsk, Russia
haska87@gmail.com

Abstract. The objective is to analyze the neutron diffusion benchmark developed by the Atomic Energy Research community for verification of best-estimate neutronics codes. The 3D benchmark of Schulz models a VVER-1000 core in steady state. The assemblies are homogeneous, represented by given diffusion theory parameters. There are seven material compositions including four enrichments, burnable absorber, control rods and a reflector. The finite element method on tetrahedron computational grids is used to solve the three-dimensional neutron problem. The software has been developed using the engineering and scientific library FEniCS. The matrix spectral problem is solved using a scalable and flexible toolkit SLEPc. The solution accuracy of the benchmark is analyzed by condensing the computational grid and varying the degree of the finite elements.

1 Introduction

The physical processes in a nuclear reactor [1] depend on the distribution of the neutron flux, whose mathematical description is based on the neutron-transport equation [2,3]. The equation, describing the process is of integro-differential form and the required distribution of neutrons flux depends on time, energy, spatial and angular variables. As a rule, simplified forms of the neutron transport equation are used for practical computations of nuclear reactors. The system of equations obtained by the multigroup diffusion approach is often used for reactor analysis [4,5] and is applied in most engineering calculation codes.

The processes in a nuclear reactor are essentially non-stationary. The stationary state of neutron flux, which is related to the critical state of the reactor, is characterized by local balance between the neutron generation and absorption. This boundary state is usually described by the solution of a spectral problem (Lambda Modes problem, λ-eigenvalue problem) provided that the fundamental eigenvalue (the maximal eigenvalue) that is called k-effective of the reactor core, is equal to unity. In this case, the stationary neutron flux is a corresponding eigenfunction. Computations of k-effective of the reactor using the Lambda Modes problem are obligatory for reactor design calculations.

I. Lirkov and S. Margenov (Eds.): LSSC 2017, LNCS 10665, pp. 435–442, 2018.
https://doi.org/10.1007/978-3-319-73441-5_47

To verify neutron diffusion codes several benchmark problems have been developed. This work is focused on the solution of the 3D benchmark problem of a VVER-1000 core in steady state. Convergence of the benchmark solution is under investigation.

2 Problem Description

Let's consider modeling neutron flux in a multi-group diffusion approximation. Neutron flux dynamics is considered within a bounded 2D or 3D domain Ω ($\boldsymbol{x} = \{x_1, \dots, x_d\} \in \Omega$, $d = 2, 3$) with a convex boundary $\partial\Omega$. The neutron transport is described by the following set of equations without taking into account delayed neutron sources:

$$\begin{aligned} \frac{1}{v_g}\frac{\partial \phi_g}{\partial t} - \nabla \cdot D_g \nabla \phi_g + \Sigma_{rg}\phi_g - \sum_{g \neq g'=1}^{G} \Sigma_{s,g' \to g}\phi_{g'} \\ = ((1-\beta)\chi_g + \beta\widetilde{\chi}_g) \sum_{g'=1}^{G} \nu\Sigma_{fg'}\phi_{g'}, \quad g = 1, 2, \dots, G. \end{aligned} \tag{1}$$

Here $\phi_g(\boldsymbol{x}, t)$ is the neutron flux of group g at point $\boldsymbol{x}$ and time t, G is the number of groups, v_g is the effective velocity of neutrons in the group g, $D_g(\boldsymbol{x})$ is the diffusion coefficient, $\Sigma_{rg}(\boldsymbol{x}, t)$ is the removal cross-section, $\Sigma_{s,g' \to g}(\boldsymbol{x}, t)$ is the scattering cross-section from group g' to group g, β is the effective fraction of delayed neutrons, χ_g, $\widetilde{\chi}_g$ is the spectra of instantaneous and delayed neutrons, $\nu\Sigma_{fg}(\boldsymbol{x}, t)$ is the generation cross-section of group g. The conditions so-called albedo-type are set at the boundary $\partial\Omega$:

$$D_g \frac{\partial \phi_g}{\partial n} + \gamma_g \phi_g = 0, \qquad g = 1, 2, \dots, G, \tag{2}$$

where n is the outer normal to the boundary $\partial\Omega$. Let's consider problem (1) with boundary conditions (2) and initial conditions

$$\phi_g(\boldsymbol{x}, 0) = \phi_g^0(\boldsymbol{x}), \quad g = 1, 2, \dots, G. \tag{3}$$

We write the boundary problem (1), (2), (3) in operator notation. We define the vector $\boldsymbol{\phi} = \{\phi_1, \phi_2, \dots, \phi_G\}$ and the matrices

$$V = (v_{gg'}), \quad v_{gg'} = \delta_{gg'} v_g^{-1},$$

$$D = (d_{gg'}), \quad d_{gg'} = -\delta_{gg'} \nabla \cdot D_g \nabla,$$

$$S = (s_{gg'}), \quad s_{gg'} = \delta_{gg'} \Sigma_{rg} - \Sigma_{s,g' \to g},$$

$$R = (r_{gg'}), \quad r_{gg'} = ((1-\beta)\chi_g + \beta\widetilde{\chi}_g)\nu\Sigma_{fg'},$$

$$g, g' = 1, 2, ..., G,$$

where

$$\delta_{gg'} = \begin{cases} 1, \, g = g', \\ 0, \, g \neq g', \end{cases}$$

is the Kronecker delta. Consider the set of vectors $\boldsymbol{\phi}$, whose components satisfy the boundary conditions (3). Using the above introduced definitions, the system of Eq. (1) takes in the form of the first-order evolutionary equation

$$V\frac{d\boldsymbol{\phi}}{dt} + (D + S)\boldsymbol{\phi} = R\boldsymbol{\phi}. \tag{4}$$

We solve the Cauchy problem for (4), when

$$\boldsymbol{\phi}(0) = \boldsymbol{\phi}^0, \tag{5}$$

where $\boldsymbol{\phi}^0 = \{\phi_1^0, \phi_2^0, \ldots, \phi_G^0\}$.

To characterize the dynamic processes in a nuclear reactor, which are described by the Cauchy problem (4), (5), solutions of some spectral problems (see [6]). Let's consider the solution of the spectral problem, called Lambda Modes problem:

$$(D + S)\boldsymbol{\varphi} = \lambda^{(k)} R\boldsymbol{\varphi}. \tag{6}$$

Problem (6) is known as the Lambda modes problem for a given configuration of the reactor core. The minimal eigenvalue is used for characterization of neutron flux, thus

$$k = \frac{1}{\lambda_1^{(k)}}$$

is the effective multiplication factor (see [2]).

3 AER Benchmark

The 3D benchmark, Schulz [7], models a VVER-1000 core in a steady state. The assemblies are homogeneous, represented by given two-group diffusion theory (see more [1,2]) parameters. There are seven material compositions including four enrichments, burnable absorber, control rods and reflector. The core height is 355 cm, covered with axial and radial reflectors. The nodes are large, with assembly lattice pitch of 24.1 cm. Figure 1 shows the radial and axial model geometry. Diffusion neutronics constants in the common notations are given in Table 1.

4 Computational Results

In the framework of a two-group model, the spectral problem (6) can be written as:

$$\begin{aligned} -\nabla \cdot D_1 \nabla \varphi_1 + \Sigma_{r1}\varphi_1 &= \lambda^{(k)}(\nu\Sigma_{f1}\varphi_1 + \nu\Sigma_{f2}\varphi_2), \\ -\nabla \cdot D_2 \nabla \varphi_2 + \Sigma_{r2}\varphi_2 - \Sigma_{s,1\to 2}\varphi_1 &= 0. \end{aligned} \tag{7}$$

The boundary conditions (2) are used at $\gamma_g = 0.5, \; g = 1, 2$.

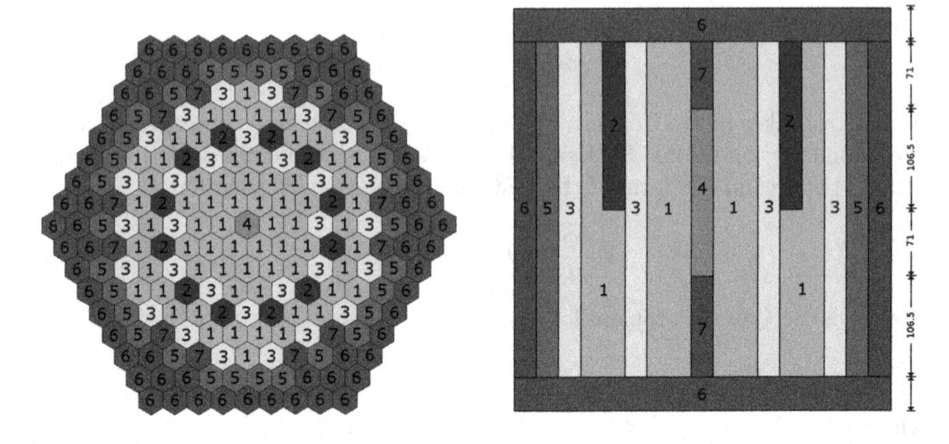

Fig. 1. Radial and axial geometry of the model.

Table 1. Diffusion neutronics constants.

Material	Group	D	Σ_r	$\Sigma_{1\to2}$	Σ_f	$\nu\Sigma_f$
1	1	1.37548	2.4135e−2	1.5946e−2	6.0130e−7	4.7663e−3
	2	0.38333	6.6002e−2		1.1231e−5	8.3980e−2
2	1	1.40950	2.4769e−2	1.4346e−2	5.9305e−7	4.7020e−3
	2	0.38756	7.4988e−2		1.1253e−5	8.4128e−2
3	1	1.37067	2.3800e−2	1.5172e−2	7.4429e−7	5.8437e−3
	2	0.38028	8.0442e−2		1.5336e−5	1.1468e−1
4	1	1.39447	2.4069e−2	1.3903e−2	7.8731e−7	6.1632e−3
	2	0.38549	9.4773e−2		1.6848e−5	1.2598e−1
5	1	1.36938	2.3697e−2	1.4855e−2	8.1014e−7	6.3396e−3
	2	0.37877	8.7681e−2		1.7381e−5	1.2998e−1
6	1	1.00000	4.0644e−2	2.4875e−2	0.0000e−0	0.0000e−0
	2	0.33333	5.2785e−2		0.0000e−0	0.0000e−0
7	1	1.36966	2.3721e−2	1.4927e−2	7.9536e−7	6.2284e−3
	2	0.37911	8.5850e−2		1.6866e−5	1.2612e−1

The method of finite elements [8] on tetrahedron calculation grids is used for the approximate solution. The software has been developed using the engineering and scientific calculation library FEniCS [9]. The SLEPc package [10] has been used for the numerical solution of the spectral problems

$$Ax = \lambda Bx. \tag{8}$$

Used Krylov-Schur algorithm with an accuracy of 10^{-15}. In the computations the following parameters are varied:

- the number of tetrahedrons per one assembly κ;
- the number of tetrahedrons in height z;
- order of finite element p.

The number of tetrahedrons per one assembly κ varies from 6 to 96 and per height z varies from 12 to 48.

The standard Lagrangian finite elements of degree $p = 1, 2, 3$ are used.

The following parameters were calculated:

- the effective multiplication factor k;
- the power distribution P per assembly with the normalization of the mean value of the core:

$$P = a(\Sigma_{f1}\varphi_1 + \Sigma_{f2}\varphi_2), \tag{9}$$

where a is normalization coefficient by a given value of the integral power.

The results are compared with those, obtained using the diffusion program CRONOS [11]. The extrapolated finite-element solution of the second-order CRONOS results is recommended as the reference solution ($k_{ref} = 1.049526$) of the Schulz benchmark.

Let's consider the following variations in the calculated parameters:

- for the effective multiplication factor, absolute deviation from the reference value k_{ref}: $\Delta k = |k - k_{ref}|$, expressed in *pcm* (percent-milli, i.e. 10^{-5});
- for power distribution per assembly P_i calculated relative deviation ε_i (expressed in %):

$$\varepsilon_i = \frac{P_i - P_i^{ref}}{P_i^{ref}},$$

where P_i^{ref} — «reference» value of power per assembly i ($i = 1, \ldots, N_e$).
- by deviations ε_i calculated integral deviation:

$$\text{RMS} = \sqrt{\frac{1}{N_e}\sum_{i=1}^{N_e}\varepsilon_i^2}, \quad \text{AVR} = \frac{1}{N_e}\sum_{i=1}^{N_e}|\varepsilon_i|, \quad \text{MAX} = \max_i |\varepsilon_i|.$$

Results of the solution of λ-spectral problem (6) for the main eigenvalue k_1 using different grids and finite elements are given in Table 2. Here N is the size of the matrices A, B (8). These data demonstrate the convergence of the computed eigenvalues with thickening of the grid κ, z and with increasing the degree p. Comparison of the reference solutions with the result of the solution at the parameters $p = 2, \kappa = 6, z = 24$ (in Table 2 highlighted in green) is shown in Fig. 2.

Table 3 shows the results obtained for the next three eigenvalues for different meshes. Vertical and horizontal cuts of power distribution for the first four eigenvalues are shown in Fig. 3.

Table 2. k_1 results for the Schulz benchmark.

p	κ	z	k_1	Δk	MAX	AVR	RMS	N
1	6	12	1.0476057	192.03	7.9382	2.5145	2.7918	18,278
1	6	24	1.0484070	111.90	7.6614	2.3465	2.5983	35,150
1	6	48	1.0486511	87.49	7.6793	2.3643	2.6092	68,894
1	24	12	1.0482940	123.20	2.2234	0.5665	0.7041	70,382
1	24	24	1.0487937	73.23	1.9377	0.4050	0.4774	135,350
1	24	48	1.0493645	16.15	1.9823	0.3980	0.4612	265,286
1	96	12	1.0483122	121.38	1.2015	0.3780	0.4857	276,146
1	96	24	1.0490651	46.09	0.2647	0.1129	0.1389	531,050
1	96	48	1.0493997	12.63	0.4554	0.1019	0.1243	1,040,858
2	6	12	1.0496463	-12.03	0.9739	0.4801	0.5581	135,350
2	6	24	1.0497290	-20.30	0.9576	0.4504	0.5314	265,286
2	6	48	1.0497379	-21.19	0.9576	0.4501	0.5307	525,158
2	24	12	1.0494978	2.82	0.3246	0.1577	0.1861	531,050
2	24	24	1.0495665	-4.05	0.2597	0.1176	0.1414	1,040,858
2	24	48	1.0495858	-5.98	0.2435	0.1117	0.1335	2,060,474
2	96	12	1.0494551	7.09	0.1786	0.0662	0.0771	2,103,650
2	96	24	1.0495265	-0.05	0.0844	0.0309	0.0377	4,123,154
2	96	48	1.0495471	-2.11	0.0573	0.0256	0.0298	8,162,162
3	6	12	1.0495750	-4.90	0.2149	0.0956	0.1153	444,962
3	6	24	1.0495782	-5.22	0.2110	0.0931	0.1125	877,898
3	6	48	1.0495771	-5.11	0.2005	0.0900	0.1082	1,743,770
3	24	12	1.0495406	-1.46	0.0430	0.0177	0.0216	1,756,982
3	24	24	1.0495382	-1.22	0.0317	0.0123	0.0146	3,466,478
3	24	48	1.0495381	-1.21	0.0286	0.0102	0.0126	6,885,470
3	96	12	1.0495357	-0.97	0.0317	0.0106	0.0137	6,982,418
3	96	24	1.0495338	-0.78	0.0211	0.0092	0.0110	13,776,122
3	96	48	1.0495336	-0.76	0.0162	0.0080	0.0100	27,363,530

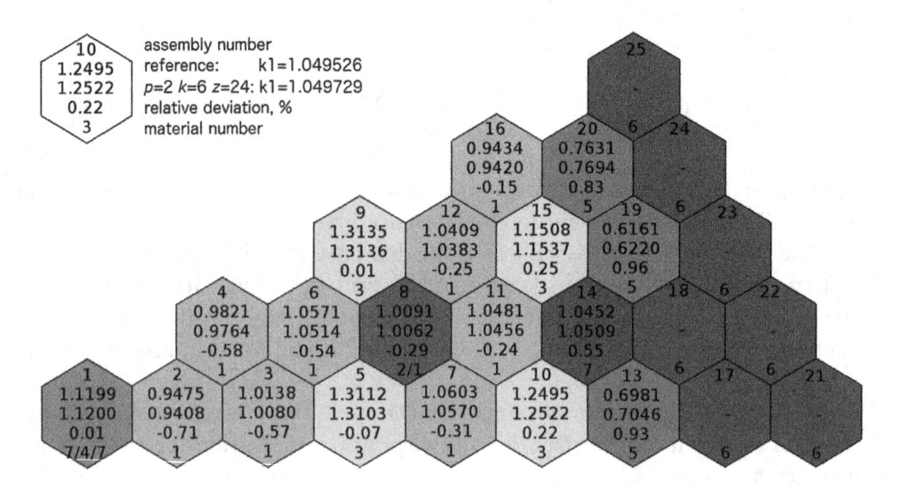

Fig. 2. Power distribution at $p = 2, \kappa = 6, z = 24$.

Table 3. The eigenvalues k_2, k_3 Đÿ k_4.

p	κ	z	k_2	k_3	k_4
1	6	12	1.0367578	1.0367505	1.0265919
1	6	24	1.0378381	1.0378355	1.0286801
1	6	48	1.0381457	1.0381434	1.0292950
1	24	12	1.0378816	1.0378762	1.0277542
1	24	24	1.0389665	1.0389631	1.0299569
1	24	48	1.0392872	1.0392857	1.0306072
1	96	12	1.0378388	1.0378365	1.0276575
1	96	24	1.0390329	1.0390303	1.0300412
1	96	48	1.0393812	1.0393798	1.0307428

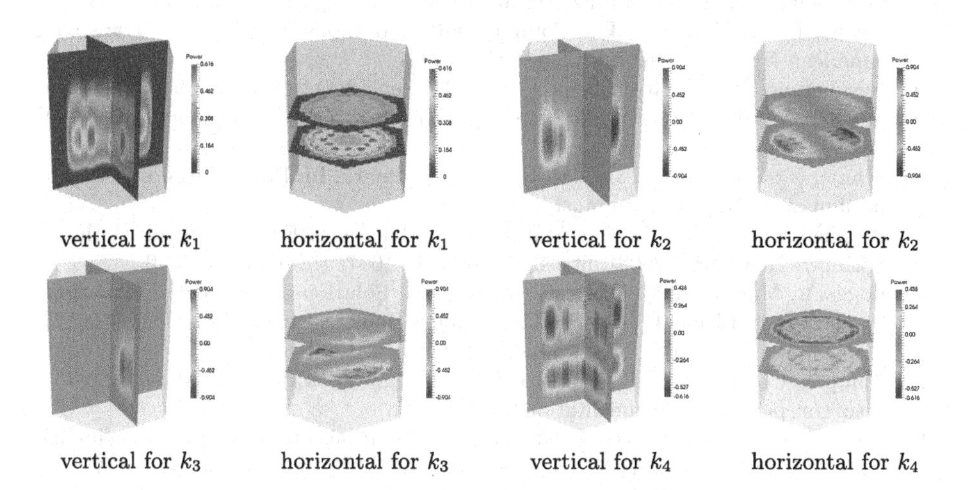

vertical for k_1 horizontal for k_1 vertical for k_2 horizontal for k_2

vertical for k_3 horizontal for k_3 vertical for k_4 horizontal for k_4

Fig. 3. Sections.

5 Summary

In this paper we have performed computational analysis of the 3D neutron diffusion benchmark of a VVER-1000 core. The software has been developed using the engineering and scientific library FEniCS and the matrix spectral problem is solved using the SLEPc package. The number of tetrahedrons per assembly κ varies from 6 to 96; the number of tetrahedrons in height z varies from 12 to 48. The finite elements of degree $p = 1, 2, 3$ are used. The results are compared with the extrapolated finite-element solution of the second-order CRONOS results recommended as the reference solution. An excellent agreement of the results was obtained for the maximum values of the parameters ($\kappa = 96$, $z = 48$ and $p = 3$). There are deviations in the results at the rounding error level. In the practice of engineering calculations it is sufficient to use the following parameters: $\kappa = 6$, $z = 24$ and $p = 2$. The results can be useful for the diffusion codes intercomparison.

Acknowledgments. The research was supported by the Government of the Russian Federation (project 14.Y26.31.0013).

References

1. Duderstadt, J.J., Hamilton, L.J.: Nuclear Reactor Analysis. Wiley, Hoboken (1976)
2. Stacey, W.M.: Nuclear Reactor Physics. Wiley, Hoboken (2007)
3. Hetrick, D.L.: Dynamics of Nuclear Reactors. University of Chicago Press, Chicago (1971)
4. Marchuk, G.I., Lebedev, V.I.: Numerical Methods in the Theory of Neutron Transport. Harwood Academic Pub, Newark (1986)
5. Lewis, E.E., Miller, W.F.: Computational Methods of Neutron Transport. American Nuclear Society, Philadelphia (1993)
6. Avvakumov, A.V., Vabishchevich, P.N., Vasilev, A.O., Strizhov, V.F.: Spectral properties of dynamic processes in a nuclear reactor. Ann. Nucl. Energy **99**, 68–79 (2017)
7. Schulz, G.: Solution of a 3D VVER-1000 benchmark. In: Proceedings of 6th Symposium of AER, Kikkonummi, Finland (1996)
8. Brenner, S.C., Scott, R.: The Mathematical Theory of Finite Element Methods. Springer, New York (2008). https://doi.org/10.1007/978-0-387-75934-0
9. Logg, A., Mardal, K.A., Wells, G.: Automated Solution of Differential Equations by the Finite Element Method: The FEniCS Book. Springer, Heidelberg (2012). https://doi.org/10.1007/978-3-642-23099-8
10. Campos, C., Roman, J.E., Romero, E., Tomas, A.: SLEPc Users Manual (2013). http://slepc.upv.es/documentation/manual.htm
11. Kolev, N.P., Lenain, R., Fedon-Magnaud, C.: Solutions of the AER 3D benchmark for VVER-1000 by CRONOS. In: Proceedings of 7th Symposium of AER on VVER Reactor Physics and Safety, Hoernitz, Germany (1997)

Precipitation Pattern Estimation with the Standardized Precipitation Index in Projected Future Climate over Bulgaria

Hristo Chervenkov(✉) and Valery Spiridonov

National Institute of Meteorology and Hydrology, Bulgarian Academy of Sciences, Tsarigradsko Shose blvd. 66, 1784 Sofia, Bulgaria
hristo.tchervenkov@meteo.bg

Abstract. The expected reduction of the precipitation over many regions in a non-stationary climate is a major concern for Europe. Climate change has the potential to increase drought risk by subjecting areas with all natural, social and economic consequences. Thus it is of great interest to analyze future drought severity projected from the climate models in order to elaborate effective mitigation strategies. In the presented work the tendencies of the total precipitation for two 30-year time periods, 2021–2050 and 2071–2100, simulated at the National Institute of Meteorology and Hydrology (NIMH-BAS) in the frame of the CECILIA project with ALADIN Climate model are used. Intending to find the precipitation sums in these periods, the tendencies are superposed over the averages for the World Meteorological Organization (WMO) reference period 1961–1990, obtained from the E-OBS gridded data set. The Standardized Precipitation Index (SPI) is selected to quantify the drought conditions and is calculated for three time-scales: SPI-3 for each season, SPI-6 for the cold and warm halves of the year and SPI-12 for the whole year. The study suggests that the magnitude of the prevailing negative tendency over the domain, which is better expressed in the second 30-year period, leads only to insignificant change of the resulting SPI distribution patterns.

Keywords: Project CECILIA · ALADIN climate
Climate projections · Precipitation change · SPI

1 Introduction

In the recent decade many studies are dedicated on the climate projections over Europe and the Mediterranean basin. The are performed with Atmosphere-Ocean General Circulation Models (AOGCMs), single or ensemble of AOGCMs-driven regional climate models, (RCMs). Despite of the overall agreement for general reduction of the precipitation amount in the middle and at the end of the 21st century, there are still many differences in the magnitude of the expected changes, annual and seasonal variability and areal distributions ([4,6,21]).

I. Lirkov and S. Margenov (Eds.): LSSC 2017, LNCS 10665, pp. 443–449, 2018.
https://doi.org/10.1007/978-3-319-73441-5_48

The Coupled Model Intercomparison Project Phase 5 (CMIP5) projections generally agree on warming in all seasons in Europe during this century, while precipitation projections are more variable across different parts of Europe and seasons. Central and Eastern Europe (C&EE) is a region where precipitation changes remain still uncertain. Even the findings of recent coordinated downscaling experiments in Europe, for example, projects PRUDENCE (Prediction of Regional Scenarios and Uncertainties for Defining European Climate Change Risks and Effects) [3] or ENSEMBLES (ENSEMBLE based predictions of climate changes and their impacts) [16] using RCM simulations of 25–50 km grid spacing are consistent with the CMIP5 projections and do not indicate any significant precipitation change over C&EE. Although regional climate change amplitudes of temperature and precipitation in Europe follow global trends, they can be also affected by changes in the large-scale circulation and regional feedback processes [15]. The comprehensive study of Stagge et al. [23] study makes use of the current RCMs from CORDEX (the Coordinated Regional Climate Downscaling Experiment, [7]) forced with CMIP5 climate projections to quantify the projected change in meteorological drought for Europe during XXI century. Meteorological drought is quantified using the SPI, which is calculated on a gridded scale at a grid spacing of 0.11° for the three projected emission pathways. Historical and future projections are analyzed with regard to mean, variance, frequency of moderate and severe droughts, the distribution of drought durations, number of drought events, and the maximum drought duration. Among others, the study reveals increasing projected drought throughout the Mediterranean, including the eastern Mediterranean. Similar approach is applied in the newly work of Osuch et al. [22]. Potential future trends in the SPI over the period 1971–2099 have been analyzed using a modified Mann-Kendall test [14,18] applied to precipitation time series derived from six ENSEMBLE RCM projections for the territory of Poland. Projections of SPI values indicate a decrease in the degree of dryness (better water availability) during the winter months and an increase in the summer period (more water scarcity) that confirm findings in previous studies. Hence only tendencies obtained an annual basis, rather than time series are available in our study, the analysis is focused only over the SPI in the selected time scales, caused by the projected precipitation change and this is the main difference in comparison with the both commented works. Actually this is a serious limitation, because, from one hand, these averages smooth over a lot of important information such as the one that characterizes the behavior of extreme SPI-values, which are usually responsible for impacts and from other, the absence of time series hampers any statistical analysis, in particular trend tests. Nevertheless such study is necessary as a step forward of obtaining overall picture of the drought conditions till the end of the century and this is the motivation of the work.

The paper is organized as follows. The methodological aspects of the study are described in the Sect. 2. The performed computations and the results are placed and commented in the Sect. 3. The concluding remarks are in the last, Sect. 4.

2 Methodology

The climate simulations with model system ALADIN Climate with focus on average precipitation and 2 m temperature were performed in the frame of the project CECILIA (Central and Eastern Europe Climate Change Impact and Vulnerability Assessment, http://www.cecilia-eu.org/), supported by European Commission's 6th Framework Programme, which key objectives were the climate change impacts and vulnerability assessment [9,12,13]. The used version of ALADIN-Climate corresponds to cycle 24 of the ARPEGE/IFS code, which physical parameterization package is derived directly from the one used in GCM ARPEGE-CLIMAT 4 [1]. The model was driven by the boundary condition of 50 km horizontal resolution coming from a "stretch mesh" version of ARPEGE-CLIMAT 4 GCM. This version of the GCM has a variable horizontal resolution being around 50 km over Southern Europe and decreasing to approximately 300 km at the antipode. The runs are performed for three time slices: 1961–1990 (present climate control run, CTL), 2021–2050 (near future run, NF), and 2071–2100 (far future run, FF) an annual basis and in grid resolution of 10 km. All future simulations were carried out using CO_2 concentrations as described by the IPCC A1B scenario [20]. In order to eliminate the systematic bias, the CTL is subtracted from the NF and FF, producing the tendencies for these two time slices.

The model domain is determined by the ALADIN climate simulations - it is centered over Bulgaria, but accommodates significant part of the Balkan peninsula and thus the performed work can be described as local to regional climate study. The SPI is a meteorological drought index, which was developed by McKee et al. [19] for monitoring drought conditions based on the conversion of precipitation data to probabilities using long-term precipitation records computed for different time scales. It normalizes precipitation anomalies over several months relative to a reference climate for a given location and time window. In this case, the SPI was normalized to the period from January 1971 to December 2010, which is strongly recommended by the Water Scarcity and Drought Expert Group. The SPI has been recommended as a key drought indicator by the World Meteorological Organization (WMO) [24] (WMO, 2006), the Lincoln Declaration on Drought [10], the DROUGHT R&SPI project [8] and, consequently, widely used worldwide in both research and operational modes (for review see [2,17]). Recently, Chervenkov et al. [2] have presented four centennial-long global gridded datasets of the SPI obtained from the authors from four different data sources for the most frequently used time windows of 1, 3, 6, and 12 months, noted traditionally as SPI-1, SPI-3, SPI-6, and SPI-12. These datasets are available free of charge at ftp://xeo.cfd.meteo.bg/SPI/ in standard netCDF format.

The well known and widely used in the climatological community gridded data-base E-OBS [11] is used for representative of the current climate. The high quality, relatively long time span (up to 65 years on daily basis) and high spatial resolution ($0.25° \times 0.25°$) makes this data-set very suitable for regional studies like the one presented in this paper.

3 Calculations and Results

First, the available on daily basis E-OBS data for the precipitation are preprocessed, obtaining the seasonal (winter: DJF, spring: MAM, summer: JJA, autumn: SON), warm (AMJJAS), cold (ONDJFM) half-year averages for the WMO-reference period 1961–1990, as well as this for the whole year. Second, the precipitation obtained with ALADIN climate tendencies for the NF and FF are mapped onto the coarser E-OBS grid. The annual tendencies are timely allocated for the aforementioned time-windows, using the proposed in [5] technique. The tendencies for each of the considered time-windows are added to these in the reference period, estimating in such a fashion the total precipitation for the both time slices.

The SPI-3, SPI-6, and SPI-12 for the considerd time-windows are calculated following the classical approach, proposed in [19]. This procedure is an equiprobability transformation and more details about the theoretical background behind the SPI can be found, for example, in [17]. The strength of the drought anomaly is classified as follows: −2 or less - extreme drought; −1.50 to −1.99 - severe drought; −1.00 to −1.49 - moderate drought; 0 to −0.99 - mild drought; 0 to 0.99 - mildly wet; 1.00 to 1.49 moderately wet; 1.50 to 1.99 severely wet; 2.00 or more - extremely wet.

The results are shown on Figs. 1 and 2.

Generally, the simulations reveals overall pattern of total precipitation, which, however, are relatively small and uneven distributed. As a whole, the climate change signal is stronger in the FF and in the SE part of Bulgaria.

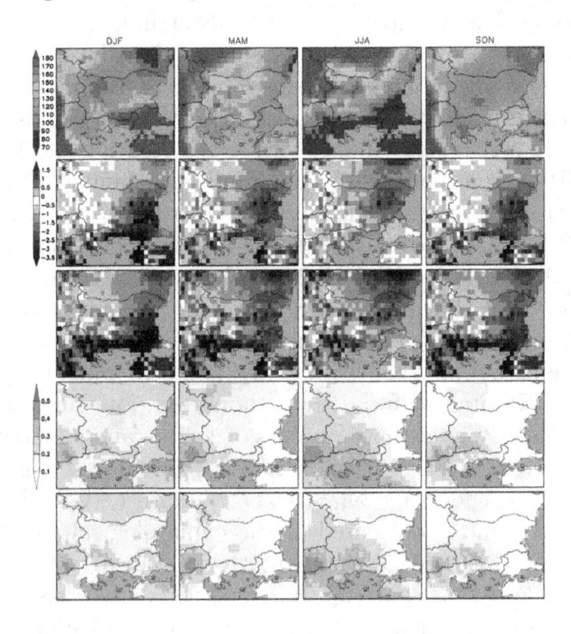

Fig. 1. Precipitation averages (unit: mm, upper legend) for the reference period on the first row, tendencies (unit: mm, midle legend) for the near on the second and far on the third row; SPI-3 (lower legend) for the near on the fourth and the far on the fifth row

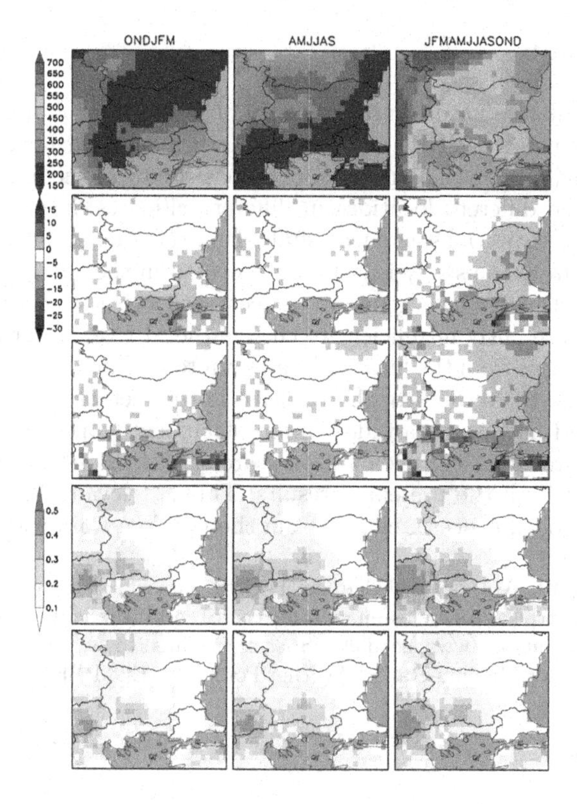

Fig. 2. Same as Fig. 1, but for the cold and warm half-year, as well as for the whole year

No drastic differences between the precipitation change fields for the seasons and half-years can be detected for the both time slices.

According the pioneering work [19], a drought event for each time scale is defined as a period in which the SPI is continuously negative and the SPI reaches a value of -1.0 or less. The drought begins when the SPI first falls below zero and ends with the positive value of SPI following a value of -1.0 or less. The SPI is standardized, both in time and space and is uniquely related to probability. The SPI is also normally distributed so it can be used to monitor wet as well as dry periods. Changes of the precipitation sum with one standard deviation σ causes change of the SPI with one unit. Consequently, if the simulated precipitation change is far below this threshold, as in the ALADIN climate output, is reasonable to expect that the SPI will remain in the same classification category. Nevertheless however, the explicit computation of the SPI is necessary in order to estimate the projected change more precisely.

The last two rows of the both figures confirms this expectation. First, the SPI is influenced relatively small. Even for the FF, were the precipitation change signal is stronger, the change generally is in order of tens. Second and most important, this change can not cause bridging of the SPI-threshold of 0, which, as pointed above, is associated with drought.

4 Conclusion

The probability-based nature (probability of observed precipitation transformed into an index) makes the SPI well suited to risk management and triggers for decision-making. The spatial and temporal consistency of the index allows comparisons between different locations in different climates, in particular estimations for climate projections, as successfully demonstrated in the presented work. Main outcome from the study is that the expected in the two time periods precipitation reductions, which generally agrees with other objective estimations (see [1] again), are fairly insufficient to change the SPI values for all time-scales and both time slices from the 'near normal' even to the 'mild drought' bin. Thus, strictly speaking, the term 'drought', at least in the meteorological sense, would be incorrect and have to be avoided in any document, considering this climate projection. Expressions as 'precipitation scarcity/reduction/shortage' are much more consistent with the presented results. This is relevant, keeping always in mind the high social concern, caused from the problem worldwide.

Acknowledgment. We acknowledge the E-OBS dataset from the EU-FP6 project ENSEMBLES (http://ensembles-eu.metoffice.com) and the data providers in the ECA&D project http://www.ecad.eu. Special thanks to I. Tsonevsky from the European Centre for Medium-Range Weather Forecasts (ECMWF) for the suggestions and his help.

References

1. Belda, M., Skalák, P., Farda, A., Halenka, T., Déqué, M., Csima, G., Bartholy, J., Torma, C., Boroneant, C., Caian, M., Spiridonov, V.: CECILIA regional climate simulations for future climate: analysis of climate change signal. Adv. Meteorol. **2015**, 1–13 (2015). https://doi.org/10.1155/2015/354727. Article ID 354727
2. Chervenkov, H., Tsonevsky, I., Slavov K.: Presentation of four centennial-long global gridded datasets of the standardized precipitation index. Int. J. Environ. Agric. Res. (IJOEAR) **2**(3), ISSN 2454-1850 (2016)
3. Christensen, J.H., Carter, T.R., Rummukainen, M., Amanatidis, G.: Evaluating the performance and utility of regional climate models: the PRUDENCE project. Clim. Change **81**(1), 16 (2007)
4. Dai, A.: Increasing drought under global warming in observations and models. Nat. Clim. Change **3**, 52–58 (2013)
5. Dèquè, M.: Frequency of precipitation and temperature extremes over France in an anthropogenic scenario: model results and statistical correction according to observed values. Glob. Planet. Change **57**(1–2), 16–26 (2007)
6. Giorgi, F., Lionello, P.: Climate change projections for the Mediterranean region. Glob. Planet. Change **63**(2–3), 90–104 (2008)
7. Giorgi, F., Gutowski Jr., W.J.: Regional dynamical downscaling and the CORDEX initiative. Annu. Rev. Environ. Resour. **40**, 467–490 (2015)
8. Gudmundsson, L., Stagge, J.H.: SCI: Standardized Climate Indices such as SPI, SRI or SPEI. R package version 1.0.1 (2014)
9. Halenka, T.: Regional climate modeling activities in CECILIA project introduction. Idöjárás **112**, 3–9 (2008)

10. Hayes, M., Svoboda, M., Wall, N., Widhalm, M.: The Lincoln declaration on drought indices: universal meteorological drought index recommended. Bull. Am. Meteorol. Soc. **92**, 485–488 (2011)
11. Haylock, M.R., Hofstra, N., Klein Tank, A.M.G., Klok, E.J., Jones, P.D., New, M.: A European daily high resolution gridded dataset of surface temperature and precipitation. J. Geophys. Res. (Atmospheres) **113**, D20119 (2008). https://doi.org/10.1029/2008JD10201
12. Huszar, P., Juda-Rezler, K., Halenka, T., Chervenkov, H., Syrakov, D., Krüger, B.C., Zanis, P., Melas, D., Katragkou, E., Reizer, M., Trapp, W., Belda, M.: Effects of climate change on ozone and particulate matter over Central and Eastern Europe. Clim. Res. **50**(1), 51–68 (2011)
13. Juda-Rezler, K., Reizer, M., Huszar, P., Krüger, B.C., Zanis, P., Syrakov, D., Katragkou, E., Trapp, W., Melas, D., Chervenkov, H., Tegoulias, I., Halenka, T.: Modelling the effects of climate change on air quality over Central and Eastern Europe: concept, evaluation and projections. Clim. Res. **53**(3), 179–203 (2012)
14. Kendall, M.G., Stuart, A.: The Advanced Theory of Statistics. Distribution Theory, vol. I. Griffin, London (1976)
15. Kjellstrom, E., Nikulin, G., Hansson, U., Strandberg, G., Ullerstig, A.: 21st century changes in the European climate: uncertainties derived from an ensemble of regional climate model simulations. Tellus, Ser. A: Dyn. Meteorol. Oceanogr. **63**(1), 24–40 (2011)
16. van der Linden, P., Mitchell, J.F.B. (eds.): ENSEMBLES: Climate Change and Its Impacts: Summary of Research and Results from the ENSEMBLES Project. Met, Office Hadley Centre, London, UK (2009)
17. Lloyd-Hughes, B., Saunders, M.A.: A drought climatology for Europe. Int. J. Climatol. **22**, 1571–1592 (2002)
18. Mann, H.B.: Nonparametric tests against trend. Econometrica **13**, 245–259 (1945)
19. McKee, T.B., Doesken, N.J., Kleist, J.: The relationship of drought frequency and duration to time scales. In: Proceedings of the 8th Conference on Applied Climatology, 17–22 January, Anaheim, CA. American Meteorological Society: Boston, MA pp. 179–184 (1993)
20. Nakicenovic, N.: Intergovernmental Panel on Climate Change: Emission Scenarios, a Special Report of Working Group III of the Intergovernmental Panel on Climate Change. Cambridge University Press, Cambridge (2000)
21. Orlowsky, B., Seneviratne, S.I.: Elusive drought: uncertainty in observed trends and short- and long-term CMIP5 projections. Hydrol. Earth Syst. Sci. **17**(5), 1765–1781 (2013)
22. Osuch, M., Romanowicz, R.J., Lawrence, D., Wong, W.K.: Trends in projections of standardized precipitation indices in a future climate in Poland. Hydrol. Earth Syst. Sci. **20**, 1947–1969 (2016)
23. Stagge, J.H., Rizzi, J., Tallaksen, L.M., Stahl, K.: Future meteorological drought: projections of regional climate models for Europe, Technical report No. 25, Future Meteorological Drought Projections of Regional Climate, DROUGHT-RSPI Project, University of Oslo, Oslo, 23 p. (2015)
24. WMO Drought monitoring and early warning: concepts, progress and future challenges. WMO-No. 1006, World Meteorological Organization. Geneva, Switzerland (2006)

Time Discretization/Linearization Approach Based on HOC Difference Schemes for Semilinear Parabolic Systems of Atmosphere Modelling

I. Dimov[1], J. Kandilarov[2(✉)], V. Todorov[1], and L. Vulkov[2]

[1] Institute of Information and Communication Technologies, BAS, Sofia, Bulgaria
ivdimov@bas.bg, venelintodorov@fmi.uni-sofia.bg
[2] Department of Mathematics, University of Ruse, Ruse, Bulgaria
{ukandilarov,lvalkov}@uni-ruse.bg

Abstract. We implement *implicit-explicit (IMEX)* linear multistep time-discretization to HOC difference schemes for weakly coupled nonlinear parabolic systems with desirable time-step restrictions and positivity preservation of the numerical solution. Numerical experiments are performed with IMEX-BDF1 (backward difference method of order one), IMEX-BDF2 (backward difference method of order two) and CN-LF (Crank-Nicolson Leap Frog) to check the properties of the methods.

1 Introduction

In our recent papers [2,3], we proposed *high-order compact (HOC)* difference schemes for semilinear parabolic systems. Our main interest are problems of atmosphere modelling with coupling in the nonlinear reaction terms [1,4],

$$\partial u_l/\partial t \;-\; K_l \triangle u_l \;+\; \mathbf{b}_l.\nabla u_l \;+\; R_l(x,y,u_1,\ldots,u_L) = f_l(x,y,t), \quad u_l|_{\partial\Omega\times R^+} = 0, \tag{1}$$

where $u_l = u_l(x,y,t)$, $l = 1,...,L$, are concentrations of L chemical species (pollutants), $K_l > 0$ are diffusion coefficients, R_l are nonlinear reaction terms, while $f_l(x,y,t)$ are source terms. We pose initial and boundary conditions

$$u_l = 0, \quad (x,y,t) \in \partial\Omega \times (0,T], \tag{2}$$

$$u_l = u_l^0(x,y), \qquad (x,y) \in \Omega, \quad l = 1,...,L. \tag{3}$$

With time-splitting by the fractional step method one has to solve subproblems that often are not consistent with the full models [4,5,8,10]. In the time-splitting approach multi-schemes cannot be used in a natural fashion, either. In this communication some *alternatives to time splitting* for air pollution models will be discussed.

We have used in [2,3] the *Crank-Nicolson/Newton* (CNN) scheme for solving the systems of ODEs arising after the space discretizations of the

I. Lirkov and S. Margenov (Eds.): LSSC 2017, LNCS 10665, pp. 450–457, 2018.
https://doi.org/10.1007/978-3-319-73441-5_49

system (1)–(3). To overcome this difficulty we implement IMEX linear multistep time-discretization to HOC schemes of (1) with desirable time-step restrictions and positivity preservation of the numerical solution. Numerical experiments are performed with IMEX-BDF1, IMEX-BDF2, and CN-LF to check the properties of the methods.

The rest of the paper is organized as following. In Sect. 2, we briefly describe HOC schemes while four time discretization methods are introduced in Sect. 3. In Sect. 4 on a 3-components practical problem [6,9] we compare the accuracy and efficiency between the four (including CNN) time-stepping methods. We conclude the paper in Sect. 5 by some conclusions.

2 Space Discretization

Let for simplicity the domain Ω be a rectangle $\Omega = [0, X] \times [0, Y]$. We introduce uniform meshes in the following way: $\overline{\omega}_{h,x} = \{x_i = ih_x,\ \ i = 0, 1, \dots, M_x,\ \ h_x = X/M_x\}$, $\overline{\omega}_{h,y} = \{y_j = jh_y,\ \ j = 0, 1, \dots, M_y,\ \ h_y = Y/M_y\}$ and then $\overline{\Omega}_h = \overline{\omega}_{h,x} \times \overline{\omega}_{h,y}$, $\overline{\Omega}_h = \Omega_h \cup \partial\Omega_h$, where Ω_h consist of all interior mesh points and $\partial\Omega_h$ - of all boundary mesh points.

We will use the index pair (i, j) to represent the mesh point (x_i, y_j) and define

$$u_{l,ij} = u_l(x_i, y_j, t), \quad R_{l,ij} = R_l(x_i, y_j, t, \mathbf{u}_{i,j}), \quad \text{etc.}$$

We order the mesh points lexicographically from left to right in x direction and from the bottom to the top in y direction. Excluding the boundary mesh points $(i, j) \in \partial\Omega_h$, for $j = 1, 2, ..., M_y - 1$ we define the following $(M_x - 1)$ dimensional vectors:

$$\mathbf{U}^h_{l,j} = \left(u^h_{l,1j}, u^h_{l,2j}, ..., u^h_{l,M_x-1j}\right), \qquad \mathbf{R}_{l,j}(\mathbf{U}^h_{1,j}, ..., \mathbf{U}^h_{L,j}) = \left(R_{l,1j}, ..., R_{l,M_x-1\,j}\right),$$
$$\mathbf{U}_l = \left(\mathbf{U}^h_{l,1}, \mathbf{U}^h_{l,2}, ..., \mathbf{U}^h_{l,M_y-1}\right)^T, \qquad \mathbf{R}_l = \left(\mathbf{R}_{l,1}, \mathbf{R}_{l,2}, ..., \mathbf{R}_{l,M_y-1}\right)^T.$$

For semidiscretization in space we use two difference schemes: central difference scheme (CDS) on a 5-point stencil and compact finite difference scheme (CFDS) on a 9-point stencil, proposed in details in [3]. After the application of CDS we obtain a system of ordinary differential equations

$$\frac{d}{dt}\mathbf{U}_l + P_l\mathbf{U}_l = \mathbf{R}_l + \mathbf{\Phi}_l, \quad t \in (0, T], \tag{4}$$

with initial conditions $\mathbf{U}_l(0)$ obtained from u^0_l for $(i, j) \in \Omega_h$ after the reordering. In (4) the matrix P_l is $(M_y - 1) \times (M_y - 1)$ block-tridiagonal matrix $P_l = tridiag(P_{l,k,k-1}, P_{l,k,k}, P_{l,k,k+1})$ and $P_{l,k,s}$, $s = k - 1, k, k + 1$ are tridiagonal matrices for $s = k$ and diagonal for $s = k \pm 1$ of order $(M_x - 1) \times (M_x - 1)$. Let for two natural numbers m and M, $m < M$ denote $m : M = m, m + 1, ..., M$ and assume that $\mathbf{p}_{k,m:M}$ is a vector with entries $\mathbf{p}_{k,m:M} = (p_{k,m}, p_{k,m+1}, ..., p_{k,M})$. The entries of $P_{l,k,s}$ are

$$P_{l,k,s} = tridiag(\mathbf{p}^{(-1,\varepsilon)}_{l,k,2:M_x-1}, \mathbf{p}^{(0,\varepsilon)}_{l,k,2:M_x}, \mathbf{p}^{(1,\varepsilon)}_{l,k,1:M_x-2}) \qquad s = k + \varepsilon,\ \ \varepsilon = 0, \pm 1, \tag{5}$$

where

$$p_{l,i,j}^{(\pm 1,0)} \pm \frac{b_{l,1}}{2h_x} - \frac{K_l}{h_x^2}, \quad p_{l,i,j}^{(0,\pm 1)} \pm \frac{b_{l,2}}{2h_y} - \frac{K_l}{h_y^2}, \quad p_{l,i,j}^{(0,0)} 2\frac{K_l}{h_x^2} + 2\frac{K_l}{h_y^2}. \tag{6}$$

The vectors Φ_l in (4) are associated with the boundary functions and also depend on time t.

Similarly, by the CFDS we obtain the following system of PDEs

$$Q_l \frac{d}{dt}\mathbf{U}_l + P_l\mathbf{U}_l = Q_l\mathbf{R}_l + \mathbf{\Phi}_l, \quad t \in (0,T], \tag{7}$$

with initial conditions $\mathbf{U}_l(0)$ obtained from u_l^0 for $(i,j) \in \Omega_h$ after the reordering. In the system (7) the matrix P_l is $(M_y-1) \times (M_y-1)$ block-tridiagonal matrix $P_l = tridiag(P_{l,k,k-1}, P_{l,k,k}, P_{l,k,k+1})$ and $P_{l,k,s}$, $s = k-1, k, k+1$ are also tridiagonal matrices of order $(M_x-1) \times (M_x-1)$. The entries of $P_{l,k,s}$ are

$$\bar{P}_{l,k,s} = tridiag(p_{l,k,2:M_x-1}^{(-1,\varepsilon)}, p_{l,k,2:M_x}^{(0,\varepsilon)}, p_{l,k,1:M_x-2}^{(1,\varepsilon)}) \qquad s = k+\varepsilon, \;\; \varepsilon = 0, \pm 1\,. \tag{8}$$

Similarly, the matrix Q_l is $(M_y-1) \times (M_y-1)$ block-tridiagonal matrix $Q_l = tridiag(Q_{l,k,k-1}, Q_{l,k,k}, Q_{l,k,k+1})$ and the entries of $Q_{l,k,s}$ are

$$Q_{l,k,s} = tridiag(q_{l,k,2:M_x-1}^{(-1,\varepsilon)},\; q_{l,k,2:M_x}^{(0,\varepsilon)},\; q_{l,k,1:M_x-2}^{(1,\varepsilon)}) \qquad s = k+\varepsilon, \;\; \varepsilon = 0, \pm 1 \tag{9}$$

with a remark that for $\varepsilon = \pm 1$ matrices $Q_{l,k,s}$ are diagonal (instead tridiagonal) matrices. The entries of P_l and Q_l are:

$$\begin{aligned}
p_{l,i,j}^{(\pm 1,-1)} &= \frac{1+\sigma^2}{\sigma}\left(-\frac{\sigma}{2}K_l \pm \frac{1}{4}(\sigma b_{l,1} \mp b_{l,2})h_x \mp \frac{b_{l,1}b_{l,2}}{8}h_x^2\right) \\
p_{i,j}^{(\pm 1,1)} &= \frac{1+\sigma^2}{\sigma}\left(-\frac{\sigma}{2}K_l \pm \frac{1}{4}(\sigma b_{l,1} \mp b_{l,2})h_x \pm \frac{b_{l,1}b_{l,2}}{8}h_x^2\right) \\
p_{i,j}^{(\pm 1,0)} &= (\sigma^2 - 5)K_l \pm \frac{5\sigma^2 - 1}{2\sigma}b_{l,2}h_x - \frac{b_{l,1}^2}{2K_l}h_x^2 \\
p_{i,j}^{(0,\pm 1)} &= (1 - 5\sigma^2)K_l \pm \frac{5-\sigma^2}{2}b_{l,1}h_x - \frac{b_{l,2}^2}{2K_l}h_x^2 \\
p_{i,j}^{(0,0)} &= 10(1+\sigma^2)K_l + \frac{b_{l,1}^2 + b_{l,2}^2}{K_l}h_x^2
\end{aligned} \tag{10}$$

and

$$\begin{aligned}
&q_{i,j}^{(\pm 1,\pm 1)} = 0,\; q_{i,j}^{(\pm 1,0)} = \tfrac{1}{4}(2 \mp \tfrac{b_{l,1}}{K_l}h_x)h_x^2, \\
&q_{i,j}^{(0,\pm 1)} = \tfrac{1}{4}(2 \mp \tfrac{b_{l,2}}{\sigma K_l}h_x)h_x^2,\; q_{i,j}^{(0,0)} = 4h_x^2.
\end{aligned} \tag{11}$$

3 Time Discretization

We rewrite the ODE systems (4) and (7) in the following common form

$$H_l \frac{\partial \mathbf{U}_l}{\partial t} + P_l\mathbf{U}_l = H_l\mathbf{R}_l(\mathbf{U}_1, ..., \mathbf{U}_L) + \mathbf{\Phi}_l, \quad l = 1, ..., L, \tag{12}$$

where

$$H_l = \begin{cases} I - \text{identity matrix}, & \text{for CDS}, \\ Q_l, & \text{for CFDS}. \end{cases} \tag{13}$$

Let $\Omega_\tau = \{t_j = j\tau,\ j = 0, 1, \ldots, N,\ \tau = T/N\}$ be uniform in time mesh. For the full discretization we consider the following variants.

- *Weight θ-discretization (θ - time-stepping)*

$$H_l \frac{\mathbf{U}_l^{j+1} - \mathbf{U}_l^j}{\tau} + P_l \mathbf{U}_l^{j,\theta} = H_l \mathbf{R}_l^{j,\theta}(\mathbf{U}_1, ..., \mathbf{U}_L) + \Phi_l^{j,\theta}, \tag{14}$$

  where $W^{j,\theta} = \theta W^{j+1} + (1-\theta) W^j$, $0 \le \theta \le 1$ for $W = \mathbf{U}_l, \mathbf{R}_l, \mathbf{\Phi}_l$.
- *IMEX - BDF1 (IMEX backward difference method of order one)*

$$H_l \frac{\mathbf{U}_l^{j+1} - \mathbf{U}_l^j}{\tau} + P_l \mathbf{U}_l^{j+1} = H_l \mathbf{R}_l(\mathbf{U}_1^j, ..., \mathbf{U}_L^j) + \mathbf{\Phi}_l^{j+1}. \tag{15}$$

- *IMEX - BDF2 (IMEX backward difference method of second order)*

$$H_l \frac{3/2\mathbf{U}_l^{j+1} - 2\mathbf{U}_l^j + 1/2\mathbf{U}_l^{j-1}}{\tau} + P_l \mathbf{U}_l^{j+1} =$$
$$2\left((H_l \mathbf{R}_l(\mathbf{U}_1^j, ..., \mathbf{U}_L^j) + \mathbf{\Phi}_l^j\right) - \left(H_l \mathbf{R}_l(\mathbf{U}_1^{j-1}, ..., \mathbf{U}_L^{j-1}) + \mathbf{\Phi}_l^{j-1}\right).$$

- *CN-LF (Crank-Nicolson Leap Frog)*

$$H_l \frac{\mathbf{U}^{j+1} - \mathbf{U}^{j-1}}{2\tau} + P_l \frac{\mathbf{U}^{j+1} + \mathbf{U}^{j-1}}{2} = (H_l \mathbf{R}_l(\mathbf{U}_1^j, ..., \mathbf{U}_L^j) + \mathbf{\Phi}_l^j.$$

One can show that IMEX-BDF2 and CN-LF are stable and second order accurate in time, whereas IMEX-BDF1 is stable but only first order accurate.

4 Numerical Study

We concentrate on the system (1)–(3) for $L = 3$ with coefficients, reaction and source terms that correspond to the atmosphere model based on Chapman's cycle, see e.g. [8,9]. While a realistic atmospheric/air-pollution model main contain dozens of reacting species [4,5], our simple model capture the basic features of the complete practical models and the methods developed in the paper are already implemented to the model of the equations solved in [5,7].

The components of the system are the nitrogen oxide (NO), nitrogen dioxide (NO_2) and ozone (O_3) denoted by u_1, u_2, u_3 respectively:

$$R_l(\mathbf{u}) = -r(\mathbf{u}),\ \ l = 1, 3,\ \ R_2(\mathbf{u}) = r(\mathbf{u}),\ \ r(\mathbf{u}) = k_1 u_1 u_3 - k_2 u_2,$$

where k_1, k_2 are the forward and backward reaction rates.

Example 1. Exact analytical solution. Here we consider a problem slightly different from the problem (1)–(3):

$$\frac{\partial u_l}{\partial t} - K\Delta u_l + \mathbf{b}_l.\nabla u_l = R_l(x,y,\mathbf{u}) + \xi_l(x,y,t), (x,y,t) \in \Omega \times (0,T]. \quad (16)$$

The functions ξ_l, $l = 1,2,3$, and the initial and boundary conditions are chosen so that the exact solution is

$$\begin{aligned} u_l &= \exp(-t)sin(\pi x)sin(\pi y), \quad l = 1,2, \quad (x,y,t) \in \overline{\Omega} \times [0,T], \\ u_3 &= 1 + \exp(-t)sin(\pi x)sin(\pi y), \quad (x,y,t) \in \overline{\Omega} \times [0,T]. \end{aligned}$$

The other parameters are as follows: $\overline{\Omega} = [0,1] \times [0,1]$, $T = 1$, $b_l = (0.1, 0.1)$, for $l = 1,2,3$, $K_1 = 1$, $K_2 = K_3 = 5$.

For the lth substance with $error_{M,l}$ we denote the error (the difference between the exact and the numerical solution) in maximum norm, obtained on the last time layer $t_N = T$ for the number of space subintervals $M_x = M_y = M$ by

$$error_{M,l} = \max_{i,j \in \bar{\Omega}_h} \| u_l(x_i, y_j, t_N) - u_l^h(i,j,N) \|.$$

The ratio between the errors obtained on two consecutive mesh refinements (usually doubling) is denoted by *ratio*:

$$ratio = ratio_{M,l/2M,l} =: error_{M,l}/error_{2M,l}.$$

In Table 1 the results of mesh refinement using CDS and CFDS combined with IMEX-BDF1 for $k_1 = 1, k_2 = 2$ and in Table 2 for $k_1 = 1000, k_2 = 2000$ are presented. The IMEX-BDF1 method is of first order with respect to time, so to observe the second order, when doubling the number of mesh points in space one must take quadruple mesh points in time. The numerical results confirm the theoretical rate of convergence, i.e. the ratio near four confirm second order for the CDS. Unfortunately, CFDS with IMEX-BDF1 gives no rise of order, as it is when Newton method is used, but the results are more precise. The CPU time for both cases is equivalent. When $k_1 = 1000, k_2 = 2000$ we need smaller time steps, so we take $N = 4096$ in all experiments.

Table 1. The error in maximum norm of pollutant u_1 for *Example* 1, obtained by CDS and CFDS, combined with IMEX-BDF1, $k_1 = 1, k_2 = 2$

		CDS and IMEX-BDF1			CFDS and IMEX-BDF1		
M	N	Error	Ratio	CPU	Error	Ratio	CPU
4	16	5.0771e−02	-	0.99	3.9589e−04	-	0.84
8	64	1.2410e−02	4.09	4.75	2.8545e−04	1.39	4.59
16	256	3.0852e−03	4.02	44.04	8.3939e−05	3.40	48.58
32	1024	7.7153e−04	4.00	618.1	2.1783e−05	3.85	662.15
64	4096	2.4076e−04	3.20	9331	5.4959e−06	3.96	9219

Table 2. The error in maximum norm of pollutant u_1 for *Example* 1, obtained by CDS and CFDS, combined with IMEX-BDF1, $k_1 = 1000, k_2 = 2000$,

		CDS and IMEX-BDF1			CFDS and IMEX-BDF1		
M	N	Error	Order	CPU	Error	Order	CPU
4	4096	3.1992e−02	-	89.59	5.5178e−04	-	100.54
8	4096	8.6821e−03	3.68	216.83	5.0463e−05	10.93	231.66
16	4096	4.0316e−03	2.15	687.37	8.8275e−05	0.57	714.90
32	4096	1.2072e−03	3.34	1427	9.0648e−05	0.97	1930
64	4096	6.4502e−04	1.87	9445	9.0796e−05	0.99	9628

Table 3. The error in maximum norm for the particle u_1, obtained by CDS combined with the IMEX-BDF2 and CN-LF, compared with implicit CDS, $k_1 = 1$, $k_2 = 2$

		implicit CDS			IMEX-BDF2			CN-LF		
M_x	N	Error	Order	CPU	Error	Order	CPU	Error	Order	CPU
4	16	4.8985e−02	-	0.91	4.9012e−02	-	0.75	4.9001e−02	-	0.71
8	64	1.2037e−02	4.07	6.45	1.2037e−02	4.07	3.97	1.2037e−02	4.07	3.33
16	256	2.9958e−03	4.01	56.71	2.9958e−03	4.01	35.99	2.9958e−03	4.01	35.25
32	1024	7.4811e−04	4.00	596.76	7.4811e−04	4.00	163.95	7.4875e−04	4.00	449.58
64	4096	1.8698e−04	4.00	35240	1.8728e−04	4.00	6384	2.5245e−04	2.97	6438

In Table 3 the results obtained using CDS combined with Newton method, IMEX-BDF2 and CN-LF for $k_1 = 1, k_2 = 2$ are presented. It is obviously, that IMEX-BDF2 and CN-LF methods need less CPU time to obtain the errors of the same order as fully implicit scheme.

Table 4 shows results obtained by CFDS combined with Newton method, IMEX-BDF2 and CN-LF for $k_1 = 1, k_2 = 2$. As it is expected all the methods are of *fourth order* in space and *second order* in time. Again IMEX-BDF2 and CN-LF methods need less CPU time (approximately 6 times for smaller mesh sizes in space) to obtain the errors of the same accuracy as fully implicit scheme. The reason is that in the IMEX schemes the nonlinear terms are evaluated on the previous time layers, which make the algebraic system of equation to be solved on each time layer linear and we need no iterations, that is the case in the Newton method.

In Table 5 results using CFDS combined with Newton method (implicit method), IMEX-BDF2 and CN-LF for $k_1 = 1000, k_2 = 2000$ are presented. The increasing of the reaction terms lead to decreasing of the stability condition (τ/h^2). For this reason we use the smaller mesh size in time, i.e. $N = 4096$. The first two methods give good results and are of fourth order of accuracy in space. Again, the CPU time for the IMEX-BDF2 method is less than in comparison with the case of implicit CFDS. The CN-LF method gives no results (NaN - not a number) and needs of smaller mesh in time.

Table 4. The error in maximum norm for the particle u_1, obtained by CFDS combined with the IMEX-BDF2 and CN-LF, compared with implicit CFDS, $k_1 = 1$, $k_2 = 2$

		Implicit CDFS			CFDS and IMEX-BDF2			CFDS and CN-LF		
M	N	Error	Order	CPU	Error	Order	CPU	Error	Order	CPU
4	16	9.4510e−04	-	0.82	9.4559e−04	-	1.03	9.4540e−04	-	0.97
8	64	6.1079e−05	15.47	6.26	6.1081e−05	15.48	3.58	6.1080e−05	15.48	3.42
16	256	3.8451e−06	15.88	60.77	3.8450e−06	15.88	33.04	3.8451e−06	15.89	36.66
32	1024	2.4074e−07	15.97	882.5	2.4062e−07	15.98	462.1	2.4074e−07	15.97	469.9
64	4096	1.5053e−08	15.99	37477	2.7121e−08	8.88	6355	1.5063e−08	15.98	6242

Table 5. The error in maximum norm for the particle u_1, obtained by CFDS combined with the IMEX-BDF2 and CN-LF, compared with implicit CFDS, $k_1 = 1000, k_2 = 2000$

Mesh sizes		Implicit CFDS			IMEX BDF2			CN-LF		
M_x	N	Error	Order	CPU	Error	Order	CPU	Error	Order	CPU
4	4096	6.2389e−04	-	95.13	6.2390e−04	-	78.42	NaN	-	-
8	4096	4.0232e−05	15.51	299.58	4.0298e−05	15.48	175.07	NaN	-	-
16	4096	2.5355e−06	15.88	947.04	2.4767e−06	16.27	192.43	NaN	-	-
32	4096	1.5880e−07	15.97	2674	1.6280e−07	15.21	1728	NaN	-	-

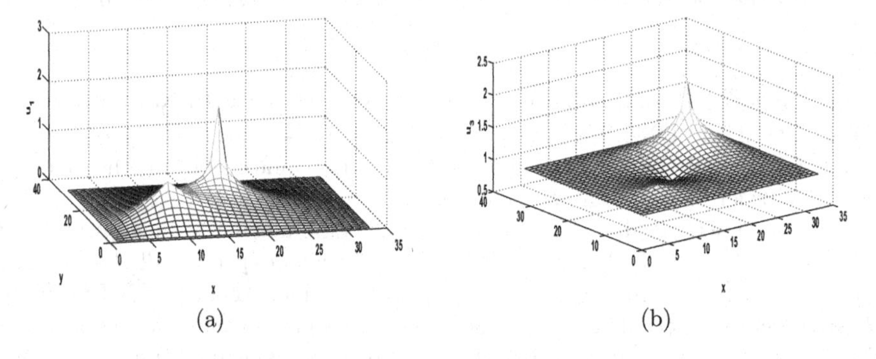

(a) (b)

Fig. 1. Numerical solution with mesh parameters $M_x = M_y = 32$, $N = 256$ for *Example* 2: (a) for NO - u_1; (b) for O_3 - u_3

Example 2. Problem with Delta - source terms. In this example we consider problem (16), where functions ξ_l now are point source terms of the form

$$\xi_l(x, y, t) = f_l(t)\delta(x - \overline{x}_l, y - \overline{y}_l), \quad l = 1, 2, 3.$$

The parameters are as follows: $\overline{\Omega} = [0, 1] \times [0, 1]$, $T = 1$, $b_l = (-0.1, 0)$, for $l = 1, 2, 3$, $K_1 = 1$, $K_2 = K_3 = 5$, $k_1 = 1000, k_2 = 2000$, $(\overline{x}_1, \overline{y}_1) = (0.5, 0.5)$, $(\overline{x}_2, \overline{y}_2) = (0.25, 0.25)$, $(\overline{x}_3, \overline{y}_3) = (0.75, 0.75)$, $f_1(t) = 7$, $f_2(t) = 11$, $f_3(t) = 13$. On Fig. 1 the numerical solution obtained by CFDS and IMEX-BDF2 with mesh parameters $M_x = M_y = 32$, $N = 256$ is presented: (a) for NO - u_1; (b) for O_3 - u_3. The influence of the point delta-sources is clearly seen.

5 Conclusion

We present here results of extensive numerical experiments that suggest that our four time-stepping methods perform well the HOC difference schemes for semilinear parabolic systems of atmosphere modeling. One future study is the analysis under which restrictions on the time and space mesh steps the full approximations conserve the positivity of the exact solutions. Also, the investigation of convergence is complicated because the luck of regularity of the solutions in the presence of concentrated sources.

Acknowledgements. This work was partially supported by the Bulgarian National Fund of Science under the grant DFNI I02-20/2014, as well as by the Program for career development of the Young scientists, BAS, Grant No. DFNP-91/04.05.2016 and by the Project 2017-FNSE-03 of the University of Ruse.

References

1. Bátkai, A., Csomós, P., Faragó, I., Horányi, A., Szépszó, G. (eds.): Mathematical Problems in Meteorological Modelling. MI, vol. 24. Springer, Cham (2016). https://doi.org/10.1007/978-3-319-40157-7
2. Dimov, I., Kandilarov, J., Todorov, V., Vulkov, L.: High-order compact difference schemes with Richardson extrapolation for semilinear parabolic systems. Applications of Mathematics in Engineering and Economics (AMEE 2016). AIP Conference Proceedings, vol. 1789, pp. 030002–030002-7 (2016). https://doi.org/10.1063/1.4968448
3. Dimov, I., Kandilarov, J., Todorov, V., Vulkov, L.: A comparison study of two high accuracy numerical methods for a parabolic system in air pollution modelling. arXiv:1701.03049
4. Dimov, I., Zlatev, Z.: Computational and Numerical Challenges in Air Polution Modelling. Elsevier Science, Amsterdam (2006)
5. Georgiev, K., Zlatev, Z.: Implementation of sparse matrix algorithms in an advection-diffusion-chemistry model. J. Comp. Appl. Math. **236**(3), 342–353 (2011)
6. Jacobsen, D.: Fundamental of Atmospheric Modelling. Cambridge University Press, Cambridge (2005)
7. Karatson, J., Kurics, T.: A preconditioned iterative solution schem for nonlinear parabolic systems arizing in air pollution modeling. Math. Modell. Anal. **18**(5), 641–653 (2013)
8. Kim, J., Cho, S.: Computation accuracy and efficiency of the time splitting method. J. Atmosph. Envir. **31**(15), 2215–2224 (1997)
9. Mamonov, A.V., Tsai, Y.-H.R.: Point source identification in non-linear advection-diffusion-reaction systems. Inverse Prob. **29**(3), 035009 (2012)
10. Ostromsky, T., Alexandrov, V., Dimov, I., Zlatev, Z.: On the performance, scalability and sensitivity analysis of a large air pollution model. Procedia Comput. Sci. **80**, 2053–2061 (2016)

Landslide Hazard, Environmental Dependencies and Computer Simulations

Nina Dobrinkova[1(✉)] and Pierluigi Maponi[2]

[1] Institute of Information and Communication Technologies, Bulgarian Academy of Sciences, 1113 Sofia, Bulgaria
ninabox2002@gmail.com

[2] School of Science and Technology, University of Camerino, 62032 Camerino, MC, Italy

Abstract. Shallow landslides triggered by weather conditions are a major hazard in most mountainous and hilly regions of Europe. An efficient system for the evaluation of landslide hazard level from weather forecast data allows the possibility to develop fast alert systems and well-organized crisis management plans. A system of this kind should estimate the effect of weather on the slope stability and translate this estimation in a easy to use hazard map. So, its main components are a model for the dynamics of soil moisture, a model for the slope stability analysis, and a proper information system able to manage the corresponding data flow. The system presented in this paper has been developed during the LANDSLIDE project and it is available at web site http://93.123.110.111/landslide/.

Keywords: Soil moisture dynamics · Landslide hazard evaluation
Early alert system

1 Introduction

Landslides are a widespread hazard in many mountainous and hilly regions. They cause significant losses as well as human victims. This is the most frequent trigger of landslides occurrence in Europe [1], and is expected to increase in the future due to climate change. Catastrophic landslides are widely distributed throughout Europe, however, with a great concentration in mountainous areas. In the last 20 years (1995–2014), they are responsible for a total of 1370 deaths and average economic loss per year is approximately 4.7 billion Euros [2]. In the United States, it was estimated in 1985 that the total dollar losses from landslides average between $1 billion and $2 billion per year. (As this is an old estimate, a more realistic number is between $2 billion and $4 billion per year, computed for year 2010 dollars) [3].

Sliding surfaces deeply located below the maximum rooting depth of trees usually involve deep regolith, weathered rock, and/or bedrock. Deep landslides normally include large slope failure and are characterized by a slowly moving mass.

I. Lirkov and S. Margenov (Eds.): LSSC 2017, LNCS 10665, pp. 458–465, 2018.
https://doi.org/10.1007/978-3-319-73441-5_50

On the other hand, a sliding surface located within the soil biomantle is called a shallow landslide. A number of predisposing factors can increase the landslide susceptibility of a slope. However, a quite usual situation is provided by slopes with high permeable soils on top and with low permeable bottom soils. Slopes of this kind can become very unstable when the top soils are filled with water like during an intense rainstorm.

Methods for landslide hazard evaluation are today mainly based on scientific literature of geomorphologic studies and of historical landslide events, see [5, 6] and the references therein for an example; these studies do not consider or underestimate the impact of climate change. Therefore, it is important that an interdisciplinary approach for landslide monitoring and prediction between the nowadays ICT (Information and Communications Technology) solutions and landslide experts has to start. This cooperation is useful in order to provide new tools that can adapt to the new conditions by correctly evaluating and predicting landslide hazards which is a fundamental prerequisite for accurate risk mapping and assessment and for the consequent implementation of appropriate prevention measures.

In particular, for shallow landslide, a physical approach allows the evaluation of the hazard level by the weather forecast data. In fact, weather forecast can be used to predict the soil moisture, that in turn can be used to evaluate landslide hazard by a slope stability analysis like Limit Equilibrium Analysis. This is a simple conceptual approach, but its practical implementation is not trivial due to the complexity of the soil system.

The above mentioned approach has been realized by the project LANDSLIDE risk assessment model for disaster prevention and mitigation (acronym: LANDSLIDE). The project consortia had jointly developed an information system [4] for landslide hazard assessment that has been tested in four hydrographic basins selected as test sites.

2 Mathematical Background and Used Formulas

The LANDSLIDE information system is a continuous-time tool for the automatic evaluation of landslide hazard from weather conditions. So, the main components in this system is the soil moisture dynamics and the slope stability analysis.

The dynamics of soil moisture is a complex phenomenon depending on atmospheric conditions, geological features of the region under study, and the corresponding land use. It can be formally described by Richards equation that arises from the Darcy law, and the mass continuity law. Richards equation is defined in terms of several physical parameters that must be chosen on the basis of the geological features of the region under study. Let B be a three-dimensional domain defining the portion of the hydrographic basin under study; let x, y, $z \in B$ be the space variables, and $t > 0$ the time variable. Richards equation can be formulated as follows:

$$\left(C(\psi) + S\frac{\theta(\psi)}{n}\right)\frac{\partial h}{\partial t} = \nabla\left(K(\psi)\nabla h\right) + W(x,y,z,t) - ET(x,y,z,t),$$
$$(x,y,z) \in B,\ t > 0, \qquad (1)$$

where $h = \psi + z$ is the hydraulic head and ψ is the pressure head, K is a diagonal matrix describing the hydraulic conductivity, which measures the ability of water to flow in the porous isotropic medium, C is the specific moisture capacity, S is the storage coefficient, n is the porosity of the soil, W is the recharge and it is related to the rate of precipitation, ET is the evapo-transpiration and it represents the loss of water due to the evaporation and transpiration of plants. Note that in (1) appears also function θ, that is the water content of the soil; it can be computed from the pressure head ψ through the Van Genuchten formula [7]. So, the soil moisture content θ can be obtained from the knowledge of the solution h of the Eq. (1). A detailed description of the Richards equation can be found in [8,9].

The two source terms W and ET appearing in Eq. (1) depend on the weather data and on the land use, respectively. These functions have a narrow support that extends beneath top surface S_T; moreover, W is computed from the precipitation data, and ET from the so called Penman-Montieth equation [9], that gives an evaluation of evapotranspiration by using vegetation data and weather data.

The spatial domain B, where Eq. (1) is defined, gives a slice of soil beneath the slope under study; so, the boundary ∂B of B is constituted by a top surface S_T, describing the soil surface, a bottom surface S_B describing the depth where the soil moisture content is analyzed, and a vertical surface S_V joining the boundary of S_T and S_B.

Appropriate initial-boundary conditions must be considered with Eq. (1) in order to define a unique solution h, see [8] for details. The resulting initial-boundary value problem constitutes the proposed model for the soil moisture dynamics.

The parameters of (1) must be set on the basis of the geomorphological features of the territory and its land cover. When this adaptation step is complete, a numerical approximation method must be used for the computation of the soil-moisture dynamics from the weather data inputs. In particular, the numerical solution of the initial-boundary value problem for (1) is computed by a finite difference method resembling the well-known Cranck-Nicolson method for diffusion problems, see [10] for details.

The slope stability analysis is the other main component of the LANDSLIDE information system. It must give an evaluation of the resistance of inclined surfaces to failure by sliding or collapsing. The factor of safety F expresses the hazard degree. It is the ratio between the forces that prevent the slope from failing and those that make the slope fail; the Mohr-Coulomb criterion [11] is used for this evaluation. Of course, when the factor of safety is greater than one the slope is stable; on the other hand, a factor of safety less than one means that the slope may be unstable. Among the various methods for the evaluation of F, the Infinite Slope Model is probably the simplest possible one. In this method, the soil surface is approximated by an infinite inclined plane. In this case the following approximation of F can be obtained:

$$F = \frac{C + (z\gamma - z_w\gamma_w)\cos^2(\beta)\tan(\phi)}{z\gamma\sin(\beta)\cos(\beta)} \tag{2}$$

where C is the effective cohesion, γ is the weight of soil, γ_w is the weight of water soil, z is the depth of failure surface, z_w is the depth of water table, β is the slope surface inclination, ϕ is the angle of internal friction, see [5] for details. When the soil moisture is known, the factor of safety F can be easily computed from formula (2) in every point of the slope under study. In the LANDSLIDE project this is used as an estimate of the landslide hazard index, and it is automatically produced by the software tool. The corresponding hazard map is delivered to the competent territorial authority.

These mathematical models have been implemented in the LANDSLIDE information system that is able to make completely automatic predictions of landslide on a day to day basis. Moreover, the statistical analysis of the time-series of these predictions provide an evaluation of the impact of climate change in a medium-long term.

3 System Architecture and Visualization

The LANDSLIDE software has its web-based platform and functionalities available for the general public on the web site http://93.123.110.111/landslide/. It has been designed in a way to be as user-friendly as possible. Users have as options all information from the meteorological stations available "on the fly" in case the data is needed to be extracted. The hazard maps for potential landslide occurrence are available after both the database with the meteo data and the UniCamerino calculation module end their processes. Figure 1 represent the general view of the system.

The system has been designed to fulfil the web-based software requirements for real time data collection from the four test sites meteorological stations via Internet connections and fills in a dedicated database every 3 h with the meteorological conditions for the specific locations. These inputs are very important for a calculations module developed by the team in University of Camerino in Italy, which combines all database information with soil structure and local specific geological conditions resulting into a hazard map with colours, red, yellow and green depicting the severity of the calculated potential for landslide occurance in the period for the specific location according to the meteorological conditions and soil specifics. The most vulnerable areas are in red, and the one which are in the safe zone are in green. General structure of the data flow and calculation processes can be seen on Fig. 2.

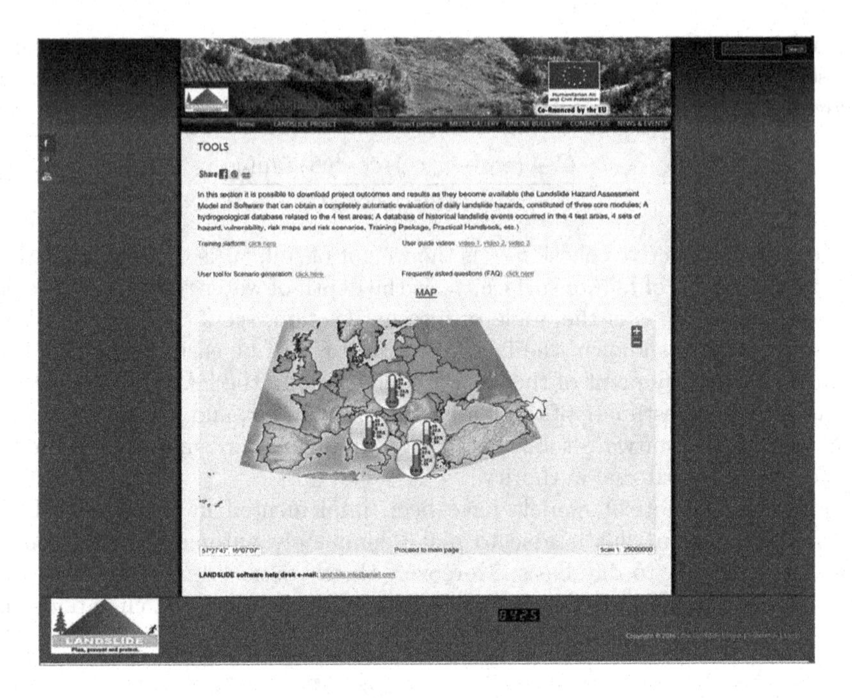

Fig. 1. LANDSLIDE platform main screen.

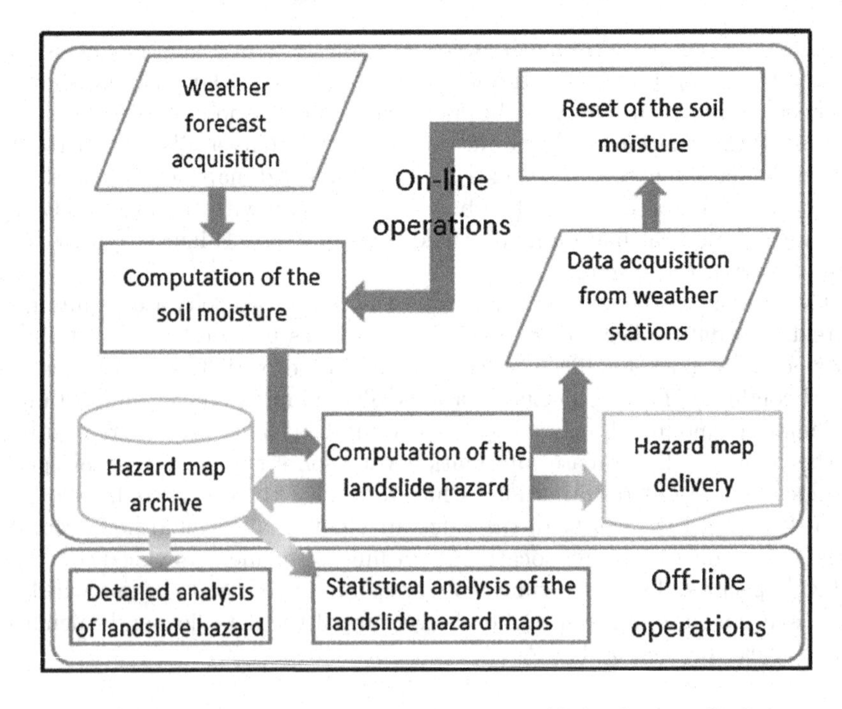

Fig. 2. LANDSLIDE system architecture. (Color figure online)

4 Operational Usage in the Four Test Sites

Four test areas have been considered in the LANDSLIDE project for the development of the information system. These areas are located on the territories of Greece, Italy, Bulgaria, and Poland. A short description of these areas is given.

The test area of Greece is located in the Peloponnesus peninsula. The responsible partner for that area is the National Observatory of Athens (NOA). Peloponnesus is a very mountainous area with over 1000000 inhabitants, including also remote villages at risk of landslides. The administration of the territory is divided between the Region of Peloponnesus with Tripoli as the capital city, and the Region of Western Greece with Patras as the capital city. The area of Panagopoula was chosen as the test site, because past landslide events have cut rail and road connection from the Greek capital Athens to Patras for weeks. Patras is the most important port of western Greece. At the moment new rail and road connections from Athens to Patras (co-financed by European Union) are being constructed under Panagopoula as the morphology of Northern Peloponnesus does not allow any alternatives. Any major landside event may have tremendous consequences for the economy of western Greece. The Panagopoula landslide area is situated in the northern part of the Prefecture of Achaia (northwestern Peloponnesus, about 15 km east of Patras). NOA has selected the pilot area very carefully. All measurements on the field have been completed and 10 GIS maps were elaborated. To correctly collect data, a meteorological station has been acquired and installed in a safe position and started transmitting data in real time in the weather meteorological stations network of NOA. According to the specifications of the LANDSLIDE protocol, two drillings have been performed by very specialized external experts.

The second test area is located in Province of Ancona, Italy. The pilot area selected for the test and implementation of the LANDSLIDE model and software in Italy has an extension of 11.69 km^2 and is located in the mid part of the Esino River basin (i.e. in the central part of the territory of the Province of Ancona). The test area was chosen as it is representative, also of the other hydrographic basins of the Province of Ancona: in effect about 30% of the whole test area is concerned by landslides (i.e. 3.55 km^2) and a landslide has recently damaged and interrupted the provincial road connecting 3 municipalities. Furthermore, several meteorological stations are located in the test area and previous geological and geotechnical studies have already been carried out. Having identified the hydrographic basin, drilling and sampling operations are being implemented according to the protocol agreed among project partners, as well as geological and geomorphological data of the area, are being collected to elaborate proper GIS maps.

The third test area used by the project model and software is the Smolyan Region in Bulgaria, and it is located at *Smolyan Lakes*. This is an area at northwest of the town of Smolyan. Its borders match the Kriva River to the west, Muneva River to the East, Cherna River to the South, and the steep rock cliff, which is located south of the peak Snejanka. The length of the landslide is 5.35 km, average width is about 1.38 km and the depth ranges 35 m to 80 m;

its total area is 7.4 km^2. Drilling and sampling operations have been performed according to the partnership protocol. Geological and geomorphological data of the area is also being collected to elaborate GIS maps. The biggest landslide in Bulgaria was registered in 1923. It is located in the western suburbs of the very town of Smolyan. It has a width of 1 km and length of 5 km. It is active, with speed of movement of 5–25 cm per year. Its depth is 80 m. The landslide is moving slowly, however if no measures will be taken it could affect the main road of the Region, disconnecting the access to Smolyan.

The fourth test area is located in Poland at the Bielsko-Biala District. The test area has been set within the Small Beskid mountain range, and field studies are taking place in an inactive sandstone quarry in Kozy. The area of the quarry has been considered to be in danger of further movements judging by the results of physicochemical studies of the material gathered in the dump. It has been found that the major part of the slope is in a state close to an equilibrium border and even small changes in the distribution of stresses, caused for example by weather conditions, may lead to the development of landslide processes. The first deformation processes within the Small Beskid area were already registered in 1968 in the quarry located in Kozy. The surface mass movements have been a continuous phenomenon, and their development have been constantly observed. In 2010 a total number of 437 landslides were registered (active, periodically active, and inactive).

5 Future Work and Conclusions

LANDSLIDE project proposes an innovative tool to predict and evaluate landslide hazards from weather conditions. More precisely, the dynamics of soil moisture and the slope stability analysis are jointly used for a quantitative evaluation of the hazard level. The main outputs of the LANDSLIDE software are the Hazard map (that indicates the probability of a landslide event in the considered areas) and the Depth map (that shows the depth at which hazard index is computed).

The project LANDSLIDE has been done in a way that the partners developing the model and the software work very closely with each other. The end user partners from Greece, Italy, Bulgaria, and Poland will have more than six months for testing and validating the final tool developed by the teams of University of Camerino and IICT-BAS.

LANDSLIDE project gives a "proof-of-concept" for a new way of landslide hazard evaluation from weather conditions. The method proposed in this project differs from the classical hazard evaluation techniques, however, we expect that a clever integration of these different assessment methods could give a very effective procedure in landslide hazard estimation.

Landslides are complex mechanical phenomena, where a large uncertainty on relevant data is present, such as geotechnical characterization of the soil and weather conditions. So, the physical models and the corresponding software tools proposed in LANDSLIDE project must be tested and improved in order to obtain an effective tool in fast alert systems and in emergency management services. However, the method proposed in the LANDSLIDE project has a high potentiality; in fact, the physical models do not depend on the particular test areas taken into account, so LANDSLIDE model and software can be easily adapted also to other territories. Finally, with similar approaches other water related events (such as forest fires and floods) can be taken into account; this would allow the creation of a platform for multi-hazard evaluation.

Acknowledgments. This work has been supported by the projects the Bulgarian National Fund 02/20 and LANDSLIDE DG ECHO/SUB/2014/693902.

References

1. Polemio, M., Petrucci, O.: Rainfall as a landslide triggering factor: an overview of recent international research. In: Bromhead, E., Dixon, N., Ibsen, M.-L. (eds.) Landslides in Research, Theory and Practice, vol. 3, pp. 1219–1226. Thomas Telford, London (2000)
2. Haque, U., Blum, P., da Silva, P.F., Andersen, P., Pilz, J., Chalov, S.R., Malet, J.-P., Auflič, M.J., Andres, N., Poyiadji, E., Lamas, P.C., Zhang, W., Peshevski, I., Pétursson, H.G., Kurt, T., Dobrev, N., García-Davalillo, J.C., Halkia, M., Ferri, S., Gaprindashvili, G., Engström, J., Keellings, D.: Fatal landslides in Europe. Landslides **13**, 1545–1554 (2016)
3. United States Geological Survey. https://www2.usgs.gov/faq/categories/9752/2607
4. LANDSLIDE project official web site. http://landslideproject.eu/
5. Van Westen, C.J., Castellanos, E., Kuriakose, S.L.: Spatial data for landslide susceptibility, hazard, and vulnerability assessment: an overview. Eng. Geol. **102**(3–4), 112–131 (2008)
6. Guzzetti, F., Reichenbach, P., Cardinali, M., Galli, M., Ardizzone, F.: Probabilistic landslide hazard assessment at the basin scale. Geomorphology **72**, 272–299 (2005)
7. Van Genuchten, M.: A closed form equation for predicting the hydraulic conductivity of unsatured soil. Soil Sci. Soc. Am. J. **44**, 892–898 (1980)
8. Pinder, G., Celia, M.: Subsurface Hydrology. Wiley Interscience, New York (2006)
9. Banta, R.: Modflow-2000. The U.S. Geological Survey Modular Ground-Water-Model, Denver, Colorado (2000)
10. Crank, J., Nicolson, P.: A practical method for numerical evaluation of solutions of partial differential equations of the heat conduction type. Proc. Camb. Philos. Soc. **43**, 50–67 (1947)
11. Duncan, J.M., Wright, S.G., Brandon, T.L.: Soil Strength and Slope Stability, 2nd edn. Wiley, Hoboken (2014)

On the Winter Wave Climate of the Western Black Sea: The Changes During the Last 115 Years

Vasko Galabov(✉) and Hristo Chervenkov

National Institute of Meteorology and Hydrology, Bulgarian Academy of Sciences, Tsarigradsko Shose blvd. 66, 1784 Sofia, Bulgaria
vasko.galabov@meteo.bg

Abstract. We present a study of the winter wave climate of the Western Black Sea with a focus on the annual maximums and the mean seasonal wave heights. We did a numerical simulation of the wave parameters in the Black Sea by the wave model SWAN for a period of 110 years. The input wind fields are from the atmospheric reanalysis ERA-CLIM. We also performed a hindcast for the period 1980–2015 using winds from the CFSR reanalysis. Extended winter (December–March) was studied. We also studied the characteristics of the pressure gradients in a larger region attempting to quantify this way the interaction of Mediterranean lows with blocking highs. No significant long term changes were found for any of the characteristics of the mean and extreme wave climate.

Keywords: Wave climate · Black Sea · Storminess · SWAN ERA-CLIM · CFSR

1 Introduction

The wave climate of the Western Black Sea was a subject of a numerous studies during the last years. In situ measurements are not available for long periods and the atmospheric reanalyses made it possible to use numerical simulations to study the wave climate. The focus of the article is on the Western Black Sea. Valchev et al. [18] constructed various proxies of the Western Black Sea storminess for the period 1948–2010 and found no significant upward or downward trends but for almost all proxies, an increasing trend until the 1980s or the 1990s and a return to average or even calm conditions in the late 2000s. Arkhipkin et al. 2014 [3] hindcasted the Black Sea wave climate for the period 1948–2010 and found that the total duration of storms and their quantity remains nearly stable with some periods of increases and decreases and explained the interdecadal changes with the influence of NAO. Akpinar et al. [1,2] studied the trends of the Black Sea mean and maximum significant wave heights and the trends of the wave energy for the period 1979–2010 based on numerical simulation using winds from the CFSR reanalysis [15]. Their conclusion is that for the Western Black Sea there are no statistically significant trends for the 31 years period for the

I. Lirkov and S. Margenov (Eds.): LSSC 2017, LNCS 10665, pp. 466–473, 2018.
https://doi.org/10.1007/978-3-319-73441-5_51

annual maximum significant wave heights and the mean annual wave energy flux. Divinsky and Kosyan [5] found an increasing trend for the period 1990–2010. In their further study [6] extended from 1979 their conclusions are not in favor of such strong trend. Polonsky et al. [13] studied the wave climate of the northern Black Sea and found signs of low frequency oscillations with a period of about 60 years consistent with the changes of the phases of the Atlantic Multidecadal Oscillation (AMO) and Pacific Decadal Oscillation (PDO). We performed a 110 years hindcast of the wave parameters based on the ERA-CLIM [16] wind data and 35 years long simulation based on CFSR in order to confirm or reject the possible trends. We focused on the winter season because as it is confirmed by all mentioned studies, it is the season with the highest seasonal mean significant wave heights and almost all relevant storms happen during the winter.

2 Data and Methodology

The numerical simulation of the wave parameters was performed by SWAN wave model [4]. Version is 40.91ABC. Detailed explanation of a model application to reconstruct historical storms in the Black Sea is available in [7]. The computational grid is regular spherical with 1/30° spatial resolution and the discretisation in spectral space is 36 directions and 31 frequencies between 0.05 and 1 Hz. The model domain covers the entire Black Sea with 451×211 computational nodes. The temporal resolution of the output is 1 h. The parameterisations of the wave energy generation and dissipation are based on Komen physics [9] with parameter $\delta = 1$ (known as "Rogers trick" [14]). The 10 m wind input data is from the ERA-CLIM reanalysis (dataset ERA-20C) and CFSR for the period 1979–2010 and from the continuation CFSv2 from 1911 to 1915. We studied the period December-January-February-March (DJFM). We focus on two parameters- DJFM mean SWH (hereafter denoted H_m) and DJFM maximum SWH (hereafter H_{max}). We selected 4 locations representative for the Western Black Sea shelf- the Bosporus and Ahtopol in the South Western shelf and Shabla and Delta in the North Western shelf (see Table 1). All points are roughly 10 km far from the coast with depth more than 50 m. The tests for significance of the trends are performed using the Mann-Kendal test with confidence at least 95% ($p = 0.01$) [8,11].

Necessary prerequisite for development of strong wind conditions in the synoptic scale is the formation and sustenance of elevated pressure gradients. Significant gradients are formed in synoptic situations when cyclone and anticyclone are located relatively close to each other, causing high-low dipole pattern. Intending to find favorable conditions for easterlies over the Western Black Sea, the gradient between the grid-cell with the highest mean sea-level pressure over arbitrary region in West Russia from one side and the location with the lowest one over the Eastern Mediterranean from the other, is computed. The computations are performed in 0.5° grid spacing for each time step of ERA-CLIM for the full 110-year time span of this dataset.

The correlations of the time series has been calculated for the period 1951–2010 with the North Atlantic Oscillation (NAO), Arctic Oscillation (AO),

Table 1. The locations where the model output was studied

Name of the point	Latitude, °	Longitude, °	Description of the location
Bosporus	41.50	29.00	North of Bosporus-south western coast
Ahtopol	42.20	28.20	Near Ahtopol- southernmost Bulgarian coast
Shabla	43.60	28.90	Near cape Shabla
Delta	44.81	30.00	North Western shelf/close to Danube delta

Atlantic Multidecadal Oscillation (AMO), Pacific Decadal Oscillation (PDO) Also East Atlantic (EA), Polar-Eurasia (POL), Scandinavian Pattern (SCA), East Atlantic/West Russia (EA/WR). All these indexes were averaged for the DJFM periods. We also obtained the correlation with the DJFM anomaly of the Northern Hemisphere temperatures (NHT) in order to look for an influence of the climate changes during the second half of the 20th century. For the period 1900–2010 we tested only with NAO, AMO, PDO, and NHT. Finally we compared the correlations obtained using the CFSR input and ERA-CLIM input for the period 1980–2010.

3 Results

The results of the SWAN model simulation for the 110 years period at two locations (Ahtopol and Delta) for H_m and H_{max} are shown in Fig. 1. For Ahtopol H_m is decreasing from 1900 to 1920 and then increasing from 1920 to 1970 with peaks during the winters of 1929 and 1931 (the winter of 1929 is with negative AO and NAO and the coldest winter in Bulgaria during the 20th century). At the location Delta H_m also reaches his peak in 1929, but its value is not as significant as at Ahtopol location. There is no decrease in H_m at the Delta location during the period 1900–1920. The period of 1900–1920 is characterized by high values of AO and NAO and only during one winter season both AO and NAO where negative. The period 1930–1970 is with decrease of AO and NAO and during 1960–1970 8 winters are with negative AO and NAO and 1969 is with extreme negative values of AO and NAO and high H_m for both locations.

After 1970 H_m decreased until the decade 1990–2000 together with increase of AO and NAO. During the period 2001–2010 there is no further decrease for the Delta location but for Ahtopol and Bosporus the decade is with lower H_m than the previous decade. H_{max} time series on the other hand does not show a pattern like the pattern of the mean and for all 4 locations there is no statistically significant correlation between these parameters of the mean and extreme wave climate. Some of the notable annual maximums occurred during winters with H_m below the average. We tested the trends of H_m and H_{max} for the 110 years period and found that there are no statistically significant trends

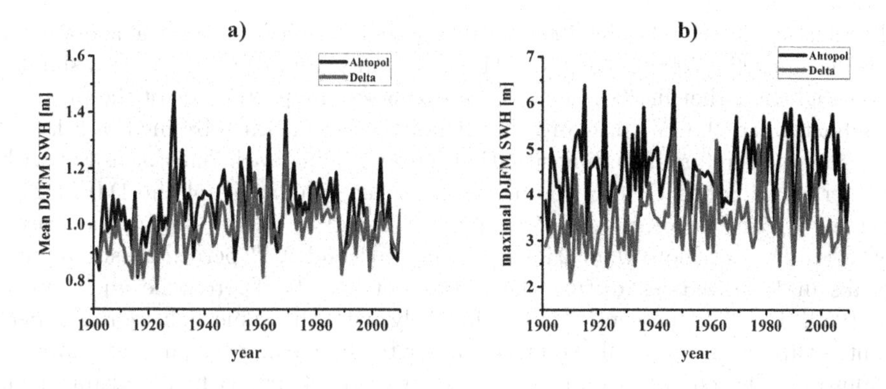

Fig. 1. (a) Mean DJFM SWH, 1901–2010 (b) maximal DJFM SWH, 1901–2010

in any of the four locations. We also did a comparison between the results of a simulation using CFSR winds and ERA-CLIM winds for the period of 1980–2010. The comparisons are shown in Fig. 2. The CFSR based simulation results in higher means and extremes by 20% (as it was shown by Van Vledder et al. [19]. CFSR winds are the most accurate from the numerous tested reanalisys datasets with some overestimation of the winds). Generally the shape of the ERA-CLIM means and CFSR means are similar. Again the time series of the extremes and the means are not similar to each other. In both cases the trends of H_{max} are rejected with confidence more than 99% ($p < 0.01$).

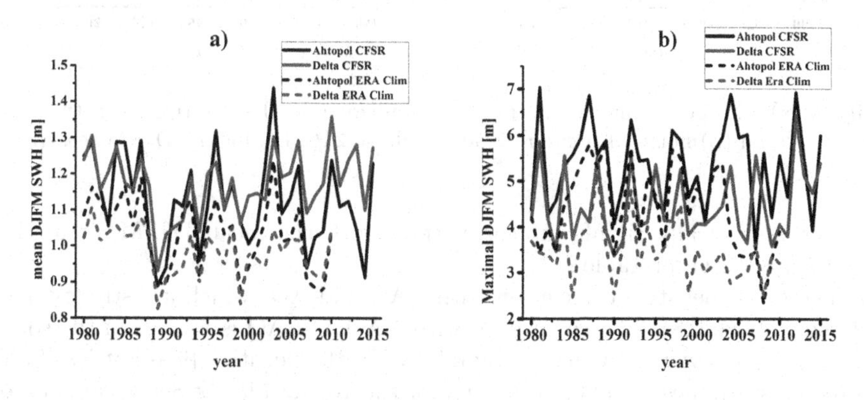

Fig. 2. (a) Mean DJFM SWH- with CFSR and ERA-CLIM; (b) maximal DJFM SWH- with CFSR and ERA-CLIM

For the two locations in the Northwestern shelf there is no statistically significant trend of H_m in CFSR and in ERA-CLIM based simulations. For the Southwestern shelf there is a statistically significant decrease in the ERA-CLIM based simulation during the period 1980–2010 with confidence of 95%. In CFSR based simulation there is a weak negative trend but not statistically significant.

The average for the decade 2001–2010 H_m is slightly lower than the average for the decade 1991–2000 in ERA-CLIM simulation and in the CFSR based simulation slightly higher in 2001–2010. Figure 3 shows the percentiles of the pressure gradient when the low pressure is north of the Black Sea (obtained in a larger region and therefore including situations also without Black Sea storms but with Easterlies in South East Europe). We show the time series of the DJFM 25th percentile, 75th percentile and 95th percentile and the hours during the season with gradient of more than 2 Pa/km. The 75th and 95th percentile show some peaks in 1929, a peak during the forties only on 95th percentile and a peak in 1969. The 25th percentile is qualitatively with the same behavior. All percentiles are with temporal evolution similar to the mean wave climate temporal evolution. The cumulative hours with high gradient are with remarkably high peak in 1969. The trends for the percentiles and the cumulative hours are not statistically significant.

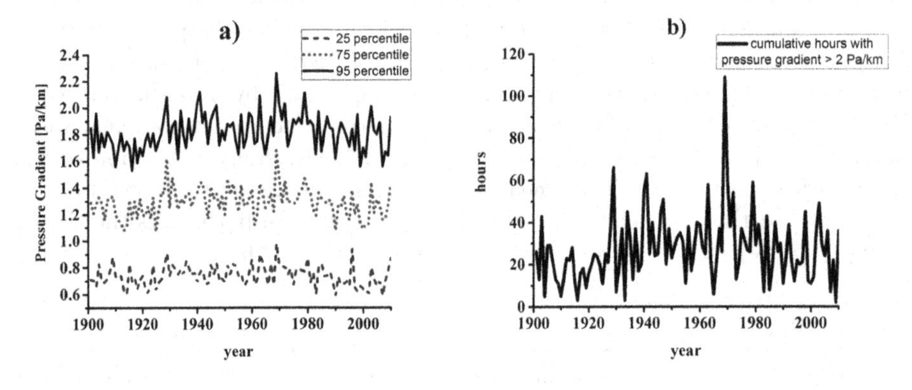

Fig. 3. (a) The percentiles of the pressure gradient (see the description in the text) DJFM season (b) hours of pressure gradient above 2 Pa/km for the DJFM season

The statistically significant correlations with the mean DJFM circulation indices are shown in Table 2.

There is a negative correlation with NAO and AO (which are strongly correlated - above 0.8). The correlation with NAO and AO is weaker in the south and increases towards the north. Link between the negative phase of NAO and increased storminess similar to the link for the Western Black Sea was found for the Mediterranean by Lionello and Sanna [10]. The positive phase of NAO leads to a decline in the frequency of the intense Mediterranean cyclones and increase of the frequency of the non-intensive cyclones that cannot lead to high waves in the Western Black Sea [17]. The correlation with the EAWR pattern is significant for the two southern locations. As it was found by Oguz et al. [12] EAWR has a significant influence on the Black Sea surface temperature and sea level anomaly and other physical characteristics and when NAO < 0 and EAWR > 0 the winters are cold with cold air outbreaks from the northern sector. When

Table 2. Statistically significant correlations for the period 1951–2010

Point	NAO	AO	NHT	SCA	EAWR	POL
Bosporus mean DJFM SWH	−0.32	-	−0.36	0.29	0.43	0.39
Ahtopol mean DJFM SWH	−0.38	−0.33	−0.36	0.36	0.33	0.33
Shabla mean DJFM SWH	−0.48	−0.49	−0.33	0.33	-	0.28
Delta mean DJFM SWH	−0.49	−0.59	−0.27	-	-	-
25th percentile of the pressure gradient	−0.58	−0.51	−0.55	0.61	-	0.35
75th percentile of the pressure gradient	−0.54	−0.51	−0.51	0.56	-	0.31
95th percentile of the pressure gradient	−0.42	−0.38	−0.34	0.36	-	-
Hours of press. gradient above 2 Pa/km DJFM	−0.32	−0.35	−0.26	0.30	-	-

NAO > 0 and EAWR > 0 this leads to strong cold air intrusions from northwest and northeast and the coldest winters in the Black Sea. When EAWR is in negative phase generally the east winds are less frequent. The correlations with SCA and POL are either because SCA is significantly correlated with AO and POL with EAWR, or because they serve as additional modulator of the influence of NAO. If we normalize the H_m for the studied locations to obtain the anomalies for the period 1951–2010 and select the winters with anomalies of H_m higher than 1.0 for Ahtopol and Bosporus- we obtain 8 cases. For 7 of these 8 cases NAO and AO are negative and in 6 out of 7 cases strongly negative. The only of these 8 cases with positive NAO and AO is with positive EAWR, POL and SCA. There are 9 winters with highly negative AO and NAO but negative anomalies of H_m for Ahtopol and Bosporus. For all these winters EAWR is negative. Obviously EAWR is the teleconnection than complements the NAO/AO influencing the winter wave climate of these locations. As for the winters with very low anomaly of H_m (below −1): 9 of them are with positive NAO and/or AO and the rest four are with negative NAO/AO and negative EAWR. For the two northern locations situation is the same - there are 8 winters with H_m anomaly above 1 and again there is only one year with positive NAO/AO and positive EAWR. Years with anomalies below −1 are 13 (for 10 of them NAO and/or AO are positive and for 3 they are both negative. All these 3 are with EAWR < 0). When we extend the period to 110 years and test again the correlations with NAO and NHT turnes out that the correlations with NAO are the same but the correlations for all locations with NHT are below 0.1 and not significant. We obtained the correlations for the period 1980–2010 using the output based on CFSR and the output based on ERA-CLIM (Table 3). The conclusions about the correlations with AO and NAO remain the same.

In the simulation with CFSR for high positive anomalies (>1) of H_m for all 6 cases for Ahtopol and Bosporus they are associated with negative NAO

Table 3. Correlation for the period 1980–2010 for the two hindcasts

Point	Wind data source	NAO	AO	NHT	SCA	EAWR	POL
Bosporus mean DJFM SWH	CFSR	−0.36	−0.36	-	0.44	-	0.43
	ERA-CLIM	−0.32	−0.30	−0.41	-	0.28	0.50
Ahtopol mean DJFM SWH	CFSR	−0.44	−0.46	-	0.47	0.24	0.39
	ERA-CLIM	−0.41	−0.44	−0.40	0.44	0.43	0.43
Shabla mean DJFM SWH	CFSR	−0.47	−0.57	-	0.52	-	-
	ERA-CLIM	−0.45	−0.56	-	0.44	-	-
Delta mean DJFM SWH	CFSR	−0.47	−0.64	-	0.47	-	-
	ERA-CLIM	−0.41	−0.58	-	0.36	-	-

and AO. For Shabla and Delta there are 8 years with high anomaly of H_m and 7 of these are with negative NAO and AO. For all locations the years with high negative anomaly of H_m are always associated with positive NAO and AO. The correlation with EAWR here may not be apparent due to the shorter period but it plays a clear role for the cases prior to 1979 when winter NAO and AO are more frequently negative NAO/AO and a lot of years with negatives are not associated with positive anomalies of the seasonal mean wave height. Obviously the period after 1980 is not long enough to reveal other influences besides NAO/AO. The H_{max} values for all locations do not correlate significantly with any index when using CFSR or ERA-CLIM.

We used a multiple regression to look for the importance of the indexes in the explanation of the variance of H_m: for Bosporus and Ahtopol NAO or AO explains about 25% of the variance and EAWR about 25% NHT. SCA and POL does not explain significant variance. For Shabla AO explains 35% of the variance and EAWR 15%. For Delta 45% of the variance is explained by AO and less than 10% by EAWR without significant role of SCA, POL and NHT. For the 25th percentile of the pressure gradient 64% of the variance is explained and NAO, SCA and POL are with equal share. For the 75th percentile 45% of the variance is explained with primary role of AO and minor of SCA and POL. For the 95th percentile only 30% of the variance are explained by AO. For the cumulative hours of high gradient below 25% of the variance is explained by AO.

4 Conclusion

We studied the Western Black Sea wave climate using numerical hindcasts based on ERA-CLIM and CFSR for 110 and 35 years respectively for the winter season. No significant trends were found of the mean and extreme wave climate. The extremes does not show any significant signs of regular changes. Significant links of the mean wave climate of the Western Black Sea with the teleconnection patterns were found with AO (alternatively NAO) increasing in the north and EAWR increasing in the south.

Acknowledgment. Deep gratitude to the organizations and institutes (ECMWF, NOAA-Climate Prediction Center, NCEP-NCAR, Unidata, MPI-M and all others), which provides free of charge software and data. Without their innovative data services and tools this study would be not possible.

References

1. Akpınar, A., Bingölbali, B.: Long-term variations of wind and wave conditions in the coastal regions of the Black Sea. Nat. Hazards **84**, 69–92 (2016)
2. Akpınar, A., Bingölbali, B., Van Vledder, G.P.: Long-term analysis of wave power potential in the Black Sea, based on 31-year SWAN simulations. Ocean Eng. **130**, 482–497 (2017)
3. Arkhipkin, V.S., Gippius, F.N., Koltermann, K.P., Surkova, G.V.: Wind waves in the Black Sea: results of a hindcast study. Nat. Hazards Earth Syst. Sci. **14**, 2883–2897 (2014)
4. Booij, N., Holthuijsen, L.H., Ris, R.C.: A third-generation wave model for coastal regions. Model description and validation. J. Geophys. Res. **104**, 7649–7666 (1999)
5. Divinsky, B.V., Kosyan, R.D.: Observed wave climate trends in the offshore Black Sea from 1990 to 2014. Oceanology **55**, 837–843 (2015)
6. Divinsky, B.V., Kosyan, R.D.: Spatiotemporal variability of the Black Sea wave climate in the last 37 years. Cont. Shelf Res. **136**, 1–19 (2017)
7. Galabov, V., Kortcheva, A., Bogatchev, A., Tsenova, B.: Investigation of the hydro-meteorological hazards along the Bulgarian coast of the Black Sea by reconstructions of historical storms. J. Environ. Prot. Ecol. **16**, 1005–1015 (2015)
8. Kendall, M.G., Stuart, A.: The Advanced Theory of Statistics: Distribution Theory, vol. I. Griffin, London (1976)
9. Komen, G.J., Hasselmann, S., Hasselmann, K.: On the existence of a fully developed wind-sea spectrum. J. Phys. Oceanogr. **14**, 1271–1285 (1984)
10. Lionello, P., Sanna, A.: Mediterranean wave climate variability and its links with NAO and Indian Monsoon. Clim. Dyn. **25**, 611–623 (2005)
11. Mann, H.B.: Nonparametric tests against trend. Econometrica **13**, 245–259 (1945)
12. Oguz, T., Dippner, J.W., Kaymaz, Z.: Climatic regulation of the Black Sea hydro-meteorological and ecological properties at interanual-to-decadal time scales. J. Mar. Syst. **60**, 235–254 (2005)
13. Polonsky, A., Evstigneev, V., Naumova, V., Voskresenskaya, E.: Low-frequency variability of storms in the northern Black Sea and associated processes in the ocean-atmosphere system. Reg. Environ. Chang. **14**, 1861–1871 (2014)
14. Rogers, W.E., Hwang, P.A., Wang, D.W.: Investigation of wave growth and decay in the SWAN model: three regional-scale applications. J. Phys. Oceanogr. **33**, 366–389 (2003)
15. Saha, S., et al.: The NCEP climate forecast system reanalysis. Bull. Am. Meteorol. Soc. **91**, 1015–1057 (2010)
16. Stickler, A., Brönnimann, S., Valente, M.A., Bethke, J., Sterin, A., Jourdain, S., Roucaute, E., Vasques, M.V., Reyes, D.A., Allan, R., Dee, D.: ERA-CLIM: historical surface and upper-air data for future reanalyses. Bull. Am. Meteorol. Soc. **95**, 1419–1430 (2014)
17. Trigo, I.F., Davies, T.D., Bigg, G.R.: Decline in Mediterranean rainfall caused by weakening of Mediterranean cyclones. Geophys. Res. Lett. **27**, 2913–2916 (2000)
18. Valchev, N.N., Trifonova, E.V., Andreeva, N.K.: Past and recent trends in the western Black Sea storminess. Nat. Hazards Earth Syst. Sci. **12**, 961–977 (2012)
19. Van Vledder, G.P., Akpınar, A.: Wave model predictions in the Black Sea: sensitivity to wind fields. Appl. Ocean Res. **53**, 161–178 (2015)

Computer Simulations of Atmospheric Composition in Urban Areas. Some Results for the City of Sofia

Ivelina Georgieva(✉), Georgi Gadzhev, Kostadin Ganev, and Nikolay Miloshev

National Institute of Geophysics, Geodesy and Geography, Bulgarian Academy of Sciences, Acad. G. Bonchev str., bl. 3, 1113 Sofia, Bulgaria
iivanova@geophys.bas.bg

Abstract. Some extensive numerical simulations of the atmospheric composition fields in the city of Sofia have been recently performed. An ensemble, comprehensive enough as to provide statistically reliable assessment of the atmospheric composition climate of Sofia—typical and extreme features of the special/temporal behavior, annual means and seasonal variations, etc. has been constructed. The simulations were carried out using the American Environment Protection Agency (US EPA) Models-3 system. As the National Centers for Environmental Prediction (NCEP) Global Analysis Data with 1 degree resolution was used as meteorological background, the system nesting capabilities were applied for downscaling the simulations to a 1 km resolution over Sofia. The national emission inventory was used as an emission input for Bulgaria, while outside the country the emissions were taken from the Netherlands Organization for Applied Scientific research (TNO) inventory. Special preprocessing procedures are created for introducing temporal profiles and speciation of the emissions. The biogenic emissions of Volatile Organic Compound (VOC) are estimated by the model Sparse Matrix Operator Kernel Emissions (SMOKE). The air pollution pattern is formed as a result of interaction of different processes, so knowing the contribution of each for different meteorological conditions and given emission spatial configuration and temporal behavior could be interesting. Different characteristics of the numerically obtained concentration fields of pollutants as well as of determining the contribution of different types of pollutants and pollution sources will be demonstrated in the present paper.

Keywords: Atmospheric composition · Numerical modeling
Integrated process rate analysis
Ensemble of numerical simulation results

1 Introduction

The atmospheric composition in urban areas is one of the primary tasks in air pollution studies. The urban air pollution climate in Sofia have not been systematically studied yet, though, of course, some air pollution modeling for the

I. Lirkov and S. Margenov (Eds.): LSSC 2017, LNCS 10665, pp. 474–482, 2018.
https://doi.org/10.1007/978-3-319-73441-5_52

city had been performed [12] and even air pollution forecast for the city is operationally going on [15]. Recently extensive studies for long enough simulation periods and good resolution of the atmospheric composition status in Bulgaria have been carried out using up-to-date modeling tools and detailed and reliable input data [5,6], but next step in studying the atmospheric composition climate is performing simulations in urban scale. The simulations aim at constructing of ensemble, comprehensive enough as to provide statistically reliable assessment of the atmospheric composition climate of the city of Sofia. Different characteristics of the numerically obtained concentration fields of pollutants as well as of determining the contribution of different types of pollutants and pollution sources will be presented in the present paper.

2 Methodology

The simulations were carried out using the US EPA Models-3 system.

- WRF [13]—Weather Research and Forecasting Model, used as meteorological pre-processor;
- CMAQ - the Community Multiscale Air Quality System [2,3], being the Chemical Transport Model (CTM) of the system, and
- SMOKE - the Sparse Matrix Operator Kernel Emissions Modelling System [4]—the emission pre-processor of Models-3 system. The large scale (background) meteorological data used in the present study is the NCEP Global Analysis Data with $1° \times 1°$ resolution. WRF and CMAQ nesting capabilities are applied for downscaling the simulations to a 1 km step for the innermost domain (Sofia). The national emission inventory is used as an emission input for Bulgaria, while outside the country the emissions are taken from the TNO inventory [17–19]. The simulations are performed for 7 years (2008 to 2014) with Two-Way Nesting mod on. Special pre-processing procedures are created for introducing temporal profiles and speciation of the emissions. The biogenic emissions of VOC are estimated by the model SMOKE [8,14,16].

At Table 1 it can be seen that the computer resource requirements for the performed WRF-SMOKE-CMAQ simulations are big enough. On the other hand the planned numerical experiments were organized in the effective HPC environment.

The calculations were implemented on the Supercomputer System "Avitohol" at the Institute of Information and Communication Technologies at the

Table 1. Computer resource requirements on 16 CPU-s for 1 day simulation

•	WRF	CMAQ and SMOKE	Total
Time (h)	3	2	5
HDD (GB)	0.5	1	1.5

Bulgarian Academy of Sciences (IICT-BAS) consists of 150 HP Cluster Platform. The CMAQ simulations were organized in separate jobs, which makes the jobs run time for 6 days real time fairly reasonable [1]. The model output storage, however, is too large. As not all the output information is so valuable for further air quality and environmental considerations a post-processing procedure and respective software were developed, in order for the output to be filtered and only the necessary information to be kept.

3 Results

The most simple and natural atmospheric composition evaluations are the surface concentrations. By averaging over the whole simulated fields of ensemble the mean annual and seasonal surface concentrations can be obtained and treated like "typical" daily concentration patterns. Plots of some of these typical annual surface concentrations for Nitrogen Dioxide (NO_2), Ozone (O_3), Sulfur dioxide (SO_2), and Fine Particulate Matter (FPRM) in 06:00 and 18:00 UTC are shown in Fig. 1.

Because the major NO_2 source in the city is the road transport (surface sources) the surface NO_2 concentrations are higher early in the morning and much smaller at noon, when the atmosphere is usually unstable, and so the turbulence transports the NO_2 aloft more intensively. The spatial distribution is significantly heterogeneous—the maximal concentrations are formed in the city centre and along the boulevard with most busy traffic and over the Thermal Power Plants (TPPs) in the city. The regions with high NO_2 concentrations also are the south part of the city. The behavior of the surface ozone is more complex. The ozone in Bulgaria is to a great extent due to transport from abroad [7,9–11]. The ozone concentrations early in the morning are smaller than at noon because less intensive transport from higher levels. The other reason, the ozone photochemistry, which explains higher O_3 concentrations at daytime and during the summer and the O_3 gaps in the regions, where the NO_2 concentrations are large. The surface concentrations of SO_2 are high also in morning hours and the regions that can be seen are in the city centre and in the bottom of the mountain Vitosha. The general SO_2 pollutant are the two TPPs in the city and the domesting heating. High concentrations can be seen also at the north-west part of the domain where is Konstinbrod. There are situated many factory and oil storage depot. The surface FPRM concentrations are bigger during the morning hours and winter months and can be seen at the city centre and near the Kostinbrod (north-west part). The main reason of this are probably the atmosphere stability in morning and so the FPRM sources are mostly the road transport which is more intensive in this time. Here also it can be seen the most busy roads in the city and the southern part of Sofia.

Determining the contribution of different types of pollutants and pollution sources to Sofia city.

The TNO emission inventory is for a 10 Selected Nomenclature for Air Pollution (SNAP) categories and allowing evaluate of the contribution of various

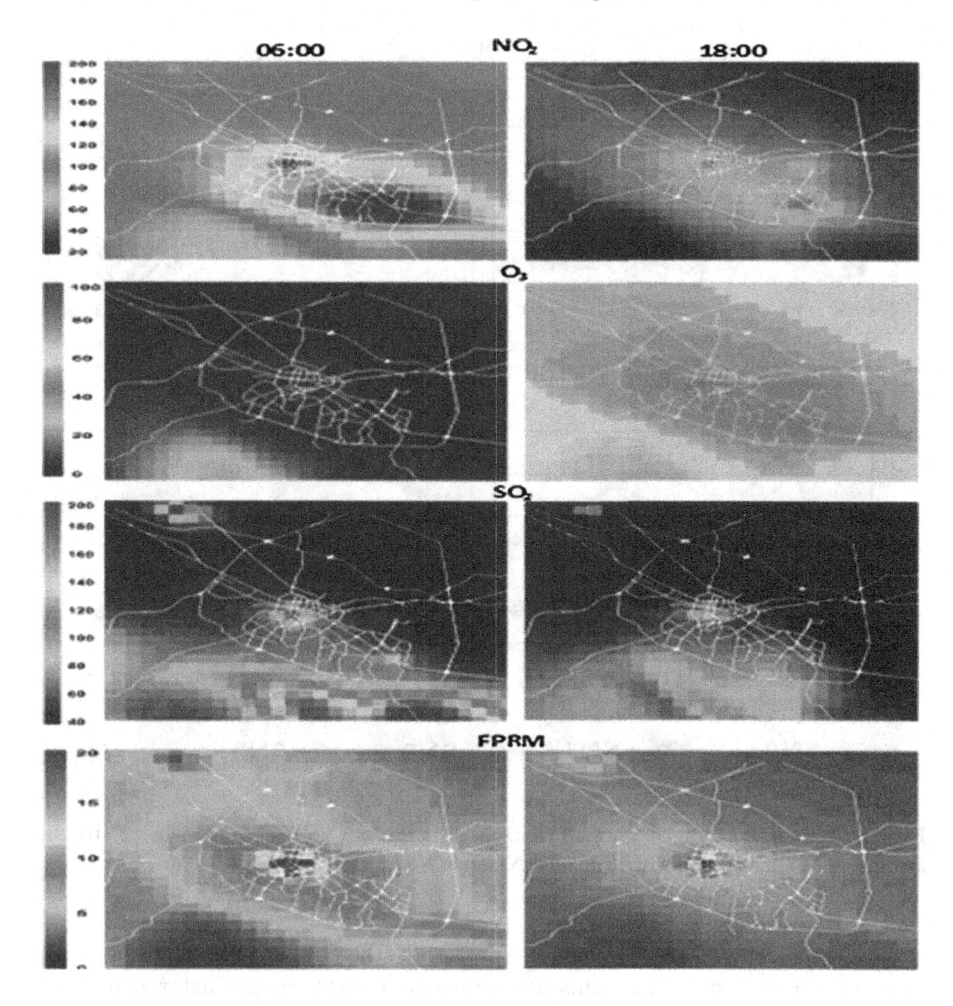

Fig. 1. Surface typical annual concentrations of NO_2, O_3, SO_2, and FPRM [μg/m^3] at 06:00 and 18:00 UTC

anthropogenic activities to the overall picture of pollution. The simulation are made for 7 years from 2008 to 2014. Five emission scenarios will be considered in the present paper: Simulations with all the emissions, simulations with the emissions of SNAP categories 1 (energetic), 2 (non-industrial combustions), 3 (industrial combustions) and 7 (road transport) for Sofia reduced by a factor of 0.8. This makes it possible to evaluate the contribution of road transport, energetic, industrial and non-industrial combustions to the atmospheric composition in the city. The relative contribution of scenario with all the reduced SNAP's was calculated for each day of this 7 year period and then by averaging the typical fields of relative contribution of this emissions to each of the compound surface concentrations were calculated for the 4 seasons and annually. Some illustrations of the emission impact evaluations will be given in the present paper only.

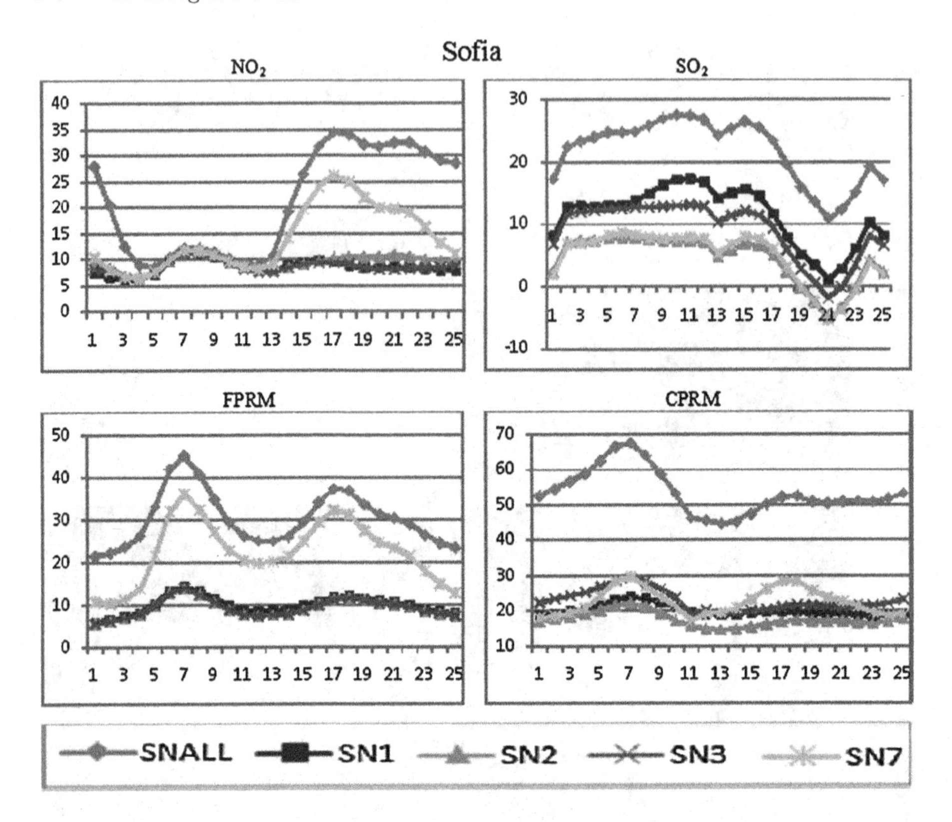

Fig. 2. Annually averaged contribution of different SNAP categories [in %] to the formation of surface concentrations of NO_2, SO_2, FPRM, and CPRM for Sofia

For all the emission categories the pattern of the contribution fields is rather complex, which reflects the emission source configuration, the heterogeneity of topography, land use and meteorological conditions. Plots of this kind can give a good qualitative impression of the spatial complexity of the emission contribution. In order to demonstrate the emission contribution behavior in a more simple and easy to comprehend way, the respective fields can be averaged over some domain, which makes it possible to follow and compare the diurnal behavior of the respective contributions for different species. Graphics of the diurnal evolution of the "typical" relative contribution of annual emissions of already mentioned SNAP categories 1, 2, 3, 7 and all to the surface concentrations of NO_2, SO_2, FPRM and CPRM for the territory of Sofia city are shown in Fig. 2.

In first place it should be noticed that the contribution of the emissions from all SNAPs to the formation of surface concentrations of NO_2, SO_2, FPRM and CPRM is positive. It could be seen that contribution of SNAP7 road transport is higher than the emissions from other SNAPs to the formation of surface concentrations of NO_2. The SNAP1 contribution to the SO_2 concentrations is bigger than other SNAPs, but the second dominant contribution is from emissions of

SNAP3. It should be noted that the SNAPALL and SNAP1 contribution to the surface SO_2 concentrations is smaller than one should expect, having in mind that the TPPs in Sofia are among the biggest sulfur sources. Probably, a significant amount of SO_2 from these sources becomes a subject of larger scale transport and so it is moved outside the domain. For FPRM and CPRM higher contribution have the emissions from SNAP7 road transport, but for CPRM in morning hours contribution of the emissions from SNAP3 is higher than this of SNAP7. The highest values of the contribution are in morning hours and in the afternoon. The smallest contributions have the emissions from SNAP2. The relative contributions for the territory of Sofia have well displayed diurnal and seasonal course. Seasonal course can be different for different pollutants and varies depending on the source of the given issue.

The contribution of the all emission categories to the surface ozone is negative, and cannot determine qualitative or quantitative differences in the contribution of emissions from different SNAP categories, which only confirms that ozone over Sofia city formed outside the city and its quantity and distribution due to transfer outside the city. It should be noticed that the warm seasons stands out with negative contributions tending to zero of all categories over the entire area except over Vitosha Mountain, where the contributions are with highest negative values.

The contribution of emissions from all categories SNAPALL is less than 100%, which means that part of the concentrations are formed from sources external to the domain D5 and are result of transfer outside the Sofia city.

4 Conclusion

The results, presented in the paper are just a first glance on the atmospheric composition status in urban areas, so very few decisive conclusions can be made at this stage of the study. Nevertheless, some of the major findings so far will be listed below:

- the behavior of the surface concentrations, averaged over the whole ensemble annually, or for the four seasons is reasonable and demonstrates effects which for most of the compounds can be explained from a point of view of the generally accepted schemes of dynamic influences (turbulent transport, atmospheric stability), local atmospheric circulations or chemical transformations;
- the surface concentrations of PM and NO_2 are probably due to the surface sources - mostly the road transport;
- the behavior of the surface ozone shows lower concentrations early in the morning than at noon. Higher O_3 concentrations at daytime and during the summer and the O_3 gaps in the regions, where the NO_2 concentrations are large (less intensive transport from higher levels, the ozone photochemistry);
- the surface concentrations of SO_2 are due to the general SO_2 pollutant - two TPPs in the city and the local domesting heating;

- Different emissions relative contribution to the concentration of different species could be rather different, varying from almost 80% to several %. The contributions of different emission categories to different species surface concentrations have different diurnal course and different meaning.
- For all of the pollutants the contribution of SNAPALL is dominant, but this contribution of emissions is less than 100%, which means that part of the concentrations are formed from sources external outside the Sofia city, and can be say that are result of transfer.
- The contribution of emissions a of all SNAPs to ozone levels in Sofia is rather small and strictly negative. This is probably due to the fact that the NOx concentrations are relatively small and they are the limitation factor for ozone formation. That means that the surface O_3 in Sofia came from outside the city.
- The contribution of all SNAP7 (road transport) to NO_2 surface concentrations is positive and reaches 50% around the most busy traffic roads. The SNAP7 has dominant contributions to the NO_2 and FPRM surface concentrations.
- SNAP1 contribution to the surface SO_2 concentrations is also big and positive and dominant with maximum around noon.
- The dominant contribution to the surface CPRM concentrations is emissions of SNAP7 (road transport) and SNAP3 (industrial combustions).

Acknowledgment. Deep gratitude to the organizations and institutes (NCEP-NCAR, Unidata, MPI-M, EMEP and to the TNO) for providing free-of-charge data and software and the high-resolution European anthropogenic emission inventory and all others.

The present work is supported by the Bulgarian National Science Fund (grant DN-04/2/13.12.2016),

EC -H2020 project 675121(project VI-SEEM),

EC-7FP grant PIRSES-GA-2013-612671 (project REQUA),

Program for career development of young scientists, BAS.

I. Georgieva is World Federation of Scientists grant holder.

References

1. Atanassov, E., Gurov, T., Karaivanova, A., Ivanovska, S., Durchova, M., Dimitrov, D.: On the parallelization approaches for intel MIC architecture. In: AIP Conference Proceedings, vol. 1773, p. 070001 (2016). https://doi.org/10.1063/1.4964983
2. Byun, D.: Dynamically consistent formulations in meteorological and air quality models for multiscale atmospheric studies part I: governing equations in a generalized coordinate system. J. Atmos. Sci. **56**, 3789–3807 (1999)
3. Byun, D., Ching, J.K.S.: Science Algorithms of the EPA Models-3 Community Multiscale Air Quality (CMAQ) Modeling system, United States Environmental Protection Agency, Office of Research and Development, Washington, D.C. 20460, EPA-600/ R-99/030 (1999)
4. CEP: Sparse Matrix Operator Kernel Emission (SMOKE) Modeling System, University of Carolina, Carolina Environmental Programs, Research Triangle Park, North Carolina (2003)

5. Gadzhev, G., Jordanov, G., Ganev, K., Prodanova, M., Syrakov, D., Miloshev, N.: Atmospheric composition studies for the Balkan region. In: Dimov, I., Dimova, S., Kolkovska, N. (eds.) NMA 2010. LNCS, vol. 6046, pp. 150–157. Springer, Heidelberg (2011). https://doi.org/10.1007/978-3-642-18466-6_17
6. Gadzhev, G., Syrakov, D., Ganev, K., Brandiyska, A., Miloshev, N., Georgiev, G., Prodanova, M.: Atmospheric composition of the Balkan region and Bulgaria. Study of the contribution of biogenic emissions. In: AIP Conference Proceedings, vol. 1404, pp. 200–209 (2011)
7. Gadzhev, G., Ganev, K., Syrakov, D., Miloshev, N., Prodanova, M.: Contribution of biogenic emissions to the atmospheric composition of the Balkan Region and Bulgaria. Int. J. Environ. Pollut. **50**(1/2/3/4), 130–139 (2012)
8. Gadzhev, G., Ganev, K., Miloshev, N., Syrakov, D., Prodanova, M.: Numerical study of the atmospheric composition in Bulgaria. Comput. Math. Appl. **65**, 402–422 (2013)
9. Gadzhev, G., Ganev, K., Syrakov, D., Prodanova, M., Miloshev, N.: Some statistical evaluations of numerically obtained atmospheric composition fields in Bulgaria. In: The Proceedings of 15th International Conference on Harmonisation within Atmospheric. Dispersion Modelling for Regulatory Purposes, Madrid, Spain, 6–9 May 2013, pp. 373–377 (2013)
10. Gadzhev, G., Ganev, K., Prodanova, M., Syrakov, D., Atanasov, E., Miloshev, N.: Multi-scale atmospheric composition modelling for Bulgaria. In: Steyn, D., Builtjes, P., Timmermans, R. (eds.) Air Pollution Modeling and Its Application XXII. NATO Science for Peace and Security Series C: Environmental Security, vol. 137, pp. 381–385. Springer, Dordrecht (2014). https://doi.org/10.1007/978-94-007-5577-2_64
11. Gadzhev, G., Ganev, K., Miloshev, N., Syrakov, D., Prodanova, M.: Calculation of some ozone pollution indeces for Bulgaria. Ecol. Saf. **8**, 384–392 (2014). ISSN 1314-7234
12. Ganev, K., Dimitrova, R., Miloshev, N.: Air flows and pollution transport in the Sofia valley under some typical background conditions. In: Proceedings of the XXVI. International Technical Meeting on Air Pollution Modelling and Its Applications, 26–30 May 2003, Istanbul, Turkey, pp. 593–594. Kluwer Academic/Plenum Publishing Corporation (2004)
13. Shamarock, W., Klemp, J.B., Dudhia, J., Gill, D.O., Barker, D.M., Duda, M.G., Huang, X., Wang, W., Powers, J.G.: A Description of the Advanced Research WRF Version 2 (2007)
14. Schwede, D., Pouliot, G., Pierce, T.: Changes to the biogenic emissions inventory system version 3 (BEIS3). In: Proceedings of 4th Annual CMAS Models-3 Users's Conference, 26–28 September 2005, Chapel Hill, NC (2005)
15. Syrakov, D., Ganev, K., Prodanova, M., Miloshev, N., Slavov, K.: Fine resolution modeling of climate change impact on future air quality over Bulgaria. In: Steyn, D., Builtjes, P., Timmermans, R. (eds.) Air Pollution Modeling and Its Application XXII. NATO Science for Peace and Security Series C: Environmental Security. Springer, Dordrecht (2012). https://doi.org/10.1007/978-94-007-5577-2_73
16. Syrakov, D., Etropolska, I., Prodanova, M., Slavov, K., Ganev, K., Miloshev, N., Ljubenov, T.: Downscaling of Bulgarian chemical weather forecast from Bulgaria region to Sofia city. In: American Institute of Physics Conference Proceedings, vol. 1561, pp. 120–132 (2013). https://doi.org/10.1063/1.4827221

17. Vestreng, V.: Emission data reported to UNECE/EMEP: evaluation of the spatial distribution of emissions. Meteorological Synthesizing Centre - West, The Norwegian Meteorological Institute, Oslo, Norway, Research Note 56, EMEP/MSC-W Note 1/2001 (2001)
18. Vestreng, V., Breivik, K., Adams, M., Wagner, A., Goodwin, J., Rozovskaya, O., Pacyna, J.M.: Inventory review 2005 (emission data reported to LRTAP convention and NEC directive), Technical report MSC-W 1/2005, EMEP (2005)
19. Visschedijk, A.J.H., Zandveld, P.Y.J., Denier van der Gon, H.A.C.: A high resolution gridded European emission database for the EU integrate project GEMS, TNO-report 2007-A-R0233/B, Apeldoorn, The Netherlands (2007)

Numerical Simulation of Deformations of Softwood Sawn Timber

Vladimir N. Glukhikh[1], Anna Yu. Okhlopkova[2], and Petr V. Sivtsev[3(✉)]

[1] Saint-Petersburg State University of Architecture and Civil Engineering, 2-nd Krasnoarmeiskaya St. 4, 190005 St. Petersburg, Russia
[2] Saint-Petersburg State Forest Technical University, Institutskiy per 5, 194021 St. Petersburg, Russia
[3] Ammosov North-Eastern Federal University, 58, Belinskogo, 677000 Yakutsk, Russia
sivkapetr@mail.ru

Abstract. Export of softwood sawn timber requires development of sawmill technology for better wood recovery. Therefore, problem of optimization of raw material cutting to obtain the maximum volume of high-quality sawn timber is of urgent priority. In this work we consider the elasticity equations that describe stress-strain state of timber. For numerical solution we approximate our system using finite element method. As the model problem we consider the deformations of the sawn timber under grown stresses depending on cutting patterns to define their board grade, which is linked with warp value. Wood parameters and inner stress model correspond to dahurian larch wood, which accounts for great part of timber export of Yakutia. The numerical simulation of the 3D problem is presented.

1 Introduction

The development of timber cutting technology is tightly linked with sawn material quality and therefore it cost. Timber cutting pattern determines inner stress distribution and hence, it deformations, which is one of the aspects of quality of the timber [1,2]. Thereby, the optimal cutting pattern, which provides the most profitable income for each wood, is in high demand. In Fig. 1 some examples of cutting pattern are presented.

In order to present the most robust way of determination of timber cutting pattern one must produce accurate mathematical model, which can predict a real timber stress-strain state. The main goal of investigation is to evaluate maximum value of bow and crook deformation as these values directly affect timber quality. These types of timber warping are explained in Fig. 2.

It is known that each type of wood has unique structure and physical parameters, which highly differ from another one. Therefore, each wood species must be described by certain unique mathematical model.

The research was supported by mega-grant of the Russian Federation Government (N 14.Y26.31.0013).

I. Lirkov and S. Margenov (Eds.): LSSC 2017, LNCS 10665, pp. 483–490, 2018.
https://doi.org/10.1007/978-3-319-73441-5_53

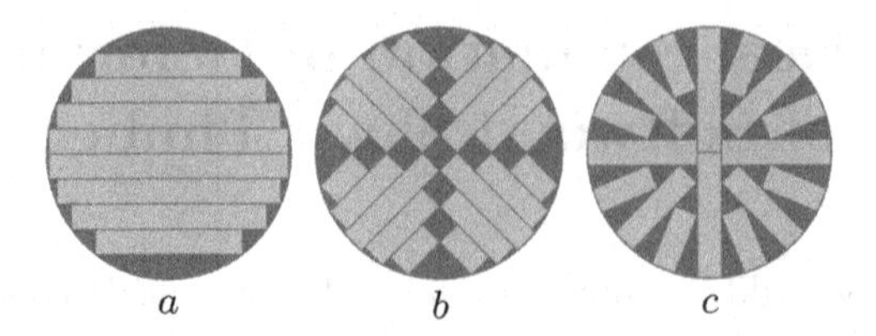

Fig. 1. Plain (a), quarter (b) and rift (c) cutting patterns

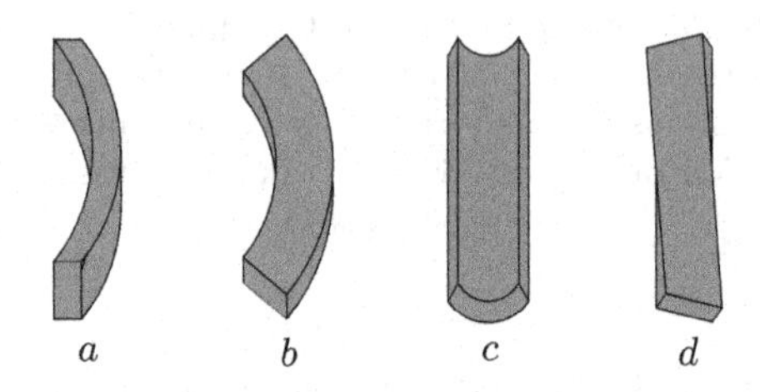

Fig. 2. Types of timber warping: bow (a), crook (b), cup (c), twist (d).

There are many applied mathematical problems related to calculation of stress-strain state of solid bodies [3–7]. In first approximation one uses models of linear elasticity that are described by Lame equations for displacement.

In this work we consider numerical simulation of mathematical model for dahurian larch wood, which accounts for great part of timber export of Yakutia. Dahurian larch refers to coniferous wood species with a noticeable core zone. Heartwood and sapwood have different physical and mechanical characteristics. That affects on the formation and distribution of internal stresses during growth and subsequent processing of wood.

The main goal is to verify empirically derived expression for inner stress components by numerical simulation. The computational algorithm is based on the finite-element approximation of displacement in space [8,9]. Numerical realization of method is performed using collection of free software FEniCS [10].

Features of computational algorithm are illustrated by calculation data for three-dimensional problem of dahurian larch sawn timber.

2 Problem Statement

Let us consider mathematical model that describes stress-strain state in computational domain Ω, which refers to sawn timber affected by inner growth stress in terms of linear elasticity problem for anisotropic material

$$\operatorname{div} \boldsymbol{\sigma}(\boldsymbol{x}) = 0, \quad \boldsymbol{x} \in \Omega, \tag{1}$$

where $\boldsymbol{x} = (x_1, x_2, x_3)$ is Cartesian coordinate vector. This coordinate system axis are parallel to the sides of sawn timber for convenience of calculations.

Then we add relation between stress tensor $\boldsymbol{\sigma}$ and deformation tensor $\boldsymbol{\varepsilon}$ for orthotropic material defined in local rotated coordinate system (r, t, a):

$$\begin{aligned}
\varepsilon_r &= \frac{\sigma_r}{E_r} - \frac{\mu_{tr}\sigma_t}{E_t} - \frac{\mu_{ar}\sigma_a}{E_a}, \\
\varepsilon_t &= -\frac{\mu_{rt}\sigma_r}{E_r} + \frac{\sigma_t}{E_t} - \frac{\mu_{at}\sigma_a}{E_a}, \\
\varepsilon_a &= -\frac{\mu_{ra}\sigma_r}{E_r} - \frac{\mu_{ta}\sigma_t}{E_t} + \frac{\sigma_a}{E_a}, \\
\gamma_{rt} &= \frac{\tau_{rt}}{G_{rt}}, \quad \gamma_{ra} = \frac{\tau_{ra}}{G_{ra}}, \quad \gamma_{ta} = \frac{\tau_{ta}}{G_{ta}},
\end{aligned}$$

where E_r, E_t, E_a are radial, tangential and longitudinal Young's moduli, G_{ij}, μ_{ij} are Shear moduli and Poisson's ratio, respectively.

Previous system of equations can be converted to the following form:

$$\boldsymbol{\sigma}(\boldsymbol{x}) = \boldsymbol{C}\boldsymbol{\varepsilon}(\boldsymbol{x})$$

in order to define equation for displacement vector. Here

$$\boldsymbol{C} = \begin{bmatrix} C_{11} & C_{12} & C_{13} & 0 & 0 & 0 \\ C_{21} & C_{22} & C_{23} & 0 & 0 & 0 \\ C_{31} & C_{32} & C_{33} & 0 & 0 & 0 \\ 0 & 0 & 0 & C_{44} & 0 & 0 \\ 0 & 0 & 0 & 0 & C_{55} & 0 \\ 0 & 0 & 0 & 0 & 0 & C_{66} \end{bmatrix}$$

is symmetric matrix with following components [11]

$$\begin{aligned}
C_{11} &= \frac{1 - \mu_{at}\mu_{ta}}{\Delta E_t E_a}, & C_{22} &= \frac{1 - \mu_{ar}\mu_{ra}}{\Delta E_r E_a}, & C_{33} &= \frac{1 - \mu_{rt}\mu_{tr}}{\Delta E_r E_t}, \\
C_{12} &= \frac{\mu_{tr} + \mu_{ar}\mu_{ta}}{\Delta E_t E_a}, & C_{13} &= \frac{\mu_{ar} + \mu_{tr}\mu_{at}}{\Delta E_t E_a}, & C_{23} &= \frac{\mu_{at} + \mu_{rt}\mu_{ar}}{\Delta E_r E_a}, \\
C_{44} &= G_{rt}, & C_{55} &= G_{ra}, & C_{66} &= G_{ta},
\end{aligned}$$

where

$$\Delta = \frac{1 - \mu_{rt}\mu_{tr} - \mu_{ra}\mu_{ar} - \mu_{ta}\mu_{at} - 2\mu_{tr}\mu_{at}\mu_{ra}}{E_r E_t E_a}.$$

In order to perform calculation using vectors from Cartesian coordinate system we need to transform deformation tensor calculated from displacement vector $\boldsymbol{u} = (u_1, u_2, u_3)$

$$\varepsilon_{i,j} = \frac{1}{2}\left(\frac{\partial u_i}{\partial x_j} + \frac{\partial u_j}{\partial x_i}\right)$$

for each point by using appropriate rotation matrix.

$$\boldsymbol{R} = \begin{bmatrix} \cos(\alpha) & -\sin(\alpha) & 0 \\ \sin(\alpha) & \cos(\alpha) & 0 \\ 0 & 0 & 1 \end{bmatrix}$$

where $\cos(\alpha) = \frac{x_1}{r}$ and $\sin(\alpha) = \frac{x_2}{r}$.

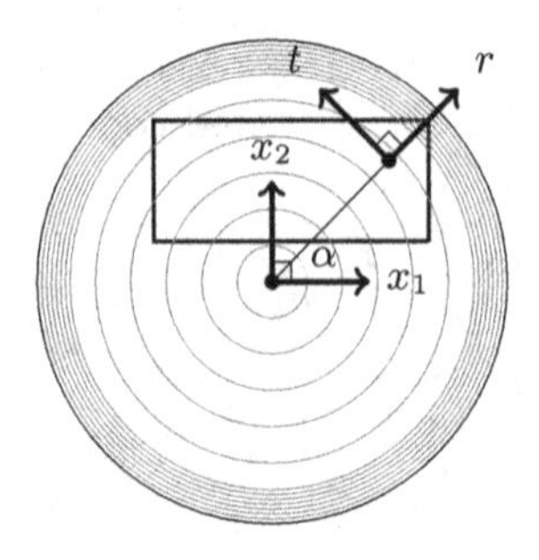

Fig. 3. Cartesian and local rotated coordinate system axis in $r - t$ plane

This rotation is shown in $r - t$ - plane in Fig. 3.

Then we add boundary conditions that correspond to surface stresses and fixation. In particular, we define Neumann boundary condition

$$(\boldsymbol{\sigma} \cdot \boldsymbol{n})(\boldsymbol{x}) = \boldsymbol{\sigma}_n(\boldsymbol{x}), \quad \boldsymbol{x} \in \Gamma_N, \tag{2}$$

that defines sawn timber inner growth stress at whole surface of the sawn timber. So we have pure Neumann problem, to solve which we use fixation of pinpoints defined by following Dirichlet boundary conditions

$$\boldsymbol{u}(\boldsymbol{x}) = \boldsymbol{0}, \quad \boldsymbol{x} \in \Gamma_D, \tag{3}$$

which defines fixation of displacement at three certain point by 1, 2 and all axis, respectively. Such fixation helps to remove random shift of body and does not constrain body transformations.

3 Finite-Element Discretisation

To get numerical solution for Eq. (1) we build finite-element approximation by space [12].

Let $L_2(\Omega)$ be Hilbert space with inner product and norm as

$$(u, v) = \int_\Omega u(\boldsymbol{x})\, v(\boldsymbol{x})\, dx, \quad ||u|| = (u, u)^{1/2},$$

and $\boldsymbol{L_2} = (L_2(\Omega))^d$ be space for vectors, where $\Omega \in \mathbb{R}^d$, and $d = 2, 3$ is dimension of the problem. Also let $H^1(\Omega)$ and $\boldsymbol{H}^1(\Omega)$ be Sobolev spaces of the first order.

Then, using boundary conditions (3), (2) we obtain the following variational formulation of problem: find $\boldsymbol{u} \in \hat{\boldsymbol{V}}$ that

$$\int_\Omega \boldsymbol{\sigma}(\boldsymbol{u})\, \boldsymbol{\varepsilon}(\boldsymbol{v}) d\boldsymbol{x} = \int_{\Gamma_N} (\boldsymbol{\sigma}_n, \boldsymbol{v}) ds, \quad \forall \boldsymbol{v} \in \hat{\boldsymbol{V}}, \tag{4}$$

where $\hat{\boldsymbol{V}}$ is test function space, which is defined as

$$\hat{\boldsymbol{V}} = \{\boldsymbol{v} \in \boldsymbol{H}^1(\Omega) : \boldsymbol{v}(\boldsymbol{x}) = 0, \quad \boldsymbol{x} \in \Gamma_D\},$$

and $\boldsymbol{V}$ is space of trial functions, shifted from test function space by the value of Dirichlet condition $\boldsymbol{u}_0$.

$$\boldsymbol{V} = \{\boldsymbol{v} \in \boldsymbol{H}^1(\Omega) : \boldsymbol{v}(\boldsymbol{x}) = \boldsymbol{u}_0, \quad \boldsymbol{x} \in \Gamma_D\}.$$

Then we define bilinear and linear forms as

$$a(\boldsymbol{u}, \boldsymbol{v}) = \int_\Omega \boldsymbol{\sigma}(\boldsymbol{u})\,\varepsilon(\boldsymbol{v})d\boldsymbol{x},$$

$$L(\boldsymbol{v}) = -\int_{\Gamma_N} (\boldsymbol{\sigma}_n, \boldsymbol{v})ds.$$

Then Eq. (4) switches to following variational formulation: find $\boldsymbol{u} \in \boldsymbol{V}$ so, that

$$a(\boldsymbol{u}, \boldsymbol{v}) = L(\boldsymbol{v}), \quad \forall \boldsymbol{v} \in \hat{\boldsymbol{V}}.$$

4 Investigation Object

The object of investigation is dahurian larch sawn timber with 156 mm × 54 mm × 5 m in size. The wood, from which timber has been sawn, is modeled as conoid. For dahurian larch we use physical parameter values presented in Table 1.

According to previous investigations [1] we present following inner stress distribution model for dahurian larch

$$\sigma_r(r) = \sigma_{ta} \ln\left(\frac{r}{R}\right),$$

$$\sigma_t(r) = \sigma_{ta}\left(1 + \ln\left(\frac{r}{R}\right)\right),$$

$$\sigma_a(r) = \sigma_0\left(1 - 7\frac{r}{R} - 8\frac{r^{14}}{R^{14}}\right).$$

Here $r = \sqrt{x_1^2 + x_2^2}$ is a radius of certain point in surface and R is wood radius which depends on x_3. For $x_3 = 0$ we have $R(0) = R_1$ and $R(5) = R_2$ at two edges of sawn timber.

Therefore, for solution of pure Neumann problem we fix a displacement of three separated points in geometry by some axis. Geometry of sawn timber, pinpoints and axis are shown in Fig. 4.

The pinpoint 1 is clamped by all three axis. The pinpoint 2 is fixed by x_2 and the pinpoint 3 locked by x_1 and x_2 axis.

Table 1. Physical parameters

Elasticity modulus			Shear modulus			Poisson coefficient		
E_r, GPa	E_t, GPa	E_a, GPa	G_{rt}, GPa	G_{ar}, GPa	G_{at}, GPa	μ_{tr}	μ_{ar}	μ_{ta}
0.7	0.35	11	0.33	1.14	0.71	0.62	0.56	0.035

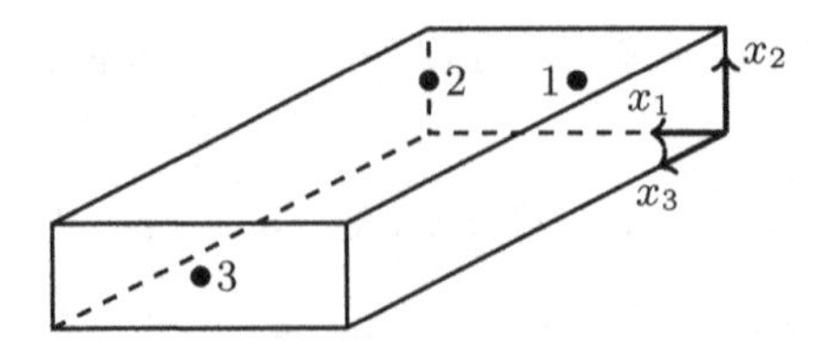

Fig. 4. Geometry and pin points of the sawn timber

5 Numerical Simulation

As an example of numerical simulation of model problem we consider sawn timber with long side parallel to longitudinal axis of the tree, so axis $a = x_3$. Location of timber inside wood in $r - t$ section is shown in Fig. 5.

For simulation we used typical values for $\sigma_0 = 2\,\text{MPa}$ and $\sigma_{ta} = 0.1\,\text{MPa}$. Calculations were performed on the mesh presented in Fig. 6 that is proved as mesh with optimal size, providing accurate results.

For these values we receive displacement distribution shown in Fig. 7.

As it is shown, for placement of sawn timber shown in Fig. 7 we have strong bow and crook deformations. These values according to GOST26002-83 standard refers to 5th quality grade, which is unlikely to be exported. Cup and twist deformations are nearly undetectable.

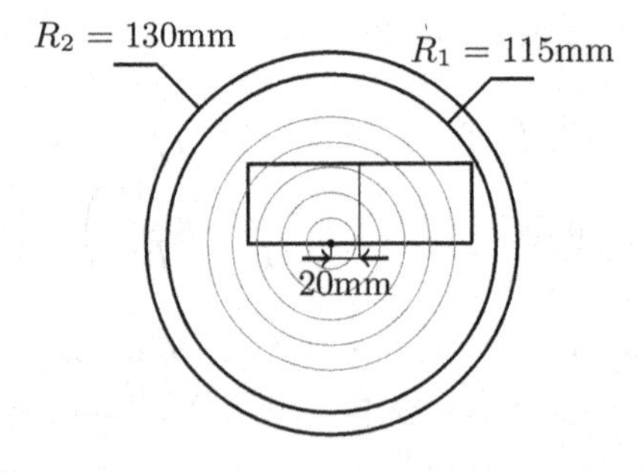

Fig. 5. Timber location inside the wood

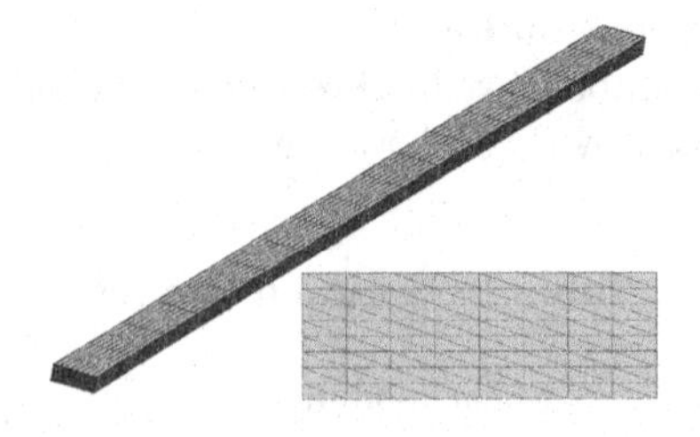

Fig. 6. Full mesh and mesh side facet (1377 vertices, 7560 elements)

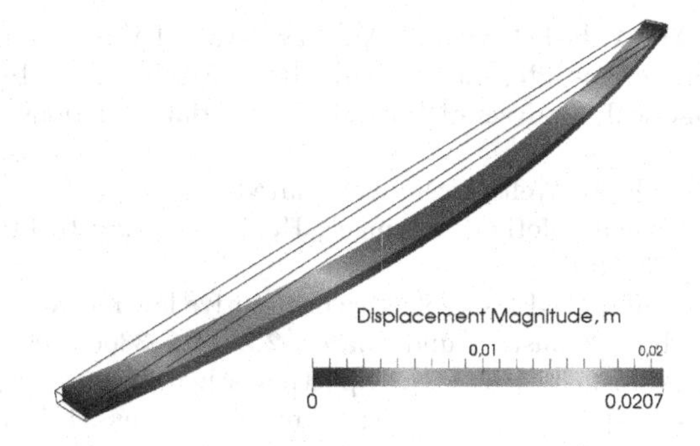

Fig. 7. Displacement magnitude distribution. Shape warped by 10× displacement vector

6 Conclusion

We accomplish numerical solution for linear elasticity problem for sawn timber, which has anisotropic elastic nature. As a result, for typical values we received adequate distribution of displacement.

In future we plan to hold investigations related directly with experimental data in order to test mathematical model in practical usage.

References

1. Glukhikh, V.N., Akopyan, A.L.: Nachal'nye naprjazhenija v drevesine: monografija. The Ministry of Education and Science of the Russian Federation, pp. 114–115 (2016)
2. Glukhikh, V.N., Khrabrova, O.Yu.: Bending of sawn wood products obtained from conventional sawing and parallel to generatix sawing. Archit. Eng. **1**(1), 4–9 (2016)
3. Vabishchevich, P.N., Vasil'eva, M.V., Kolesov, A.E.: Shema rasshheplenija dlja zadach porouprugosti i termouprugosti. Zhurnal vychislitel'noj matematiki i matematicheskoj fiziki **54**(8), 1345–1355 (2014)
4. Kolesov, A.E., Vabishchevich, P.N., Vasil'eva, M.V.: Splitting schemes for poroelasticity and thermoelasticity problems. Comput. Math. Appl. **67**(12), 2185–2198 (2014)
5. Kolesov, A.E., Vabishchevich, P.N., Vasilyeva, M.V., Gornov, V.F.: Splitting scheme for poroelasticity and thermoelasticity problems. In: Dimov, I., Faragó, I., Vulkov, L. (eds.) FDM 2014. LNCS, vol. 9045, pp. 241–248. Springer, Cham (2015). https://doi.org/10.1007/978-3-319-20239-6_25
6. Sivtsev, P.V., Vabishchevich, P.N., Vasilyeva, M.V.: Numerical simulation of thermoelasticity problems on high performance computing systems. In: Dimov, I., Faragó, I., Vulkov, L. (eds.) FDM 2014. LNCS, vol. 9045, pp. 364–370. Springer, Cham (2015). https://doi.org/10.1007/978-3-319-20239-6_40
7. Sivtsev, P.V., Kolesov, A.E., Sirditov, I.K., Stepanov, S.P.: The numerical solution of thermoporoelastoplasticity problems. In: AIP Conference Proceedings, vol. 1773, p. 110010 (2016)

8. Afanas'eva, N.M., Vabishchevich, P.N., Vasil'eva, M.V.: Unconditionally stable schemes for convection-diffusion problems. Russ. Math. **57**(3), 1–11 (2013)
9. Lui, S.H.: Numerical Analysis of Partial Differential Equations. Wiley, Hoboken (2012)
10. Logg, A., Mardal, K.A., Wells, G.N.: Automated Solution of Differential Equations by the Finite Element Method. Springer, Heidelberg (2012). https://doi.org/10.1007/978-3-642-23099-8
11. Oudjene, M., Khelifa, M.: Elasto-plastic constitutive law for wood behaviour under compressive loadings. Constr. Build. Mater. **23**, 3359–3366 (2009)
12. Kolesov, A.E., Vabishchevich, P.N.: Splitting schemes with respect to physical processes for double-porosity poroelasticity problems. Russ. J. Numer. Anal. Math. Model. **32**(2), 99–113 (2017)

Large Scale Computations in Fluid Dynamics

Valentin A. Gushchin[1,2(✉)]

[1] Institute for Computer Aided Design Russian Academy of Sciences, Moscow, Russia
gushchin@icad.org.ru

[2] Moscow Institute of Physics and Technology, State University, Dolgoprudny, Russia

Abstract. Many phenomena in the nature may be considered in the frame of the incompressible fluid flows. Such flows are described by the Navier-Stokes equations. As usually we have deal with the flows with large gradients of hydrodynamic parameters (flows with a free surface, stratified fluid flows, separated flows, etc.). For direct numerical simulation of such flows finite difference schemes should possess by the following properties: high order of accuracy, minimum scheme viscosity, dispersion and monotonicity. The Splitting on the physical factors Method for Incompressible Fluid flows (SMIF) with hybrid explicit finite difference scheme based on Modified Central Difference Scheme (MCDS) and Modified Upwind Difference Scheme (MUDS) with special switch condition depending on the velocity sign and the signs of the first and second differences of transferred functions has been developed. This method has been successfully applied for the flows with a free surface including regimes with broken surface wave, for 3D separated homogeneous and stratified fluid flows around a sphere and a circular cylinder including transitional regimes. The air, heat and mass transfer in the clean rooms for the pharmaceutical industry is considered. The parallelization of the algorithm has been made and applied on the massive parallel computers with a distributed memory. Some examples of calculated problems will be discussed.

Keywords: Direct numerical simulation · Viscous fluid flows
Sphere · Cylinder · Clean rooms

1 Introduction

Unsteady 3D separated and wavy fluid flows around a moving blunt body are very wide spread phenomena in the nature. The understanding of such flows is very important both from theoretical and from practical points of view. The aim of the present paper is the demonstration of the opportunities of the mathematical modeling of the separated flows of the homogeneous and the stratified fluid around blunt bodies on the basis of the Navier-Stokes equations in the Boussinesq approximation. In the homogeneous water case only the wakes of

I. Lirkov and S. Margenov (Eds.): LSSC 2017, LNCS 10665, pp. 491–498, 2018.
https://doi.org/10.1007/978-3-319-73441-5_54

the obstacles are observed. In the stratified water along with wakes the internal waves are generated.

The other problem which also may be considered in the frame of incompressible fluid flows is clean room problem. For the designing and construction of the clean rooms for aerospace, microelectronic, pharmaceutical, chemical and food processing industries, one can ensure minimum contamination and optimal heat conditions. As usually we have deal with the flows with large gradients of hydrodynamic parameters (flows with a free surface, stratified fluid flows, separated flows, etc.). For direct numerical simulation of such flows finite difference schemes should possess by the following properties: high order of accuracy, minimum scheme viscosity, dispersion and monotonicity. The Splitting on the physical factors Method for Incompressible Fluid flows (SMIF) with hybrid explicit finite difference scheme based on Modified Central Difference Scheme (MCDS) and Modified Upwind Difference Scheme (MUDS) with special switch condition depending on the velocity sign and the signs of the first and second differences of transferred functions has been developed and proved [3,4]. This method was successfully applied for the flows with a free surface including regimes with broken surface wave, for 3D separated homogeneous and stratified fluid flows around a sphere and a circular cylinder including transitional regimes [1,2,5,6], for the air, heat and mass transfer in the clean rooms [7]. The parallelization of the algorithm has been made and applied on the massive parallel computers with a distributed memory such as PARAM 10000 (based on Ultra Sparc II processors), MPS (based on Intel Xeon processors), PARAM PADMA.RU (based on IBM Power 5 processors). The code was parallelized by using a domain decomposition method in radial direction for a sphere and a circular cylinder. The number of grid points varied from $(120 \times 60 \times 120)$ till $(240 \times 240 \times 72)$ per each of five (or six) unknown functions.

2 Equations and Boundary Conditions

Let $\rho(x,y,z) = 1 - x/2A + S(x,y,z)$ is the non-dimensional (by ρ_0) density of the linearly stratified fluid where x, y, z are the Cartesian coordinates; z, x, y are the streamwise, lift and lateral directions non-dimensionalized by $d/2$, d is a diameter of the moving body; $A = \Lambda/d$ is the scale ratio, Λ is the buoyancy scale, which is related to the buoyancy frequency N and period T_b $(N = 2\pi/T_b, N^2 = g/\Lambda)$; g is the scalar of the gravitational acceleration; S is a dimensionless perturbation of salinity. The density stratified viscous fluid flows have been simulated on the basis of the Navier-Stokes equations in the Boussinesq approximation (1)–(3) (including the diffusion equation (1) for the stratified component (salt)) with four dimensionless parameters: Froude number $Fr = U/(N \cdot d)$, Reynolds number $Re = U \cdot d/\nu$, the scale ratio $A \gg 1$, Schmidt number $Sc = \nu/\kappa = 709.22$.

$$\frac{\partial S}{\partial t} + (\mathbf{v} \cdot \nabla) S = \frac{2}{Sc \cdot Re} \triangle S + \frac{v_x}{2A} \tag{1}$$

$$\frac{\partial \mathbf{v}}{\partial t} + (\mathbf{v} \cdot \nabla) \mathbf{v} = -\nabla p + \frac{2}{Re} \triangle \mathbf{v} + \frac{A}{2Fr^2} S \frac{\mathbf{g}}{g} \tag{2}$$

$$\nabla \cdot \mathbf{v} = 0 \tag{3}$$

In (1)–(3) $\mathbf{v} = (v_x, v_y, v_z)$ is the velocity vector (non-dimensionalized by U), p is a perturbation of pressure (non-dimensionalized by $\rho_0 U^2$), U is the scalar of the body velocity, ν is the kinematical viscosity, κ is the salt diffusion coefficient, t is time (non-dimensionalized by $t_0 = d/(2 \cdot U) = 1/(2 \cdot Fr \cdot N)$). It is convenient to introduce one more non-dimensional time $T = t \cdot t_0 / T_b = [t/(2 \cdot Fr \cdot N)] \cdot N/2\pi = t/(4\pi \cdot Fr)$. In the case of the spherical coordinate system R, θ, φ ($x = R \sin\theta \cos\varphi, y = R \sin\theta \sin\varphi, z = R\cos\theta, \mathbf{v} = (v_R, v_\theta, v_\varphi)$) the following boundary conditions have been used on the sphere surface:

$$v_R = v_\theta = v_\varphi = 0, \left.\frac{\partial p}{\partial R}\right|_{R=d/2} = \left.\left(\frac{\partial S}{\partial R} - \frac{1}{2A}\frac{\partial x}{\partial R}\right)\right|_{R=d/2} = 0 \tag{4}$$

On the external boundary of the O-type grid the following boundary conditions have been used: (1) for $z < 0 : v_R = \cos\theta, v_\theta = -\sin\theta, v_\varphi = 0, S = 0$; (2) for $z \geq 0 : v_R = \cos\theta, v_\theta = -\sin\theta, \frac{\partial v_\varphi}{\partial R} = 0, \frac{\partial S}{\partial z} = 0$.

3 Numerical Method SMIF

For solving of the Navier-Stokes equations (1)–(3) the Splitting on physical factors Method for Incompressible Fluid flows (SMIF) has been used [3,4]. Let the velocity, the perturbation of pressure and the perturbation of salinity are known at some moment $t_n = n \cdot \tau$, where τ is time step and n is the number of time-steps. Then the calculation of the unknown functions at the next time level $t_{n+1} = (n+1) \cdot \tau$ for Eqs. (1)–(3) can be presented in the following four-step form:

$$\frac{S^{n+1} - S^n}{\tau} = -\left(\mathbf{v}^n \cdot \nabla\right) S^n + \frac{2}{Sc \cdot Re} \triangle S^n + \frac{{v_x}^n}{2A} \tag{5}$$

$$\frac{\tilde{\mathbf{v}} - \mathbf{v}^n}{\tau} = -\left(\mathbf{v}^n \cdot \nabla\right) \mathbf{v}^n + \frac{2}{Re} \triangle \mathbf{v}^n + \frac{A}{2Fr^2} S^{n+1} \frac{\mathbf{g}}{g} \tag{6}$$

$$\tau \triangle p = \nabla \cdot \tilde{\mathbf{v}} \tag{7}$$

$$\frac{\mathbf{v}^{n+1} - \tilde{\mathbf{v}}}{\tau} = -\nabla p \tag{8}$$

In order to understand the finite-difference scheme for the convective terms of the Eqs. (5)–(6) let us consider the linear model equation and a finite-difference approximation of this equation:

$$f_t + u f_x = 0$$

where f - is unknown function, $u = \text{const}$.

$$\frac{f_i^{n+1} - f_i^n}{\tau} + u \frac{\tilde{f}_{i+1/2}^n - \tilde{f}_{i-1/2}^n}{h} \tag{9}$$

Let us investigate the class of the difference schemes which can be written in the form of the two-parameter family and depends on the parameters α and β as follows:

$$\begin{aligned} \tilde{f}^n_{i+1/2} &= \alpha \begin{pmatrix} f^n_{i-1} \\ f^n_{i+2} \end{pmatrix} + (1-\alpha-\beta) \begin{pmatrix} f^n_{i} \\ f^n_{i+1} \end{pmatrix} + \beta \begin{pmatrix} f^n_{i+1} \\ f^n_{i} \end{pmatrix}, \begin{matrix} u \geq 0 \\ u < 0 \end{matrix} \\ \tilde{f}^n_{i-1/2} &= \alpha \begin{pmatrix} f^n_{i-2} \\ f^n_{i+1} \end{pmatrix} + (1-\alpha-\beta) \begin{pmatrix} f^n_{i-1} \\ f^n_{i} \end{pmatrix} + \beta \begin{pmatrix} f^n_{i} \\ f^n_{i-1} \end{pmatrix}, \begin{matrix} u \geq 0 \\ u < 0 \end{matrix} \end{aligned} \tag{10}$$

In this case the first differential approximation for Eq. (9) has the form

$$f_t + uf_x = \frac{Ch^2}{2\tau}\left[1 + 2(\alpha - \beta) - C\right] f_{xx} + \frac{Ch^3}{3!\tau}\text{sign}u \left[C^2 - 6\alpha - 1\right] f_{xxx} \dots \tag{11}$$

where $C = \frac{|u|\tau}{h}$ is the Courant number. If we put $\alpha = \beta = 0$ in (10) we'll obtain usual first order monotonic scheme which is stable when $0 \leq C \leq 1$. It is known that it is impossible to construct a homogeneous monotonic difference scheme of higher order than the first order of the approximation for Eq. (9). A monotonic scheme of higher order can therefore only be constructed either on the basis of second-order homogeneous schemes using smoothing operators, or on the basis of the hybrid schemes using different switch conditions from one scheme to another (depending on the nature of the solution), possibly with the use of smoothing. Here we use the hybrid monotonic difference scheme [4] with the following switch condition. Let $\Delta f_{i+1} = f_{i+1} - f_i$, $\Delta^2 f_i = \Delta f_i - \Delta f_{i-1}$. If $(u \cdot \Delta f \cdot \Delta^2 f) \leq 0$ the Modified Central Difference Scheme (MCDS) with $\alpha = 0, \beta = 0.5\,(1 - C)$ is used and if $(u \cdot \Delta f \cdot \Delta^2 f) > 0$ the Modified Upwind Difference Scheme (MUDS) with $\alpha = -0.5\,(1 - C)\,, \beta = 0$ is used. It was shown [4] that the areas of monotonicity of pointed schemes have non-zero intersection. So the constructed hybrid scheme with MCDS and MUDS has second order of accuracy with respect to the time and spatial variables, zero scheme viscosity and monotonous. It is stable when the Courant criterion $0 \leq C \leq 1$ is satisfied. Moreover it was shown [4] that this hybrid scheme comes nearest to the third order schemes. The generalization of the considered finite-difference scheme for 2D and 3D problems is easily performed.

4 The Visualization Techniques

For the visualization of the 3D vortex structures in the fluid flows the isosurfaces of β, λ_2 (Figs. 1 and 2) and the streamwise component of vorticity ω_z (Fig. 3, $\omega = rot\mathbf{v}$) have been drawing, where β is the imaginary part of the complex-conjugate eigenvalues of the velocity gradient tensor G [8] (Fig. 1b), λ_2 is the second eigenvalue of the $\mathbf{S}^2 + \boldsymbol{\Omega}^2$ tensor, where S and Ω are the symmetric and antisymmetric parts of G [9] (Fig. 1a, c).

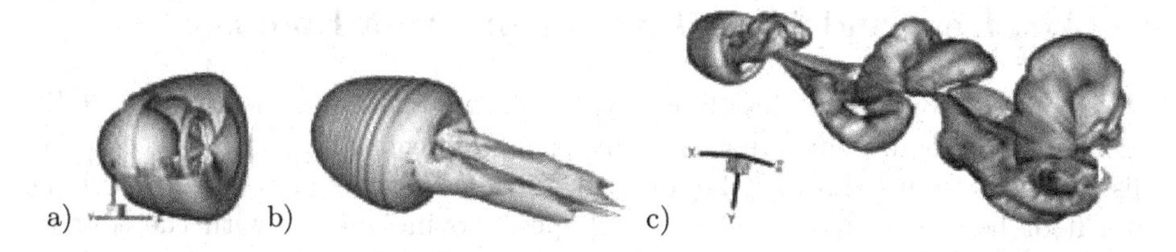

Fig. 1. Vortex structures of the sphere wake for $Fr > 10$: a–c - $Re = 200, 250, 350$; (a) $\lambda_2 = -10^{-6}$ and -0.16; (b) $\beta = 0.04$; (c) $\lambda_2 = -2 \cdot 10^{-5}$

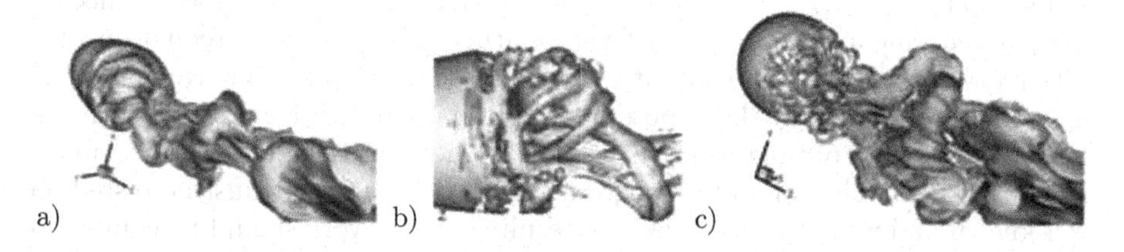

Fig. 2. Vortex structures of the sphere wake for $Fr > 10$: a–c - $Re = 10^4, 10^4, 5 \cdot 10^5$; (a) $\lambda_2 = -10^{-4}$; (b) $\beta = 1$; (c) $\lambda_2 = -10^{-4}$

5 Some Fluid Flow Regimes Around a Circular Cylinder

Let us consider the homogeneous viscous fluid flow regimes around a circular cylinder. For $Re > 40$ the periodical formation of vortex tubes is simulated in the wake. For $Re = 191$ 2D-3D transition is observed in the wake. It means that for $Re > 191$ there is a periodicity of the flow along the circular cylinder axis. For $191 < Re \leq 300$ and $300 \leq Re \leq 400$ the periodicity scales are equal to $3.5d \leq \lambda \leq 4d$ (mode A) and $0.8d \leq \lambda \leq 1.0d$ (mode B) correspondingly (Fig. 3). Owing to our investigations it was found that the values of the maximum phase difference along the circular cylinder axis are approximately equal to $0.1 - 0.2T_f$ (for mode A) and $0.015 - 0.030T_f$ (for mode B), where the time T_f is the period of the flow [2].

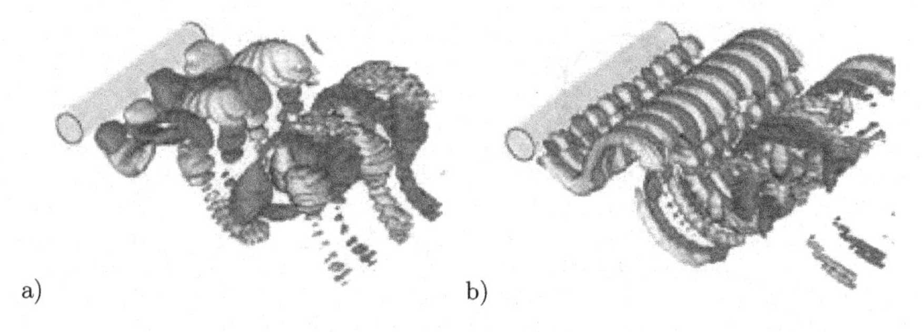

Fig. 3. The isosurfaces of two opposite streamwise components of the vorticity in the wake of the circular cylinder: (a–b) $Re = 230, 320$; $\lambda/d = 3.75, 0.83$

6 Air, Heat and Mass-Transfer in Clean Rooms

Let's consider some examples of air, heat and mass-transfer in clean rooms (CR) problem. The cleanrooms are three-dimensional objects with complex geometry, distributed systems through the air inlet and outlet vents or perforated ceiling and floor hosted equipment complex shapes, moving robots, with the sources of particles of different sizes and various laws of their motion. The air, heat fluxes and movement of particles are three-dimensional and turbulent in terms of gas dynamics. For mathematical modelling of these processes it is necessary to develop the adequate mathematical models that correctly describe the mechanisms of movement. To solve the relevant mathematical problems (equation with initial and boundary conditions), it is necessary to develop an effective numerical algorithms, which will make it possible to obtain quantitative results (within a reasonable amount of time and acceptable accuracy) in the form of tables, curves or flow pictures. The computer programs or applied packages must be tested on well-known and well-studied tasks, the results of their work should be compared with the calculations of other authors, with the data of physical experiments and measurements. Some such methodical and test calculations have been done for example in [1–4,6].

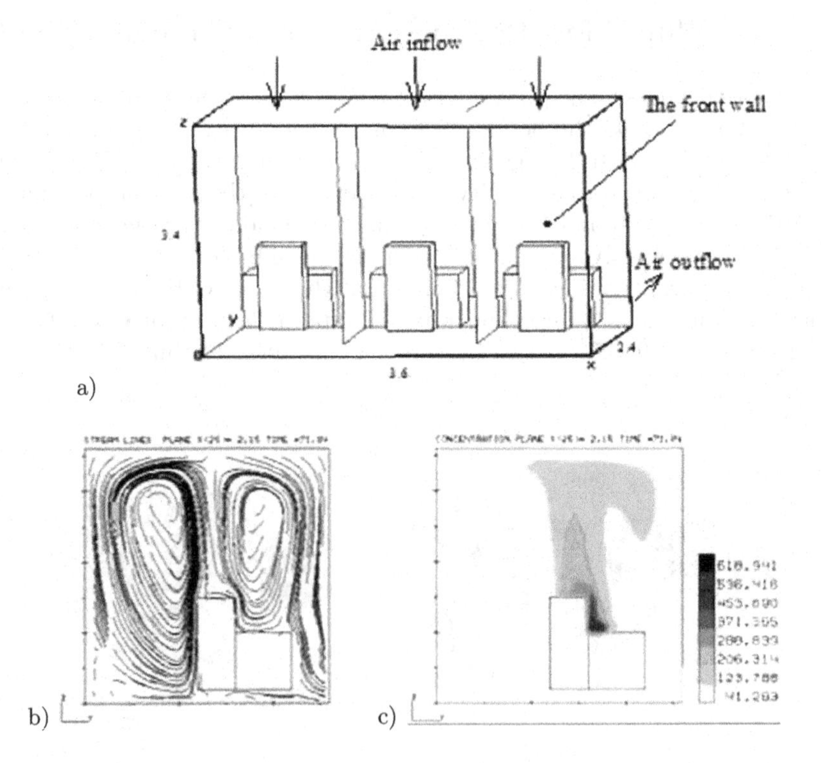

Fig. 4. (a) CR configuration; (b) stream lines; (c) the picture for concentration of powder substance

The following example demonstrates how to use the applied package CRAG-Clean Room Air-dynamic Guide (based on SMIF) at schematic design phase, i.e. when the designer wants to get answers for some questions of interest and select the optimal solution before to start designing. Let us consider the example of designing a weight room where one intends to weigh some powder substances for the pharmaceutical industry. Room configuration, consisting of three compartments with the tables and operators is shown in Fig. 4. Air is blown into the room through the ceiling with a speed of 0.02 m/c and is discharged through the bars with a height of 0.5 m at the bottom of the front wall of the room. Supply air temperature is 22 ° C. The main question was the following: is it possible the powder substance from a table of one of the compartments to fall into other compartments. If the temperature in the room and all objects within it (including operators), equal to the temperature of inlet air, then the situation will prove to be quite acceptable. However, if we take into account that the body temperature of the operator is equal to 36 ° C then the vertical convective flow can significantly change the situation.

7 Conclusion

The brief description of the Splitting on the physical factors Method for Incompressible Fluid flows (SMIF) with hybrid explicit finite difference scheme based on Modified Central Difference Scheme (MCDS) and Modified Upwind Difference Scheme (MUDS) with special switch condition depending on the velocity sign and the signs of the first and second differences of transferred functions has been given. This method was successfully applied for the investigation of flows with a free surface including regimes with broken surface wave, for 3D separated homogeneous and stratified fluid flows around a sphere and a circular cylinder including transitional regimes. Some examples of numerical calculations and visualizations for 3D flows around a sphere and a circular cylinder were described. The 3D visualisation used here gave us the possibility for more careful analysis of the transitional regimes for the fluid flows around a sphere and a 3D circular cylinder and the further understanding and refining of classification of the flow regimes. The mathematical modelling of the air, heat and mass transfer in the weight room for the pharmaceutical industry shows the necessity of taking into account the body temperature of operator.

Acknowledgments. This work has been partly supported by Russian Science Foundation (grant No. 17-11-01286).

References

1. Gushchin, V.A., Matyushin, P.V.: Numerical simulation and visualization of vortical structure transformation in the flow past a sphere at an increasing degree of stratification. J. Comput. Math. Math. Phys. Russ. **51**(2), 251–263 (2011)
2. Gushchin, V.A., Kostomarov, A.V., Matyushin, P.V.: 3D visualization of the separated fluid flows. J. Vis. Jpn. **7**(2), 143–150 (2004)

3. Gushchin, V.A., Konshin, V.N.: Computational aspects of the splitting method for incompressible flow with free surface. J. Comput. Fluids **21**(3), 345–353 (1992)
4. Gushchin, V.A.: Family of quasi-monotonic finite-difference schemes of the second-order of approximation. Math. Models Comput. Simul. **8**, 487–496 (2016). https://doi.org/10.1134/S2070048216050094
5. Gushchin, V., Matyushin, P.: The theory and applications of the SMIF method for correct mathematical modeling of the incompressible fluid flows. In: Dimov, I., Faragó, I., Vulkov, L. (eds.) FDM 2014. LNCS, vol. 9045, pp. 209–216. Springer, Cham (2015). https://doi.org/10.1007/978-3-319-20239-6_21
6. Gushchin, V.A., Kostomarov, A.V., Matyushin, P.V., Pavlyukova, E.R.: Direct numerical simulation of the transitional separated fluid flows around a sphere and a circular cylinder. J. Wind Eng. Ind. Aerodyn. **90**, 341–358 (2002)
7. Chafle, G., Gushchin, V.A., Narayanan, P.S.: Parallel computing of industrial aerodynamics problems: clean rooms. In: Schiano, P., Ecer, A., Periaux, J., Satofuka, N. (eds.) Parallel Computational Fluid Dynamics: Algorithms and Results Using Advanced Computers, pp. 305–311. Elsevier Science B.V., Amsterdam (1997)
8. Chong, M.S., Perry, A.E., Cantwell, B.J.: A general classification of three dimensional flow fields. Phys. Fluids **A2**(5), 765–777 (1990)
9. Jeong, J., Hussain, F.: On the identification of a vortex. J. Fluid Mech. **285**, 69–94 (1995)

A Domain Decomposition Multilevel Preconditioner for Interpolation with Radial Basis Functions

Gundolf Haase[1(✉)], Dirk Martin[2], Patrick Schiffmann[2], and Günter Offner[2]

[1] Institute for Mathematics and Scientific Computing, University of Graz, Graz, Austria
gundolf.haase@uni-graz.at
[2] AVL List GmbH, Graz, Austria
http://imsc.uni-graz.at/haasegu

Abstract. We present the reasonableness of the extension of a two-level domain decomposition method to a multilevel method as a preconditioner for interpolation with radial basis functions (RBF) on distributed memory systems. The arising subproblems are efficiently solved using the FGP algorithm, a method that is well-suited for shared memory settings.

1 Introduction

Many scientific and industrial simulation problems involve interpolation over numerous data sites as a subtask. Interpolation with radial basis functions (RBF) is an interpolation method applicable with (almost) no conditions on the distribution of the data sites and in arbitrary dimensions, that features favourable smoothness and convergence properties. Nevertheless, a straightforward application is hindered by the ill-conditioning of the resulting dense systems of linear equations arising from the interpolation task [2,3,15].

We use RBF interpolation for the deformation of computational domains in $\mathbb{R}^3$ as proposed in [4]. The originating simulation tasks require the utilisation of distributed memory systems, thus a distributed solution for the interpolation task is demanded. We pursue our approach in [12] by extending the presented two-level method to a multilevel method. As the initial approach, our method is based on [2,13]. The arising subproblems are efficiently solved by a preconditioned Krylov subspace projection method suitable for shared memory settings [8,9], the FGP algorithm.

The remaining paper is organized as follows. Section 2 gives an introduction to RBF interpolation. Section 3 features preconditioning methods for shared memory and distributed memory settings. This includes the extension of a two-level method to a multilevel method. We present numerical results in Sect. 4 and annotate some conclusions in Sect. 5.

I. Lirkov and S. Margenov (Eds.): LSSC 2017, LNCS 10665, pp. 499–506, 2018.
https://doi.org/10.1007/978-3-319-73441-5_55

2 Interpolation with Radial Basis Functions

For detailed discussions of the analysis on RBF interpolation we refer to the books [6,18]. A comprehensive view over aspects of theory and application is given in [10].

The general setting for RBF interpolation is a given set of data sites $X = \{x_i\}_{i=1}^N$ from a domain $\Omega \subseteq \mathbb{R}^d$ and a set of associated real function values $f_i = f(x_i)$. Sought is an approximating function $s : \Omega \to \mathbb{R}$ by interpolation $s|_X = f|_X$ of the form

$$s(x) = \sum_{i=1}^{N} \lambda_i \phi(\|x - x_i\|) + p(x), \tag{1}$$

with real coefficients λ_i, a radial function Φ, and a polynomial term $p \in \mathbb{P}_d^{k-1}$. In general, a function $\Phi : \mathbb{R}^d \to \mathbb{R}$ is called radial if there exists a univariate function $\phi : [0, \infty) \to \mathbb{R}$ such that $\Phi(\boldsymbol{x}) = \phi(\|\boldsymbol{x}\|_2)$. The polynomial term p is required for the existence and uniqueness of a solution, the required degree depends on the choice of the basis function ϕ. In case the choice of the basis function requires a polynomial term, the given set of points X has to be unisolvent with respect to polynomials of the corresponding degree. A set $X \subset \mathbb{R}^d$ is called unisolvent for $\mathbb{P}_d^{k-1}$, if $p|_X = 0 \Rightarrow p \equiv 0$ for polynomials $p \in \mathbb{P}_d^{k-1}$.

Let $p_j, j = 1, \dots, M$ be a basis of $\mathbb{P}_d^{k-1}$. The interpolant s can be determined by inserting the given data sites and function values in (1). Demanding a side condition for the coefficients of the required polynomial term leads to a dense system of linear equations:

$$\sum_{i=1}^{N} \lambda_i \phi(\|x_i - x_k\|) + \sum_{j=1}^{M} \pi_j p_j(x_k) = f(x_k), \qquad 1 \le k \le N, \tag{2}$$

$$\sum_{i=1}^{N} \lambda_i p_l(x_i) = 0, \qquad 1 \le l \le M, \tag{3}$$

or, in short notation

$$A\boldsymbol{x} = \boldsymbol{b} := \begin{pmatrix} \Phi & \Pi \\ \Pi^\top & 0 \end{pmatrix} \begin{pmatrix} \boldsymbol{\lambda} \\ \boldsymbol{\pi} \end{pmatrix} = \begin{pmatrix} \boldsymbol{f} \\ \boldsymbol{0} \end{pmatrix}. \tag{4}$$

We restrict our numerical examples to the choice of the multiquadric basis function

$$\phi(x) = \sqrt{x^2 + c^2}$$

with a real scaling parameter c to allow comparability with earlier results [12,14]. The multiquadric basis function requires a constant polynomial term.

3 Preconditioning RBF Interpolation

The system (4) is w.l.o.g. dense and ill-conditioned. Therefore, the numerical solution requires iterative methods with a suitable preconditioning, if the system size surpasses certain limits.

3.1 The FGP Algorithm

The FGP algorithm [8,9,11] is a preconditioned Krylov subspace projection method based on the semi-inner product

$$\langle s,t\rangle_\phi = \varrho \boldsymbol{\lambda}^\top \Phi \boldsymbol{\mu}$$

with $s(x) = \sum_{i=1}^N \lambda_i \phi(\|x - x_i\|) + \sum_{j=1}^M \alpha_j p_j(x)$ for $\lambda_i, \alpha_j \in \mathbb{R}$ and $t(x) = \sum_{i=1}^N \mu_i \phi(\|x - x_i\|) + \sum_{j=1}^M \beta_j p_j(x)$ for $\mu_i, \beta_j \in \mathbb{R}$, induced by the radial basis function ϕ. Φ denotes the associated kernel to ϕ. The factor $\varrho \in \{-1, 1\}$ depends on the choice of the basis function.

The included preconditioning is based on an approximation of the Lagrange basis of the spanned function space, which can be obtained by solving the interpolation tasks

$$\hat{u}_k(x_i) = \sum_{\ell=1}^N \zeta_{k,\ell}\phi(\|x_i - x_\ell\|) + \sum_{j=1}^M \pi_{k,j} p_j(x_i) := \delta_{ik}, \quad \text{for } i,k = 1,\ldots,N,$$

where δ_{ik} denotes the Kronecker-delta. As an approximation, the functions $\hat{u}_k$, $k = 1,\ldots,N$ are computed on subsets of X, that contain not more than q interpolation centers. Generally the relation $q \ll N$ holds. These subsets are called Lagrange-sets ($\mathcal{L}$-sets). The $\mathcal{L}$-sets are chosen such that they span subsets of X in various scales.

The FGP algorithm is well-suited for shared memory settings. We applied this method effectively on CPUs and accellerators [12,14]. Nevertheless, the distribution of the $\mathcal{L}$-sets impedes the application for distributed memory settings.

3.2 Domain Decompostion Methods

Beatson et al. present a list of four building blocks that they consider the 'essential ingredients for a domain decomposition interpolatory fitter' [2]:

1. A space subdivision method.
2. A solution method for small interpolation subproblems.
3. A fast evaluation of the RBF interpolations occuring at various spaces.
4. An outer iteration.

For the construction of a domain decomposition method (DDM), the given set of interpolation centers X is subdivided into the overlapping subdomains $X_j, j = 1,\ldots,D$ such that $X = \bigcup_{j=1}^D X_j$ and $X_i \cap X_j \neq \emptyset$. Associated with this subdivision is a set of non-overlapping subdomains $\tilde{X}_j, j = 1,\ldots,D$ such that $X = \bigcup_{j=1}^D \tilde{X}_j$, where $\tilde{X}_j \subset X_j$ and $\tilde{X}_i \cap \tilde{X}_j = \emptyset$. The cardinalities of the subsets are denoted by $N_j = |X_j|$ and $\tilde{N}_j = \left|\tilde{X}_j\right|$ respectively.

Let $\mathfrak{R}_s \in \mathbb{R}^{N \times N_s}$ denote the restriction matrix projecting a vector $\boldsymbol{x}$ from a domain X onto a vector $\boldsymbol{x}_s = \mathfrak{R}_s \boldsymbol{x}$ on the subdomain X_s and $\tilde{\mathfrak{R}}_s \in \mathbb{R}^{N \times \tilde{N}_s}$

denote the restriction matrix which restricts a vector $\boldsymbol{x}$ from domain X onto a vector $\tilde{\boldsymbol{x}}_s = \tilde{\mathfrak{R}}_s x$ on the subdomain $\tilde{X}_s$.

For the non-overlapping subdivision a vector over the domain X can be composed by applying the transposed mapping operations $\tilde{\mathfrak{R}}_s^\top$ on the local vectors $\boldsymbol{x}_s$

$$\boldsymbol{x} = \sum_{s=1}^{D} \tilde{\mathfrak{R}}_s^\top \boldsymbol{x}_s.$$

Beatson et al. present a two-level method [2] that also employs a coarse grid correction. The selection of the set of the coarse grid interpolation centers X_C is denoted as $\mathfrak{R}_C \in \mathbb{R}^{N \times N_C}$, where $N_C := |X_C|$. The coarse grid nodes are chosen such that $X_C \cap X_s \neq \emptyset$ for $s = 1, \ldots, D$.

Retaining the notation from (4), we write $A_s, \boldsymbol{x}_s, \boldsymbol{b}_s$ for the systems over subdomains. The restriction matrices $\mathfrak{R}_s$ (and $\tilde{\mathfrak{R}}_s$) are extended by zero blocks to the matrices R_s (and $\tilde{R}_s$)

$$R_s := \begin{pmatrix} \mathfrak{R}_s & 0 \\ 0 & 0 \end{pmatrix} \in \mathbb{R}^{(N+M)\times(N_s+M)} \text{ and } \tilde{R}_s := \begin{pmatrix} \tilde{\mathfrak{R}}_s & 0 \\ 0 & 0 \end{pmatrix} \in \mathbb{R}^{(N+M)\times(\tilde{N}_s+M)}.$$

We denote the projection of a vector in $\mathbb{R}^N$ onto $\left(\mathbb{P}_d^{k-1}\right)^\perp$ as $\mathfrak{P} \in \mathbb{R}^{N\times N}$. As for the restriction matrices we extend the projection in order to spare the polynomial term to an operator

$$P := \begin{pmatrix} \mathfrak{P} & 0 \\ 0 & 0 \end{pmatrix} \in \mathbb{R}^{(N+M)\times(N+M)}.$$

A notable difference to the extended restriction matrices defined above is the matrix $\hat{R}$, since we need to carry the update for the polynomial part from the coarse grid solution:

$$R_C := \begin{pmatrix} \mathfrak{R}_C & 0 \\ 0 & 0 \end{pmatrix} \in \mathbb{R}^{(N+M)\times(N_C+M)} \text{ and } \hat{R}_C := \begin{pmatrix} \mathfrak{R}_C & 0 \\ 0 & I \end{pmatrix} \in \mathbb{R}^{(N+M)\times(N_C+M)}.$$

We can now write the iteration step as follows:

$$\boldsymbol{x}^{(i+1/2)} = \boldsymbol{x}^{(i)} + P \sum_{j=1}^{D} \tilde{R}_j^\top A_j^{-1} R_j \left(\boldsymbol{f} - A\boldsymbol{x}^{(i)}\right)$$

$$\boldsymbol{x}^{(i+1)} = \boldsymbol{x}^{(i+1/2)} + \hat{R}_C^\top A_C^{-1} R_C \left(\boldsymbol{f} - A\boldsymbol{x}^{(i+1/2)}\right).$$

The presented algorithm can be regarded as a two-level Schwarz method that is multiplicative between the levels and a restricted additive Schwarz method (RASM) within the fine level [16]. A notable difference to general Schwarz methods is the required correction of the fine-level coefficients that assures (3).

It is common for applications of Schwarz methods to replace the correction over the subdomains with approximate solvers [16, Sect. 1.2]. Ling and Kansa proposed to use approximate solvers for a DDM for radial basis function interpolation [13].

The theory of preconditioned Krylov subspace projection methods requires the preconditioning operator to be constant for all iterations. In general, this does not hold if the approximated solution is achieved by an iterative method itself. As shown in Sect. 4, our numerical tests show only a slight increase of the required iteration numbers, backing up the statements about the applicability of approximated solutions over subdomains in [13, 16].

In [12], we implemented the two-level block Jacobi approach using the data division that was predetermined by our application, in particular finite volume discretizations with one cell-layer overlap. For the generation of the coarse representation we chose a coarsening factor of 1/8. We approximately solve the arising subproblems applying the FGP algorithm. Both, the outer iteration of the DDM and the inner iteration of the FGP algoritm require the computation of a matrix-vector product with a dense system matrix. We approximate this matrix-vector product with a multipole method [1, 7].

In order to extend this approach to a general multilevel method we define subsets $X_l, l = 1, \dots, L$ of the given set X, such that $X_l \subset X_{l+1}$ for $l = 1, \dots, L-1$ and $X_L \equiv X$. Each level $l > 1$ is subdivided into overlapping subdomains $X_{j,l}, j = 1, \dots, D_l$ and the associated non-overlapping subdivision $\tilde{X}_{j,l}, j = 1, \dots, D_l$. The notation of the cardinalities is similar to the notation above $N_{j,l} = |X_{j,l}|$ and $\tilde{N}_{j,l} = \left|\tilde{X}_{j,l}\right|$. The notation of the restriction matrices is also extended by the current level in a similar manner.

For the RASM within the levels $l > 1$, we again require a projection of the composed coefficient vector onto the subspace $\left(\mathbb{P}_d^{k-1}\right)^{\perp}$.

The extension to a general multilevel method on L levels now has the form

$$
\begin{aligned}
\boldsymbol{x}^{(i+1/L)} &= \boldsymbol{x}^{(i)} + P \sum_{j=1}^{D_L} \tilde{R}_{j,L}^{\top} A_{j,L}^{-1} R_{j,L} \left(\boldsymbol{f} - A\boldsymbol{x}^{(i)}\right) \\
\boldsymbol{x}^{(i+2/L)} &= \boldsymbol{x}^{(i+1/L)} + P \sum_{j=1}^{D_{L-1}} \tilde{R}_{j,L-1}^{\top} A_{j,L-1}^{-1} R_{j,L-1} \left(\boldsymbol{f} - A\boldsymbol{x}^{(i+1/L)}\right) \\
&\vdots \\
\boldsymbol{x}^{(i+1)} &= \boldsymbol{x}^{(i+(L-1)/L)} + \hat{R}_1^{\top} A_1^{-1} R_1 \left(\boldsymbol{f} - A\boldsymbol{x}^{(i+(L-1)/L)}\right).
\end{aligned}
$$

For the extension of our implementation, the predetermined data division was replaced with an octree as hierarchical data structure in $\mathbb{R}^3$. Each nonempty box represents a subdomain, the overlap is constructed from contributions of the neighbor boxes. Each refinement level of the octree represents a level of the multilevel method. As for the two-level implementation we employ the FGP algorithm as local solution method. The computation of the involved matrix-vector products is approximated with a multipole method in a straightforward manner.

4 Numerical Examples

In this section we present iteration counts for non-preconditioned Krylov subspace projection methods compared to our implementation of the FGP algorithm and the presented multilevel method and timing measurements for our distributed multilevel implementation. We imposed test function F3 [5] as boundary conditions on N nodes distributed over a sphere.

Table 1 compares the iteration counts for the CG method, the GMRES method, the FGP algorithm with $|\mathcal{L}\text{-set}| = 50$, and variations of our multilevel approach for solving the system to a L_2-error less than 10^{-5}. The value of the overlap is given relatively to the box size of the corresponding octree box. The relative tolerance depicts the relative reduction of the residual at the approximate solution of the subdomain correction. We additionally vary the number of levels.

Table 1. Iteration count for various iterative methods.

				Multilevel DDM							
Overlap				1/8				1/16			
Rel. tol.				1E−8		1E−2		1E−8		1E−2	
Levels				2	3	2	3	2	3	2	3
N	CG	GMRES	FGP								
2^8	88	120	5	–	–	–	–	–	–	–	–
2^{10}	148	257	6	7	–	8	–	–	–	–	–
2^{12}	118	155	7	4	8	4	8	6	–	6	–
2^{14}	122	132	7	3	4	4	4	4	8	4	9
2^{16}	115	121	8	3	3	4	4	3	4	3	4
2^{18}	141	338	9	3	3	3	4	3	4	4	4

The near constant iteration numbers indicate a sound preconditioning effect as soon as a sufficient overlap is created. However, the method fails to converge if no sufficient overlap is created.

We present timing results for two test systems in Table 2. The first test system is equipped with two Intel®Xeon®E5-2640 v3 CPUs. We used an overlap of 1/8 box size, a multilevel tree depth of $\max(1, \lceil \log_8(p) \rceil)$ for p cores, and a relative tolerance of 10^{-2} for subsystems and an absolute tolerance of 10^{-6} for the total system. The column labeled 'sequential' lists the timings for the sequential code, the remaining columns state the timings for the MPI code running on the respective number of cores.

The second test system is a cluster of 16 Nvidia®Tegra®X1 SoC nodes, each equipped with 4-Plus-1 quad-core ARM Cortex A-57 CPUs [17]. We retained the settings as above except a multilevel tree depth of $\max(1, \lceil \log_8(p) \rceil) + 1$ for p cores.

Table 2. Time in sec. per iteration on the two test systems.

Intel Xeon system						
N	Cores					
	Sequential	1	2	4	8	16
2^{10}	4.50E−2	6.28E−2	3.31E−2	2.11E−2	1.37E−2	6.83E−3
2^{12}	3.24E−1	3.40E−1	1.74E−1	1.03E−1	6.74E−2	3.83E−2
2^{14}	2.69E+0	2.71E+0	1.42E+0	7.74E−1	4.75E−1	2.15E−1
2^{16}	1.53E+1	2.11E+1	1.10E+1	6.38E+0	4.08E+0	1.52E+0
2^{18}	–	1.37E+2	7.19E+1	3.82E+1	2.47E+1	1.11E+1
2^{20}	–	–	–	3.07E+2	1.87E+2	7.86E+1
2^{22}	–	–	–	–	–	6.49E+2
ARM system						
N	Cores					
	4	8	16	24	32	48
2^{14}	2.64E+0	1.82E+0	1.66E+0	1.52E+0	–	–
2^{16}	1.79E+1	1.14E+1	7.38E+0	6.79E+0	8.17E+0	7.58E+0
2^{18}	1.37E+2	7.18E+1	3.49E+1	3.25E+1	3.50E+1	3.20E+1
2^{20}	–	–	1.59E+2	1.30E+2	1.17E+2	9.66E+1

The measured timings imply a quasilinear time complexity on both systems for the sequential and the distributed implementation. The method scales as long as an adequate workload remains per node.

5 Conclusion

We showed the applicability of our multilevel domain decomposition method using an octree as hierarchical structure. The near constant iteration numbers back up the reasonableness of the choice of the coarsening factor 1/8 for this approach. Again, an approximate solution over the subdomains using the FGP algorithm proved useful. However, the area of application of our multilevel approach starts at $\approx 2^{10}$ boundary nodes. This clearly falls short of the threshold where distributed systems are demanded by our simulation applications.

We identified bottlenecks in the current communication design and the data division layout in our implementation that have to be addressed to further improve the scalability of our method. The optimization of the load balancing on heterogeneous systems utilizing different accelerators will also have an impact on the future data division layout.

Acknowledgements. This project has received funding from the European Union's Horizon 2020 research and innovation programme under grant agreement No. 671697.

References

1. Beatson, R.K., Greengard, L.: A short course on fast multipole methods. In: Wavelets, Multilevel Methods and Elliptic PDEs, pp. 1–37, Oxford University Press (1997)
2. Beatson, R.K., Light, W., Billings, S.: Fast solution of the radial basis function interpolation equations: domain decomposition methods SIAM. J. Sci. Comput. **22**(5), 1717–1740 (2001)
3. Beatson, R., Levesley, J., Mouat, C.: Better bases for radial basis function interpolation problems. Comput. Appl. Math. **236**, 434–446 (2011)
4. de Boer, A., van der Schoot, M.S., Bijl, H.: Mesh deformation based on radial basis function interpolation. Comput. Struct. **85**(11–14), 784–795 (2007)
5. Bozzini, M.T., Rossini, M.F.: Testing methods for 3D scattered data interpolation. Multivariate Approximation and Interpolation with Applications. (Almunecar 2001). Acad. Cienc. Exact. Fıs. Quım. Nat. **20**, 111–135 (2002)
6. Buhmann, M.: Radial Basis Functions. Theory and Implementations. Cambridge Monographs on Applied and Computational Mathematics. Cambridge University Press, New York (2003)
7. Cherrie, J.B., Beatson, R.K., Newsam, G.N.: Fast evaluation of radial basis functions: methods for generalized multiquadrics in $\mathbb{R}^n$ SIAM. J. Sci. Comput. **23**(5), 1549–1571 (2001)
8. Faul, A.C., Powell, M.J.D.: Krylov Subspace Methods for Radial Basis Function Interpolation. University of Cambridge, DAMP, Cambridge (1999)
9. Faul, A.C., Goodsell, G., Powell, M.J.D.: A Krylov subspace algorithm for multiquadric interpolation in many dimensions. IMA J. Numer. Anal. **25**(1), 1–24 (2005)
10. Fasshauer, G.: Meshfree Approximation Methods with MATLAB. World Scientific, Singapore (2007)
11. Gumerov, N., Duraiswami, R.: Fast radial basis function interpolation via preconditioned Krylov iteration. SIAM J. Sci. Comput. **29**(5), 1876–1899 (2007)
12. Haase, G., Martin, D., Offner, G.: Towards RBF interpolation on heterogeneous HPC systems. In: Lirkov, I., Margenov, S.D., Waśniewski, J. (eds.) LSSC 2015. LNCS, vol. 9374, pp. 182–190. Springer, Cham (2015). https://doi.org/10.1007/978-3-319-26520-9_19
13. Ling, L., Kansa, E.J.: Preconditioning for radial basis functions with domain decomposition methods. Math. Comput. Model. **40**(13), 1413–1427 (2004)
14. Martin, D., Haase, G.: Interpolation with radial basis functions on GPGPUs using CUDA. Technical report SFB-Report 2014-04, SFB MOBIS, University of Graz (2014)
15. Powell, M.J.D.: Some algorithms for thin plate spline interpolation to functions of two variables. Adv. Comput. Math. **4**, 303–319 (1993)
16. Smith, B.F., Bjørstad, P.E., Gropp, W.D.: Domain Decomposition: Parallel Multilevel Methods for Elliptic Partial Differential Equations. Cambridge University Press, New York (1996)
17. Rajovic, N., Paul, M., Gelado, I., Puzovic, N., Ramirez, A., Valero, M.: Supercomputing with commodity CPUs: are mobile SoCs ready for HPC? In: Proceedings of the International Conference on High Performance Computing, Networking, Storage and Analysis (2013)
18. Wendland, H.: Scatterred Data Approximation. Cambridge Monographs on Applied and Computational Mathematics. Cambridge University Press, New York (2010)

Sampling in *In Silico* Biomolecular Studies: Single-Stage Experiments vs Multiscale Approaches

Nevena Ilieva[1(✉)], Jiaojiao Liu[2], Xubiao Peng[3], Jianfeng He[2], Antti Niemi[2,4], Peicho Petkov[5], and Leandar Litov[5]

[1] Institute of Information and Communication Technologies, Bulgarian Academy of Sciences, Sofia, Bulgaria
nevena.ilieva@parallel.bas.bg
[2] School of Physics, Beijing Institute of Technology, Beijing 100081, People's Republic of China
[3] University of British Columbia, Vancouver, Canada
[4] Nordita, 106 91 Stockholm, Sweden
[5] Physics Faculty, Sofia University "St. Kl. Ohridsky", Sofia, Bulgaria

Abstract. *In silico* studies of biological molecules face the problem of sampling quality due to the systems size (in atom numbers), the time scale of the investigated processes and the admissible computational time step. Advances in hardware alone are incapable of resolving this problem and the efforts are oriented towards sampling techniques enhancements, multilevel system representations and development of multistage and multiscale methods through synergistic protocols from complementary approaches. We combine a mean field approach with all atom molecular dynamics (MD), to develop a multistage algorithm that can model protein folding and dynamics over very long time periods with atomic-level precision. We compare the quality of conformation-space sampling for villin headpiece (PDB ID 2F4K) with a 125 µs long folding simulation performed on the dedicated supercomputer ANTON.

Keywords: Biological molecules · Molecular modeling
Molecular dynamics · Mean field theory · Protein folding
Numerical simulations

1 Introduction

All-atom molecular dynamics (MD) [6] provides microscopic data about the time evolution of every single atom of the investigated system—in the case of life sciences, biological molecules or biomolecular complexes, in solvent environment. It produces a discrete and piecewise linear trajectory of each atom, as a solution of the discretized equations of motion (Newton's equations). It is an approximation to the solution of continuum Newton's equation. The time step size is determined by the requirements for physical adequacy of the approximation and

I. Lirkov and S. Margenov (Eds.): LSSC 2017, LNCS 10665, pp. 507–515, 2018.
https://doi.org/10.1007/978-3-319-73441-5_56

for stability of the computational procedure. Thus, the ratio between the time step and the characteristic atomic motion time scale, $\epsilon \sim \Delta t/\tau$, should be small. The fastest molecular motions—the covalent bond oscillations—have duration of a few femtoseconds that limits Δt to 1–2 femtoseconds. Such a short time step makes an all-atom approach to protein dynamics an extreme computational challenge [7,21,22]. Many biological processes take minutes and even hours, but even processes of a few seconds are complicated to simulate as the performance of the most powerful supercomputers is limited to at most microseconds of time evolution per day. For example, the folding time of myoglobin – a protein, containing 154 amino acid residues–is around 2.5 s [9]. A powerful dedicated supercomputer like ANTON [17] can produce at most several microseconds of *in vitro* folding trajectory per day for much shorter proteins. One has to keep in mind also the inaccuracy of the existing force fields, despite the efforts that are concentrated in this direction [7]. Their limitations tend to essentially affect a folding trajectory no later than around ten microseconds [17]. On the software site, in biomolecular simulations weak scaling is largely irrelevant, but one rather seeks strong-scaling software engineering techniques or ensemble simulation techniques [16]. Some possibilities for performance improvement are provided by multilevel representations of the system under consideration—geometry or dynamics-based coarse-graining methods, which allow for a larger time step and thus up to 2–3 orders of magnitude fold speed-up compared to all-atom approaches, though with loss in accuracy [7,20].

We build on complementarity of deterministic and stochastic approaches to develop a computationally effective multiscale method for *in silico* investigation of structure and dynamics of proteins (see [13] and the references therein). We have demonstrated the subatomic precision of the reconstruction protocol on the example of different types of proteins. Here we address the sampling efficiency of the newly developed method. As an example, we use villin headpiece. We elaborate on the findings in [10], where the folding and unfolding process of a chicken villin headpiece subdomain was studied within the soliton model of protein folding [3,15]. The reference is a 125 µs trajectory [12], obtained at ANTON and provided to us by DESRES (D.E. Shaw Research, New York, NY 10036, USA). We scrutinize the generalized bond and torsion angle distributions in two particular domains of the protein—the first α-helix (residues 3–10) and the loop between the first two helices (residues 11–13).

2 Methods and Computations

2.1 Molecular Dynamics and Monte Carlo Methods

In MD, the time evolution is obtained by solving the classical (Newtonian) equations of motion for all atoms in the system—protein, as well as solvent. Following the ergodic hypothesis, one identifies the time average of an observable with the ensemble average. In the case of conservative forces, the total energy is preserved, which – together with the fixed volume and number of particles – makes the MD averaging equivalent to that over a microcanonical ensemble

$$\bar{A}_{MD} \simeq \langle A \rangle_{\text{microcanonical}} \tag{1}$$

This ensemble reproduces the Hamiltonian nature of the dynamics but is not convenient for comparisons with the experiment. To this end, MD simulations in other ensembles are performed, by coupling the system to a thermal bath, changing the box size, etc. (see, e.g., [19]).

Prior to MD, computer simulations of many-body systems with a very large number of accessible states were carried out within the Monte Carlo approach. The MC-based methods (e.g., the MCMC Metropolis method [14]) are stochastic in nature and rely on importance-weighted random walk in the configuration space.

A complete cycle (all N atoms of the system are randomly moved) defines an MC step, which is a convenient unit for describing the time evolution of the system. The MC-based methods generate average over a canonical ensemble

$$\langle A \rangle \simeq \langle A \rangle_{\text{canonical}}. \tag{2}$$

Thus, MD methods are precise but limited by the physical model restrictions and the computation bottlenecks. MC methods are not suitable for studying time-dependent phenomena or momentum-dependent properties but give better coverage of the configuration space through tunneling between energetically separated regions. Our approach for modeling protein folding and dynamics builds on complementarity of MD and MC-based methods [5,8].

2.2 Soliton Formalism in Protein Dynamics

A mean-field approach is often the best choice when the problem description in fundamental-level constituents is tedious or with large uncertainties in the involved parameters. In case of proteins, a mean-field approach proved constructive in describing the protein folding dynamics in terms of a complete set or order parameters—generalized bond and torsion angles $\kappa_i \in [0, \pi)$ and $\tau_i \in [-\pi, \pi)$, in the Frenet frames associated with the backbone Cα atoms

$$\kappa_{i+1,i} \equiv \kappa_i \;= \arccos(\mathbf{t}_{i+1} \cdot \mathbf{t}_i) \tag{3}$$

$$\tau_{i+1,i} \equiv \tau_i \;= \text{sign}[(\mathbf{b}_{i-1} \times \mathbf{b}_i) \cdot \mathbf{t}_i] \arccos(\mathbf{b}_{i+1} \cdot \mathbf{b}_i), \tag{4}$$

where $(\mathbf{t}_i, \mathbf{b}_i, \mathbf{n}_i)$ are the unit backbone tangent, binormal and normal vectors, i numbers the aminoacid residues, $i = 1, \ldots, N$ [3,8,15].

The difference in κ values between consecutive residues $\Delta\kappa_i = |\kappa_{i+1} - \kappa_i|/\pi$ is small, as seen from the crystallographic data in Protein Data Bank (PDB) [2] and from MD simulation data. Thus, $\Delta\kappa_i$ can be employed as an expansion parameter instead of ϵ. An additional advantage of the expansion in powers of $\Delta\kappa_i$ is its independence of any particular time scale, so this description should be valid over arbitrary time periods.

Fig. 1. PDB structures employed in the study: 2F4K, in purple and grey, compared to 1YRF, in orange (left panel); Initial conformation for the blind MC procedure: experimental structure, in red, interlaced with the 8-soliton approximation, in grey (right panel). (Color figure online)

In the limit of small $\Delta\kappa_i$ the Landau free energy can be written as

$$E(\kappa,\tau) = \sum_{i=1}^{N-1} \Delta\kappa_i^2 + \sum_{i=1}^{N} \left\{ \lambda\,(\kappa_i^2 - m^2)^2 + \frac{d}{2}\,\kappa_i^2\tau_i^2 - \; b\kappa_i^2\tau_i - a\tau_i + \frac{c}{2}\tau_i^2 \right\} + \mathcal{O}(\Delta\kappa_i^4), \tag{5}$$

where (λ, m, a, b, c, d) are parameters (see, e.g. [15]). These parameters should be computed from first principles for any particular protein, but at present this computations are practically impossible. A viable solution of this problem is provided by training a minimum energy configuration of (5) to model the corresponding PDB backbone. We use the code *Propro*, available also on-line at http://www.folding-protein.org.

In (5), a rather straightforward generalization of the Hamiltonian of the discrete nonlinear Schrödinger equation (DNLS) can be identified [3]. It was shown that for certain parameter values it supports a kink-like solution – dark DNLS soliton.

2.3 Side Chains in Mean Field Approach

The mean field theory (5) builds on the Cα coordinates, while the role of the side-chain orientation is accounted for implicitly by the interactions in (5). The side-chain coordinates are reconstructed from the minimal-energy Cα structure, e.g., with *Pulchra* [18], with a subsequent MD run for equilibration and energy

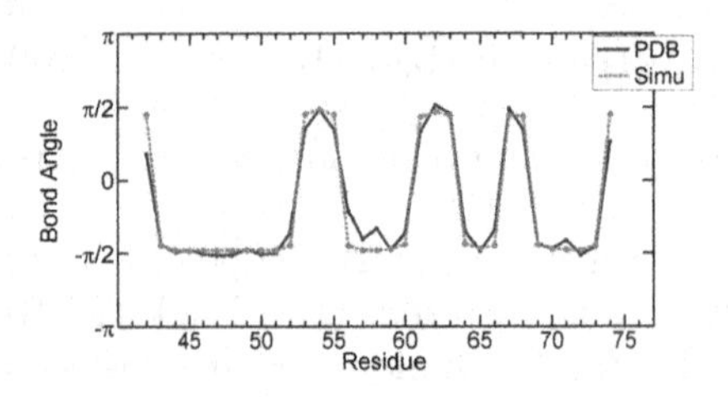

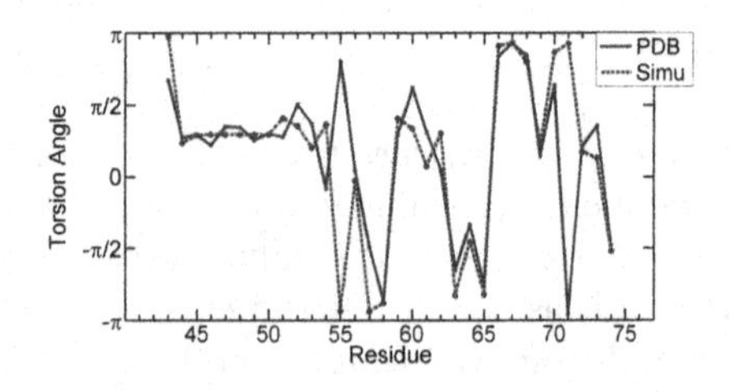

Fig. 2. Generalized backbone bond (left) and torsion (right) angles for the experimental structure (frame 000125 from DESRES trajectory) and the eight-soliton approximation.

minimization [17]. The generated minimal-energy $C\alpha$ structures are first investigated for steric clashes with a pre-selected threshold $R_0 = 1.6$ Å for any pair of *not* covalently bonded atoms. Recall, that the covalent bond distance between C, N and O atoms is at most ∼1.54 Å. We only proceed to the stability analysis with mean field structures, which comply with this criterion.

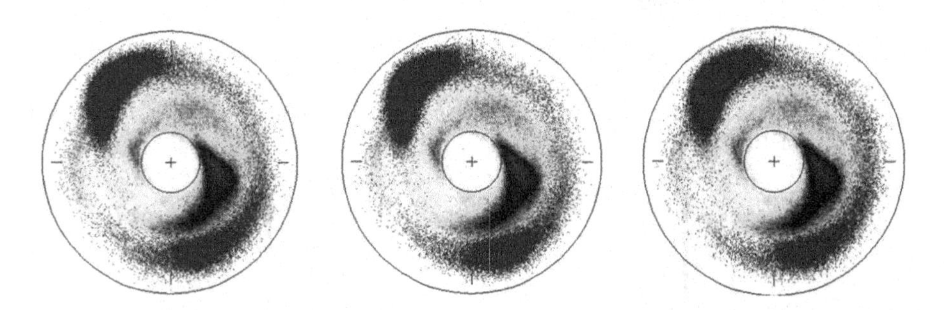

Fig. 3. Distribution of the generalized backbone bond angle κ for the first helix in the first, second and third sets of 10000 frames.

2.4 Villin Headpiece Domain: Structures Employed

We exemplify our consideration on a widely studied protein—the villin headpiece protein. Villin is a medium-sized actin-binding protein consisting of an N-terminal six-repeat core and a C-terminal cap (headpiece, denoted VHP67). The C-terminal subdomain, denoted VHP35 [1], is a bundle of three α-helices forming a well-packed hydrophobic core. The size (35 amino acid residues) and the high helical content imply fast folding that has been also experimentally confirmed: wild type folds in 16 µs [11] and with its small size is a perfect playground for MD protein studies, sometimes even called "the hydrogen atom of MD". In the unfolded state, 34% of the residues were engaged in native secondary structure, 7%—in non-native, the rest were unstructured. We employ two different structures: 2F4K [11] and 1YRF [4], which differ by two residues in the third helix, with a backbone RMSD of only 0.664 Å. The difference affects binding behavior but is largely irrelevant for the folding dynamics.

3 Results and Discussion

The MD trajectory provided by DESRES [12] was 120 µs long, produced at 360 K, with a time step of 2.5 fs, the data was recorded every 200 ps, thus giving rise to some 628000 frames, describing 34 consecutive folding and unfolding events with an average duration of 2.8 µs, resp. 0.9 μs. The C_α–RMSD between the center of the most populated cluster in the simulation and the experimental structure was 1.3 Å. In Fig. 3 it is given the distribution of the generalized backbone bond angles for the first helix. Each of the panels corresponds approximately to one folding–unfolding cycle and the three of them exhibit no statistically significant variations.

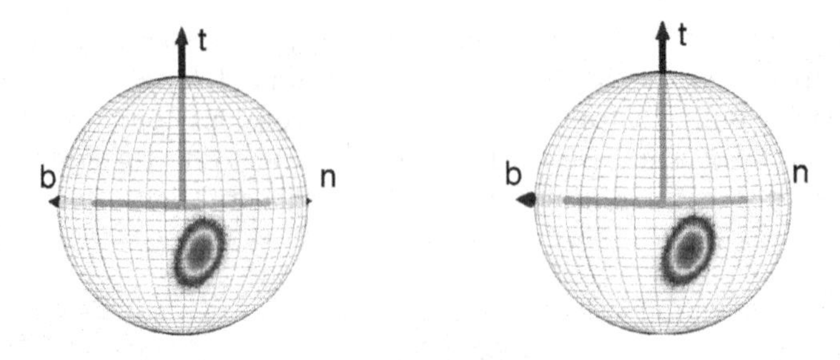

Fig. 4. Side-chain orientation in terms of the η angle for the first helix in the first and second sets of 10000 frames.

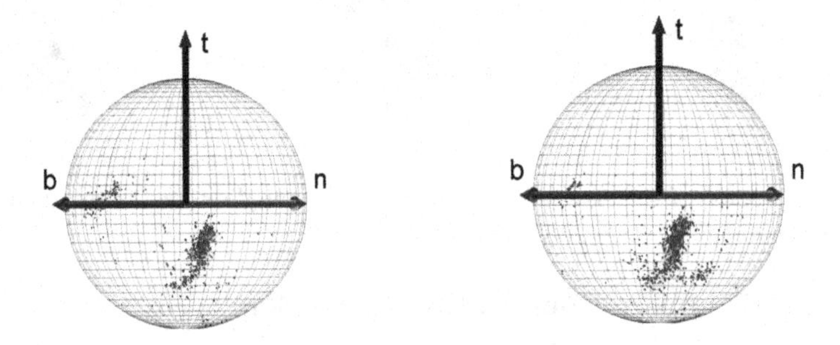

Fig. 5. Side-chain orientation for the first helix in MC-generated states for VHP (PDB ID 1YRF): backfolding structure (left panel) vs misfolding structure (right panel).

We use the structure from frame 000125 for training the energy function (5); the eight-soliton approximation obtained that way is presented in Fig. 1 (right panel) and in Fig. 2, in terms of generalized bond/torsion angle comparison. It was then used for a blind MC procedure [8,13], producing a scattered cluster distribution without clear convergence to the biological fold (figure not shown).

In Fig. 4 the side-chain orientation for the residues in the first helix is given, in terms of the η angle (see [5] for definition). Figure 5 plots the same data for the MC generated states from the two-soliton approximation from [10]. The left plot refers to structures kept below melting temperature ($T_m = 361\,K$) in the heating phase, so correctly folding back, while the right refers to overheated and consequently misfolding MC structures, in both cases—after 15×10^5 MC steps. The statistics is obviously insufficient but already capable of capturing the important structural change in the misfolded conformations. This is even better seen in the histograms in Figs. 6 and 7, where the bond- and torsion-angle distributions are shown for the residues that were originally in a helical conformation. Recall that for regular secondary motifs the $\{\kappa, \tau\}$ values are $\{\pi/2, 1\}$ for α-helix, resp. $\{1, \pi\}$ for β-strand.

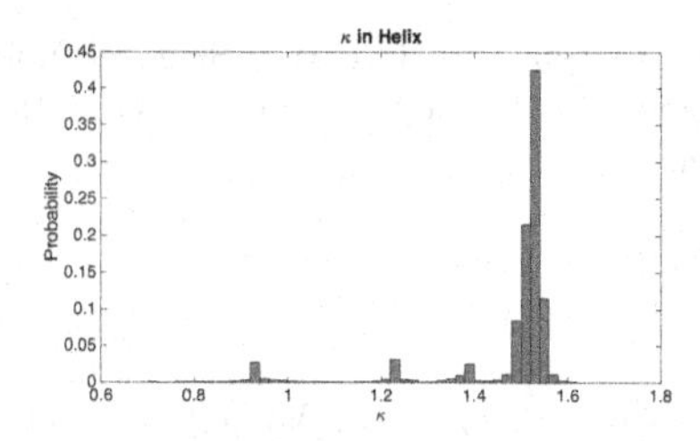

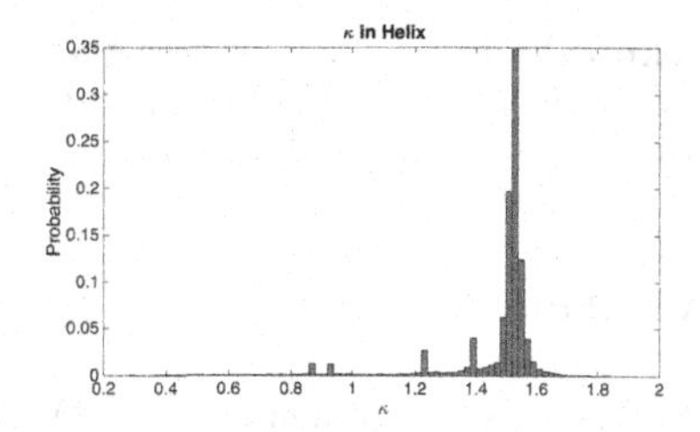

Fig. 6. Generalized bond-angle distribution for the residues in helical PDB conformation in MC-generated states for VHP (PDB ID 1YRF): backfolding structure (left panel) vs misfolding structure (right panel).

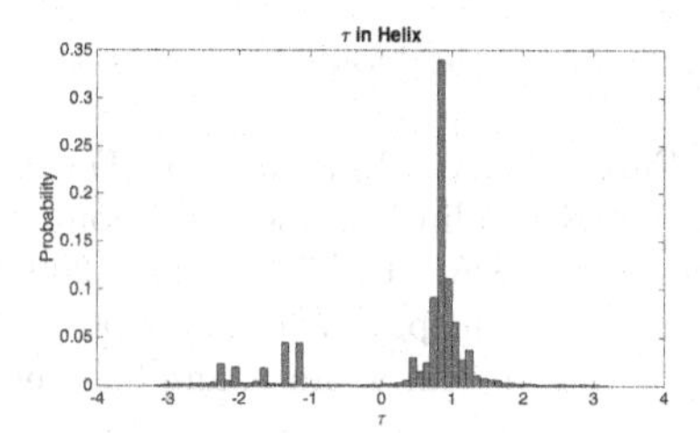

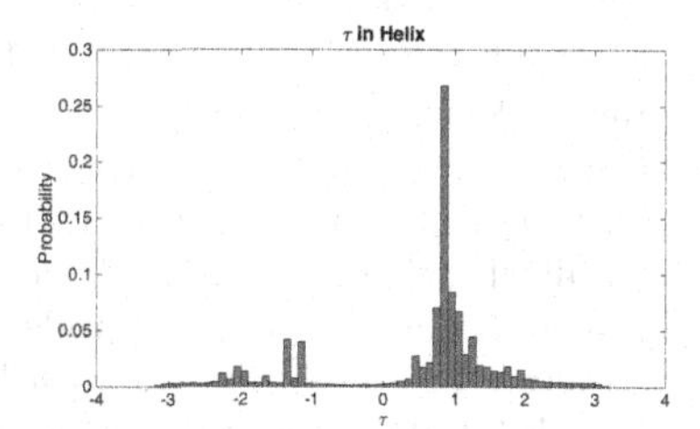

Fig. 7. Distribution of generalized torsion angles for the residues in helical PDB conformation in MC-generated states for VHP (PDB ID 1YRF): backfolding structure (left panel) vs misfolding structure (right panel).

4 Conclusions

Several important conclusions follow from the above considerations. The extensive single-stage all-atom MD simulations exhibit impressively stable patterns in a high number repeated folding–unfolding cycles. The aforementioned simulations still cannot guarantee sufficiently complete conformation space sampling as additional steps are required to identify all possible (quasi)stable folds.

The multi-stage approach we advocate here provides qualitatively comparable sampling of the conformation space; quantitative improvement can be achieved through adjustment of the MC procedure parameters. The analysis of the side-chain orientation provides adequate statistically relevant information about the backbone dynamics. A blind initial state choice could not be justified: the trial conformation was neither completely extended, nor close in any sense to the biological fold that probably gave higher weight to random effect influences. Acquisition of better statistics in the MC-based procedure is needed for biologically reliable results.

In both cases we observe visible similarity in the analyzed distribution, which on the one hand justifies the folding protocol, but on the other hand, does not provide sufficient evidence for better exploration of the conformation space in the long single-stage simulations as compared to the single-cycle ones, be they part of single or multi-stage procedures.

Acknowledgments. This research was supported in part by Bulgarian Science Fund (Grant DNTS-CN-01/9/2014) and Intergovernmental S&T Cooperation Project at the Ministry of Science and Technology of P.R. China.

References

1. Bazari, W.L., Matsudaira, P., Wallek, M., Smeal, T., Jakes, R., Ahmed, Y.: Villin sequence and peptide map identify six homologous domains. Proc. Natl. Acad. Sci. U.S.A. **85**, 4986–4990 (1988)
2. Berman, H.M., Westbrook, J., Feng, Z., Gilliland, G., Bhat, T.N., Weissig, H., Shindyalov, I.N., Bourne, P.E.: The protein data bank. Nucleic Acids Res. **28**(1), 235–242 (2000)
3. Chernodub, M., Hu, S., Niemi, A.J.: Topological solitons and folded proteins. Phys. Rev. E **82**, 011916 (2010)
4. Chiu, T.K., Kubelka, J., Herbst-Irmer, R., Eaton, W.A., Hofrichter, J., Davies, D.R.: High-resolution x-ray crystal structures of the villin headpiece subdomain, an ultrafast folding protein. Proc. Natl. Acad. Sci. U.S.A. **102**, 7517–7522 (2005)
5. Dai, J., Niemi, A., He, J., Sieradzan, A., Ilieva, N.: Bloch spin waves and emergent structure in protein folding with hiv envelope glycoprotein as an example. Phys. Rev. E **93**, 032409 (2016)
6. Frenkel, D., Smit, B.: Understanding Molecular Simulation. Academic Press, Cambridge (2001)
7. Gelman, H., Gruebele, M.: Fast protein folding kinetics. Q. Rev. Biophys. **47**(2), 95–142 (2014)
8. Ilieva, N., Liu, J., Marinova, R., Petkov, P., Litov, L., He, J., Niemi, A.J.: Are there folding pathways in the functional stages of intrinsically disordered proteins? In: AIP Conference Proceedings, vol. 1773, no. 1, p. 110008 (2016)
9. Jennings, P., Wright, P.: Formation of a molten globule intermediate early in the kinetic folding pathway of apomyoglobin. Science **262**, 892–896 (1993)
10. Krokhotin, A., Lundgren, M., Niemi, A.J., Peng, X.: Soliton driven relaxation dynamics and protein collapse in the villin headpiece. J. Phys.: Condens. Matter **25**(32), 325103 (2013)
11. Kubelka, J., Chiu, T., Davies, D., Eaton, W., Hofrichter, J.: Sub-microsecond protein folding. J. Mol. Biol. **359**, 546–553 (2006)
12. Lindorff-Larsen, K., Piana, S., Dror, R.O., Shaw, D.E.: How fast-folding proteins fold. Science **334**(6055), 517–520 (2011)
13. Liu, J., Dai, J., He, J., Niemi, A.J., Ilieva, N.: Multistage modeling of protein dynamics with monomeric Myc oncoprotein as an example. Phys. Rev. E **95**, 032406 (2017)
14. Metropolis, N., Rosenbluth, A., Rosenbluth, M., Teller, A., Teller, E.: Equation of state calculations by fast computing machines. J. Chem. Phys. **21**(6), 1087–1092 (1953)
15. Niemi, A.J.: Gauge fields, strings, solitons, anomalies, and the speed of life. Theoret. Math. Phys. **181**(1), 1235–1262 (2014)
16. Páll, S., Abraham, M.J., Kutzner, C., Hess, B., Lindahl, E.: Tackling exascale software challenges in molecular dynamics simulations with GROMACS. In: Markidis, S., Laure, E. (eds.) EASC 2014. LNCS, vol. 8759, pp. 3–27. Springer, Cham (2015). https://doi.org/10.1007/978-3-319-15976-8_1

17. Raval, A., Piana, S., Eastwood, M.P., Dror, R.O., Shaw, D.E.: Refinement of protein structure homology models via long, all-atom molecular dynamics simulations. Proteins: Struct. Funct. Bioinf. **80**(8), 2071–2079 (2012)
18. Rotkiewicz, P., Skolnick, J.: Fast procedure for reconstruction of full-atom protein models from reduced representations. J. Comput. Chem. **29**(9), 1460–1465 (2008)
19. Rovere, M.: Lecture notes on Monte Carlo and molecular dynamics simulations. School of Neutron Scattering "F. P. Ricci", Santa Margherita di Pula, 22 September–3 October 2008. http://webusers.fis.uniroma3.it/~rovere/FISLIQ/lect-notes.pdf
20. Saunders, M., Voth, G.: Coarse-graining methods for computational biology. Annu. Rev. Biophys. **42**, 73–93 (2013)
21. Shaw, D.E., et al.: Anton, a special-purpose machine for molecular dynamics simulation. SIGARCH Comput. Archit. News **35**(2), 1–12 (2007)
22. Shaw, D.E., et al.: Millisecond-scale molecular dynamics simulations on Anton. In: Proceedings of the Conference on High Performance Computing Networking, Storage and Analysis, SC 2009, pp. 39:1–39:11. ACM, New York (2009)

Cultural Heritage RC Structures Environmentally Degradated: Optimal Seismic Upgrading by Tention-Ties Under Shear Effects

A. Liolios[1], K. Liolios[2(✉)], A. Moropoulou[3], K. Georgiev[2], and I. Georgiev[2,4]

[1] Division of Structural Engineering, Department of Civil Engineering, Democritus University of Thrace, 67100 Xanthi, Greece
aliolios@civil.duth.gr

[2] Institute of Information and Communication Technologies, Bulgarian Academy of Sciences, Acad. G. Bonchev str. Bl. 25A, 1113 Sofia, Bulgaria
kostisliolios@gmail.com, georgiev@parallel.bas.bg

[3] School of Chemical Engineering, National Technical University of Athens, Iroon Polytechniou Street 9, Zografou Campus, Athens, Greece
amoropul@central.ntua.gr

[4] Institute of Mathematics and Informatics, Bulgarian Academy of Sciences, Acad. G. Bonchev str. Bl. 8, 1113 Sofia, Bulgaria
ivan.georgiev@math.bas.bg

Abstract. A computational approach is presented for the seismic response of existing Cultural Heritage industrial reinforced concrete (RC) structures, which have been degraded due to extreme actions (environmental, seismic etc.) and are to be seismically upgraded by using cable elements (tension-ties). Emphasis is given to shear effects, which are common for old RC buildings not designed according to new (after 2000) seismic codes concerning Civil Engineering praxis. The unilateral behavior of the cable-elements and the non-linear behavior of the RC structural elements are strictly taken into account and result to inequality constitutive conditions. For the numerical treatment of the system of partial differential relations (PDE), a double discretization, in space by the Finite Element Method and in time by a direct incremental approach, is used. So, in each time-step, a non-convex linear complementarity problem is solved. The decision for the optimal cable-strengthening scheme under seismic sequences is obtained on the basis of computed damage indices, as shown in a numerical example.

1 Introduction

Recent built Cultural Heritage includes, besides the usual historic monumental structures (churches, monasteries, old masonry buildings etc.), also existing industrial buildings of reinforced concrete (RC), e.g. old factory premises [1]. As concerns the seismic behavior of such RC structures, it often arises in Civil Engineering praxis the need for seismic upgrading [2–4]. Moreover, for such old RC buildings, the flexural failure of RC structural members can be

I. Lirkov and S. Margenov (Eds.): LSSC 2017, LNCS 10665, pp. 516–526, 2018.
https://doi.org/10.1007/978-3-319-73441-5_57

strongly influenced by shear effects, as building damages due to recent earthquakes have shown. This holds mainly because old RC structures have been designed and constructed without taking into account the requirements imposed by the recent seismic codes, see e.g. [4–6]. Eurocode EC8 [5] is now valid for all the seismically active countries which are members of the European Union. The seismic upgrading for Cultural Heritage RC structures must be realized by using materials and methods in the context of the Sustainable Construction [2,7,12]. So, the use of cable-like members (tension-ties) can be considered as an alternative strengthening method in comparison to other traditional methods (e.g. RC mantles) [7–11]. These cable-members (ties) can undertake tension but buckle and become slack and structurally ineffective when subjected to a sufficiently large compressive force. Thus the governing problem conditions take an equality as well as an inequality form and the problem becomes a highly nonlinear one [7–11,13,14]. In the present study, a computational approach is presented for the seismic analysis of Cultural Heritage existing industrial RC frame-buildings, which can appear shear insufficiencies and are to be strengthened by cable elements. Damage indices [7–9] are computed, on the one hand for the seismic assessment of such historic industrial structures and on the other hand for the choice of the optimal cable-bracing strengthening version. Finally, in a Civil Engineering application, a simple typical Cultural Heritage example is presented for the case of multiple earthquakes excitation.

2 The Computational Approach

2.1 Mathematical Formulation of the Problem

The formulation of the seismic problem concerning Cultural Heritage RC structures strengthened by ties includes inequality conditions due to unilateral behavior of the tie-elements. For the so-resulted system of the governing partial differential equations (PDE), a double discretization, in space and time, is applied as usually in structural dynamics [15,16]. Details of the developed numerical approaches given in [7–10] are briefly summarized herein. First, the structural system is discretized in space by using frame finite elements [15,16]. Pin-jointed bar elements are used for the cable-elements. The behavior of these elements includes loosening, elastoplastic or/and elastoplastic-softening-fracturing and unloading - reloading effects. Non-linear behavior is considered as lumped at the two ends of the RC frame elements, where plastic hinges can be developed. All these non-linear characteristics, concerning the ends of frame elements and the cable constitutive law, can be expressed mathematically by the subdifferential relation:

$$s_i(d_i) \in \hat{\partial} S_i(d_i). \tag{1}$$

Here s_i and d_i are the (tensile) force (in [kN]) and the deformation (elongation) (in [m]), respectively, of the i-th cable element, $\hat{\partial}$ is the generalized gradient and S_i is the superpotential function, see [13,14].

Next, dynamic equilibrium for the assembled structural system with cables is expressed by the matrix relation:

$$\boldsymbol{M}\ddot{\boldsymbol{u}} + \boldsymbol{C}(\dot{\boldsymbol{u}}) + \boldsymbol{K}(\boldsymbol{u}) = \boldsymbol{p} + \boldsymbol{A}\boldsymbol{s}. \tag{2}$$

Here $\boldsymbol{u}$ and $\boldsymbol{p}$ are the displacement and load vectors, respectively, and $\boldsymbol{s}$ is the cable stress vector. $\boldsymbol{M}$ is the mass matrix and $\boldsymbol{A}$ is a transformation matrix. The damping and stiffness terms, $\boldsymbol{C}(\dot{\boldsymbol{u}})$ and $\boldsymbol{K}(\boldsymbol{u})$, respectively, concern the general non-linear case. Dots over symbols denote derivatives with respect to time. For the case of ground seismic excitation $\boldsymbol{x}_g$, the loading history term $\boldsymbol{p}$ becomes $\boldsymbol{p} = -\boldsymbol{M}\boldsymbol{r}\ddot{\boldsymbol{x}}_g$, where $\boldsymbol{r}$ is the vector of stereo-static displacements [15,16].

The above relations (1)–(2), combined with the initial conditions, consist the problem formulation, where, for given $\boldsymbol{p}$ and/or $\ddot{\boldsymbol{x}}_g$, the vectors $\boldsymbol{u}$ and $\boldsymbol{s}$ have to be computed. From the strict mathematical point of view, using (1) and (2), we can formulate the problem as a dynamic hemivariational inequality one by following [13,14] and investigate it.

2.2 Numerical Treatment of the Problem

In Civil Engineering practical cases, based on experimental investigations, the above constitutive relations (1) are piecewise linearized. Thus, simplified stress-deformation constitutive diagrams are used. Details of the numerical solution of the problem are given in [7–11]. In each time-step a relevant non-convex linear complementarity problem of the following matrix form is solved:

$$\boldsymbol{v} \geq \boldsymbol{0}, \quad \boldsymbol{D}\boldsymbol{v} + \boldsymbol{a} \leq \boldsymbol{0}, \quad \boldsymbol{v}^T.(\boldsymbol{D}\boldsymbol{v} + \boldsymbol{a}) = \boldsymbol{0}. \tag{3}$$

So, the nonlinear Response Time-History (RTH) can be computed for a given seismic ground excitation.

An alternative approach for treating numerically the problem is the incremental one. Thus, the matrix incremental dynamic equilibrium is expressed by the relation:

$$\boldsymbol{M}\boldsymbol{\Delta}\ddot{\boldsymbol{u}} + \boldsymbol{C}\boldsymbol{\Delta}\dot{\boldsymbol{u}} + \boldsymbol{K_T}\boldsymbol{\Delta}\boldsymbol{u} = -\boldsymbol{M}\boldsymbol{\Delta}\ddot{\boldsymbol{u}}_g + \boldsymbol{A}\boldsymbol{\Delta}\boldsymbol{s}. \tag{4}$$

where $\boldsymbol{C}$ and $\boldsymbol{K_T}(\boldsymbol{u})$ are the damping and the tangent stiffness matrix, respectively, and $\ddot{\boldsymbol{u}}_g$ is the ground seismic acceleration. On such incremental approaches the structural analysis software Ruaumoko [15] is based, which uses the finite element method and is applied hereafter. The applicability and the effectiveness of Ruaumoko have been computationally confirmed by using experimental results in [8,9] for flexural-critical RC frames and in [19] for shear-critical RC frames.

2.3 Consideration of Effects Due to Shear and to Concrete Cracking and Confinement

During a seismic excitation, the RC structural elements can appear flexural, shear or combined flexural-shear failure mode. Considering shear effects, the RC structural elements are classified [3] according to their shear span-to-depth ratio:

$$a_s = \frac{L_s}{d} = \frac{M}{Vd}, \tag{5}$$

where $L_s = M/V$ is the *shear length* corresponding to the considered member end, M the bending moment, V the shear force and d the section depth (height). In the case of RC frame columns or beams, the shear length L_s is equal to one-half of the clear length of the examined structural element according to EC8 [5,6]. Due to shear effects, the RC structural elements appear one of the three following behaviours [3] according to their shear span-to-depth ratio:

1. For $a_s \geq 7.0$, bending prevails, and in this case bending failure occurs before any shear failure, no matter if shear reinforcement exists or not.
2. For $2.0 \leq a_s \leq 7.0$, the failure mode depends on the shear reinforcement of the web.
3. For $a_s \leq 2.0$, the case of short R/C columns is considered, where a special design procedure must be followed so that an explosive cleavage failure of the short column is avoided.

As had been remarked for bridge structural systems [17], in general the shear effects can reduce the available plastic curvature ϕ_p of critical sections. This is shown in the Fig. 1 of a typical moment-curvature ($M - \varphi$) diagram. Due to shear effects in plastic hinges, the flexural ductile behaviour can be modified to a flexural-brittle one. Thus a shear failure, and a so-caused reduction of the curvature ductility $\mu_\varphi = \varphi_u/\varphi_y$, can be eventually appeared or not. This must be checked at every time-step of the seismic excitation.

Under a seismic excitation, the RC structural elements have a linear-elastic behavior until one or two of their critical end-regions, after cracking, enter to the yielding state and plastic hinges are appeared. These effects of cracking on columns and beams are estimated by applying the guidelines of Eurocode 8, part 3, see [4,5,18] and Greek Retrofitting Code [6]. So, the effective flexural stiffness $E_c I_{eff}$ is given by the EC8-formula:

$$E_c I_{eff} = \frac{M_y L_s}{3\theta_y}. \tag{6}$$

Here M_y and θ_y are the flexural moment and the chord rotation at yield, respectively, which are calculated by EC8 formulas given in [4–6], and L_S is the *shear length* corresponding to the member end. On the other hand, the available cyclic shear strength V_R (in MN), corresponding to the considered member end, is decreased with the incremental demand plastic chord rotation $\theta_{p,d}$ according

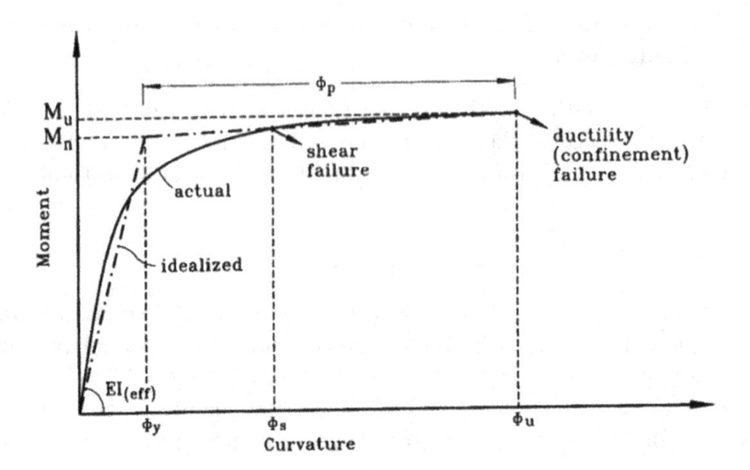

Fig. 1. Reduction of available plastic curvature and flexural ductility due to shear effects [17].

to the following semi-empirical experimental expression given by Eq. (A.12) of Eurocode EC8 [5]:

$$V_R = \frac{1}{\gamma_{el}} \left[\frac{(h-x)\lambda_1}{2L_s} + (1 - 0.05\lambda_2)[0.16\lambda_3(1 - 0.16\lambda_4)A_c\sqrt{f_{cm}/CF} + V_W] \right] \tag{7}$$

Here, γ_{el} is a safety factor [18] that is taken equal to 1.15 for primary seismic structural elements (due to scattering of the experimental values) and is taken 1.00 for secondary seismic members. x is the compression zone depth (in meters) that is known by the sectional analysis, CF is the Confidence Factor according to Table 3.1 of EC8 [5,18]; V_W is the contribution of the transverse reinforcement to shear strength, taken as being equal to $V_W = \rho_W b_W z f_{yw,m}/CF$ for cross-section with rectangular web of width b_W; ρ_W is the transverse reinforcement ratio that is given by $\rho_W = (A_{SW} l_W)/(h_C b_C s_h)$, where l_W is the total length of the stirrups, A_{SW} is the steel section area of the stirrup, h_C and b_C the dimensions of the confined core of the section and s_h is the centerline spacing of stirrups.

The other parameters in (7) are [18]:

$$\lambda_1 = \min(N, 0.55A_c f_{cm}/CF) \tag{7a}$$

$$\lambda_2 = \min(5, \mu_\Delta^p) \tag{7b}$$

$$\lambda_3 = \max(0.5, 100\rho_{tot}) \tag{7c}$$

$$\lambda_4 = \min(5, a_s) \tag{7d}$$

where N is the axial force in [MN], that is positive for compression, while when the axial force is tensional then N is taken zero; $A_C = b_W d$ for rectangular sections with b_W as width of compression zone and d is the depth of the tension

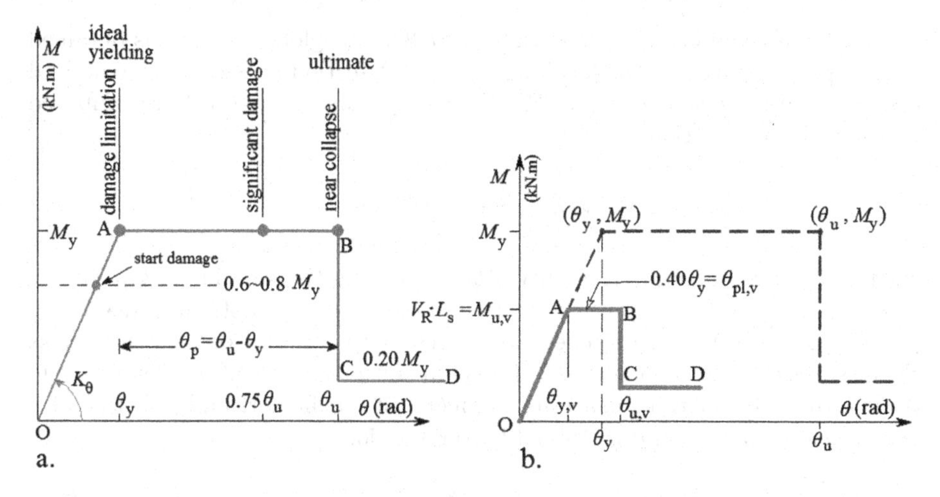

Fig. 2. Moment-Chord Rotation Diagram (a) for ductile failure and (b) for shear failure [18].

reinforcement in meters; f_{cm} is the concrete compressive strength (mean value) in [MPa]; $\mu_{\Delta}^{p} = \theta_p/\theta_y$; ρ_{tot} is the total longitudinal reinforcement ratio (tensional, compression and intermediate); and a_s is the contemporary shear ratio.

In order to define the final elastic-plastic diagram of Moment-Chord Rotation $(M - \theta)$ of the critical section at the considered member end, it must be checked which type of failure precedes, the flexure-failure or the shear-failure [18,19]. Thus, using the known value of the shear strength V_R from (7), the moment $M_{u,v}$ at the critical section due to V_R is calculated as

$$M_{u,v} = L_s V_R \tag{8}$$

When $M_{u,v} > M_y$, i.e. when $M_{u,v}$ is greater than the flexural yielding moment M_y, then the flexural failure precedes the shear one. In that case, the final elastic-plastic diagram of Moment-Chord Rotation $(M - \theta)$ is given by Fig. 2(a).

On the contrary, when $M_{u,v} < M_y$, i.e. when $M_{u,v}$ is smaller than the flexural yielding moment M_y, then the shear failure precedes the flexural one. In this case, the final elastic-plastic diagram of Moment-Chord Rotation $(M - \theta)$ diagram of the considered member end is given as the curve OABCD of Fig. 2(b) according to Sect. 7.2.4.2 of KANEPE [6].

After the above check, the final (corrected) values of M_y and θ_y are used in (6).

2.4 Investigations for the Optimal Cable-Strengthening Versions by Damage Indices

The decision about a possible strengthening for an existing structural system of RC Cultural Heritage, damaged by a seismic event, can be taken after an assessment realization [2–4]. Here the assessment is based on a relevant evaluation of suitable damage indices. After Park/Ang [7,20], the global damage is obtained as a weighted average of the local damage at the section ends of each structural element or at each cable element. First the local damage index DI_L is computed by the following relation:

$$DI_L = \frac{\mu_m}{\mu_u} + \frac{\beta}{F_y d_u} E_T \tag{9}$$

where μ_m is the maximum ductility attained during the load history; μ_u is the ultimate ductility capacity of the section or element; β is a strength degrading parameter; F_y is the yield generalized force of the section or element; E_T is the dissipated hysteretic energy; and d_u is the ultimate generalized deformation. As known [2–4], ductility concerns the metelastic behavior, and in terms of a generalized deformation d is defined by the relation:

$$\mu = \frac{d}{d_y} \tag{10}$$

where d_y denotes the yield generalized deformation and it holds $d \geq d_y$ Next, the dissipated energy E_T is chosen as the weighting function and the global damage index DI_G is computed by using the following relation:

$$DI_G = \frac{\sum_{i=1}^{n} DI_{Li} E_i}{\sum_{i=1}^{n} E_i} \tag{11}$$

where $DI_L i$ is the local damage index after Park/Ang at location i; E_i is the energy dissipated at location i; and n is the number of locations at which the local damage is computed. Finally, for the choice of the optimal system of the strengthening ties, various virtual such systems are proposed and the minimum value according to (11) is sought.

3 Numerical Example

3.1 Description of the Considered Cultural Heritage RC Structural System

The RC frame of Fig. 3, denoted as System F0, is element of an old industrial Cultural Heritage RC building and will be strengthened by a ties-system. The frame is of concrete class C 16/20, it was designed according to old Greek building codes (before 1985), and so is characterized as a flexure-shear critical one [19]. The beams are of rectangular section 30/60 (width/height, in cm) and have a total vertical distributed load 30 KN/m (each beam). The columns have section dimensions, in cm: 40/40.

After its seismic assessment by damage indices, system F0 is to be strengthened by ties. The cable-bracings systems B, C and D, shown in Fig. 4, are proposed and the optimal one will be chosen. The cable elements have a cross-sectional area $F_c = 18\,\text{cm}^2$ and they are of steel class S220 with yield strain

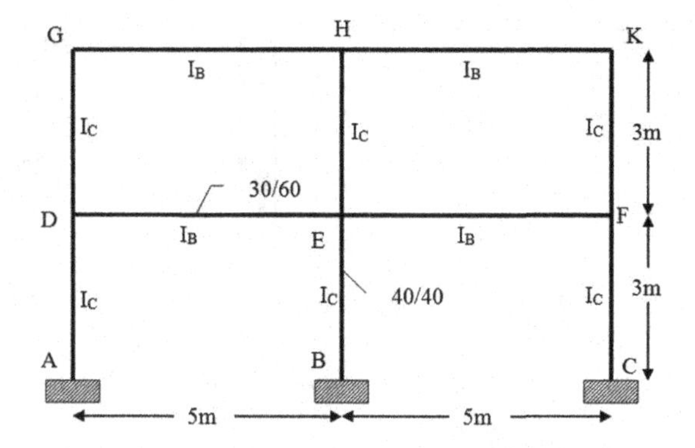

Fig. 3. System F0: the industrial RC frame without cable-strengthening

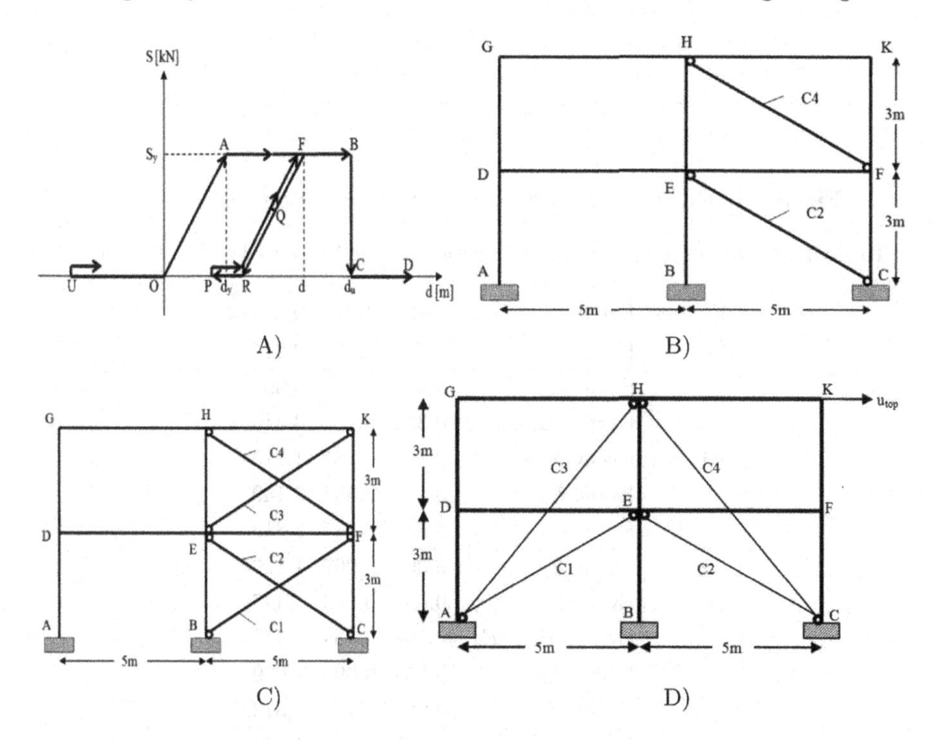

Fig. 4. (A) The constitutive law of the cable-elements, (B) the F1 two-ties-system, (C) the F2 four-ties-system X, (D) the F3 four-ties-system inverted V.

$\epsilon_y = 0.11\%$, fracture strain $\epsilon_f = 2\%$ and elasticity modulus $E_c = 200GPa$. The cable constitutive law, concerning the unilateral (slackness), hysteretic, fracturing, unloading-reloading etc. behavior, is depicted in Fig. 4(A). All the systems F0, F1, F2 and F3 are considered to be subjected to the Coalinga multiple ground seismic excitation shown in Fig. 5 and described in Table 2, see also [7].

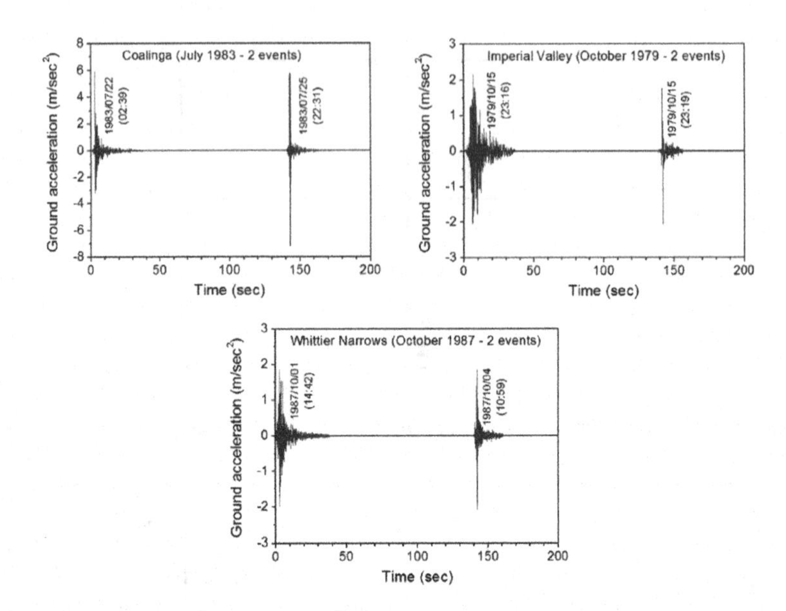

Fig. 5. Ground acceleration records of simulated seismic sequences [7].

Table 1. Representative response quantities for the Systems F0, F1, F2 and F3

Frames	Events	DI_G	DI_L	$u_{top}[cm]$
F0	Event E_1	0.134	0.179	2.227
	Event E_2	0.301	0.474	3.398
	Event $(E_1 + E_2)$	0.334	0.481	3.410
F1	Event E_1	0.133	0.185	1.715
	Event E_2	0.256	0.354	3.149
	Event $(E_1 + E_2)$	0.317	0.385	3.813
F2	Event E_1	0.068	0.007	1.126
	Event E_2	0.097	0.136	1.447
	Event $(E_1 + E_2)$	0.108	0.154	1.471
F3	Event E_1	0.054	0.009	1.069
	Event E_2	0.082	0.128	1.313
	Event $(E_1 + E_2)$	0.085	0.137	1.314

3.2 Representative Results

Representative results of the numerical investigation are presented in next Table 1. In column 2, Event E_1 corresponds to Coalinga seismic event of 0.165 normalized PGA and Event E_2 to 0.200 normalized PGA. The sequence of events E_1 and E_2 is denoted as Event (E_1+E_2). In column 3 the Global Damage Indices and in column 4 the Local Damage Index DI_L for the bending moment at the

Table 2. Multiple earthquakes data

No	Seismic sequence	Date (Time)	Magn. (M_L)	Rec. $PGA(g)$	Norm. $PGA(g)$
1	Coalinga	1983/07/22 (02:39)	6.0	0.605	0.165
		1983/07/25 (22:31)	5.3	0.733	0.200
2	Imperial Valley	1979/10/15 (23:16)	6.6	0.221	0.200
		1979/10/15 (23:19)	5.2	0.211	0.191
3	Whittier Narrows	1987/10/01 (14:42)	5.9	0.204	0.192
		1987/10/04 (10:59)	5.3	0.212	0.200

left fixed support of the frames are given. Finally, in the column 5, the maximum horizontal top displacement u_{top} (absolute value) is given.

As the table values show, multiple earthquakes generally increase, in an accumulative way, the response quantities. Based on the values of the horizontal top displacement $u_{top} = u_2^{(A)}$, it can be concluded that the optimal global strengthening version is the system F3 of Fig. 4(D).

4 Concluding Remarks

As the results of the numerical example have shown, the herein presented computational approach can be effectively used for the numerical investigation of the seismic inelastic behaviour of Cultural Heritage old industrial RC frame-systems, which are strengthened by cable elements and subjected to earthquake sequences. The flexure-shear critical characteristics, which these structures can appear, are taken into account. Further, the optimal cable-bracing scheme can be selected among investigated alternative ones by using computed damage indices.

Acknowledgments. This work was partially supported by the Bulgarian Academy of Sciences through the "Young Scientists" Grant No. DFNP-97/2016 for K. Liolios and K. Georgiev, and by the Bulgarian NSF Grant No. DFNI I-02/9 for I. Georgiev.

References

1. Asteris, P.G., Plevris, V.: Handbook of Research on Seismic Assessment and Rehabilitation of Historic Structures. IGI Global, Hershey (2015)
2. Leftheris, B.P., Stavroulaki, M.E., Sapounaki, A.C., Stavroulakis, G.E.: Computational Mechanics for Heritage Structures. WIT Press, Southampton (2006)
3. Penelis, Ge.G., Penelis, Gr.G.: Concrete Buildings in Seismic Regions. CRC Press, Boca Raton (2014)
4. Fardis, M.N.: Seismic Design, Assessment and Retrofitting of Concrete Buildings: Based on EN-Eurocode 8. Springer, Berlin (2009). https://doi.org/10.1007/978-1-4020-9842-0
5. EC8: Eurocode 8, Design of structures for earthquake resistance, Part 3: Assessment and Retrofitting of buildings, (EC8-part3), EN 1998–3, Brussels (2004)

6. KANEPE: Hellenic Code of Retrofitting of Reinforced Concrete Buildings. Organiz. Seismic Design and Protection (OASP). Athens, Greece (in Greek and in English translation) (2012)
7. Liolios, A., Moropoulou, A., Liolios, A., Georgiev, K., Georgiev, I.: A computational approach for the seismic sequences induced response of cultural heritage structures upgraded by ties. In: Margenov, S., Angelova, G., Agre, G. (eds.) Innovative Approaches and Solutions in Advanced Intelligent Systems. SCI, vol. 648, pp. 47–58. Springer, Cham (2016). https://doi.org/10.1007/978-3-319-32207-0_4
8. Liolios, A.: A computational investigation for the seismic response of RC structures strengthened by cable elements. In: Papadrakakis, M. et al. (eds.) Proceedings of the COMPDYN 2015: Computational Methods in Structural Dynamics and Earthquake Engineering, vol. II, pp. 3997–4010 (2015)
9. Liolios, A., Chalioris, C.: Reinforced concrete frames strengthened by cable elements under multiple earthquakes: A computational approach simulating experimental results. In: Proceedings of 8th GRACM International Congress on Computational Mechanics, Volos (2015)
10. Liolios, A., Chalioris, K., Liolios, A., Radev, S., Liolios, K.: A computational approach for the earthquake response of cable-braced reinforced concrete structures under environmental actions. In: Lirkov, I., Margenov, S., Waśniewski, J. (eds.) LSSC 2011. LNCS, vol. 7116, pp. 590–597. Springer, Heidelberg (2012). https://doi.org/10.1007/978-3-642-29843-1_67
11. Liolios, A., Elenas, A., Liolios, A., Radev, S., Georgiev, K., Georgiev, I.: Tall RC buildings environmentally degradated and strengthened by cables under multiple earthquakes: a numerical approach. In: Dimov, I., Fidanova, S., Lirkov, I. (eds.) NMA 2014. LNCS, vol. 8962, pp. 187–195. Springer, Cham (2015). https://doi.org/10.1007/978-3-319-15585-2_21
12. Moropoulou, A., Bakolas, A., Spyrakos, C., Mouzakis, H., Karoglou, M., Labropoulos, K., Delegou, E., Diamandidou, D., Katsiotis, N.: NDT investigation of Holy Sepulchre complex structures. In: Radonjanin, V., Crews, K. (eds.) Proceedings of Structural Faults and Repair 2012. Proceedings in CD-ROM (2012)
13. Panagiotopoulos, P.D.: Hemivariational Inequalities. Applications in Mechanics and Engineering. Springer, New York (1993). https://doi.org/10.1007/978-3-642-51677-1
14. Mistakidis, E.S., Stavroulakis, G.E.: Nonconvex Optimization in Mechanics. Smooth and Nonsmooth Algorithms, Heuristic and Engineering Applications. Kluwer, London (1998)
15. Carr, A.J.: RUAUMOKO - inelastic dynamic analysis program. Department Civil Engineering, University of Canterbury, Christchurch, New Zealand (2008)
16. Chopra, A.K.: Dynamics of Structures: Theory and Applications to Earthquake Engineering. Pearson Prentice Hall, New York (2007)
17. Priestley, M.J.N., Seible, F.C., Calvi, G.M.: Seismic Design and Retrofit of Bridges. Wiley, Hoboken (1996)
18. Makarios, T.: Modelling of characteristics of inelastic members of reinforced concrete structures in seismic nonlinear analysis. In: Padovani, G., Occhino, M. (eds.) Focus on Nonlinear Analysis Research, pp. 1–41. Nova Publishers, New York (2013)
19. Liolios, A., Chalioris, C.: Reinforced concrete frames strengthened by tention-tie elements under cyclic loading: a computational approach. In: Papadrakakis, M., et al. (eds.) Proceedings of COMPDYN 2017: Computational Methods in Structural Dynamics and Earthquake Engineering, paper C18195 (2017)
20. Park, Y.J., Ang, A.H.S.: Mechanistic seismic damage model for reinforced concrete. J. Struct. Div. ASCE **111**(4), 722–739 (1985)

New Approach to Identifying Solitary Wave Solutions of Modified Kawahara Equation

Tchavdar T. Marinov[1] and Rossitza S. Marinova[2(✉)]

[1] Department of Natural Sciences, Southern University at New Orleans, 6801 Press Drive, New Orleans, LA 70126, USA
tmarinov@suno.edu

[2] Department of Mathematical and Physical Sciences, Concordia University of Edmonton, 7128 Ada Boulevard, Edmonton, AB T5B 4E4, Canada
rossitza.marinova@concordia.ab.ca

Abstract. Stationary localized waves are considered in the frame moving to the right. The original ill–posed problem has a non–unique solution. To cope with this issue, the bifurcation problem is reformulated into a problem for identification of an unknown coefficient from over-posed boundary data in which the trivial solution is excluded. This approach to solving the modified fifth order Kawahara equation is original allowing identification of the non–trivial solutions. The numerical solutions are compared with known analytical solution. The convergence of the difference scheme is illustrated with numerical examples.

1 Introduction

The modified Kawahara equation was formulated first by Kawahara in [9]. This nonlinear equation has attracted attention due to the numerous applications in science and engineering. It plays an important role in the theory of fluid mechanics, optical fibers, biology, solid state physics, chemical kinematics, chemical physics, and geochemistry. Lots of methods deal with obtaining exact and approximate analytic solutions of the equation, see Biswas in [3], Yusufoĝlu and Bekir in [11], Jabbari and Kheiri in [7], Araruna et al. in [1]. Bekir et al. in [2] studied variable-coefficient modified Kawahara equations using solitary wave ansatz method to find dark optical as well as bright optical soliton solutions. Other authors study solitonic solutions by using a variety of powerful methods such as the variational iteration method (VIM) proposed by He in [6], Jin in [8], mesh-free method based on the collocation with radial basis functions used by Zarebnia and Aghili in [12].

Since not all solutions of the modified Kawahara equation are analytic, there is a need of studying the equation numerically. This work is dealing with numerical approach to solving the Modified Kawahara equation.

I. Lirkov and S. Margenov (Eds.): LSSC 2017, LNCS 10665, pp. 527–535, 2018.
https://doi.org/10.1007/978-3-319-73441-5_58

1.1 Localized Waves Problem

The modified Kawahara equation is a non–linear partial differential equation in the form

$$u_t + u^2 u_x + p u_{xxx} - q u_{xxxxx} = 0, \tag{1}$$

where $p > 0$ and $q > 0$ are given constants.

The motivation of this work is to devise an approach for obtaining new solitonic solutions of the stationary waves in the moving frame $\xi = x - ct$. After integrating with respect to ξ and taking into account the localized character of the investigated solutions, the Eq. (1) reduces to the following nonlinear ordinary differential equation

$$\mathcal{L}(u) = -cu + \frac{u^3}{3} + p u_{\xi\xi} - q u_{\xi\xi\xi\xi} = 0. \tag{2}$$

1.2 Asymptotic Behavior of the Tails

We are looking for solutions of the Eq. (2) with $u \to 0$ when $\xi \to \infty$. Then $u^3 \ll u$ in the tails and the linearized version of the Eq. (2) coincides with its linear part or reduces to

$$q u_{\xi\xi\xi\xi} - p u_{\xi\xi} + cu = 0. \tag{3}$$

Equation (3) possesses harmonic solutions of the type $e^{k\xi}$. The corresponding dispersion relation reads

$$qk^4 - pk^2 + c = 0 \quad \Longrightarrow \quad k^2 = \frac{1}{2q}\left[p \pm \sqrt{p^2 - 4cq}\right]. \tag{4}$$

Equation (4) shows that the asymptotic tails of the localized wave can be either monotonic, purely oscillatory or damped oscillatory, depending on whether k^2 is real positive, real negative or complex.

2 Introducing Unknown Coefficient

The problem is ill-posed according to the definition of Hadamard (see Hadamard [5]) because Eq. (2) possesses two solutions for the given boundary conditions $u(\xi) \to 0$ when $\xi \to \pm\infty$. In addition, the trivial solution is a very strong attractor. This brings challenges while constructing schemes based on finite difference and finite element methods for solving the problem.

To tackle the solitary-wave identification problem, we introduce a coefficient and transform the original problem to a problem for finding the coefficient in similar way as in Marinov et al. in [10].

Let $u(\xi) \not\equiv 0$ be a solution of Eq. (2) with $u(\xi) \to 0$ when $\xi \to \pm\infty$. Obviously, the function $u(-\xi)$ is also a solution, i.e. the solution is an even function, namely $u(\xi) = u(-\xi)$, and only even functions can be solutions to these equations.

Therefore, the problem can be considered on the half-line. In this case, the boundary conditions at $\xi = 0$ are

$$u(0) = \chi, \quad u'(0) = 0, \quad u'''(0) = 0, \tag{5}$$

where $\chi \neq 0$ is an unknown constant.

It is convenient to scale the function $u(\xi)$ by the unknown constant χ by introducing a new unknown function $w(\xi)$ as $u(\xi) = \chi w(\xi)$.

The equation for coefficient identification becomes

$$\mathcal{A}(w, \chi) \equiv -cw + \chi^2 \frac{w^3}{3} + pw_{\xi\xi} - qw_{\xi\xi\xi\xi} = 0, \tag{6}$$

with boundary conditions

$$w(0) = 1, \quad w'(0) = 0, \quad w'''(0) = 0, \tag{7}$$

$$w(\xi) \to 0 \text{ when } \xi, \to \infty, \quad w'(\xi) \to 0 \text{ when } \xi \to \infty. \tag{8}$$

Thus, the difficulties connected with the unknown constant χ in the boundary condition (5) are circumvented. Yet, it is a problem of unknown coefficient χ from over-posed boundary data. Under certain conditions it is possible to find a constant χ such that the Eq. (6) has a solution $w(\xi)$ and this solution also satisfies the boundary conditions (7)–(8). In such a case we say that the pair (w, χ) constitutes a solution to the problem (6)–(8). This is an inverse problem for identification of coefficient from over-posed data.

3 Solving the Coefficient Identification Problem

We use the Method of Variational Imbedding (MVI) proposed by Christo I. Christov, see Christov [4], for solving the coefficient identification problem.

First, we rewrite Eq. (6) as a system of ordinary differential equations

$$w' - \alpha = 0, \quad \alpha' - \beta = 0, \quad \beta' - \gamma = 0, \tag{9}$$

$$-cw + \chi^2 \frac{w^3}{3} + p\beta - q\gamma' = 0. \tag{10}$$

The respective boundary conditions at $\xi = 0$ become

$$w(0) = 1, \;\; \alpha(0) = \gamma(0) = 0. \tag{11}$$

Since $w \to 0$ when $\xi \to \infty$, all derivatives of the function w are approaching 0 and corresponding boundary conditions at infinity are

$$w(\infty) = \alpha(\infty) = \beta(\infty) = \gamma(\infty) = 0. \tag{12}$$

Following MVI, the original problem is replaced by a problem for minimizing a functional. In this case the functional is

$$\mathcal{I}(w,\chi) = \int_0^\infty \Big[(w' - \alpha)^2 + (\alpha' - \beta)^2 + (\beta' - \gamma)^2 + \left(-cw + \chi^2 \frac{w^3}{3} + p\beta - q\gamma' \right)^2 \Big] \mathrm{d}x \to \min, \tag{13}$$

where w, α, β, and γ must satisfy the conditions (11) and (12), and $\chi \neq 0$ is an unknown constant. The functional $\mathcal{I}(w,\chi)$ is a quadratic and homogeneous function of its arguments w and χ; hence, it attains its minimum if and only if all arguments are zero. In this sense there is one-to-one correspondence between the solution of the original problem (6)–(8) and the minimization problem (13).

3.1 Linearization

The Euler-Lagrange equations for this functional possess cubic nonlinearity with respect to the function w. It means that for solving the Euler-Lagrange equations numerically, one has to linearize the said equation. Alternatively, one can linearize the integrand in (13) considering the function w as known (say, from the previous iteration) when they appear as coefficients.

Following the latter approach for linearization, we introduce $\mu(\xi) = w^3(\xi)$ and consider the problem for minimization of the following functional

$$\mathcal{I}(w,\chi) = \int_0^\infty (w' - \alpha)^2 + (\alpha' - \beta)^2 + (\beta' - \gamma)^2 + \left(-cw + \chi^2 \frac{\mu}{3} + p\beta - q\gamma' \right)^2 \mathrm{d}x \to \min. \tag{14}$$

3.2 Imbedded Boundary-Value Problem

Necessary conditions for minimization of the functional in (14) are derived from the Euler-Lagrange equations for the functions $w(\xi)$, $\alpha(\xi)$, $\beta(\xi)$, and $\gamma(\xi)$, and for the constant χ. The equations for the functions $w(\xi)$, $\alpha(\xi)$, $\beta(\xi)$, and $\gamma(\xi)$ are

$$\frac{d}{d\xi}(w' - \alpha) + c(-cw + \chi^2 \frac{\mu}{3} + p\beta - q\gamma') = 0, \tag{15}$$

$$\frac{d}{d\xi}(\alpha' - \beta) + (w' - \alpha) = 0, \qquad \frac{d}{d\xi}(\beta' - \gamma) + (\alpha' - \beta) = 0, \tag{16}$$

$$-q\frac{d}{d\xi}(-cw + \chi^2 \frac{\mu}{3} + p\beta - q\gamma') + (\beta' - \gamma) = 0, \tag{17}$$

The system (15)–(17) is of the eight order. The solution of this system may satisfy four conditions at each boundary point. There exist exact number of boundary conditions for $\xi \to \infty$, see (12). We make use of the so-called natural boundary condition for minimization of the functional at $\xi = 0$, which is nothing else than the original Eq. (6), i.e.

$$-cw + \chi^2 \frac{\mu}{3} + p\beta - q\gamma' = 0. \tag{18}$$

The problem is coupled with the equation for χ. We rewrite the functional (13) in the form

$$\mathcal{I} = \frac{\chi^4}{9} \int_0^\infty w^6 \mathrm{d}x + \frac{2\chi^2}{3} \int_0^\infty w^3 \left(-cw + p\beta - q\gamma'\right) \mathrm{d}x \tag{19}$$

$$+ \int_0^\infty \left[\left(-cw + p\beta - q\gamma'\right)^2 + \left(w' - \alpha\right)^2 + \left(\alpha' - \beta\right)^2 + \left(\beta' - \gamma\right)^2\right] \mathrm{d}x.$$

After some algebraic manipulations, the equation for χ^2 reduces to

$$\chi^2 = -3 \left(\int_0^\infty w^3 \left(-cw + p\beta - q\gamma'\right) \mathrm{d}x \right) / \left(\int_0^\infty w^6 \mathrm{d}x \right). \tag{20}$$

4 Finite Difference Method

4.1 Grid and Approximations

The grid is chosen to be uniform allowing approximation of all operators with central differences. The grid spacing is defined as $h = \frac{\xi_\infty}{n-2}$, where n is the total number of grid points and ξ_∞ is a sufficient large number used to approximate infinity. Then the grid points are defined as follows: $\xi_i = (i - 1.5)h$ for $i = 1, \ldots, n$. The grid point $\xi = 0$ is the mid-point $\xi_{\frac{3}{2}}$, which is important for the numerical procedure. Let us also introduce the notation $w_i = w(\xi_i)$, $\alpha_i = \alpha(\xi_i)$, $\beta_i = \beta(\xi_i)$, $\gamma_i = \gamma(\xi_i)$, for $i = 1, \ldots, n$. We employ symmetric central differences for approximating the differential equations (15)–(17) and obtain

$$\frac{w_{i-1} - 2w_i + w_{i+1}}{h^2} - \frac{-\alpha_{i-1} + \alpha_{i+1}}{2h} - cq\frac{-\gamma_{i-1} + \gamma_{i+1}}{2h} + p\beta_i - c^2 w_i = -c\chi^2 \frac{\mu_i}{3} + O(h^2), \tag{21}$$

$$\frac{\alpha_{i-1} - 2\alpha_i + \alpha_{i+1}}{h^2} - \frac{-\beta_{i-1} + \beta_{i+1}}{2h} + \frac{-w_{i-1} + w_{i+1}}{2h} - \alpha_i = 0 + O(h^2), \tag{22}$$

$$\frac{\beta_{i-1} - 2\beta_i + \beta_{i+1}}{h^2} - \frac{-\gamma_{i-1} + \gamma_{i+1}}{2h} + \frac{-\alpha_{i-1} + \alpha_{i+1}}{2h} - \beta_i = 0 + O(h^2), \tag{23}$$

$$q^2 \frac{\gamma_{i-1} - 2\gamma_i + \gamma_{i+1}}{h^2} + (1 - pq)\frac{-\beta_{i-1} + \beta_{i+1}}{2h} + cq\frac{-w_{i-1} + w_{i+1}}{2h} - \gamma_i = q\chi^2 \frac{-\mu_{i-1} + \mu_{i+1}}{6h} + O(h^2). \tag{24}$$

for $i = 2, \ldots, n - 1$.

The grid allows approximating the boundary conditions with the second order as well. Namely, at $\xi = 0$, the boundary conditions become

$$w(0) = 1 \Longrightarrow w_1 + w_2 = 2 + O(h^2), \tag{25}$$
$$\alpha(0) = 0 \Longrightarrow \alpha_1 + \alpha_2 = 0 + O(h^2), \tag{26}$$
$$\gamma(0) = 0, \Longrightarrow \gamma_1 + \gamma_2 = 0 + O(h^2). \tag{27}$$

Finally, the equation $-cw + \chi^2 \frac{\mu}{3} + p\beta - q\gamma' = 0$ is approximated as

$$p\frac{\beta_1 + \beta_2}{2} - q\frac{\gamma_2 - \gamma_1}{h} = c\frac{w_1 + w_2}{2} - \frac{\chi^2}{3}\left(\frac{w_1 + w_2}{2}\right)^3 + O(h^2). \tag{28}$$

The boundary conditions at $\xi = \xi_\infty$ are approximated as

$$w(\xi) \to 0, \quad \xi \to \infty \Longrightarrow w_n + w_{n-1} = 0 + O(h^2), \tag{29}$$
$$\alpha(\xi) \to 0, \quad \xi \to \infty \Longrightarrow \alpha_n + \alpha_{n-1} = 0 + O(h^2), \tag{30}$$
$$\beta(\xi) \to 0, \quad \xi \to \infty \Longrightarrow \beta_n + \beta_{n-1} = 0 + O(h^2), \tag{31}$$
$$\gamma(\xi) \to 0, \quad \xi \to \infty \Longrightarrow \gamma_n + \gamma_{n-1} = 0 + O(h^2). \tag{32}$$

The linearization proposed in Subsect. 3.1 allows to invert the matrix of the linear system only once, at the beginning of the iteration process, thus, save computational time.

4.2 Estimation of χ

We use the 'extended midpoint rule' for approximating the integrals in the Eq. (20) for χ

$$\chi^2 = -3\left(\sum_{k=2}^{n-1} \bar{w}_k^3 \left(-c\bar{w}_k + p\bar{\beta}_k - q\frac{-\gamma_{k-1} + \gamma_k}{h}\right)\right) / \left(\sum_{k=2}^{n-1} \bar{w}_k^6\right) + O(h^2), \tag{33}$$

where $\bar{w}_k$ and $\bar{\beta}_k$ are the midpoint values of the functions. This approach secures the second order approximation.

4.3 Iterative Algorithm

Step I. Solve the eight-order boundary value problem (15)–(17) for the function w with given χ and μ.

Step II. If the difference between two consecutive approximations of w is smaller than ε, then proceed to **Step III**; otherwise, μ is replaced by the newly approximated value of w^3 and then go to **Step I**;

Step III. With the newly computed w, the coefficient χ is evaluated from formula (33). If the following criterion is satisfied

Table 1. Number of iterations, l^2 norm of $w - w_{\text{exact}}$, the estimated values of the coefficients χ and the rates of convergence.

h	Iterations	$\|\|w - w_{\text{exact}}\|\|_{l^2}$	Rate$_w$	χ	$\|\chi - \chi_{\text{exact}}\|$	Rate$_\chi$
0.1	56	0.00886880	—	2.373540	0.001832	—
0.05	55	0.00220981	2.0048	2.372175	0.000467	1.9719
0.025	50	0.00055174	2.0019	2.371826	0.000117	1.9969
0.0125	44	0.00013789	2.0005	2.371738	0.000029	2.0124

$$\frac{\max |w^{n+1} - w^n|}{h \max |w^{n+1}|} < \varepsilon \quad \text{and} \quad \frac{\max |\chi^{n+1} - \chi^n|}{h \max |\chi^{n+1}|} < \varepsilon, \tag{34}$$

then the iterations are terminated. Otherwise the index of iteration is incremented $n := n + 1$ and the algorithm is returned to step **Step I**.

The chosen tolerance in (34) is $\varepsilon \leq 10^{-10}$.

5 Numerical Results

In order to verify the performance of the described numerical scheme we started the numerical experiments with an analytical solution (see Jin [8])

$$u_{\text{an}}(\xi) = -\frac{3p}{\sqrt{10q}} \operatorname{sech}^2\left(\frac{\sqrt{p}}{2\sqrt{5q}}\xi\right), \text{ where } c = \frac{4p^2}{25q}. \tag{35}$$

Choosing $p = 5$ and $q = 4$ in the Eq. (35), we obtain the following value for the coefficient: $\chi_{\text{an}} = u(0) = \frac{3\sqrt{10}}{4} \approx 2.371708245126285\ldots$.

The value of the approximated numerically infinity ξ_∞ is chosen as $\xi_\infty = 30$ for the presented here results. The discretization error term is $O(h^2)$, and the total error of approximation is $O(h^2)$. The results from the numerical calculations clearly demonstrate these error orders.

Figure 1 shows the point-wise numerical errors for the function w calculated with four different spacings: $h = 0.1; 0.05; 0.025; 0.0125$. It can be seen that if the spacing decreases twice, the errors decrease approximately four times. This shows that the order of convergence is 2.

The rates of convergence are calculated as

$$\text{rate}_w = \log_2 \frac{||w_{2h} - w_{\text{exact}}||}{||w_h - w_{\text{exact}}||}, \qquad \text{rate}_\chi = \log_2 \frac{||\chi_{2h} - \chi_{\text{exact}}||}{||\chi_h - \chi_{\text{exact}}||}. \tag{36}$$

The l^2 norm of the difference between the numerical solution w and the analytical one with four different spacings $h = 0.1; 0.05; 0.025; 0.0125$ are given in Table 1 along with the rates of convergence. The estimated values of the coefficients χ and the rates of convergence using the same four grid spacings are also given in Table 1. The numerical experiments clearly confirm the second order of convergence of the numerical solution to the exact solution.

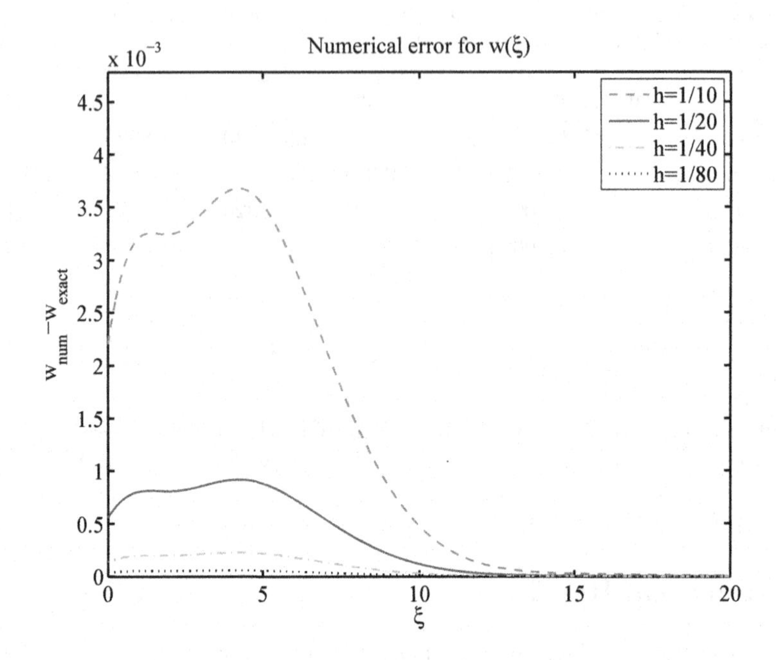

Fig. 1. The difference between numerical and exact values of $w(\xi)$.

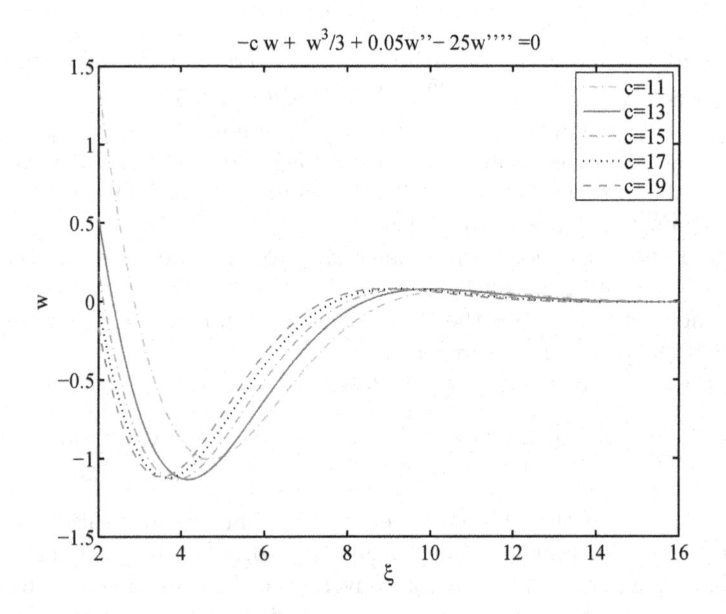

Fig. 2. Solitary wave solutions with oscillatory tails for $p = 0.05$, $q = 25$.

5.1 Solitary Waves with Oscillatory Tails

As we already mentioned in Subsect. 1.2, for certain values of the coefficients we can expect solitary-wave solutions with oscillatory tails. In this subsection we keep the coefficients $p = 0.05$ and $q = 25$ and change the values of c to $c = 11; 13; 15; 17; 19$. The shape of the obtained solution is given in Fig. 2. For this calculations we use $\xi_\infty = 60$. The oscillatory behavior is more and more visible as c is increased.

Conclusions. In this paper, a new approach has been applied to solve the modified Kawahara equation. Numerical results have been presented to show the efficiency of the method. We can conclude from the results that the methods provide high accuracy for the modified Kawahara equation. Therefore, the method proved to be an efficient method for solving the modified Kawahara equation. Next work will involve identifying solutions for different sets of parameters.

References

1. Araruna, F., Capistrano-Filho, R., Doronin, G.: Energy decay for the modified Kawahara equation posed in a bounded domain. J. Math. Anal. Appl. **385**, 743–756 (2012)
2. Bekir, A., Güner, Ö., Bilgil, H.: Optical soliton solutions for the variable coefficient modified Kawahara equation. Optik **126**, 2518–2522 (2015)
3. Biswas, A.: Solitary wave solution for the generalized Kawahara equation. Appl. Math. Lett. **22**, 208–210 (2009)
4. Christov, C.I.: A method for identifying homoclinic trajectories. In: Proceedings of the 14th Spring Conference, Union of Bulgarian Mathematicians, pp. 571–577 (1985)
5. Hadamard, J.: Le Probleme de Cauchy et les Equations aux Derivatives Partielles Lineares Hyperboliques. Hermann, Paris (1932)
6. He, J.H.: Variational iteration method for delay differential equations. Commun. Nonlinear Sci. Numer. Simul. **2**(4), 235–236 (1997)
7. Jabbari, A., Kheiri, H.: New exact traveling wave solutions for the Kawahara and Modified Kawahara Equations by using Modified Tanh-Coth Method. Acta Universitatis Apulensis **23**, 21–38 (2010)
8. Jin, L.: Application of variational iteration method and homotopy perturbation method to the modified Kawahara equation. Math. Comput. Model. **49**, 573–578 (2009)
9. Kawahara, T.: Oscillatory solitary waves in dispersive media. J. Phys. Soc. Jpn. **33**, 260–264 (1972)
10. Marinov, T.T., Christov, C.I., Marinova, R.S.: Novel numerical approach to solitary-wave solutions identification of Boussinesq and Korteweg-de Vries equations. Int. J. Bifurcat. Chaos **15**(2), 557–565 (2005)
11. Yusufoğlu, E., Bekir, A.: Symbolic computation and new families of exact travelling solutions for the Kawahara and modified Kawahara equations. Comput. Math. Appl. **55**, 1113–1121 (2008)
12. Zarebnia, M., Aghili, M.: A new approach for numerical solution of the modified Kawahara equation. J. Nonlinear Anal. Appl. **2**, 48–59 (2016)

Sequential Variational Data Assimilation Algorithms at the Splitting Stages of a Numerical Atmospheric Chemistry Model

Alexey Penenko[1,2](✉), Vladimir Penenko[1,2], Elena Tsvetova[1], Anastasia Grishina[1,2], and Pavel Antokhin[3]

[1] Institute of Computational Mathematics and Mathematical Geophysics SB RAS, pr. Akademika Lavrentjeva 6, 630090 Novosibirsk, Russia
a.penenko@yandex.ru

[2] Novosibirsk State University, Pirogova str. 1, 630090 Novosibirsk, Russia

[3] V.E. Zuev Institute of Atmospheric Optics SB RAS, Academician Zuev sq. 1, 634055 Tomsk, Russia

Abstract. A variational data assimilation algorithm is studied numerically. *In situ* concentration measurement data are assimilated into transport and transformation model of atmospheric chemistry. The algorithm is based on decomposition and splitting methods with solution of variational data assimilation problems for separate splitting stages. A direct algorithm without iterations is used for the linear transport stage. An iterative gradient algorithm is applied for data assimilation at the nonlinear chemical transformation stage. In a realistic numerical experiment, the contributions of data assimilation algorithms for the different splitting stages are compared.

Keywords: Variational data assimilation · Splitting
Advection-diffusion-reaction models

1 Introduction

Forecasting of chemical composition in the atmosphere is a challenge for researchers due to the nonlinear, multi-scale, multi-dimensional and dynamic character of the corresponding transport and transformation processes [1]. An accurate forecast needs information about various model parameters like emissions data, meteorological conditions, reaction rates. Some of them are difficult to obtain. Data assimilation algorithms can be applied to improve the model forecasts by adjusting the model parameters in the course of modeling with the use of incoming measurement data of the adequate accuracy.

Computational issues with large-scale models can be solved by using the splitting technique which admits the efficient parallelization. In this paper we study an approach to data assimilation when variational data assimilation is carried out quasi-independently within the separate splitting stages with shared

I. Lirkov and S. Margenov (Eds.): LSSC 2017, LNCS 10665, pp. 536–543, 2018.
https://doi.org/10.1007/978-3-319-73441-5_59

measurement data [2,3]. Data assimilation at the transport stage is fulfilled with a direct algorithm. An iterative gradient algorithm is applied in the nonlinear transformations stage. We have developed the consistent numerical schemes for direct and adjoint problems related to both stages. The approach allows us to assimilate data in large-scale problems as well. The iterative algorithm for the nonlinear stage is time-consuming. That is why it is important to determine if its contribution is significant for the performance of the algorithm in a realistic scenario.

2 Data Assimilation Problem

A horizontally homogeneous scenario is considered in a domain $\Omega_T = \Omega \times [0,T]$, $\Omega = [0,Z]$, $T > 0$, bounded by $\partial\Omega_T = \partial\Omega \times [0,T]$. An advection-diffusion-reaction model for $l = 1,\ldots,N_c$ is

$$\begin{aligned} \frac{\partial \varphi_l}{\partial t} &+ div\,(w\,\varphi_l - \mu\ grad\varphi_l) + P_l(t,\boldsymbol{\varphi})\varphi_l \\ &= \Pi_l(t,\boldsymbol{\varphi}) + f_l + r_l \quad (z,t) \in \Omega_T, \end{aligned} \tag{1}$$

$$\mu\frac{\partial \varphi_l}{\partial \boldsymbol{n}} + \beta\varphi_l = g_l, \quad (z,t) \in \partial\Omega_T, \tag{2}$$

$$\varphi_l = \varphi_l^0, \quad z \in \Omega,\ t = 0, \tag{3}$$

where N_c is the number of considered substances, $\varphi_l(z,t)$ denotes the concentration of the lth substance at a point $(z,t) \in \Omega_T$, $\boldsymbol{\varphi}$ is the vector of φ_l for $l = 1,\ldots,N_c$. The functions $w(z,t)$, $\mu(z,t)$ correspond to the "wind speed" and diffusion coefficient, $\boldsymbol{n}$ is the boundary outer normal direction, $f_l(z,t)$, $g_l(z,t)$, $\varphi_l^0(z)$ are *a priori* sources function and initial conditions, r_l is a control function (uncertainty), that is introduced in the model structure to assimilate the data. Destruction and production operator elements $P_l, \Pi_l : [0,T] \times \mathbb{R}_+^{N_c} \to \mathbb{R}_+$ are defined by the chemical kinetics system. We suppose all the functions and model parameters to be smooth enough for the solutions to exist and the further transformations to make sense.

Direct problem: Given $\boldsymbol{f}$, $\boldsymbol{g}$, $\boldsymbol{r}$, $\boldsymbol{\varphi}^0$, determine $\boldsymbol{\varphi}$ from (1)–(3).

Exact solution $\boldsymbol{\varphi}^*$ is the solution of the direct problem corresponding to the emission function $\boldsymbol{r}^*$ that is "unknown" in a data assimilation scenario. Instead, there is a set of available measurements:

$$\xi_m = \left\{(z_M^m, t_M^m, l_M^m, I_m, \sigma_M^m)\right\}, m = 1, ..., N_M,$$

where z_M^m is the height of the measurement, t_M^m is the time moment of the measurement, l_M^m is the number of the substance measured,

$$I_m = (1 + (\sigma_M^m)\,\eta)\varphi^*_{l_M^m}\,(z_M^m, t_M^m)\,,$$

is the measured concentration, $\eta \in N(0,1)$ is a normally distributed variable, $\sigma_M^m > 0$.

Data assimilation problem: For any $t^* \in [0,T]$ determine $\boldsymbol{\varphi}^*$ for $t \geq t^*$ with (1)–(3), the functions $\boldsymbol{f}$, $\boldsymbol{g}$, $\boldsymbol{\varphi}^0$ and $\{\xi_m | t_M^m \leq t^*\}$.

3 Weakly Coupled Data Assimilation Scheme for the Split Model

For the numerical solution, a uniform temporal grid $\omega_t = \{t^j\}_{j=1}^{N_t}$ on $[0,T]$ with step size Δt and N_t points and a uniform spatial grid ω_z with N_z grid points on Ω are defined. Let all the measurement points be on the grid: $\{(z_M^m, t_M^m)\}_{m=1}^{N_M} \subset \omega_z \times \omega_t$. For every time step j the measurement operator can be defined:

$$H^j \varphi = \left\{\varphi_{l_M^m}\left(z_M^m, t_M^m\right) | t_M^m = t^j\right\}, \; I^j = \left\{I_m | t_M^m = t^j\right\}.$$

An additive-averaged splitting scheme [4, p. 341] is used in the intervals $[t^{j-1}, t^j]$. The splitting is done with respect to a physical process: advection-diffusion (corresponding variables are marked with index z) and transformation (marked with index c) with *a priori* sources partition $\boldsymbol{f} = \boldsymbol{f}_c + \boldsymbol{f}_z$. The details will be presented further.

Variational data assimilation algorithms estimate the solution of a data assimilation problem as the minimum of the functional with the constraints imposed by the model. The functional usually combines measurement data misfit with a norm of the control variable:

$$J^j\left(\boldsymbol{\varphi}, \boldsymbol{r}\right) = \alpha \left\|H^j\boldsymbol{\varphi} - I^j\right\|_{s^j}^2 + \left\|\boldsymbol{r}\right\|^2, \tag{4}$$

where $\|.\|$ is the Euclidean norm over the space of the real grid functions on $\omega_z \times \{1,\ldots,N_c\}$, $\|.\|_{s^j}$ is the weighted Euclidean norm in $\mathbb{R}^{N_M^j}$, where N_M^j is the number of measurements at t^j, α is the assimilation (regularization) parameter, which defines whether the solution will be closer to the direct model solution for $\boldsymbol{r} = 0$, $(\alpha \to 0)$ or better reproduce the measurements $(\alpha \to \infty)$. At the time step t^j only the control variable $\boldsymbol{r}^j$ for the step is updated.

In the weakly-coupled (or fine-grained) approach [2,3], the same data are assimilated quasi-independently to different parts of the model and the results are merged afterwards. Given the solution $\boldsymbol{\varphi}^{j-1}$ on the previous time step, the functional

$$J_f^j\left(\left\{\boldsymbol{\varphi}_\beta^j, \boldsymbol{r}_\beta^j\right\}_{\beta\in\{z,c\}}\right) = \sum_{\beta\in\{z,c\}} J^j\left(\boldsymbol{\varphi}_\beta^j, \boldsymbol{r}_\beta^j\right)$$

is minimized with the split model acting as a constraint. Here $\left\{\boldsymbol{\varphi}_\beta^j, \boldsymbol{r}_\beta^j\right\}_{\beta\in\{z,c\}}$ are state and control functions corresponding to the splitting stages. The next step is approximated with

$$\boldsymbol{\varphi}^j = \sum_{\beta\in\{z,c\}} \gamma_\beta \boldsymbol{\varphi}_\beta^j, \; \sum_{\beta\in\{z,c\}} \gamma_\beta = 1, \; \gamma_\beta \geq 0, \; \beta \in \{z,c\}. \tag{5}$$

The advantage of the scheme is that all the optimization problems can be solved in parallel.

3.1 Advection-Diffusion Stage

At the stage, the different substances can be treated in parallel. Let $l \in \{1, \ldots, N_c\}$,

$$\gamma_z \frac{\partial (\varphi_z)_l}{\partial t} + \frac{\partial}{\partial z}\left(w (\varphi_z)_l - \mu \frac{\partial}{\partial z} (\varphi_z)_l \right)$$

$$= (f_z)_l + (r_z)_l, \ (z,t) \in \Omega \times \left[t^{j-1}, t^j\right], \tag{6}$$

$$\mu \frac{\partial (\varphi_z)_l}{\partial \boldsymbol{n}} + \beta (\varphi_z)_l = g_l, \ (z,t) \in \partial\Omega \times \left[t^{j-1}, t^j\right], \tag{7}$$

$$(\varphi_z)_l = \varphi_l, \ z \in \Omega, \ t = t^{j-1}. \tag{8}$$

The model (6)–(8) is approximated with a diagonally dominant tridiagonal matrix system:

$$- a_i \phi_{i+1}^j + b_i \phi_i^j - c_i \phi_{i-1}^j = \phi_i^{j-1} + \Delta t \ r_i^j + \Delta t \ f_i^j, \ i = 1, \ldots, N_z, \tag{9}$$

$$a_i \geq 0, \ b_i > 0, \ c_i \geq 0, \ i = 1, \ldots, N_z, \ c_0 = a_{N_z} = 0, \tag{10}$$

where $\phi_i^j = (\varphi_z)_l \left(z_i, t^j\right)$, $r_i^j = (r_z)_l \left(z_i, t^j\right)$, $f_i^j = (f_z)_l \left(z_i, t^j\right)$. The conditional minimization problem for (4) with the constraints (9)–(10) has the following Lagrange function:

$$\bar{J}^j \left(\phi^j, r^j, \psi^j\right) \Delta t = \left(\alpha \sum_{i=1}^{N_z} \left(\frac{\phi_i^j - I_i^j}{s_i} \right)^2 M_i^j + \sum_{i=1}^{N_z} \left(r_i^j\right)^2 \right) \Delta t$$

$$+ \sum_{i=1}^{N_z} \left(- a_i \phi_{i+1}^j + b_i \phi_i^j - c_i \phi_{i-1}^j - \phi_i^{j-1} - \Delta t r_i^j - \Delta t f_i^j \right) \psi_i^j, \tag{11}$$

where ψ_i^j is the adjoint function (Lagrange multiplier). If there is a measurement data at the point $\left(z_i, t^j\right)$ for the substance l, then M_i^j is equal to 1, and equal to 0 otherwise. If $M_i^j = 1$, then I_i^j is equal to the measurement data, and equal to 0 otherwise. If $M_i^j = 1$, then s_i is equal to the measurement weight, and equal to 1 otherwise. The stationary point of (11) can be found by the solution of the following tridiagonal matrix equation with the direct Gaussian elimination method [2,3]:

$$- A_i \Phi_{i+1}^j + B_i \Phi_i^j - C_i \Phi_{i-1}^j = F_i^j, i = 1, ..., N, \quad C_0 = A_N = 0,$$

$$A_i = \begin{pmatrix} a_i & 0 \\ 0 & c_{i+1} \end{pmatrix}, B_i = \begin{pmatrix} b_i & -\frac{\Delta t}{2} \\ \frac{2\alpha M_i \Delta t}{s_i^2} & b_i \end{pmatrix}, C_i = \begin{pmatrix} c_i & 0 \\ 0 & a_{i-1} \end{pmatrix},$$

$$\Phi_i^j = \begin{pmatrix} \phi_i^j \\ \psi_i^j \end{pmatrix}, F_i^{j+1} = \begin{pmatrix} \phi_i^{j-1} + \Delta t f_i^j \\ \frac{2\alpha M_i \Delta t}{s_i^2} I_i^j \end{pmatrix}.$$

3.2 Chemical Reaction Stage

At the stage, the different height levels can be treated in parallel. Let $z \in \omega_z$,

$$\gamma_c \frac{\partial (\varphi_c)_l}{\partial t} + P_l (t, \varphi_c) (\varphi_c)_l = \Pi_l (t, \varphi_c) + (f_c)_l + (r_c)_l, \quad (12)$$

$$t \in \left[t^{j-1}, t^j\right], \; l = 1, ..., N_c, \quad (13)$$

$$(\varphi_c)_l = \varphi_l, \; t = t^{j-1}, \; l = 1, ..., N_c. \quad (14)$$

Due to the stiffness, the equation for the chemical reaction stage is calculated on the temporal grid $\bar{\omega}_t$ that is $\bar{N}_t$ times finer than temporal grid ω_t, i.e. each interval $\left[t^{j-1}, t^j\right]$ is divided into $\bar{N}_t - 1$ sub-intervals with the lengths $\overline{\Delta t}$ and grid points $\left\{\bar{t}^j\right\}_{j=1}^{\bar{N}_t}$. Let $\bar{t}^1 = t^{j-1}$ and $\bar{t}^{\bar{N}_t} = t^j$. To solve (12)–(14), an unconditionally monotonic scheme is used that is equivalent to the part of QSSA scheme [5]. For $l = 1, ..., N_c$:

$$\phi_l^{j+1} = \phi_l^j \exp(-P_l(\bar{t}^j, \phi^j)\overline{\Delta t}) + g\left(P_l(\bar{t}^j, \phi^j)\right)\left(\Pi_l(\bar{t}^j, \phi^j) + f_l + r_l\right), \quad (15)$$

$$g(P) = \int_0^{\overline{\Delta t}} \exp(-P(\overline{\Delta t} - \xi))d\xi = \frac{1 - \exp(-P\overline{\Delta t})}{P\overline{\Delta t}}, \quad j = 1, \ldots, \bar{N}_t - 1, \quad (16)$$

$$\phi_l^1 = \varphi_l(z, \bar{t}^1), \quad (17)$$

where $\phi_l^j = (\varphi_c)_l (z, \bar{t}^j)$, $r_l = (r_c)_l (z, t^j)$, $f_l = (f_c)_l (z, t^j)$. The concentration measurement data for t^j is interpreted as the mean value of the state function on the fine grid. The control function $\boldsymbol{r}$ on the fine grid is supposed to be constant. Hence the target functional for this stage is

$$J(\boldsymbol{r}, \boldsymbol{\phi}) = \alpha \sum_{l=1}^{N_c} \left(\frac{1}{\bar{N}_t} \sum_{j=1}^{\bar{N}_t} \phi_l^j - I_l\right)^2 \frac{M_l}{s_l^2} + \sum_{l=1}^{N_c} r_l^2, \quad (18)$$

where $\boldsymbol{\phi} = \boldsymbol{\phi}(\boldsymbol{r})$ is the solution of (15)–(17). If there is a measurement data for the substance l at the point $\left(z, t^j\right)$, then M_l is equal to 1, and equal to 0 otherwise. If $M_l = 1$, then I_l is equal to the measurement data, and equal to 0 otherwise. If $M_l = 1$, then s_l is equal to the measurement weight, and equal to 1 otherwise. With the help of adjoint equations, the gradient of functional (18) can be derived

$$\nabla_{\mathbf{r}} J(\mathbf{r}, \boldsymbol{\phi}(\boldsymbol{r})) = 2\alpha \sum_{j=1}^{\bar{N}_t - 1} \left(\mathbf{R}^j\right)^* \psi^j + 2\boldsymbol{r},$$

where ψ is the solution of the adjoint problem

$$\psi^{j-1} = \left(\mathbf{W}^j\right)^* \psi^j + \boldsymbol{h}^j,\ j = 1, ..., \bar{N}_t - 1, \quad \psi^{\bar{N}_t - 1} = \boldsymbol{h}^{\bar{N}_t},$$

$$\mathbf{W}^j = \underset{l=1,...,N_c}{diag} \exp(-P_l\left(\bar{t}^j, \phi^j\right) \overline{\Delta t}) + \left\{S_l\left(\bar{t}^j, \phi^j, \boldsymbol{r}^j\right)\right\}_{l=1}^{N_c},$$

$$S_l\left(t, \varphi, \boldsymbol{r}\right) = g\left(P_l(t, \varphi)\right) \nabla \Pi_l\left(t, \varphi\right)$$

$$+ \left(\phi_l\left(-\exp(-P_l(t, \varphi)\overline{\Delta t})\overline{\Delta t}\right) + g'\left(P_l(t, \varphi)\right)\left(\Pi_l(t, \varphi) + r_l\right)\right) \nabla P_l\left(t, \varphi\right),$$

$$\mathbf{R}^j = \underset{l=1,...,N_c}{diag} g\left(P_l\left(t^j, \phi^j\right)\right), \quad h_l^j = \frac{2M_l}{s_l^2 \bar{N}_t}\left(\frac{1}{\bar{N}_t}\sum_{j=1}^{\bar{N}_t} \phi_l^j - I_l\right).$$

Here $\underset{l=1,...,N_c}{diag} A_c$ denotes the diagonal matrix with A_c on diagonal and $\{S_l\}_{l=1}^{N_c}$ is the matrix with rows S_l. The optimization problem is then solved with a conjugate gradient method. Negative values of control functions r_β can lead to negative concentrations. At the both stages the negative concentrations are set to zero.

4 Numerical Study of Splitting Stages Contributions to the Data Assimilation Result

For the numerical experiment, a realistic scenario corresponding to a background region in Western Siberia on August 8, 2013, has been chosen. The turbulence diffusion coefficient $\mu\left(z, t\right)$ was calculated from *k-theory* [9] with the results of the ground heat flow measurements, friction tension and by the vertical wind profiles. The grid parameters were $N_z = 30$, $Z = 3000\,\mathrm{m}$, $T = 75600\,\mathrm{s}$, $N_t = 281$, $\bar{N}_t = 10^3$. The measurements are regular in space and time and were defined by the temporal $\Delta t_m = 5400\,\mathrm{s}$ and the spatial $\Delta z_m = 500\,\mathrm{m}$ gaps between measurement points, $\sigma_M^m = 0$, $\alpha = 10^{100}$ (the measurements are reproduced as accurate as possible), $s_i = s_l = 1$, $\boldsymbol{f} = 0$. The atmospheric chemistry transformation model [6–8] contained 20 reactions and 22 reacting species. The reaction rates have been taken from [6]. They depended on the time of day.

The following configurations were compared: NoDA (background solution): direct problem solution that is driven by initial and boundary conditions only; Exact solution: direct problem solution that is driven by both initial and boundary conditions and by the "unknown" source of NO "on the ground" ($z = 0$). The emission rate was chosen to provide a unitary relative difference between "exact" and background solutions; DA (data assimilation solution): The result of the data assimilation algorithm. The initial and boundary conditions are the same as for the NoDA and "exact" solutions. The source is absent as in the NoDA case, but there are NO_2 and O_3 concentration measurements available for the data assimilation system. In Fig. 1 the results are presented for O_3 which measurement data is assimilated and for NO that is emitted in the exact solution case. It shows that the data assimilation algorithm is able to estimate unmeasured substances concentrations.

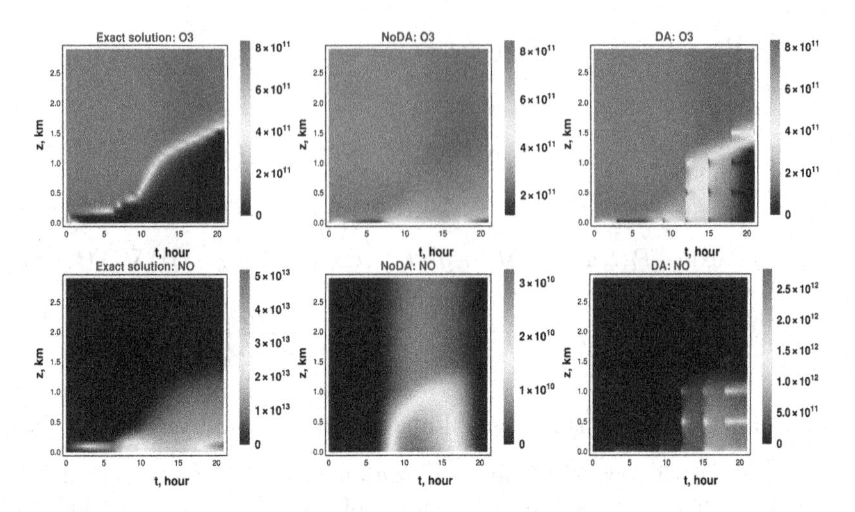

Fig. 1. Comparison of the exact solution (left column), solution without data assimilation (central column) and the data assimilation results (right column) for O_3 which concentration data is assimilated (upper row) and NO that is emitted (lower row).

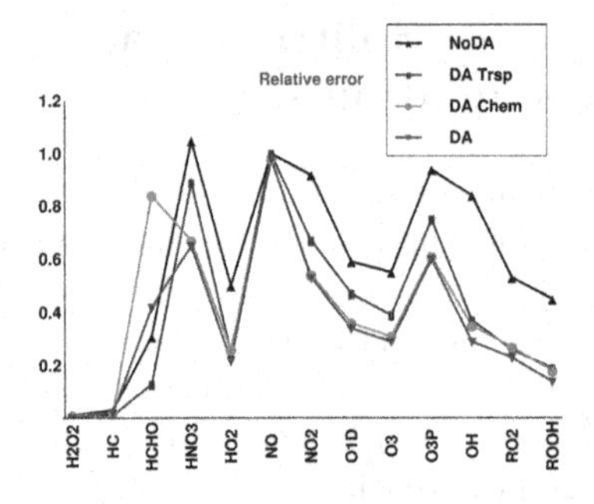

Fig. 2. Relative errors for different configurations: NoDA corresponds to the solution without data assimilation; DA is the result of data assimilation; in DA Trsp, the data are assimilated at the transport stage only; in DA Chem, the data are assimilated at the reactions stage only.

To estimate the impact of data assimilation at different splitting stages, we have considered the configurations when data were assimilated at the advection-diffusion stage only (DA Trsp) and when they were assimilated at the reaction stage only (DA Chem). The result is presented in Fig. 2. It can be seen that the data assimilation at the reaction stage was able to improve the result for the most substances compared to data assimilation at the transport stage only. For $HCHO$, the transport stage data assimilation was able to stabilize the solution. According to Fig. 2, the relative error for the unmeasured driving variable NO is almost the same as for the NoDA case. But in Fig. 1 we can see that NO is qualitatively reconstructed, albeit the reconstructed concentration levels have different magnitudes with respect to the exact solution.

5 Conclusions

Combination of splitting, direct and iterative variational data assimilation schemes at the splitting stages allows constructing scalable algorithms for assimilation of *in situ* concentration measurements to the advection-diffusion-reaction models of atmospheric chemistry. In the realistic numerical experiment considered, the contribution of data assimilation at the reaction stage was more significant than the contribution of data assimilation at the advection-diffusion stage.

Acknowledgments. The work is partially supported by the Presidium of RAS under the Programs I.33P and II.2P/I.3-3; by the projects MK-8214.2016.1, RFBR 17-01-00137 and 17-05-00374. The SB RAS Siberian Supercomputer Center is gratefully acknowledged for providing supercomputer facilities.

References

1. Elbern, H., Strunk, A., Schmidt, H., Talagrand, O.: Emission rate and chemical state estimation by 4-dimensional variational inversion. Atmos. Chem. Phys. Discuss. **7**(1), 1725–1783 (2007). https://doi.org/10.5194/acpd-7-1725-2007
2. Penenko, A., Penenko, V., Nuterman, R., Baklanov, A., Mahura, A.: Direct variational data assimilation algorithm for atmospheric chemistry data with transport and transformation model. In: Romanovskii, O.A. (ed.) 21st International Symposium Atmospheric and Ocean Optics: Atmospheric Physics. SPIE - The International Society for Optical Engineering, November 2015. https://doi.org/10.1117/12.2206008
3. Penenko, A.V., Penenko, V.V., Tsvetova, E.A.: Sequential data assimilation algorithms for air quality monitoring models based on a weak-constraint variational principle. Numer. Anal. Appl. **9**(4), 312–325 (2016). https://doi.org/10.1134/s1995423916040054
4. Samarskii, A.A., Vabishchevich, P.N.: Computational Heat Transfer, Mathematical Modelling. Wiley, Hoboken (1996)
5. Hesstvedt, E., Hov, O., Isaksen, I.S.: Quasi-steady-state approximations in air pollution modeling: comparison of two numerical schemes for oxidant prediction. Int. J. Chem. Kinet. **10**(9), 971–994 (1978). https://doi.org/10.1002/kin.550100907
6. Gery, M.W., Whitten, G.Z., Killus, J.P., Dodge, M.C.: A photochemical kinetics mechanism for urban and regional scale computer modeling. J. Geophys. Res. **94**(D10), 12925 (1989). https://doi.org/10.1029/jd094id10p12925
7. Stockwell, W.R.: Simulation of a reacting pollutant puff using anadaptive grid algorithm by R.K. Srivastava et al. J. Geophys. Res. **107**(D22) (2002). https://doi.org/10.1029/2002jd002164
8. Byun, D., Schere, K.L.: Review of the governing equations, computational algorithms, and other components of the models-3 community multiscale air quality (CMAQ) modeling system. Appl. Mech. Rev. **59**(2), 51 (2006). https://doi.org/10.1115/1.2128636
9. Troen, I.B., Mahrt, L.: A simple model of the atmospheric boundary layer; sensitivity to surface evaporation. Bound.-Layer Meteorol. **37**(1–2), 129–148 (1986). https://doi.org/10.1007/bf00122760

Computational Modelling of the Full Length hIFN-γ Homodimer

Peicho Petkov[1], Elena Lilkova[2(✉)], Nevena Ilieva[2], Genoveva Nacheva[3], Ivan Ivanov[3], and Leandar Litov[1]

[1] Faculty of Physics, Sofia University "St. Kliment Ohridsky", Sofia, Bulgaria
[2] Institute of Information and Communication Technologies, Bulgarian Academy of Sciences, Sofia, Bulgaria
elilkova@parallel.bas.bg
[3] Institute of Molecular Biology, Bulgarian Academy of Sciences, Sofia, Bulgaria

Abstract. Human interferon gamma (hIFN-γ) is an important signalling molecule, which plays a key role in the formation and modulation of immune response. The controversial conclusions concerning the function of hIFN-γ C-termini as well as the lack of structural information about this domain motivated us to perform molecular dynamics simulations in order to model the structure of the hIFN-γ C-terminal part. The simulations were carried out with the CHARMM22 force field, starting from a fully extended conformation of the C-termini. They showed unambiguously that the C-termini tend to approach the globular part of the protein, so that the whole hIFN-γ molecule adopts a more compact conformation. The energetic favourability of the more compact conformations of the whole cytokine was also confirmed by means of free energy perturbation simulations.

1 Introduction

Interferon gamma (IFN-γ) is an important cytokine with multiple biological effects (for a review, see [17]). Human interferon gamma (hIFN-γ) is organized in a non-covalent homodimer in which the two monomers are associated in antiparallel orientation [16]. Each monomer is a 17 kDa protein consisting of 144 amino acids (aa), organized in six α-helices denoted A to F (comprising 62% of the molecule), linked by short unstructured regions. Additionally, the prot ein contains a long unstructured C-terminal domain composed of 21 aa.

The existing structure–functional studies, do not shed much light on the role of the hIFN-γ 21 aa-long C-terminal tail for its biological functions. The reason for this is the lack of X-ray diffraction data about the unstructured C-terminal domain [16]. The hIFN-γ C-terminal region is highly positively charged, thus it is highly susceptible to proteases. Two domains enriched in basic amino acids are distinguished in this region. The first one (denoted D1) encompasses amino acids Arg^{129}-Lys-Arg-Ser^{132} and the second one (D2) includes the sequence Arg^{137}-Gly-Arg-Arg^{140}. Functionally, D1 appears to be more important than

I. Lirkov and S. Margenov (Eds.): LSSC 2017, LNCS 10665, pp. 544–551, 2018.
https://doi.org/10.1007/978-3-319-73441-5_60

the D2 domain because the deletion of the latter results in a significant increase in hIFN-γ activity [13], whereas the complete removal of the unstructured C-terminal region deprives the cytokine of biological activity [4].

The controversial conclusions about the function of the hIFN-γ C-termini as well as the lack of a reasonable model explaining their role in the receptor–ligand interaction prompted us to model their structure by means of molecular dynamics (MD) and free energy perturbation (FEP) simulations.

2 Computations and Analysis

2.1 Initial Structure

We proceeded from the crystallographic structural data obtained with a recombinant hIFN-γ homodimer in complex with the extracellular part of its receptor hIFN-γR1 with PDB ID 1FG9 [16]. There, the coordinates of the hIFN-γ atoms are resolved up to the 126-th amino acid, while the last 18 residues of each monomer are missing. There is experimental evidence that the C-termini of hIFN-γ are very flexible and do not adopt any particular rigid conformation, so as first, using the molecular manipulation and visualization program PyMOL [15], we designed these missing regions in a completely extended conformation with randomly perturbed backbone dihedral angles. The structure so obtained was further minimized and equilibrated with GROMACS [8]. The final equilibrated structure of hIFN-γ with reconstructed C-termini was used as input for the folding simulations (Fig. 1).

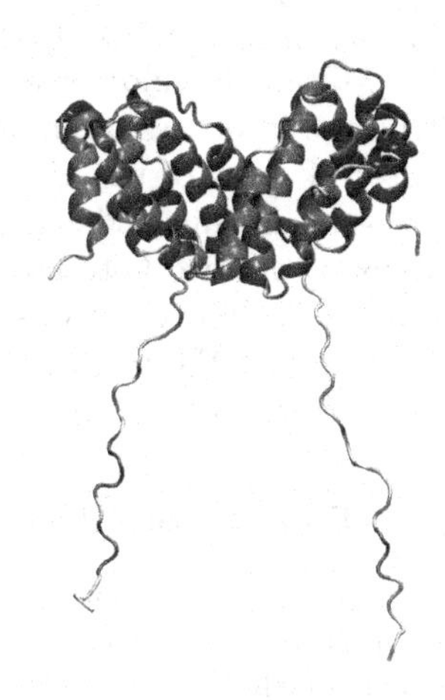

Fig. 1. Structure of hIFN-γ with reconstructed C-termini, used as starting structure for the folding simulations.

2.2 Molecular Dynamics Folding Simulations

The minimized and equilibrated starting structure was solvated in a simulation box with dimensions 100 × 120 × 130 Å^3 with imposed periodic boundary conditions. 18 chlorine ions were added to neutralize the net charge of the system. A 200 ns molecular dynamics simulation was performed to fold the reconstructed C-termini. The molecular modelling package NAMD 2.7 [14] was used. Topology was build with the CHARMM22 force field [12], combined with the modified TIP3P water model for the solvent [12]. Constraints were imposed on all bonds using the SHAKE algorithm [2]. We used the impulse-based Verlet-I/r-RESPA multiple time-stepping integration scheme [18] with an integration time step of 2 fs: short-range non-bonded interactions were calculated at every time step, and full electrostatics evaluation was performed at every 2 time steps. The PME method [5] was applied for calculation of the electrostatic interactions. The PME cut-off distance for the direct summation was 12 Å and 128 FFT grid points were used in each direction in the reciprocal space. Van der Waals interactions were truncated at a cut-off distance of 12 Å with a switching function which starts to smoothly switch them off at a distance of 10 Å. The system was simulated in the canonical ensemble at a temperature of 310 K. The constant temperature was maintained by the Langevin thermostat [7] with a damping coefficient of 5/ps. The conformation of the system was written to the trajectory at intervals of 500 ps. In what follows, this simulation is quoted as "folding simulation".

2.3 Cluster Analysis

The folding trajectory was subjected to cluster analysis, in order to analyse the simulation and to capture the most probable and representative conformations of the C-terminal region. First, the trajectory was least-square fitted to the starting structure in order to remove all rotational and translational movements of the center of mass of the globular part of the hIFN-γ molecule (aa 1–122). Then a root mean square deviation (RMSD) matrix of the positions of the backbone atoms of the two C-termini (aa 123–144) was constructed between all pairs of frames. The clustering was performed with the program g_cluster of GROMACS [8] using GROMOS clustering algorithm [3] with a cut-off of 10.0 Å based on the RMSD matrix.

2.4 Free Energy Perturbation Simulations

The comparison of the solvation free energy of different structures can be used as an estimate of their energetic favourability and correspondingly as an indication of their probability. Therefore, we performed free energy perturbation (FEP) simulations for calculation of the solvation free energy of the minimized and equilibrated starting structure with the C-terminal amino acids in extended conformation and the centroids of the three biggest clusters from the MD folding simulation.

The FEP simulations were performed with the package NAMD 2.7 [14]. The solvation free energy was calculated by decoupling the protein from bulk water, gradually switching off the interactions of the atoms of the protein and the solvent. The resulting free energy change is the free energy of solvation. In all FEP simulations the protein was described by the CHARMM22 force field with the modified TIP3P water model for the solvent [12]. FEP simulations were performed in the isothermal-isobaric ensemble under the following conditions: (i) temperature of 310 K, maintained by a Langevin thermostat with a damping coefficient of 5/ps; and (ii) pressure of 1 atm, maintained by a Nosé-Hoover Langevin piston [6] with an oscillation period of 100 fs and damping time scale of 150 fs. All bonds were constrained via SHAKE [2]. The integration time step was 2 fs. The Van der Waals interactions were smoothly turned off from 10 Å and cut at 11 Å. Electrostatic interactions were calculated with the PME algorithm [5] with a cut-off for the direct summation of 11 Å and a grid spacing in the reciprocal space of 1.3 Å.

Due to the soft-core potential, separate simulations were run for the forward ($\lambda = 0 \rightarrow \lambda = 1$) and the reverse ($\lambda = 1 \rightarrow \lambda = 0$) transformations. The electrostatic interaction was fully decoupled for appearing particles for values of $\lambda < 0.5$ and for disappearing particles for values of $\lambda > 0.5$. The Lennard-Jones potential was shifted for small values of the coupling parameter by 5.0 Å. The forward and reverse transformation simulations consisted of 67 windows, distributed according to the protocol described in Table 1. In each window a short minimization of 500 steps was first performed, followed by 7000 steps of equilibration and 2000 steps of data collection for ensemble averaging.

Table 1. FEP protocol for calculating free energy of solvation.

Forward simulation			Reverse simulation		
λ_{start}	λ_{stop}	$\Delta\lambda$	λ_{start}	λ_{stop}	$\Delta\lambda$
0.000	0.001	0.001	1.000	0.999	−0.001
0.001	0.005	0.004	0.999	0.995	−0.004
0.005	0.010	0.005	0.995	0.990	−0.005
0.010	0.060	0.010	0.990	0.700	−0.010
0.060	0.100	0.020	0.700	0.600	−0.020
0.100	0.600	0.025	0.600	0.100	−0.025
0.600	0.700	0.020	0.100	0.060	−0.020
0.700	0.990	0.010	0.060	0.010	−0.010
0.990	0.995	0.005	0.010	0.005	−0.005
0.995	0.999	0.004	0.005	0.001	−0.004
0.999	1.000	0.001	0.001	0.000	−0.001

3 Results and Discussion

3.1 Folding Simulations

In the folding simulations the C-termini did not remain in extended conformation. The folding starts within 15–20 ns, the C-termini tending to approach the globular part of the protein. During the rest of the simulation they rearrange and reorient around the globule. The root mean square difference of the C-terminal C_α positions (aa 123–144) with respect to the center of mass position of the globule (aa 1–122) for the two folding simulations is presented with the black line on Fig. 2a.

During the folding simulation, the structure of the globule is preserved quite stable within the whole 200 ns. In this force field a short α-helical turn in the C-terminal tails, including residues $Ala^{124} - Gly^{128}$ is formed.

The fluctuations of the positions are relatively small. This behaviour is also seen from the results of the cluster analysis. We found that the 400 conformations of the folding trajectory formed 11 clusters. The largest cluster includes about the half of the conformations (grey blocks on Fig. 2a). The structure of the centroid of the biggest cluster from the folding simulation is presented on Fig. 2b. The small fluctuations in the positions of the atoms in the folding simulations are most probably due to the chosen temperature control scheme. In addition to the stochastic force applied, the Langevin thermostat also applies a random friction which damps the fluctuations of the atomic positions.

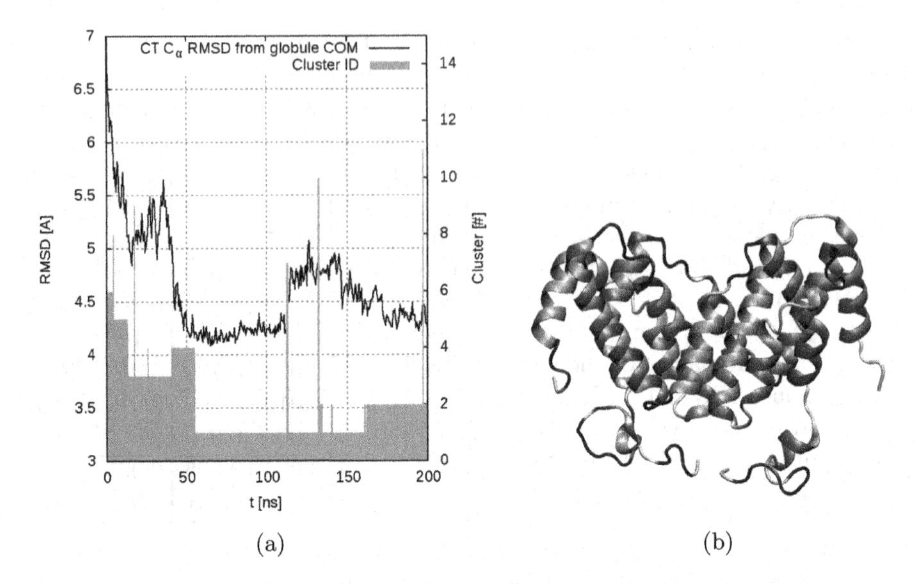

Fig. 2. (a) Correlation between cluster population (grey blocks) and RMS distance of the C-termini C_α atoms to the center of mass of the globule (black solid line); (b) Conformation of the centroid of the largest cluster from the folding simulation.

3.2 Solvation Free Energies

The cluster analysis gives no clear answer which is the most plausible and energetically favourable structure amongst the obtained most populated ones. The latter can be determined by calculating and comparing their solvation free energies. Unfortunately, at present it is not possible to calculate very accurately free energies of large macromolecules from MD simulations in reasonable time. However, one can still use MD simulations to obtain a rough estimation of the relative free energies and the respective energetic favourability of protein structures.

In order to asses the conformations favourability, we performed several free energy perturbation simulations to calculate the relative solvation free energies of the centroids of the biggest clusters with respect to the extended starting conformation.

The FEP simulation protocol employed in this study (see Sect. 2.4 and Table 1) was designed to consist of rather short simulations in each window on purpose. The aim was to avoid significant conformational changes that occur as the protein-solvent interactions are scaled. They are especially tangible in the C-terminal domains of hIFN-γ because of the higher charge density and its scaled screening by the solvent. The limited sampling was compensated for by performing three independent FEP simulations for each structure using different starting seeds for velocities generation.

The FEP simulations were analysed with the ParseFEP plug-in of VMD [10] using the SOS free-energy estimator [11]. The results of all simulations are summarized in Table 2. ParseFEP estimated the statistical error of each FEP

Table 2. Free energies of solvation of the extended starting structure and the centoids of the three biggest clusters of the folding simulation.

Structure	ΔG_{solv}^{Run1}	ΔG_{solv}^{Run2}	ΔG_{solv}^{Run3}	ΔG_{solv}^{AVE}	STD	$\Delta\Delta G_{solv}^{rel}$
Extended	−2314.93	−2229.41	−2303.25	−2282.53	46.37	0.00
C_1	−2464.60	−2400.70	−2431.62	−2432.31	31.96	−149.78
C_2	−2423.81	−2454.19	−2353.64	−2410.55	51.57	−128.02
C_3	−2379.25	−2362.20	−2330.03	−2357.16	24.99	−74.63

simulation to be at most 3 kcal/mol. However, the standard deviation of the three independent runs for each structure ranges between 30 and 60 kcal/mol.

It is important to note that the solvation free energies in columns 2–5 of Table 2 are not absolute. They include the solvation free energy of the respective conformation and a bias, introduced by the PME algorithm that is used for calculating the long-range electrostatic interactions. The latter is proportional to the square of the charge difference between the initial and the final state of the FEP transformation and reciprocal to the volume of the simulation box [9]. In this context the individual values of the solvation free energy of each structure are not conclusive by themselves. However, since we simulated different conformations of the same protein in identical initial conditions, (simulation box and ionic concentration), the PME related bias is the same in all FEP simulations. Hence, the difference of the free energies between two structures should be bias-free. The relative solvation free energies of the centroids of the biggest clusters with respect to the extended starting structure are given in the last column of Table 2.

As expected, the two biggest clusters which encompass about 72% of the folding simulation are characterized by lower solvation energies than the starting extended structure. These also exhibit the most compact conformations. The solvation free energy decreases with the SASA, the radius of gyration and the RMS distance between the C-terminal C_α positions from the globular center of mass (Fig. 3).

At present, there is neither structural model for the conformation of the C-terminal domain of hIFN-γ nor consensus about the role of this part of the molecule for the biological activity of the cytokine. In a previous study of the role of hIFN-γ C-termini we have applied a continuum model to investigate the electrostatic interactions in hIFN-γ both as a free homodimer and as a homodimer associated with the hIFN-γR1 receptor [1]. A large number of titratable groups with a dramatic shift in their pK values were registered and this effect was less expressed in the hIFN-γ bound to the receptor. Based on these results we concluded that the hIFN-γ C-terminal domain was capable of adopting a limited number of conformations to decrease the unfavourable electrostatic effects of the two basic C-terminal domains D1 and D2 at the time of binding to the hIFN-γ receptor. These conformational changes favour the reduction of the exposure of the titratable groups of the globular part to the solvent, thus promoting its transition to a more stable conformation.

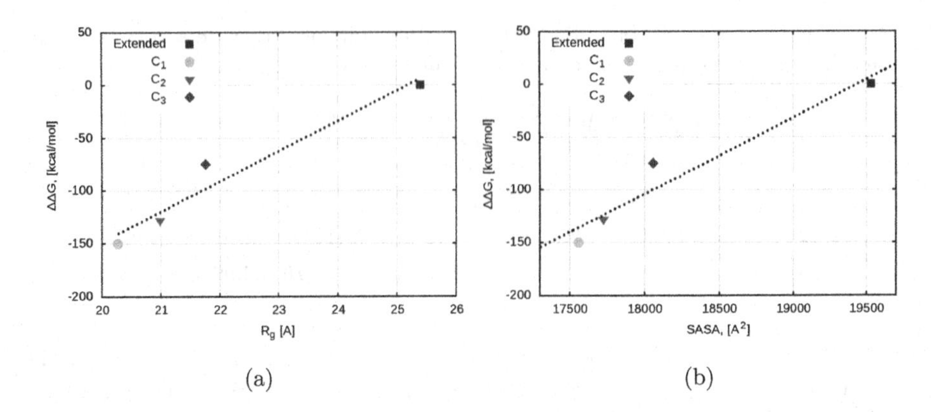

Fig. 3. Correlation between the relative solvation free energy of the structures and (a) gyration radius, and (b) solvent accessible surface area.

In the present study we provide further evidence in favour of the relative compactness of the whole hIFN-γ molecule. The next step is to investigate in detail the influence of the level of compactness of the cytokine on the dynamic of the receptor binding and thus, on the signal transduction pathway.

4 Conclusions

We investigated the structure of hIFN-γ C-termini by means of MD simulations. The last 22 residues tend to adopt compact conformations near the globular part of the cytokine. The tendency for compactness was also studied by a series of FEP simulations for calculating the solvation free energies of representative for the folding simulations conformations. The results from the solvation free-energy analysis confirm the energetic favourability of the more compact conformation. The obtained folded and equilibrated structure of the full length hIFN-γ homodimer is useful for further simulations of the interaction of the cytokine with other macromolecules, in particular with the extracellular domain of its receptor.

Acknowledgements. EL and NI acknowledge financial support under the Programme for Young Scientists' Career Development at the Bulgarian Academy of Sciences (Grant DFNP-99/04.05.2016).

References

1. Altobelli, G., Nacheva, G., Todorova, K., Ivanov, I., Karshikoff, A.: Role of the C-terminal chain in human interferonγ stability: an electrostatic study. Proteins: Struct. Funct. Bioinform. **43**(2), 125–133 (2001)
2. Ciccotti, G., Ryckaert, J.: Molecular dynamics simulation of rigid molecules. Comput. Phys. Rep. **4**(6), 346–392 (1986)

3. Daura, X., Gademann, K., Jaun, B., Seebach, D., van Gunsteren, W.F., Mark, A.E.: Peptide folding: when simulation meets experiment. Angew. Chem. Int. Ed. **38**(1–2), 236–240 (1999)
4. Dobeli, H., Gentz, R., Jucker, W., Garotta, G., Hartmann, D., Hochuli, E.: Role of the carboxy-terminal sequence on the biological activity of human immune interferon (ifn-γ). J. Biotechnol. **7**(3), 199–216 (1988)
5. Essmann, U., Perera, L., Berkowitz, M.L., Darden, T., Lee, H., Pedersen, L.G.: A smooth particle mesh Ewald method. J. Chem. Phys. **103**(19), 8577–8593 (1995)
6. Feller, S.E., Zhang, Y., Pastor, R.W., Brooks, B.R.: Constant pressure molecular dynamics simulation: the Langevin piston method. J. Chem. Phys. **103**(11), 4613–4621 (1995)
7. Grest, G.S., Kremer, K.: Molecular dynamics simulation for polymers in the presence of a heat bath. Phys. Rev. A **33**, 3628–3631 (1986)
8. Hess, B., Kutzner, C., van der Spoel, D., Lindahl, E.: Gromacs 4: algorithms for highly efficient, load-balanced, and scalable molecular simulation. J. Chem. Theory Comput. **4**(3), 435–447 (2008)
9. Hummer, G., Pratt, L., Garcia, A.: Ion sizes and finite-size corrections for ionic-solvation free energies. J. Chem. Phys. **107**, 9275–9277 (1997)
10. Liu, P., Dehez, F., Cai, W., Chipot, C.: A toolkit for the analysis of free-energy perturbation calculations. J. Chem. Theory Comput. **8**(8), 2606–2616 (2012)
11. Lu, N., Kofke, D.A., Woolf, T.B.: Improving the efficiency and reliability of free energy perturbation calculations using overlap sampling methods. J. Comput. Chem. **25**, 28–39 (2004)
12. MacKerell Jr., A.D., Bashford, D., Bellott, M., et al.: All-atom empirical potential for molecular modeling and dynamics studies of proteins. J. Phys. Chem. B **102**(18), 3586–3616 (1998)
13. Nacheva, G., Todorova, K., Boyanova, M., Berzal-Herranz, A., Karshikoff, A., Ivanov, I.: Human interferon gamma: significance of the C-terminal flexible domain for its biological activity. Arch. Biochem. Biophys. **413**(1), 91–98 (2003)
14. Phillips, J.C., Braun, R., Wang, W., Gumbart, J., Tajkhorshid, E., Villa, E., Chipot, C., Skeel, R.D., Kalé, L., Schulten, K.: Scalable molecular dynamics with NAMD. J. Comput. Chem. **26**(16), 1781–1802 (2005)
15. Schrodinger, LLC, The PyMOL molecular graphics system, version 1.3r1 (2010). http://www.pymol.org
16. Thiel, D., le Du, M.H., Walter, R., D'Arcy, A., Chène, C., Fountoulakis, M., Garotta, G., Winkler, F., Ealick, S.: Observation of an unexpected third receptor molecule in the crystal structure of human interferon-γ receptor complex. Structure **8**, 927–936 (2000)
17. Tsanev, R., Ivanov, I.: Immune Interferon: Properties and Clinical Application. CRC Press LLC, Boca Raton (2001)
18. Tuckerman, M., Berne, B.J., Martyna, G.J.: Reversible multiple time scale molecular dynamics. J. Chem. Phys **97**(3), 1990–2001 (1992)

Using Advanced Mathematical Tools in Complex Studies Related to Climate Changes and High Pollution Levels

Zahari Zlatev[1], Ivan Dimov[2], Krassimir Georgiev[2](✉), and Radim Blaheta[3]

[1] Department of Environmental Science, Aarhus University, Roskilde, Denmark
zz@envs.au.dk

[2] Institute of Information and Communication Technologies, Bulgarian Academy of Sciences, Sofia, Bulgaria
ivdimov@bas.bg, georgiev@parallel.bas.bg

[3] Institute of Geonics AS CR, Studentska 1768, 70800 Ostrava-Poruba, Czech Republic
radim.blaheta@ugn.cas.cz

Abstract. **UNI-DEM** is a large-scale environmental model described by a non-linear system of partial differential equations (PDEs) and used in many studies of air pollution levels in different European countries. The discretization of **UNI-DEM** leads to a long series of huge computational tasks, because it is necessary to run the discretized model with many different scenarios during long time-periods of many consecutive years. Therefore, both the storage requirements and the computational work are enormous. We had to resolve **four** difficult problems in the efforts to perform successfully the required simulations. More precisely, we had to do the following:

(a) to implement fast numerical methods,
(b) to select suitable splitting procedures,
(c) to exploit efficiently the cache memories of the available high-speed computers
(d) to parallelize the computer codes.

We use several runs over sixteen consecutive years and with fourteen scenarios. Our main purpose will be to show the long-range transport of potentially dangerous air pollutants to Bulgaria.

1 Description of the Unified Danish Eulerian Model (UNI-DEM)

The development of the Danish Eulerian Model (UNI-DEM) started in the beginning of the 1980-ies. The model has been improved many times in an attempt to increase its reliability and accuracy [2,15,18]. It was used in many air pollution studies in different European countries as well as in evaluating the impact of the climatic changes, and first and foremost of the global warming, on some high pollution levels, which might be dangerous for plants, animals and human health

I. Lirkov and S. Margenov (Eds.): LSSC 2017, LNCS 10665, pp. 552–559, 2018.
https://doi.org/10.1007/978-3-319-73441-5_61

[1,3–15,17,19–22]. UNI-DEM is described mathematically by the following system of s non-linear partial differential equations (PDEs):

$$\begin{aligned}\frac{\partial c_i}{\partial t} = & -u\frac{\partial(c_i)}{\partial x} - v\frac{\partial(c_i)}{\partial y} - w\frac{\partial(c_i)}{\partial z} \\ & + \frac{\partial}{\partial x}\left(K_x\frac{\partial c_i}{\partial x}\right) + \frac{\partial}{\partial y}\left(K_y\frac{\partial c_i}{\partial y}\right) + \frac{\partial}{\partial z}\left(K_z\frac{\partial c_i}{\partial z}\right) \\ & + Q_i(t,x,y,z,c_1,c_2,\dots c_s) - (k_{1i}+k_{2i})c_i + E_i(t,x,y,z), \quad i=1,2,\dots s,\end{aligned} \tag{1}$$

where: $c_i = c_i(t,x,y,z)$ are concentrations of chemical species, $u = u(t,x,y,z)$, $v = v(t,x,y,z)$ and $w = w(t,x,y,z)$ are wind velocities along the three coordinate axes, $K_x = K_x(t,x,y,z)$, $K_y = K_y(t,x,y,z)$, and $K_z = K_z(t,x,y,z)$ are diffusivity coefficients, the non-linear terms Q_i are describing the chemical reactions, k_{1i} and k_{2i} are dry and wet deposition coefficients and the E_i are terms describing the emission sources. We assume that the spatial domain is a parallelepiped P, defined by $x \in [a_1,b_1]$, $y \in [a_2,b_2]$, and $z \in [a_3,b_3]$. The time interval is $t \in [a,b]$ and the initial values $c_i(a,x,y,z)$ are given. We shall furthermore assume that some boundary conditions are always available (much more details about these issues can be found in [15,18]).

2 Difficulties in the Numerical Treatment of UNI-DEM

The direct discretization of the derivatives in (1) leads to huge computational tasks. Assume that the numbers of grid points used in relation with the intervals $[a_1,b_1]$, $[a_2,b_2]$, $[a_3,b_3]$, and $[a,b]$ are N_1, N_2, N_3, and N respectively. Then the system of PDEs (1) will be transformed into a system of ordinary differential equations (ODEs), which contains $s \times N_1 \times N_2 \times N_3$ equations, which have to be treated during N time–steps. A typical run of UNI–DEM over a period with meteorological and emission data for one year will lead to the solution of systems of PDEs containing $56 \times 480 \times 480 \times 10 = 128024000$ equations during 213120 time–steps. Note too, that we have to run UNI-DEM over long sequences of consecutive years and with many different scenarios. It is clear that it is impossible to run the model directly. Therefore, we had to resolve several difficult tasks: (a) to select fast and reliable numerical methods, (b) to parallelize the computations and to use powerful high-speed computers and (c) to exploit efficiently the cache memories of the available computers. It turned out, however, that even the successful solution of these three tasks was not enough, especially when we had to carry out climatic investigations and sensitivity studies. It was necessary to resolve an extra task: (d) to implement some splitting procedure. This task is extremely important, because it allows us to divide the complex problem containing different operators in several sub-problems. This simplifies the solution of the previous three tasks (a)–(c), because it is possible to optimize further the computational process by applying different numerical methods, different parallel techniques and different ways of exploiting the cache memories in connection with the obtained during the splitting sub-models. Therefore, it is appropriate to start with the crucial choice of an **optimal** splitting procedure.

3 Introduction of an Optimal Splitting Procedure

The following splitting procedure is in some sense the best one:

$$\frac{\partial c_i^{[1]}}{\partial t} = -u\frac{\partial c_i^{[1]}}{\partial x} - v\frac{\partial c_i^{[1]}}{\partial y} + \frac{\partial}{\partial x}\left(K_x\frac{\partial c_i^{[1]}}{\partial x}\right) + \frac{\partial}{\partial y}\left(K_y\frac{\partial c_i^{[1]}}{\partial y}\right),$$
$$i = 1, 2, \ldots s, \tag{2}$$

$$\frac{\partial c_i^{[2]}}{\partial t} = +Q_i(t, x, y, z, c_1^{[2]}, c_2^{[2]}, \ldots c_s^{[2]}) - (k_{1s} + k_{2s})c_i^{[2]} + E_i(t, x, y, z),$$
$$i = 1, 2, \ldots s, \tag{3}$$

$$\frac{\partial c_i^{[3]}}{\partial t} = -w\frac{\partial c_i^{[3]}}{\partial z} + \frac{\partial}{\partial z}\left(K_z\frac{\partial c_i^{[3]}}{\partial z}\right),$$
$$i = 1, 2, \ldots s. \tag{4}$$

The first sub-model describes the combination of the horizontal transport (the advection) and the horizontal diffusion. The second one deals with the chemical reactions, the deposition terms and the emission sources. The last sub-model describes the vertical exchange. **This splitting procedure is optimal in the sense that it does not require extra boundary conditions**, i.e. the same boundary conditions that are necessary for the original system of PDEs defined by (1) are also needed in the treatment of the three smaller systems of PDEs (2)–(4). More precisely, the boundary conditions on the vertical sides of parallelepiped P are needed in (2), the boundary conditions on the bottom and on the top of P are used in (4) and no boundary conditions are necessary in the treatment of (3), which does not contain spatial derivatives and can easily be rewritten as a system of ODEs. We must emphasize here that there are many other splitting procedures, which lead to simpler, and even much simpler, sub-problems, but require introduction of artificial boundary conditions and that might cause some problems.

4 Choice of Numerical Methods

Consider the sub-problems (2)–(4) and discretize the spatial derivatives by some numerical scheme (it must be reiterated here that **different** numerical schemes can be applied in the different sub-models and this is one of the great advantages of using splitting techniques: for each sub-model one may and must select the most suitable algorithm). Then the systems of PDEs (2)–(4) are transformed into three systems of ODEs:

$$\frac{\partial g^{[1]}}{\partial t} = f^{[1]}\left(t, g^{[1]}\right), \quad \frac{\partial g^{[2]}}{\partial t} = f^{[2]}\left(t, g^{[2]}\right), \quad \frac{\partial g^{[3]}}{\partial t} = f^{[3]}\left(t, g^{[3]}\right), \tag{5}$$

$$g^{[m]}(t) \in \mathbb{R}^{s \times N_1 \times N_2 \times N_3},\ m = 1, 2, 3$$

The components of the unknown functions $g^{[m]}(t) \in \mathbb{R}^{s \times N_1 \times N_2 \times N_3}$, $m = 1, 2, 3$ are approximations at time t of the concentrations at all spatial grid-points and for all species. The components of functions $f^{[m]}(t, g^m) \in \mathbb{R}^{s \times N_1 \times N_2 \times N_3}$, $m = 1, 2, 3$ depend both on quantities involved in the right-hand-side of (2)–(4) and on the numerical algorithms used in the discretization of the spatial derivatives.

Various methods (finite elements, finite differences, a pseudo-spectral algorithm, schemes based on wavelets, etc.) can be used in the discretization of the spatial derivative in (2) and (4). Simple linear finite elements are rather efficient. Predictor-corrector methods with several different correctors [14] selected in an attempt to enhance the numerical stability of the equations are used in the solution of the first system in (5). The second system in (5) is very ill-conditioned and stiff [18]. Different experiments show that the L-stable Backward Differentiation Formula is performing rather well. The θ-method with $\theta = 0.75$ is currently used in the treatment of the third system in (5). Much more information about the numerical methods can be found in [2,14–16,18].

5 Achieving Parallelization

The possibility of choosing different numerical algorithms for handling the different mathematical operators in the sub-models is not the only advantage of using splitting procedures. The division of the system of PDEs (1) into the sub-problems (2)–(4) leads to natural parallelism that can easily be exploited on different computers. It is easily seen that the following conclusions are true. The system of PDEs (2) contains $s \times N_3$ independent sub-systems and the number of equations in each sub-system is $N_1 \times N_2$. The system of PDEs (3) contains $N_1 \times N_2 \times N_3$ independent sub-systems and the number of equations in each sub-system is s. The system of PDEs (4) contains $s \times N_1 \times N_2$ independent sub-systems and the number of equations in each sub-system is N_3. It should be noted here that these tasks could successfully be run on different types of parallel computers [2,15,18].

6 Exploiting the Cache Memory of the Available Computer

Exploiting the natural parallelism is relatively easy, but by far not sufficient. It is also necessary to exploit efficiently the cache memory. This is not an easy problem, because the sizes of the parallel tasks are different. The numbers of equations per parallel task are $N_1 \times N_2 = 480 \times 480 = 230400$, $s = 56$, and $N_3 = 10$ respectively for the sub-problems (2), (3), and (4). In the first case, the tasks are very large, while these are too small in the other two cases. Several small tasks in the last two cases have to be united in larger tasks, called **chunks**. A series of experiments have to be used in order to find the optimal size of the chunks on the available computer. The first case is much more complicated. It is necessary to find additional parallelism within each large parallel task.

7 Numerical Demonstrations

Results obtained over a time-period of one year performed on an IBM BLUE GENE computer (some details about this computer could be found in http://internetofthingsagenda.techtarget.com/definition/Blue-Gene and [8]) are given in Fig. 1. It is seen that the optimal length of the chunks is 16 and the speed-up rates for the advection and chemical parts are nearly optimal for this length of the chunks, the total speed-up is rather satisfactory although the overhead part is running sequentially. UNI-DEM is run as a two-dimensional model in this experiment.

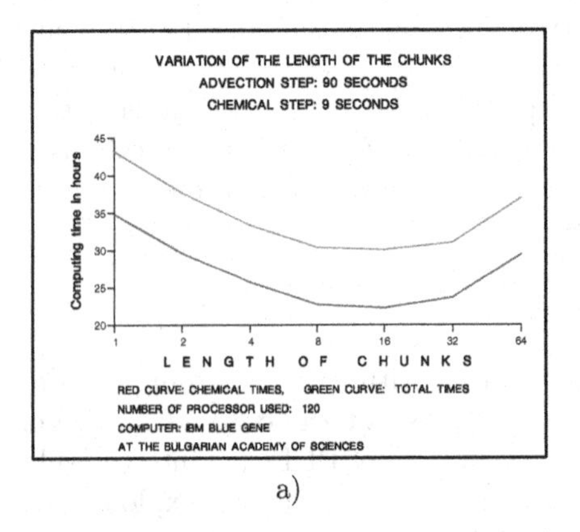

a)

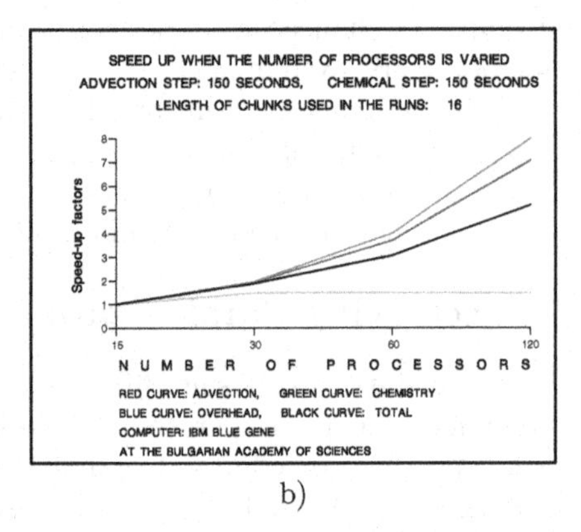

b)

Fig. 1. (a) Dependence of the computing time on the length of chunks, and (b) the speed-up achieved with the optimal length of the chunks (Color figure online)

8 Long-Range Transport of Air Pollution and Climatic Changes

The fact that the climatic changes might lead to substantial increases of some pollution levels is illustrated in Figs. 2 and 3 for Bulgaria. The results in Fig. 2 indicate that the numbers of *bad days* (days in which the 8-h averages of the ozone concentrations exceed at least once the critical level of 60 ppb; "*bad days*" can have damaging effects on some groups of human beings, e.g. people who suffer from asthmatic diseases, etc.) are (*a*) often greater than the EU-critical level of 25 days and (*b*) the climatic changes can lead to additional increases of up to 78%. The same tendency (the climatic changes lead to increased number of bad days) holds not only for year 2000, but also for a period of 16 consecutive years; see Fig. 3. The meteorological data and the emission fields were obtained from the Meteorological Synthesizing Centre – WEST at the Norwegian Meteorological Institute in the frame of EMEP (the European Monitoring and Evaluation Program), www.emep.int.

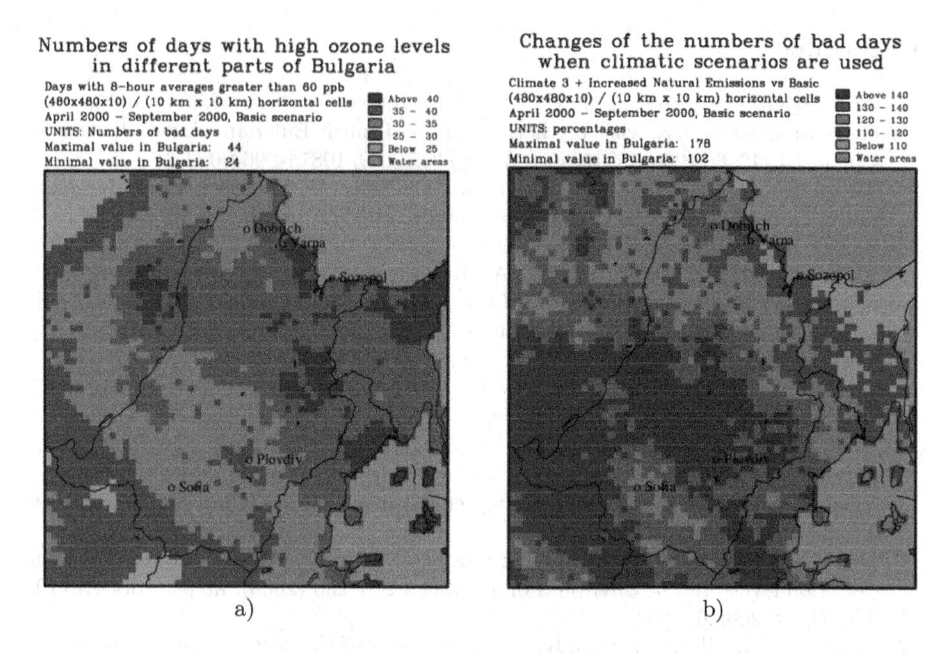

Fig. 2. (a) Distribution of the numbers of bad days in Bulgaria for year 2000, and (b) the corresponding increases (in percent) when one of the climatic scenario is used. (Color figure online)

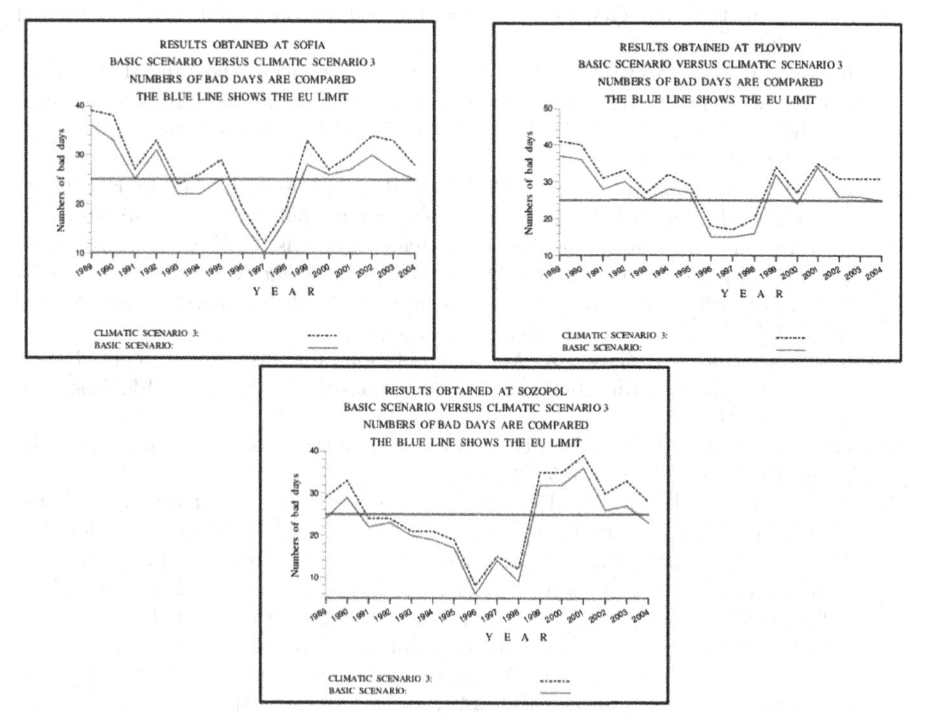

Fig. 3. Variation of the numbers of bad days in three different Bulgarian sites during a time-period of 16 years. (Color figure online)

References

1. Abdalmogith, S., Harrison, P.M., Zlatev, Z.: Intercomparison of inorganic aerosol concentrations in UK with predictions of the Danish Eulerian Model. J. Atmos. Chem. **54**, 43–66 (2006). https://doi.org/10.1007/s10874-006-9012-3
2. Alexandrov, V., Owczarz, W., Thomsen, P.G., Zlatev, Z.: Parallel runs of large air pollution models on a grid of SUN computers. Math. Comput. Simul. **65**, 557–577 (2004). https://doi.org/10.1016/j.matcom.2004.01.022
3. Ambelas Skjøth, C., Bastrup-Birk, A., Brandt, J., Zlatev, Z.: Studying variations of pollution levels in a given region of Europe during a long time-period. Syst. Anal. Model. Simul. **37**, 297–311 (2000). http://dl.acm.org/citation.cfm?id=345435
4. Bastrup-Birk, A., Brandt, J., Uria, I., Zlatev, Z.: Studying cumulative ozone exposures in Europe during a 7-year period. J. Geophys. Res. **102**, 23917–23935 (1997)
5. Csomós, P., Cuciureanu, R., Dimitriu, G., Dimov, I., Doroshenko, A., Faragó, I., Georgiev, K., Havasi, Á., Horváth, R., Margenov, S., Ostromsky, T., Prusov, V., Syrakov, D., Zlatev, Z.: Impact of Climate Changes on Pollution Levels in Europe, Final report for a NATO Linkage Project (Grant 980505) (2006). http://www.cs.elte.hu/~faragois/NATO.pdf
6. Dimov, I., Geernaert, G., Zlatev, Z.: Impact of future climate changes on high pollution levels. Int. J. Environ. Pollut. **32**(2), 200–230 (2008). https://doi.org/10.1504/IJEP.2008.017103
7. Geernaert, G., Zlatev, Z.: Studying the influence of the biogenic emissions on the AOT40 levels in Europe. Int. J. Environ. Pollut. **23**(1–2), 29–41 (2004). https://doi.org/10.1504/IJEP.2004.005485
8. Georgiev, K., Zlatev, Z.: Studying air pollution levels in he Balkan peninsula area by using an IBM Blue Gene/P computer. Int. J. Environ. Pollut. **46**(1/2), 97–114 (2011). ISSN 0957-4352
9. Harrison, R.M., Zlatev, Z., Ottley, C.J.: A comparison of the predictions of an Eulerian atmospheric transport chemistry model with experimental measurements over the North Sea. Atmos. Environ. **28**, 497–516 (1994). https://doi.org/10.1016/1352-2310(94)90127-9
10. Hass, H., van Loon, M., Kessler, C., Stern, R., Mathijsen, J., Sauter, F., Zlatev, Z., Langner, J., Foltescu, V., Schaap, M.: Aerosol modelling: results and intercomparison from European regional-scale modelling systems. GSF-National Research Center for Environment and Health, International Scientific Secretariat, EUROTRAC, Münich (2004). http://www.trumf.fu-berlin.de/veranstaltungen/events/glream/GLOREAM_PMmodel-comparison.pdf
11. Havasi, Á., Bozó, L., Zlatev, Z.: Model simulation on transboundary contribution to the atmospheric sulfur concentration and deposition in Hungary. Időjárás **105**, 135–144 (2001)
12. Havasi, Á., Zlatev, Z.: Trends of Hungarian air pollution levels on a long time-scale. Atmos. Environ. **36**, 4145–4156 (2002)
13. Roemer, M., Beekman, M., Bergsröm, R., Boersen, G., Feldmann, H., Flatøy, F., Honore, C., Langner, J., Jonson, J.E., Matthijsen, J., Memmesheimer, M., Simpson, D., Smeets, P., Solberg, S., Stevenson, D., Zandveld, P., Zlatev, Z.: Ozone trends according to ten dispersion models. GSF - National Research Center for Environment and Health, International Scientific Secretariat, EUROTRAC, Münich (2004). http://www.mep.tno.nl/eurotrac/EUROTRAC-trends.pdf
14. Zlatev, Z.: Application of predictor-corrector schemes with several correctors in solving air pollution problems. BIT **24**, 700–715 (1984). https://doi.org/10.1007/BF01934925

15. Zlatev, Z.: Computer Treatment of Large Air Pollution Models. Environmental and Technology Library, vol. 2. Kluwer Academic Publishers, Dordrecht/Boston/London (1995). https://doi.org/10.1007/978-94-011-0311-4
16. Zlatev, Z.: Partitioning ODE systems with an application to air pollution models. Comput. Math. Appl. **42**, 817–832 (2001). http://www.sciencedirect.com/science/article/pii/S0898122101002012
17. Zlatev, Z.: Impact of future climate changes on high ozone levels in European suburban areas. Clim. Chang. **101**, 447–483 (2010). https://doi.org/10.1007/s10584-009-9699-7
18. Zlatev, Z., Dimov, I.: Computational and Numerical Challenges in Environmental Modelling. Studies in Computational Mathematics, vol. 13. Elsevier, Amsterdam (2006). https://books.google.dk/books?isbn=0080462480
19. Zlatev, Z., Dimov, I., Georgiev, K.: Relations between climatic changes and high pollution levels in Bulgaria. Open J. Appl. Sci. **6**, 386–401 (2016). https://doi.org/10.4236/ojapps.2016.67040
20. Zlatev, Z., Georgiev, K., Dimov, I.: Influence of climatic changes on pollution levels in the Balkan Peninsula. Comput. Math. Appl. **65**, 544–562 (2013). https://doi.org/10.1016/j.camwa.2012.07.006
21. Zlatev, Z., Moseholm, L.: Impact of climate changes on pollution levels in Denmark. Ecol. Model. **217**, 305–319 (2008). https://doi.org/10.1016/j.ecolmodel.2008.06.030
22. Zlatev, Z., Havasi, Á., Faragó, I.: Influence of climatic changes on pollution levels in Hungary and surrounding countries. Atmosphere **2**, 201–221 (2011). https://doi.org/10.3390/atmos2030201

Large-Scale Numerical Computations for Sustainable Energy Production and Storage

Parallel Aggregation Based on Compatible Weighted Matching for AMG

Ambra Abdullahi[1(✉)], Pasqua D'Ambra[2], Daniela di Serafino[3], and Salvatore Filippone[4]

[1] Università degli Studi di Roma "Tor Vergata", Roma, Italy
ambra.abdullahi@uniroma2.it
[2] Istituto per le Applicazioni del Calcolo "Mauro Picone", CNR, Naples, Italy
pasqua.dambra@cnr.it
[3] Università degli Studi della Campania "Luigi Vanvitelli", Caserta, Italy
daniela.diserafino@unicampania.it
[4] Cranfield University, Cranfield, UK
salvatore.filippone@cranfield.ac.uk

Abstract. We focus on the extension of the MLD2P4 package of parallel Algebraic MultiGrid (AMG) preconditioners, with the objective of improving its robustness and efficiency when dealing with sparse linear systems arising from anisotropic PDE problems on general meshes. We present a parallel implementation of a new coarsening algorithm for symmetric positive definite matrices, which is based on a weighted matching approach. We discuss preliminary results obtained by combining this coarsening strategy with the AMG components available in MLD2P4, on linear systems arising from applications considered in the Horizon 2020 Project "Energy oriented Centre of Excellence for computing applications" (EoCoE).

Keywords: AMG · Parallel aggregation · Weighted matching

1 Introduction: AMG Based on Compatible Weighted Matching

Algebraic MultiGrid (AMG) methods have proved to be very promising in preconditioning general sparse linear systems of equations when no information on the origin of the problem is available. Their goal is to define automatic coarsening processes depending only on the coefficient matrix, without using any a priori information or characterization of the *algebraically smooth error*, that is, the error not reduced by the relaxation method. Recent theoretical developments provide general approaches to the construction of coarse spaces for AMG that has optimal convergence, i.e., convergence independent of the problem size, in the case of general linear systems [14, 22]. However, despite these theoretical developments, almost all currently available AMG methods and software

I. Lirkov and S. Margenov (Eds.): LSSC 2017, LNCS 10665, pp. 563–571, 2018.
https://doi.org/10.1007/978-3-319-73441-5_62

rely on heuristics to drive the coarsening process among variables; for example the *strength of connection* heuristics is derived from a characterization of the algebraically smooth vectors that is theoretically well understood only for M-matrices.

A new approach for coarsening sparse symmetric positive definite (s.p.d.) matrices has been proposed in recent papers [10,11], following some theoretical and algorithmic developments [5,14]: *coarsening based on compatible weighted matching*. It defines a pairwise aggregation of unknowns where each pair is the result of a maximum weight matching in the matrix adjacency graph. Specifically, the aggregation scheme uses a maximum product matching in order to enhance the diagonal dominance of a matrix representing the hierarchical complement of the resulting coarse matrix, thereby improving the convergence properties of a corresponding compatible relaxation scheme. The matched nodes are aggregated to form coarse index spaces, and piecewise constant or smoothed interpolation operators are applied for the construction of a multigrid hierarchy. No reference is made to any a priori knowledge on the matrix origin and/or any definition of strength of connection; however, information about the smooth error may be used to define edge weights assigned to the original matrix graph. More aggressive coarsening can be obtained by combining multiple steps of the pairwise aggregation.

This method has been implemented in a C-language software framework called *BootCMatch: Bootstrap AMG based on Compatible Weighted Matching*, which incorporates the coarsening method in an adaptive algorithm that computes a composite AMG for general s.p.d. matrices with a prescribed convergence rate through a bootstrap process [10,12]. The main computational kernel of this coarsening approach is the computation of a maximum product matching, accounting for the largest fraction of the time needed to build the AMG preconditioner.

Let $G = (V, E, C)$ be a weighted undirected graph associated with a symmetric matrix A, where V is the set of vertices, E the set of edges and $C = (c_{ij})_{i,j=1,\ldots,n}$ is the real positive matrix of edge weights. A *matching* in G is a subset of edges $M \subseteq E$ such that no two edges share a vertex. The number of edges in M is the cardinality of the matching; a *maximum cardinality matching* contains the largest possible number of edges. A *maximum product weighted matching* is a matching maximizing the product of the weights of the edges in the matching. By a simple weight transformation it is possible to formulate the computation of a maximum product weighted matching in terms of a general maximum weight matching, that is, a matching which maximizes the sum of the weights for each edge in the matching; this is also called an *assignment problem*. Such problems are widely studied in combinatorial optimization and are often used in sparse direct linear solvers for reordering and scaling of matrices [13,17].

Three algorithms for maximum product weighted matching are used in BootCMatch:

MC64: the algorithm implemented in the MC64 routine of the HSL library (http://www.hsl.rl.ac.uk). It works on bipartite graphs, i.e. graphs where the vertex set is partitioned into two subsets V_r and V_c (for example, the rows and columns indices of A) and $(i, j) \in E$ connects $i \in V_r$ and $j \in V_c$; it finds optimal matchings with a worst-case computational complexity $\mathcal{O}(n(n + nnz) \log n)$, where n is the matrix dimension and nnz the number of nonzero entries.

Half-approximate: the algorithm described in [19]. It is a greedy algorithm capable of finding, with complexity $\mathcal{O}(nnz)$, a matching whose total weight is at least half the optimal weight and whose cardinality is at least half the maximum cardinality.

Auction-type: a version of the near-optimal auction algorithm, based on a notion of approximate optimality called ε-complementary slackness, first proposed by Bertsekas [3] and implemented in the SPRAL Library as described in [17]. The original auction algorithm has a worst case computational complexity $\mathcal{O}(n \cdot nnz \log(cn))$, in the case of integer weights, where $c = \max_{ij} |c_{ij}|$; however, its SPRAL version reduces the cost per iteration and the average number of iterations, producing a near-optimal matching at a much lower cost than that of the (optimal) MC64.

Here we discuss a parallel version of the coarsening algorithm described above and its integration into the library of parallel AMG preconditioners MLD2P4, to obtain preconditioners that are robust and efficient on large and sparse linear systems arising from anisotropic elliptic PDE problems discretized on general meshes. The algorithm is implemented with a decoupled approach, where each parallel process performs matching-based coarsening on the part of the matrix owned by the process itself. This requires an extension of the MLD2P4 software framework, in order to efficiently interface the BootCMatch functions implementing the coarsening and to combine the new functionality with the other AMG components of MLD2P4. Computational experiments have been performed on linear systems coming from applications in the Horizon 2020 Project "Energy oriented Centre of Excellence for computing applications" (EoCoE, http://www.eocoe.eu).

2 Interfacing MLD2P4 with BootCMatch

MLD2P4 is a package of parallel AMG and domain decomposition preconditioners, designed to provide scalable and easy-to-use preconditioners [6–9], which has been successfully exploited in different applications (see, e.g., [2,4]). It is based on the PSBLAS computational framework [16] and can be used with the PSBLAS Krylov solvers. MLD2P4 is written in Fortran 2003 using an object-oriented approach [15]; thanks to its layered software architecture, its classes and functionalities can be easily extended and reused. It is freely available from https://github.com/sfilippone/mld2p4-2.

Two classes are used to represent AMG preconditioners in MLD2P4: the *base preconditioner* and the *multilevel preconditioner*. The multilevel preconditioner

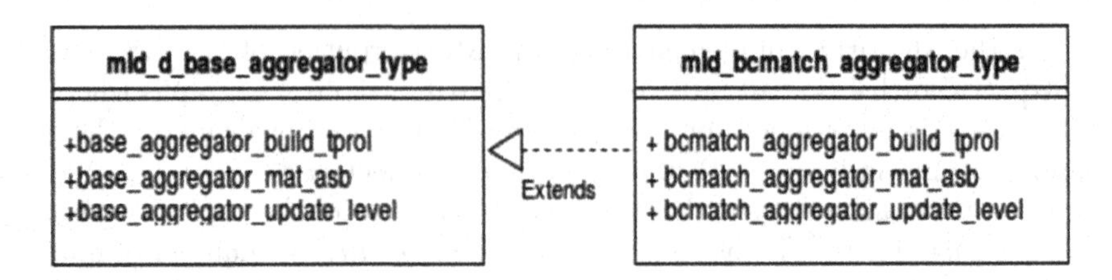

Fig. 1. Extending the aggregator class in MLD2P4 (only the relevant methods are displayed)

holds the coarse matrices of the AMG hierarchy, the corresponding smoothers (stored as base preconditioners), and prolongation and restriction operators. An *aggregator* object, encapsulated in the base preconditioner structure, hosts the aggregation method for building the AMG hierarchy. The basic aggregation available in MLD2P4 is the decoupled smoothed aggregation algorithm presented in [20], where the final prolongator is obtained by applying a suitable smoother to a tentative prolongator. In particular, the method *base_aggregator_build_tprol* builds the tentative prolongator and stores other information about it and the corresponding restriction operator.

The MLD2P4 *basic_aggregator_type* class can be extended to define new aggregation strategies, possibly adding attributes and methods, or overriding the basic ones. In order to interface the aggregation algorithm implemented in BootCMatch, the class *bcmatch_aggregator_type* was defined, extending *basic_aggregator_type* and providing the *bcmatch_aggregator_build_tprol* method, which ovverrides the basic method and performs matching-based aggregation; a schematic picture of the new class is given in Fig. 1.

When implementing the interface, we had to take into account that MLD2P4 is written in Fortran while BootCMatch is written in C, and MLD2P4 and BootCMatch use different data structures for storing sparse matrices and vectors. Fortran 2003 provides a standardized way for creating procedures, variables and types that are interoperable with C. We exploited this feature by creating in MLD2P4 two derived types that are interoperable with C and correspond to the *bcm_CSRMatrix* and *bcm_vector* structures used in BootCMatch.

Furthermore, BootCMatch is sequential and thus needed a strategy to be applied in the parallel MLD2P4 environment. Therefore, BootCMatch was interfaced with the decoupled parallel aggregation scheme available in MLD2P4. This was implemented by passing to the aggregator the local part of the matrix held by the current process, so that the aggregation algorithm could be run on that submatrix, obtaining a local tentative prolongator. Furthermore, a subroutine called *mld_base_map_to_tprol* was developed to create a global tentative prolongator from the local ones. Of course, the resulting parallel matching-based aggregation is generally different from the sequential one.

The *base_aggregator_mat_asb* method performs several actions: it applies a smoother to the tentative prolongator (if required), computes the restriction operator as $R = P^T$, and creates the coarse-level matrix A_c using the Galerking approach, i.e. $A_c = RAP$. This method was overridden to take into account that the BootCMatch tentative prolongator has a more general form than the basic tentative prolongator implemented in MLD2P4. The *base_aggregator_update_level* method was also overriden by *bcmatch_aggregator_ update_level*, which takes care of projecting on the next coarse level information needed by the matching-based aggregation.

3 Computational Experiments

In order to illustrate the behaviour of the parallel AMG preconditioners described in the previous sections, we show the results obtained on linear systems derived from two applications in the framework of the Horizon 2020 EoCoE Project.

A first set of linear systems comes from a groundwater modelling application developed at the Jülich Supercomputing Centre (JSC); it deals with the numerical simulation of the filtration of 3D incompressible single-phase flows through anisotropic porous media. The linear systems arise from the discretization of an elliptic equation with no-flow boundary conditions, modelling the pressure field, which is obtained by combining the continuity equation with Darcy's law. The discretization is performed by a cell-centered finite volume scheme (two-point flux approximation) on a Cartesian grid [1]. The systems considered in this work have s.p.d. matrices with dimension 10^6 and 6940000 nonzero entries distributed over seven diagonals. The anisotropic permeability tensor in the elliptic equation is randomly generated from a lognormal distribution with mean 1 and three standard deviation values, i.e., 1, 2 and 3, corresponding to the three systems M1-Filt, M2-Filt and M3-Filt. The systems were generated by using a Matlab code implementing the basics of reservoir simulations and can be regarded as simplified samples of systems arising in ParFlow, an integrated parallel watershed computational model for simulating surface and subsurface fluid flow, used at JSC.

A second set of linear systems comes from computational fluid dynamics simulations for wind farm design and management, carried out at the Barcelona Supercomputing Center (BSC) by using Alya, a multi-physics simulation code targeted at HPC applications [21]. The systems arise in the numerical solution of Reynolds-Averaged Navier-Stokes equations coupled with a modified $k - \varepsilon$ model; the space discretization is obtained by using stabilized finite elements, while the time integration is performed by combining a backward Euler scheme with a fractional step method, which splits the computation of the velocity and pressure fields and thus requires the solution of two linear systems at each time step. Here we show the results concerning two systems associated with the pressure field, which are representative of systems arising in simulations with discretization meshes of different sizes. The systems, denoted by M1-Alya and M2-Alya, have s.p.d. matrices of dimensions 790856 and 2224476, with 20905216 and 58897774 nonzeros, respectively, corresponding to up to 27 entries per row.

All the systems were preconditioned by using an AMG Kcycle [18] with decoupled unsmoothed double-pairwise matching-based aggregation. One forward/backward Gauss-Seidel sweep was used as pre/post-smoother with systems M1-Filt, M2-Filt and M3-Filt; one block-Jacobi sweep, with ILU(0) factorization of the blocks, was applied as pre/post-smoother with M1-Alya and M2-Alya. UMFPACK (http://faculty.cse.tamu.edu/davis/suitesparse.html) was used to solve the coarsest-level systems, through the interface provided by MLD2P4. The experiments were performed by using the three matching algorithms described in Sect. 1. The truncated Flexible Conjugate Gradient method FCG(1) [18], implemented in PSBLAS, was chosen to solve all the systems, according to the variability introduced in the preconditioner by the Kcycle. The zero vector was chosen as starting guess and the iterations were stopped when the 2-norm of the residual achieved a reduction by a factor of 10^{-6}. A generalized row-block distribution of the matrices was used, computed by using the Metis graph partitioner (http://glaros.dtc.umn.edu/gkhome/metis/metis/overview).

The experiments were carried out on the yoda linux cluster, operated by the Naples Branch of the CNR Institute for High-Performance Computing and Networking. Its compute nodes consist of 2 Intel Sandy Bridge E5-2670 8-core processors and 192 GB of RAM, connected via Infiniband. Given the size and the sparsity of the linear systems, at most 64 cores, running as many parallel processes, were used; 4 cores per node were considered, according to the memory bandwidth requirements of the linear systems. PSBLAS 3.4 and MLD2P4 2.2, installed on the top of MVAPICH 2.2, were used together with a development version of BootCMatch and the version of UMFPACK available in SuiteSparse 4.5.3. The codes were compiled with the GNU 4.9.1 compiler suite.

We first discuss the results for the systems M1-Filt, M2-Filt and M3-Filt. In Table 1 we report the number of iterations performed by FCG with the preconditioners based on the different matching algorithms, as the number of processes, np, varies. In Table 2 we report the corresponding execution times (in seconds) and speedup values. The times include the construction of the preconditioner and the solution of the preconditioned linear system. In general the preconditioned FCG solver shows reasonable algorithmic scalability, i.e. the number of iterations

Table 1. M1-Filt, M2-Filt and M3-Filt: number of iterations

	M1-Filt			M2-Filt			M3-Filt		
np	MC64	Half-app	Auction	MC64	Half-app	Auction	MC64	Half-app	Auction
1	13	11	12	18	29	18	46	58	52
2	15	11	14	20	35	21	79	68	65
4	15	11	13	20	32	21	62	83	64
8	15	11	13	20	31	22	69	77	71
16	13	11	15	19	29	19	75	69	101
32	15	11	15	21	37	21	79	68	82
64	15	11	13	26	32	21	86	76	78

Table 2. M1-Filt, M2-Filt and M3-Filt: execution time and speedup

	M1-Filt						M2-Filt						M3-Filt					
	MC64		Half-app		Auction		MC64		Half-app		Auction		MC64		Half-app		Auction	
np	time	sp	time	sp	time	sp	time	sp	time	sp	time	sp	time	sp	time	sp	time	sp
1	18.58	1.0	8.72	1.0	8.90	1.0	19.54	1.0	12.46	1.0	9.73	1.0	25.15	1.0	18.11	1.0	15.61	1.0
2	9.67	1.9	4.43	2.0	4.71	1.9	11.16	1.8	6.61	1.9	5.58	1.7	16.24	1.5	10.07	1.8	9.48	1.6
4	5.30	3.5	2.68	3.2	2.75	3.2	5.55	3.5	4.05	3.1	3.14	3.1	8.29	3.0	7.17	2.5	5.63	2.8
8	2.66	7.0	1.42	6.1	1.32	6.7	2.79	7.0	2.02	6.2	1.63	6.0	4.41	5.7	3.66	5.0	3.24	4.8
16	1.05	17.7	0.71	12.2	0.73	12.2	1.18	16.6	1.07	11.7	0.77	12.6	1.88	13.4	1.43	12.6	2.05	7.6
32	0.67	27.9	0.52	16.8	0.52	17.2	0.75	26.2	0.82	15.3	0.60	16.1	1.25	20.2	1.16	15.7	1.07	14.6
64	0.43	43.0	0.43	20.4	0.39	22.9	0.51	38.5	0.60	20.7	0.40	24.1	0.82	30.6	0.82	22.2	0.83	18.7

for all systems does not vary too much with the number of processes. A larger variability in the iterations can be observed with M3-Filt, due to the higher anisotropy of this problem and its interaction with the decoupled aggregation strategy. With a more in-depth analysis (not shown here for the sake of space) we find the coarsening ratio between consecutive AMG levels ranges between 3.6 and 3.8 for the MC64 and auction algorithms, while it is between 3.0 and 3.3 for the half-approximation one, except with 64 processes where it reduces to 2.8 for M2-Filt and M3-Filt. None of the three matching algorithms produces preconditioners that are clearly superior in reducing the number of FCG iterations; indeed, for these systems there is no advantage in using the optimal matching algorithm implemented in MC64, since the non-optimal ones appear very competitive. The times corresponding to the half-approximation and auction algorithms are generally smaller, mainly because the time needed to build the AMG hierarchies is reduced. The speedup decreases as the anisotropy of the problem grows due to the larger number of FCG iterations and the well-known memory-bound nature of our computations. The largest speedups are obtained with MC64, because of the larger time required by MC64 on a single core.

In the case of M1-Alya and M2-Alya, only MC64 is able to produce an effective coarsening (with a ratio greater than 3.8), hence avoiding a significant increase of the number of iterations when the number of processes grows. Therefore, in Table 3, we only report results (iterations, time and speedup) for this case. The preconditioned solver shows good algorithmic scalability in general; the number of iterations on a single core is much smaller because in this case the AMG smoother reduces to an ILU factorization. A speedup of 42.8 is achieved for M1-Alya, which reduces to 29.3 for the larger matrix M2-Alya; in our opinion, this can be considered satisfactory. Note that we did not achieve convergence to the desired accuracy by using the basic classical smoothed aggregation algorithm available in MLD2P4.

In conclusion, the results discussed here show the potential of parallel matching-based aggregation and provide an encouraging basis for further work in this direction, including the application of non-decoupled parallel matching algorithms and the design of ad-hoc coarsening strategies for some classes of smoothers, according to the principles of compatible relaxation [14,22].

Table 3. M1-Alya and M2-Alya: number of iterations, execution time and speedup, using weighted matching based on MC64.

	M1-Alya			M2-Alya		
np	it	time	sp	it	time	sp
1	8	51.22	1.0	8	76.00	1.0
2	39	25.39	2.0	40	81.90	0.9
4	40	11.69	4.4	44	39.26	1.9
8	50	5.96	8.6	43	19.24	3.9
16	57	2.89	17.7	48	10.84	7.0
32	84	2.18	23.5	52	5.25	14.5
64	58	1.20	42.8	48	2.60	29.2

Acknowledgments. This work has received funding from the European Union's Horizon 2020 Research and Innovation Programme under Grant Agreement No. 676629 (Project EoCoE). The authors wish to thank Herbert Howen (BSC, Barcelona), Stefan Kollet and Wendy Sharples (JSC, Jülich) for making available the test problems.

References

1. Aarnes, J.E., Gimse, T., Lie, K.-A.: An introduction to the numerics of flow in porous media using matlab. In: Hasle, G., Lie, K.-A., Quak, E. (eds.) Geometric Modelling, Numerical Simulation, and Optimization, pp. 265–306. Springer, Heidelberg (2007). https://doi.org/10.1007/978-3-540-68783-2_9
2. Aprovitola, A., D'Ambra, P., Denaro, F.M., di Serafino, D., Filippone, S.: SParCLES: enabling large eddy simulations with parallel sparse matrix computation tools. Comput. Math. Appl. **70**, 2688–2700 (2015)
3. Bertsekas, D.P.: The auction algorithm: a distributed relaxation method for the assignment problem. Ann. Oper. Res. **14**, 105–123 (1988)
4. Borzì, A., De Simone, V., di Serafino, D.: Parallel algebraic multilevel Schwarz preconditioners for a class of elliptic PDE systems. Comput. Vis. Sci. **16**, 1–14 (2013)
5. Brannick, J., Chen, Y., Kraus, J., Zikatanov, L.: Algebraic multilevel preconditioners for the graph Laplacian based on matching in graphs. SIAM J. Numer. Anal. **51**, 1805–1827 (2013)
6. Buttari, A., D'Ambra, P., di Serafino, D., Filippone, S.: Extending PSBLAS to build parallel Schwarz preconditioners. In: Dongarra, J., Madsen, K., Waśniewski, J. (eds.) PARA 2004. LNCS, vol. 3732, pp. 593–602. Springer, Heidelberg (2006). https://doi.org/10.1007/11558958_71
7. Buttari, A., D'Ambra, P., di Serafino, D., Filippone, S.: 2LEV-D2P4: a package of high-performance preconditioners for scientific and engineering applications. Appl. Algebra Eng. Commun. Comput. **18**, 223–239 (2007)
8. D'Ambra, P., di Serafino, D., Filippone, S.: On the development of PSBLAS-based parallel two-level Schwarz preconditioners. Appl. Numer. Math. **57**, 1181–1196 (2007)

9. D'Ambra, P., di Serafino, D., Filippone S.: MLD2P4: a package of parallel algebraic multilevel domain decomposition preconditioners in Fortran 95. ACM Trans. Math. Softw. **37**(3), Article No. 30 (2010). https://github.com/sfilippone/mld2p4-2
10. D'Ambra, P., Vassilevski, P.S.: Adaptive AMG with coarsening based on compatible weighted matching. Comput. Vis. Sci. **16**, 59–76 (2013)
11. D'Ambra, P., Vassilevski, P.S.: αAMG based on weighted matching for systems of elliptic PDEs arising from displacement and mixed methods. In: Russo, G., Capasso, V., Nicosia, G., Romano, V. (eds.) ECMI 2014. MI, vol. 22, pp. 1013–1020. Springer, Cham (2016). https://doi.org/10.1007/978-3-319-23413-7_142
12. D'Ambra, P., Filippone, S., Vassilevski, P.S.: BootCMatch: a software package for bootstrap AMG based on graph weighted matching. Submitted
13. Duff, I.S., Koster, J.: On algorithms for permuting large entries to the diagonal of a sparse matrix. SIAM J. Matrix Anal. Appl. **22**, 973–996 (2001)
14. Falgout, R.D., Vassilevski, P.S.: On generalizing the algebraic multigrid framework. SIAM J. Numer. Anal. **42**, 1669–1693 (2004)
15. Filippone, S., Buttari, A.: Object-oriented techniques for sparse matrix computations in Fortran 2003. ACM Trans. Math. Softw. **38**(4), Article No. 23 (2012)
16. Filippone, S., Colajanni, M.: PSBLAS: a library for parallel linear algebra computation on sparse matrices. ACM Trans. Math. Softw. **26**, 527–550 (2000)
17. Hogg, J., Scott, J.: On the use of suboptimal matchings for scaling and ordering sparse symmetric matrices. Numer. Linear Algebra Appl. **22**, 648–663 (2015)
18. Notay, Y., Vassilevski, P.S.: Recursive Krylov-based multigrid cycles. Numer. Linear Algebra Appl. **15**, 473–487 (2008)
19. Preis, R.: Linear time 1/2-approximation algorithm for maximum weighted matching in general graphs. In: Meinel, C., Tison, S. (eds.) STACS 1999. LNCS, vol. 1563, pp. 259–269. Springer, Heidelberg (1999). https://doi.org/10.1007/3-540-49116-3_24
20. Vaněk, P., Mandel, J., Brezina, M.: Algebraic multigrid by smoothed aggregation for second and fourth order elliptic problems. Computing **56**, 179–196 (1996)
21. Vásquez, M., Houzeaux, G., Koric, S., Artigues, A., Aguado-Sierra, J., Arís, R., Mira, D., Calmet, H., Cucchietti, F., Owen, H., Taha, A., Burness, E.D., Cela, J.M., Valero, M.: Alya: multiphysics engineering simulation toward exascale. J. Comput. Sci. **14**, 15–27 (2016)
22. Xu, J., Zikatanov, L.: Algebraic multigrid methods. Acta Numerica **26**, 591–721 (2017). https://doi.org/10.1017/S0962492917000083

Efficient Solution Techniques for Multi-phase Flow in Porous Media

Henrik Büsing(✉)

E.ON Energy Research Center,
Institute for Applied Geophysics and Geothermal Energy,
RWTH Aachen University, Aachen, Germany
hbuesing@eonerc.rwth-aachen.de
http://www.eonerc.rwth-aachen.de/GGE

Abstract. Multi-phase flow in porous media is relevant for many applications, e.g. geothermal energy production, groundwater remediation, CO_2 sequestration, enhanced oil recovery or nuclear waste storage. The arising non-linear partial differential equations are highly non-linear and thus often solved in a fully implicit way. We present a Schur complement reduction method relying on algebraic multigrid methods for solving the arising linear systems. This method is compared to a classical Constrained Pressure Residual (CPR-AMG) approach. It turns out that the new method is competitive to the classical approach with the advantage that it relies only on scalable algebraic multigrid (AMG) and not on incomplete LU (ILU) preconditioning. Scaling results are presented for both methods on the Jülich high performance computer JUQUEEN.

Keywords: Multi-phase flow · Fully implicit newton method
Constrained pressure residual (CPR-AMG)
Algebraic multigrid (AMG) · Schur complement reduction
High performance computing (HPC)

1 Introduction

Many application areas rely on the solution of multi-phase flow equations in porous media: (i) geothermal energy is produced from hot subsurface steam reservoirs, where the two phases are water in liquid and in gaseous form. For the prediction of reservoir performance, it is necessary to simulate the behavior of the phases over time. (ii) CO_2 sequestration is considered as one option to mitigate anthropogenic effects on climate. Here, CO_2 is injected in saline sandstone aquifers in depths greater than 800 m to keep the CO_2 in supercritical state. The two phases are CO_2 and brine. For assessing the safety of the CO_2 injection, the propagation of the CO_2 plume in the subsurface needs to be modelled. (iii) non-aqueous phase liquids (NAPLs) may contaminate groundwater reservoirs. A remediation strategy often involves the injection of hot steam into the reservoir. The viscosity of the present NAPLs is thus reduced and they may be transported away from the reservoir. In this case the present phases are

I. Lirkov and S. Margenov (Eds.): LSSC 2017, LNCS 10665, pp. 572–579, 2018.
https://doi.org/10.1007/978-3-319-73441-5_63

groundwater, steam and the NAPL. (iv) for enhanced oil recovery, water or CO_2 is injected into an oil reservoir to produce a bigger percentage of oil. Here, water and oil, and possibly CO_2, are the occurring phases. (v) finally, in nuclear waste storage the containers with the nuclear waste may be attacked by saline waters leading to corrosion and subsequently to the generation of H_2. Here, the saline waters and the generated gas are the two phases. The prediction of the safety of the nuclear waste repository relies on simulations for the gas migration.

In all these cases, mass balance equations for the two (or more) phases are solved. These equations are highly non-linear and strongly coupled. One standard solution approach relies on a fully implicit formulation using the implicit Euler method in time and a two-point flux approximation in space. After discretization, the non-linear algebraic equations are solved with Newton's method. Consequently, in every time step, in every Newton step, a linear system needs to be solved. We concentrate on this computing kernel and present efficient methods for the solution of the linear systems.

One key aspect in our study is the comparability of the used methods. To ensure this, we use PETSc (the Portable Extensible Toolkit for Scientific computation) [1,2] as the simulation framework. This allows us to interchange the linear solver on a command line basis and ensures comparability of the methods as well as easy reproducibility of the results.

We present a weak scaling test for the Schur complement reduction method and CPR-AMG on JUQUEEN solving a problem from CO_2 geo-sequestration. The efficiency of both methods in terms of matrix-vector products is demonstrated.

The paper is organized as follows: In Sect. 2, we present the equations for multi-phase flow in porous media. Afterwards the numerical method for the solution of the coupled systems is presented in Sect. 3. The results of a weak scaling test on the high performance computer JUQUEEN are presented in Sect. 4. Finally, we discuss our findings and give an outlook how the presented methods might be improved. Appendix A lists all the solver options for easy reproduction.

2 Mathematical Model

The multi-phase flow equations rely on a modified version of Darcy's law [7]. The volumetric flow rates v_α for phase $\alpha \in \{w, n\}$ based on relative permeabilities (cf. [3]) are given by

$$\begin{aligned} \boldsymbol{v}_w &= -\frac{k_{rw}}{\mu_w} \mathbb{K} (\nabla p_w - \rho_w \boldsymbol{g}), \\ \boldsymbol{v}_n &= -\frac{k_{rn}}{\mu_n} \mathbb{K} (\nabla p_n - \rho_n \boldsymbol{g}). \end{aligned} \tag{1}$$

Here, $\mathbb{K}$ is the absolute permeability, $k_{r\alpha}$ are the relative permeabilities of phase α, $\boldsymbol{g} = (0, 0, -g)^T$ is gravity, p_α is the phase pressure of phase α and ρ_α and μ_α are the phase densities and dynamic viscosities.

The mass balance equations for the two phases are given by

$$\phi \frac{\partial(\rho_w S_w)}{\partial t} + \operatorname{div}(\rho_w \boldsymbol{v}_w) = q_w$$
$$\phi \frac{\partial(\rho_n S_n)}{\partial t} + \operatorname{div}(\rho_n \boldsymbol{v}_n) = q_n, \tag{2}$$

with porosity ϕ, sources and sinks q_α and saturations S_α. These equations are supplemented by two algebraic constraints for the saturations and capillary pressure.

$$S_w + S_n = 1 \tag{3}$$

and

$$p_c = p_n - p_w. \tag{4}$$

Insertion of the Darcy velocities (1) as well as the algebraic constraints (3) and (4) into (2) and choosing water pressure p_w and gas saturation S_n as primary variables we end up with

$$\phi \frac{\partial(\rho_w(1-S_n))}{\partial t} - \operatorname{div}\left(\rho_w \frac{k_{rw}}{\mu_w} \mathbb{K}\left(\nabla p_w - \rho_w \boldsymbol{g}\right)\right) = q_w$$
$$\phi \frac{\partial(\rho_n S_n)}{\partial t} - \operatorname{div}\left(\rho_n \frac{k_{rn}}{\mu_n} \mathbb{K}\left(\nabla(p_c + p_w) - \rho_n \boldsymbol{g}\right)\right) = q_n. \tag{5}$$

The two equations are strongly coupled through the relative permeabilities and the capillary pressure. Common relationships for these two quantitites are the models of [4,8]. We choose the Brooks-Corey approach for capillary pressure

$$p_c = p_d S_e^{-1/\sigma}, \tag{6}$$

which is often used with the [6] model for relative permeabilities

$$k_{rw} = S_e^{\frac{2+3\sigma}{\sigma}} \tag{7}$$
$$k_{rn} = (1-S_e)^2 \left(1 - S_e^{\frac{2+\sigma}{\sigma}}\right). \tag{8}$$

The effective saturation S_e is defined as

$$S_e = \frac{S_w - S_{wr}}{1 - S_{wr} - S_{nr}}. \tag{9}$$

The equations of system (5) are hyperbolic or degenerate parabolic in the case of high capillary pressures and pose a challenge for algebraic multigrid methods. A common approach is to divide both equations of system (5) by the corresponding phase densities and sum them up. In the incompressible case, this leads to an elliptic equation for the pressure. This pressure equation may now be solved efficiently by algebraic multigrid methods. As second equation, the mass balance of either wetting or non-wetting phase may be chosen. We use the non-wetting phase equation.

In the next section, we discuss how the equations are discretized in time and space and how the resulting non-linear algebraic system is solved.

3 Numerical Method

Equation system (5) is discretized in time by the implicit Euler method and in space by a two-point flux approximation method. The mass balance of phase α then reads

$$\begin{aligned}
&\phi \frac{(\rho_\alpha S_\alpha)_i^{n+1} - (\rho_\alpha S_\alpha)_i^n}{\Delta t} V_i \\
&- \sum_j \left(\rho_\alpha \frac{k_{r\alpha}}{\mu_\alpha}\right)_{ij}^{n+1} \left(\frac{p_{w,j} - p_{w,i}}{d_i + d_j} - \rho_{ij} g_{ij}\right)^{n+1} A_{ij} \\
&- q_{\alpha,i}^{n+1} V_i = 0. \qquad (10)
\end{aligned}$$

Here, A_{ij} denotes the area between cell i and j; V_i is the volume of cell i; d_i is the distance to the interface separating cells i and j with area A_{ij}; n and $n+1$ are time indices indicating the solution at step n and $n+1$. After discretization in space and time the non-linear algebraic equations are solved with Newton's method:

$$\frac{\partial F}{\partial u} \Delta u = -F(u).$$

Here, $F = (F_1, F_2)^T$ represents the first and second equation of system (5) and $u = (p_w, S_n)^T$ is the vector of unknowns. In the incompressible case, using $G_1 = \rho_w^{-1} F_1 + \rho_n^{-1} F_2$ as first (pressure) equation and $G_2 = F_2$ as second equation the corresponding Jacobian $J = \begin{pmatrix} J_{11} & J_{12} \\ J_{21} & J_{22} \end{pmatrix}$ has the form

$$J_{11} = \frac{\partial G_1}{\partial p_w} = -\operatorname{div}\left(\left(\frac{k_{rw}(S_w)}{\mu_w} + \frac{k_{rn}(S_w)}{\mu_n}\right) \mathbb{K} \nabla \operatorname{id}\right) \qquad (11)$$

$$\begin{aligned}
J_{12} = \frac{\partial G_1}{\partial S_n} &= -\operatorname{div}\left(\frac{\mathrm{d}k_{rw}(S_w)}{\mathrm{d}S_n} \frac{\mathbb{K}}{\mu_w} (\nabla p_w - \rho_w \boldsymbol{g}) \ldots\right. \\
&\left. - \frac{\mathrm{d}k_{rn}(S_w)}{\mathrm{d}S_n} \frac{\mathbb{K}}{\mu_n} (\nabla(p_c + p_w) - \rho_n \boldsymbol{g}) + k_{rn}(S_w) \frac{\mathbb{K}}{\mu_n} \nabla \frac{\mathrm{d}p_c}{\mathrm{d}S_n}\right) \qquad (12)
\end{aligned}$$

$$J_{21} = \frac{\partial G_2}{\partial p_w} = -\operatorname{div}\left(\frac{\rho_n}{\mu_n} k_{rn}(S_w) \mathbb{K} \nabla \operatorname{id}\right) \qquad (13)$$

$$\begin{aligned}
J_{22} = \frac{\partial G_2}{\partial S_n} &= -\operatorname{div}\left(\frac{\rho_n}{\mu_n} \frac{\mathrm{d}k_{rn}(S_w)}{\mathrm{d}S_n} \mathbb{K} (\nabla(p_c + p_w) - \rho_n \boldsymbol{g}) \ldots\right. \\
&\left. + \frac{\rho_n}{\mu_n} k_{rn}(S_w) \mathbb{K} \nabla \frac{\mathrm{d}p_c}{\mathrm{d}S_n}\right), \qquad (14)
\end{aligned}$$

with id being the identity.

The CPR-AMG method [5,9] now solves the pressure equation (J_{11} block) and updates the global residuals. Afterwards, an ILU preconditioned flexible [11] GMRES [12] solves the full system and deals with the hyperbolic contributions. For the pressure solve we use BoomerAMG from the Hypre package (cf. [10])

and for the outer solve we use a block Jacobi method with ILU(0) on the blocks as preconditioner.

The Schur complement reduction method decomposes the Jacobian (cf. [1]) into

$$
\begin{aligned}
J^{-1} &= \left[\begin{pmatrix} I & 0 \\ J_{21}J_{11}^{-1} & I \end{pmatrix} \begin{pmatrix} J_{11} & 0 \\ 0 & S \end{pmatrix} \begin{pmatrix} I & J_{11}^{-1}J_{12} \\ 0 & I \end{pmatrix} \right]^{-1} \\
&= \begin{pmatrix} I & -J_{11}^{-1}J_{12} \\ 0 & I \end{pmatrix} \begin{pmatrix} J_{11}^{-1} & 0 \\ 0 & S^{-1} \end{pmatrix} \begin{pmatrix} I & 0 \\ -J_{21}J_{11}^{-1} & I \end{pmatrix}
\end{aligned}
$$

with Schur complement $S = J_{22} - J_{21}J_{11}^{-1}J_{12}$. Note, that an actual inverse is never formed, but rather the solution of a linear system is computed. Instead of solving the full $2N \times 2N$ system, two $N \times N$ systems are solved, reducing the computational effort significantly. The solution method relies heavily on a good preconditioner for the Schur complement S. We use J_{22} to construct the preconditioner for S. For the solution of the linear systems we again use BoomerAMG from the Hypre package. In the next section we compare the Schur complement method with the CPR-AMG method in a weak scaling test.

4 Simulations

We show an example from CO_2 sequestration for benchmarking the Schur complement method and the CPR-AMG method on JUQUEEN, an IBM Blue-Gene/Q with 28 672 nodes. Each node consists of an IBM PowerPC A2 running at 1.6 GHz with 16 GB of memory. This example mimics a typical CO_2 injection scenario, where CO_2 is injected with a rate of $0.2\,\mathrm{kg\,s^{-1}}$ into a domain with extensions of $600\,\mathrm{m} \times 1\,\mathrm{m} \times 100\,\mathrm{m}$ in a depth of 800 m.

We prescribe a porosity $\phi = 0.2$, permeability $\mathbb{K} = 10^{-12}\,\mathrm{m}^2$, pore size distribution index $\sigma = 2$ and an entry pressure $p_d = 10^3\,\mathrm{Pa}$ as rock properties. Constant fluid parameters are used for wetting density $\rho_w = 1\,000\,\mathrm{kg\,m^{-3}}$ and viscosity $\mu_w = 10^{-3}\,\mathrm{Pa\,s}$, as well as non-wetting density $\rho_n = 454.4\,\mathrm{kg\,m^{-3}}$ and viscosity $\mu_n = 3.29 \cdot 10^{-5}\,\mathrm{Pa\,s}$.

The domain is iteratively refined in x- and z-direction for the weak scaling test, starting on level one with 393 216 cells ($1536 \times 1 \times 256$) up to level four with 25 165 824 cells ($12288 \times 1 \times 2048$). We start with one node (i.e. 2×8 cores) on level one and end with 64 nodes (i.e. 1 024 cores) on level four. One MPI process per core is used. The distributed array communication data structure is created with PETSc's DMDACreate and PETSc determines the partitioning across processes using MPI.

We execute five time steps with a fixed small time step size of $\Delta t = 100\,\mathrm{s}$. In practice, a refinement of the space discretization would also need a subsequent refinement for the time step size, and consequently more time steps to reach a certain time.

Figure 1 shows the efficiency over the number of cores. Ideally, the solution time stays the same for all the refinement levels. Actually, we see a decrease in

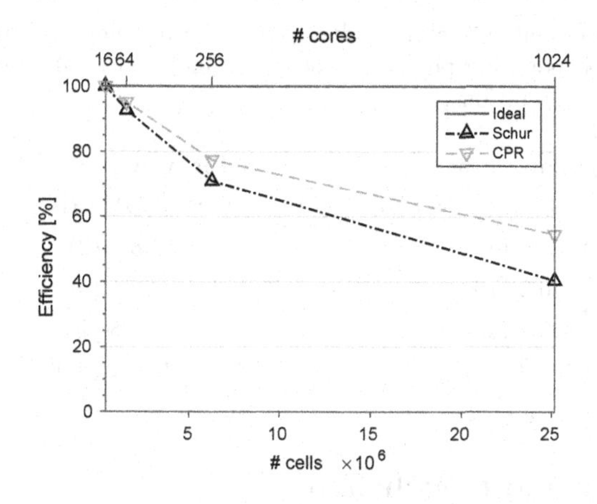

Fig. 1. Weak scaling test on JUQUEEN with up to 1 024 cores using PETSc 3.7.4.

Table 1. Total solution time (TT), Jacobian computation time (FJ) and linear solve time (LS) in seconds (and percentage of total time) for Schur complement method and CPR-AMG.

	Schur complement			CPR-AMG		
Level	TT [s]	FJ [s (%)]	LS [s (%)]	TT [s]	FJ [s (%)]	LS [s (%)]
1	367.6	173.1 (47)	26.6 (7)	354.5	173.2 (49)	15.5 (4)
2	396.4	173.8 (44)	54.8 (14)	372.6	173.6 (47)	28.3 (8)
3	517.7	173.6 (34)	145.4 (28)	457.6	173.6 (38)	73.4 (16)
4	909.2	173.6 (19)	507.8 (56)	650.5	173.6 (27)	254.4 (39)

efficiency down to 40 % for the Schur complement method and down to 55 % for the CPR-AMG method. The reason for this are longer solution times for the iterative solver. Our application code is in parts compute-bound, and in other parts memory-bandwidth bound. An example of the former would be the calculation of the Jacobian matrix, and for the latter the solution of the linear systems.

Table 1 shows the total computation time, the time spent for the calculation of the Jacobian and the time needed for the linear system solve. The time for the solution of the Jacobian is totally scalable, taking for all refinement levels the same amount of time. In contrast, the solution time for the linear system drastically increases. Table 2 shows the number of matrix-vector-products as a proxy for the amount of linear iterations as well as the time spent for extracting the submatrices. Most of the time spent in the linear solve is not due to an increase in linear iterations, but rather due to the time spent for extracting submatrices.

Table 2. Number of matrix-vector-products (MVP), time spent to extract submatrices (SubMatrix) in seconds (and percentage of total time) and total number of nonlinear (NI) and linear (LI) iterations.

	Schur complement				CPR-AMG			
Level	MVP	SubMatrix [s (%)]	NI	LI	MVP	SubMatrix [s (%)]	NI	LI
1	64	16.4 (4)	5	70	31	8.2 (2)	5	21
2	74	40.2 (10)	5	106	36	20.1 (5)	5	26
3	79	129.5 (25)	5	115	38	64.8 (14)	5	28
4	84	490.2 (54)	5	126	41	245.2 (38)	5	31

5 Summary and Conclusion

We presented a constrained pressure residual (CPR-AMG) method and compared it to a Schur complement reduction method. Both methods solve the linear systems arising in the solution of multi-phase flow in porous media in a very efficient way. Nevertheless, they both do not scale ideally in the presented weak scaling test. Interestingly, this is not due to a drastic increase in linear iterations due to refinement, but rather due to an increase in time for the extraction of sub-matrices. To finally give an answer to the question, which method performs better, this kernel needs to be improved. In view of these results, the author contacted the PETSc developers and MatGetSubMatrix has been improved and now scales optimally. The use of PETSc as computation framework for both methods allows for a fair comparison between the two methods. Additionally, it becomes very simple to improve these methods. The Schur complement reduction method could be improved by a better preconditioner for the Schur complement. The CPR-AMG method would benefit, e.g. from an additional solve on the saturation field.

Acknowledgements. The research leading to these results has received funding from the European Union's Horizon2020 Research and Innovation Program under grant agreement No. 640573 (Project DESCRAMBLE) and No. 676629 (Project EoCoE).

We gratefully acknowledge the very helpful discussions with the PETSc developer team.

A PETSc Command Line Options

In this appendix we present the different command line options used to invoke the (non)linear solvers in PETSc.

A.1 Schur Complement Reduction

```
-snes_type newtonls -snes_linesearch_type basic -snes_stol 1e-6
-snes_atol 1e-6 -snes_rtol 1e-6 -snes_max_it 10 -ksp_max_it 100
-ksp_rtol 1e-6 -ksp_atol 1e-6 -ksp_type fgmres -pc_type fieldsplit
-pc_fieldsplit_type schur -pc_fieldsplit_schur_precondition a11
-fieldsplit_0_ksp_type gmres -fieldsplit_1_ksp_type gmres
-fieldsplit_0_pc_type hypre -fieldsplit_0_pc_hypre_type boomeramg
-fieldsplit_1_pc_type hypre -fieldsplit_1_pc_hypre_type boomeramg
-fieldsplit_0_ksp_max_it 10 -fieldsplit_1_ksp_max_it 10
```

A.2 Constrained Pressure Residual (CPR-AMG)

```
-snes_type newtonls -snes_linesearch_type basic -snes_stol 1e-6
-snes_atol 1e-6 -snes_rtol 1e-6 -snes_max_it 10 -ksp_max_it 100
-ksp_rtol 1e-6 -ksp_atol 1e-6 -ksp_type fgmres -pc_type composite
-pc_composite_type multiplicative -pc_composite_pcs fieldsplit,bjacobi
-sub_0_ksp_type fgmres -sub_0_pc_fieldsplit_type additive
-sub_0_fieldsplit_0_ksp_type gmres -sub_0_fieldsplit_1_ksp_type preonly
-sub_0_fieldsplit_0_pc_type hypre -sub_0_fieldsplit_0_pc_hypre_type boomeramg
-sub_0_fieldsplit_1_pc_type none -sub_0_fieldsplit_1_sub_pc_type none
-sub_0_fieldsplit_1_ksp_max_it 10 -sub_1_sub_pc_type ilu
```

References

1. Balay, S., Abhyankar, S., Adams, M.F., Brown, J., Brune, P., Buschelman, K., Dalcin, L., Eijkhout, V., Gropp, W.D., Kaushik, D., Knepley, M.G., McInnes, L.C., Rupp, K., Smith, B.F., Zampini, S., Zhang, H., Zhang, H.: PETSc users manual. Technical report ANL-95/11 - Revision 3.7, Argonne National Laboratory (2016). http://www.mcs.anl.gov/petsc
2. Balay, S., Gropp, W.D., McInnes, L.C., Smith, B.F.: Efficient management of parallelism in object oriented numerical software libraries. In: Arge, E., Bruaset, A.M., Langtangen, H.P. (eds.) Modern Software Tools in Scientific Computing, pp. 163–202. Birkhäuser Press (1997)
3. Bear, J.: Dynamics of Fluids in Porous Media. Elsevier, New York (1972)
4. Brooks, R.J., Corey, A.T.: Hydraulic Properties of Porous Media, vol. 3. Colorado State University Hydrology Paper, Fort Collins (1964)
5. Bui, Q.M., Elman, H.C., Moulton, J.: Algebraic multigrid preconditioners for multiphase flow in porous media. arXiv preprint arXiv:1611.00127 (2016)
6. Burdine, N.T.: Relative permeability calculations from pore-size distribution data. Pet. Trans. AIME **198**, 71–77 (1953)
7. Darcy, H.: Les fontaines publiques de la ville de Dijon. Dalmont, Paris (1856)
8. van Genuchten, M.T.: A closed-form equation for predicting the hydraulic conductivity of unsaturated soils. Soil Sci. Soc. Am. **44**, 892–898 (1980)
9. Lacroix, S., Vassilevski, Y., Wheeler, J., Wheeler, M.: Iterative solution methods for modeling multiphase flow in porous media fully implicitly. SIAM J. Sci. Comput. **25**(3), 905–926 (2003)
10. Lawrence Livermore National Laboratory: HYPRE: High performance preconditioners (2016). http://www.llnl.gov/CASC/hypre/
11. Saad, Y.: A flexible inner-outer preconditioned GMRES algorithm. SIAM J. Sci. Comput. **14**(2), 461–469 (1993)
12. Saad, Y., Schultz, M.H.: GMRES: a generalized minimal residual algorithm for solving nonsymmetric linear systems. SIAM J. Sci. Stat. Comput. **7**(3), 856–869 (1986)

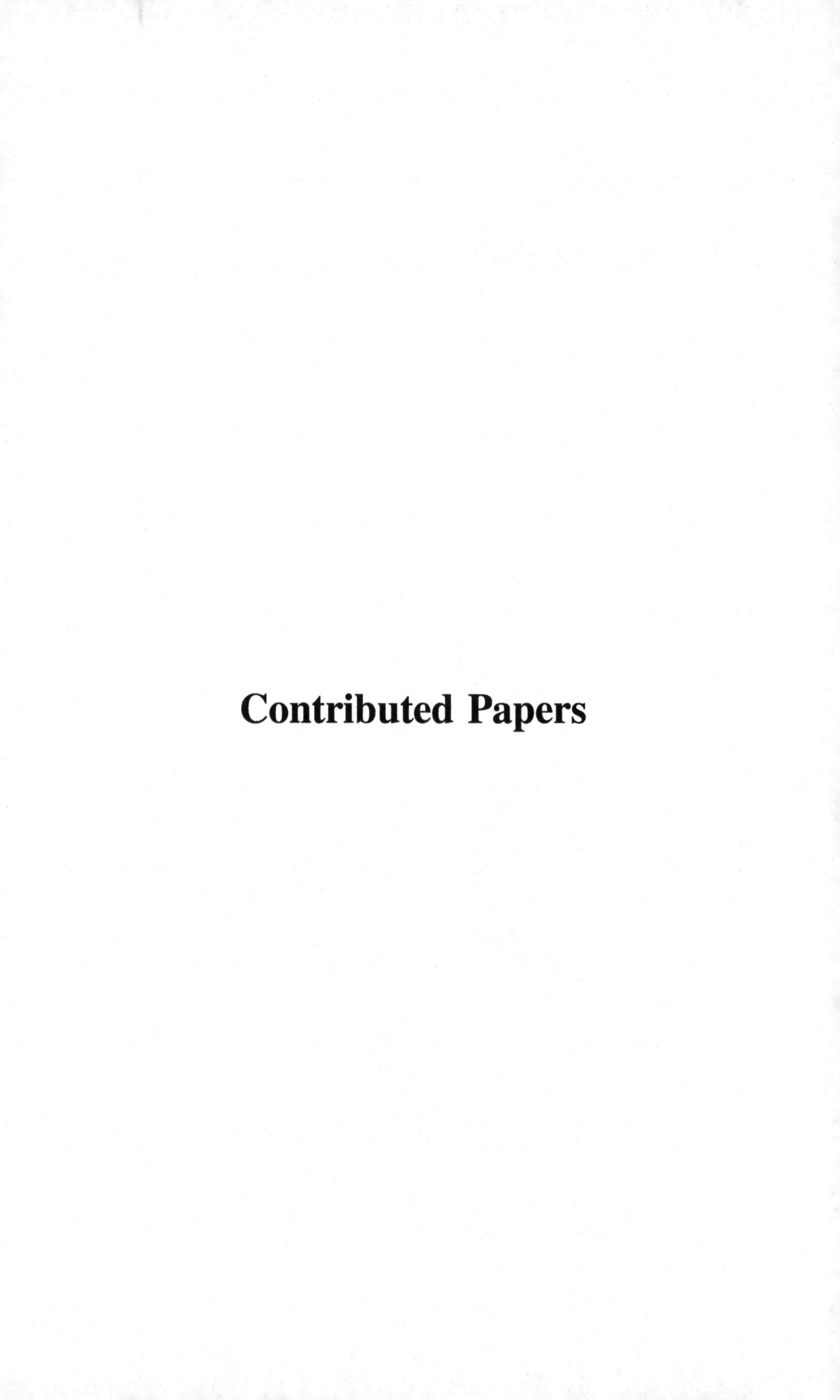

Contributed Papers

A Unified Numerical Approach for a Large Class of Nonlinear Black-Scholes Models

Miglena N. Koleva(✉) and Lubin G. Vulkov

University of Ruse, 8 Studentska str., 7017 Ruse, Bulgaria
{mkoleva,lvalkov}@uni-ruse.bg

Abstract. In this paper, we consider a class of non-linear models in mathematical finance, where the volatility depends on the second spatial derivative of the option value. We study the convergence and realization of the constructed, on a fitted non-uniform meshes, implicit difference schemes. We implement various Picard and Newton iterative processes. Numerical experiments are discussed.

1 Introduction

The main factor, which governs the models for pricing derivatives is the *volatility* of the stock price [2,9,10].

We consider the non-linear Black-Scholes equation in which the volatility ($\sigma(\cdot) > 0$) is assumed to be a function of the underlying asset S, the time t and the *sign* of Greek Gamma ($\Gamma = V_{SS}$) of the option value $V(S,t)$. After time inversion, the model can be formulated as a Cauchy problem for the equation

$$V_t = \frac{1}{2}\sigma^2\left(S, t, \operatorname{sign}\left(V_{SS}\right)\right) S^2 V_{SS} + (r-q) S V_S - rV, \quad S > 0, \quad 0 < t \leq T. \quad (1)$$

Here $r > 0$ is the interest rate, $q \geq 0$ is the dividend yield rate, T is the maturity.

The motivation for solving the non-linear equation (1) arises from more realistic option pricing models in which the volatility fluctuates and one can take into account nontrivial transaction costs, uncertain volatility model, feedback and illiquid market effects due to large traders choosing given stock-trading strategies, etc. [2,10].

There are a number of papers on the numerical solution of Black-Scholes equations with different non-linear volatility, governing European option (cf., for example [2,6,7,9]).

We develop a full implicit numerical method for solving a class of non-linear Black-Scholes models for (1). The numerical scheme is constructed on non-uniform fitted mesh. Then we present the corresponding non-linear system of algebraic equations, in a form similar to the multi-dimensional absolute value equation and solve it by different Picard-like and Newton-like methods.

Further for computations it is necessary to restrict the underlying stock price in a finite region $(0, S_{\max})$, where $S_{\max}$ is a sufficiently large positive number. Then, we define the payoff and boundary conditions for (1) as follows

$$V(S,0) = g_1(S),\ S \in (0, S_{\max}),\ \ V(0,t) = g_2(t),\ V(S_{\max},t) = g_3(t),\ t \in (0,T]. \quad (2)$$

I. Lirkov and S. Margenov (Eds.): LSSC 2017, LNCS 10665, pp. 583–591, 2018.
https://doi.org/10.1007/978-3-319-73441-5_64

Here g_1, g_2, and g_3 are given non-negative functions, satisfying the compatibility conditions $g_1(0) = g_2(0)$ and $g_1(S_{\max}) = g_3(0)$.

The choices of g_1, g_2, and g_3 depend on the type of the option. Popular European options are Vanilla and Butterfly Spread for which we have

$$g_1 = \begin{cases} (S-K)^+ & \text{for Vanilla call,} \\ (K-S)^+ & \text{for Vanilla put,} \\ (S-K_1)^+ - 2(S-K)^+ + (S+K_2)^+ & \text{for Butterfly Spread,} \end{cases}$$

$$g_2 = \begin{cases} 0 & \text{for Vanilla call,} \\ Ke^{-rt} & \text{for Vanilla put,} \\ 0 & \text{for Butterfly Spread,} \end{cases} \qquad g_3 = \begin{cases} S_{\max} - Ke^{-rt} & \text{for Vanilla call,} \\ 0 & \text{for Vanilla put,} \\ 0 & \text{Butterfly Spread,} \end{cases}$$

where $v^+ = \max\{v, 0\}$ (and $v^- = \max\{-v, 0\}$), K, K_1, and K_2 denote the strike prices of the options.

Applying the substitution $U(S,t) = V(S,t) + g_2(t) + S[g_3(t) - g_2(t)]/S_{\max}$ in (1) and (2), we obtain a transformed problem with zero Dirichlet boundary conditions, initial function $U(S,0)$ and right-hand side:

$$f(S,t) = \left(\frac{S}{S_{\max}} - 1\right) g_2' - \frac{S}{S_{\max}} g_3' + q\frac{S}{S_{\max}}(g_2 - g_3) - rg_2.$$

Some well known volatility terms of (1), depending on $\Gamma = V_{SS} = U_{SS}$.

• *Leland model* (LM):

$$\sigma^2(\text{sign}(\Gamma)) = \sigma_0^2 (1 + \text{Le} \times \text{sign}(\Gamma)), \quad \text{Le} = \sqrt{\frac{2}{\pi}} \left(\frac{\kappa}{\sigma_0 \sqrt{\delta t}}\right),$$

where σ_0^2 represents the historical volatility, $0 < \text{Le} < 1$ is the Leland number, δt denotes the transaction frequency and κ denotes transaction cost measure.

• *Boyle and Vorst model* (BVM):

$$\sigma^2(\text{sign}(\Gamma)) = \sigma_0^2 \left(1 + \text{Le}\sqrt{\frac{\pi}{2}} \times \text{sign}(\Gamma)\right),$$

• *Hoggard, Whalley, and Wilmott model* (HWWM). The volatility term in the case of long and short position is represented by

$$\sigma^2(\text{sign}(\Gamma)) = \begin{cases} \sigma_0^2(1 - \varkappa \times \text{sign}(\Gamma)), \text{ long position,} & \varkappa = \frac{\kappa}{\sigma_0}\sqrt{\frac{8}{\pi \delta t}}, \\ \sigma_0^2(1 + \varkappa \times \text{sign}(\Gamma)), \text{ short position.} \end{cases}$$

• *Uncertain volatility model* (UVM). In this model, the volatility vary between two known values, depending on the sign of Gamma Greek of the option. The prices obtained under a no-arbitrage analysis are no longer unique. All that can be computed are the best case and worst case prices, for a specified long or short position [9]. The uncertain volatility is given by worst/best case for an investor with a long position in the option:

$$\sigma^2\left(\mathrm{sign}(\Gamma)\right) = \begin{cases} \sigma_2^2, \, \Gamma \le 0, \\ \sigma_1^2, \, \Gamma > 0, \end{cases} \qquad \sigma^2\left(\mathrm{sign}(\Gamma)\right) = \begin{cases} \sigma_2^2, \, \Gamma > 0, \\ \sigma_1^2, \, \Gamma \le 0, \end{cases} \tag{3}$$

where $\sigma_1 \le \sigma(\mathrm{sign}(\Gamma)) \le \sigma_2$. Prices for investors with short positions can be obtained by the negative of the solutions (1)–(3).

The above considerations motivate us to study the modified model (1) and (2) for $0 < S \le S_{\max}$, $0 < t \le T$ in the form

$$U_t - \frac{1}{2}\left(\alpha U_{SS} + \beta\left|U_{SS}\right|\right) S^2 - (r-q)SU_S + rU = f(S,t), \tag{4}$$

where $0 < \alpha < 1$, $|\beta| < 1$, $\alpha + \beta > 0$ are given by

$$\alpha = \begin{cases} \sigma_0^2, & \mathsf{LM, BVM,} \\ & \mathsf{HWWM;} \\ (\sigma_2^2+\sigma_1^2)/2, & \mathsf{UVM.} \end{cases} \qquad \beta = \begin{cases} \sigma_0^2 Le, & \mathsf{LM}, \\ \sigma_0^2 Le\sqrt{\pi/2}, & \mathsf{BVM}, \\ -\sigma_0^2 \varkappa & \mathsf{HWWM}\text{: long position,} \\ \sigma_0^2 \varkappa & \mathsf{HWWM}\text{: short position,} \\ (\sigma_1^2-\sigma_2^2)/2, & \mathsf{UVM}\text{: worst case long,} \\ (\sigma_2^2-\sigma_1^2)/2, & \mathsf{UVM}\text{: best case long.} \end{cases}$$

In the next section we construct a difference scheme on fitted mesh. In Sect. 3 we discuss numerical methods for solving non-linear system of equations. Numerical results are given in Sect. 4. We finish with concluding remarks.

2 Finite Difference Approximations

We now construct finite difference discretization of (4). We divide $(0, S_{\max})$ into M sub-intervals (S_i, S_{i+1}), $i = 0, 1, \dots, M-1$, satisfying $0 = S_0 < S_1 < S_2 < \dots < S_M = S_{\max}$. Denote by $h_i = S_{i+1} - S_i$, $i = 0, 1, \dots, M-1$ and $\hbar_0 = h_0/2$, $\hbar_i = (h_{i-1} + h_i)/2$, $i = 1, 2, \dots, M-1$, $\hbar_M = h_{M-1}/2$.

We consider a smooth non-uniform grid, cf. in 't Hout et al. [4] - uniform inside the region $[S_l, S_r] = [m_1 K, m_2 K]$, $0 < m_1 < 1$, $m_2 > 1$ (such that to include the points $S = K_1$ and $S = K_2$) and non-uniform outside, with stretching parameter $c = K/10$:

$$S_i := \phi(\xi_i) = \begin{cases} S_l + c\sinh(\xi_i), & \xi_{\min} \le \xi_i < 0, \\ S_l + c\xi_i, & 0 \le \xi_i \le \xi_{\mathrm{int}}, \\ S_r + c\sinh(\xi_i - \xi_{\mathrm{int}}), & \xi_{\mathrm{int}} \le \xi_i < \xi_{\max}. \end{cases}$$

The uniform partition of $[\xi_{\min}, \xi_{\max}]$ is defined by $\xi_{\min} = \xi_0 < \dots < \xi_M = \xi_{\max}$: $\xi_{\min} = \sinh^{-1}(-S_l/c)$, $\xi_{\mathrm{int}} = (S_r - S_l)/c$, $\xi_{\max} = \xi_{\mathrm{int}} + \sinh^{-1}\left((S_{\max} - S_r)/c\right)$.

Similarly, we divide $[0, T]$ into sub-intervals with mesh nodes $\{t_n\}_{n=0}^N$ satisfying $0 = t_0 < t_1 < \dots < t_N = T$ and let $\triangle t_n = t_{n+1} - t_n$, $n = 0, 1, \dots, N-1$.

The numerical solution at point (S_i, t^n) is denoted by U_i^n. For the finite difference approximations of the first and second derivatives we use the notations

$$(U_S)_i^n = \frac{U_{i+1}^n - U_i^n}{h_i}, \quad (U_{\bar{S}})_i^n = \frac{U_i^n - U_{i-1}^n}{h_{i-1}}, \quad (U_{\mathring{S}})_i^n = \frac{h_{i-1}(U_S)_i^n + h_i(U_{\bar{S}})_i^n}{2\hbar_i},$$
$$(U_{\bar{S}S})_i^n = \frac{(U_S)_i^n - (U_{\bar{S}})_i^n}{\hbar_i}.$$

Applying central difference approximation for the second derivative and upwind scheme, combined with 'maximal use of central differencing' [9] for the convection term, we construct the full implicit discretization of (4)

$$\begin{aligned}\frac{U_i^{n+1} - U_i^n}{\Delta t_n} &- \frac{1}{2}\left(\alpha(U_{\bar{S}S})_i^{n+1} + \beta|(U_{\bar{S}S})_i^{n+1}|\right) S_i^2 \\ &- (r-q)^+ S_i[\chi_i^+ (U_S)_i^{n+1} + (1-\chi_i^+)(U_{\mathring{S}})_i^{n+1}] \\ &+ (r-q)^- S_i[\chi_i^- (U_{\bar{S}})_i^{n+1} + (1-\chi_i^-)(U_{\mathring{S}})_i^{n+1}] + rU_i^{n+1} = f_i^{n+1},\end{aligned} \tag{5}$$

for $f_i^{n+1} = f(S_i, t_{n+1})$, $n = 0, 1, \dots, N-1$, $i = 2, 3, \dots, M-1$ and

$$\chi_i^+ = \begin{cases} 0, & h_i < \dfrac{\sigma_{\min}^2 S_i}{(r-q)^+}, \\ 1, & \text{otherwise}, \end{cases} \qquad \chi_i^- = \begin{cases} 0, & h_{i-1} < \dfrac{\sigma_{\min}^2 S_i}{(r-q)^-}, \\ 1, & \text{otherwise}, \end{cases}$$

where $\sigma_{\min}^2 = \alpha - |\beta|$. In practice, $\chi_i^+ = 1$ or $\chi_i^- = 1$ only in a few grid nodes close to $S = 0$. The finite difference scheme is completed by

$$\begin{aligned} &U_0^n = U_M^n = 0, \quad n = 1, \dots, N, \\ &U_i^0 = g_1(S_i) - g_2(0) - S_i[g_3(0) - g_2(0)]/S_{\max}, \quad i = 0, \dots, M. \end{aligned} \tag{6}$$

Theorem 1. *The solution of the full implicit discretization* (5) *and* (6) *converges unconditionally to the viscosity solution of the non-linear PDE* (4) *as* $\Delta t_n \to 0$ *and* $|h| \to 0$, *where* $|h| = \max_{1 \le i \le M} h_i$.

Proof (outline). Following similar arguments as in [9], we prove that the solution, obtained by the numerical scheme (5) and (6) is monotone (in the sense of [9, Definition 1]), stable in maximal discrete norm and consistent. Then the statement of the theorem follows from the theory in [1].

3 Solution of the Non-linear Algebraic Systems

Let $U^n = (U_0^n, U_1^n, \dots, U_M^n)^T$, then for any $n = 0, 1, \dots, N-1$, $i = 1, \dots, M-1$ the scheme (5) and (6) can be presented in the equivalent matrix-vector form:

$$A_1 U^{n+1} - \beta|A_2 U^{n+1}| = F(U^n), \quad n = 0, 1, \dots, N-1, \tag{7}$$

where $F(U^n) = [F_1^n, F_2^n, \dots, F_{M-1}^n]^T$, $F_i^n = U_i^n/\Delta t_n + f_i^{n+1}$,

$$A_1 = \text{tridiag}\{-\lambda_i, \mu_i, -\gamma_i\},\ A_2 = \text{tridiag}\{-\widetilde{\lambda}_i, \widetilde{\mu}_i, -\widetilde{\gamma}_i\},\ i = 1, 2, \ldots, M-1,$$

$$\lambda_i = \frac{\alpha S_i^2}{2\hbar_i h_{i-1}} - (r-q)^+ S_i \frac{(1-\chi_i^+)h_i}{2\hbar_i h_{i-1}} + (r-q)^- S_i \left[\frac{\chi_i^-}{h_{i-1}} + \frac{(1-\chi_i^-)h_i}{2\hbar_i h_{i-1}}\right],$$
$$\gamma_i = \frac{\alpha S_i^2}{2\hbar_i h_i} - (r-q)^- S_i \frac{(1-\chi_i^-)h_{i-1}}{2\hbar_i h_i} + (r-q)^+ S_i \left[\frac{\chi_i^+}{h_i} + \frac{(1-\chi_i^+)h_{i-1}}{2\hbar_i h_i}\right],$$
$$\mu_i = \frac{1}{\Delta t_n} + \lambda_i + \gamma_i + r,\quad \widetilde{\lambda}_i = \frac{S_i^2}{2\hbar_i h_{i-1}},\quad \widetilde{\gamma}_i = \frac{S_i^2}{2\hbar_i h_i},\quad \widetilde{\mu}_i = \widetilde{\lambda}_i + \widetilde{\gamma}_i.$$

The system (7) can be written as $M-1$-dimensional absolute value equation, substituting $Y^n = A_2 U^n$, $n = 0, 1, \ldots, N$ and dividing (7) by β. Then to solve the resulting non-linear system of algebraic equation, we can apply methods, developed for single or multi-dimensional absolute value equation $Ax - |x| = b$, $A \in \mathbb{R}^{n\times n}$, $x, b \in \mathbb{R}^n$ [3,5,8,12].

We prefer to keep the sparse structure of the coefficient matrix. That is why we consider the present form of (7) and adapt these methods appropriately.

At each time level $n = 0, 1, \ldots, N-1$, the numerical solution U^{n+1} is computed by the following iterative methods.

Picard iterative method (PIM):

$$A_1 U^{(k+1)} = F(U^n) + \beta |A_2 U^{(k)}|,\quad k = 0, 1, 2, \ldots$$

The results will be compared with Newton-like methods. Let denote

$$D(U^{(k)}) = \text{diag}(\text{sign}(A_2 U^{(k)})),\quad \mathcal{F}(U^{(k)}) = A_1 U^{(k)} - \beta|A_2 U^{(k)}| - F(U^n).$$

Generalized Newton method (GNM):

$$U^{(k+1)} = (A_1 - \beta D(U^{(k)})A_2)^{-1} F(U^n),\quad k = 0, 1, 2, \ldots$$

Improved generalized Newton method 1 (IGNM1):

$$U^{(k+1/2)} = (A_1 - \beta D(U^{(k)})A_2)^{-1} F(U^n),$$
$$U^{(k+1)} = U^{(k+1/2)} - (A_1 - \beta D^{(k)} A_2)^{-1} \mathcal{F}(U^{n+1/2}),\qquad k = 0, 1, 2, \ldots$$

Improved generalized Newton method 2 (IGNM2). Let $\|\cdot\|$ is the maximal norm.

$$U^{(k+1)} = U^{(k)} + (1 - a^{(k)}) d^{(k)},\quad k = 0, 1, 2, \ldots$$
$$a^{(k)} = \frac{\|\mathcal{F}(U^{(k)} + d^{(k)})\|}{\|2\mathcal{F}(U^{(k)} + d^{(k)}) - \mathcal{F}(U^{(k)})\|},\quad d^{(k)} = -(A_1 - \beta D(U^{(k)})A_2)^{-1} \mathcal{F}(U^{(k)}),$$

The iteration processes are terminated when reaching the desired tolerance $\max_i |U_i^{(k)} - U_i^{(k+1)}| (\max_i \{1, |U_i^{(k+1)}|\})^{-1} < \text{tol}$. Then, $U^{n+1} := U^{(k+1)}$.

Theorem 2. *At each time level $n = 0, 1, \ldots, N-1$ and at each iteration $k = 0, 1, \ldots$, the coefficient matrices A_1 and $A_1 - \beta D(U^{(k)})A_2$ are M-matrices.*

Proof (outline). From the definition of χ_i^+, χ_i^- follows that $\lambda_i \geq 0$, $\mu_i > 0$, $\gamma_i \geq 0$, $\lambda_i \pm |\beta|\widetilde{\lambda}_i \geq 0$, $\mu_i \pm |\beta|\widetilde{\mu}_i > 0$, $\gamma_i \pm |\beta|\widetilde{\gamma}_i \geq 0$, $i = 0, \ldots, M$. Next, it s clear that A_1 and $A_1 - \beta D(U^{(k)})A_2$ are strictly diagonally dominant and irreducible. Therefore, they are M-matrices [11].

On the base of the results in [9, Theorem 1] can be proved that the generalized Newton iteration method converges to the unique solution to equation (4) for any given initial iterate U^0.

4 Numerical Simulations

In this section we test the accuracy, the order of convergence and the efficiency of the proposed methods. In order to minimize oscillations in the spatial error, the payoff is averaged in the neighbouring nodes of the strikes K, K_1, K_2 (depending on the option) $g_1(S_i) = (S_{i+1/2} - S_{i-1/2})^{-1} \int_{S_{i-1/2}}^{S_{i+1/2}} g_1(\widetilde{S})d\widetilde{S}$.

We consider the more challenging butterfly option for Leland model. The model parameters are: $\sigma_0 = 0.2$, $Le = 0.5$, $S_{\max} = 120$, $K = 50$, $K_1 = 40$, $K_2 = 60$. Thus, for computations we set $[S_l, S_r] = [K/2, 3K/2]$.

We give the values of the solution at strike point K at maturity $T = 1$. The notation diff stands for the error (absolute value of the difference) between the solution on two consecutive refinement grids. The order of convergence CR is calculated as $\log_2$ from the ratio between two consecutive values of diff. For the iteration process we choose tol $= 1.e - 8$. We compute the solution by the full implicit scheme (7), $r = 0.04$, $q = 0$ and $r = 0.01$, $q = 0.05$ with fixed time step $\triangle t = \min\limits_i h_i^2$ in order to verify the order of convergence both in space and time. In Tables 1 and 3, we give the value of the solution $V(K, T)$, computed by generalized Newton algorithms GNM, IGNA1, IGNM2 for different meshes, quantity diff, CR and CPU time (in seconds) and averaging number of iterations (iter) k at each time level. Similar results, obtained by Picard iteration process PIM are listed in Tables 2 and 4. We observe that with all Newton methods, we obtain one and the same values $V(K, 1)$ (up to 8 digits after decimal point) for one and the same meshes. The values of $V(K, 1)$, obtained by Picard iterations differ insignificantly with the one obtained by Newton methods. The convergence is first order in time and second order in space at strike point K for all methods, but the efficiency (see CPU times) is different. We observe better performance of Picard iteration approach, although for more coarse meshes this algorithm accomplishes more iterations, than the other ones. In our opinion, the reason for the best efficiency of PIM is due to the facts that this is a one stage algorithm (in contrast to IGNA1 and IGNA2), does not require formation of the matrix $D(U^{(k)})$ at each iteration and the coefficient matrix A_1 of the corresponding linear system, is one and the same at each iteration (in contrast to GNM, IGNA1 and IGNA2).

On Fig. 1 we plot evolution graphics of the option value V and Greek gamma Γ, computed with PIM. We observe that no oscillations appear.

Table 1. Leland model, $r = 0.04$, $q = 0$, generalized Newton methods

M	$V(K,1)$	diff	CR	GNM		IGNA1		IGNA2	
				CPU	iter	CPU	iter	CPU	iter
400	5.092484			0.02	3.07	0.03	3.07	0.03	3.07
800	5.074317	1.81677e−2		0.07	2.39	0.11	2.63	0.08	2.38
1600	5.069853	4.46377e−3	2.025	0.37	2.00	0.65	2.35	0.43	2.00
3200	5.068745	1.10766e−3	2.011	2.87	2.00	4.71	2.06	3.54	2.00
6400	5.068469	2.76064e−4	2.004	21.56	2.00	34.61	2.00	26.92	2.00
12800	5.068400	6.89876e−5	2.001	170.29	2.00	271.77	2.00	206.26	2.00

Table 2. Leland model, $r = 0.04$, $q = 0$, Picard iterations

M	$V(K,1)$	diff	CR	CPU	iter
400	5.092481			0.01	7.39
800	5.074311	1.81695e−2		0.03	4.68
1600	5.069849	4.46211e−3	2.026	0.15	3.26
3200	5.068739	1.11012e−3	2.007	0.87	2.21
6400	5.068467	2.71810e−4	2.030	5.58	2.03
12800	5.068400	6.71843e−5	2.016	40.62	2.00

Table 3. Leland model, $r = 0.01$, $q = 0.05$, generalized Newton methods

M	$V(K,1)$	diff	CR	GNM		IGNA1		IGNA2	
				CPU	iter	CPU	iter	CPU	iter
400	5.142095			0.02	3.11	0.03	3.11	0.03	3.11
800	5.124640	1.74550e−2		0.07	2.49	0.12	2.77	0.09	2.49
1600	5.120367	4.27309e−3	2.030	0.37	2.00	0.68	2.40	0.44	2.00
3200	5.119308	1.05873e−3	2.013	2.81	2.00	4.72	2.06	3.55	2.00
6400	5.119044	2.63732e−4	2.005	21.58	2.00	34.85	2.00	27.02	2.00
12800	5.118978	6.58951e−5	2.001	171.82	2.00	275.70	2.00	206.81	2.00

Table 4. Leland model, $r = 0.01$, $q = 0.05$, Picard iterations

M	$V(K,1)$	diff	CR	CPU	iter
400	5.142093			0.01	7.39
800	5.124634	1.74591e−2		0.03	4.70
1600	5.120363	4.27084e−3	2.031	0.15	3.34
3200	5.119301	1.06270e−3	2.007	0.88	2.20
6400	5.119042	2.58842e−4	2.038	5.87	2.03
12800	5.118978	6.38831e−5	2.019	41.78	2.00

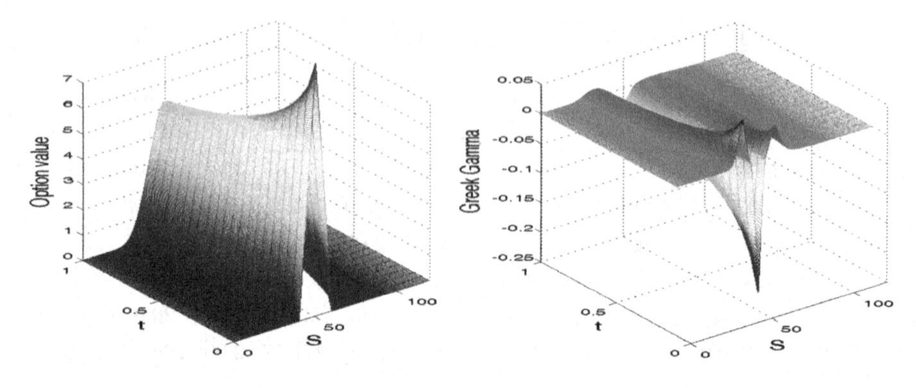

Fig. 1. Evolution graphics of V and Γ, $M = 800$, $\triangle t = 0.05$, $p = 0.02$, $q = 0.01$

5 Conclusions

We propose a full implicit numerical method for solving a class of non-linear Black-Scholes models with non-zero dividend and the volatility term, which depends on second derivative of the option value. The numerical scheme is constructed on non-uniform fitted mesh. For the approximation of the convection term we apply an upwind method, combined with 'maximal use of central differencing' [9], where a second order (on non-uniform mesh) discretization is implemented. The generated non-linear system of algebraic equations is presented in the form, similar to the multi-dimensional absolute value equation and solved by different Picard-like and Newton-like methods, which are adapted for the particular case. We observe much better performance of the Picard iteration method.

Acknowledgements. This research was supported by the Bulgarian National Fund of Science under Project "Advanced Analytical and Numerical Methods for Nonlinear Differential Equations with Applications in Finance and Environmental Pollution"-2017.

References

1. Barles, G.: Convergence of numerical schemes for degenerate parabolic equations arising in finance. In: Rogers, L.C.G., Talay, D. (eds.) Numerical Methods in Finance. Cambridge University Press, Cambridge (1997)
2. Ehrhardt, M.: Nonlinear Models in Mathematical Finance: New Research Trends in Option Pricing. Nova Science Publishers, New York (2008)
3. Feng, J., Liu, S.: An improved generalized Newton method for absolute value equations. SpringerPlus **5**(1), 10–42 (2016)
4. Haentjens, T., In't Hout, K.J.: Alternating direction implicit finite difference schemes for the Heston-Hull-White PDE. J. Comput. Finan. **16**(1), 83–110 (2012)
5. Haghani, F.K.: On generalized Traub's method for absolute value equations. J. Optim. Theory Appl. **166**(2), 619–625 (2015)

6. Koleva, M.N., Vulkov, L.G.: On splitting-based numerical methods for nonlinear models of European options. Int. J. Comput. Math. **3**(5), 781–796 (2016)
7. Lesmana, D.C., Wang, S.: An upwind finite difference method for a nonlinear Black-Scholes equation governing European option valuation under transaction costs. Appl. Math. Comput. **219**(16), 8811–8828 (2013)
8. Mangasarian, O.L.: A generalized Newton method for absolute value equations. Optim. Lett. **3**(1), 101–108 (2009)
9. Pooley, D., Forsythy, P., Vetzalz, K.: Numerical convergence properties of option pricing PDEs with uncertain volatility. IMA J. Numer. Anal. **23**(2), 241–267 (2003)
10. Ševčovič, D.: Nonlinear parabolic equations arising in mathematical finance. In: Ehrhardt, M., Günther, M., ter Maten, E.J.W. (eds.) Novel Methods in Computational Finance. MI, vol. 25, pp. 3–15. Springer, Cham (2017). https://doi.org/10.1007/978-3-319-61282-9_1
11. Varga, R.S.: Matrix Iterative Analysis. Prentice-Hall, Engelwood Cliffs (1962)
12. Yong, L.: An iterative method for absolute value equations problem. Information **16**(1), 7–12 (2013). International Information Institute (Tokyo)

Beta-Function B-splines and Subdivision Schemes, a Preliminary Study

Arne Lakså(✉)

University of Tromsø - The Arctic University of Norway, Narvik, Norway
arne.laksa@uit.no

Abstract. This paper is an initial study of subdivision schemes in connection with blending technics such as Expo Rational B-splines, see [1]. The study is done on curves, but surfaces are a natural next step. It turns out that blending two second degree polynomial curves, which interpolates three points, generate a 4-point subdivision scheme for Catmull Rom Splines, see [2]. It can be shown that different subdivision schemes can be developed from a blending spline construction using different types of local curves and/or blending functions. We will show examples, as circular arcs, and discuss some problems and properties.

Keywords: Spline · Subdivision · Blending · Curve

1 Introduction

How we define curves and surfaces is important in geometric modeling. Spline curves of different types, polynomial B-splines, Hermite splines, Expo-Rational B-splines and other related spline types are the most common description of curves today. They are typically defined by a knot vector, a control polygon, and a degree/order. The knot vector and the degree defines the basis functions. To calculate points and derivatives on the curve we compute the basis functions at a given parameter value and use this as weights in an affine combination of the control points, [3]. However, if we only want to display a curve, we can just refine the control polygon into a sequence of control points that, in the limit, converge to a curve. By doing this, freedom from a closed-form mathematical expression is achieved, and a wide variety of curve types can be expressed. The curves are commonly called subdivision curves as the methods are based upon the binary subdivision of the uniform B-spline curves, [4]. The same can be done with curves that are interpolating the control points, interpolatory subdivision, [5].

In the following sections we will look at subdivision schemes for curves, we will define B-functions and look at blending spline curves, and finally we will see examples of subdivision derived from blending.

I. Lirkov and S. Margenov (Eds.): LSSC 2017, LNCS 10665, pp. 592–599, 2018.
https://doi.org/10.1007/978-3-319-73441-5_65

2 4-Point Subdivision Scheme for Catmull Rom Splines

A 3th degree polynomial curve $c(t)$ that interpolates the 4 points, P_0, P_1, P_2, P_3 at the parameter values $t = \{-1, 0, 1, 2\}$ can be made using the Lagrange polynomials:

$$c(t) = \frac{t(t-1)(t-2)}{-6}P_0 + \frac{(t+1)(t-1)(t-2)}{2}P_1 + \frac{(t+1)t(t-2)}{-2}P_2 + \frac{(t+1)t(t-1)}{-6}P_3.$$

If we calculate the curve at the average parameter value, $t = \frac{-1+0+1+2}{4} = \frac{1}{2}$, we get

$$c\left(\frac{1}{2}\right) = -\frac{1}{16}P_0 + \frac{9}{16}P_1 + \frac{9}{16}P_2 - \frac{1}{16}P_3, \tag{1}$$

which is a point between the points P_1 and P_2. This 4-point subdivision scheme for Catmull Rom spline curves is also called the Dubuc-Deslaurier scheme, see [6]. We insert points between all two neighboring points to get a new refined point set. We repeat this procedure until the point set is large enough. It can be shown that this converges towards a curve with continuity between C^1 and C^2, see [2].

3 4-Point Subdivision Scheme from Blending

We now interpolate the points P_0, P_1, P_2 with a 2nd degree polynomial curve $c_1(t)$ at the parameter values $t = \{-1, 0, 1\}$, we then interpolate the three points P_1, P_2, P_3 with another 2nd degree polynomial curve $c_2(t)$ at the parameter values $t = \{0, 1, 2\}$. The result is

$$c_1(t) = \frac{t(t-1)}{2}P_0 + \frac{(t+1)(t-1)}{-1}P_1 + \frac{(t+1)t}{2}P_2,$$

and

$$c_2(t) = \frac{(t-1)(t-2)}{2}P_1 + \frac{t(t-2)}{-1}P_2 + \frac{t(t-1)}{2}P_3.$$

We compute the curves at the average parameter value $t = \frac{1}{2}$,

$$c_1\left(\frac{1}{2}\right) = -\frac{1}{8}P_0 + \frac{3}{4}P_1 + \frac{3}{8}P_2,$$

and

$$c_2\left(\frac{1}{2}\right) = \frac{3}{8}P_1 + \frac{3}{4}P_2 - \frac{1}{8}P_3.$$

We blend these two curves at $t = \frac{1}{2}$ using the weight $\frac{1}{2}$,

$$\begin{aligned} c\left(\frac{1}{2}\right) &= \frac{1}{2}\left(-\frac{1}{8}P_0 + \frac{3}{4}P_1 + \frac{3}{8}P_2\right) + \left(1 - \frac{1}{2}\right)\left(\frac{3}{8}P_1 + \frac{3}{4}P_2 - \frac{1}{8}P_3\right) \\ &= -\frac{1}{16}P_0 + \frac{9}{16}P_1 + \frac{9}{16}P_2 - \frac{1}{16}P_3. \end{aligned} \tag{2}$$

What we see is that (1) and (2) give the same result. Thus, it follows that this method also gives a 4-point subdivision scheme for Catmull Rom spline curves. Note that we used $\frac{1}{2}$ as weight in the blending. We can use other weights, which will be discussed later.

This gives us an idea that we can look at the use of other types of blending than the linear, and we might use other types of local curves in the blending than polynomial based curves. This will be investigated in the following sections.

4 B-functions and Blending

B-function is short for blending function. It is typically used in blending two or more functions, either vector or point valued functions, where we blend either points or curves or tensor product/triangular surfaces etc. The definition first formulated in [7], is:

Definition 1. *A B-function is:*

D1 *a permutation function* $B : I \to I \quad (I = [0, 1] \subset \mathbb{R})$,
D2 *where* $B(0) = 0$,
D3 *and* $B(1) = 1$,
D4 *and that is monotone, i.e.* $B'(t) \geq 0, \quad t \in I$.
D5 *A B-function is called symmetric if,* $B(t) + B(1-t) = 1, \quad t \in I$.

This symmetry is a point symmetry (around the point (0.5 0.5)).

Figure 1 shows four example of B-functions plotted together with their derivatives. From the figure we can observe that for all four B functions are $b(0) = 0$ and $b(1) = 1$. All B-functions in Fig. 1 are also monotone and symmetric.

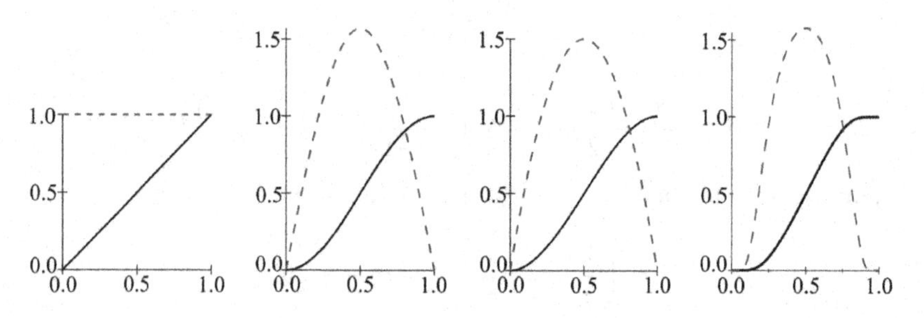

Fig. 1. B-functions (solid) and their derivatives (dashed). From left hand side is a linear, trigonometric, polynomial, and finally an Expo-Rational B-function.

Definition 2. *The order of B-functions is a property that plays an important role. It is the number of subsequent derivatives which is zero at start and end. The order of a B-function is short for the Hermite order of a B-function, denoted S, it is for a symmetric B-function determined by*

$$B^{(j)}(0) = B^{(j)}(1) = 0, \quad j = 1, 2, ..., S. \tag{3}$$

For a non-symmetric B-function we have to differ between the start and the end,

$$\begin{aligned} B^{(j)}(0) &= 0, \quad j = 1, 2, ..., S_0. \\ B^{(j)}(1) &= 0, \quad j = 1, 2, ..., S_1. \end{aligned} \tag{4}$$

This property is also called the Hermite property of a B-function.

The Hermite property of a B-function determines the continuity of a blending type curve constructed by using the given B-function. If the order is one, the curve is at least C^1-continuous.

5 Curves Constructed with a B-function

Given n control points $c_0, c_2, ..., c_{n-1}$ and a knot vector $\bar{t} = \{t_0, t_1, ..., t_{n+d}\}$. A B-spline curve of degree d defined by these control points and knot vector $\bar{t}$ is

$$C(t) = \sum_{i=0}^{n-1} c_i \; b_{d,i}(t),$$

where the basis functions $b_{d,i}(t)$ are defined by the knot-vector, $\bar{t}$, and the degree d. The recursive definition of the basis functions is:

$$b_{d,i}(t) = w_{d,i}(t) \; b_{d-1,i}(t) + (1 - w_{d,i+1}(t)) \; b_{d-1,i+1}(t) \tag{5}$$

where

$$w_{d,i}(t) = \frac{t - t_i}{t_{i+d} - t_i}, \tag{6}$$

terminating with

$$b_{0,i}(t) = \begin{cases} 1, \text{ if } & t_i \leq t < t_{i+1}, \\ 0, \text{ otherwise}, \end{cases} \tag{7}$$

The B-spline can be changed by deforming (6), by a reparametrisation. We add a B-function to (6), and thus get $B \circ w_{d,i}(t)$. The result is a non polynomial B-spline, i.e. a Generalized Expo-Rational B-spline (GERBS) curve, see [8].

If the order is 2 (degree 1 in the polynomial case) we get:

$$b_{1,i}(t) = B \circ w_{1,i}(t) \; b_{0,i}(t) + (1 - B \circ w_{1,i+1}(t)) \; b_{0,i+1}(t) \tag{8}$$

The result is that the continuity of the B-spline increase with the order of the B-function. In the ERBS case, when we use an Expo-Rational B-function, is the continuity actually infinite, described in [9].

6 GERBS, the Extended Blending Spline

Notice that a second order B-spline curve is piecewise linear. This does not change by adding a B function to the B-spline basis as shown in expression (8). However, adding a B function gives some new properties that allow us to

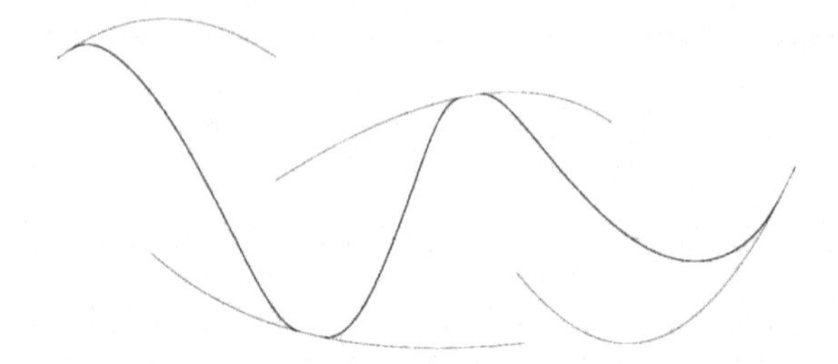

Fig. 2. An Expo-rational B-spline with four local curves. The local curves are interpolating the GERBS-curve at their respective knot values, with both position and all derivatives.

replace the control points with local (control) curves. We now get a GERBS blending splines,

$$c(t) = \sum_{i=0}^{n-1} c_i(t)\ b_{1,i}(t)$$

where $c_i(t)$ are local curves defined over the two knot intervals, $[t_{i-1}, t_{i+1}]$. This is described in [9,10].

In Fig. 2 there is an example of an GERBS-curve and the local curves it is made from. Since the order is two, and the B-function is symmetric, we can simplify the equation for each knot interval, i.e.

$$c(t) = (1 - B(t))\ c_i\,(t) + B(t)\ c_{i+1}\,(t)\,, \tag{9}$$

where $B(t) = B \circ w_{1,k}(t)$, and it follows that we reparameterize each interval to now be $[0, 1]$.

There is also another property described in definition 2 and connected to blending and the order of the B-function. It is the Hermite interpolation property. It follows that the local curves interpolates the global curve with the position and subsequent derivatives up to the order of the B-function used. This because, if we calculate the derivation of (9) we get,

$$c^{(j)}(t) = c_i^{(j)}(t) + B(t)\ (c_{i+1}^{(j)}(t) - c_i^{(j)}(t)) + \sum_{i=0}^{j} \binom{j}{i} B^{(i)}(t)\ (c_{i+1}^{(j-i)}(t) - c_i^{(j-1)}(t))$$

Since $B(t_i) = 0$ and $B^{(j)}(t_i) = 0,\ j = 1, ..., S$ is $c^{(j)}(t_i) = c_i^{(j)}\,(t)\,,\ j = 0, ..., S$. and since $B(t_{i+1}) = 1$ and $B^{(j)}(t_{i+1}) = 0,\ j = 1, ..., S$ is $c^{(j)}(t_{i+1}) = c_i^{(j)}\,(t)\,,\ j = 0, ..., S$. Where S is the order of the B-function. This Hermite interpolation is illustrated in Fig. 2.

If the local curve interpolate all three subsequent control points, then the order of the Hermite interpolation will increase by one. This is illustrated in Fig. 3. Here, the B-function is the linear function $B(t) = t$, but still the result is C^1-continuity. This is called the expanded Hermite interpolation property and is described in [7].

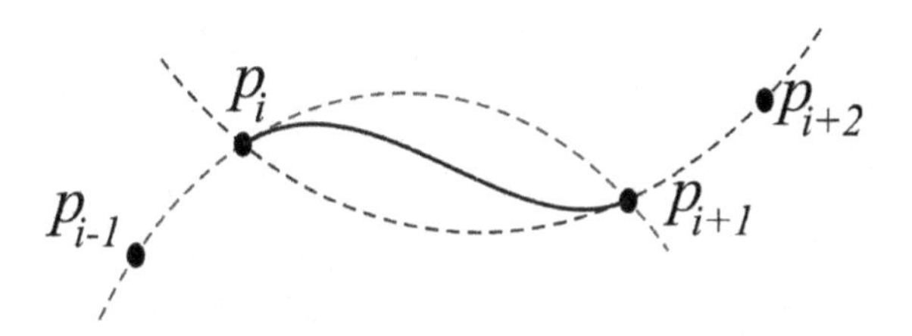

Fig. 3. Example of circle splines, see [11]. Between the point p_i and p_{i+1} there are two circular arcs blended to one curve. As we can see, the curve segment between the points p_i and p_{i+1} is only defined by the four points: p_{i-1}, p_i, p_{i+1}, and p_{i+2}.

7 Subdivision and Blending Using a Regularized Beta Function

As we can see in Sect. 3, we can derive the 4-point Dubuc-Deslaurier subdivision scheme for Catmull Rom splines from blending of two curves, where each interpolates their respective three control points. The scheme is constructed using two 2nd degree polynomial curves which each interpolates their respective three consecutive points and that these curves are blended with a linear B-function $B(t) = t$.

To compare the curves made using different B-functions and/or different types of local curves, we use a common initial point set. Thus we can show how the different schemes change the shape of the curve. The initial point set is:

$$P = \{(0,0),(1,1),(2,-1),(3,1),(4,0)\}\,, \tag{10}$$

and in addition is a first point $F = (-1,0)$ and a last point $L = (5,0)$. The points F and L are used to give the first derivative at start and end of the curve.

In Fig. 4 it is the given point set used to make a curve with the ordinary 4-point Dubuc-Deslaurier subdivision scheme. The small cubes that can be seen in the figure are the initial points. The cubes outside the curve are the first point F and last point L to decide the derivatives at start and end, and the recursive refinement of these "derivative" points.

What happens if we want to use another B-function, one with an order greater than 0. Remember that the B-functions are symmetric and it follows that all symmetric B-functions are 0.5 at $t = 0.5$. To use another B-function means that we have to insert points other places than at $t = 0.5$. In the next plot, Fig. 5, there is the same point set as in Fig. 4 used. The curve is made with the same local curves, but an order 1 regularized beta-function B-function is used, $B(t) = 3t^2 - 2t^3$. We now insert new points at $t = \frac{1}{3}$ and $t = \frac{2}{3}$, giving $B\left(\frac{1}{3}\right) = \frac{7}{27}$ and $B\left(\frac{2}{3}\right) = \frac{20}{27}$. As we can see, the curve is more "curved" over all, but the maximum curvature is smaller.

In the third plot, Fig. 6, the local curves are circular arcs that each interpolates their respective three consecutive points. In this example we have used the linear, order 0 B-function, $B(t) = t$. The figure shows that the resulting curve is

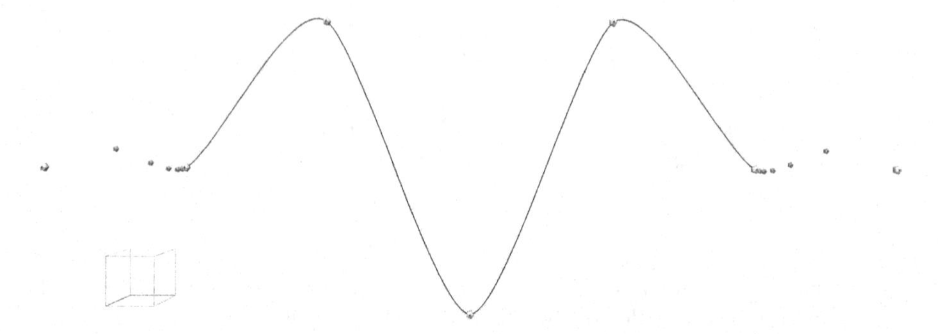

Fig. 4. The figure shows a curve made from the point set described in expression (10). The scheme is the ordinary 4-point Dubuc-Deslaurier subdivision scheme for Catmull Rom splines that provides a C^1-continuous curve.

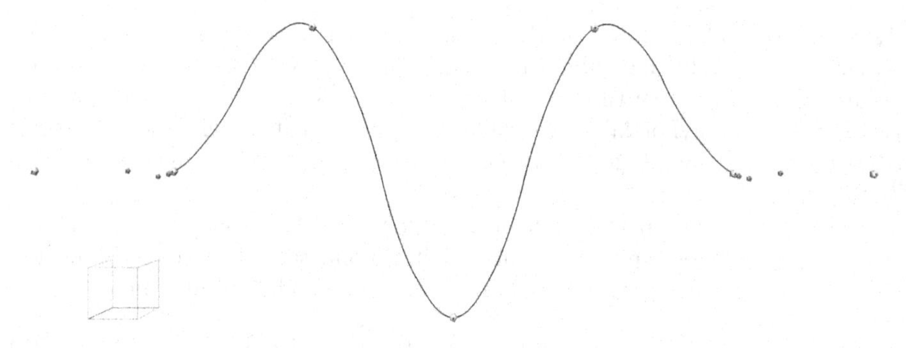

Fig. 5. The figure shows a curve made from the point set described in expression (10). Here the B-function, $B(t) = 3t^2 - 2t^3$, is of order 1, this gives a C^2-continuous curve.

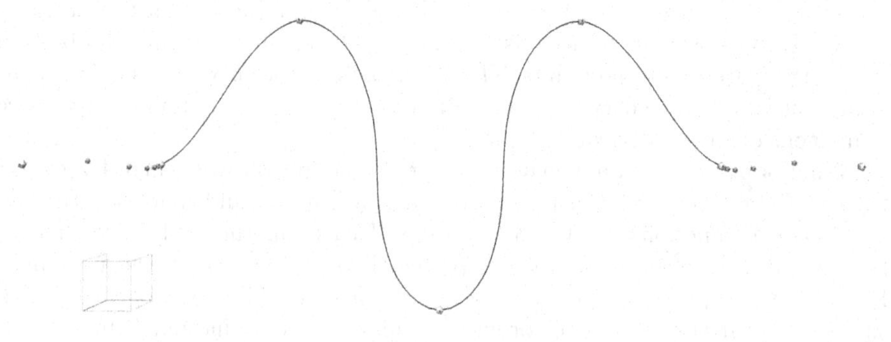

Fig. 6. The figure shows a curve made from the point set described in expression (10). Here the B-function is the linear one, $B(t) = t$, but the local curves are circular arcs.

more curved than the curves in the two previous examples. Another observation is that, in the first example - Fig. 4, has the tangent vector at the original points the same direction as the line from the original point before the given original point to the original point after the given original point. This is a typical property of the 4-point subdivision scheme for Catmull Rom splines. What we can see is that the same is the case for the curve in Fig. 5. It appears that changing B function does not affect this property. In the last example, Fig. 6, this is obvious not the case.

8 Concluding Remarks

This study is only initial. The properties has to be studied in more detail and more testing on other schemes must be done. It is also important to compare with other schemes and find relationship with them.

It is too early to conclude whether blending is a fruitful approach to developing subdivision schemes. In order to do that, the next step is to move on to surface subdivision schemes. They are widely used in computer graphics, computer games and virtual reality applications in general.

References

1. Lakså, A.: Basic properties of expo-rational B-splines and practical use in Computer Aided Geometric Design. In: unipubavhandlinger, vol. 606. Unipub, Oslo (2007)
2. Catmull, E., Rom, R.: A class of local interpolating splines. In: Barnhill, R.E., Riesenfeld, R.F. (eds.) Computer Aided Geometric Design, vol. 30, pp. 317–326. Academic Press, New York (1974)
3. Farin, G.: Curves and Surfaces for CAGD. Morgan Kaufmann Publishers Inc., San Francisco (2002)
4. Rehan, K., Siddiqi, S.S.: A family of ternary subdivision schemes for curves. App. Math. Comput. **270**, 114–123 (2015)
5. Herbst, B.M., Hunter, K.M., Rossouw, E.: Subdivision of curves and surfaces: an overview. ResearchGate.net (2007). https://www.researchgate.net/publication/228948724_Subdivision_of_curves_and_surfaces_An_overview
6. Deslauriers, G., Dubuc, S.: Symmetric iterative interpolation processes. Constr. Approximation **5**(1), 49–68 (1989)
7. Lakså, A.: Non polynomial B-splines. In: 41th International Conference Applications of Mathematics in Engineering and Economics AMEE 2013, vol. 1690, p. 030001. American Institute of Physics (AIP) (2015)
8. Dechevsky, L.T., Bang, B., Lakså, A.: Generalized expo-rational B-splines. Int. J. Pure Appl. Math. **57**, 833–872 (2009)
9. Dechevsky, L.T., Lakså, A., Bang, B.: Expo-rational B-splines. Int. J. Pure Appl. Math. **27**, 319–369 (2006)
10. Lakså, A., Bang, B., Dechevsky, L.T.: Exploring expo-rational B-splines for curves and surfaces. In: Dæhlen, M., Mørken, K., Schumaker, L. (eds.) Mathematical Methods for Curves and Surfaces, pp. 253–262. Nashboro Press (2005)
11. Wenz, H.: Interpolation of curve data by blended generalized circles. Comput. Aided Geom. Des. **13**, 673–680 (1996)

Conjugate Gradient Method for Identification of a Spacewise Heat Source

V. I. Vasil'ev, V. V. Popov, and A. M. Kardashevsky(✉)

North-Eastern Federal University, Yakutsk, Russia
kardam123@gmail.com

Abstract. In this research we study a problem of identifying of the right-hand side in a parabolic equation dependent on spatial variables in multidimensional domain. For numerical solution of the set inverse initial-boundary problem we use a conjugate gradient method with purely implicit time approximation. The results of the computational experiment performed on model problems with quasi-real solutions, including those with noise in input data are being discussed.

Keywords: Inverse problem · Identification of spacewise heat source
Parabolic partial differential equation · Implicit difference scheme
Conjugate gradient method

1 Introduction

Numerous up-to-date problems of applied science and technology lead to inverse problems. Theoretical study, development of efficient numerical methods of various inverse problems of mathematical physics are summarized in monographs of Samarsky and Vabishchevich [12], Kabanikhin [8], Lions and Magenes [9], Isakov [3]. Inverse tasks of source identification in heat conduction equation are of great interest in terms of their applications. It is notable that Cannon [1], Isakov [4] studied different aspects of correctness of the inverse problems setting including problems of identification of spatially distributed source for parabolic equations.

In their works [5,6] Johansson and Lesnik suggest an iterative algorithm of sustainable iterative source renewal using a boundary element method.

In work [2] a finite element method is suggested, and the work [7] – suggests a finite difference method. In the suggested methods a correct direct problem is being solved at each iteration. In his works [14,15] Vabishchevich has developed, proved and numerically implemented an iterative method that was earlier proposed in his monograph [12] and, thus, built the second iterative process of Picard type.

It should be noted that in work [13] there has been suggested an iterative method with minimal residuals for numerical solution of retrospective inverse problem of heat conduction, and in work [16] we can see a conjugate gradient method.

I. Lirkov and S. Margenov (Eds.): LSSC 2017, LNCS 10665, pp. 600–607, 2018.
https://doi.org/10.1007/978-3-319-73441-5_66

In the present research work we suggest an application of iterative method of conjugate gradients for numerical solution of a source identification problem, which depends on spatial variables in a parabolic equation. The method is quite simple and effective. We provide solution examples for model test problems for one-dimensional and two-dimensional cases. Besides, we present calculation results under perturbed condition.

2 Problem Setting

Let us assume that $\boldsymbol{x} = (x_1, x_2, \ldots, x_p) \in \Omega = \prod_{\alpha=1}^{p}[0, l_\alpha]$ is a constrained parallelepiped from R^p. Let us put a direct problem for linear parabolic equation with variable coefficients. We need to find a function $u(\boldsymbol{x}, t)$, $\boldsymbol{x} = (x_1, x_2, \ldots, x_p) \in \Omega$, $0 < t \leqslant T$, $T > 0$, that satisfies a parabolic equation of second order

$$\frac{\partial u}{\partial t} - \sum_{\alpha=1}^{p} \frac{\partial}{\partial x_\alpha}\left(k_\alpha(\mathbf{x}, t)\frac{\partial u}{\partial x_\alpha}\right) = f(\boldsymbol{x}), \quad \boldsymbol{x} \in \Omega, \quad 0 < t \leqslant T \tag{1}$$

with following boundary and initial conditions:

$$u(\boldsymbol{x}, t) = 0, \quad \boldsymbol{x} \in \partial\Omega, \quad 0 < t \leqslant T, \tag{2}$$

$$u(\boldsymbol{x}, 0) = u_0(\boldsymbol{x}), \quad \boldsymbol{x} \in \overline{\Omega}. \tag{3}$$

If coefficients of Eq. (1) have quite smooth functions and satisfy the conditions $0 < c_1 \leqslant k_\alpha(\mathbf{x}, t) \leqslant c_2 < \infty$, $\alpha = 1, 2, \ldots, p$, the right-hand side and initial condition are similarly smooth and bounded functions. Besides, if an initial condition goes to zero at domain boundary Ω, then the direct problem (1)–(3) is formulated correctly.

Further we will consider an inverse problem when a function $f(\boldsymbol{x})$ is unknown and depends only on spatial variables in Eq. (1). Let us give the following additional condition

$$u(\boldsymbol{x}, T) = \varphi(\boldsymbol{x}), \quad \boldsymbol{x} \in \overline{\Omega}. \tag{4}$$

The set inverse problem of finding a pair $u(\boldsymbol{x}, t)$, $\varphi(\boldsymbol{x})$ under conditions (1)–(3) and additional condition (4) is correct if there are existence and a unique solvability that can be observed in the above-cited works. In the present research we focus on the issues of numerical solution of inverse problems, skipping theoretical questions of convergence of approximate solutions to exact ones.

3 Difference Scheme of the Problem

We conduct a numerical solution of a parabolic problem (1)–(3) using the finite difference method [11]. For that we introduce a uniformly spaced grid in the domain Ω in each direction x_α with $h_\alpha, \alpha = 1, \ldots, p$ steps and designate a set of internal nodes by ω

$$\omega_\alpha = \left\{x_\alpha = i_\alpha h_\alpha,\ i_\alpha = 1, 2, \ldots, N_\alpha - 1,\ N_\alpha h_\alpha = l_\alpha\right\}, \quad \alpha = 1, \ldots, p,$$

$$\omega = \prod_{\alpha=1}^{p} \omega_\alpha, \quad \overline{\omega} = \omega \cup \partial\omega.$$

Let us introduce Hilbert space of grid functions $y, v \in H = L_2(\omega)$ where a scalar product and norm are defined as follows:

$$(y, w) \equiv \sum_{\boldsymbol{x}\in\omega} y(\boldsymbol{x})\, w(\boldsymbol{x}) \prod_{\alpha=1}^{p} h_\alpha, \quad \boldsymbol{x} \in \omega, \quad \|y\| \equiv \sqrt{(y, y)}.$$

In the assumption of sufficient coefficient smoothness $k_\alpha(\boldsymbol{x}, t)$ for all internal nodes $\boldsymbol{x} \in \omega$ a finite difference scheme for multidimensional elliptic operator $A(t)$ is written in the following way:

$$A(t) = \sum_{\alpha=1}^{p} A_\alpha(t), \quad \boldsymbol{x} \in \omega, \tag{5}$$

where $A_\alpha(t)$ is a difference scheme for elliptic part of differential operator of an original problem (1)–(2) on direction $\alpha = 1, \ldots, p$ and is written in the following way:

$$A(t)_\alpha y = -(a_\alpha(\boldsymbol{x}) y_{\overline{x}_\alpha})_{x_\alpha}, \quad \alpha = 1, \ldots, p, \qquad \boldsymbol{x} \in \omega. \tag{6}$$

Here we use the formulas for coefficients according to integro-interpolation method of discrete construction [11]:

$$a_\alpha(\boldsymbol{x}, t) = k_\alpha(x_1, \ldots, x_\alpha + 0.5h_\alpha, \ldots, x_p, t), \quad \alpha = 1, \ldots, p, \qquad \boldsymbol{x} \in \omega.$$

In the set of grid functions vanishing (going to zero) at the set of boundary nodes $\partial\omega$, a constructed operator $A(t)$ is self-conjugated, positively definite and bounded

$$A(t) = A^*(t), \quad 8c_1 \sum_{\alpha=1}^{p} \frac{1}{l_\alpha^2} \leqslant \|A(t)\| \leqslant 4c_2 \sum_{\alpha=1}^{p} \frac{1}{h_\alpha^2}.$$

Let us denote a difference solution by y_n at time $t^n = n\tau$, where $\tau > 0$ is a time step, while $N\tau = T$. When using a fully implicit two-level scheme a transition to a new time level in the problem is implemented in accordance with

$$\frac{y^n - y^{n-1}}{\tau} + A^n y^n = f(\boldsymbol{x}), \quad \boldsymbol{x} \in \omega, \quad n = 1, 2, \ldots, N, \tag{7}$$

$$y_0 = u_0(\boldsymbol{x}), \quad \boldsymbol{x} \in \overline{\omega}. \tag{8}$$

It is known that fully implicit difference scheme (7)–(8) is certainly stable on the right-hand side and initial condition. And for its solution due to a positivity of the operator $A(t) > 0$ the following a priori estimate is true:

$$\|y^N\| < \|u_0\| + T\|f\|. \tag{9}$$

Here for inverse problem solution of identification of a space-distributed source $f(\boldsymbol{x})$, $\boldsymbol{x} \in \overline{\omega}$ we use an additional condition in the form of the finite difference scheme (7)

$$y^N(\boldsymbol{x}) = \varphi(\boldsymbol{x}), \ \boldsymbol{x} \in \overline{\omega}. \tag{10}$$

4 An Inverse Problem: A Solution Algorithm

Let us give the problem a corresponding operator formulation. Thus, at the finite moment of time from (7)–(8) for the given $y^0 = u_0(\boldsymbol{x})$ we will obtain:

$$y^N = \varphi = \mathcal{A}u_0 + \mathcal{B}f, \quad \boldsymbol{x} \in \overline{\omega}, \quad \mathcal{A} = \prod_{n=1}^{N} S^n, \quad \mathcal{B} = \sum_{n=1}^{N} \tau \mathcal{C}^n, \tag{11}$$

where S^n is a transitional operator from one $(n-1)$ – time layer to the next n-th time layer:

$$S^n = (I + \tau A^n)^{-1}, \tag{12}$$

and operator $\mathcal{C}^n$ is determined according to the formula

$$\mathcal{C}^n = \prod_{j=n}^{N} (I + \tau A^j)^{-1}. \tag{13}$$

Therefore, an approximate solution of the inverse problem taking into account the conditions (1)–(4) with the help of fully implicit difference scheme (7)–(8) and (10) leads to solving of the following finite difference operator equation:

$$\mathcal{B}f(\boldsymbol{x}) = \phi(\boldsymbol{x}) - \mathcal{A}y^n, \quad \boldsymbol{x} \in \overline{\omega}. \tag{14}$$

Since A is selfadjoint, the operator $\mathcal{C}^n$ is self-adjoint, and hence the operator $\mathcal{B}$ in Eq. (14) is also selfadjoint. A unique solvability of the operator Eq. (14) follows from the positivity of $\mathcal{B}$ operator.

To decrease a number of iterations we use an iterative method of conjugate gradients [10,11]. It should be noted that it is also appropriate to use the iteration method of conjugate gradients in the systems of linear algebraic equations at each time step.

5 Numerical Calculations

Let us have calculation examples using the proposed iterative method for test examples with quasi-real solutions.

Example 1. Identification of the right-hand side given as Gaussian distribution (Example 2 from [5]):

$$k(x,t) = 1, \quad x \in [0,l], \quad 0 < t \leqslant T, \quad u_0(x) = 10\sin(\pi x/l), \quad x \in [0,l],$$

$$f(x) = e^{(x-x_0)^2/(2\sigma^2)}/(\sigma\sqrt{2\pi}), \quad x \in [0,l],$$

Here

$$x_0 = l/2, \quad \sigma = 0.1, \quad l = 1, \quad T = 1.$$

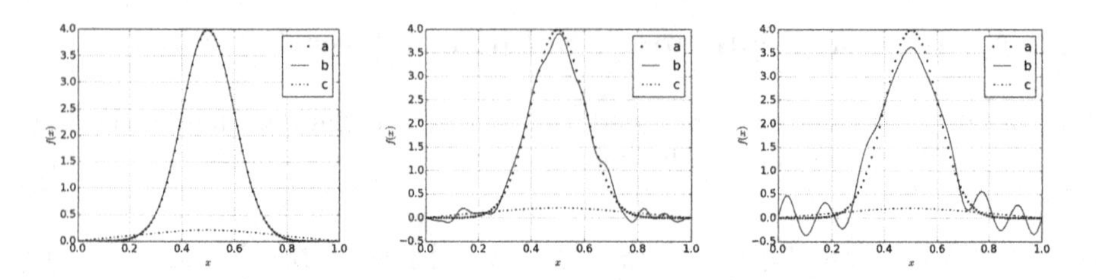

Fig. 1. Identification of a smooth right-hand side (Example 1).

Calculations have been conducted on a spatial grid consisting of 101 nodes, and a time nod consisting of 100 layers.

Results of the inverse problem solution for renewal of the right-hand side of the parabolic equation are illustrated in Example 1. On the left figure we see diagrams: a is a precise value diagram of the right-hand side, b is right-hand side calculated by iterative method, c is diagram of quasi-real additional condition $\varphi(x)$. For identification of the right-hand side $\varepsilon = 10^{-6}$ it was required to conduct $k = 5$ iterations. Mean-root square relative error for the right-hand side definition calculated by the formula

$$R = \frac{\|f_k(\mathbf{x}) - f(\mathbf{x})\|}{\|f(\mathbf{x})\|} \tag{15}$$

is equal to $R = 0.000305$. In the center and on the right of the same figure we see solution results of the problem under certain conditions of an additional assumption $\varphi(x)$ given with an error $\delta = 1\%,\ 5\%$ accordingly. In the experiments a disturbance of the additional condition was as follows:

$$\varphi_\delta(x) = \varphi(x) + \delta\sigma(x) \max_{x\in\Omega} \varphi(x)/2, \quad x \in \Omega, \tag{16}$$

where $\sigma(x)$ is a random variable uniformly distributed on the interval $(-1, 1)$. In order to distinguish smoother solution, let us take K-times used three-point formula as a smoothing operator:

$$\varphi_i^{k+1} = (\varphi_{i+1}^k + 4\varphi_i^k + \varphi_{i-1}^k)/6, \ i = 1, \ldots, M-1, \ k = 0, \ldots, K-1; \tag{17}$$

$$\tilde{\varphi}_i = \varphi_i^K, \ i = 1, \ldots, M-1 \tag{18}$$

of grid functions vanishing at boundary nodes. In the center there are similar diagrams under perturbation of 1% ($\varepsilon = 10^{-3}, k = 19, K = 30, R = 0.0723$), and at right 5% ($\varepsilon = 10^{-3}, k = 15, K = 60, R = 0.1682$).

Example 2. It differs from the previous example by an additional condition introduced in the form of discontinuous function [5]

$$f(x) = \begin{cases} 1, \, x \in [0.3l, 0.7l]; \\ 0, \, \mathit{else}. \end{cases} \tag{19}$$

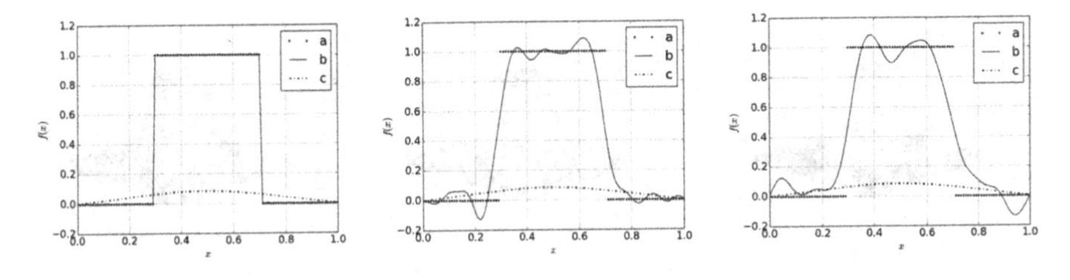

Fig. 2. Identification of discontinuous right-hand side (Example 2).

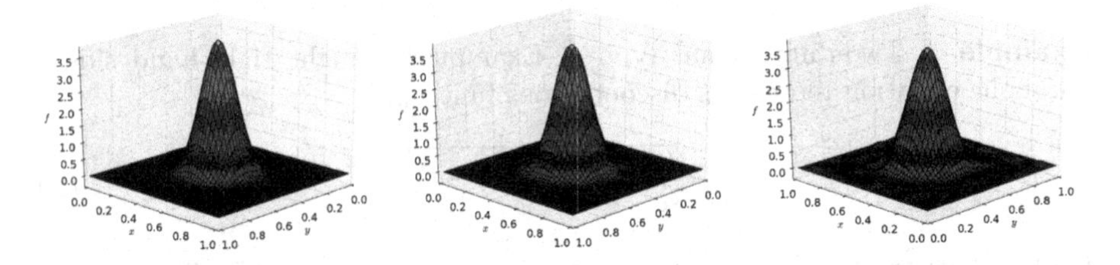

Fig. 3. Identification of the 2-dimensional smooth space-wise heat source.

Figure 2 shows a solution of the problem identifying the right-hand side under the same input data and parameters of the spatiotemporal grid as in the previous example, under given additional condition in the form of (19). On the left diagram 2 the line b is a right-hand side calculated using the proposed iterative method under the given additional condition (19), here ($\varepsilon = 10^{-6}, k = 19, R = 0.00423$). The right-hand side obtained under condition of 1% perturbation of the additional condition is represented by a curve b in the central figure ($\varepsilon = 10^{-3}, k = 5, K = 30, R = 0.2047$). On the right figure calculation results for 5% perturbation of the additional condition ($\varepsilon = 10^{-3}, k = 6, K = 60, R = 0.22482$) are presented.

Example 3. Two-dimensional type of Example 1:

$$\begin{aligned} k_\alpha(\boldsymbol{x}, t) &= 1, \quad \alpha = 1, 2, \quad \boldsymbol{x} \in \Omega, \quad 0 < t \leqslant T, \\ u_0(\boldsymbol{x}) &= 10 \sin(\pi x_1 / l_1) \sin(\pi x_2 / l_2), \quad \boldsymbol{x} \in \Omega, \\ f(\boldsymbol{x}) &= e^{(\boldsymbol{x} - \boldsymbol{x}_0)^2 / (2\sigma^2)} / (\sigma \sqrt{2\pi}), \boldsymbol{x} \in \Omega, \end{aligned}$$

Here

$$\boldsymbol{x}_0 = (l_1/2, l_2/2), \quad \sigma = 0.1, \quad l_1 = l_2 = 1, \quad T = 1.$$

Parameters of a spatiotemporal grid are equal to $N_1 = N_2 = N = 100$. Figure 3 presents diagrams of the right-hand side (on the left) found by the suggested iterative method of the right-hand side for 1% perturbation of the additional condition (see central figure: $\varepsilon = 10^{-3}, k = 7, K = 60, R = 0.042$) and for 5% perturbation of the additional condition (right figure: $\varepsilon = 10^{-3}, k = 5, K = 60, R = 0.089$).

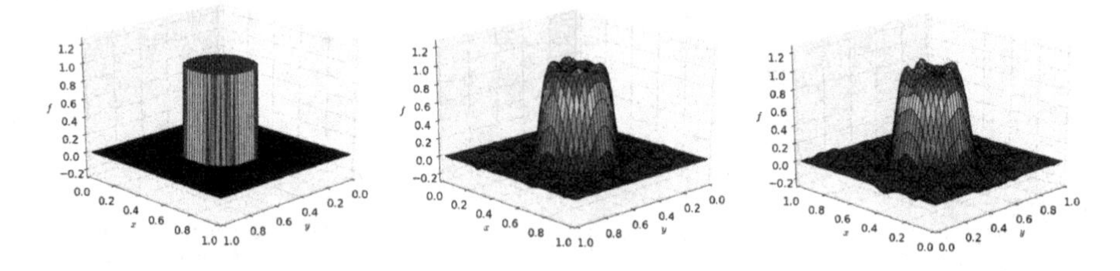

Fig. 4. Identification of the 2-dimensional discontinuous space-wise heat source.

Example 4. Two-dimensional type of Example 2. Sought right-hand side of parabolic equation represents discontinuous function

$$f(\boldsymbol{x}) = \begin{cases} 1, \sqrt{(x_1 - l/2)^2 + (x_2 - l/2)^2} < 0.2; \\ 0, \textit{else}. \end{cases} \tag{20}$$

On Fig. 4 we present the solution results of the inverse problem for identification of the right-hand side at the same input data and parameters of the spatiotemporal grid as in the previous example under additional condition in the form of (20). On the left diagram there is the sought right-hand side. The right-hand side found by the proposed iterative method for 1% perturbation of the additional condition (20) is shown on the central figure($\varepsilon = 10^{-3}, k = 10, K = 30, R = 0.2476$). On the right figure the counter results for 5% perturbation of the additional condition are presented ($\varepsilon = 10^{-3}, k = 7, K = 60, R = 0.2974$).

6 Conclusion

In quasi-real computational experiment, we calculate additional data for the inverse problem using the results of numerical solution for a direct problem. Calculated data show high rate of convergence of the considered iterative processes. The quoted results of the computational experiment confirmed high efficiency of the iterative method of conjugate gradients for numerical solution of the inverse problem for identification of the right-hand side in parabolic equation, dependent on spatial variables only. Such good results have been obtained even for 5% perturbation of the input data due to application of conjugate gradients at each iteration for the right-hand side at solving of the systems of algebraic equations resulting on each time layer.

Acknowledgments. The authors express their sincere gratitude to Professor P.N. Vabishchevich for constructive comments and fruitful discussions. The research was supported by the Government of the Russian Federation (project № 14.Y26.31.0013).

References

1. Cannon, J.R.: Determination of an unknown heat source from overspecified boundary data. SIAM J. Numer. Anal. **5**, 275–286 (1968)
2. D'haeyer, S., Johansson, B.T., Slodicka, M.: Reconstruction of a spacewise-dependent heat source in a time-dependent heat diffusion process. IMA J. Appl. Math. **79**(1), 33–53 (2014)
3. Isakov, V.: Inverse Source Problems: Mathematical Surveys and Monographs, vol. 34. American Mathematical Society (AMS), Providence (1990)
4. Isakov, V.: Inverse parabolic problems with the final overdetermination. Commun. Pure Appl. Math. **44**(2), 185–209 (1991)
5. Johansson, T., Lesnic, D.: Determination of a spacewise dependent heat source. J. Comput. Appl. Math. **209**(1), 66–80 (2007)
6. Johansson, B.T., Lesnic, D.: A variational method for identifying a spacewise-dependent heat source. IMA J. Appl. Math. **72**, 748–760 (2007)
7. Erdem, A., Lesnic, D., Hasanov, A.: Identification of a spacewise dependent heat source. Appl. Math. Model. **37**, 10231–10244 (2013)
8. Kabanikhin, S.I.: Inverse and Ill-Posed Problems: Theory and Applications. De Gruyter, Berlin (2011)
9. Lions, J.-L., Magenes, E.: Non-Homogeneous Boundary Value Problems and Applications, vol. I. Springer, Berlin (1972). https://doi.org/10.1007/978-3-642-65217-2
10. Saad, Y.: Iterative Methods for Sparse Linear Systems. SIAM, Philadelphia (2003)
11. Samarskii, A.A.: The Theory of Difference Schemes. Marcel Dekker, New York (2001)
12. Samarskii, A.A., Vabishchevich, P.N.: Numerical Methods for Solving Inverse Problems of Mathematical Physics. De Gruyter, Berlin (2007)
13. Samarskii, A.A., Vabishchevich, P.N., Vasil'ev, V.I.: Iterative solution of a retrospective inverse problem of heat conduction. Mat. Model. **9**(5), 119–127 (1997)
14. Vabishchevich, P.N.: Additive Operator-Difference Schemes. Walter de Gruyter GmbH, Berlin (2014)
15. Vabishchevich, P.N.: Iterative computational identification of a space-wise dependent source in parabolic equation. Inverse Probl. Sci. Eng. **25**(8), 1168–1190 (2017)
16. Vasil'ev, V.I., Kardashevsky, A.M.: Iterative solution of the retrospective inverse problem for a parabolic equation using the conjugate gradient method. In: Dimov, I., Faragó, I., Vulkov, L. (eds.) Numerical Analysis and Its Applications. LNCS, vol. 10187, pp. 698–705. Springer, Heidelberg (2017). https://doi.org/10.1007/978-3-319-57099-0_80

Author Index